RANGER SCHOOL

PLANT GROWTH AND DEVELOPMENT

McGRAW-HILL SERIES IN ORGANISMIC BIOLOGY

Consulting Editors

Gardiner: *The Biology of Invertebrates*
Kramer: *Plant and Soil Water Relationships: A Modern Synthesis*
Leopold and Kriedemann: *Plant Growth and Development*
Patten and Carlson: *Foundations of Embryology*
Phillips: *Introduction to the Biochemistry and Physiology of Plant Growth Hormones*
Price: *Molecular Approaches to Plant Physiology*

PLANT GROWTH AND DEVELOPMENT

A. CARL LEOPOLD
Department of Horticulture
Purdue University
Lafayette, Indiana

PAUL E. KRIEDEMANN
C.S.I.R.O. Division of Horticultural Research
Merbein, Victoria, Australia

SECOND EDITION

McGRAW-HILL BOOK COMPANY
New York St. Louis San Francisco Auckland Düsseldorf Johannesburg
Kuala Lumpur London Mexico Montreal New Delhi
Panama Paris São Paulo Singapore Sydney Tokyo Toronto

Library of Congress Cataloging in Publication Data

Leopold, Aldo Carl, date
Plant growth and development.

(McGraw-Hill series in organismic biology)
Bibliography: p.
Includes index.
1. Growth (Plants) 2. Plant physiology.
I. Kriedemann, Paul E., joint author. II. Title.
[DNLM: 1. Plants—Growth and development. QK731
L587p]
QK731.L44 1975 582′.03 74-20970
ISBN 0-07-037200-4

PLANT GROWTH AND DEVELOPMENT

3 4 5 6 7 8 9 0 V H V H 7 9 8 7 6

This book was set in Century Schoolbook by Black Dot, Inc.
The editors were William J. Willey, Andrea Stryker-Rodda, and Carol First;
the designer was Nicholas Krenitsky;
the production supervisor was Leroy A. Young.
New drawings were done by J & R Services, Inc.
Von Hoffmann Press, Inc., was printer and binder.

In its early common naïve state science . . . imagined
that we could observe things in themselves,
as they would behave in our absence.
Naturalists . . . are now beginning to realize
that even the most objective of their observations
are steeped in the conventions they adopted at the outset . . .
so that when they reach the end of their analyses
they cannot tell with any certainty
whether the structure they have made
is the essence of the matter they are studying,
or the reflection of their own thought.

Pierre Teilhard de Chardin
The Phenomenon of Man

CONTENTS

PREFACE TO THE SECOND EDITION

In preparing this second edition, we are struck by two special characteristics in the manner of change in our science.

First, the rate of change is astonishing. We had thought that the passage of 10 years would have left at least half the information on plant growth still usable, but on the contrary, we find that both the bibliography and the selection of figures show nearly a 75 percent turnover since 10 years ago. This would indicate a half-life of biological information of as little as 5 years. Some data of J. Margolis (*Science,* 155:1213-1219, 1967) indicate that citations of scientific papers show a 50 percent falloff in about 5 or 6 years, and data of B. C. Brookes (*J. Documentation,* 26:283-294, 1970) indicate that the checkout rate of library books in the science area shows about a 50 percent falloff in 6 or 7 years. If a professional physiologist were to stop reading the literature in, say, 1975, he would then have lost half his scientific background in about 1980 or 1982. Therefore, in compiling the references for this edition, we have been particularly careful to include very recent literature whenever possible, as well as the classic articles in the field.

The second special characteristic we have noted is the channeling of research efforts into special areas. One might use a metaphor of research rain falling on the scientific field, quickly forming channels in which the water cuts into new ground, but leaving other equally interesting parts of the field unchanged. In the subject area of photosynthesis, for example, such a fast-cutting channel is the C_4 type of photosynthesis; in auxin it is the mechanism of action at the plasmalemma and cell-wall level; in ethylene it is the regulation of biosynthesis. Areas of little change might include the relations between molecular structure and activity of any of the hormones, and the entire field of photoperiodism and vernalization. We have attempted to achieve a balanced approach, following the new channels without neglecting the more basic older ones.

As coauthors, we have each taken primary responsibility for certain sections: P. K. for Parts I and IV, C. L. for Parts II, III, and V, each of us working within the areas of his own specializations. The organization of this edition follows the same general outline as the first edition, but we added a new chapter on ethylene and have substantially expanded and revised the material in certain sections to reflect the new developments and approaches mentioned above.

We wish to express special thanks to Dr. W. K. Purves for reading the entire manuscript after the first draft, and to F. B. Abeles, J. H. Cherry, and C. Y. Tsai for reading specific parts of the manuscript. Thanks are also due to A. S. Crafts, D. J. Morre, J. V. Possingham, B. R. Loveys, and R. K. deal Fuente for supplying photographs.

A. Carl Leopold
Paul E. Kriedemann

PREFACE TO THE FIRST EDITION

As a treatise on plant physiology, this book has three directions of special emphasis: first, to develop the student's reliance on experiments in forming generalizations about his science; second, to depict science as a complex of imperfect approximations derived by the scientific method; and third, to reorganize the subject to make it more nearly representative of modern plant physiology in the laboratory and in the field.

I should like to explain the reasons for this approach. As a teacher, I am disturbed to find young professional plant physiologists who have so completely embraced the generalizations which have been tutored into them and which they have cherished and repeated for their examinations in the best tradition of scholasticism, that they do not really have a working knowledge of the science that they must revise and help to improve. Comprehension of the scientific generalization without knowledge of underlying specific experimental facts seems to me to be a poor tool for hewing out new science. Therefore, in this book I have drawn on experimental evidence insofar as possible to illustrate the concepts being discussed.

I am also disturbed to find that a large proportion of research workers do not realize when they enter this profession that the scientific method is an imperfect game—one that involves imperfect deductions from imperfect experiments; and that the nature of the game must lead to disagreement—sometimes quiet and rational, sometimes not. In using a case-history approach, I hope that the student will recognize the inconsistencies among the "facts" accumulated so far, and that this will make him a little more sanguine about the sharp disagreements between equally dependable scientists and will teach him to relish the exchange of criticism that sharpens good science and shrinks poor science.

The book is centered about the workings of the growing plant, without organized coverage of biochemistry and nutrition. This assumes a logical pedagogical division of the subject of plant physiology into a section on growth and development and another section on nutrition and metabolism. This is also done with the hope of maintaining an appropriate level of interest in the functions of the living plant, in the face of a current tendency toward preoccupation with grindates, supernates, and simulated life activities in test tubes without sufficiently clear relationship to the growth of the whole plant.

In short, it is hoped that the presentation here will paint an approximate picture of the status of this science today and at the same time will make the student think in terms of experimental units of information. The best experimental units incorporate the weaknesses of the human minds that conceived them and carried them out, and no generalization from these units is more dependable than the underlying experiments.

The biological puzzle is a dynamic and exciting one, including some pieces which are changing shape even as we try to fit them together, some that need to be polished and changed before they will be able to fit into their places, and some which have no apparent relevance to the rest of the pieces at hand. The challenge for the scientist is to fit together an improved picture, using his ingenuity to devise more precise reconstructions of the working mechanisms inside the living organism.

A. Carl Leopold

PART 1
ASSIMILATION AND GROWTH

PHOTOSYNTHESIS
TRANSLOCATION
THE DYNAMICS OF GROWTH

The emergence of higher plants during the course of our planet's evolutionary history is one manifestation of selective pressures for efficient forms of growth and development. Multicellular organisms embody this characteristic, but their continued survival and eventual success has necessitated close coordination of anatomical form and physiological function. This coordination is discernible at three distinct levels: in cellular and subcellular processes, within tissue systems, and at the macroscopic, or whole-plant, level.

The combined activities of molecular biologists, biochemists, and electron microscopists have made it possible to ascribe biochemical function to anatomical form at the ultrastructural level. For example, the long recognized compartmentation of biochemical events within cells and tissues was once described by mathematical analogies but is now visualized in terms of subcellular components. Each organelle, or subcellular complex, possesses a distinctive enzyme complement and is encompassed by differentially permeable membranes. Their metabolic function varies accordingly.

Tissue specialization has also assumed increasing significance during the course of evolution because multicellular organisms require a vascular system which combines both supportive and sensory roles. With centers of growth becoming further removed from sites of assimilation, some form of substrate translocation was clearly necessitated, but a plant's vascular network also enables it to transmit regulatory agents or signals. These help to coordinate the whole organism in response to more localized environmental cues.

This environment-plant interaction is fundamental to growth and development and is amenable to mathematical analysis at the macroscopic level. Such quantitative descriptions emphasize the dynamic nature of plant growth and help us develop concepts of how multicellular organisms manage to pursue a gene-controlled strategy aimed at survival. As our understanding of plant growth and development improves, these stratagems take on biochemical or biophysical dimensions and we are able to build up a framework of molecular ecology, the regulation of plant growth by biochemical and biophysical forces with a spectrum of activity that ranges from subcellular levels to the macroscopic.

1 PHOTO-SYNTHESIS

Plants very probably draw through their leaves some part of their nourishment from the air . . . may not light, also, by freely entering surfaces of leaves and flowers, contribute much to ennobling the principles of vegetables?

Stephen Hales
"Vegetable Staticks," 1727

These early concepts foreshadowed our own view of photosynthesis as the most significant chemical event in the biosphere, a process which has made possible the evolution and maintenance of terrestrial life as we know it. The amount of solar energy that is usefully absorbed for plant growth represents only a tiny fraction of the total energy incident upon the earth, and yet this absorption winds the mainspring for all the earth's biological machinery.

THE GLOBAL SCALE

The true magnitude of global photosynthesis is difficult to assess, partly because of the extreme variability in rates (both marine and terrestrial). Land surfaces have been credited with fixing 20 to 30 billion metric tons of CO_2 per year, while the oceans are thought to assimilate about 40 billion (Bolin, 1970). Global estimates vary between 10 and 100 billion tons, with a more generous estimate of 150 billion tons by Rabinowitch (1951). Estimates of the relative magnitudes of terrestrial and marine photosynthesis also seem somewhat speculative. Rabinowitch (1951) credits the oceans with 90 percent of the global total, although more recent figures (Ryther, 1970; Woodwell, 1970) suggest that only one-third of the CO_2 fixed by the earth's ecosystems can be attributed to marine organisms.

The oceans have a universally recognized role as CO_2-buffering systems for the earth's atmosphere, and since the middle of the nineteenth century our atmospheric CO_2 concentration has increased by only 10 percent (Woodwell, 1970). The increasing release due to the consumption of fossil fuels and the reduction in CO_2-fixing capacity of terrestrial vegetation due to deforestation and related practices (Pearman and Garratt,

1972) have accentuated the upward trend in atmospheric CO_2 concentration, which reached 1 ppm/year between 1958 and 1968 (Sawyer, 1972). Counterbalancing this trend is the enormous buffering capacity of the oceans where CO_2 enters into an equilibrium with dissolved bicarbonate ions. Release of gaseous CO_2 is then associated with the precipitation of equimolar quantities of insoluble carbonate.

$$2HCO_3^- \rightleftharpoons CO_2 \uparrow\downarrow + CO_3^{=} \downarrow + H_2O$$

This equilibrium, although temperature-dependent, is modified by other solutes in the ocean. The overall outcome is that the CO_2 content of the oceans is well buffered and thus regulates the atmospheric concentration.

Photosynthesis has also played a key role in evolution on a global scale. About 2 billion years ago, certain marine organisms developed which were capable of photosynthetic evolution of oxygen. Before this time, photosynthesis presumably occurred without oxygen production, much as it does in present-day green and purple bacteria (Van Neil, 1949). The entry of oxygen into what had been a reducing atmosphere eventually gave rise to sufficient oxygen and ozone in the upper atmosphere to attenuate the ultraviolet component in the solar radiation. Organisms were then able to emerge from their protective layer of seawater, and the stage was set for the evolution of land plants. The presence of oxygen in the atmosphere also provided an opportunity for the utilization of the more efficient aerobic types of respiration.

TERRESTRIAL PHOTOSYNTHESIS

Community photosynthesis is a summation of the photosynthetic activities of individual plants and their component leaves, plus other photosynthetic surfaces. This terrestrial leaf community is of course more than just a mass of chlorophyll-containing tissues; it is a beautifully structured array of organs highly effective in the absorption of solar radiation and atmospheric CO_2.

The advent of infrared gas analyzers as physiological tools has made it possible to study CO_2 exchange by single leaves, whole plants, and indeed whole communities of plants (Chap. 3). Rates of CO_2 uptake per unit leaf surface can vary, even under favorable conditions, over almost two orders of magnitude. Some broad comparisons are shown in Table 1-1. Bear in mind the generalized nature of these groupings, which are in no way exhaustive and are simply intended to give some general idea of observed photosynthetic rates in various categories of plants. The real-life situation is more complex than Table 1-1 suggests because of variation due to environmental factors and differing physiological states.

Two distinct categories of plants emerge: (1) Those in which the CO_2-fixation system depends almost entirely on the enzyme ribulose 1,5-diphosphate (RuDP)† carboxylase and which

†A list of chemical and other abbreviations used in this book, chemical names of compounds with common names, and prefixes for powers of 10 for use with metric units of measure will be found in Appendix Tables A-1 to A-3.

Table 1-1 Comparative rates of maximum photosynthesis vary over almost two orders of magnitude.

C_3 photosynthesizers: Rate, mg CO_2/dm-h†	
Slow-growing perennials (desert species, orchids, some Crassulaceae)	1–10
Evergreen woody plants (tropical and subtropical trees and shrubs)	5–15
Deciduous woody plants (horticultural plants, trees)	15–30
Rapidly growing agronomic plants (wheat, soybeans, sugar beet, cotton, sunflower)	20–40
C_4 photosynthesizers:	
Tropical grasses and other plants with carboxylic acid-fixation pathway (sugarcane, corn, *Amaranthus, Atriplex*)	50–90

†These general values are based on the projected area of the leaf and take no account of differing stomatal distribution between the upper and lower surfaces of the leaves.

have been described as possessing the reductive pentose phosphate or Calvin cycle of CO_2 fixation. They include the great majority of plant species and are commonly referred to as C_3 plants. (2) Those which possess an additional photosynthetic system. Members of this group have almost universally higher rates of photosynthesis and represent a more recent evolutionary advance, thought to be the outcome of environmental pressures for short-term rapid growth. Associated with their high rates of photosynthesis, compared with the C_3 plants, is an enhanced turnover of assimilate (Hofstra and Nelson, 1969), low resistance to the inward diffusion of CO_2 (El-Sharkawy and Hesketh, 1965), a virtual absence of photorespiratory CO_2 evolution, CO_2 compensation points close to zero (Forrester et al., 1966), and specialized chloroplast structures in mesophyll vs. vascular-bundle sheath cells (Laetsch, 1968; Bisalputra et al., 1969). Their CO_2 fixation system is principally due to the enzyme phosphoenolpyruvate (PEP) carboxylase with subsequent entry of fixed carbon into a C_4 dicarboxylic acid cycle. To differentiate this plant category from the first group they are referred to as C_4 plants.

Physiological differences between the C_3 and C_4 plants will be reiterated later in this chapter.

THE PHOTOSYNTHETIC SITE

The chloroplast is a highly functional organelle which houses the complete machinery for photosynthesis. It is a self-replicating body (see Fig. 1-1) with its own complement of nucleic acids and ribosomes, which manufacture distinctive proteins (Boardman et al., 1965). These properties indicate a large measure of self-reliance, although the nucleus of the cell can have a controlling influence on chloroplast biochemistry, e.g., in chlorophyll synthesis.

The chloroplast is motile in the cytoplasm and generally capable of altering its position in response to light. Under low light intensities chloroplasts arrange themselves into positions which maximize light interception, and under high intensities they arrange themselves in positions which minimize light interception. How this locomotion is achieved is largely unknown, although the role of phytochrome in regulating chloroplast movement is well established (Chap. 15).

The initial sightings of chloroplasts were probably made before 1700 by the Dutch microscopist Anton van Leeuwenhoek, although chloroplasts were first described by von Mohl in 1837, who also noted the presence of starch in them; however, he did not seize on the relationship between chloroplasts, photosynthesis, and starch. Sachs (1862) claims to have been the first to appreciate this point.

The detection of specific chemical events related to energy input was delayed by lack of suitable instrumentation until the classic work of Hill (1937), who demonstrated a light-driven evolution of oxygen by leaf material suspended in water with ferric oxalate. The utilization of light-acquired chemical energy for CO_2 reduction and a pathway for carbon fixation was subsequently established by 1954 from the work of Calvin and associates at Berkeley.

Chloroplasts are more than passive inhabitants of cells. It is quite possible that higher plants derived their chloroplasts from the early invasion of a photosynthetic organism something like a blue-green alga (Cohen, 1970). This unique organelle confers the CO_2-fixing ability upon plants by virtue of its ability to utilize light for the generation of ATP and $NADPH_2$ necessary for CO_2 fixation. Numerous kinds of cells, both plant and animal, are able to fix CO_2 provided they have a supply of ATP and $NADPH_2$; the photosynthetic cell thus specifically equips the green plant for its autotrophic existence.

Chloroplast Structure

The chloroplast (Fig. 1-3), a lens-shaped organelle completely encompassed by an outer membrane, is considerably smaller than the nucleus of the cell. Chloroplasts occur with great abundance in the leaf; Kirk and Tilney-Bassett (1967) calculate a density of one-half million chloroplasts per square millimeter of leaf surface. Their existence and essential structure were known well before the end of the nineteenth century. The remarkable description of grana in 1883 by Arthur

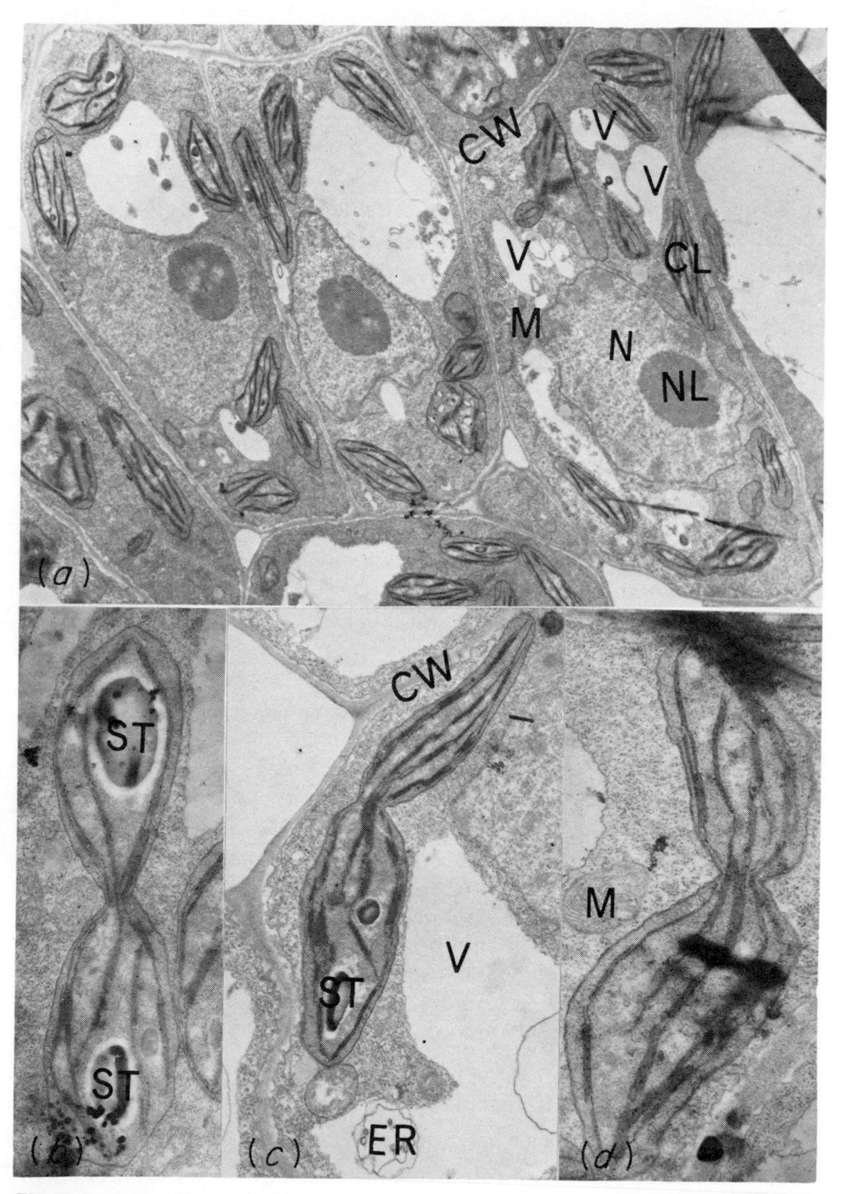

Fig. 1-1 Spinach chloroplasts form a central constriction in mature plastids which then produce two daughter organelles with a full complement of structural components. (*a*) Cells in leaf 10 of a 16-day-old spinach plant (X9000). (*b*) to (*d*) Chloroplasts with central constrictions (X13,200, 9,000, and 16,800, respectively). All preparations were stained with lead hydroxide and were from the base of young leaves about 1 cm long. CW = cell wall, V = vacuole, CL = chloroplast, M = mitochondrion, N = nucleus, ST = starch grain, ER = endoplasmic reticulum (Possingham and Saurer, 1969).

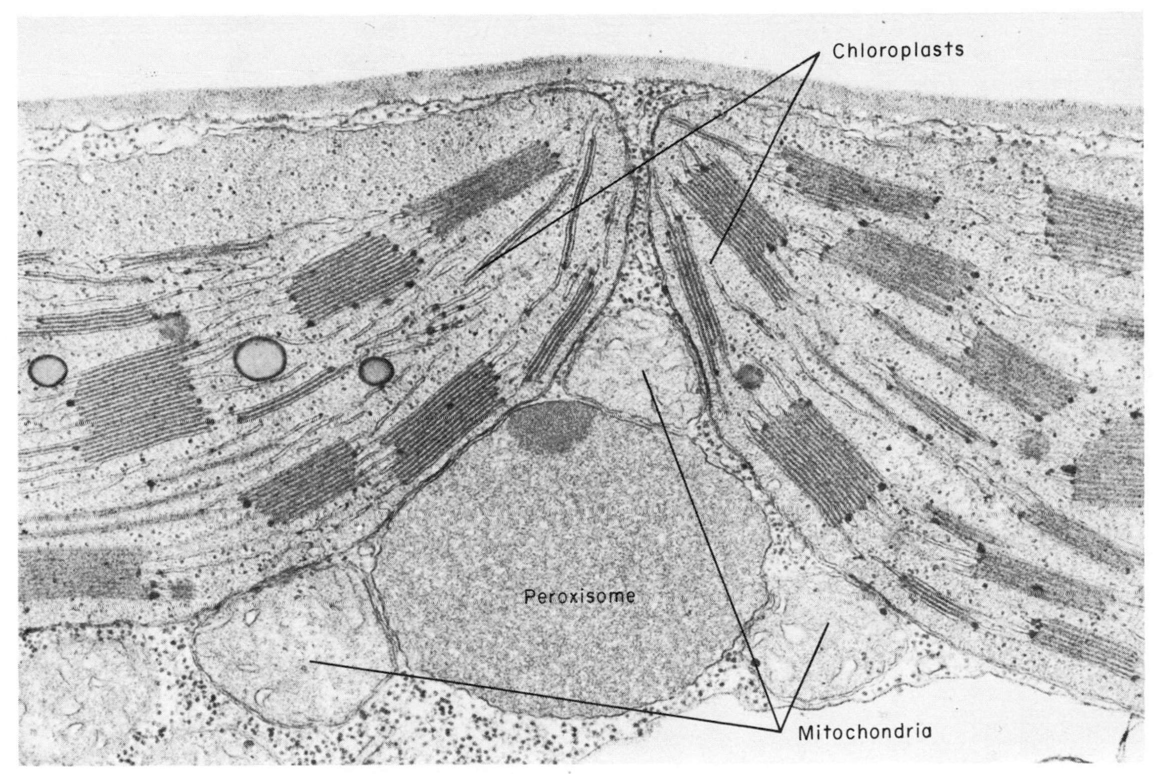

Fig. 1-2 Sites of CO_2 fixation are often juxtaposed against regions of CO_2 generation. Organelles such as peroxisomes and mitochondria can be appressed to chloroplasts as illustrated in this mesophyll cell from tobacco leaf (X35,000). Compare with Fig. 1-18. (Courtesy of *Prof. Eldon Newcomb, University of Wisconsin.*)

Meyer testifies to careful observation on the part of early microscopists.

The ground substance of chloroplasts (Fig. 1-3), known as *stroma,* is a slightly electron-dense granular matrix. Embedded in the stroma is a large number of membrane-bounded flattened sacs, or lamellae. Menke (1962) coined the term *thylakoids* (saclike) for these structures. Between 2 and 100 thylakoids may be stacked up to form a granum (Fig. 1-3*a*). The three-dimensional spatial relationships of the thylakoids have yet to be established with certainty; e.g., it is not really known whether they represent ramifications from one "large" membrane-bounded sac or are, in fact, an assemblage of many distinct sacs.

A notable variation in this generalized picture of higher-plant chloroplasts is provided by the two types of chloroplasts in C_4 plants (Fig. 1-3). In corn, for example, the chloroplasts in parenchyma cells of the vascular bundle sheath often have no grana, and single thylakoids extend over the full length of the plastid. Chloroplasts in the mesophyll tissue of a C_4 plant leaf (Fig. 1-3*a*) adhere to the general structural pattern described earlier, although they tend not to form starch grains between the stroma lamellae as in the C_3 plants (see also Woo et al., 1971).

These same mesophyll chloroplasts in C_4 plants, such as corn, have additional distinctive features. They are equipped with a network of fine tubules forming a reticulum underneath the outer en-

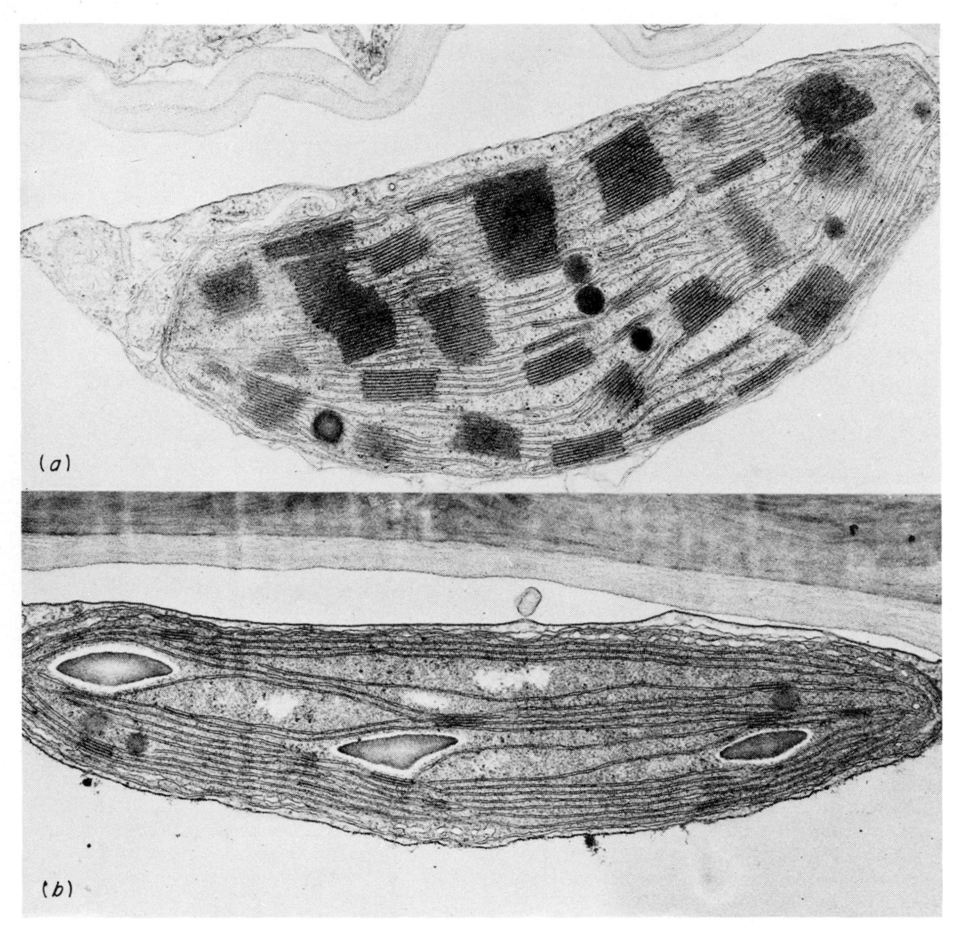

Fig. 1-3 Chloroplasts of C_4 plants show structural specialization according to location within the leaf. Bundle-sheath cells (*b*) have chloroplasts devoid of pronounced grana but with abundant starch grains. Mesophyll cells (*a*), on the other hand, contain more typical plastids (compare Fig. 1-2) but starch is absent (Andersen et al., 1972). X20,000. (*Courtesy of Dr. David Bishop C.S.I.R.O. Division of Food Research, Sydney, Australia.*)

velope of the chloroplast. Laetsch (1968) and Rosado-Alberio et al. (1968) have described these features and have speculated on their involvement in the translocation of newly formed photosynthate.

The surface of the thylakoid is covered with a mosaic of particles in a regular arrangement, whose size varies according to the region of the thylakoid (grana vs. stroma). Electron micrographs give the impression that each particle consists of four subunits. These particles, known as *quantosomes,* were originally thought to occur inside the thylakoid, but more recent opinion favors a surface location (Park, 1965). Quantosomes were initially heralded as the fundamental photosynthetic unit, although their size (150 Å) seems inadequate to house the necessary components for complete photosynthesis (Rabinowitch and Govindjee, 1969).

Although there is some divergence of opinion regarding the origin of mature chloroplasts within the cell, the present weight of evidence favors the view that they do not arise *de novo* but come from preexisting plastids during both vegetative and sexual reproduction. The antecedent to the chloroplast, i.e., the proplastid, is a small unpigmented (or pale-green) body which occurs in rapidly dividing but undifferentiated meristimatic tissues. These proplastids multiply freely and so keep pace with cell division. The proplastid eventually gives rise to a mature plastid appropriate to the tissue being formed (chloroplasts in the leaf, leucoplasts in a tuber).

The mature chloroplasts of both algae and higher plants are still capable of division. Early observations on algal cells combined with more recent evidence from higher plants (Possingham and Saurer, 1969) (Fig. 1-1) support the concept of cytological continuity of chloroplasts. In reproducing cells of both lower and higher plants, plastid populations can be maintained either by proplastid division (with subsequent differentiation) or by the division of mature chloroplasts (see Fig. 1-1).

The synthesis of chlorophyll is closely associated with the conversion of vesicles into disks and lamellae, and while electron micrographs do not permit detection of chlorophyll pigment, the formation of the pigment is closely associated with the organization of the disk and lamellar structures. Since the chlorophyll is structurally a part of the lamella, it is generally accepted that they are formed together. Upon exposure of etiolated *Euglena* cells to light, there is a gradual accumulation of chlorophyll for about 8 h before detectable photosynthesis begins (Fig. 1-4). This light effect is also evident in higher plants; there appears to be a diurnal periodicity of chlorophyll synthesis, with relatively rapid synthesis occurring during the daylight hours (Sironval, 1963).

The formation of the lamella and its associated chlorophyll has been found by numerous workers to require protein synthesis (Aoki and Hase, 1964; Hudock et al., 1964). Applications of poisons of protein synthesis, e.g., chloramphenicol, can prevent lamellar development and chlorophyll formation. Similar blockage is obtained with actinomycin D, indicating that RNA synthesis is also involved in chloroplast development (Bogorad, 1967).

Substantial amounts of DNA are found in chloroplasts, and although this material represents only 2 to 14 percent of the total DNA within a cell, labeling experiments with the green alga *Chlamydomonas* indicate that during the day most of the DNA synthesis takes place in the chloroplasts. These experiments were performed with synchronized cultures of the alga in which the mitotic activity occurs exclusively during the night (Chiang and Sueoka, 1967). Chloroplastic DNA is distinctive in having a slightly different density (Tewari and Wildman, 1966), being singularly free of histones (von Wettstein, 1967), and having somewhat different base ratios than nuclear DNA (Green and Gordon, 1966).

RNA is present in the chloroplast, especially as ribosomal RNA (rRNA) with density characteristics more closely resembling bacterial ribosomes than cytoplasmic ribosomes (Boardman, 1967). Chloroplast ribosomes have been isolated from tobacco leaves as polyribosomes and shown to be very effective in carrying out protein synthesis (Boardman et al., 1965; Chen and Wildman, 1967). Thus the chloroplast is endowed with its own set of genetic instructions and the ability to translate them into RNA and proteins.

A rapid turnover in chloroplast components was demonstrated by Sironval (1963), who incorporated radioactive α-aminolevulinic acid into the chlorophyll of soybeans and followed the de-

Fig. 1-4 If *Euglena* cells are held at 100 fc in mannitol (to prevent divisions), some chlorophyll formation occurs before photosynthesis begins. Oxygen evolution showed a lag time of 8 h (Stern et al., 1964).

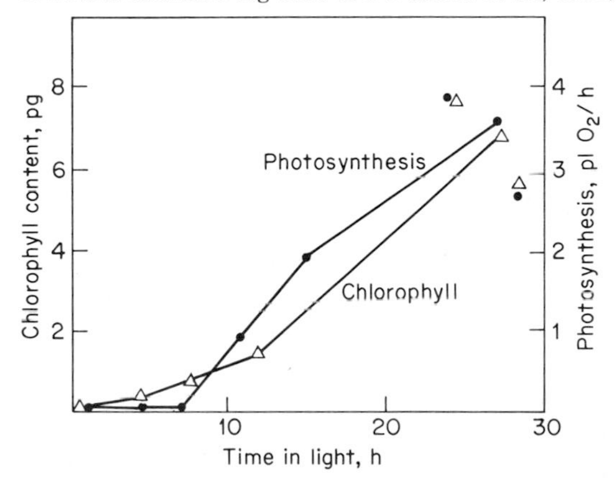

cline in radioactivity as a measure of turnover; he concluded that the pigment-protein complex responsible for energy absorption has a half-life of about 5 days. Further indirect evidence may be taken from the rate of deterioration of leaf components in the dark (Shaw et al., 1965), which suggests that half of the chlorophyll content of wheat leaves will disappear in 5 days. Similar rates of disappearance are evident for protein and RNA components of the leaves as well. Bogorad (1967), for example, has noted a rapid decay of RNA polymerase in chloroplasts.

Taken collectively, these data support the view that maintenance of photosynthetic effectiveness in chloroplasts does require extensive renewal of its components.

Like all biological entities, chloroplasts do not last forever but eventually degenerate. Chromoplasts are formed in some instances, and this sequence can be associated with the normal senescence of maturing fruits. Spurr and Harris (1968) have shown this transformation in the epidermal (peel) tissues of *Capsicum* as the fruit ripens, and Thomson (1966) has described similar changes in Valencia oranges. The familiar golden color of ripened oranges is due to the synthesis of carotenoids which replace chlorophyll as the predominant pigment (Eilati et al., 1969). This loss of chlorophyll is associated with degeneration of chloroplasts into chromoplasts. The physiological role of these secondary organelles has not been documented.

Chloroplast Function

Light reactions A plant's utilization of light depends upon a specific molecule absorbing radiant energy in a usable way. The porphyrin protein of the chlorophyll molecule (its chromophore) is capable of trapping radiant energy and then transferring energy to other molecules. Understandably, the chemical differences between the various forms of chlorophyll equip them with different light-absorbing properties. Chlorophyll *a* and chlorophyll *b*, for example, differ in the nature of their participation in the two photosystems described below.

Within the chloroplast, chlorophyll is confined to the lamellae. All plants have chlorophyll *a*, and most plants also have chlorophyll *b*, *c*, or *d*. The ratio between chlorophylls *a* and *b* varies according to both plant and environmental conditions. C_4 plants show *a*/*b* ratios of 3.9 vs. 2.8 for C_3 plants (Black et al., 1969; Black and Mayne, 1970; Chang and Troughton, 1972); a given species will have a relatively greater *a*/*b* ratio if grown under high light intensities (Rabinowitch and Govindjee, 1969; Reger and Krauss, 1970). Chlorophyll *a* exists in more than one form, and the variant known as P700 (named from its absorption peak) plays a key role in photochemical reactions. P700 is very rapidly bleached and represents only 0.1 percent of the total chlorophyll *a* present. A second form of chlorophyll *a* (P690) is also fundamental to the light reaction of photosynthesis. These two forms of chlorophyll are key pigments in the photochemical reactions of photosynthesis and represent the chief energy traps for photosystems I and II, respectively (see Fig. 1-5).

The absorption of photons by a pigment assembly known as photosystem II catalyzes the removal of electrons from water molecules. These energized electrons then move down a series of acceptors and donors until they eventually get to NADP and contribute to its reduction to NADPH. Along this downhill sequence, some energy is siphoned off into ATP formation. Electrons which were initially boosted against the electrochemical gradient above photosystem II following photon absorption are thought to be accepted by a substance called Q (it takes this name from its quenching effect on fluorescence). Electrons subsequently energized within photosystem I reach the ferredoxin-reducing substance (FRS). The downhill run in their energy status again promotes NADP reduction into NADPH.

The chemical identity of these various acceptors and donors is not known in complete detail, and their very existence is often hypothetical and based on the outcome of experiments which combine artificial electron acceptors with compounds that inhibit electron flow at specific

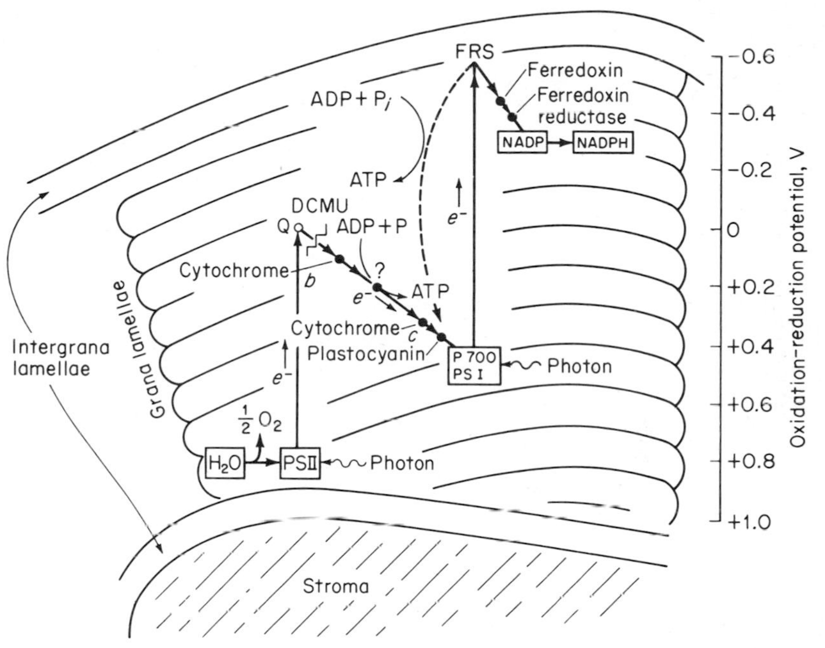

Fig. 1-5 There are two photoreductive steps in photosynthesis, photosystem II, which photolyzes water, and photosystem I, which reduces ferredoxin. By their action, ATP and $NADPH_2$ generated within the grana are then utilized for CO_2 fixation within the chloroplast stroma. (*Adapted from Levine, 1969.*)

points in the overall sequence (see diuron† (DCMU), in Fig. 1-5).

This picture of photosynthesis, which depends on the cooperation of two photochemical reactions, has its origin in the *Emerson enhancement effect.* Emerson and Lewis (1943) reported that quantum efficiency declines abruptly at wavelengths above 680 nm even though chlorophyll a absorption in vivo still occurs within this band. About 10 years later Emerson et al. (1957) discovered that photosynthetic efficiency around 680 to 700 nm is improved substantially if light of a shorter wavelength is superimposed. Of particular interest was their finding that the photosynthetic rate under simultaneous illumination with these two beams of light is greater than the sum of photosynthetic rates when short- and long-wavelength light are supplied separately; hence the term enhancement (Fig. 1-6). This effect is now explicable in terms of two light reactions, proposed almost simultaneously by a number of investigators between 1960 and 1961. Unlike the process in higher plants, bacterial photosynthesis appears to be driven by a single light reaction (photosystem I in purple bacteria), and as a consequence it shows no enhancement effect.

Chloroplasts in higher plants are beautifully structured organelles; specific events tend to be associated with specific organized particles, or at least with separable fragments. Boardman and Anderson (1964) for example, demonstrated the existence of two separate categories of particles within the spinach chloroplast. The first had a chlorophyll *a*/*b* ratio of 2 and was primarily concerned with photosystem II (the one involved in the photolysis of H_2O). Their second particle had a chlorophyll *a*/*b* ratio of 6 and was responsible for photosystem I (a major site of ATP and $NADPH_2$ generation).

†For chemical names see appendix Table A-2.

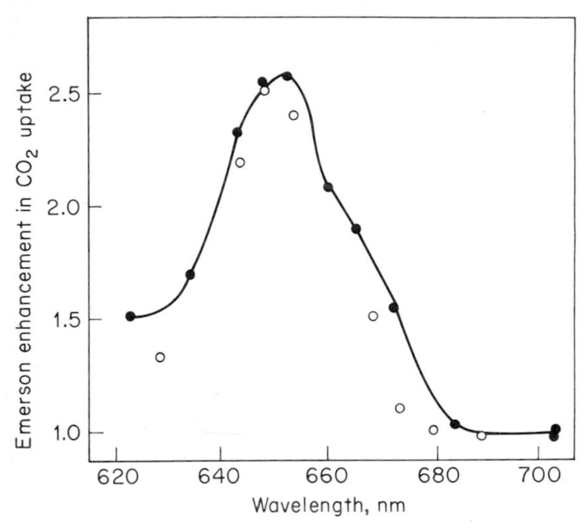

Fig. 1-6 The Emerson enhancement of photosynthesis depends upon the wavelength of the supplementary light; 650 nm superimposed on a steady supply at 700 nm produces the maximal enhancement in *Mimulus* (solid circles) and in *Solidago* (open circles) (Björkman, 1968).

Other forms of chlorophyll, as well as the accessory pigments of the chloroplast, are capable of deriving excitation energy from incident light, but for this energy to be useful in the photolysis of water and subsequent photochemical events, it apparently has to be channeled via these special forms of chlorophyll *a*. The chlorophyll molecule is equipped with highly mobile pi electrons, which allow a very rapid transfer of energy. Nevertheless, because trapped energy has to be channeled through common reducing centers (photosystems I and II), these can impose an upper limit on the overall rate of photosynthesis and can reduce the photosynthetic efficiency of the leaf, especially at high (saturating) light intensities.

Although there are divergent opinions regarding details of electron donors and acceptors during the initial stages of the light reactions following chlorophyll *a* excitation (Fig. 1-5), there is a general agreement that the production of ATP and $NADPH_2$ occurs in or on the thylakoids, whereas the conversion of CO_2 to carbohydrate takes place in the stroma of the chloroplast. An exchange of ATP or ADP and $NADPH_2$ or NADPH between thylakoids and stroma is therefore necessitated.

These suggested locations for the light and dark reactions of photosynthesis are based upon experiments with sonicated chloroplast preparations. If the crude preparation is centrifuged (145,000*g* for 30 min) to yield a pellet and straw-colored supernatant, the chlorophylls and carotenoids plus electron-transfer components are with the thylakoid fragments that constitute the pellet. The supernatant, on the other hand, contains soluble chloroplast protein which accounts for almost all of the RuDP carboxylase activity of the preparation. The supernatant is then able to fix CO_2 into phosphorylated intermediates if ATP and $NADPH_2$ are supplied, either exogenously or by preincubation with functional thylakoids.

Table 1-2 illustrates this point with data from experiments on CO_2 fixation by chlorophyll-free extracts vs. complete chloroplast systems from spinach. The combined photosystems I and II (Fig. 1-5) generate ATP by a light-induced electron cascade through a series of carriers. This process of photophosphorylation was recognized by Arnon (see Arnon, 1960, and literature cited),

Table 1-2 Assimilatory power from functional chloroplasts can drive subsequent CO_2 fixation in a chlorophyll-free preparation. This assimilatory power, that is, $NADPH_2$ and ATP, was generated in vitro by incubating a preparation of broken chloroplasts from spinach leaves under illumination but in the absence of bicarbonate. Chloroplasts were then removed by centrifugation, and the chlorophyll-free supernatant was retained. Subsequent dark fixation of $NaH^{14}CO_3$ by the remaining chlorophyll-free preparation was substantially increased by this pretreatment (Arnon, 1960).

Treatment	*$^{14}CO_2$ fixed, cpm*
A. Complete chloroplast systems	
In light	200,000
In dark	20,000
B. Chlorophyll-free preparation	
In dark (ATP + $NADPH_2$ from chloroplast system)	134,000
In dark (No supplement)	9,000

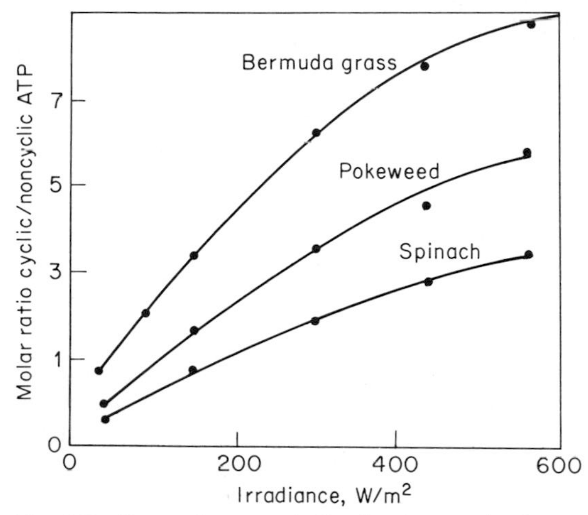

Fig. 1-7 Burmuda grass (a C_4 plant) not only shows a superior rate of photosynthesis compared with C_3 plants such as pokeweed and spinach, but it also generates a larger proportion of ATP via cyclic photophosphorylation (Chen et al., 1969).

who distinguished two forms, cyclic and noncyclic. It now appears (Rabinowitch and Govindjee, 1969) that noncyclic photophosphorylation can be identified with the main sequence of electron transfer shown in Fig. 1-5 and thus accompanies the Hill reaction, which involves the photolysis of H_2O and release of O_2. By contrast, cyclic photophosphorylation is the simple outcome of a back reaction in photosystem I indicated by dashed lines in the figure, and can proceed (even enhanced) when noncyclic photophosphorylation is inhibited with DCMU.

The question now arises whether the ATP and $NADPH_2$ generated photochemically are sufficient to drive the CO_2-fixing system. Bauchop and Elsden (1960) suggest that 6 mol of ATP is needed for each mole of CO_2 assimilated, 3 for reducing the CO_2 to the carbohydrate level, and another 3 to meet the energy demands of growth. Even if we restrict the present argument to the CO_2-assimilation phase, the question becomes especially relevant for C_4 plants because of their higher energy requirements for CO_2 fixation.

Bassham and Calvin (1957) estimated that 3 mol of ATP and 2 mol of $NADPH_2$ participate in the fixation of 1 mol of CO_2 by C_3 plants. But for C_4 plants, Hatch and Slack (1970) suggest that 5 mol of ATP plus 2 mol of $NADPH_2$ are required. Clearly, then, C_4 plants must have a higher capacity for photophosphorylation (presumably cyclic) to sustain their CO_2-fixing system. Chen et al. (1969) have furnished evidence on this point (see Fig. 1-7) by comparing the photophosphorylation performance of Bermuda grass (a C_4 plant) with that of spinach (a C_3 plant). The superiority of Bermuda grass chloroplasts for ATP formation in cyclic photophosphorylation is accentuated at the higher light intensities. This difference (Fig. 1-7) between the C_3 and C_4 plants in their response to light intensity resembles the comparative light response curves for photosynthesis (Fig. 1-8) in C_3 and C_4 plants.

This argument for a higher cyclic photophosphorylation capacity in C_4 plants is further developed by Black and Mayne (1970), who show that C_4 plants not only have a higher chlorophyll *a*/*b* ratio but that the P700 component of chloro-

Fig. 1-8 Bermuda grass (a C_4 plant) shows a greater rate of photosynthesis than orchard grass (a C_3 plant) and a higher light requirement for saturation (Chen et al., 1969).

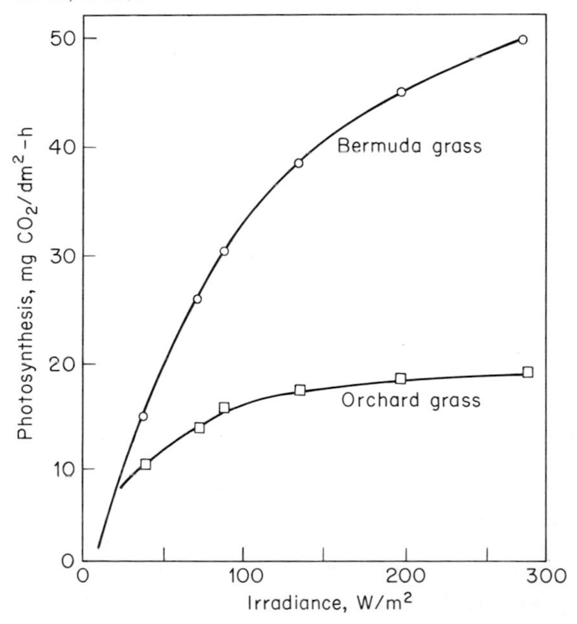

phyll *a* is also greater. Photosystem I, and thus cyclic photophosphorylation, seems to be abundantly supplied with the raw materials for ATP generation in these C_4 plants.

The absence of grana in bundle-sheath plastids has further implications. It could relate to their high light requirements for photophosphorylation, and since granal stacks are considered to be the primary locus of photosystem II within the chloroplast (Homann and Schmid, 1967), it is particularly noteworthy that bundle-sheath chloroplasts from leaves of *Sorghum bicolor,* which lose their grana during ontogeny, also lose photosystem II activity (Downton and Pyliotis, 1971). The two photosystems can become unlinked in these plastids as they mature, and their ability to photoreduce NADP diminishes accordingly (Bishop et al., 1971).

The efficiency of light utilization In a thermodynamic sense, photosynthesis is an unlikely event because a weak oxidant (CO_2) and a weak reductant (H_2O) must be converted into a strong oxidant (O_2) and a strong reductant (carbohydrate).

Van Niel has formulated photosynthesis with the photolysis of water as its primary photochemical event:

$$H_2O \rightarrow H^+ + OH^-$$

Subsequently,

$$4\ OH^- \rightarrow 2H_2O + O_2$$

The initial photolysis would therefore have to occur four times to allow evolution of one O_2 or to produce enough oxygen to match the assimilation of one CO_2 and maintain molar equivalence of O_2 and CO_2. The overall quantum requirement for photosynthesis would then be at least 4. If the primary photolysis of each H_2O required 2 quanta, photosynthesis as a whole would need 8 quanta. This value has been confirmed experimentally.

A controversy over the true quantum yield of photosynthesis that developed between Emerson and Lewis (1943) and Warburg and coworkers (1922 and later papers) led to extensive measurements of photosynthetic efficiency by scores of investigators from 1922 on. This intense and competitive interest had unexpected consequences: detailed observation on photosynthetic yield at specific wavelengths revealed a decrease in photosynthesis toward the red end of the spectrum that was more acute than the decline in absorption. If the light-absorbing pigments were equally effective for photosynthesis at all absorbed wavelengths, shouldn't the action spectrum coincide with the absorption spectrum?

However, this departure at the red end is more than offset if supplementary red light of a shorter wavelength is also provided. Photosynthesis now increases by an amount in excess of that due simply to the added energy; i.e., there has been an enhancement effect. This Emerson enhancement effect has a clearly defined action spectrum (Fig. 1-6) and implies that the two energy sinks within the photosynthetic apparatus are acting in a sort of booster reaction. The proposed scheme which embodies two such reactions was shown in Fig. 1-5.

If in fact these two light reactions do occur in such a way that the quantum energy of two photons is consumed in the photolysis of water, the overall quantum requirement for O_2 release can be calculated on this basis: it must be 8.

The consistent demonstration of a lower quantum requirement by Warburg and coworkers is especially difficult to reconcile with this calculation (8 quanta), which relies implicitly on Van Neil's formulation. The discrepancy led Warburg to adopt a different concept for photosynthesis, a complex between chlorophyll and carbonic acid (the photolyte) being regarded as the primary reactive unit. By computing photosynthetic yield purely on the basis of energy absorbed by this photolyte, Warburg et al. (1969) claim to have obtained photochemical equivalence: their quantum requirement for splitting this photolyte was always 1 and photosynthesis was calculated to be perfect!

Action spectra Measurements of quantum efficiency at a number of wavelengths easily become an action spectrum, but to measure quantum yield it is necessary to relate the number of molecules of CO_2 converted or O_2 generated to the number of photons absorbed; i.e., moles of reactant are divided by the einsteins of absorbed energy (an einstein is simply Avogadro's number, 6.02×10^{23}, of photons), so that

$$\text{Quantum yield } \Phi = \frac{\text{moles of } CO_2 \text{ converted}}{\text{einsteins absorbed}}$$

A light source or filter combination is required to give as nearly monochromatic light as possible, and then the measurements of photosynthesis are made at a similar photon flux throughout the spectrum. Meaningful estimates of Φ are best obtained at subsaturating intensities of monochromatic light because the effectiveness or incident energy is consistent over the linear phase of the light-response curve (Fig. 1-9*a*). Similarly, it is important to relate photosynthetic rate to absorbed energy (and not simply to incident energy) because the degree of absorption of incident energy is also wavelength-dependent (Fig. 1-9*b*).

Notwithstanding all these precautions, the measurement of Φ reveals an overall discrepancy between the leaf's absorption spectrum and the action spectrum for photosynthesis (compare Fig. 1-9*b*). This discrepancy of more absorption in the blue end of the spectrum per unit of photosynthetic activity is due largely to the presence of accessory pigments which in effect extend the wavelength range of energy that can be used by the plant. The excitation energy from each pigment system must still be routed via the common reductive center in the photosynthetic apparatus, and this immediately imposes a limitation on light utilization. With increases in light intensity above a certain level, photosynthetic efficiency inevitably declines (see Fig. 1-10). C_4 plants tend to be at an advantage here because they can generally utilize higher light intensities than C_3 plants (Fig. 1-11).

The decline in photosynthetic efficiency at higher light intensities is especially evident in continuous light (Fig. 1-10) but occurs in intermittent light. Emerson and Arnold investigated this problem in 1932.

Their findings are amenable to the concept of a photosynthetic unit in which excitation energy derived from a broad spectrum is fed to a common reduction center; the capacity of this center can then become rate-limiting. The actual size of this *photosynthetic unit,* in terms of the number of

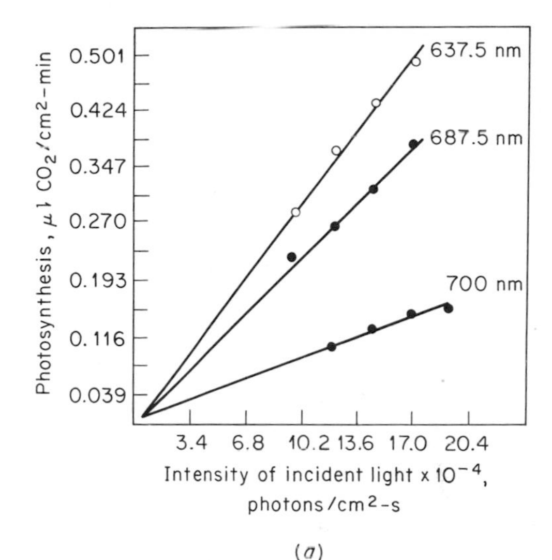

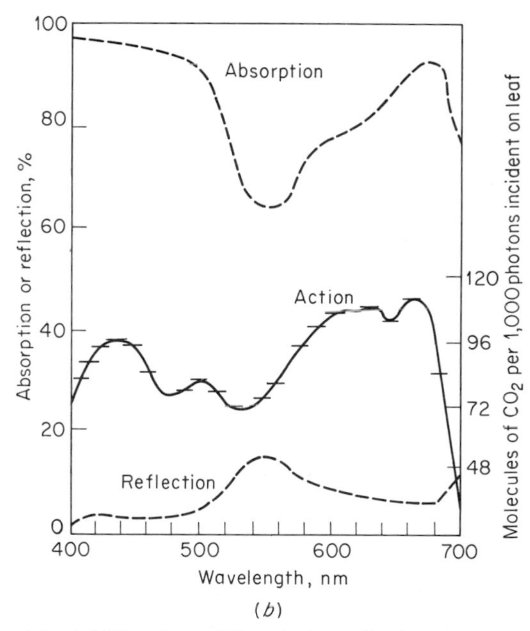

Fig. 1-9 (*a*) The slope of the photosynthetic response to increase in light yields a value for photosynthetic efficiency in terms of molecules of CO_2 consumed per 1,000 photons. (*b*) The measurements are repeated at various wavelengths to give an action spectrum (Balegh and Biddulph, 1970).

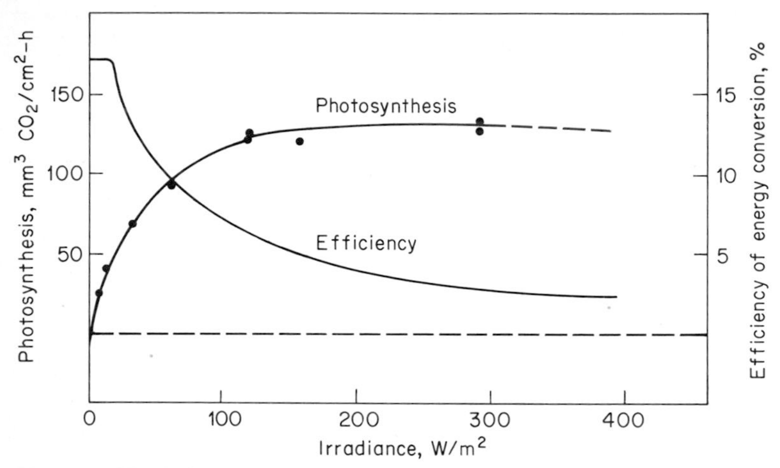

Fig 1-10 The light-response curve for photosynthesis is matched by a decline in efficiency as the photosynthetic response becomes curvilinear (Gaastra, 1958).

chlorophyll molecules per operative center, can be derived from laboratory measurements. Assuming a quantum requirement of 8, and further assuming that each chlorophyll molecule can act independently, a brief but intense flash of light which activates all the chlorophyll molecules should generate one molecule of O_2 for every eight molecules of chlorophyll. However, Emerson and Arnold's (1932) experiments showed only one O_2 from every 2,500 chlorophyll molecules. Apparently the chlorophyll molecules were acting in units of 2,500/8, that is, about 300.

Light saturation generally occurs when about 1 chlorophyll molecule in every 300 is absorbing energy. That additional energy is *not* usefully absorbed is the basic problem: once a given chlorophyll molecule in a photosynthetic unit becomes excited due to absorbed energy, no other chlorophyll (pigment) molecule in that same unit can usefully absorb any more until the initial charge is put to photosynthetic use or lost in fluorescence. Bonner (1962) has developed the argument that this system of a common reduction center for the utilization of absorbed energy imposes the final upper limit on photosynthetic yield. In his view, the light-harvesting antennae of the chloroplast offer greater potential for energy capture than can be utilized at reaction sites within the plastid. These organelles show peak efficiency in weak light (Chap. 15).

The *dark reactions,* where ATP and $NADPH_2$ drive CO_2 reduction, can impose an additional, and often rate-limiting, restriction on overall photosynthetic yield, especially at high light intensity. This was one reason for the use of light flashes, rather than continuous light, in experiments where the size of a photosynthetic unit was being estimated. In Fig. 1-10, for example, the departure from linearity of the light-response curve for photosynthesis shows that the dark enzymatic reactions are becoming overtaxed. Their capacity eventually imposes a ceiling on photosynthesis so that increased light intensity no longer elicits a response in the rate of photosynthesis.

Induction phenomena in photosynthesis find a related explanation: an organism's photosynthetic potential is not fully realized the moment it is illuminated; instead there is a brief time lag from the start of illumination to the attainment of steady CO_2 fixation. During this *induction period* the CO_2 assimilating machinery is being primed by the ATP and $NADPH_2$ generated in the light reactions. Oxygen will have been released, due to the photolysis of water, but without the concomitant fixation of CO_2; i.e., the dark reactions responsible for accruing intermediates of the carbon pathway have lagged behind the photochemical events (Rabinowitch, 1956).

These carbon-fixation systems can very easily become a study in themselves, but the main features of CO_2 assimilation are described in the next section.

In summary, the energy for CO_2 fixation is derived from a pair of closely coupled light reactions located on or in the internal membrane system of the chloroplast. The photochemical efficiency of the initial photolysis of water is such that two photons are required to split each H_2O molecule, but other "frictional" losses within the system raise the overall quantum requirement for complete photosynthesis. Efficiency is lost at elevated light intensities because excitation energy from a variety of pigments must be channeled through a common reduction center which then represents a bottleneck to energy utilization.

CO_2 assimilation The overall photosynthetic reaction is frequently stated as:

$$CO_2 + H_2O \xrightarrow{nh\nu} CH_2O + O_2$$

This is a misleading oversimplification both because it gives no hint of the wide array of compounds generated during the CO_2 reduction cycle and because it does not differentiate between the two categories of plants which employ the C_3 and C_4 fixation pathways.

The availability of long-lived ^{14}C-labeled compounds after World War II (Calvin et al., 1949) gave a new impetus to work on the chemical identity of CO_2-fixation products. Indeed, it is now difficult to imagine how details of metabolic pathways could have been established without the combination of these materials and paper-partition chromatography. Success was spectacular, and the research efforts of the Berkeley group resulted in a Nobel prize for Melvin Calvin in 1961.

The CO_2-fixation cycles in C_3 and C_4 plants have been amply illustrated (see for example, Bassham and Calvin, 1957; Stiller, 1962; Bassham, 1964; Hatch and Slack, 1970). Our prime concern here is not with the finite details of chemical transformations but the central significance of CO_2 assimilation as a starting point for all autotrophic growth. We also note the variety of intermediates in the CO_2-fixation sequence which can act as starting compounds for the synthesis of more complicated molecules.

C_3 plants Armed with the wisdom of retrospect (and a copy of Calvin's cycle), we can understand why the first labeling experiments, which sometimes lasted for hours, gave no clue to the initial fixation products. Sachs had asserted in 1862 that starch was the direct product of photosynthesis, but this knowledge was of limited use in ascertaining the pathway leading to its formation.

When more sophisticated experiments at Berkeley used algal suspensions (Bassham, 1965) and brief exposures to $^{14}CO_2$, the current picture of the CO_2-fixation pathway began to emerge. Typically, in 2 s, the labeled carbon is found principally in 3-phosphoglyceric acid; after 7 s label is located also in sugar monophosphate, diphosphates, and 3-phosphoglyceric acid; and after 60 s radioactivity appeared in a variety of sugar phosphates, plus phosphoenolpyruvic

Fig. 1-11 Sugarcane (a C_4 plant) shows both a higher rate of photosynthesis and a greater light requirement for saturation compared to C_3 plants (Hesketh and Moss, 1963).

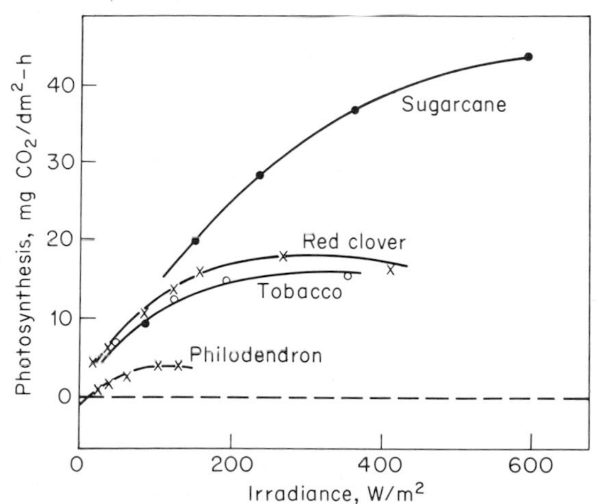

acid, carboxylic and amino acids, and sugars. The preponderance of phosphorylated intermediates in the early fixation products is demonstrated in Table 1-3, which shows some data on isolated chloroplasts.

Kinetic experiments involving brief exposures to $^{14}CO_2$ pointed to phosphoglyceric acid as the initial fixation product.

Subsequent kinetic experiments using a brief pulse of $^{14}CO_2$ followed by a longer chase with $^{12}CO_2$ indicated that some metabolites have a turnover time of only 5 s. The importance of short-term labeling experiments is self-evident.

More detailed information on the chemical steps of the carbon-fixation cycle came from the labeling patterns of various intermediates. Specific molecules were degraded carbon by carbon, and the position of the ^{14}C within the molecule helped analysts decide how that molecule could have been synthesized (Bassham and Calvin, 1957).

The proposed CO_2-fixation pathway (Bassham and Calvin, 1957) provoked intense interest, and, almost inevitably, some difficulties began to emerge (Stiller, 1962). First, cell-free chloroplast preparations capable of assimilating CO_2 showed fixation rates, on a chlorophyll basis, that were far below the rates achieved by intact tissues. Second, the enzyme thought to be primarily responsible for CO_2 fixation (RuDP carboxylase) showed a K_m well in excess of the CO_2 concentration expected to occur within the leaf. Moreover, the amount of enzyme present was inadequate to account for assimilation rates observed on intact material.

With improvement in the techniques for isolating chloroplasts, higher fixation rates were achieved. For example, nonaqueous methods of preparation minimized the loss of water-soluble enzymes (Stocking, 1959), and this favored greater physiological activity in the preparation. On the other hand, in aqueous preparations, the intact chloroplasts can act as an osmometer and will easily rupture in hypotonic media. The osmotic potential of the extracting medium must therefore be adjusted to that of the chloroplasts to avoid excessive buildup in turgor (see, for example, Shephard et al., 1968). Taking suitable precautions, Gibbs et al. (1967) reported chloroplast preparations that assimilate CO_2 at 235 mol per milligram of chlorophyll per hour compared to 180 to 200 for an intact spinach leaf. Similarly Shephard et al. (1968) were able to prepare fully active chloroplasts from *Acetabularia* capable (Table 1-4) of as much photosynthesis per unit of chlorophyll as intact leaves. This was the final proof, then, that chloroplasts are self-sufficient photosynthetic units. Bidwell (1972) confirms this assertion.

However, the problem of enzyme activity remains. We are confronted by the paradox of a leaf whose *overall* K_m for CO_2 is comparable to normal atmospheric CO_2 concentrations, having a CO_2-fixing enzyme whose K_m in vitro is higher by almost two orders of magnitude, as illustrated in Table 1-5.

Clearly, the RuDP enzyme has a higher affinity for CO_2 than for HCO_3^-. Cooper et al. (1969) have even provided evidence that CO_2 is the more likely substrate in vivo. But, irrespective of the

Table 1-3 $^{14}CO_2$ fixation products change their distribution from phosphoglyceric acid toward sugar phosphates during the first few minutes of exposure to $^{14}CO_2$ (Arnon, 1960).

Illumination, min	Total $^{14}CO_2$ fixed, 10^{-3} cpm	Total ^{14}C fixed	
		Phosphoglyceric acid, %	Sugar phosphates, %
1	108	100	0
2	196	75	24
5	450	54	38
10	600	40	57
20	750	13	86

Table 1-4 $^{14}CO_2$-fixation products in isolated chloroplasts resemble those in whole cells, indicating the self-contained nature of the plastids (Shephard et al., 1968).

Photosynthesis in $^{14}CO_2$ for 10 min	*Acetabularia mediterrania*	
	Whole cells	Chloroplasts
^{14}C, nCi/mg chlorophyll:		
Alcohol-soluble	264	202
Alcohol-insoluble	303	285
Alcohol-soluble compounds, % of total soluble radioactivity:		
Sugar phosphates	8.3	5.0
Phosphoglyceric acid	5.9	4.5
Glycolic acid	9.4	11.0
Malic acid	2.6	3.1
Sucrose	31.6	20.8
Glucose	14.8	17.6
Aspartic acid	8.9	12.8
Glutamic acid	5.9	2.7
Alanine	12.6	22.4

precise nature of the carbon source for this enzyme, we must conclude that either RuDP carboxylase behaves very differently in vitro, where all substrates except CO_2 are nonlimiting, and in vivo, where other photosynthetic intermediates influence activity; or else there is some kind of CO_2-concentrating mechanism at the photosynthetic site.

While we do have reason for retaining the first possibility (Bahr and Jensen, 1974), we also have evidence for a CO_2-concentrating mechanism: carbonic anhydrase catalyzes the reversible hydration of carbon dioxide:

$$\underbrace{CO_2 + H_2O}_{} \rightleftharpoons \underbrace{HCO_3^- + H^+}_{}$$
$$\leftrightharpoons H_2CO_3 \leftrightharpoons$$

Enns (1967) has shown that the rate of CO_2 diffusion can be increased by this enzyme—a hundredfold in a model system. Carbonic anhydrase is known to occur in plant cells, although its exact location differs between C_3 and C_4 plants (Everson and Slack, 1968). Everson and Slack (1968) and Bidwell et al. (1969) have each provided evidence suggesting its participation in CO_2 fixation.

The wide occurrence of carbonic anhydrase in higher plants is in itself suggestive evidence for a physiological role, and it is still conceivable that some equilibration between HCO_3^- and CO_2 which could occur as part of the photosynthetic reaction in the cell is facilitated by the enzyme.

C4 plants The problem of reconciling observed rates of CO_2 uptake in intact leaves with measured levels of enzyme activity (especially RuDP carboxylase) is even more accentuated in this group because tropical grasses (and some dicotyledonous genera) show particularly high rates of photosynthesis. The situation is further complicated by the nature of $^{14}CO_2$-fixation products in these plants. Kortschak et al. (1965) demonstrated that organic acids, especially malate, account for most of the fixed $^{14}CO_2$ in short-term experiments on sugarcane leaves. Even under mildly varying environmental conditions, and irrespective of sugarcane variety, most of the

Table 1-5 The affinity of carbon-fixing enzyme systems depends upon the nature of the substrate. RuDP carboxylase shows a distinct preference for CO_2.

Enzyme	Substrate	K_m†	Reference
RuDP carboxylase	CO_2	0.45 mM	Hatch and Slack, (1970)
	HCO_3^-	20.00 mM	Hatch and Slack, (1970)
PEP carboxylase	HCO_3^-	0.40 mM	Hatch and Slack, (1970)
Overall CO_2 assimilation by intact leaf	CO_2	300 ppm	Goldsworthy (1968)
CO_2 solubility in H_2O at 300 ppm		0.008 mM	Hatch and Slack (1970)

†The K_m value for the enzyme indicates the substrate concentration yielding one-half the saturated rate. A high affinity between enzyme and substrate leads to a low K_m. Within the leaf, the likely CO_2 concentration in the aqueous phase is still far below the K_m for any of the CO_2-fixing systems.

$^{14}CO_2$ assimilated in 7 to 15 s was fixed as malic and aspartic acids (Kortschak and Hartt, 1966).

It seemed curious that an organic acid should be an early fixation product; this was something more typical of dark CO_2 fixation, as in crassulacean plants (Beevers et al., 1966). Calvin's work had in a sense established a kind of precedent, and a three-carbon compound, or at least something more closely related to a sugar, had become regarded as a typical fixation product for a photosynthesizing organ. In the Crassulaceae, PEP carboxylase is operative for CO_2 fixation; if this were also the case for sugarcane leaves, some formation of ^{14}C-labeled oxalacetate was to be expected, but Kortschak and coworkers had found none.

The anomaly was resolved by a research group at the David North Plant Research Laboratory in Brisbane, Australia, where the work of Hatch and Slack (1966 and subsequent papers) clarified the issue. They used high-specific-activity $^{14}CO_2$, and their exposure periods were particularly short, down to 1s; but, more important, the extract was treated with 2,4-dinitrophenylhydrazine. Any oxalacetate in the extract was thereby converted into a more stable derivative. Oxalacetate then made its appearance as an early fixation product and confirmed the suspected operation of PEP carboxylase. The fixed carbon thus makes its way eventually to sugar via carbon atom 1 of a four-carbon organic acid (see Fig. 1-12).

Further work by Slack and Hatch (1967) established that PEP carboxylase activity in these C_4 plants is 60 times more active (per unit of chlorophyll) than in typical C_3 plants, e.g., wheat, oats, or sugar beet (see Table 1-6). Moreover, this enzyme adopts a different form according to species (Ting and Osmond, 1973*b*), while its metabolic function also shows species dependence. In C_4 plants the enzyme makes malate and/or aspartate as an intermediate, whereas the enzyme in C_3 plants produces malate and/or aspartate as a photosynthetic product (Ting and Osmond, 1973*a*). These related observations underlined the central significance of PEP carboxylase for CO_2 assimilation by C_4 plants.

Not only is the C_4 fixation pathway highly characteristic of the tropical grasses examined to date and of a limited number of dicotyledonous genera, but it seems to be further restricted to certain chloroplasts within these plants. Earlier in this chapter (Fig. 1-3) we referred to the two classes of chloroplasts in C_4 plants; those with numerous well-developed grana but no starch grains, which occur in the mesophyll cells, and those largely devoid of grana but capable of manufacturing starch grains, which occur in the parenchyma cells of the vascular bundle sheath. Making use of the slight difference in density between these two types of plastids, Slack et al. (1969) achieved reasonable separation and were able to study the enzyme complements of these differing chloroplasts. Table 1-6 illustrates the point; C_4 activity occurs primarily in the mesophyll tissue, while C_3 activity, gauged by the higher level of RuDP carboxylase, is essentially a feature of the bundle-sheath plastids.

The integration of these functionally different chloroplasts toward a more efficient assimilation and transport of carbon by the leaf as a whole is discussed further by Hatch (1972). Reactions of the C_4 pathway are shown in simplified form in Fig. 1-13.

CO_2 is fixed initially in mesophyll chloroplasts, where PEP serves as acceptor. The reaction sequence leads to the four-carbon acids malate and aspartate. One or other of these acids [depending upon species (Downton, 1970)] migrates to the bundle sheath, where the four-carbon acid is decarboxylated to yield CO_2 and a three-carbon compound. PEP in the mesophyll is regenerated from the three-carbon compound while the CO_2 released internally is refixed through the action of RuDP carboxylase. The great abundance of specific forms of PEP carboxylase in the mesophyll (Ting and Osmond, 1973*a,b*) in combination with its high affinity for CO_2 (Hatch, 1972) equips the C_4 plant with an impressive system for CO_2 assimilation. This division of labor between chloroplasts in the bundle sheath and those in mesophyll cells was anticipated almost a century ago by Haberlandt (1884). What is now known to be an intricate complex of biochemical and photochemical events has been described more recently as *cooperative photosynthesis* (Karpilov, 1970).

Fig. 1-12 Brief exposure of sugarcane leaves to radioactive CO_2 reveals that malate is one of the first major photosynthetic products (Hatch and Slack, 1966).

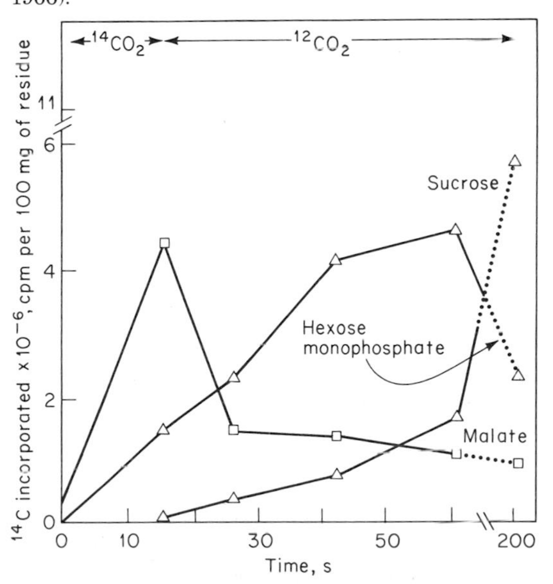

Table 1-6 The general pattern of enzyme activity as related to CO_2 assimilation varies between C_3 and C_4 plants (Hatch and Slack, 1970).

Enzyme	Relative activity	
	C_4	C_3
PEP carboxylase (mesophyll)	60	1†
RuDP carboxylase (bundle sheath)	1	1‡
Carbonic anhydrase	1	8
Glycolate oxidase	1	15

†Huang and Beevers (1972) have confirmed such compartmentation for this enzyme. In C_4 plants *Atriplex rosea* and *Sorghum sudanense*, PEP carboxylase was largely confined to mesophyll tissue, whereas RuDP carboxylase was located principally in bundle-sheath cells.

‡Relative activities of RuDP carboxylase in C_3 and C_4 plants was originally published (Hatch and Slack, 1970) as a ratio of 1:10, but the low level in C_4 plants was subsequently found to be an artifact of the extraction method. C_3 and C_4 plants are now regarded as distinctive with respect to the location of this enzyme. In C_4 plants, mesophyll cells are devoid of RuDP carboxylase. This enzyme is wholly located in the bundle sheath (Hatch et al., 1971; Slack et al., 1969).

As stated in Table 1-5, RuDP carboxylase has a low affinity for CO_2 (at least in vitro), and concentrations within a leaf would be totally inadequate to sustain maximum fixation if CO_2 levels depended entirely on inward diffusion. C_3 plants might circumvent this problem through the action of carbonic anhydrase (Graham et al., 1971), but in C_4 plants the mechanism seems clear; reactions within the mesophyll serve to concentrate CO_2 for subsequent fixation by RuDP carboxylase within the bundle-sheath plastids. As shown in Fig. 1-13, CO_2 fixed by the action of PEP carboxylase is subsequently transported in the form of a four-carbon acid to the bundle sheath. CO_2 is in effect concentrated at this internal site, and direct evidence is available that a

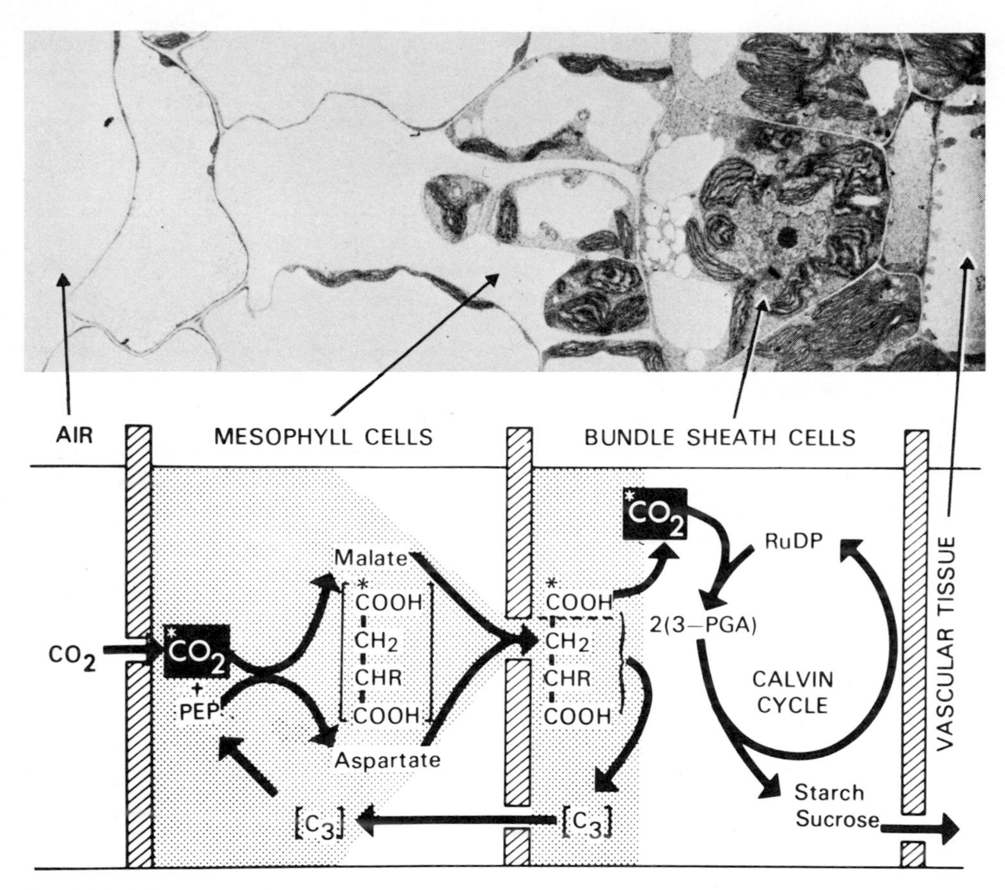

Fig. 1-13 The successful operation of the C_4 pathway of photosynthesis rests on a combination of leaf anatomy and biochemistry. The chloroplasts of mesophyll and bundle sheath cells make distinctive contributions to CO_2 assimilation (Hatch, 1972).

substantial pool of CO_2 does exist within C_4 leaves under illumination. These internal levels of CO_2 are much greater than could be expected by physical diffusion from ambient air and do in fact seem adequate to overcome the low substrate affinity of RuDP carboxylase (Hatch, 1972).

GAS EXCHANGE BY A LEAF

Atmospheric CO_2 reaches the chloroplast by diffusion, and the rate of its diffusion depends upon both the steepness of the CO_2 concentration gradient and the resistance offered by the leaf. However, the source of photosynthetic CO_2 can be either internal or atmospheric, and this complicates our analysis. As light energizes a fixation of CO_2, there is often a concomitant enhancement of respiratory CO_2 production, perhaps in part associated with the mass-action effect of CO_2 removal. The CO_2 exchange in photosynthesis is thus a combination of internally recycled CO_2 and a net flux of CO_2 from the outside air. The derivation of these fluxes and resistances allows a better understanding of the rate-limiting processes for photosynthesis, e.g., stomatal aperture vs. internal restraints; hence our interest.

This present section will therefore cover gas exchange by a single leaf and will refer to the environmental factors that influence this exchange. The response of a whole community of leaves to the environment is considered in Chap. 3.

Photorespiration

Despite its photosynthetic origin, O_2 causes a substantial inhibition of photosynthesis. Warburg (1920) reported such an effect for *Chlorella*, and McAlister and Meyers (1940) subsequently confirmed this same response in wheat. Both O_2 evolution and CO_2 assimilation were reduced in the presence of atmospheric O_2, and the phenomenon became known as the *Warburg effect* (its counterpart, the *Pasteur effect* was first recognized in 1876 as an inhibition of sugar breakdown by O_2). The Warburg effect in using a spinach-chloroplast preparation is illustrated in Fig. 1-14. Oxygen causes greatest inhibition when CO_2 levels are low and light levels are saturating. In fact, as shown in Fig. 1-14, the inhibition can be relieved by a sufficiently high HCO_3^- concentration.

Since the O_2 inhibition is maximal when photosynthesis is light-saturated, it is conceivable that the O_2 is interfering with the generation of reducing power ($NADPH_2$), or, as Warburg (1920) suggested, the O_2 may be reoxidizing a primary photochemical product and thus competing with CO_2 for this product's reducing power. However, Ellyard and Gibbs (1969) demonstrated that generation of reducing power is unaffected by O_2 and concluded that for most environmental conditions this process would not be rate-limiting.

To understand this Warburg effect, some related observations on the gas exchange of whole leaves must now be considered. Krotkov and coworkers at Kingston, Canada, repeatedly noticed that respiratory CO_2 evolution is higher in the light than in darkness and that there is a postillumination burst of CO_2 output, especially at higher oxygen concentrations (Figs. 1-15 and 1-16). Krotkov (1963) introduced the term *photorespiration* to differentiate between these apparently separate forms of CO_2 evolution. The distinction between dark respiration and photorespiration also shows up as a differential sensitivity to O_2, temperature, and metabolic inhibitors and in the specific activity of respiratory substrates following photosynthesis in $^{14}CO_2$.

As well as stimulating CO_2 evolution in the light, O_2 also caused a depression of net photosynthesis (Forrester et al., 1966), so that additional reasons for the reduction in photosynthesis suggest themselves:

Fig. 1-14 Photosynthetic assimilation of $^{14}CO_2$ by chloroplasts is inhibited by oxygen; the effect is counteracted by higher bicarbonate concentrations (Ellyard and Gibbs, 1969).

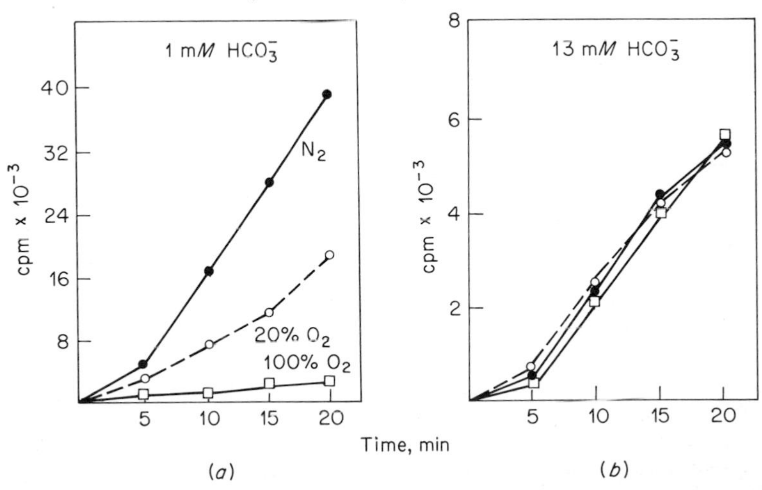

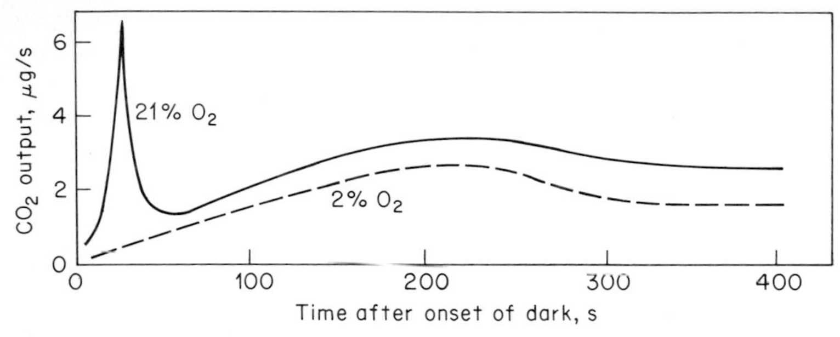

Fig. 1-15 The postillumination burst of dark CO_2 output by detached tobacco leaves is stimulated by oxygen (Tregunna et al., 1966).

1 The CO_2 evolved in photorespiration lowers the concentration gradient which draws CO_2 into the leaf.

2 If the precursor for the CO_2 evolved comes from the Calvin cycle, this will lower the concentration of potential CO_2 acceptor molecules.

3 The photorespiratory CO_2, by recycling into the CO_2-fixing system, competes with CO_2 derived from the atmosphere.

4 O_2 causes a direct and indeed competitive inhibition of RuDP carboxylase.

A reduction in net photosynthesis then becomes apparent.

A key finding in this photorespiratory jigsaw was the observation that glycolate synthesis is enhanced by O_2 (Gibbs, 1970). Glycolate had previously been recognized as a product of photosynthesis, and was commonly observed to be excreted into the medium by chloroplasts or algae, especially in the presence of certain inhibitors such as isoniazid (see Fig. 1-17). Despite the stimulated release of glycolic acid referred to in Fig. 1-17, CO_2 assimilation by the *Chlorella* is unaffected by the poison.

Although chloroplasts produce glycolate, they are virtually unable to metabolize ^{14}C glycolate supplied exogenously; i.e., they do not possess glycolate oxidase (Jackson and Volk, 1970). This observation has now to be reconciled with the very low levels of glycolate found in leaves: if the chloroplast is synthesizing this acid, something must be happening to it outside of the chloroplast.

The *peroxisome* offers an explanation. (see Fig. 1-2). This organelle occurs in the cytoplasm, conceivably appressed to the chloroplast, and it contains glycolate oxidase, glyoxylate reductase, catalase, and some malate dehydrogenase. Tolbert and coworkers (Tolbert et al., 1969) found peroxisomes in all leaf homogenates tested, although photorespiratory enzymes were particularly low in C_4 plants.

While peroxisomes account for oxidation of glycolate and the associated O_2 requirement for photorespiration, the light-induced evolution of CO_2 is still not explained because ^{14}C glycolate supplied to peroxisomes does not result in $^{14}CO_2$ evolution (Jackson and Volk, 1970). A direct decarboxylation of glyoxylate could occur within the chloroplast, but CO_2 is also generated within the cytoplasm. It seems that the peroxisome generates ^{14}C glycine from added ^{14}C glycolate, and upon entry into the cytoplasm of the cell two molecules of glycine react to produce serine plus CO_2. Hence the previously observed effect of *Chlorella*, that inhibition of transaminase activity causes a buildup of glycolate in the medium, is explained. The sequence of reactions envisaged for higher plants is summarized schematically in Fig. 1-18. Although there is doubt about the initial steps leading to glycolate synthesis (Gibbs, 1970), the overall picture is sufficiently clear to explain in general terms the overall interaction between photosynthesis and photorespiration. Since the oxidation of glycolate

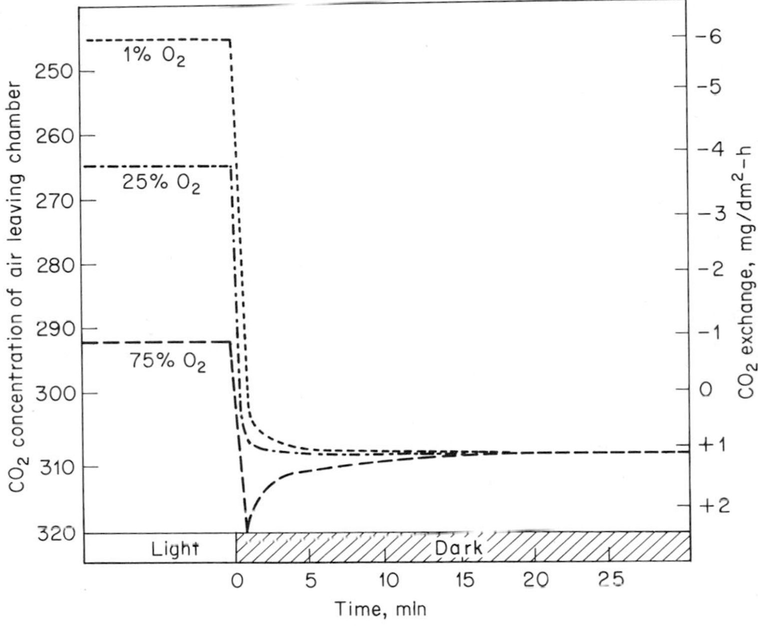

Fig. 1-16 Oxygen reduces photosynthesis and stimulates the postillumination burst in CO_2. Both effects relate to photorespiration (Egle and Fock, 1967).

to glyoxylate has an absolute requirement for O_2 which is not met by O_2 produced during photosynthesis (Gibbs, 1970), we have an explanation of the O_2 stimulation of photorespiration and associated reduction in photosynthesis.

During photosynthesis by algae and by C_3 plants, up to 50 percent of the carbon fixed may have to pass through the glycolate cycle en route to sucrose, so that photorespiration must be considered as more than simply an addendum to photosynthesis. Hatch and Slack (1970) adopt the view that photorespiration is a metabolic adjunct to the Calvin cycle. Since the oxidation of glycolate is not coupled to ATP formation, this process of photorespiration seems doubly wasteful, an unlikely situation for a higher plant. A positive function for photorespiration remains obscure.

In an illuminated leaf, we can visualize two opposing processes: photosynthesis consuming CO_2 and photorespiration generating CO_2. Their relative rates and proximity within the cell will govern the steady-state CO_2 concentration inside the leaf. As the CO_2 level declines due to photosynthesis, CO_2 will become rate-limiting. But at the same time, photorespiration will be encouraged, especially at normal O_2 ten-

Fig. 1-17 Glycolic acid secretion by *Chlorella* is stimulated by isoniazid, an inhibitor of transaminase activity (Whittingham and Pritchard, 1963).

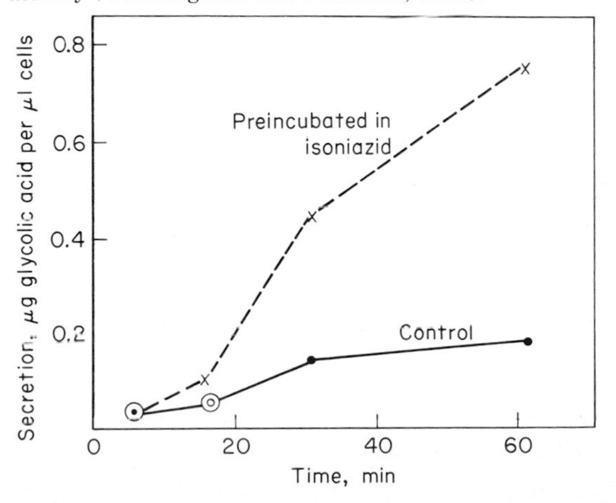

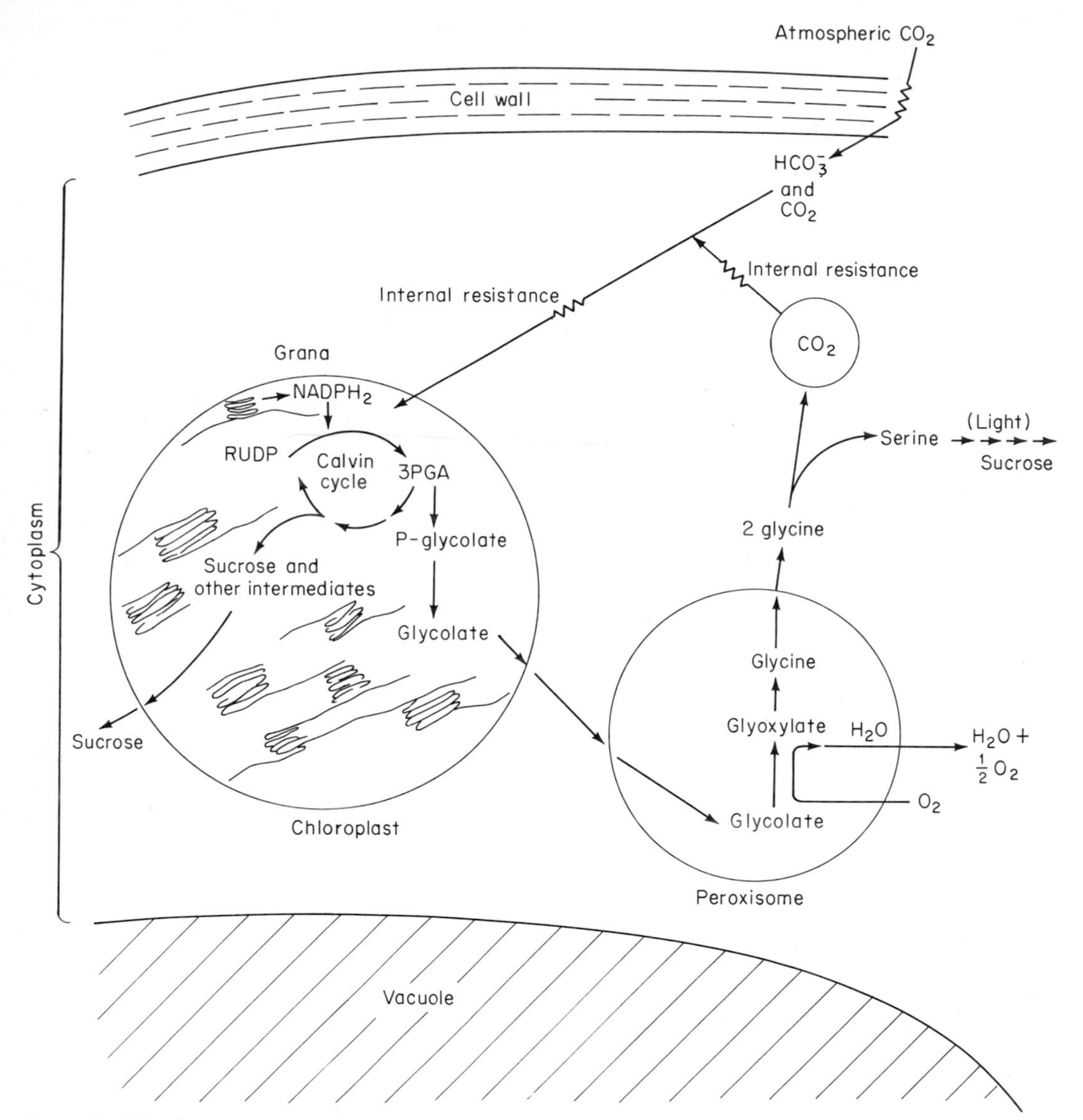

Fig. 1-18 The participation of chloroplasts and peroxisomes in formation and dissimilation of substrates during photorespiration (highly schematic).

sions or above. Since photorespiration depends upon current photosynthesis for its glycolate substrate, the traffic in CO_2 (consumption vs. generation) will achieve a balance for any given set of environmental conditions.

This predicted situation is borne out in experiments on intact leaves held in an illuminated assimilation chamber. As the air in the chamber is recirculated over the leaf and through a CO_2 analyzer, the level declines to a steady-state value termed *compensation point* Γ. Miller and Burr (1935) first made such an observation, and the parameter Γ has been subsequently analyzed in great detail by Meidner and other physiologists at Reading. Although highly sensitive to temperature and moisture stress, as well as other plant factors (see Jackson and Volk, 1970, and literature cited), Γ shows a surprising consistency at about 50 ppm CO_2 for C_3 plants held in a mild environment. Admittedly there is some variation between different genera of C_3 plants, but, more remarkably, C_4 plants show Γ values consistently close to *zero*. This observation is in line with the virtual absence of photorespiration in C_4 plants. The ability of C_4 plants to exhaustively remove CO_2 from ambient air makes them powerful scavengers of CO_2.

Diffusive resistances Water evaporates from a leaf because of a vapor-pressure gradient from the leaf to the air, but the rate of evaporation is less than from a free water surface because of the diffusive resistance offered by the leaf. Renner applied physical principles in analyzing this situation as early as 1910 and showed how to derive the magnitude of boundary-layer resistance. Subsequently the subject has been given more detailed attention, and accurate mathematical descriptions of both H_2O and CO_2 fluxes have been made (see, for example, Raschke, 1956; Gaastra, 1959).

If we know the concentration of H_2O vapor in the air around a leaf and the rate at which it is transpiring we can calculate the total diffusive resistance of water loss, commonly termed Σ_r:

$$\Sigma_r = \frac{A - B,\ \mu\text{g/cm}^3}{\text{transpiration},\ \mu\text{g/cm}^2\text{-s}}$$

where A is water-vapor concentration in leaf and B is water-vapor concentration in air. This resistance (Σ_r) will then have the curious dimension of seconds per centimeter. Conductance would of course have the reciprocal dimension, (centimeters per second), and sometimes this term is easier to grasp.

In the equation above, water vapor is presumed to emerge from the substomatal cavity of a leaf, and we assume that the atmosphere there is saturated with water vapor. The water-vapor concentration inside the leaf would then be equivalent to the saturation vapor pressure at the particular leaf temperature. Under strongly desiccating conditions or where solute accumulation on the evaporation surface of the substomatal cavity is likely, as with some arid-zone species (Whiteman and Koller, 1964) internal saturation is not maintained (see Chap. 17). For present purposes we simplify the issue and take the case of a well-watered mesophyte.

Gas exchange is shown diagrammatically in Fig. 1-19. The term Σ_r will be the straight summation of boundary-layer resistance r_a and stomatal resistance r_s. Since these two resistances are in series, they can be summed (as in electrical theory), but to isolate the term r_c (cutiular resistance) in Fig. 1-19 we would need to deal with reciprocal resistances because these two resistances are in parallel (as with the electrical analog). To simplify the calculations, the term *leaf resistance* r_l is commonly derived to represent the summation of r_s and r_c, so that Σ_r becomes $= r_a + r_l$. Boundary-layer resistance can in turn be calculated from measurements of evaporation of filter-paper replicas which have been moistened and then held in the normal position in the leaf chamber.

CO_2 entering the leaf will have to overcome these same resistances to reach the substomatal cavity, but the CO_2 molecules are then confronted by additional impedances, both physical and biochemical, before being fixed within the chloroplasts. To measure the total diffusive resistance to CO_2 entry Σ_r' we would need to know the CO_2 concentration gradient and the rate of CO_2 assimilation. The external CO_2 concentration is easily measured, as is the rate of CO_2 uptake, but what is the CO_2 concentration at the site of photosynthesis? Gaastra (1959) presumed that it would be very low, and for the purpose of calculating Σ_r' took a value of zero. However, we know that photorespiration will generate a certain CO_2 tension, and furthermore the fixation of CO_2 in a chemical reaction will necessitate something greater than zero substrate.

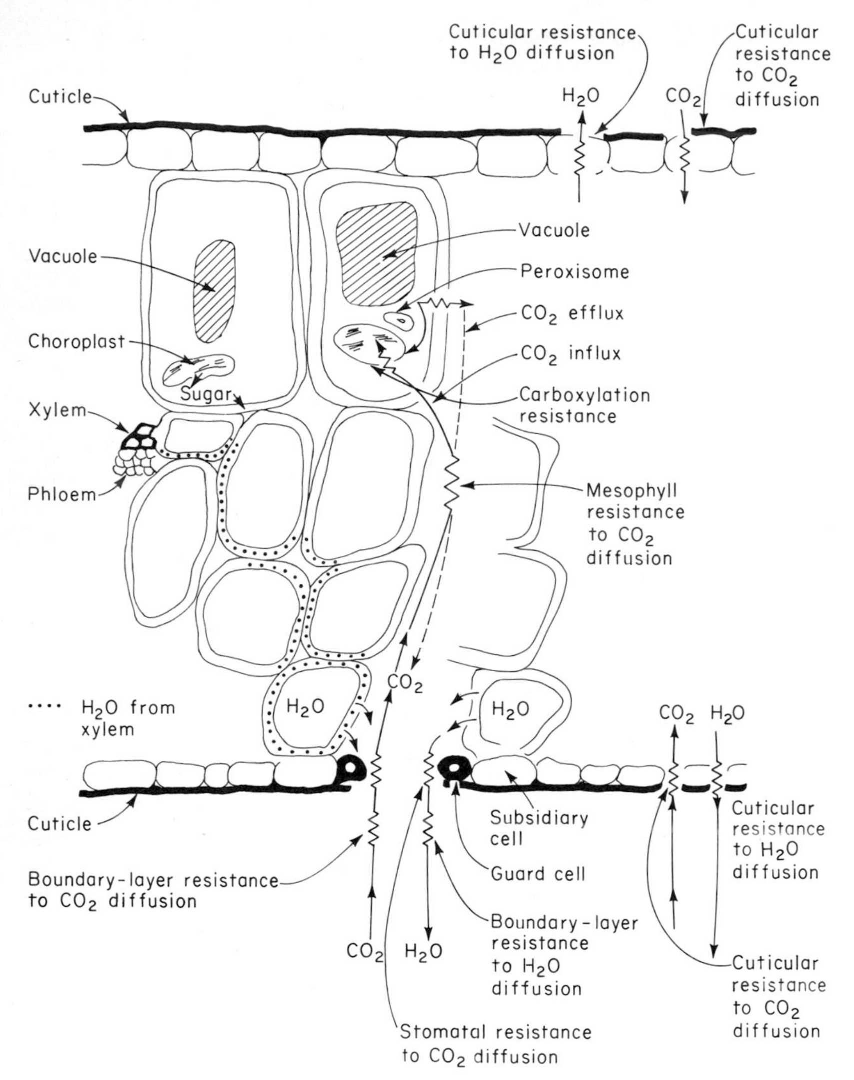

Fig. 1-19 A highly schematic view of component resistances to H_2O and CO_2 exchange.

To minimize photorespiration, it has become more common to measure photosynthesis in O_2-free air and then take the internal CO_2 concentration as zero. Holmgren et al. (1965) devised the interesting tactic of measuring the photosynthetic response to CO_2 in O_2-free air and at light saturation and derive Σ_r' from the slope of the response. Low Σ_r' will allow a steeper response and vice versa (Holmgren et al., 1965). These two approaches to the calculation of Σ_r' in O_2-free air can yield comparable results (Kriedemann, 1971). Σ_r' is then the summation of r_a' and r_l' plus the internal resistances mentioned previously.

Since CO_2 entering the substomatal cavity has to traverse the same pathway as the escaping H_2O molecules, the corresponding resistances for CO_2 diffusion (r_a' and r_l') can be derived by correcting for the difference in diffusion coefficients D between H_2O and CO_2 molecules. The ratio D_{H_2O}/D_{CO_2} is 0.258/0.165 cm²/s, so that r_a' and $r_l' = 1.56r_a$ and r_l. To allow for some degree of turbulent transfer, as opposed to true diffusion in the boundary layer, Gale and Poljakoff-Mayber (1968) prefer the ratio of 1.35 when deriving r_a'. These corrections (and more particularly r_l to r_l') can be made with greater confidence at lower transpiration rates because there is less likelihood of mutual interference between CO_2 molecules entering and H_2O molecules emerging from the stomatal pore (Parkinson and Penman, 1970).

Having estimated r_a' and r_l', we are now in a position to examine these *internal* resistances to CO_2 diffusion according to the relationship

$$\Sigma_r' = r_a' + r_l' + r_m.$$

Residual resistance r_r can be considered as wholly internal; it is synonymous with the term mesophyll resistance r_m, which occurs widely in the literature (for example, Gaastra, 1959). Meidner (1969) and others have deplored the use of r_m in the present context because it was originally intended (Penman, 1942, cited by Meidner, 1969) for the resistance to CO_2 diffusion in the vapor phase within the leaf. In a sense we are begging the question here and adopting r_r as the collective term which embodies the physical and biochemical resistance to CO_2 assimilation within the leaf.

The diffusive resistances shown in Table 1-7 indicate that photosynthesis is more likely to be limited by internal processes r_r than by stomatal capacity. The use of antitranspirants as an aid to increased water-use efficiency was anticipated on that basis (Slatyer and Bierhuizen, 1964). According to diffusion theory, r_l could be increased by chemical means; assuming that r_r is not affected proportionately, one would expect transpiration to decline more than photosynthesis. Some success has been predicted for such an application (Parkinson, 1970), but in most cases the treatment required to limit transpiration significantly has had undesirable side effects (Gale and Hagan, 1966; Kriedemann and Neales, 1963).

Since water-vapor loss is governed by Σ_r and CO_2 uptake by Σ_r', the water-use efficiency of a leaf, under given environmental conditions, can be specified by the ratio Σ_r/Σ_r'. This quantity varies (0.21 to 0.34) between different classes of plants (Holmgren et al., 1965) and shows some promise as a means of predicting potentially efficient plants for arid regions.

Table 1-7 The magnitudes of stomatal vs. internal resistances and absolute rates of photosynthesis are species-dependent. Control systems for gas exchange vary accordingly (El-Sharkawy and Hesketh, 1965).

Species	Photosynthesis, mg CO_2/dm²-h	Transpiration, g H_2O/dm²-h	Stomatal resistance r_s, s/cm	Residual (internal) resistance† r_r, s/cm
Corn	63 ± 2	3.3 ± 0.2	1.5	1.0
Sunflower	50 ± 1	3.5 ± 0.2	1.7	1.5
Cotton	38 ± 1	3.1 ± 1.0	2.0	2.9
Oats	31 ± 3	3.6 ± 1.6	1.7	4.1
Hibiscus	23 ± 5	3.1 ± 0.7	1.9	7.3
Tobacco	21 ± 5	2.3 ± 1.0	3.6	5.5
Soybean	25 ± 1	2.3 ± 0.6	2.7	5.5
Thespesia	18 ± 5	2.8 ± 0.5	1.1	9.7

†Often quoted as r_m, that is, mesophyll resistance.

Equipped with reliable estimates of the various diffusive resistances shown in Fig. 1-19 and Table 1-7, one can formulate a mathematical model to describe and perhaps predict photosynthetic rates under stated environmental conditions (see, for example, Brown, 1969). By utilizing this general form of model making, Lake (1967) successfully predicted the higher respiratory rates now known to occur in the light.

Further apparent differences between C_3 and C_4 plants also became evident when their gas-exchange rates were analyzed in terms of diffusive resistances. As mentioned in connection with Table 1-7, the photosynthetic activity of C_3 plants is commonly limited by r_r, especially in the slower-growing woody perennials (Jarvis and Jarvis, 1964; Holmgren et al., 1965). In C_4 plants, however, internal resistance r_r is so low that it is comparable in magnitude to stomatal diffusive resistance r_s. For all practical purposes, then, stomatal aperture regulates both transpiration and photosynthesis. We would therefore expect no improvement in water-use efficiency in such a plant by controlling stomatal aperture, a prognosis borne out by experience (Gale and Hagan, 1966).

Changes in leaf efficiency which are associated with *age* can also be diagnosed by separating stomatal from internal controls in terms of diffusive resistance. Leaf photosynthesis often achieves a peak at about the time the lamina reaches full size and then undergoes a decline. Hardwick et al. (1968) suggest that the subsequent decline is partly determined by an increase in mesophyll resistance (Fig. 1-20) although r_l can also increase. A loss in stomatal responsiveness by aging leaves has already been described (Cowan and Milthorpe, 1968).

An analysis of photosynthetic responses to these internal and to other environmental factors, in terms of diffusive resistances, becomes the topic of our next section.

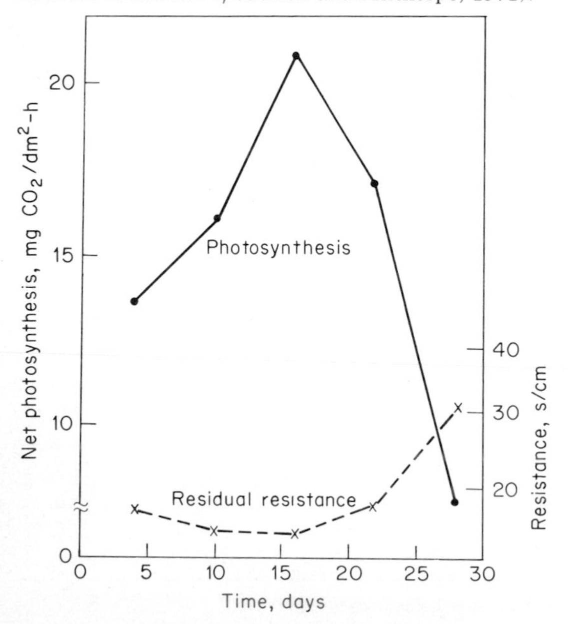

Fig. 1-20 Wheat leaves show an increase in photosynthetic activity during the course of their expansion and reach peak rates about 15 days after emergence. Subsequent decline with age correlates with increased residual resistance r_r (Osmon and Milthorpe, 1971).

CONTROL SYSTEMS FOR PHOTOSYNTHESIS

For convenience, factors regulating photosynthesis can be divided into two broad categories, internal and environmental. Internal factors refer to systems which provide some measure of self-regulation, whereas environmental factors refer to those which can be manipulated from outside the leaf, e.g., light, CO_2 concentration, and water supply. Under environmental we should also include those factors which tend to influence the rate or the end products of photosynthetic reactions, including temperature, O_2 concentration, nutrient supply, exogenous chemicals, light quality and intensity, and even day length.

Inevitably, these two groupings (internal and environmental) tend to be artificial because, in nature, so many factors interact continuously to dictate plant performance. In the interests of clarity there must be a compromise, and so we shall deal individually with some major control systems in the present section, and then in Part 4 under Ecological Physiology, endeavor to cover their interactions. At present we con-

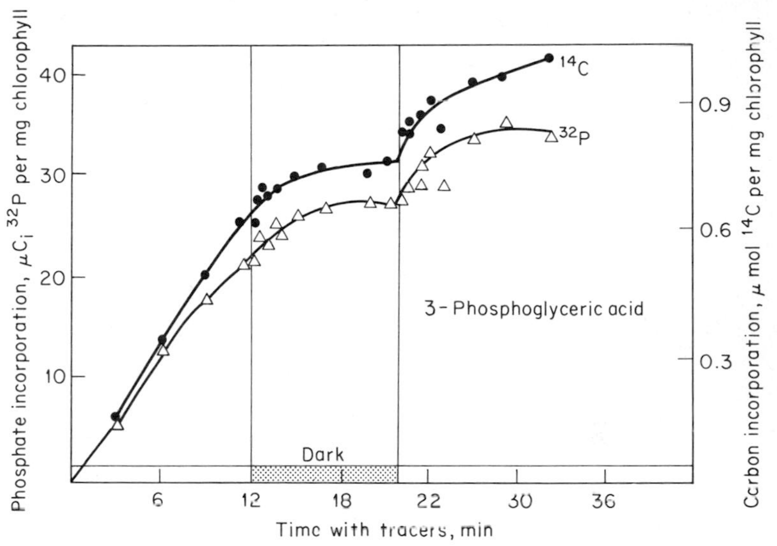

Fig. 1-21 The initial fixation product during C_3 photosynthesis (phosphoglyceric acid) incorporates both ^{14}C and ^{32}P during the light. Formation of labeled PGA declines immediately after transfer to darkness (Bassham and Jensen, 1967).

sider factors which offer physiological possibilities for controlling photosynthesis.

Internal Regulation

Photosynthetic enzyme systems Enzymes can govern not only the rate but sometimes also the direction of metabolic events. This is especially true for photosynthesis since gas-exchange measurements on single leaves show that r_r (including carboxylation resistance) can be rate-limiting, especially where low rates of photosynthesis are observed. In other words, the CO_2-fixing system within the leaf is often the chief impediment to photosynthesis.

Given that r_r (and thus the carboxylation systems) can be rate-limiting, we must still explain the inordinately low test-tube activity of the CO_2-fixing enzyme. Enzyme assays on cell-free systems suggest that the activity of the major carboxylating system in C_3 plants (RuDP carboxylase) is really insufficient to support even the limited rate of CO_2 assimilation observed in the intact leaf (Stiller, 1962). Extrapolation from the cell-free systems to the intact leaf is beset with difficulty, but some lines of evidence point to a higher in vivo activity compared with that assayed in the test tube. Light activation of RuDP carboxylase is a case in point. Figure 1-21 demonstrates, for example, a sudden fall in the activity of the carboxylation system in darkness, despite an adequate level of substrate and cofactors.

This key photosynthetic enzyme (RuDP carboxylase) is also sensitive to end-product inhibition and has a sigmoidal saturation curve for substrates. These features would allow fine control over CO_2 assimilation (Bassham and Jensen, 1967).

Another critical enzyme for the photosynthetic apparatus is the NADP-linked glyceraldehyde-3-phosphate dehydrogenase, which is also located in the chloroplast. This enzyme shows a rapid increase in activity following illumination (5 to 10 min.), and the light-response curve for enzyme activity resembles that for photosynthetic activity (Fig. 1-22) (Ziegler et al., 1965; Müller and Ziegler, 1969). Biochemists are especially interested in this system because the counterpart of the NADP-linked enzyme, NAD-linked glyc-

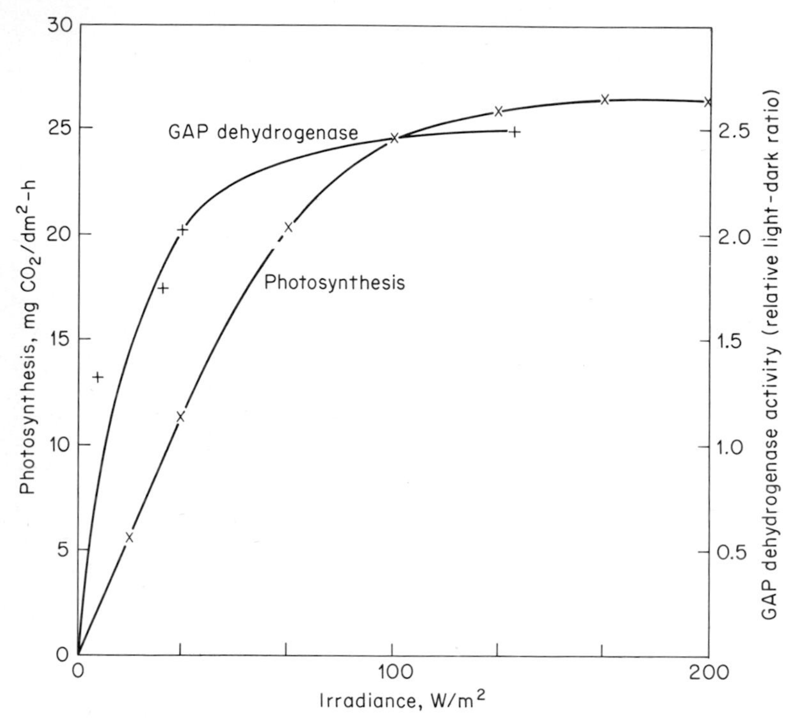

Fig. 1-22 The light-response curve for glyceraldehyde-3-phosphate dehydrogenase activity in leaves of *Vicia faba* shows a close similarity to the overall photosynthesis activity (*Sinapis alba*) (Ziegler et al., 1965).

eraldehyde-3-phosphate dehydrogenase, shows a reciprocal change during this light activation.

Enzymes of the C_4 fixation pathway may also exert a regulatory role as well as simply catalyzing reactions because the primary CO_2-fixing system, PEP carboxylase, is inhibited by its own reaction product, oxaloacetate. Similarly, another enzyme of the C_4 pathway, pyruvate P_i dikinase, can be selectively inhibited in either the forward or reverse direction by appropriate reactants, and, furthermore, this enzyme is light-sensitive. Its activity increases within minutes of exposure to light with a response pattern resembling that for photosynthesis at least up to 200 fc (Hatch and Slack, 1970).

Greater enzyme activity at higher light intensity has obvious merit in enabling a leaf to absorb more light usefully, but the mechanism of the initial activation of the enzyme due to light remains obscure. Graham et al. (1968) have demonstrated that phytochrome seems to be the primary photoreceptor for regulating synthesis of Calvin cycle enzymes, but the mechanism behind direct light activation of enzymes seems more elusive. Nevertheless, some encouraging results are at hand: Wildner and Criddle (cited by Preiss and Kosuge, 1970) succeeded in extracting a *light-activating factor* from the alga *Rhodospirillum,* and so apparently the mechanisms of light activation may ultimately find a firm chemical footing.

Leaf resistance Since CO_2 reaches the chloroplasts via the open stomata, stomatal resistance can help regulate photosynthesis. Figure 1-23 demonstrates how both H_2O and CO_2 exchange are under stomatal control. In C_4 plants, r_r is generally of the same order of magnitude as stomatal resistance r_s and does not in itself represent the major limitation for photosynthesis (see corn-leaf data in Table 1-7). For these C_4 plants, leaf resistance (controlled primarily by stomatal aperture) does appear to regulate photosynthesis. The situation is different for C_3 plants, and although leaf resistance appears to exert comparable control over both H_2O and CO_2 exchange, in Figure 1-23 there is a greater reduction in transpiration than in photosynthesis as the stomata begin to close; i.e., the ratio of transpiration to photosynthesis is reduced. We conclude that internal resistances (including carboxylation efficiency) offer greater limitation to CO_2 fixation than stomatal resistance in such C_3 plants.

Such environmental factors as light intensity and quality, photoperiod, CO_2 concentration, atmospheric humidity, and soil moisture can therefore affect photosynthesis via stomatal resistance. These factors are discussed in more detail in Chap. 3 and again in Part 4.

Demand for photosynthate The central issue here is whether a plant grows faster because it photosynthesizes more vigorously or whether controlling factors call for rapid growth and then photosynthesis responds accordingly. From the type of evidence reviewed by Neales and Incoll (1968) we gain the impression that in some circumstances photosynthesis is a consequence rather than a cause of increased growth rate.

By analogy with the law of mass action, the greater the demand imposed upon photosynthesis the greater the yield (up to a certain point). The converse should also apply. There are numerous instances of this type of compensatory response (see the review in Neales and Incoll, 1968). If the area of leaf surface supporting a given crop is reduced, CO_2 assimilation per leaf will increase; alternatively, the demand for assimilates can be lowered by crop thinning or meristem removal, and then the photosynthetic rate declines.

We have already referred to end-product inhibition as one control system for carboxylating enzymes, but is such a system likely to operate in an intact leaf? The plant provides a system for transporting products away from their site of synthesis, but some degree of compartmentation of photosynthate will occur even in the absence of this translocation. The impeded removal of photosynthate from a leaf, e.g., by petiole girdling or by holding detached leaves in sucrose solution, can suppress current photosynthesis and at the same time lead to a buildup of sugars in the leaf (Hartt, 1963). However, it is difficult to attribute the reduction in photosynthesis to solute accumulation per se since leaf resistance was not measured, and in any event we do not know at what threshold level for solute concentration the chloroplast photosynthesis will start to decline. Moreover, the photosynthate stored in the chloroplast is often starch, and it is hard to imagine how this insoluble product could have a negative feedback effect unless the starch grains disrupted the chloroplast's structure so much that its photosynthetic apparatus was impaired.

In view of the close hormonal control of photosynthesis which is known to occur (next section), it is insufficient to attribute responses in photosynthesis following phloem interruption or removal of a sink simply to the reduced demand for

Fig 1-23 Both transpiration and photosynthesis show close dependence upon leaf resistance when measured under stable laboratory conditions (Kriedemann, 1971).

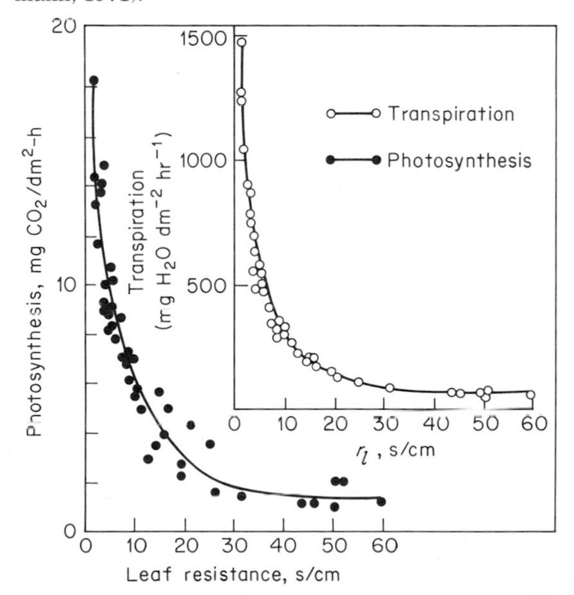

assimilates. Although the apical meristem or fruit represents a sink for assimilates, it is also a source of hormones which might be regulating events at the supply point for the photosynthate. Such a possibility gains strength from experiments reported by Bidwell and Turner (1966), where the application of indole-3-acetic acid (IAA) seemed to substitute for the detached sink, at least so far as the plant's photosynthetic machinery was concerned. This was also the explanation favored by Sweet and Wareing (1966) for their results with pine seedlings, where the removal of the apical meristem lowered photosynthesis and partial defoliation caused an increase.

Similarly, in experiments where the phloem is severed and distal leaves show lowered photosynthesis, we cannot ignore the possibility that the leaf's normal hormone supply has been interrupted due to the inactivated phloem of the petiole. This point assumes more importance in view of the demonstrations (1) that the supply of cytokinins to the leaves may be derived from the root system (see Skene and Kerridge, 1967) and (2) that cytokinins can stimulate gas exchange by leaves (Meidner, 1967; Treharne et al., 1970).

Regulation by hormones The range of physiological events which are under hormonal control continues to expand, and recent experiments indicate that photosynthesis, too, may be regulated by the plant's hormone systems. Leaf photosynthesis may be regulated by changes in either r_l or r_r, and since both parameters are under hormonal influence, the effects of hormones on leaf photosynthesis can be usefully analyzed in terms of diffusive resistances.

Two research groups in Aberystwyth (Wales) have reported clear instances where gibberellin and cytokinins have elicited increases in photosynthesis. The treated plants also showed a higher carboxylating activity, due apparently to enzyme activation rather than *de novo* synthesis (Treharne et al., 1970). Earlier experiments reported by Wareing et al. (1968) showed an increase in photosynthesis following partial defoliation, which was again matched by a greater activity of carboxylating enzymes. Neales et al. (1971) report comparable effects in defoliated bean plants. It is nevertheless difficult to attribute the enhanced assimilation in remaining leaves quantitatively to higher enzymatic activity in leaf extracts unless component resistances to CO_2 fixation are derived in vivo. The need for such data was recognized by Neales et al. (1971) and is heightened by other observations on stomatal responses to applied cytokinins. Meidner (1967), for example, reports that application of 3 μ molar kinetin causes a 12 percent increase in photosynthesis within 1 h, but this increase is associated with a lower leaf resistance (increased stomatal aperture) and reduced CO_2 compensation point. These observations immediately raise a question: Which photosynthetic control (r_l or r_r) is altered by the cytokinin? Since the available evidence cited above suggests that both resistances are under hormonal influence, the question must remain open.

Genetic controls A plant's photosynthetic activity and its responsiveness to environmental conditions are both species-dependent. Furthermore, a plant's ability to utilize the aerial environment (leaf arrangement and light interception) as well as its capacity to adapt to new situations, such as changing from high to low light intensity, can also be genetically determined. Evolutionary pressures can therefore lead to ecological races that are photosynthetically distinct (see review by Hiesey and Milner, 1965).

Broadly speaking, genetic control of photosynthesis can be exerted on both the CO_2-fixing system and the CO_2-transport system of the leaf. These influences would show up in Table 1-7 for example, as variation in r_r and r_s, respectively. We can relate the carboxylating efficiencies of the various plants listed in Table 1-7 to their r_r value, but the limits on such a relationship were discussed earlier.

Biochemical capacity to fix CO_2 is a function of the enzyme complement of the chloroplasts, which in turn is under the control of this organelle's own genes. Since the chloroplasts can transmit their own genetic information independently, here is an opportunity for genetically based variation in photosynthesis. Extensive studies on diverse members of the genus *Sorghum* (Downes, 1971) show that consistent differences do occur but can be readily offset by environmental influences. Similarly, the photosynthetic activity of hybrid and inbred lines of corn has been compared by Elmore and coworkers in Arizona, and again consistent differences exist, but any relationship with productivity was difficult to prove (Elmore et al., 1967). This picture of genetically controlled differences between closely related varieties or species is of little encouragement to plant breeders who might wish to produce material with an intrinsically higher photosynthetic capacity and correspondingly elevated yields. The problems are accentuated in the red kidney bean, for example, where the genetic mechanism controlling varietal differences shows some dominance for *low* photosynthetic efficiency (Izhar and Wallace, 1967).

Under *natural* conditions, genetic control of photosynthesis *is* expressed, and the emergence of ecological races within a given genus does reflect the plant's ability to make adaptive alterations in its photosynthetic apparatus. Probably the most impressive instance has been the emergence of C_4 plants as a physiologically distinct group that cuts across a large number of taxonomically unrelated genera. While evolutionary pressures for rapid growth over relatively short periods, as encountered in monsoonal climates, have produced the genetically distinct C_4 plant, adaptation to situations where growth is severely restricted, e.g., by light, can also have a genetic foundation. Ecological races of *Solidago* illustrate some remarkable adaptations of this type, which have been intensively studied in both Sweden and California. Björkman and Holmgren, who report their findings in a series of papers dating from 1963, first compared the sun and shade races of *Solidago* and then the hybrid clones of an F_1 population from a cross between the two races. The progeny showed a wide variation in photosynthetic behavior, parental characteristics being recombined in many different ways. Since differences in photochemical efficiency, RuDP carboxylase activity, and stomatal and mesophyll diffusive resistance all contributed to the divergence between the original parents, it is not surprising that the photosynthetic responses of the progeny showed highly heterozygous control. The ecological advantages or disadvantages of these various features (photochemical efficiency, etc.) are outlined in Part 4.

Leaf Age Photosynthetic activity has already begun to decline well ahead of visual indications of impending senescence (Fig. 11-16). The newly expanding leaf generally shows a sigmoidal growth curve, and maximum photosynthetic activity (area basis) is reached as the leaf achieves full size. For a deciduous perennial like the grapevine, this occurs 30 to 40 days after the lamina first unfolds although leaves on herbaceous annuals achieve peak rates at an earlier age (Fig. 1-20 and 11-16). The steady increase in photosynthetic activity as the leaf expands is not simply a consequence of increased chlorophyll content because assimilation number also increases; instead, changes in CO_2 compensation point plus alterations in internal anatomy and diffusive resistance seem related to the enhanced CO_2 uptake (Kriedemann et al., 1970).

Curiously enough, the emergence of the leaf as an exporting organ generally follows a pattern similar to that of changes in photosynthesis, maximum export occurring at about the time the leaf attains full size (Thrower, 1964; Hale and Weaver, 1962). The question immediately arises: Does the developmental pattern for photosynthesis govern the onset of translocation, or is the transport system the primary regulator? On the basis of our present understanding we could argue in either direction.

Assimilation rate, along with other anabolic processes which depend upon chloroplast function, declines as leaves age (Hernandez - Gil and Schaedle, 1973). There are associated increases in mesophyll resistance (Fig. 1-20), while stomatal response also becomes sluggish (Chap. 17), so that an increase in either r_s' or r_r could impede CO_2 exchange. Mechanisms underlying these changes are discussed in Chap. 11.

Environmental Factors

Light intensity The availability of sunlight is often the most significant determinant of plant productivity in agricultural situations, and the response of individual leaves to light intensity is relevant to the performance of both individual plants and their communities. We shall be dealing with the "instantaneous" response to light and for the moment disregard adaptive response to changes in light intensity that can occur within the leaf, e.g., chloroplast rearrangement, alteration in photochemical efficiency, or modified anatomy.

Since the photosynthetic activity of a leaf depends upon a continuous supply of raw materials, in particular light and CO_2, Blackman's (1905) principle of limiting factors should operate; i.e., "when a process is conditioned as to its rapidity by a number of separate factors, the rate of the process is limited by the pace of the slowest factors."

Fig. 1-24 The hypothetical response of photosynthesis to light intensity and CO_2 concentration would follow this pattern if Blackman's principle operated ideally (Blackman, 1905). The slope of the curve reflects the efficiency or quantum yield Φ.

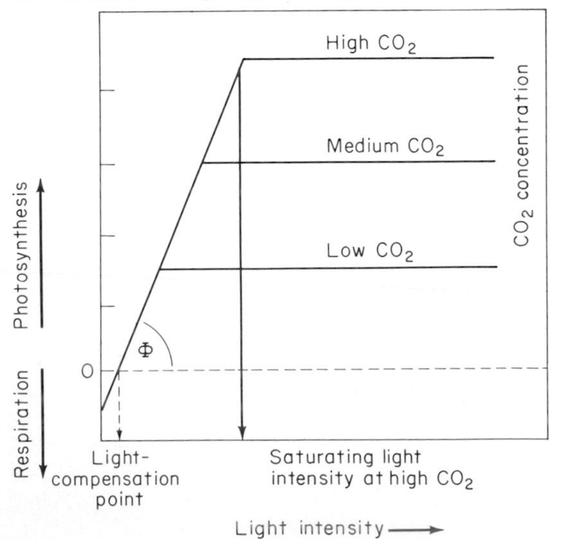

Accordingly, the interaction of light and CO_2 in determining photosynthesis should produce a family of curves, as in Fig. 1-24. The slope of the response curve will be governed by the leaf's photochemical activity (represented by Φ in Fig. 1-24), whereas the availability of CO_2 will dictate the light-saturated rate of photosynthesis; hence the stepwise increments in Fig. 1-24.

Regardless of its validity, real plants do not behave precisely as shown in Fig. 1-24. The light-response curve does not show a sharp inflexion point but a gradual transition from the linear increase at lower light intensities to a plateau at light saturation (see Fig. 1-25 and preceding light-response curves). Such a departure immediately tells us something about the photosynthesizing leaf, because an especially thin algal suspension does behave more like the theoretical picture in Fig. 1-24 (van den Honert, 1930). By contrast, a thick algal suspension, or for that matter a thick leaf, will show a greater departure from the form predicted by the principle of limiting factors (see Figs. 15-5 and 15-7 for a comparison of thick and thin leaves). Apparently, then, the gradient in light intensity within the experimental system, especially the optically thick one, has something to do with the discrepancy between the response curve predicted from Blackman's principle and that actually observed. In a thick algal suspension or thick leaf, when chloroplasts closest to the light source are light-saturated, those more remote from the light source will be considerably less than light-saturated. Such a heterogeneity within a leaf will result in a rounded light-intensity curve.

The variation in supply to chloroplasts in various positions within the leaf of the other major reactant for photosynthesis, CO_2, would also be expected to result in a rounded CO_2-concentration curve, and Fig. 1-25 confirms this point.

Photochemical efficiency Since the leaf's quantum efficiency Φ determines the maximum slope of the light-response curve (Fig. 1-24), the shapes of such light-response curves provide

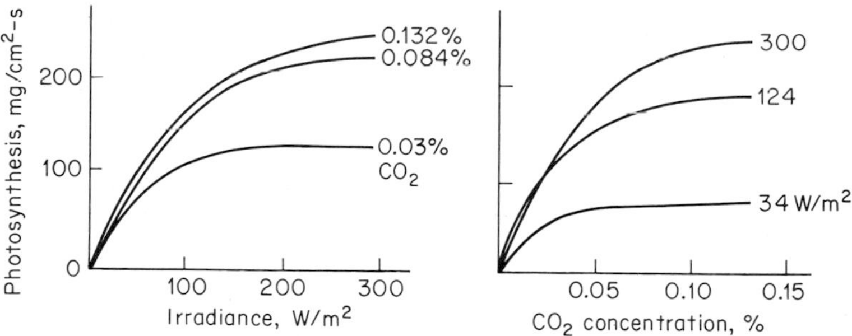

Fig. 1-25 The photosynthetic response of sugar beet leaves departs from the hypothetical relationship to light intensity and CO_2 as predicted by Blackman's principle, due to gradients in both light intensity and CO_2 concentration within the leaf (Gaastra, 1959).

a useful means for comparing leaves of different species or different physiological adaptation. In Gaastra's (1959) experiments on sugarbeet leaves, for example, maximum values for Φ were achieved at low light intensity, where photosynthesis was still showing a linear response to increases in light (up to 50W/m^2 in Fig. 1-25). Those leaves attained a quantum efficiency of 0.087 mol of CO_2 per einstein absorbed, which agrees well with the maximum quantum efficiencies ($\Phi \approx 0.1$) obtained by other workers (Wassink, 1946).

By developing such an approach, Björkman and Holmgren (1963) revealed that shade-adapted races of *Solidago* have a higher photochemical efficiency than those adapted to exposed situations (Chap. 15). Björkman (1966, 1968) then analyzed this difference in greater detail and was able to attribute differences in leaf photosynthesis to disparate photochemical and RuDP carboxylase activities. In an ecological situation where light is limiting, the shade-adapted *Solidago* and other shade plants will be at a distinct advantage, and data in Fig. 15-5 explain why. Near the light-compensation point their photosynthetic rate will actually be higher than for the corresponding exposed races, although this advantage is clearly lost at high light intensity, where the CO_2-fixing capacity of the leaf imposes a greater limitation on photosynthesis in shade leaves. Given that a leaf's carboxylating capacity is often the chief limitation for photosynthesis and that it is also the prime determinant of how much light a leaf can usefully absorb, we now have an explanation for this commonly observed correlation between saturating light intensity and photosynthetic capacity; i.e., leaves with an inherently high capacity also show higher values for light saturation.

Light compensation and adaptation Another consistent feature of the light-response curve for photosynthesis, in addition to its general form, is that the curve does not pass through the origin of the graph; i.e., zero net photosynthesis does not occur at zero light intensity. Instead, a certain level of light is required to bring the photosynthetic machinery to a point where it can offset the respiratory processes generating CO_2. At this equilibrium point, where no net flux of CO_2 occurs, the incident intensity is referred to as the *light-compensation point* and is frequently of the order of 100 to 200 fc for sun leaves but as low as 10 fc in shade-adapted foliage.

The light-compensation point of a leaf is primarily a reflection of its dark-respiration rate, and differences between species can be explained largely on that basis. Unless some specific light adaptation is involved, photochemical efficiency tends to be comparable between plants (Gaastra, 1962), and at these low light intensities photorespiration is not sufficiently activated to put the leaf at a significant disadvantage (even in C_3 plants). While there are inherent differences between leaves in their rates of dark respiration, and hence light-compensation points, these parameters still show remarkable flexibility and are often responsible for a leaf's ability to adapt from exposed to shaded habitats (Chap. 15).

C_3 and C_4 plants Differences between the plants with C_3 and C_4 types of photosynthesis can be observed in their light-response curves. Figure 1-11 shows some broad scale comparisons. These light-response curves are sufficiently detailed for us to see that the light-saturation rate for the C_4 plant, sugarcane, has not been reached even at the highest light intensity used (virtually full sunlight) but the initial slopes of the curves (except for the *Philodendron*) are similar. The C_3 species, tobacco and clover, are saturated at about one-third full sunlight. C_3 and C_4 plants therefore apparently do not differ in photochemical efficiency but are enormously different in CO_2-fixing capacity.

The high internal resistance to CO_2 diffusion in some C_3 plants has already been cited (r_r in Table 1-7), and since it seems to be the major bottleneck to photosynthesis in this group, a sufficiently high CO_2 concentration should minimize this photosynthetic difference between C_3 and C_4 plants.

Carbon dioxide The prevailing level of CO_2 in the earth's atmosphere is generally far below the level which would allow the full expression of a leaf's photosynthetic activity. It seems anomalous that evolution should have produced photosynthetic systems with CO_2 optima three or fivefold higher than the usual levels occurring in nature. The photosynthetic response curves of Fig. 1-25 illustrate the increase in photosynthesis elicited by CO_2 concentrations well above normal ambient concentrations. Obviously, the availability of light will influence the leaf's ability to respond to added CO_2, but at intensities approaching full sunlight photosynthesis is often enhanced by concentrations up to 1,000 ppm.

Species differences in photosynthetic capacity can become more easily discerned at high light intensities because the CO_2-diffusion process then becomes limiting (Gaastra, 1962) and differences in the diffusive resistance will also be exaggerated. In this connection, wide-ranging differences between various classes of organisms have been reported by Steeman-Nielsen (1952), who found that the unicellular alga *Hormidium flaccidum* is saturated at 0.05 percent CO_2; the higher terrestrial plant *Triticum sativum* achieves, maximum photosynthesis at 0.15 percent; while the aquatic phanerogam *Myriophyllum spicatum* is not replete, photosynthetically, until 1.1 percent CO_2 is provided. As stated by the author, "22 times as much CO_2 is thus necessary to effect saturation in the aquatic higher plant as in the terrestrial unicellular alga. As already shown, diffusion can without difficulty explain this."

Diffusive resistance If the diffusion of CO_2 is mainly responsible for species differences, we should be able to express these quantitatively; the CO_2 response curve for photosynthesis provides the means. The steepness of this curve has little to do with photochemical efficiency, unlike its counterpart the light-response curve; instead the total diffusive resistance for CO_2 movement to the fixation site Σ_r' determines the slope. When the experiment on CO_2 concentrations is done in an O_2-free atmosphere (Fig. 1-26), we can essentially disregard photorespiration and assume that the CO_2 concentration at the photosynthetic site is held very close to zero. In that case the response line for photosynthesis in an O_2-free atmosphere would pass close to the origin because we must expect zero photosynthesis at approximately zero CO_2 outside the leaf (not precisely zero since dark respiration will be generating CO_2). In nature, O_2 is of course present in the atmosphere, and we must anticipate some photorespiratory activity. The effective CO_2 concentration at the chloroplast will then be some positive number, at least in a C_3 plant, and the photosynthesis–CO_2-response line will be displaced to the right by a proportional amount (see Fig. 1-26). The intercept on the abscissa will now approximate the internal CO_2 concentration at the fixation site. In a C_4 plant, the photosynthesis-response curve still passes close to the origin, despite the presence of O_2 in the air (Hesketh, 1963).

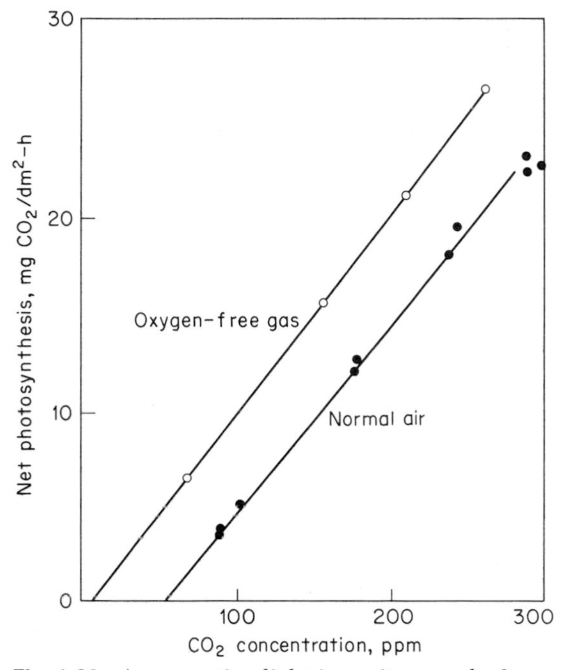

Fig. 1-26 At saturating light intensity, pear-leaf photosynthesis shows a linear response to CO_2 concentration (at least up to ambient levels). The absence of oxygen in the gas stream stimulates CO_2 fixation because photorespiration is suppressed (Kriedemann and Canterford, 1971).

Curiously enough, the apparent CO_2-compensation point for the leaf shown in Fig. 1-26 was 62 ppm and approximates the intercept point on the abscissa (60 ppm CO_2). Such coincidence is not achieved for every measurement (Troughton and Slatyer, 1969) but occurs often enough to suggest that either method (CO_2-response curve for photosynthesis or Γ determination) is a fair indication of the effective level of CO_2 inside the leaf.

CO_2 fertilization Since photosynthesis shows such an impressive response to CO_2 concentration, there is the prospect of enhancement for practical gain. Clearly the nature of this "fertilizer" limits its use. Being a gas, it needs to be contained, and greenhouse or growth-cabinet application seems the most realistic. Since CO_2 uptake involves an enzymatic fixation, there will be a considerable temperature component in the photosynthetic response to CO_2. This is to be anticipated from the data in Fig. 1-25 and is also borne out in practice (Chap. 3).

Replenishing the air in a greenhouse and then maintaining reasonable turbulance will have a favorable effect on growth (Went, 1957) due principally to the improved CO_2 supply (the effect is to reduce r_a'; see Fig. 1-19).

This enhanced CO_2 supply, especially in CO_2 enrichment experiments, does not necessarily improve yield. Excessive levels, say above 1,000 ppm, are phytotoxic (Gaastra, 1959), but even below this injurious level, the long-term productivity of a plant population does not match the short-term response in the laboratory for two reasons: (1) the higher CO_2 tension can lead to stomatal closure and thereby affect the improved availability of CO_2 from the atmosphere; (2) plants can show morphogenic responses, e.g., reduced leaf area, at levels approaching 1,000 ppm, so that productivity by the whole plant or a population of plants will not parallel the photosynthetic behavior of a single leaf (see Table 3-2).

SUMMARY

Individual leaves can be viewed in photosynthetic terms as autonomous organs which are themselves physically robust but still offer a well-buffered aqueous medium for the delicate operations of chloroplasts. These organelles are provided with energy and raw materials within the limits set by environmental conditions and subject to regulation by internal feedback. Adaptations in leaf morphology, plus physiological specialization within their photosynthetic tissues, favor light interception and CO_2 absorption with close control over evaporative losses. The emergence of highly efficient C_4 plants, with their unique coordination of biochemical and anatomical systems, emphasizes such adaptation.

Since assimilatory organs are entirely responsible for sustaining plant growth, they must embody the machinery for formation and distribution of fixation products. Photosynthetic tissue within the leaf is therefore coupled to its vascular system, which then fulfils the dual functions of distributing assimilate and integrating the performance of individual leaves into a single organism. The operation of this remarkable network is discussed in Chap. 2.

GENERAL REFERENCES

Black, C. C. 1973. Photosynthetic carbon fixation in relation to net CO_2 uptake. *Annu. Rev. Plant Physiol.* 24:253–286.

Gibbs, M. 1971. "Structure and Function of Chloroplasts." Springer-Verlag, Berlin. 286 pp.

Hatch, M. D., C. B. Osmond, and R. O. Slatyer (eds.). 1971. "Photosynthesis and Photorespiration." Wiley-Interscience, New York. 565 pp.

Kirk, J. T. O. 1970. Biochemical aspects of chloroplast development. *Annu. Rev. Plant Physiol.,* 21:11–42.

Preiss, J., and T. Kosuge. 1970. Regulation of enzyme activity in photosynthetic systems. *Annu. Rev. Plant Physiol.,* 21:433–466.

San Pietro, A., F. A. Greer, and T. J. Army (eds.). 1967. "Harvesting the Sun." Academic Press, New York. 342 pp.

Zelitch, I. 1971. "Photosynthesis, Photorespiration and Plant Productivity." Academic Press, New York. 347 pp.

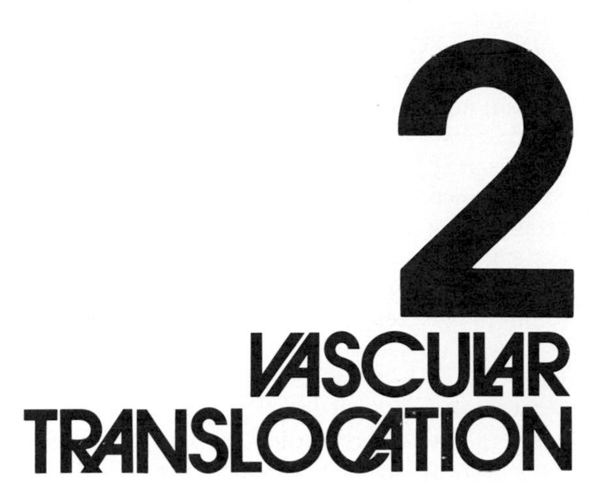

2 VASCULAR TRANSLOCATION

INTRODUCTION

A system for transporting both organic and inorganic substrates was essential for the evolution of large multicellular organisms. Photosynthetic products, as well as the raw materials for maintainance and growth, must be shifted over enormous distances, and the traffic must be sufficient to provide logistic support for rapid growth and development.

Consider the engineering problems involved in the design of such a translocating device:

1 It must be capable of moving carbohydrates up to the stem apex, where growth is most active, and down to the lower stem and roots; in short, it must be bidirectional.

2 It must be able to do this in spite of the fact that water will have to move from root to leaf in the transpiration stream, and so it must be insulated from the water system.

3 It must be able to handle solutions of difficult osmotic qualities, the molarities of carbohydrates in the translocation stream being high enough to plasmolyze most plant cells if they are in contact with them.

4 It must be capable of mending breaks in the system so that the removal of a leaf or part of a stem does not destroy the translocation system. If the translocation system utilized flow under pressure, any break could destroy this internal pressure, with lethal consequences unless there were a means of plugging the break.

These requirements demand a great deal of the translocation-system design; how the design meets the specifications is fascinating, though imperfectly understood.

There are two major translocating systems: phloem, which moves solutes in either acropetal or basipetal directions, and xylem, which moves water and minerals in the direction of the transpiration stream. The phloem translocation system has several other qualities. It can handle organic and inorganic materials, carbohydrates, nitrogenous organic materials, and ions. It supplies roots with carbon skeletons for synthesis of amino acids and related compounds. Classically, the phloem system has been identified as the locus of translocation of photosynthates and the xylem the locus of inorganic nutrients, but this separation is far from accurate. Either system can effectively translocate photosynthates, organic nitrogenous substances, plant hormones, or nutrients. In addition, some cross-transfer of water and solutes occurs between the two systems.

Whatever the fluids transported, the motive forces for phloem and xylem movement are totally different. Xylem solutes move passively as bulk flow in response to a water potential gradient. This gradient is usually a simple consequence of evaporative losses from leaves, but it can be accentuated in some halophytic species by salt secretion onto foliar surfaces. The precise distribution of mineral nutrients within aerial organs involves some complex transport processes, but their ascent up the xylem toward sites of deposition is passive and relatively uncomplicated.

By contrast, phloem transport is not simply a consequence of gradients in water potential, and attempts by physiologists to explain how the system works have involved more controversy than any other field of plant physiology. Encounters between proponents of the rival points of view often resemble those chemical reactions which generate more heat than light!

Since some knowledge of vascular anatomy is basic to an understanding of both xylem and phloem transport, our next section will deal with anatomical features of plants that are relevant to the problems of absorbing (loading), transporting, and then unloading water or solutes.

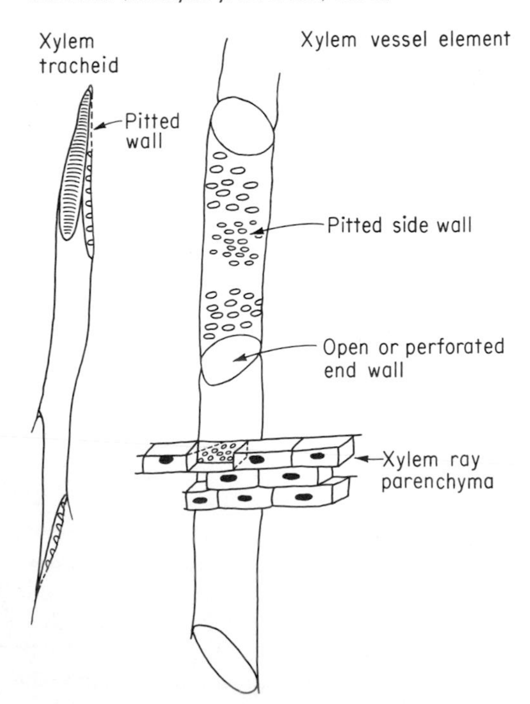

Fig. 2-1 Conducting cells in the xylem, showing a tracheid (*left*), with its pitted walls and ends, and a vessel element (*right*), with its pitted side walls but open ends. The nucleated ray parenchyma cells are shown with their pitted connections with the vessel element. (*Modified from Esau, 1953.*)

PATHWAYS OF MOVEMENT

Xylem Morphology

Anatomists know a great deal about xylem because component cells have hard and usually sculptured walls, which stay intact during tissue preparation. Xylem cells constitute an apparently inert plumbing system, and water plus solutes can pass along these conduits with a minimum of metabolic involvement. Two types of conducting elements occur (Fig. 2-1), tracheids and vessels. Tracheids have no open perforations in their walls and tend to be longer and thinner than vessels. Tracheids are formed from a single cell and are generally about 1 cm long and about 0.001 to 0.002 cm wide. Vessels, on the other hand, are formed from chains of cells by breakdown of cross walls. They are not necessarily much wider than tracheids (exceptions do occur, e.g., oak trees and grapevines) but can run for over 1 m between cross walls. Vessels offer much less resistance to longitudinal flow than tracheids. Gymnosperms contain only tracheids; angiosperm timber contains both vessels and tracheids.

Passage of water between tracheids is facilitated by pit pairs with thin primary pit membranes in the walls between adjacent or superimposed cells. Vessels, on the other hand, have open perforations in end and side walls, so that water can move more freely both longitudinally and laterally. Lateral movement of materials within the stem is facilitated by medullary rays, which pierce laterally through the xylem tissues of woody plants and continue into the phloem. These ray cells carry large numbers of pits, which match pores in the xylem tracheids and are generally oriented radially, in line with their function of aiding lateral movement. Solute exchange between xylem tracheids and ray cells can occur through these paired pits.

The tracheids of gymnosperms can be closed by a valvelike torus in the tracheid pit, whereas closure in angiosperms is accomplished by invasion from adjacent ray cells. These ingrowths, or *tyloses*, along with accumulation of oils, gums, resins, and tannins in older wood, plug the older vessels. Concurrently, tracheal closure renders the wood less accessible to fungal invasion and strengthens the woody stem.

Under conditions of high evaporative demand, a strong transpiration pull is developed in the xylem. This tension, which builds up within vessels and tracheids, emphasizes their need for thick walls. Structurally weaker cells would collapse and stop the flow of water. Since heartwood vessels are blocked, this tension (and consequently water flow) is largely restricted to the relatively narrow sapwood.

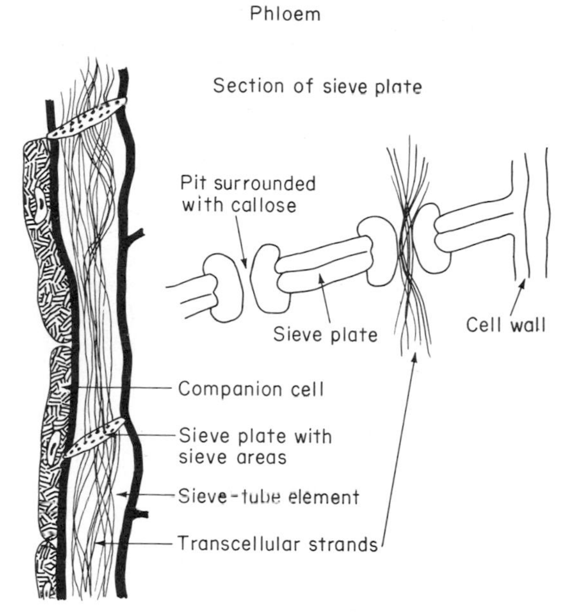

Fig. 2-2 A diagram of phloem cells representing a sieve-tube element with transcellular strands and sieve plates at each end. The nucleated companion cells with dense cytoplasm and an abundance of organelles are shown appressed to the sieve-tube element. An enlarged detail of the sieve plate in cross section illustrates the pits which connect adjacent elements and shows how fibrillar material might occupy a pore.

Phloem Morphology

Critical features of phloem are not readily discernible because the principal conducting elements (sieve tubes) are relatively small, have thin walls, and are not easily preserved for structural analysis. In particular, internal components are readily disrupted during processing, so that the anatomist's view of a preserved sieve tube does not necessarily coincide with living and functional tissue. Fortunately, the advent of electron microscopy has greatly facilitated studies of phloem morphology and contrasts sharply with the perceptive but far less sophisticated work of Hartig (1860). The newer technique has led to improved methods of fixation and staining, so that formerly obscure details can now be viewed under greater magnification and with better resolution.

In overall view (Fig. 2-2) the phloem amounts to a longitudinal pipe system comprising vertical arrays of sieve elements. Successive elements in functional phloem are separated by perforated sieve plates, which allow direct connection between adjacent cells and transmission of solutes. Inevitably, the sieve plate offers some hindrance to flow. This restriction becomes accentuated with time because each pore in the sieve areas becomes encased in a doughnut-shaped deposit of callose which enlarges with age. Callose, an amorphous glucan featuring $\beta(1,3)$ linkages, is widely distributed throughout the plant kingdom. The steady deposition of this material means that a pore can be virtually eliminated by the time phloem is mature. The effectiveness of individual sieve tubes therefore appears short-lived, but growing plants, for example, produce a renewed supply in recurring rings of phloem in the bark complementary to the xylem rings which eventually constitute the wood.

In addition to being replaced, sieve tubes can also be rejuvenated, as occurs in some perennials when they emerge from dormancy. In *Tilia americana,* for example, the vascular system is able to hydrolyze callose deposits formed previously and thereby render plugged sieve plates functional once more (Evert and Murmanis, 1965). This helps meet the urgent need for substrates throughout the tree during the resurgence of growth in spring. In ultrastructural terms (Fig. 2-3), a sieve tube contains a fairly homogeneous liquid phase in which organelle structures have been largely decomposed. The nucleus has disappeared, leaving one or several nucleoli intact. The tonoplast probably disappears, although mitochondria are sometimes detectable; but one conspicuous feature in fresh material is a highly ordered array of strands running the length of each sieve tube and passing from cell to cell across intervening sieve plates. The presence and functional significance of these strands was first commented upon by Thaine (1961) from his observations at Leeds (England). Canny, then at Cambridge, developed a translocation theory on this anatomical foundation (Canny, 1962*b*). Nevertheless, the functional significance and even the existence of these strands has been hotly debated (Esau et al., 1963). The existence of this fibrillar material has been reaffirmed (Jarvis and Thaine, 1971), and it now represents an important plank in some theories on translocation.

This fibrillar material, including fragments of disintegrated organelles, seems to stretch the entire length of successive sieve tubes (Fig. 2-2), but the material may collapse into a plug—suddenly, following tissue disruption, or gradually with age. In this form, what was once fibrillar material now constitutes a slime plug and conveys the impression of occluding sieve pores. Conversely, the very young sieve tubes in *Cucurbita* phloem appear to have localized aggregations of a similar material, slime bodies, which become dispersed into threads through the sieve tube at about the time the sieve plate develops perforations (Evert et al., 1966).

Microfibrillar material within functional sieve tubes (Fig. 2-2) appears as filaments with contractile possibilities (MacRobbie, 1971). These phloem filaments which contain the so-called P-protein of functional phloem (Walker and Thaine, 1971) are characterized by ATPase activity (Gilder and Cronshaw, 1973). While it may possess some contractile capabilities, Williamson (1972) suggests that P protein is not really analogous to contractile proteins in other biological systems, e.g., muscle. The possible involvement of phloem filaments in generating the forces necessary to drive translocation is in doubt but remains a provocative issue.

The enucleate sieve tube is often closely associated with *companion cells,* which contain prominent nuclei. When these cells are present in the phloem, they are closely appressed to the sieve tubes (Fig. 2-3) and characteristically contain dense cytoplasm with abundant mitochondria, ribosomes, and endoplasmic reticulum. Such features, plus a wealth of plasmodesmatal connections with the sieve tubes (Evert et al., 1971), suggest that companion cells might provide for the metabolic requirements of sieve tubes.

Nonvascular Movement

Cell walls have long been regarded as a pathway of movement for water or solutes derived from either phloem or xylem (Strugger, 1938–1939). They might be viewed as the final corridor of free movement for water or solutes before they are siezed metabolically by assimilative cells. For example, Gaff et al., (1964) were able to demonstrate in a particularly elegant way the degree to which water can become distributed within a leaf of *Helxine* simply through the cell walls. These workers supplied transpiring leaves with colloidal gold small enough in particle size (<100 Å) to penetrate into cell walls. Sites of gold deposition, and hence the

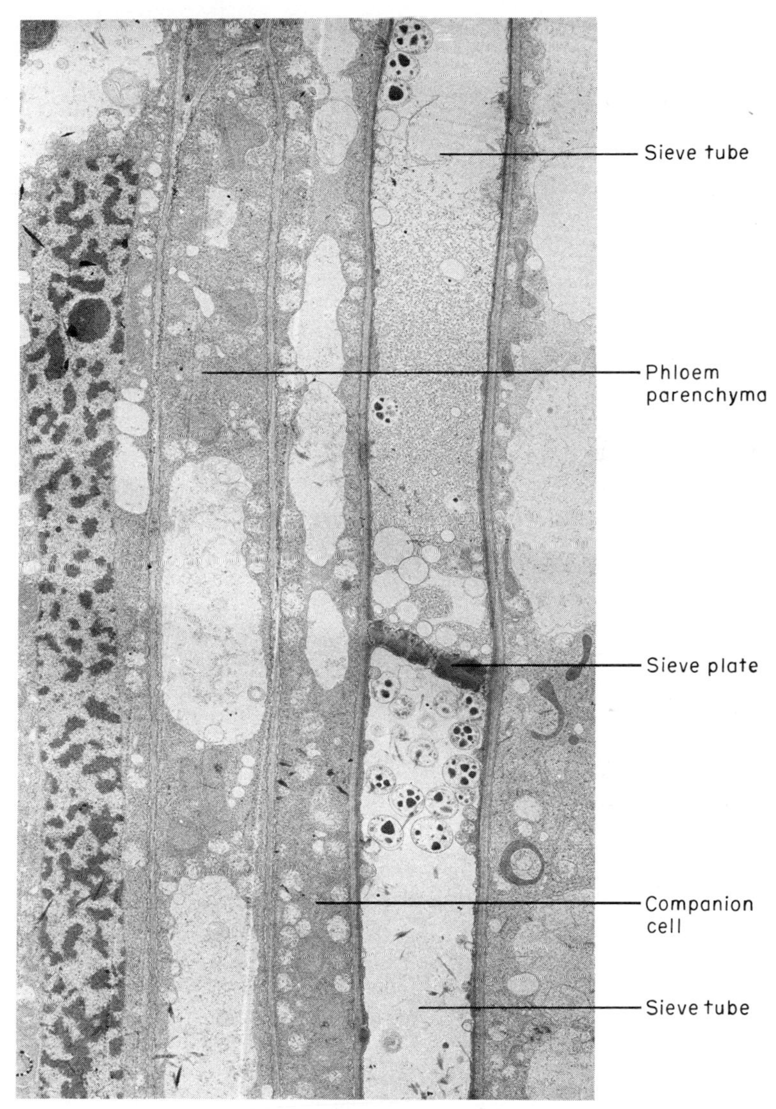

Fig. 2-3 The phloem of a young wheat leaf includes sieve tubes ensheathed by phloem parenchyma and companion cells; X1800. (*Courtesy of J. Kuo. Botany Department, Monash University, Australia.*)

pathway for water movement, were then established by electron microscopy (Fig. 2-4).

Cell walls appear to show a similar involvement in short-distance translocation of sugar into both storage tissue and developing fruits. Such substrate arrives via the phloem from sites of photosynthesis or from an exogenous supply, and the cell walls of accumulating cells then make temporary storage possible (Hawker, 1965, for sugarcane) or offer a pathway of movement (Kriedemann, 1969, with developing grapes; Lüttge and Weigl, 1965, with corn root).

Regardless of their possible role in transport and storage, the walls of even highly vacuolated

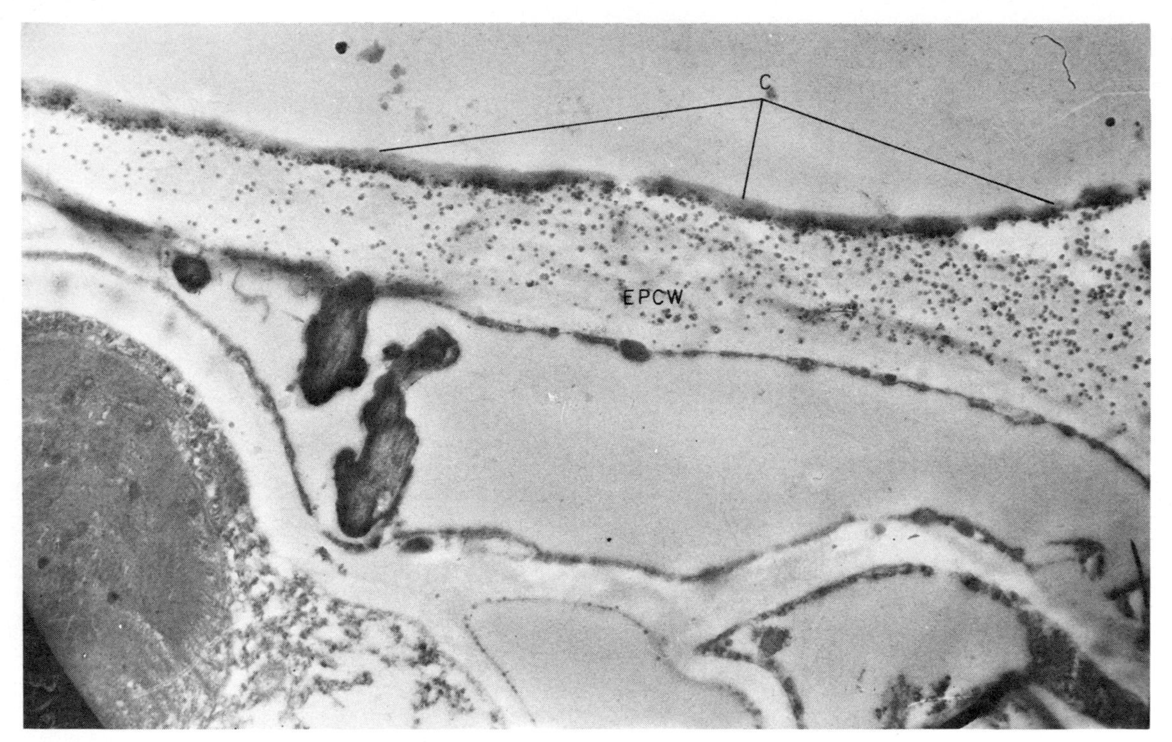

Fig. 2-4 The free movement of water in the cell wall is demonstrated by the movement of colloidal gold, supplied in the transpirational stream of *Helxine soleirolii* cuttings, into the cell-wall regions. Note the abundant gold colloidal particles in the epidermal cell wall (*EpCW*), especially along the interface with the cuticle (*C*); X20,000 (Gaff et al., 1964).

cells should not be simply regarded as inert structural members that afford free passage to solute molecules about to gain entry into accumulating cells. Certain enzymes located in (or at least bound to) the walls foster metabolic processes related to solute exchange. A good example is sucrose hydrolysis due to cell-wall invertase before accumulation by sugarcane parenchyma tissue or beanpod endocarp (Sacher, 1966*b*). This question of solute uptake and cell-wall involvement will be dealt with more fully later in this chapter.

SOLUTIONS IN TRANSIT

Xylem Sap

The solution inside the xylem usually contains only 0.1 to 0.4 percent solutes and as such is considerably more dilute than phloem sap. About one-third of xylem solutes are inorganic materials; the remaining organic materials comprise sugars, organic acids, alkaloids, lactones, and substantial amounts of organic nitrogenous materials.

Xylem sap can be extracted by applying vacuum to a section of stem and then collecting the exudate. Sap flow is encouraged by repeated excision of small portions at the distal end. This process is continued until the entire stem is consumed. A 20-cm apple stem, for example, yields 3 to 4 ml of sap. Using this method, Bollard (1953) followed changes in the composition of xylem sap from apple branches during the year. He found a great increase in inorganic salts during blooming which paralleled a high level of organic nitrogen in phloem sap (Fig. 2-5). By contrast, xylem sugar falls to a minimum value, virtually zero, during flowering but then rises in late summer and winter (Anderssen, 1929, with pear branches).

Nitrogenous constituents are represented by complicated mixtures of amino acids and amides, amino acids being the most common nitrogen compounds. A small group of trees has citrulline as the principal form of nitrogen in xylem sap, a curious amino acid which is not a normal component of plant proteins but is an intermediate in the urea cycle of animals. A few species contain principally the ureides allantoin and allantoic acid. Their role in plants is as much a puzzle as that of citrulline, but they are recognized as products of purine metabolism in animals (see Fig. 6-5). These amides, citrulline and the ureides, are especially rich in nitrogen on a molar basis and are therefore well suited to a role in nitrogen translocation.

Besides the carbohydrates and nitrogenous compounds of xylem sap, there are numerous inorganic nutrients, especially sulfur, potassium, magnesium, calcium, phosphorus, and iron. Of these, the metals are almost entirely in the form of chelates (Stewart, 1963). Some metals, e.g., iron, may be chelated with organic acids (Tiffin and Brown, 1962) or sometimes with sugars, e.g., iron with fructose or calcium with lactose (Charley et al., 1963; Charley and Saltman, 1963). It is even conceivable that one beneficial effect of organic soils may be the provision of generous amounts of natural chelating agents which facilitate cation movement through the xylem without precipitation. Among the metals, potassium is apparently not chelated; of the anions, sulfur apparently moves as sulfate, and phosphorus may move as phosphorylcholine (Bollard, 1960) or other organic forms such as sugar phosphates, phospholipids, or nucleic acids and related compounds. A wide array of phosphate esters also occurs in phloem (Bieleski, 1969), but transport there is of a different character entirely.

Fig. 2-5 Seasonal changes in nutrients in the xylem sap of apple trees. The content of potassium, phosphorus, and magnesium rises during early summer (October and November in South Africa), though the rise in potassium content is much the most pronounced (Bollard, 1958).

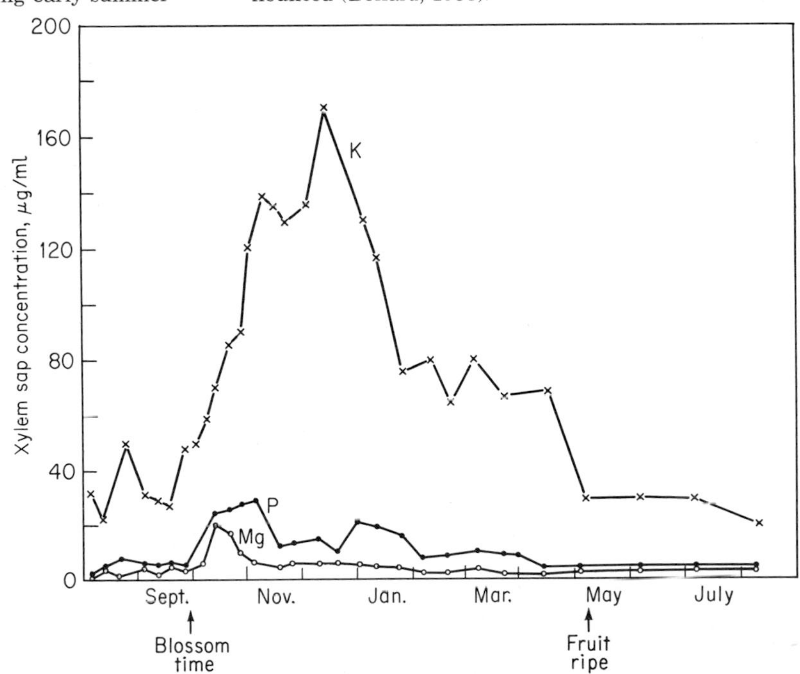

Seasonal changes in xylem nutrients are easily followed on woody perennials, and Bollard (1958) (Fig. 2-5) made a detailed analysis on apple trees. Potassium showed greatest variation with the level reaching a peak in early summer. Phosphorus and magnesium also showed maxima during this period, but absolute levels were of a different order of magnitude. Nutrient levels were relatively low during the latter part of the growing season and subsequent dormant period.

Ion concentration in the xylem appears unrelated to transpiration rate even though foliar transpiration obviously governs the volume of sap which flows through the xylem.

Large-scale movement in xylem conduits and subsequent distribution into foliage or other aerial organs is readily demonstrated with colored or fluorescent dyes, e.g., eosin, but more detailed examination is made difficult by migration of the tracer during fixation or other steps preparative to microscopy. Crowdy and Tanton (1970) overcame this problem by supplying lead ethylenediaminetetraacetic acid (EDTA) chelate to roots or leaves of wheat plants (virtually any plant should be suitable), allowing transpiration to occur for a period, and then precipitating the lead in situ with H_2S. Its location was then determined by light and electron microscopy (Fig. 2-6). Bulk movement of water occurred within the lumen of xylem when lead EDTA chelate was supplied to the roots. Outside the xylem, deposits were confined to cell walls, being particularly dense within the middle lamella.

The speed of solute movement in the xylem is greatly influenced by the rate of transpiration because xylem-sap nutrients move largely *en masse* in the transpiration stream. Speeds up to 900 cm/h have been reported, but under more moderate conditions values are an order of magnitude lower [see Sheriff (1972) for detailed measurements of sap flow using the magnetohydrodynamic technique].

Phloem Sap

Solute concentration is sufficiently high in sieve tubes for a positive pressure to be maintained even in transpiring plants. Consequently, phloem exudation will occur following decapitation or bark incision. Such an exudate is liable to contamination from damaged cells in neighboring tissues, but this problem is overcome by using an aphid stylet, a technique developed by an entomologist (Mittler, 1958) who anesthetized aphids after they had embedded their stylets and then severed them from their embedded mouthparts. When aphids feed undisturbed, they secrete honeydew, but this material has been preferentially depleted in nitrogenous compounds by the insect and is not representative of phloem sap (Esau, 1961). Severed mouth parts, on the other hand, provide an accurate index of phloem composition, and exudation can continue for many hours. Probably the only criticism that can be leveled against this technique is that some localized pressure drop must occur and could encourage water movement from elsewhere; nonetheless, this effect is small and presumably influences only concentration and not composition of solutes. Samples obtained in this way are representative for single sieve tubes because Zimmermann (1961) established by examination of microsections that each aphid stylet was embedded in a single sieve element.

Phloem sap contains sugars at a concentration between 10 and 25 percent, weight per volume, while amino acids and amides generally occur to the extent of 0.03 to 0.4 percent, weight per volume. As the chief translocated sugar, sucrose offers the plant a wealth of metabolic advantages over other soluble forms of carbohydrate (Arnold, 1968). It occurs almost universally and accounts for the bulk of carbohydrate, although higher sugars of the oligosaccharide series can accompany sucrose and have been identified in numerous trees (Zimmermann, 1957*a*). There are certain exceptions to this generalization, most notably the apple, where sucrose is replaced by sorbitol as the primary form of photosynthate in transit. Although nitrogenous materials are minor constituents quantitatively they show greater diversity than sugars. Mittler (1958) chromatographically identified 10 to 12 different amino acids and showed substantial variation ac-

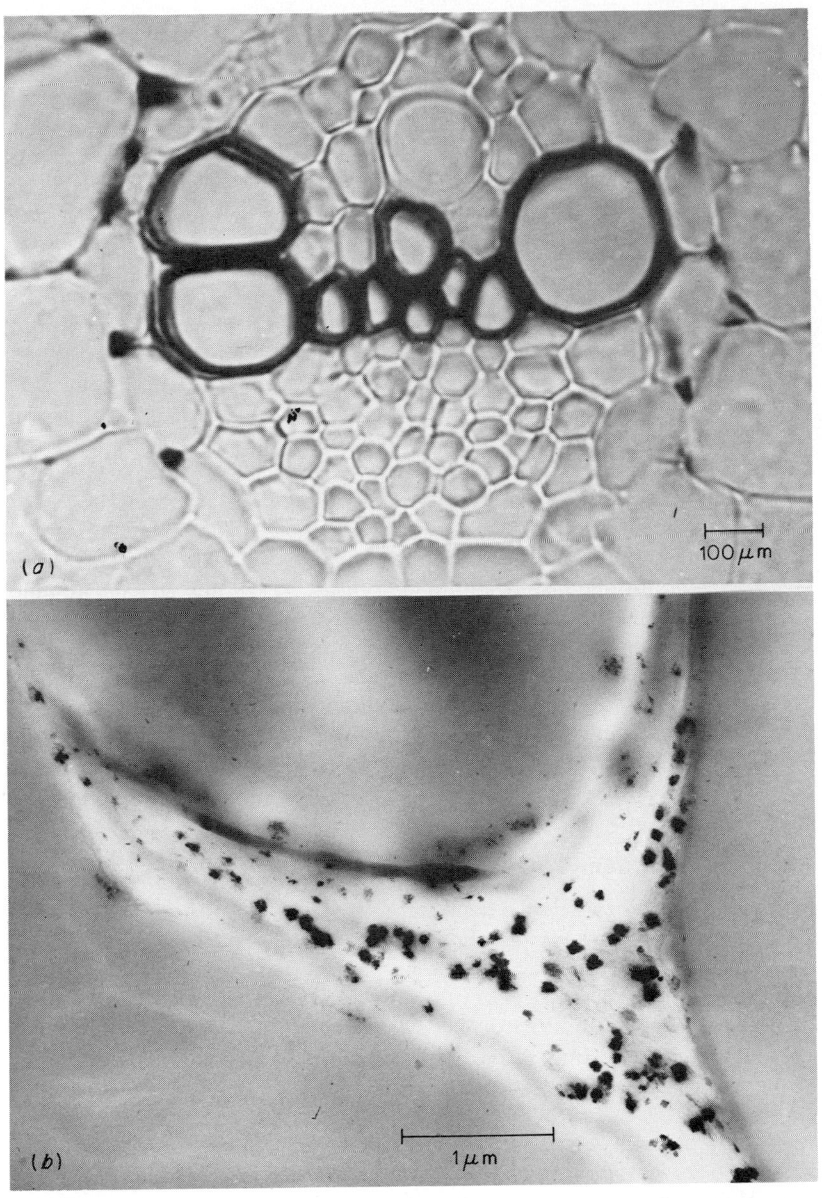

Fig. 2-6 The movement of water through the vascular strands of intact wheat seedlings can be traced using lead EDTA chelate supplied to the roots and precipitating the lead with H_2S to permit determination of its precise location. Lead deposits in the xylem vessels and intercellular spaces mark the movement of water (*a*) up the plant (X100), and (*b*) in the cell walls of the leaves (X4,000) (Crowdy and Tanton, 1970).

cording to season. Peak values were obtained in spring (0.2 percent), and a marked decline occurred after leaf growth was completed (0.03 percent), but levels rose again with senescence in autumn (0.13 percent).

Phloem exudate from aphid stylets, bark incisions, or cut stumps also contains materials which reflect the complexity of metabolic activities in the vascular system. Becker and Ziegler (1973) reported cyclic AMP in sieve-tube sap from the secondary phloem of *Robinia pseudoacacia,* while Hall et al. (1972) reported a diverse range of both organic and inorganic substances in exudate from bark incisions on activity growing *Ricinus.* Bioassays suggested the existence of auxin, gibberellins, and cytokinins; while adenosine triphosphate (ATP) was also present. ATP occurred at a concentration between 0.40 and 0.60 m*M*, in keeping with values found by other workers in exudate from aphid stylets.

Protein nitrogen is also a trace component of phloem sap (Ziegler, 1956), whereas phloem filaments are specifically proteinaceous and commonly referred to as P protein (Walker and Thaine, 1971); hence our earlier distinction between the composition of translocated materials and the phloem tissue itself.

Exchange between Phloem and Xylem

There is abundant evidence for a transfer of materials between these two conducting systems, and the radially oriented ray tissues are well suited to this purpose. Lateral movement is difficult to measure quantitatively, and few real measurements on intact material have been made. Hartig suggested as early as 1861 (cited by Biddulph et al., 1958) that soluble materials move basipetally out of leaves, descending via the bark to storage sites in wood rays. Mason and Maskell (1928*a*,*b*) also suggested the possibility of circulation but without an intervening storage period. They observed that the ratio of nitrogen, phosphorus, and potassium to carbohydrate moving down their cotton plants was greater than that required for growth and postulated that minerals were reascending in the xylem. Working on a more localized level, Webb and Gorham (1965) described (by chemical analysis and autoradiography) the radial movement of ^{14}C-labeled photosynthate in stems, petioles, and hypocotyls of squash. Such movement occurs along the entire length of the phloem but especially in young tissues. Although stachyose is the common translocated sugar, radial movement is principally as sucrose; presumably, then, sucrose passes through lateral walls of sieve tubes more readily than stachyose. Since the phloem is ultimately responsible for supplying substrate to every nonphotosynthetic cell in the plant, radial movement must be regarded as an integral part of translocation. Velocities of radial movement are of the same order as protoplasmic streaming (Webb and Gorham, 1965), and since protoplasmic material extends laterally out of sieve tubes (Evert and Murmanis, 1965; Evert et al., 1971), protoplasmic streaming offers a plausible explanation for radial translocation.

The extent of lateral movement is nevertheless strictly limited, and Zimmermann (1961) estimated for trees that only 1 cm of lateral movement occurs for every 50 cm of longitudinal translocation. A herbaceous plant like tobacco shows comparable behavior, judging from Porter's (1966) calculation that only 12 percent of sugars traveling along the midrib show lateral movement, a proportion which was not readily altered.

Grapevine shoots, with their low resistance to longitudinal flow, seem anomalous in the present context because lateral movement can be encouraged to a substantial degree. Hale and Weaver (1962) demonstrated, for example, that labeled assimilates will move radially in response to the demands of a developing cluster, while Hardy and Possingham (1969) were able to demonstrate preferential movement of certain solutes around cinctured regions of vine shoots. The abundance of medullary rays in this species would be a contributing factor.

Irrespective of the possible significance of lateral movement in whole-plant performance, solute exchange between phloem and xylem does occur and adds further difficulty to interpreting tracer profiles and confirming the reality of bidirectional tracer movement. This problem assumes special significance during protracted experiments, as discussed in the next section.

MECHANISMS OF ORGANIC TRANSLOCATION

Solutes entering the root system of a transpiring plant may ascend via the xylem in a passive and relatively uncomplicated way. By contrast, the mechanism of phloem translocation is still hotly debated. Although plants move prodigious amounts of substrates over long distances in support of growth and accumulation at remote sites, the exact mechanism remains obscure.

Following synthesis in a chloroplast, carbon skeletons must emerge from this organelle and thence, from the host cell, reach an appropriate minor vein, undergo a loading step into the translocation system, and then head off in the required direction. After traveling what might be an appreciable distance, the solute must be unloaded and either accumulated against a concentration gradient, e.g., sugar-storage tissue of a fruit or stalk, or incorporated into some polymeric substance, e.g., cellulose wall or starch grain. The original substrate molecule is liable to chemical transformation as part of its loading and unloading and during passage along the phloem because some substrate is siphoned off to meet metabolic requirements of conducting tissues en route. Having then been deposited at some sink, assimilates are not necessarily at rest, because mobilization will occur if they are called upon to support a resurgence of growth elsewhere in the plant. The above sequence of events may then be repeated.

To give these problems some perspective in quantitative terms the kinetics of translocation should be discussed before attempting to understand how the system works.

Kinetics and Motive Forces: Short-Distance Translocation

Translocation amounts to transferring "dry" matter from points of synthesis to centers of growth or accumulation. If the material is moving as a solution, the rate of dry-matter transfer is a function of solution concentration and velocity of movement. Mass transfer M with dimensions of grams of dry weight per square centimeter per hour will equal velocity multiplied by solution concentration, i.e.,

$$M\ (\text{g dry wt/cm}^2\text{-h}) = V\ (\text{cm/h})\ C\ (\text{g/cm}^3)$$

In this context M is synonymous with *rate of translocation.* Confusion can arise if the two different parameters of *velocity* and *rate* are used indiscriminately. Some authors express rates as a specific mass transfer; this term has dimensions of grams of dry weight per square centimeter of phloem per hour and makes useful comparisons between different systems possible.

In considering dry-matter transfer due to translocation, Canny (1971) suggests a maximum value for M of about 4 g per square centimeter of phloem per hour. If we assume that sieve tubes represent two-thirds of phloem cross-sectional area, M becomes 6 g per square centimeter of sieve tube per hour. If the solution transported is 25 percent sucrose, V must be 40 cm/h. As shown later in this discussion, a velocity of this order is reasonable. However, to achieve maximum rate at minimum concentration (say 5 percent sucrose), V would have to be 200 cm/h. A plant can therefore achieve maximum translocation rate despite wide variation in velocity, a factor to be borne in mind when trying to reconcile divergent values for the velocity of translocation reported in the literature.

Despite wide variations in solute concentration and apparent velocity, translocation rates can be surprisingly consistent. Canny (1960) assembled reliable data from a number of different sources (Table 2-1).

The next question concerns *speed.* How fast do solute molecules travel along the phloem? By discussing solute molecules rather than solution, we are, for the moment, avoiding the question of mechanism, but results at our disposal rely on following tagged molecules.

Early experiments utilized dyes or other detectable exogenous materials, and this approach can still be highly productive in demonstrating possible pathways for solute movement (see, for

Table 2-1 The rate of translocation, calculated on a common unit of grams dry weight per square centimeter of phloem per hour, shows similarities for the stems of various plants and markedly lower rates for the petioles (adapted from Canny, 1960).

Plant system	Specific mass transfer, g dry wt per cm^2 of phloem per hour	Reference
Stems:		
Solanum tuber stem	4.5	Dixon and Ball (1922)
Dioscorea tuber stem	4.4	Mason and Lewin (1924)
Solanum tuber stem	2.1	Crafts (1933)
Kigelia fruit peduncle	2.6	Clements (1940)
Cucurbita fruit peduncle	3.3	Crafts and Lorenz (1944)
Cucurbita fruit peduncle	4.8	Colwell, cited in Crafts and Lorenz (1944)
Mean	3.6	
Petioles:		
Phaseolus petiole	0.56	Birch-Hirschfeld (1920)
Phaseolus petiole	0.7	Crafts (1931)
Tropaeolum petiole	0.7	Crafts (1931)

example, Biddulph and Cory, 1965). However, foreign molecules are likely to move according to their own diffusive characteristics and in the direction of their own concentration gradient. A better approach is to use a tagged molecule that simulates normal photosynthetic products. This is commonly achieved by supplying $^{14}CO_2$ to a photosynthesizing leaf and studying the profile of the radioactivity in conducting tissue between source and sink. Such a distance profile for ^{14}C sucrose moving along a willow stem is shown in Fig. 2-7. Either the slope or rate of advance of this profile can serve as an index of translocation velocity.

Even assuming that the ^{14}C profile accurately reflects all solutes of the same chemical identity that are in transit, at least three problems must be considered before computing velocity from radioactivity profiles:

1 Some account must be taken of delays in loading labeled assimilates (or exogenous solutes) into the translocation system.

2 Profile shape, and in particular the location of an advancing front, depends on sensitivity in the radioactivity-detecting system, so that larger doses of label at the source will give higher *apparent* rates of translocation.

3 If application of the tagged molecule is continued for a long time, the advancing profile of radioactivity will reflect not only the normal rate of transfer but also the rate at which labeled molecules are being loaded into the translocating system.

In view of all these difficulties it may seem surprising that so much energy has been devoted to this form of experimentation, but the shortcomings listed represent the wisdom of retrospect. Tracer experiments were often completed long before their limitations were appreciated.

Measurements of translocation velocity that do not suffer from the shortcomings of ^{14}C tracer experiments have been described by Moorby et al. (1963), whose data are particularly relevant because they were able to obtain sets of radioactivity profiles for photosynthate transport on only one plant. This could not normally be achieved with ^{14}C because its emission is too soft for accurate assay on intact living specimens. Moorby and his colleagues used the more energetic isotope of carbon, ^{11}C, whose half-life is only 20 min. Some data from one particular set of experiments are reproduced in Figs. 2-8 and 2-9. Soybean leaves were exposed to $^{11}CO_2$, and radioactivity was monitored at certain distances from 10 to 30 cm below the labeled leaf. At each level, count rate rose sigmoidally with time, and extrapolation of the linear portion of each curve toward the abscissa suggests intercepts at 9, 14, and 30 min. Velocities were therefore 66, 77, and 60 cm/h, respectively.

The use of $^{11}CO_2$ for translocation studies was especially elegant because this isotope permits sequential experiments on the same plant, thus achieving replication without adding the variability inherent in large populations of individuals.

One reasonably consistent feature in all tracer experiments is the shape of the advancing front. Instead of having an abrupt front, the profile is logarithmic, and a long trail precedes the main front, much as hot water gradually reaches an outlet remote from the storage tank. This effect is particularly evident in Fig. 2-10, where data from Mortimer (1965) have been plotted arithmetically. Although low levels of activity can be detected well ahead of the main front, this does not constitute a velocity measurement. Nelson et al. (1959*a, b*) recorded particularly high speeds of translocation for such a blip of advancing activity in soybean, but this speed (2,000 or even 5,000 cm/h) does not represent the bulk of assimilates and it is conceivable that labeled gas migrated via intercellular spaces (compare ^{15}O movement in barley and rice, Barber et al., 1962). Whittle (1971) provides an additional analysis of these radioactivity-distance profiles. She distinguishes between the advancing front, which is strictly linear on a log radioactivity–distance plot, and a region behind the front,

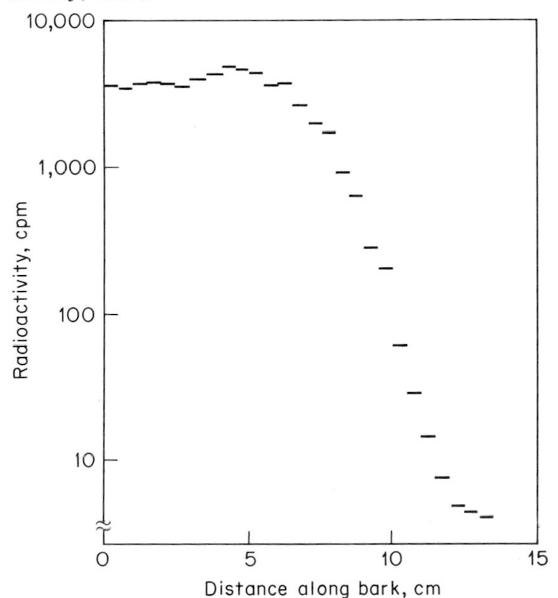

Fig. 2-7 Translocation along a willow stem can be followed as the radioactivity distribution in the stem after fixation of radioactive CO_2. The advancing profile shows a roughly logarithmic decline of radioactivity at the front after 5 h of translocation (Canny, 1961).

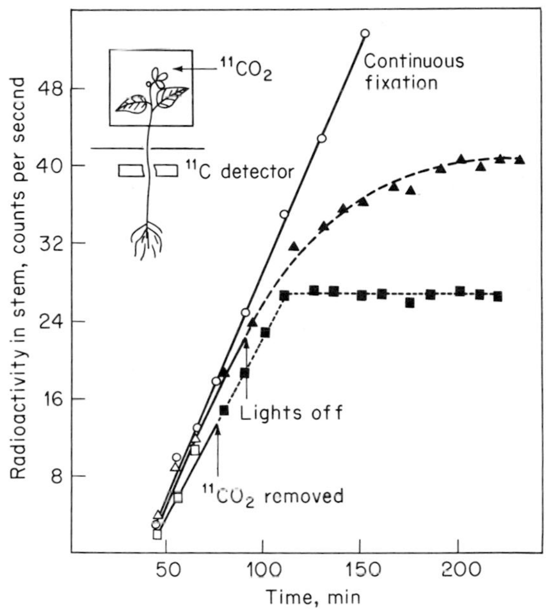

Fig. 2-8 Translocation of photosynthate in soybean plants following fixation of $^{11}CO_2$ by the illuminated leaves, showing the rise in radioactivity at the base of the stem (22 cm) under conditions of continuous fixation of $^{11}CO_2$ or when the lights were turned off at 90 min or when the radioactive CO_2 was replaced with nonradioactive CO_2 at 78 min. Either darkening or removal of the radioactive CO_2 depressed the flow of radioactivity along the stem (Moorby et al., 1963).

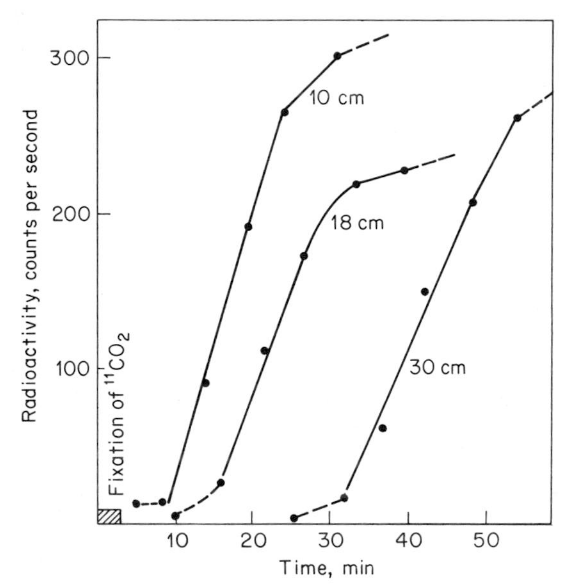

Fig. 2-9 After 3 min of photosynthesis in $^{11}CO_2$, the arrival curves for radio-activity at various positions down the stem of soybean can be used to estimate the translocation velocities. Extrapolation of these three curves yield velocity values between 60 and 75 cm/h for translocation (Moorby et al., 1963).

which persists in the stems for many hours and changes in form with time.

The diffusion analog Any freely diffusible substance, solute or solvent, should move as a function of its gradient in physicochemical potential. In a steady-state condition, this passive movement from regions of high to lower concentration will be expected to follow Fick's first law, which describes movement as a function of diffusion constant times concentration gradient, i.e.,

$$\frac{dc}{dt} = D \frac{dc}{dx}$$

where c is concentration, t is time, x is distance, and D is the diffusion constant.

Plotting solute distribution as it moves along phloem should give a straight line in a semilog plot, and this would be expected whether the phloem contents were undergoing bulk flow or not. Mason and Maskell's (1928*a,b*) now classic work on carbohydrate movement in cotton established this principle, which has subsequently been confirmed repeatedly in tracer experiments. The diffusion coefficient for sugar in water $D = 5 \times 10^{-6}$ cm²/s may be compared with Mason and Maskell's K value of 0.07 cm²/s. This value is drawn from their equation $M = K\, dS/dx$, which is formally equivalent to Fick's first law, where S is sucrose concentration and x distance. Sugar movement through Mason and Maskell's cotton plants therefore behaves according to diffusion kinetics in this respect, but the speed of movement is 30,000 times greater than would be expected by simple diffusion; hence their concept of *activated diffusion* to account for sugar movement in the plant.

Protoplasmic streaming Some measure of *activation* due to protoplasmic streaming was suggested by De Vries as early as 1885. This idea, which was supported by Curtis (1935), had the merit of helping explain bidirectional translocation and would account for inhibitory effects of metabolic poisons (cyclosis is very sensitive to oxygen deficiency or metabolic inhibitors).

However, speeds of cyclosis (as gauged from movement of cell inclusions) tend to be lower than translocation velocities. Cyclosis is generally rated at 1 to 5 cm/h, whereas translocated solutes move at 5 to 100 cm/h. Furthermore, the amounts of solutes that could be moved via this system are very limited; no more than one-quarter of the material in a cuboid cell would be transported in a given direction. This system would therefore have to be 4 times as efficient as any other mechanism involving only longitudinal movement in one direction—a discouraging requirement.

Transcellular strands Cyclosis, in the strict sense, has been reported for companion cells (see Thaine, 1964) but does not appear to occur in mature sieve tubes, thereby weakening the original protoplasmic-streaming theory still further. However transcellular movement of materials has been described by Thaine (1961); for some this provides compelling evidence and for others only controversial evidence for a protoplasmic theory of translocation (Canny, 1962*b*; Thaine, 1964). Thaine defined transcellular streaming "as the movement of the particulate and fluid constituents of cytoplasm through linear files of longitudinally oriented plant cells. . . . This phenomenon is not confined to sieve tubes, but it is suggested that some of the unusual properties of sieve tube cytoplasm are due to their extreme specialization as transcellular streaming pathways."

The very existence of these *strands,* as they became called, has aroused a heated debate. Some authors felt that Thaine was seeing diffraction lines (Esau et al., 1963), but Thaine had seen particles along the strands, and, furthermore, the strands came into and out of focus as the objective was moved *focused* up and down. Strands are also observed under the interference microscope, and this strengthens the case for their existence. Additional support comes from Jarvis and Thaine (1971), with electron micrographs showing strands in a sieve element from *Cucurbita pepo.* The plant material had been quick-frozen in Freon-12 held at the temperature of liquid nitrogen. This method removes any doubt about the reality of structures seen under the electron microscope because disruption from chemical fixatives or ice-crystal formation was avoided. We can therefore be confident that sieve plates have well-defined strands passing through their pores and that in this particular tissue pores of the sieve plate may be completely blocked.

Fig. 2-10 ^{14}C-labeled photosynthate was generated within sugar beet leaves by exposing the leaf blade to $^{14}CO_2$ for about 1 min. The distribution of label down consecutive 2-cm sections of midrib and petiole shows a declining profile which advances with time (Mortimer, 1965).

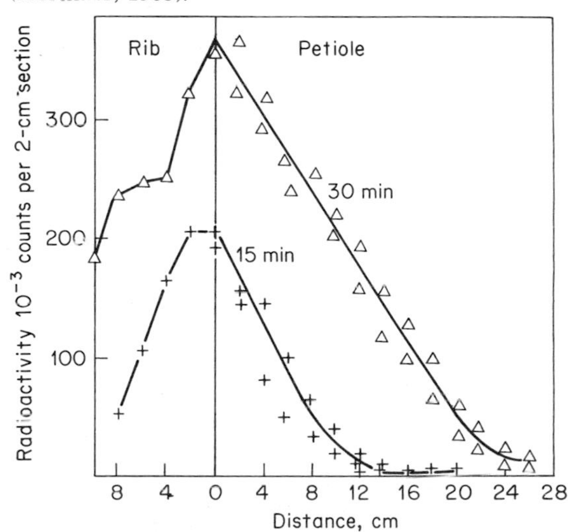

Assuming that strands exist, how can they contribute to translocation? Particulate material moves readily, as observed for cell inclusions by all observers of streaming and for foreign particles by Worley (1966), although some uncertainty exists whether these particles move in, on, or even independent of the protoplasmic strands; i.e., are the strands solid or tubular? Dissolved materials present more difficulty. Canny (1962*b*) imagines that solute diffusion initiates phloem transport. Suppose sugar is removed from a sieve element into some sink; then sugar concentration in this element will fall, and a wave of readjustment should pass along the sieve tube. A greater rate of sugar utilization favors a steeper gradient and hence a higher rate of translocation. Canny's wave of readjustment relies upon transcellular streaming and will occur much faster than could be sustained by normal diffusion. The transcellular strands might affect only sugar transport by stirring sieve-tube contents, thereby hastening the movement of sugar from one end, where it is being introduced, to the other, where it is being withdrawn. An alternative and intriguing explanation for the motive force, originally proposed by Thaine (1969), has been developed by Aikman and Anderson (1971). Working on the proposition that transcellular strands are microtubules (Wooding, 1969), Aikman and Anderson envisage peristaltic contractions running as a series of waves along transcellular strands driving a longitudinal flow of solution. The strands are known to be proteinaceous and of characteristic types (see Walker and Thaine, 1971), so that rhythmic contraction could very well occur. How biochemical energy could be transduced into a physical force within the sieve tube is open to speculation; perhaps it is analogous to muscle contraction.

Electroosmosis Within this context of contractile protein and transcellular strands, one further possibility for motive force must be discussed, that of electroosmosis. This concept was proposed independently by Fensom (1957) and Spanner (1958). Briefly, these workers proposed that sieve plates could act as a sort of pumping station due to electroosmotic forces. Fensom suggests this force stems from a differential permeability of sieve plants to H^+ and HCO_3^- ions, while Spanner envisages a circulation of K^+ ions between sieve tubes and companion cells, via sieve plates. Ions would move in response to this potential gradient across the sieve plate, and water molecules would be swept along, thereby generating a pressure gradient in the downstream direction. Differential translocation of dissociated vs. nonionized solutes in *Heracleum* [for example, K^+, tritated water (THO), and ^{14}C; see Fensom and Davidson (1970)] could suggest an electroosmotic component, but the forces involved are probably too modest to entirely account for translocation.

Kinetics and Motive Forces: Long-Distance Translocation

Tall trees must move enormous amounts of organic and inorganic material over large distances. This requirement introduces a fresh set of problems compared with phloem transport in leaves, petioles, and stems of herbaceous plants. Close proximity of source and sink generates a steeper concentration gradient, which in turn favors a higher rate of translocation. An importing fruit might be within a few centimeters of source leaves, but a tree's root system could well be 50 to 100 m from its crown; is the same translocation mechanism likely to hold in both situations?

The diffusion analog, together with related theories discussed earlier, can account for the various features of phloem transport over short distances, including bidirectional movement, selective transport of different solutes, sensitivity to metabolic inhibitors, and response to concentration gradient. The existence of transcellular strands is implicit in these theories, but are such strands continuous through the length of a tree? Mature sieve elements of secondary phloem in woody plants have been examined in great detail (see, for example, Evert and Derr, 1964), and while as many as five strands have been observed to traverse a single sieve plate, long-distance continuity is not established. Despite this observation, the general impression gained by Evert and Derr was that sieve-plate pores are generally open, a key requirement for an overall-flow system.

Münch's pressure-flow theory This famous theory was advanced in 1930 by Münch (following a preliminary report in 1926), prompted, it seems, by Pfeffer's results on osmotic control of water movement. The familiar diagram illustrating its mechanism is reproduced in Fig. 2-11. Two reservoirs, *A* and *B* represent phloem in leaf and root regions, respectively. These source and sink regions are connected by a narrow pipe, which represents sieve tubes. Increases in leaf carbohydrate (*A*) lift osmotic potential, and hence turgor pressure, whereupon solution will flow toward *B*, provided a turgor gradient is maintained. Removal of water and solutes at *B*, that is at the sink, sustains this gradient.

Fig. 2-11 Diagrammatic model the Münch pressure-flow system, showing a closed system (the phloem), in which a rise in the sugar content at *A* (the leaves) would create a gradient in concentration such that there would be a mass flow of solution toward point *B* (the stem apex, root, or fruit). The increase in osmotic value would lead to the entry of water at *A* and its loss at *B*, and the water would then return through the xylem system (right to left).

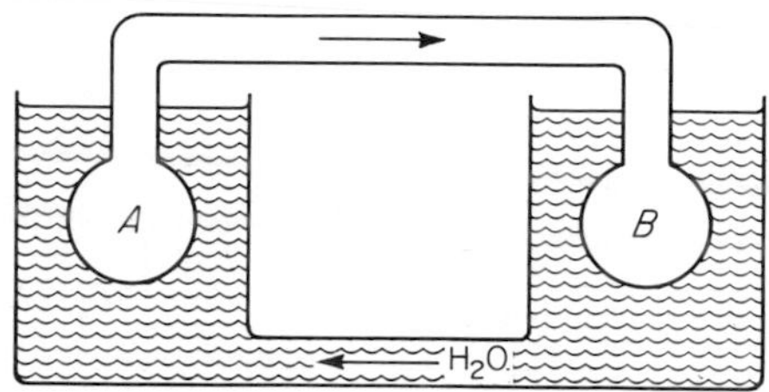

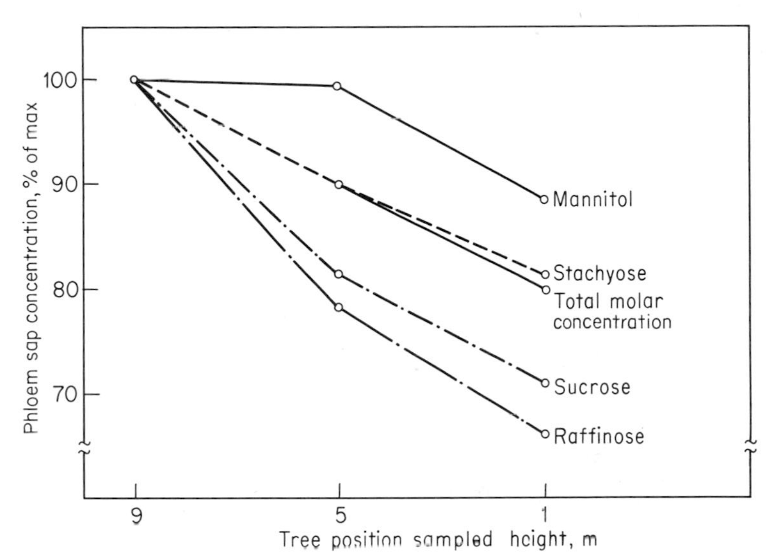

Fig. 2-12 There are gradients in sugar content down the stem of a white ash tree in July, when phloem translocation is active, and the gradient is evident not only as total sugar content of the phloem sap but also as the content of three separate sugars and mannitol (Zimmermann, 1957*b*).

If pressure flow is to apply in nature, several physical requirements must be met:

1 When translocation occurs, there must be a decreasing gradient in solute concentration away from the leaves.

2 Lateral walls of sieve tubes must have restricted permeability to these solutes.

3 Longitudinal conductivity must be high, i.e., a low resistance to flow, as would be favored by unplugged sieve pores.

4 Water must be able to return from sink to source (presumably via the xylem).

Experimental support for these points is as follows.

1 There is a gradient in molar concentration in the direction of transport, as illustrated in Fig. 2-12. Over a distance of 8 m down the trunk of white ash, Zimmermann (1957*b*) found a 20 percent drop in sap molarity. This drop applied to all sugars as well as to their overall concentration. Is this gradient strictly a consequence of leaf activity? If so, the gradient should disappear in the absence of leaves. This did in fact happen when leaves fell off in autumn. However, as trees approach dormancy, both leaf activity and substrate requirements by the root system are likely to diminish, thereby altering translocation patterns. More convincing evidence in favor of pressure flow was Zimmermann's observation that the same loss in gradient occurs when leaves are removed mechanically during the growing season (Fig. 2-13).

Nitrogenous materials in phloem sap apparently do not show such a gradient; they are occasionally more abundant toward the base than higher up the tree. Phosphate content can be similarly distributed, but, as with nitrogenous materials, this picture is complicated by an exchange with xylem contents which is relatively fast compared with carbohydrate. Consequently a concentration gradient for inorganic ions based on phloem exudate does not necessarily imply flow in that direction. This effect also confounds interpretation of multiple tracer experiments where mixtures of labeled sugar, inorganic solutes, and tritiated water are applied simultaneously to a translocating system, as in experiments on stem segments of willow reported

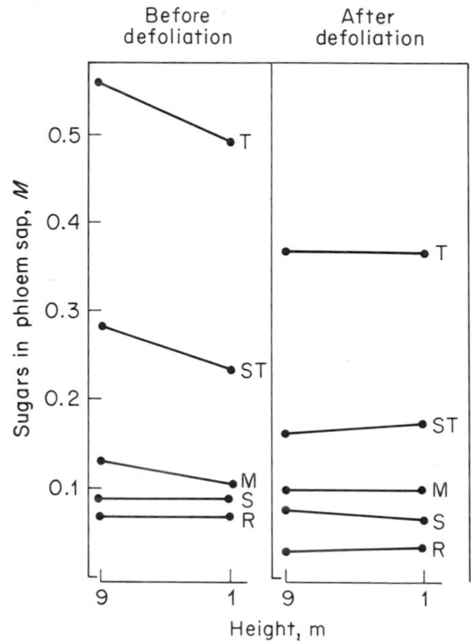

Fig. 2-13 Mass-flow theories of translocation imply that a concentration gradient should exist along the direction of flow. Sugar-concentration gradients down the trunk of white ash confirm this prediction. Their disappearance after defoliation suggests that the gradient is a direct consequence of assimilatory activity by the leaves. M = mannitol, S = sucrose, R = raffinose, ST = stachyose, T = total concentration (Zimmermann, 1958).

by Peel et al. (1969). These authors demonstrated different rates of solute movement which were in turn independent of THO translocation (Table 2-2). Nevertheless, high lateral permeability, especially to THO, weakens the argument made against a mass flow of solution.

2 On this point of lateral movement, Zimmermann (1957*b*), again using white ash, found that exudate becomes more dilute after an initial cut is made into the phloem. If sugars could move as freely as H_2O through lateral walls of sieve tubes, there would not be any dilution. Furthermore, after cutting into a tree's phloem, there is a sharp loss in turgor above and below the cut but no measurable change lateral to the cut. Presumably, osmotically active solutes (mainly sugars) have not moved across the trunk. This assemblage of observations, representative of many others, gives the overall impression that sieve tubes display good longitudinal permeability.

3 One of the most contentious issues in the massflow theory is the level of resistance offered by sieve pores to bulk movement of solution. The existence of callose plugs (Figs. 2-2 and 2-3) has been emphasized repeatedly and tends to militate against a mass-flow concept because excessively high pressures would have to be generated in order to sustain estimated rates of flow. Weatherley and Johnson (1968) have estimated the pressure differential that would have to exist for laminar flow through willow sieve tubes and calculate that a pressure gradient of about 0.25 atm/m would be required to sustain a velocity of 100 cm/h for a 10 percent sucrose solution. This calculation ignores sieve plates. The required gradient is more than doubled when they are taken into account but could still account for transport over a distance of, say, 25 m. For translocation to occur by pressure flow over greater distances (100 m, as in tall trees) either sieve pores would have to be totally unrestricted or some special mechanism for rapid solute transfer across the sieve plate would have to be invoked.

Although sieve pores are not totally unobstructed in nature, the situation favoring mass flow can be retrieved. First, Eschrich (1965), in reviewing the physiology of callose formation, asserts that callose cylinders, once imagined to be normal features of functional phloem, can now be regarded as fixation artifacts. Callose deposition and subsequent removal is in fact surprisingly fast. It can be induced by a wide variety of stimuli, including ultrasound, foreign chemicals, and heat (see McNairn and Currier, 1968, and literature cited).

Second, Evert and Derr (1964), after examining slime substance and strands in secondary phloem of six woody perennials, have decided that sieve pores are mainly open and that intervacuolar continuity exists between cells. These observations favor pressure flow, notwithstanding the strands that do run from cell to cell through sieve-plate pores (Worley, 1973).

4 The fourth requirement to be met by a pressure-flow system, as outlined previously, is that water can return via the xylem. Developing fruits, especially those on long pedicels, provide suitable objects for this type of experimentation because for all practical purposes, they are wholly dependent upon the parent tree for substrate and can easily be measured for weight and volume.

In a provocative paper, Clements (1940) reports measurements of the growth of sausage-tree fruits; he calculated that the deposition of dry matter in the fruit would require an enormous flow of water in a mass-flow system, as much as 2 to 5 l of water per fruit per day. These calculations would lead one to expect a veritable fountain of water returning in the xylem from the fruit, but severing the xylem pedicels of young fruits did not reveal a water flow. This challenge to the mass-flow theory can be tentatively accounted for on the basis of two items: (1) the calculated water movement was based on phloem-sap concentrations of 0.1 to 0.03 *M*, and in fact the sap is usually 5 or 10 times more concentrated; and (2) during the first half of the fruit growth period, when enlargement is proceeding most rapidly, the gain in dry weight is almost exactly matched by increases in volume, suggesting that a return flow of water would be expected only during the last half of the fruit growth period.

Concerning this earlier phase of fruit growth, Canny (1973), with active cooperation from Clements, undertook further measurements on sausage trees growing at the University of Hawaii. It was established that the spectacular growth in fruit size which occurs soon after pollination is not solely attributable to translocation per se. Fruits on a single stalk can gain up to 8 kg (fresh weight) in 6 weeks, but at that stage the fruit is 88 percent water, and a specific mass transfer of less than 2 g dry wt per cm^2 of phloem per hour is enough to account for fruit growth (compare with other translocation data in Table 2-1).

The inward flow of water must be substantial, and Clements' original paper does emphasize the need for a return flow in the xylem if mass flow were to operate in the phloem. This aspect of translocation has received little attention, but there is one oblique reference to such flow in a crop plant—the peanut.

This plant presents an unusual case with respect to the flow of xylem sap out of the fruits. The underground peanut fruit is particularly susceptible to deficiency of some inorganic nutrients unless they can be taken up by the fruit. Harris (1949) has noted that supplying inorganic nutrients to the roots will not satisfy the nutritive needs of the fruit, an observation consistent with the presumption that xylem flow is out of the fruit and that nutrient translocated in the xylem is available only if taken up by the fruit itself.

A final criterion for the feasibility of pressure flow is the demonstration that sap actually flows. Ziegler and Vieweg (1961) used an ingenious approach to provide this evidence. They pulled bundles of sieve tubes out of *Heracleum* petioles for short spans, leaving bundles still attached at each end. Heat was applied via a light beam to a localized area of phloem. They then measured the progress of heated material and thus estimated a velocity for sap movement. Their measuring system (thermocouples) was so sensitive that elevating sap temperature 1°C was sufficient to show movement; the velocity noted was 35 cm/h.

Table 2-2 When three radioactive components are supplied to a strip of willow bark and the movement of the labels along the bark is sampled through an aphid stylet, the specific activities of ^{14}C from sucrose and ^{32}P from phosphate are detected long before the ^{3}H from water (Peel et al., 1969).

Time from tracer application, h	Specific activity in stylet exudate 10^{-2} cpm/μl		
	^{3}H	^{14}C	^{32}P
1	0	7	6
2	0	13	26
3	0	10	37
4	0	8	38
5	0	9	43
6	0	9	47
7	0	11	85
8	0	12	91

Speed of translocation in large trees has been a subject of some debate, and results of Huber et al. (1937) are widely quoted. These German workers reported an apparent velocity of between 1.5 and 4.5 m/h for red oak (*Quercus borealis* L.). Their results are based on a sucrose-concentration wave, which apparently descended the trunk and which they followed by collecting exudate from excisions at different positions down the trunk. As Zimmermann (1969) points out, "there are reasons why one must regard this 'wave' with some scepticism. It is possible that the wave does indicate phloem transport, but is distorted by the effect of xylem tensions upon the concentration of sieve tube exudate. Another reason to throw doubt upon the concept of the moving concentration wave is its rather high apparent velocity. . . ." Canny et al. (1968) have seized upon this second point, and taking a velocity of 4 m/h at a concentration of 0.18 g/cm^3 (as reported for exudate), they estimated the oak tree must have covered 4 acres if it was to generate sufficient photosynthate to account for this high value.

Wide reference to these early data has done the pressure-flow hypothesis some disservice in the eyes of opposing protagonists, but this detraction has been rectified by the more elegant work of Zimmermann (1969) on a moving wave of concentration *ratio.* Unlike red oak, white ash does not translocate principally sucrose. This sugar accounts for only one-fifth or one-tenth of total carbohydrate in transit. The translocated sugar is a mixture of stachyose, raffinose, sucrose, and D-mannitol. Over many years, Zimmermann and workers noticed a slight but consistent diurnal variation in the relative levels of these different sugars supplied by the foliage to the trunk. This feature of white ash enabled Zimmermann to repeat the hitherto unconfirmed work of Huber, but in a less equivocal way, because this *ratio* wave would not be distorted by water potential in the neighboring phloem even though absolute sugar concentration could be influenced. Figure 2-14 illustrates the downward advance of a raffinose-stachyose ratio wave from 8 A.M. to 4 P.M. Data on sucrose + raffinose/stachyose show a wave which moves down at a velocity of 30 to 70 cm/h. In white ash, solute concentration of phloem exudate varies from 0.2 to 0.5 g/cm^3, so that specific mass transfer in sieve tubes must be between 6 and 18 g/cm^2-h. The values are of the same order as data compiled in Table 2-1.

Fig. 2-14 The ratio of raffinose to stachyose in the phloem sap of white ash shows a gradient down the tree in the afternoon, when translocation is most active, compared with early morning, when it is relatively inactive (Zimmermann, 1969).

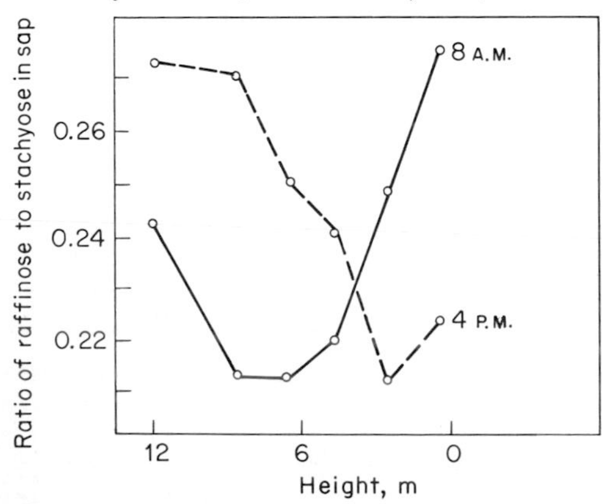

Zimmermann's data (Fig. 2-14) therefore strengthen substantially the case for pressure flow in trees by demonstrating a bulk flow of solution.

Modifications of Pressure Flow

Having established that sap can actually flow, i.e., solutes can move *en masse* as a solution, we must now reconcile this contention with additional observations on environmental and metabolic influences. Contributions from three independent laboratories impinge on this problem, and, predictably, three new versions of Münch's original pressure-flow theory have emerged. Eschrich et al. (1972) developed a theory of *volume flow;* Cataldo et al. (1972*a,b*) present a scheme for *solution flow;* while Qureshi and Spanner (1973*a,b,c*) prefer *active mass flow* to accommodate observations on their particular system. Each of these theories contributes to our appreciation of possible mechanisms for translocation and will serve as introduction to the *bimodal theory* of sucrose translocation presented by Fensom (1972), which incorporates features of both mass flow and activated diffusion.

Volume flow From their studies at Göttingen (Germany), Eschrich et al. (1972) acknowledge the evidence for mass flow of a solution but pose a question about its driving force. Is it an overall pressure gradient, as conceived in Münch's original theory? Because a turgor gradient along the entire length of the translocating system has not been established, and because partly occluded sieve plates would offer substantial resistance, the idea of an overall gradient in pressure becomes less tenable. The problem would take on a different perspective if the pressure differences which act as a driving force were more localized, even at the level of individual sieve tubes. Could sufficient pressure be generated at that level to drive observed rates of flow? Accordingly, Eschrich et al. (1972) mathematically analyzed the flow of solutions along tubular semipermeable membranes. Solution flow in closed (turgid) tubes and in open tubes without turgor was considered. In summary, their theoretical model (based on nonequilibrium thermodynamics) could account for experimental observations purely on the basis of hydrostatic and osmotic pressure differences across the semipermeable membranes of these tubes. A hydrostatic pressure gradient along the direction of solution flow was not required.

Solution flow An essential requirement of any mass-flow theory is that different solutes should move at the same speed because they are supposedly moving under physical influences as a solution. Marshall and Wardlaw (1973) make such an observation for the simultaneous translocation of ^{14}C-labeled assimilates and ^{32}P-labeled phosphate in wheat plants, but both these components are solutes. If material is really moving as a solution, both solutes and solvent should also travel concurrently. Experimental observations on ^{14}C assimilate and tritiated water (THO) show a discrepancy (see Fig. 2-15), and this becomes a persuasive argument against the mass flow of a solution. The situation could be retrieved by firm evidence that THO shows a lateral leakage not shared by solute molecules. [Peel et al. (1969) had in fact demonstrated such a leakage in their bark strips from willow.] Accordingly, Cataldo et al. (1972*a,b*), working at Ohio State University, developed a mathematical model for reversible exchange of THO between the lumen of sieve tubes and surrounding phloem tissue in their sugar beet leaf system. [The original mass-flow model of Horwitz (1958) bears some analogy.]

Cataldo and coworkers put their model to a practical test by studying the movement of THO and ^{14}C sucrose supplied simultaneously to a primary vein from a small glass capillary. Distribution profiles down the petiole of a fed leaf are shown in Fig. 2-15. The THO profile typically lags behind the sucrose profile, but if mass flow or solution flow is really operating, both solute and solvent should travel concurrently. In Fig. 2-15, THO showed an apparent velocity of 35 cm/h, whereas sucrose moved at 72 cm/h. This discrepancy is explained by the model in terms of lateral movement by THO. Cataldo et al. (1972*a,b*) support their argument on the basis of mathematical prediction, plus more experimental evidence. Insertion of a cold block over 8 cm of the petiole displaced the profiles and led to an increase in tracer accumulation below the block. THO showed an apparent accumulation 5 times greater than for sucrose, which is highly suggestive of lateral leakage on the part of THO and coincides with Peel's earlier observations on willow bark. Presumably, then, both THO and sucrose were moving more or less concurrently and within the same phloem conduits, as demanded by theories depending upon a bulk flow of solution.

Active Mass flow In studying the translocation characteristics of long stolons of *Saxifraga*

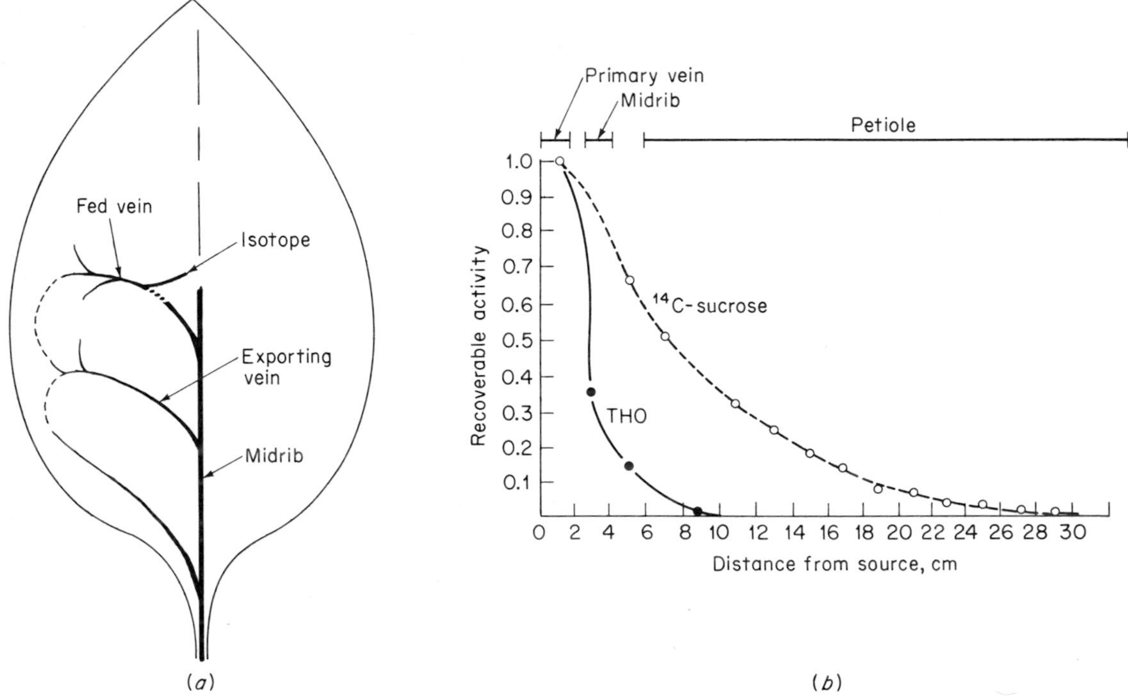

Fig. 2-15 The translocation of radioactive sucrose and water out of the leaf of beet may occur at different rates. (*a*) ^{14}C-sucrose and THO were supplied to a vein of the leaf. (*b*) The profile of the two isotopes was determined after 25 min. Apparent velocities were 72 and 35 cm/h (Cataldo et al., 1972*a*).

sarmentosa, Qureshi and Spanner (1973*a,b,c*), working at Bedford College, London, noted some general characteristics of mass flow but with a certain metabolic involvement. The movement of both ^{137}Cs and naturally assimilated ^{14}C was strongly inhibited by anoxia (N_2 atmosphere), although the effect was reversible fairly readily, while local darkness also had an adverse effect on translocation (Fig. 2-16). Since callose formation did not appear to be involved in this response, their results tended to support the early contention of Mason and Phillis (1936) that oxygen is required to maintain some "special state of the cytoplasm" which enables translocation to occur. This special state certainly appears operative in squash plants, where Sij and Swanson (1973) have demonstrated a rapid decline in petiolar translocation under anaerobic conditions.

Geiger and Christy (1971) made similar observations with their test system, where anoxia at the sink leaf of a small sugar beet plant caused a rapid fall in the import of labeled assimilate. This system did not differentiate between unloading at the sink and translocation toward the sink, but Qureshi and Spanner (1937*b*) went on to analyze this metabolic component still further. They studied the movement of both ^{14}C-labeled sucrose and ^{14}C-labeled natural assimilate between parent and daughter plants of *Saxifraga,* similar to those illustrated in Fig. 2-16. As predicted by mass-flow theories, long-distance transport was strictly unidirectional and was dependent on source-sink relationships, i.e., reversible according to illumination of parent or daughter plant (see Fig. 2-17). However, their subsequent experiments (Qureshi and Spanner, 1973*c*), showed that 5 m*M* 2,4-

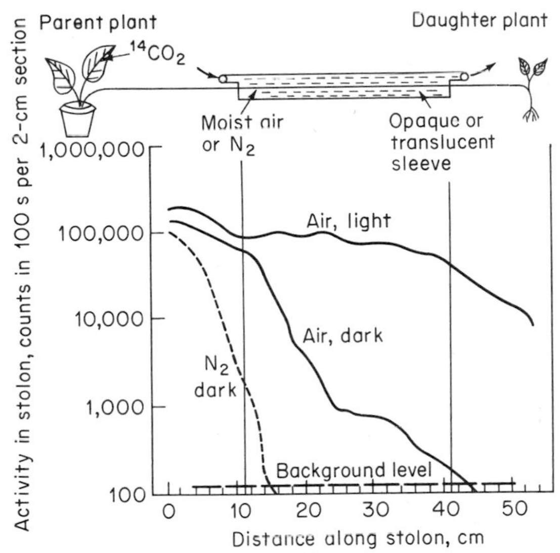

Fig. 2-16 The movement of ^{14}C-labeled natural assimilates down the stolon of *Saxifraga sarmentosa* (experimental system illustrated) is disrupted by either darkness or an atmosphere of N_2. These distance profiles (4 h after the start of feeding) combined with other data on 2,4-dinitrophenol imply some metabolic involvement in this system's translocation. (*Adapted from Qureshi and Spanner, 1973a.*)

dinitrophenol (DNP) applied to the center of a 30-cm stolon gives a strong but reversible inhibition of phloem translocation for both ^{137}Cs and ^{14}C assimilates. The DNP effect seems localized in the sieve tubes, but callose formation is not involved.

Taken collectively, the results of Qureshi and Spanner (1973*a,b,c*) are in conflict with the original Münch hypothesis because the translocating system appears to involve a metabolic component. On the other hand, the overall movement of assimilate is unidirectional, and bidirectional translocation of applied tracers is highly localized, so that the diffusion analog is an unsatisfactory alternative explanation for this system. Qureshi and Spanner reconciled these conflicts in their metabolically operated form of mass flow. The metabolic component was regarded as being synonymous with protein contractility or potassium electroosmosis.

Bimodal theory Instead of making a clear distinction between some form of mass flow and a mechanism which relies on activated diffusion, Fensom (1971) proposes a composite theory of translocation which relates to isolated but living phloem strands of *Heracleum*. Sucrose translocation is thought to occur in two chief modes (although subsidiary modes also exist).

The first mode is based upon microperistaltic movement along strands or fibrils of contractile lipoprotein, thought to extend axially through the sieve tube (evidence from light and electron microscopy). Aikman and Anderson (1971) analyzed this peristaltic wave in some detail and concluded that such a model would be quantitatively acceptable in terms of solution velocity, driving pressure, and rate of energy dissipated.

The second mode of translocation (Fensom, 1972) is conceived of as a mass flow of solution around the contractile microfibrillar material (MFM, or P protein, discussed previously). This mass flow is relatively slow and is stopped by callose formation, but it is also stopped if the MFM becomes inoperative. [How contractile

Fig. 2-17 The direction of translocation along the stolon of *Saxifraga sarmentosa* reflects a source-sink relationship. Normal translocation is from parent to daughter plant (Fig. 2-16), but providing labeled photosynthate to the daughter plant can show translocation in the opposite direction if the parent plant is darkened. Photosynthetic fixation was 1 h, translocation 3 h (Qureshi and Spanner, 1973).

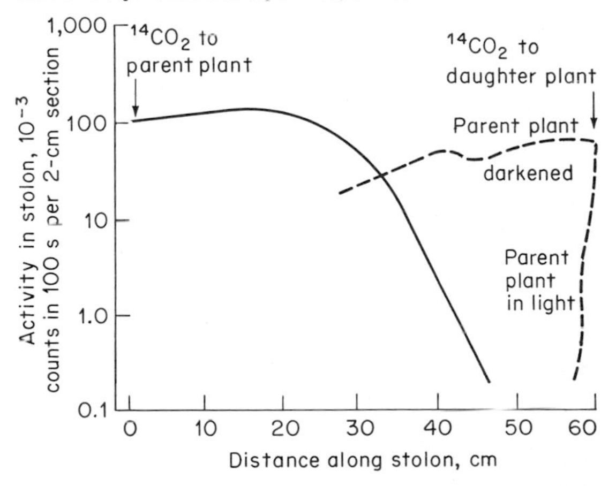

material contributes to mass flow is not apparent, but this metabolic involvement suggested by Fensom (1972) is reminiscent of Qureshi and Spanner's view mentioned above, and of Kursanov's (1961) experiments on DNP inhibition.] Microfibrillar material could also participate directly in the movement of sucrose as pulses traveling at about 400 cm/h. This mode of translocation would be stopped by cold blocks but not by callose because these proteinaceous strands can stretch between successive sieve tubes. Fensom (1972) estimates that anywhere between 50 and 350 microfibrillar strands can pass through a single sieve-plate pore.

A third, though minor, mode of translocation is also indicated as a surface-layer component which operates at high speed. This third mode is demonstrated by tracers, particularly in the form of amino acids, which travel along the surface of living and nonliving tissue. Nelson (1962) reported extraordinarily high speeds of apparent movement (7,000 cm/h) during his ^{14}C tracer experiments, which Fensom (1972) now suggests to be reflections of his third mode.

This bimodal or trimodal system of translocation may offer distinct advantages to the plant. The component of surface movement, albeit small, could allow rapid control of subsequent mechanisms by activating the system (surface charge, enzyme activity, conformational change of P protein). Loading of translocate and pulse flow may then operate from source to sink. Mass flow follows if one direction of the pulse flow predominates. Sieve plates with open pores would act as pumps but at the same time provide the plant with a margin of safety by forming callose plugs in response to acute stress or physical damage. Companion cells would still fulfil their traditional supporting role by supplying energy to sieve tubes and by participating in loading and unloading of assimilates.

A composite view Plant physiologists agree that photosynthetic products (principally sucrose) move through the phloem from areas of synthesis to sites of consumption, but there the agreement usually ends. One school favors the diffusion analog, and a second prefers mass flow in one form or another, i.e., with or without metabolic involvement.

Promoting one view at the expense of another amounts to a misrepresentation of the real situation. Sharp divergence of opinion often stems from imperfect experiments on dissimilar plants with a wide range of techniques and a variety of solutes. A plant's translocation system can undoubtedly deal with a miscellany of substances, including both ionic and nonionic solutes and even polymeric materials up to the size of virus particles. Mechanisms which sustain their movement might eventually prove to have equivalent subtleties: mass flow may well account for the movement of sucrose in solution, but virus particles do not necessarily depend upon the same mechanism for their distribution through the plant.

The prospect of developing a totally comprehensive theory which accounts for all forms of translocation and which is acceptable to all protagonists certainly seems remote.

Loading and Unloading Substrates

Regardless of motive forces within sieve tubes, translocation cannot occur unless it is preceded by solute uptake. The concentration of phloem sap is commonly greater than in surrounding tissues of stem or leaf, and this feature, coupled with the specificity of translocated sugar, is highly suggestive of active accumulation. This loading step might well be an intrinsic part of the mechanism of translocation. The general question of phloem entry then has two parts: (1) the biochemical mechanism behind accumulation and (2) the anatomical arrangements within the tissues involved.

First, the biochemistry. A wide range of living systems including microorganisms, leaf disks, storage tissue, stem-pith parenchyma, root tips, and excised vascular bundles can accumulate sugar against a concentration gradient. The accumulating system can be highly specific with regard to substrate, and when sucrose is supplied, the molecule can be accumulated intact [as in castor bean cotyledons, Kriedemann and Beevers (1967*b*)] or following hydrolysis [as in sugar cane storage tissue, Hawker and Hatch (1965), and bean-pod endocarp, Sacher (1966*b*)]. In terms of accumulating capacity, vascular tissue is far superior to its surrounding parenchyma. This has been demonstrated by Bieleski (1966), who utilized a number of species. In addition, phloem-containing tissues were also able to absorb solutes at a rate 4 to 35 times faster than comparable parenchyma. Bieleski then demonstrated an aging phenomenon on his material following aeration in dilute $CaCl_2$ or $CaSO_4$ for some hours. Fresh tissue had an accumulating mechanism of low affinity but high capacity, while aging allowed the development of a second system of high affinity and low capacity. The consequences for phosphate uptake are demonstrated in Fig. 2-18. Taken collectively, these data (plus a wealth of related information on other tissues) assure us that vascular bundles have the biochemical means of loading solutes into the translocation system. The pathway of loading and subsequent unloading is our next concern.

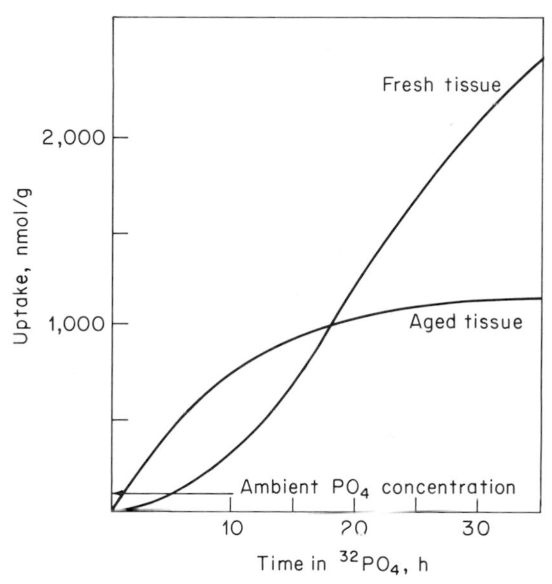

Fig. 2-18 Vascular bundles excised from celery stalks were incubated in ^{32}P-labeled phosphate solutions (10^{-4} *m* KH_2PO_4). Tissue which had been aged (held for 20 to 24 h in 10^{-4} $CaCl_2$ or $CaSO_4$) developed different uptake characteristics, although both samples achieved internal concentrations well in excess of external levels of phosphate (arrow in Figure). Aged tissue achieved a greater affinity for phosphate, but its accumulation system was of lower capacity (Bieleski, 1966).

Sugar beet provides suitable material for this exercise, and Esau (1967) has made a detailed study of vascular systems in leaves, with particular reference to minor veins. These minor veins are deeply embedded in the leaf mesophyll, with photosynthetic tissue above and below. Repeated branching of minor veins minimizes the distance between sites of sugar synthesis and those of loading into the translocation stream. The leaf's overall vascular distribution is therefore well adapted to the task of scavenging assimilates and supplying photosynthesizing cells with water, the xylem elements lying alongside the sieve tubes at these vein endings.

In the minor veins of sugar beet leaves there are companion cells and phloem parenchyma, which are particularly rich in organelles. Geiger and Cataldo (1969) found that sucrose uptake by these minor veins was 50 to 80 times the rate established by Bieleski on excised phloem, a difference probably related to these organelle-rich cells of minor veins. In such cells the surface-to-volume ratio would be conducive to sieve-tube loading (Geiger et al., 1971).

Further anatomical refinements like those evident in *transfer cells* aid the actual loading process. These highly specialized cells (Figs. 2-19 and 2-20) are appressed to the sieve tubes (Fig. 2-19) and presumably enhance solute movement from adjacent parenchyma into phloem conduits. Gunning and Pate, and their colleagues at Belfast (Gunning and Briarty, 1968; Gunning and Pate, 1969) have described these structures in almost 1,000 species of higher plants. Transfer cells are equipped with protuberant wall structures, which increase their surface-to-volume ratio by an order of magnitude compared with smooth-walled cells. This wall modification, shown in Fig. 2-20, suggests a superior capacity for solute exchange. Numerous branched plasmodesmata traverse the wall between transfer cells and associated sieve tubes and amplify this capacity.

If specialized cells of this type are of general significance for phloem loading, we should ex-

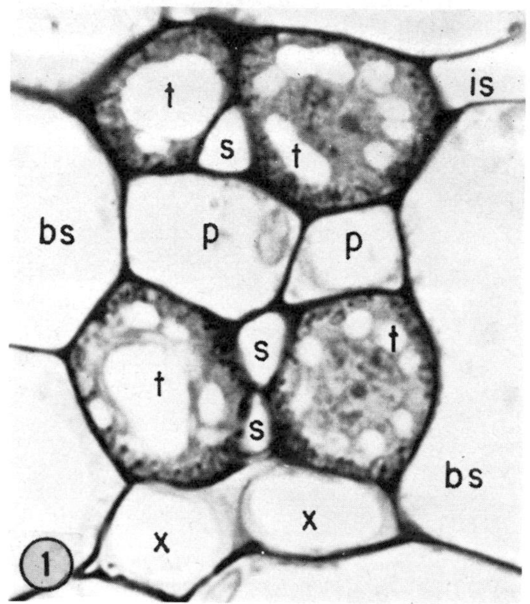

Fig. 2-19 Sieve elements (*s*) within a minor vein of *Pisum arvense* (shown here in transverse section X2000) are ensheathed by phloem parenchyma cells (*p* and *t*). Those cells with dense contents (*t*) have been termed transfer cells. Their location with respect to xylem elements (*x*) and bundle-sheath tissue (*bs*) suggests some role in facilitating solute exchange (Gunning et al., 1968).

pect them not only in actively photosynthesizing organs but also in storage structures or senescing leaves, where substrates are being mobilized and recovered for use elsewhere in the plant. Such cells do in fact occur in these other situations (Pate and Gunning, 1969) and in a specially modified form. This observation further emphasizes their key function in loading and unloading and for solute exchange between phloem and xylem. Transfer cells also participate in the xylem system by aiding movement of materials from roots to leaves. Yeung and Peterson (1972) ascribe such a function to transfer cells in the rosette plant *Hieracium floribundum*, where these specialized cells occur in the plant's rhizome and especially in association with the xylem of foliar traces. O'Brien et al. (1970) also note the exclusive association of transfer cells with tracheidal elements in certain parts of the wheat seedling and go on to suggest their involvement in accumulating nitrogenous compounds from tracheids.

Physiology of Control Systems

A process, like translocation, which consumes metabolic energy, particularly during both loading and unloading phases (Coulson et al., 1972), will obviously respond to environmental factors or chemical regulators (either stimulants or inhibitors). Apart from the indirect consequences of light, CO_2, and temperature (via their effects on photosynthesis and hence substrate availability), there are direct influences of such factors on translocation.

Fig. 2-20 This single transfer cell, shown at higher magnification (X12,000), reveals a great abundance of wall protuberances which in effect increase the surface area of its plasma membrane by at least an order of magnitude compared with a smooth-walled cell (Gunning and Briarty, 1968).

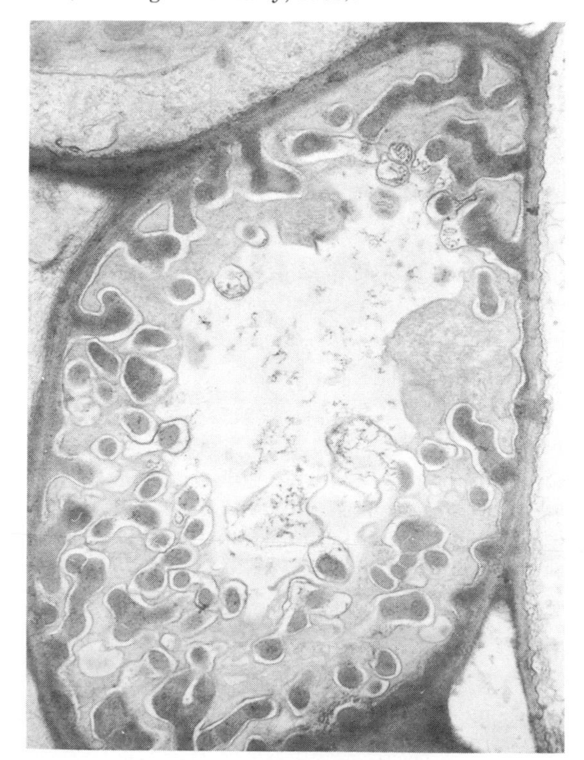

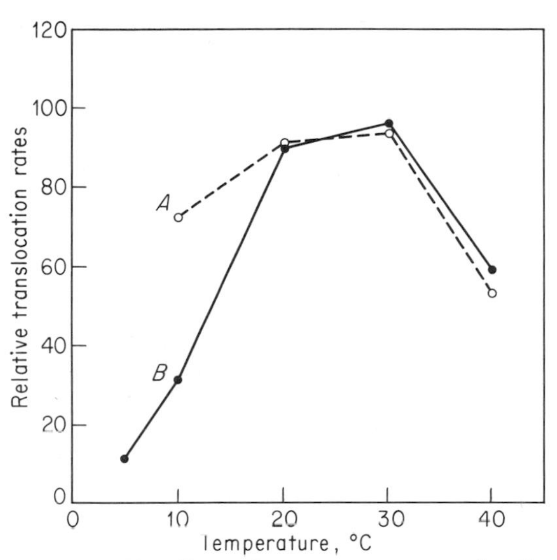

Fig. 2-21 The effects of temperature on translocation of carbohydrate from bean leaves. In curve *A* the temperature treatment is restricted to the petiole of the leaf (Swanson and Böhning, 1951), and in curve *B* the entire plant is given the temperature treatment (Hewitt and Curtis, 1948).

Temperature There has been considerable disagreement regarding temperature effects. Some early evidence suggested a Q_{10} of less than unity, i.e., translocation proceeding better at lower temperature, an anomalous situation for most living systems. Several participants to the argument then provided conclusive evidence that Q_{10} was after all closer to 2, an optimum for translocation being reached between 20 and 30°C. Data taken from such experiments (Fig. 2-21) show a similar pattern irrespective of whether the temperature treatment is given to the whole plant or just the petiole (Swanson and Böhning, 1951; Hewitt and Curtis, 1948).

In view of the basic dissimilarity between phloem and xylem, these systems are also likely to differ with respect to their temperature sensitivity. Swanson and Whitney (1953) have addressed themselves to this problem by making a definitive study of temperature effects on ion ascent and descent. They enclosed sections of stem or petiole in temperature-controlled jackets and followed the movement of different radioactive ions applied either to a leaf above the jacketed section or to the roots below. Comparisons of phosphate, potassium, cesium, and calcium in the downward stream from a leaf showed the same temperature sensitivity as carbohydrate moving in the phloem. A marked peak was obtained at 25 to 35°C (Fig. 2-22). Calcium, which is relatively immobile, was insensitive to the temperature differences. Swanson and Whitney (1953)

Fig. 2-22 The effects of temperature on translocation of cesium, phosphorus, and potassium out of bean leaves are very great. Calcium translocation is poor and quite insensitive to temperature. The temperature treatments were restricted to the petioles of treated leaves; about 5 μCi was applied to each leaf. The calcium values are multiplied by 10 (Swanson and Whitney, 1953).

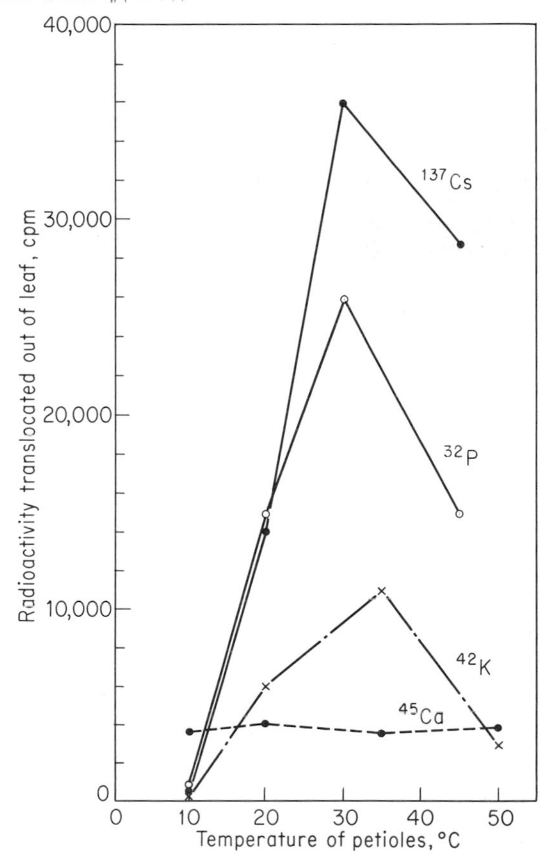

then made a direct comparison of ^{32}P movement in phloem and xylem. Translocation out of the leaf was highly sensitive to temperature, whereas ^{32}P ascending the stem from the root system was insensitive, as would be expected for material moving principally via the xylem. The same premise would lead one to expect that stem girdling would not alter upward translocation in the xylem, a prediction also borne out by other results of Swanson and Whitney.

Localized reduction in stem or petiole temperature to between 0 and 5°C generally reduces translocation of either ^{32}P- or ^{14}C-labeled assimilates, but such effects can be transient. Such a response is demonstrated in Fig. 2-23 as an abrupt reduction in movement down the petiole due to chilling but with subsequent recovery despite low temperature. Such low-temperature inhibition is time-dependent (Coulson et al., 1972). The readjustment that occurs at 1°C (Fig. 2-23) could derive from sugar buildup on the supply side of the chilled zone, resulting in a steeper concentration gradient through that portion of the petiole. In all probability, sieve-tube permeability was also altered by temperature (Giaquinta and Geiger, 1973), so that lateral transfer may have occurred (as discussed previously). This effect will complicate interpretation of data like those in Fig. 2-23.

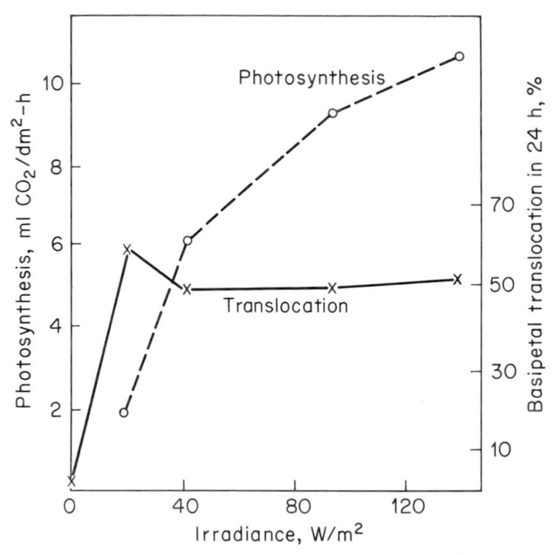

Fig. 2-24 Translocation of ^{14}C-labeled assimilates in sugarcane leaves is saturated by a much lower light intensity than photosynthesis is (Hartt, 1965).

Fig. 2-23 Translocation in sugar beet plants is acutely sensitive to temperature changes but shows some adjustment with time. $^{14}CO_2$ was supplied to a mature leaf and radioactivity measured in a young expanding leaf; petiole temperature was controlled along a 2-cm region (Swanson and Geiger, 1967).

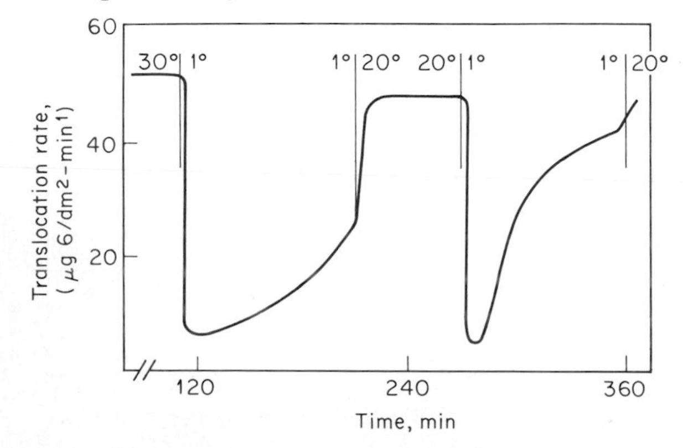

Light The movement of assimilates out of a leaf can depend upon radiant energy. Hartt et al. (1964) found with sugarcane leaves, for example, that the export of ^{14}C-labeled assimilates from a source is reduced by almost 50 percent if light is excluded following $^{14}CO_2$ assimilation. Not only is photosynthesis reduced at low intensities, but the proportion of fixed carbon leaving the leaf is also reduced. The response of basipetal translocation to light intensity is shown in Fig. 2-24. The saturating intensity for translocation is substantially lower than the anticipated value for photosynthesis and gave rise to Hartt's (1965) idea of a *phototranslocation* process, which would operate in conjunction with photosynthesis. The highly characteristic anatomy of leaves on C_4 plants, where vascular bundles are ensheathed by a cylinder of chlorophyllous tissue, makes them likely candidates for such a phototranslocation process. Sunlight trapped by bundle-sheath plastids would provide a readily available energy source for loading photosynthate into the vascular system. Hofstra and Nelson's (1969) data, which show a much faster turnover of recent assimilate in leaves of C_4 plants (corn and *Sorghum*) compared with a number of C_3 plants, supports this view. Some of their data are shown in Fig. 2-25. Moss and Rasmussen (1969) made a similar distinction between leaves of corn (C_4) and sugar beet, which differed by a factor of 3 in their rates of ^{14}C export.

Certain effects of light on ^{14}C translocation are reminiscent of similar effects on the movement of a flowering stimulus. The subject is discussed more fully in Chap. 12, but it is worth mentioning here that translocation of the flowering stimulus is impeded by darkening a portion of a leaf blade, in much the same way that translocation is slowed down. The two systems were therefore thought to have similar requirements for movement, although independent movement has been demonstrated for *Pharbitis nil* (King et al., 1968).

Light intensity can influence growth patterns in whole plants and will therefore modify the distribution of assimilates. Root and bud growth are commonly inhibited by low intensities, and this can lead to a reduction in assimilate flow to the root system (see, for example, Nelson, 1964, with conifers). On the other hand, leaf expansion in soybeans can respond positively to increased light intensity, with subsequent repercussions on assimilate import. Predictably, any restraint on leaf expansion will then lower this demand (Thrower, 1964).

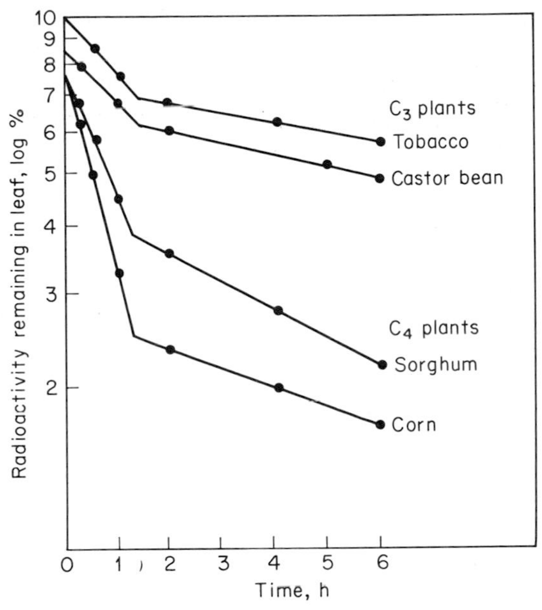

Fig. 2-25 Turnover of ^{14}C-labeled photosynthate in leaves is markedly faster in the C_4 plants sorghum and corn than in the C_3 plants tobacco and castor bean. The C_4 plants exported at least 70 percent in the 6 h following assimilation, compared with about 45 percent for the C_3 plants (Hofstra and Nelson, 1969).

Moisture stress Whole-plant growth and the distribution pattern for assimilates are both affected by moisture stress. Extension growth, which relies in some measure on positive turgor, shows an early response, whereas photosynthesis and substrate accumulation do continue despite some decrease in water potential (Iljin, 1957). If tissue enlargement has ceased, despite continued photosynthesis, assimilate concentration within the leaves will inevitably increase; Hartt (1967) has indeed observed such an increase in sugar concentration in the stem of sugarcane plants under stress. Translocation must have continued despite the reduction in growth. Presumably, then, translocation shows a different response to moisture stress compared with growth and photosynthesis, although it is necessary to differentiate between the different phases of translocation.

Assimilate transport out of a leaf is especially sensitive to moisture stress (Plaut and Reinhold, 1965), whereas movement within the vascular system is not as severely affected (Nelson, 1963). Loading assimilates into the translocation system is apparently the critical phase. This contention is supported by Wardlaw's observations on wheat and ryegrass (Wardlaw, 1967, 1969) that translocation velocity is virtually unaffected by low water potential provided the importing organs (inflorescence) are not subject to the same stress as the remaining plant parts.

Metabolic controls The loading and unloading steps in translocation require metabolic energy. If ATP is involved, uncoupling agents such as 2,4-dinitrophenol should have some selective effect. Such an inhibition has been demonstrated for the substrate-loading phase of sucrose ^{14}C translocation in castor bean seedlings (Kriedemann and Beevers, 1967*a*), whereas the movement of previously accumulated label was not so affected by the inhibitor.

There are large uncertainties about metabolic involvement in the actual movement of substrates within the phloem conduits, but the withdrawal of materials by an importing organ is known to be linked to respiration. Some results from Geiger and Christy (1971) are especially pertinent. In their sugar beet system (described in connection with temperature effects) translocation to a sink leaf was severely reduced by anoxia. This depression in translocation was almost immediate and appeared related to the subsequent decline in both respiratory activity and ATP level. The data are generally consistent with the idea that "ATP supplies energy to drive an unloading process in the sink."

Apart from their indirect effects via growth, nutrient elements also influence translocation. The role of boron, for example, had become a highly contentious issue with an earlier suggestion (Gauch and Dugger, 1953) that a boron-sugar complex is involved in movement through the phloem. This view has been largely abandoned following work of Neales (1959) suggesting that the boron effect is mediated largely by growth responses.

Key roles have been attributed from time to time to other elements, including copper, manganese, zinc, and molybdenum, but in every instance any effect on translocation is confounded by growth reactions. Potassium, however, does retain some special features in this connection. Not only is there a wealth of indirect evidence which implicates this element in sugar transport, but Amir and Reinhold (1971) have demonstrated with bean plants a clear depression of translocation under conditions of even mild potassium deficiency and showed that light is required for the expression of this effect.

Under the present heading of metabolic controls on translocation we must also consider the hormones of natural or synthetic origin. It is no coincidence that strong localized sinks (fruits, expanding leaves, terminal meristems, and the like) are also sites of active hormone synthesis; moreover, the application of exogenous substances will also influence assimilate distribution (Shindy et al., 1973). Questions immediately arise: What constitutes this attractive property of hormones with respect to assimilate movement? What interrelations exist between hormone transport and solute translocation?

In physiological test systems such as decapitated internodes or pedicels, applied auxin cytokinin or gibberellin are able to attract photosynthate (Letham, 1967). Apparently the applied hormone has either stimulated synthetic processes at the site of application, thereby creating a sink, or there has been a direct influence on translocation due to an activation of transcellular strands within the phloem. The need to make this distinction becomes more apparent after comparing results from long- and short-duration experiments.

Long-term experiments on test systems (24 to 36 h duration) provide adequate time for hormone-stimulated syntheses, but such experiments tend to show a lack of specificity in hormone-directed translocation. In this situation the hormone could simply arrest tissue deterioration, compared with controls, and hence its translocating activity is sustained. With intact or at least a self-supporting system such as excised bean leaves, this problem of deterioration is minimized. Nakata and Leopold (1967) were able to demonstrate that labeled sucrose (^{32}P to a lesser extent) is attracted toward localized sites of benzyladenine application. Instead of generating a new sink (as with bean leaves) the attracting capacity of existing organs can be enhanced; orange fruitlets on intact trees become intensified sinks after treatment with kinetin (Kriedemann, 1968) while grapevine shoots show a similar response to benzyladenine (Quinlan and Weaver, 1969).

Short-term experiments offer a stronger possibility of demonstrating direct hormone effects on translocation because the prospect of new growth is reduced. Accordingly, Mullins (1970) was able to demonstrate enhanced photosynthate accumulation in bean internodes due to a hormone mixture [IAA and gibberellic acid (GA) plus a synthetic kinin] within 2.5 h. However, it was still not possible to divorce this effect from hormone-stimulated growth because protein and RNA synthesis was also enhanced within the same interval.

DISTRIBUTION PATTERNS

Solutes Derived from the Root System

Water derived from the soil may reach all parts of the plant with relative ease, but nutrients do not. When a nutrient solution is supplied to the root system, sites of entry and upward distribution patterns for solute molecules can diverge greatly from those for water flow.

Extending regions of young roots, together with root-hair zones, are responsible for most water absorption, whereas the region immediately behind the tip is particularly active in nutrient absorption. This pattern does of course need further qualification with respect to zones of nutrient absorption and differentiation between solutes. Such information has been provided by Rovira and Bowen (1968), who examined phosphate, chloride, and sulfate uptake by seminal roots on wheat seedlings. These authors developed an ingenious method of supplying labeled solutes and then scanning the roots in much the same way as in a radiochromatogram. They reported two major absorption sites for each anion, one 4 to 7 mm from the root apex and another 20 to 30 mm behind the apex. With regard to the plant's capacity to differentiate between solutes, Rovira and Bowen (1968) found that phosphate is absorbed over the whole section (7 cm behind the apex) whereas chloride and sulfate absorption is more active near the root tip. These observations on wheat seedlings emphasize the clear distinction roots make between water and solute molecules.

When radioactive phosphate is supplied to an intact plant via the roots, activity accumulates in leaves according to age and position rather than in a more uniform pattern, as would be expected for water. Biddulph (1951) established such a picture for bean plants by measuring the amounts of phosphorus accumulated in specific leaves after 4 days. His data show how phosphate accumulated almost exponentially in successively higher leaves. Obviously the nutrient is not moving passively with the transpiration stream. Instead, there must be some mechanism behind the selective distibution of this nutrient.

Patterns of nutrient distribution show many variations. Some collect preferentially in young leaves, e.g., phosphorus; some collect in older leaves, e.g., iron; others collect along leaf veins (iron) or in hydathodes (cobalt). Old leaves tend not to accumulate phosphorus but mobilize this element from within their own tissues and export it in large amounts (a feature associated with senescence). Young leaves, on the other hand, show preferential uptake of certain nutrients and even appear to operate under the direction of the plant's apex. Biswas and Sen (1959) produced some evidence to this effect when they demonstrated that preferential accumulation of labeled phosphorus and sulfate from roots to young leaves is erased if the plant's apex is removed; decapitated plants develop equal distribution of label in all leaves. The distribution of nutrients from the root system is therefore dependent upon an upward flow of xylem sap in the first instance, followed by a directing mechanism which appears analogous in some ways to the hormone-induced movement of assimilates described earlier.

Past arguments over whether upward translocation of ions occurs in xylem or phloem were resolved after the advent of isotope tracer methods. For example, Stout and Hoagland

(1939) established that radioactive phosphate applied to the roots of *Salix* shows up specifically in the xylem within those regions of the stem where wood and bark are separated. If wood and bark are in immediate contact, some label migrates laterally into the bark. Instead of feeding roots with ^{32}P and looking for xylem transport up the stem, an alternative approach, designed to show phloem movement, would be to feed label to exporting leaves and check which vascular system is concerned with movement out of tissues. Biddulph and Markle (1944) adopted this approach, supplying radioactive phosphate to a leaf of cotton and separating bark from wood in an intermediate section of stem. This case involved downward translocation, and the isotope was abundant throughout the bark but occurred in xylem elements only near the point of application. The data imply that translocation out of the leaf is restricted to the phloem (with subsequent movement either up or down the stem also occurring in the phloem) whereas the movement of solutes from the root zone tends to occur via the xylem.

DISTRIBUTION OF NEW ASSIMILATES

Overall patterns between areas of supply (leaves) and points of demand (growing organs) will be limited by vascular connections, so that importing organs do not necessarily obtain assimilates from the closest leaves. Some strange anomalies exist in distribution patterns from particular leaves, due to complex vascular connections. In a vegetative herbaceous plant, lower leaves tend to supply assimilates to the roots, upper leaves meet the requirements of the terminal meristem, and leaves in an intermediate position might well supply in either direction (Milthorpe and Moorby, 1969). This generalized pattern is of course not rigid and may change dramatically if the plant enters its reproductive phase or suffers partial defoliation. By supplying $^{14}CO_2$ to selected leaves and then making autoradiographs it is possible to build up a picture of distribution patterns over the course of development. A typical sequence, like that described for grapevines (Hale and Weaver, 1962) or soybeans (Thrower, 1962) follows: During early growth, a leaf requires carbohydrate for its development and will show a net import of assimilates until the laminar surface is one-third to one-half full size. At that stage assimilate export starts, with substrates directed toward the apex. As the plant continues to grow and this particular leaf becomes separated from the apex by additional leaves, the root system assumes increasing prominence as a sink. Ultimately our source leaf turns senescent, and both organic and inorganic reserves become mobilized and are exported. During this last phase current photosynthate may again be imported (Thaine et al., 1959, on soybeans) but to a very minor extent.

With the onset of reproduction, this generalized pattern of assimilate distribution undergoes a major alteration. From work on a variety of plants such as peas, grapes, tomato, and cereals (Wardlaw, 1968), developing fruits show an overwhelming demand for assimilates from adjacent leaves; flowers themselves may have originally been weaker sinks by comparison (as in grapevines, Hale and Weaver, 1962). During reproductive development, roots and buds operate under difficulty because their demands for photosynthate are overshadowed by the requirements of the fruit.

Genetic factors which manifest themselves in the form a plant adopts, also influence the distribution of assimilates, but are subject to hormonal modification. Lovell (1971) describes such a situation from his comparative studies on two tall and two dwarf varieties of pea *(Pisum sativum)*. $^{14}CO_2$ was supplied to leaves on all four varieties, and assimilate distribution after 24 h was compared. ^{14}C translocation to upper shoots predominated in tall varieties; the dwarf varieties behaved similarly only following treatment with gibberellic acid. Total CO_2 fixation was not affected by gibberellic acid; only the distribution of ^{14}C assimilates responded to hormonal treatment.

MOBILIZATION AND RETRANSLOCATION

Centers of Supply and Demand

Growing organs can develop at the expense of either current photosynthate or stored reserves. In the first instance, a supply of substrate is at hand, and the importing organ has simply to generate demand. By contrast, where mobilization is involved, the question of how importing organs instigate the dissolution and retranslocation of stored material, even from remote sites, becomes a key issue.†

A plant's capacity to recycle raw materials can be viewed against a background of evolutionary advance. Single-celled organisms had no such requirement, but higher plants with their coordinated growth (apical dominance) and structural specialization (storage organs) have an obvious need. Pathogens have taken this principle a step further as an aid to their exploitation of higher plants, and as far as the host plant is concerned, infection lesions are distressingly attractive to the plant's own nutrients.

The large accumulation of carbohydrates or fats in such storage organs as fruits and tubers is considered to be a consequence of the activities of enzymes, which convert the soluble translocated carbohydrates into insoluble forms, thus creating a gradient for further accumulation (see the review by Wanner, 1958). Accumulation can be generally correlated with the activity of enzymes which synthesize such relatively insoluble materials as starches, fats, and proteins; conversely, the export of materials from storage organs can be correlated with the activity of hydrolyzing enzymes such as amylase.

Distribution Patterns

During our earlier discussions on translocation patterns for ions moving from roots to leaves, we noted that older leaves accumulate much less phosphate than younger ones (Fig. 2-26); it is significant that older leaves are also known to export large amounts of carbohydrate, nitrogen, potassium, and phosphorus as they age. Koontz and Biddulph (1957) demonstrated this feature by the simple device of applying equal amounts of isotope to leaves of different ages. Their results (Fig. 2-26) show the relatively larger amounts of ^{32}P translocated out of the older primary leaf of bean seedlings, compared with the amounts translocated out of the first and second trifoliate leaf. Notably, the direction of transport varied with leaf location, and, by analogy to the distribution of current photosynthate discussed earlier, the lower leaf exported large amounts of ^{32}P to the root system rather than to the apical tissues.

This ability of leaves to export previously accumulated substrate has enormous significance for plant growth and development. The export of materials from older leaves or the retranslocation of these materials supplies the growing

†The term *mobilization*, as used here, refers to processes associated with the conversion of stored assimilates, e.g., starch, back into soluble and easily translocated material such as sugar.

Fig. 2-26 The translocation of ^{32}P out of bean leaves varies with age. NaH_2PO_4 was applied to different leaves, and the translocation into three separate divisions of the plant (*A, B,* and *C*) was measured after 24 h. Note the relatively large amounts translocated out of the oldest leaf (Koontz and Biddulph, 1957).

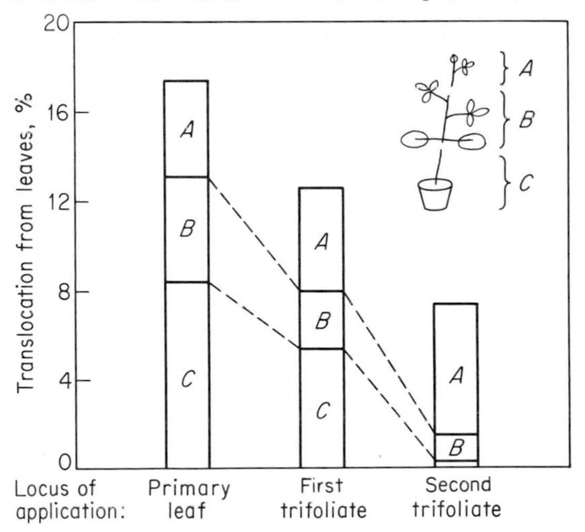

parts of the plant with a large proportion of its inorganic nutrients during later stages of growth. When small grains such as wheat and oats have reached 25 percent of full size, the net uptake of nitrogen and phosphorus by the plant is 90 percent completed, and so for the last 75 percent of their growth, the grains must be supplied with these elements principally by retranslocation from within the plant. Petrie et al. (1939) have shown how wheat leaves approaching senescence lose up to 85 percent of the nitrogen and almost 90 percent of the phosphorus originally present. This material is exported to other organs.

The magnitude of retranslocation is also illustrated by some data of Williams (1938), who plotted the distribution of nitrogen in oat plants as a function of age. Figure 2-27 illustrates how extensive amounts of nitrogen are exported from the roots and leaves during the accumulation of nitrogen into the inflorescence. At a slightly later stage, the stems also provide an apparent source of nitrogen for the inflorescence.

There is a marked selectivity between ions for retranslocation. In some interesting experiments on the ability of different ions to be retranslocated Biddulph et al. (1958) compared the distribution patterns of the relatively mobile phosphorus, the relatively immobile calcium, and the intermediately mobile sulfate. Solutions of these tracers were applied briefly to the roots of bean plants, and then the distribution patterns in the leaves were followed at intervals by autoradiography. Striking differences in retranslocation were found; phosphorus accumulated in the primary leaf up to 6 h after exposure, and then the radioactivity in this leaf declined steadily as the ion was retranslocated toward the growing point and young, developing leaves. In direct contrast, calcium accumulated in the primary leaf for about 6 h after exposure, and then no detectable decline in the radioactivity of the primary leaf was observed; i.e., no apparent retranslocation of the calcium occurred. The sulfate was retranslocated to an extent comparable with the phosphorus.

Fig. 2-27 The distribution of nitrogen components in the oat plant during growth, showing the loss of nitrogen from the leaves and stem into the inflorescence in the last 50 days (Williams, 1938).

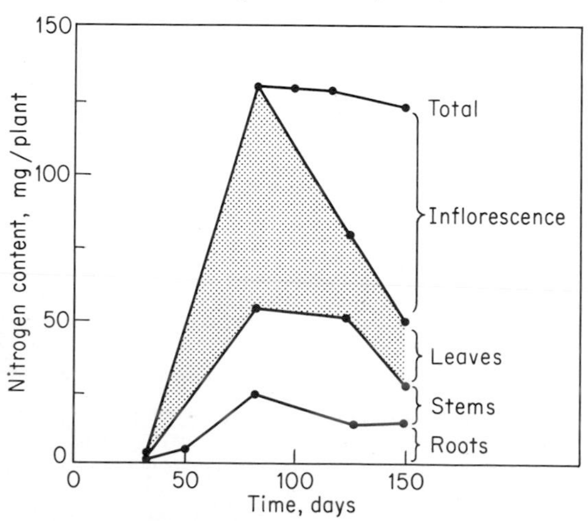

Internal Factors

In some early work, Mothes (1928) observed that the loss of nitrogenous materials from aging tobacco leaves can be prevented by topping the plant. After obtaining more elaborate data on the subject, Petrie et al. (1939) reasserted that topping greatly reduces the export of nitrogenous materials, and they described this as a sink effect, the fruits forming a large depot in which these nutrients accumulate.

A developing organ amounts to more than simply a recipient for substrate; it also exerts a positive influence over the fate of substrates. This principle was nicely demonstrated by Marrè (1948*a,b*, 1949) in his experiments on *Calonyction*. In this plant (Fig. 2-28) the receptacle and peduncle become much enlarged and fleshy during fruit growth, with a relatively high concentration of starch in the peduncle itself. If ovaries or stamens are removed, this starch is degraded and disappears from the young fruit. When Marrè took a cutting with two developing fruits and aborted the embryo of one, its stored nutrients were quickly lost. The intact fruit was characterized by a positive force which led to the

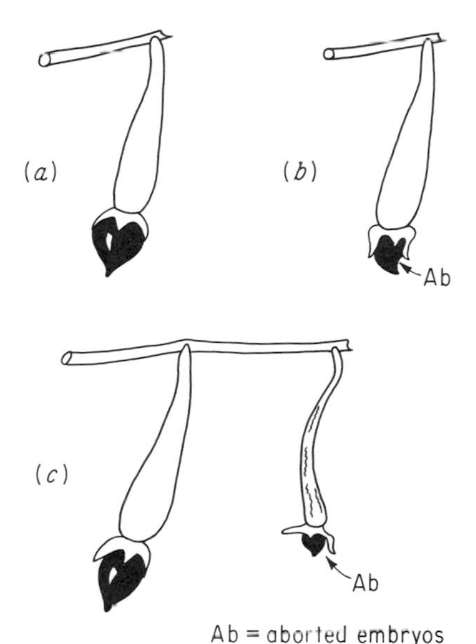

Fig. 2-28 The influence of developing fruits over mobilization of substrates in adjacent tissue is demonstrated on these cuttings of *Calonyction*. (*a*) An intact fruit; (*b*) an aborted fruit on a cutting by itself does not show degeneration of the fleshy peduncle; (*c*) an aborted fruit on a cutting with an intact fruit shows an exhaustion of the fleshy peduncle (Marrè 1948*b*).

mobilization of stored material (starch) within the neighboring peduncle and the subsequent translocation of this sugar into the intact fruit.

Some environmental factors are also known to modify mobilizing activities. Vickery et al. (1937) noted, for example, that tobacco leaves held in darkness lose their nitrogenous constituents rapidly, whereas leaves given daily light periods are relatively stable. Perhaps light can fortify the capacity of a leaf to attract materials which have been mobilized within other regions. Nutrient deficiencies also increase the export of nitrogenous material from leaves (Williams, 1955). The impoverished plant presumably experiences an accentuated need to recycle its limited supply of nutrients.

Heat treatment, on the other hand, results in loss of importing ability but leads to greater mobilization of local reserves. Engelbrecht and Mothes (1960) have shown a migration of labeled amino acids from treated to control regions of a tobacco leaf which was associated with yellowing in the treated area.

In contrast to the effects of heat, Engelbrecht and Mothes (1960) found that kinetin applied to a portion of a tobacco leaf encourages the movement of metabolites to that region. If an excised leaf is treated with kinetin in a localized area, this area remains green while the untreated portion turns yellow. Radioactive amino acids applied to one part of a leaf move into the kinetin-treated area. This effect is found whether isotopes of carbon or of nitrogen are used in the amino acids.

The cytokinin effect is further illustrated by following the nitrogenous constituents of tobacco leaves after kinetin treatment. It is evident from Fig. 2-29 that severed leaves eventually lose large amounts of their protein nitrogen and that the soluble degradation products accumulate in the midrib of the leaf. If kinetin has been

Fig. 2-29 The changes in soluble and insoluble nitrogen fractions of tobacco leaves after excision. (*Left*) The initial distribution in left and right halves of the blade and in the midrib. (*Middle*) After 9 days, considerable loss of the protein nitrogen (*shaded bars*) has occurred in the blades, and solubles (*unshaded bars*) have accumulated in the midrib. (*Right*) After 9 days, when one side of the leaf has been treated with kinetin, there is an increase in both soluble and protein nitrogen in the treated side at the expense of the untreated parts (Mothes et al., 1959).

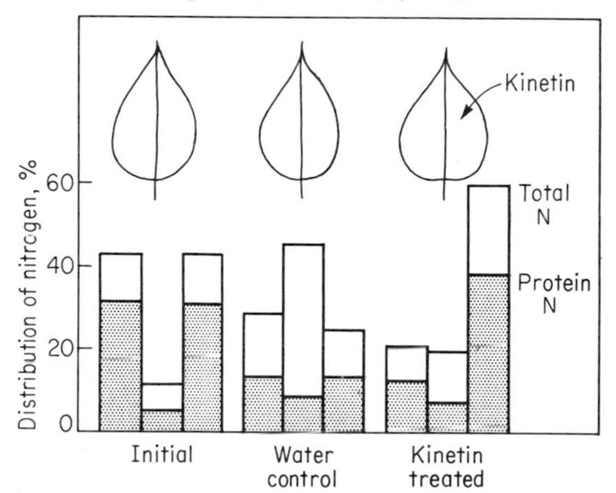

applied to one-half of the leaf blade, an accumulation of nitrogen occurs in the treated leaf part rather than the midrib. Measurements of the treated and untreated sides showed that kinetin treatment results in an enhancement of chlorophyll content, protein synthesis, and nucleic acid content of the leaf. As the protein was degraded in the leaf, the soluble products actually accumulated in the midrib, even though there was no further depot to which they could be translocated. This effect suggests that nitrogenous materials are actively exported from the senescing leaf.

SUMMARY

The emergence of translocation systems in higher plants was probably a major factor in the evolution of higher plants. These systems have sufficient capacity in some plants to permit the development of large storage organs such as fleshy fruits, tubers, bulbs, and corms. Furthermore, translocation has become coupled to some aspects of plant development, playing a major role in the regulation of leaf senescence and correlative effects of apical dominance. These features of the plant's translocation system represent a potent force in the integration of growth and development at the whole-plant level.

We have clear evidence that certain parts of the plant can instigate the mobilization of stored resources within remote organs, with subsequent translocation to these sites of rapid growth or development. Applications of kinetin and some related compounds can lead to the attraction of nutrients in a manner suggesting mobilization due to apices, flowers, fruit, and other organs. The intriguing possibility arises that correlative movement of materials within plants may be regulated to some degree by growth substances.

GENERAL REFERENCES

Canny, M. J. 1971. Translocation: Mechanisms and kinetics. *Annu. Rev. Plant Physiol.,* 22: 237–260.

Canny, M. J. 1973. "Phloem Translocation." Cambridge University Press, London. 301 pp.

Crafts, A. S., and C. E. Crisp. 1971. "Phloem Transport in Plants." W. H. Freeman and Company, San Francisco. 481 pp.

Eschrich, W. 1970. Biochemistry and fine structure of phloem in relation to transport. *Annu. Rev. Plant Physiol.,* 21:193–214

Läuchli, A. 1972. Translocation of inorganic solutes. *Annu. Rev. Plant Physiol.,* 23:197–218.

Milthorpe, F. L., and J. Moorby. 1969. Vascular transport and its significance in plant growth. *Annu. Rev. Plant Physiol.,* 20:117–138.

Pate, J. S., and B. E. S. Gunning. 1972. Transfer cells. *Annu. Rev. Plant Physiol.,* 23:173–196.

3 THE DYNAMICS OF GROWTH

Although species vary enormously in their assimilative abilities (C_4 vs. C_3, sun vs. shade plants), growth controls inherent in the plant still exert powerful effects over their general performance. Physical inputs sustain growth, but biological regulation dictates the pattern of its utilization and ultimate expression. If we are to understand the nature of this regulation at a whole-plant level and appreciate the interactions between plants and their environment, we need more detailed measurements than simply final yield. *Growth analysis* and mathematical models of growth and development provide such parameters.

GROWTH OF INDIVIDUAL PLANTS

Basic Concepts of Growth Analysis

For individual cells or organs, growth is potentially unlimited and begins as an exponential pattern (curve A in Fig. 3-1). However, mutual interactions within an individual impose limitations on growth, the actual growth curve falls away in a sigmoidal manner, and curve B results (Fig. 3-1). By analogy, microorganisms also show sigmoidal growth due to eventual limitations of space or nutrients or accumulation of end products. The overall growth of any organ or organism includes an early exponential phase (Fig. 3-1, curve B). Such parameters as volume, weight, surface area, height, cell number, and even protein content all show a similar pattern.

Growth of a higher plant, during its exponential phase, is analogous to accumulation of capital at continuous compound interest. The embryo represents initial capital, while photosynthetic efficiency determines the interest rate. Growth may be regarded as an increase in fresh weight (as in an etiolated seedling) or an accumulation of dry weight (as in a green seedling). Either system will follow the law of compound interest during its early exponential phase.

If the plant's initial weight is W_0 and the rate of compound interest is r, the total weight after a certain time t will be W_t, where $W_t = W_0(1+r)^t$. This fundamental equation describes many natural phenomena where a quantity's rate of

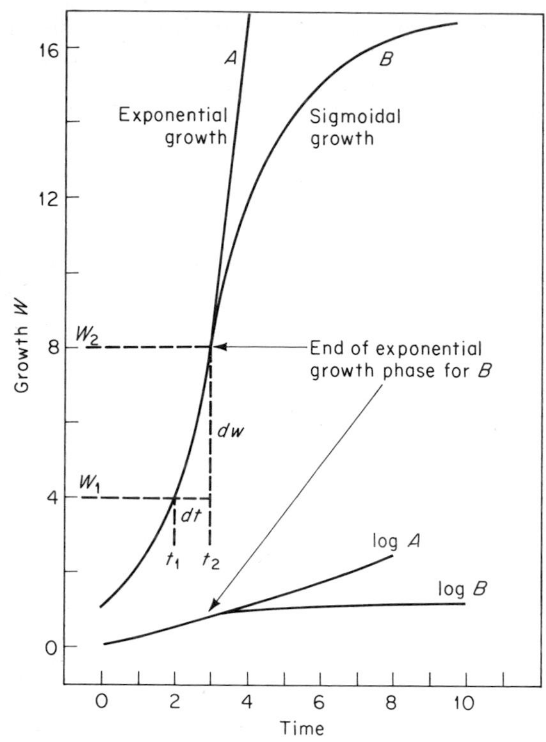

Fig. 3-1 Early vegetative growth in herbaceous plants tends to be exponential (curve *A*) although a sigmoidal pattern is more characteristic of its entire life span (curve *B*). Relative growth rate can be derived from the slope of a logorithmic plot or from the integral of $\frac{1}{W}\frac{dW}{dt}$ over the time interval $t_1 - t_2$.

increase varies according to the magnitude of the quantity itself. This relationship between growth and time is more conveniently used in the form $\ln W_t = \ln W_0 + rt \ln e$, where ln stands for the logarithm to the base *e*. This logarithmic form is simply the equation of a straight line on a semilogarithmic plot (see Fig. 3-1). The slope of this line is determined by *r*, that is, the rate of interest, and thus represents the plant's capacity to add to its own dry weight. Blackman (1919) used the term *efficiency index* to denote this same concept. It has now found wide acceptance as *relative growth rate* (RGR). In the equation above, the term *e* is a constant (2.7182) derived from the binomial theorem (and named after the mathematician Euler) and since $\ln e = 1$, then $\ln W_t = \ln W_0 + rt$. Relative growth rate, that is, *r*, or efficiency index, can now be determined from plant-growth data by sampling on two occasions t_1 and t_2 and substituting into the expression

$$\text{RGR} = \frac{\ln W_2 - \ln W_1}{t_2 - t_1}$$

W_1 and W_2 refer to whole-plant dry weight on two successive occasions t_1 and t_2. An equation of the same form emerges from integrating an expression for the rate of weight increase per unit weight already present over a discrete time interval, i.e.,

$$\frac{1}{W}\frac{dw}{dt}$$

as illustrated in Fig. 3-1.

If growth is really exponential, RGR should be constant and a semilogarithmic plot of dry weight vs. time should yield a straight line (Fig. 3-1). This holds true for early growth in many herbaceous plants, but RGR always declines later in their life cycle.

When Blackman (1919) expounded the compound-interest law of plant growth and developed this concept of efficiency index, other physiologists in England (Kidd et al., 1920) criticized his view and tended to favor the *substance quotient* of previous German workers (Noll and his pupils), which simply represented the ratio of final weight to seed weight divided by time. This view was reversed that same year (West et al., 1920). Four concepts were developed for use in growth analysis.

1 Relative growth rate (RGR)

$$R = \frac{1}{W}\frac{dw}{dt} = \frac{\ln W_2 - \ln W_1}{t_2 - t_1}$$

i.e., the increase in weight per unit of original weight over a time interval *t*.

2 Leaf-area ratio (LAR)

$$A = \frac{L}{W} = \frac{L_1 + L_2}{W_1 + W_2}$$

i.e., the ratio of leaf area *L* to whole-plant dry weight.

3 Unit leaf rate (or net assimilation rate, NAR)

$$E = \frac{W_2 - W_1}{L_2 - L_1} (\ln L_2 - \ln L_1)$$

i.e., the rate of increase in dry weight per unit leaf area, assuming that both dry weight and leaf area are increasing exponentially. Photosynthetic tissues other than leaves would of course be taken into account.

4 Relative leaf-growth rate (RLGR)

$$R_L = \ln L_2 - \ln L_1$$

analogous to relative growth rate of the whole plant.

If leaf area and dry weight are both increasing exponentially, the following relationships can be derived:

$$R = \frac{dW/dt}{W} \qquad A = \frac{L}{W} \qquad \text{and} \qquad E = \frac{dW/dt}{L}$$

Therefore $R = AE$ since

$$\frac{dW/dt}{W} = \frac{L}{W} \frac{dW/dt}{L}$$

These indexes of West et al. (1920) aimed at resolving plant growth into those attributes which govern its rate of dry-matter increase on the basis of simple experimental data such as leaf area and dry weight. Unit leaf rate E is a key determination for the assessment of plant performance since it expresses growth in terms of foliar surface. Its significance was initially recognized by Gregory (1917, 1926), who is credited with being the first to derive the growth index which subsequent authors refer to as unit leaf rate (Williams, 1946).

Gregory had used the term *net assimilation rate* (NAR), which has the more obvious connotation with photosynthesis, and calculated his index as

$$\text{NAR} = \frac{\ln L_2 - \ln L_1}{t_2 - t_1} \frac{W_2 - W_1}{L_2 - L_1}$$

i.e., the rate of increase in whole-plant dry weight per unit leaf area.

In Gregory's nomenclature, which must take priority for historical reasons, relative growth rate then becomes the product of net assimilation rate and leaf-area ratio, or RGR = NAR × LAR. Williams (1946) made a detailed analysis of such growth parameters and their mathematical validity, and discussed the relative merits of basing unit leaf rate on foliar area, dry weight, or protein content. He emphasized that Gregory's (1926) formula does not give an accurate estimate of mean NAR over an extended period of growth unless plant dry weight and leaf area have been linearly related.

Even during a plant's early phase of exponential growth, W and L are not necessarily related in this way, but any error introduced into the NAR calculation is minimized by shortening the time interval between successive harvests to a few days. During later phases of development, departure from linearity becomes appreciable, so that NAR values based on Gregory's formula must be regarded as approximations.

Evans and Hughes (1962) and Whitehead and Myerscough (1962), confronted by this same problem, developed modified formulas which take into account changes in the L/W relationship as a plant develops. Ondok (1971) and Ondok and Kvet (1971) also offer methods for exponential vs. nonexponential growth. Their approach is particularly suited to the analysis of long-term seasonal trends rather than short-term fluctuations due to climatic factors.

NAR expresses a plant's capacity to increase dry weight in terms of the area of its assimilatory surface. The term therefore represents photosynthetic efficiency in the overall sense, and in conjunction with LAR and RGR it can be used to analyze the response of plant growth to environmental conditions.

Growth in a higher plant is obviously a complex phenomenon, and a simple mathematical formulation can do little more than embody some of the factors dictating plant performance. While formulations such as NAR are useful, they must be employed with circumspection. As Watson (1952) explains, NAR does not measure real photosynthesis since it represents the net result of photosynthetic gain over respiratory loss and may therefore vary according to the magnitude of respiration. For example, if total respiration of an entire plant is expressed in terms of leaf area, this term is likely to increase with age; but LAR will decrease (an older plant is not as "leafy"), so that NAR could fall irrespective of change in photosynthetic activity.

A second precaution concerns species comparisons. Since NAR gives no direct indication of respiratory losses, this index does not necessarily serve as a direct measure of inherent photosynthetic capacities.

The third limitation is of some importance for agronomists. While NAR indicates a plant's efficiency at producing dry matter, economic yield is subject to additional controls and is not necessarily related to photosynthetic efficiency. Watson (1956), for example, found no positive association between NAR and yield in five varieties of sugar beet and three of potato. Similarly, Thorne (1960) reported growth data for sugar beet, potato, and barley, where NAR for sugar beet is consistently higher than for barley and yet final productivity showed the opposite relationship.

Fig. 3-2 In adapting to reduced light intensity, mung bean plants show a decline in photosynthetic activity (NAR) but an increase in leaf area (LAR) such that growth is sustained (RGR) (Monsi et al., 1962).

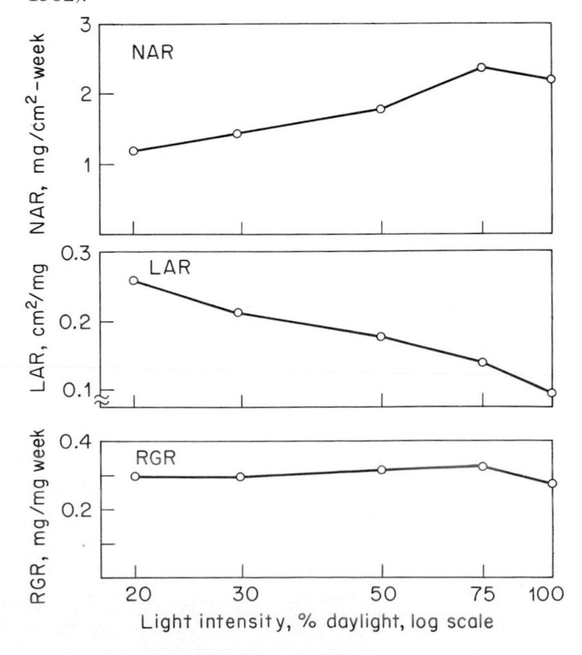

Despite these limitations, the concepts of growth analysis have proved highly effective in studying a plant's reaction to many environmental conditions. Evidence has steadily accrued since the 1920s which demonstrates variation in efficiency of plant growth according to age and environmental conditions, and consistent differences between species have also been recorded. Plants appear to have a genetically determined potential for growth and utilize certain strategies to overcome environmental factors or at least reconcile themselves with them. The mechanisms they adopt probably stem from selective pressures during their evolution and are amenable to mathematical analysis.

Environmental Factors

Light Before the widespread availability of controlled-environment facilities for plant-growth studies, the factor most amenable to modification was light intensity. A population of plants could be shaded to different degrees, and the reaction of individuals to altered light intensity was gauged by growth analysis.

Species as diverse as *Citrus* and mung beans *(Phaseolus aureus)* show the remarkable capacity to modify their leaf morphology so that growth is sustained despite severe reductions in light intensity. Such ability is evident in Fig. 3-2. Shading the mung beans reduced NAR, but these plants had grown decidedly leafy and LAR was more than doubled by an 80 percent reduction in sunlight. Increased foliar surface compensated in this case for decreased NAR at low light, so that RGR was virtually unaffected.

A woody perennial such as *Citrus* can be equally adaptive. Monselise (1951) grew sweet lime seedlings outdoors in Israel and varied light intensity by supporting loosely woven jute cloth over individual plots of seedlings. In deepest shade (30 percent full sun) NAR was still 3.573 compared with 3.549 (g/m^2-day) in exposed plots, and since shaded plants developed greater foliar surface, RGR actually *increased* from 1.523 in

Table 3-1 Under low light intensities, both a sun plant *(Helianthus)* and a shade plant *(Impatiens)* show marked reduction in the NAR and some increase in LAR; in the shade plant, the rise in LAR is sufficient to maintain relatively good RGR, whereas in the sun plant this is not true.

Daylight, %	NAR mg/cm²-week	NAR %	LAR, cm²/g	RGR g/g-week	RGR %
Helianthus annuus (Blackman and Wilson, 1951)					
100	7.97	100	82	0.66	100
24	2.86	36	140	0.42	64
12	1.34	17	170	0.23	35
Impatiens parviflora (Evans and Hughes, 1961)					
100	6.05	100	132	0.80	100
24	3.28	54	239	0.78	98
12	2.00	33	315	0.63	79

open plots to 2.055 g per 100 g per day under shaded conditions.

More typically, RGR does decline following reductions in solar radiation, especially at higher latitudes. Blackman and coworkers assembled data for 22 herbaceous species which demonstrated how the level of solar radiation in southern England limited the NAR of all species and in many instances restricted RGR. Blackman and Black (1959) were then led to postulate that for any region where growth is not restrained by unfavorable temperature, water supply, or nutrition, productivity per unit leaf area is most commonly limited by solar radiation. As their name might imply, sunflowers can certainly utilize higher levels of solar radiation than occur in southern England, and Warren-Wilson (1966) provides some evidence in support of Blackman's generalization that light limits their growth at higher latitudes. Under the long, clear days of inland Australia, NAR reached 2.0 g/dm²-week, virtually twice previous values recorded in England. Warren-Wilson (1966) estimated that maximum rates of photosynthesis in his sunflowers must have been around 50 to 65 mg CO_2/dm²-h (impressive values by any standard; cf. Table 1-1).

Helianthus annuus, a sun plant, responds favorably to greater insolation, and since available sunlight in southern England is already limiting growth, further reductions cause a decline in NAR. This decline is offset to some degree by increased LAR, but this increase is insufficient to offset the photosynthetic loss and RGR falls. *Impatiens parviflora*, on the other hand, is a shade plant and can exploit lower insolation more effectively than *Helianthus*. NAR is sustained despite a reduction in light intensity, and, in association with rising LAR, the plant's relative growth rate is maintained at a reasonable level down to about 100 cal/cm²-day, which would represent 10 to 20 percent full sun. The contrast between these two species is summarized in Table 3-1. Mechanisms which enable such photosynthetic adaptation to reduced light intensity are discussed in Chaps. 1 and 15 (see Figs. 15-4 and 15-6).

Temperature A plant's reaction to temperature is undoubtedly complex and species-dependent, but growth parameters can define some temperature effects in terms of altered efficiency. Goodall (1945), for example, obtained data on tomatoes growing under greenhouse conditions and calculated a multiple regression of NAR on light intensity and duration, humidity, and day and night temperature. The relationship was highly significant, but light intensity was the only factor shown to exert a separate effect.

Since high respiratory losses would lower the dry-weight gain from photosynthesis, NAR should be negatively correlated with temperature. Gregory (1926) found such a relationship with night temperature during the growth of barley, but a positive correlation existed with day temperature. Since higher light intensities would have been associated with Gregory's higher day temperatures, it is difficult to ascertain whether NAR was specifically dependent on day temperature.

The temperature optimum for photosynthesis will have immediate consequences for NAR, but temperature effects on leaf expansion have additive effects for plant growth as a whole. The sooner leaves expand, the sooner they become contributing organs, so that conditions favoring higher RLGR should also encourage RGR. Blackman and Black (1959) made such an observation during their ecological studies on the plant environment. In their experience, the higher temperatures of tropics and subtropics led to greater LAR and therefore enhanced RGR. Similarly, Beinhart (1962) found that clover growth is favored by increased temperature (10 to 30°C) because the rate of leaf production is increased.

The consequences of leaf growth rather than NAR were further emphasized by Warren-Wilson (1966) in his experiments on rape, sunflower, and maize grown as widely spaced populations under controlled conditions. The optimum temperature for NAR was lower than that for RGR (some other feature of plant growth was entering, namely, LAR). The temperature optimum for RGR coincided with high RLGR (Fig. 3-3); there was not a greater *weight* of leaves on the plant, but their areas were substantially greater and could be expressed in terms of leaf-area–leaf-weight ratios (Fig. 3-3). In other words, plants showing higher RGR had rapidly enlarging and thinner leaves. Photosynthetic effectiveness was clearly independent of leaf thickness.

Since temperature effects on whole-plant growth can be mediated via an effect on leaf expansion, RLGR becomes an important pa-

Fig. 3-3 At higher temperatures, leaves tend to be thinner (with higher leaf area-to-weight ratios). Growth rates for leaves (RLGR) and for the whole plant (RGR) tend to decline at supraoptimal temperatures (Warren-Wilson, 1966).

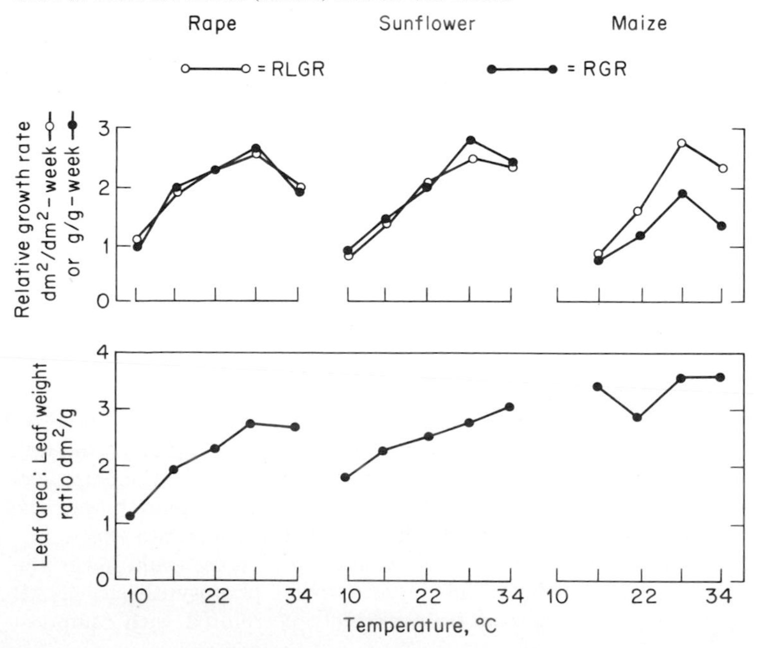

rameter for analyzing temperature response. Milthorpe and coworkers have made extensive studies on the expansion of the leaf surface, and the first paper in this series (Milthorpe, 1959) refers to the influence of temperature. Some relevant data on cucumber are shown in Fig. 3-4. The coincidence between NAR and RLGR with respect to temperature-response curves suggests that both photosynthesis and leaf expansion are optimized at 24°C. This temperature effect held for both levels of irradiance shown in Fig. 3-4.

Carbon dioxide Since photosynthesis shows an impressive response to increased CO_2 levels (Figs. 1-25 and 1-26), whole-plant growth should be similarly affected; this expectation is amply supported by data obtained from growth-analysis experiments. Understandably, the nature of this raw material limits its application to enclosed situations, so that information on growth responses to CO_2 is based upon experiments in greenhouses or light chambers. CO_2 effects under natural conditions must be viewed by inference from micrometeorological work (see the section on Community Photosynthesis, below).

The Agricultural Research Council unit of Flower Crop Physiology at Reading, England, maintains an extensive facility for controlled-environment studies, where Hughes and coworkers have analyzed chrysanthemum growth in great detail. The response of this species to both light intensity and CO_2 concentration was summarized by Hughes and Cockshull, (1971). There were significant effects of both light and CO_2 on dry weight and a large positive interaction between them. NAR, which serves as a direct index of photosynthetic activity, was also responsible and especially at high irradiance, but LAR showed an intriguing result: leaf expansion was not enhanced by elevated CO_2, and LAR showed an overall decline.

Plants would naturally require less foliar surface at elevated CO_2 concentration to achieve a given photosynthetic rate, but their adaptation in the form of lower LAR was unexpected. Tognoni et al. (1967) have described a similar type of response to CO_2 enrichment by bean and tomato plants maintained in solution culture

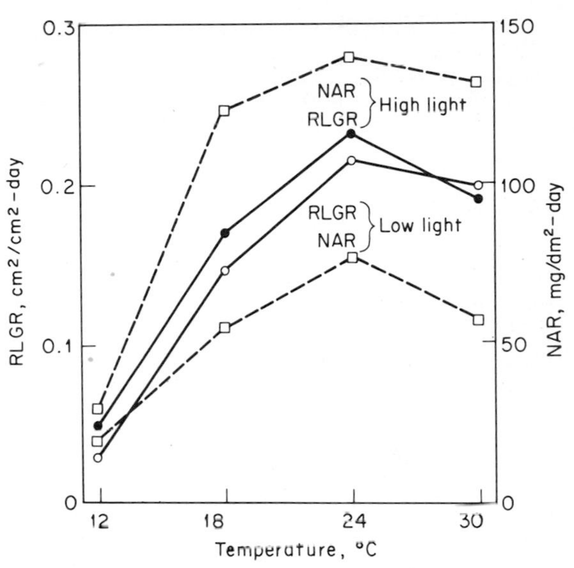

Fig. 3-4 Cucumber foliage shows an optimum at 24°C for both assimilatory activity (NAR) and the expansion of assimilating surface (RLGR) irrespective of irradiance level (28 W/m², *open symbols*; 56 W/m², *closed symbols*) (Milthorpe, 1959).

(Table 3-2). In this case, LAR declined, notwithstanding increases in both NAR and RGR. Presumably growth rate would have been further enhanced if LAR had not diminished. Ford and Thorne (1967) demonstrated a similar effect on barley, sugar beet, kale, and corn. In a teleological sense, it seems that a plant which is approaching its productivity ceiling balances the expansion of leaves against their photosynthetic effectiveness.

Water supply Stomatal opening and cell enlargement both depend on turgor, so that a restriction in water supply is likely to affect both photosynthesis and leaf expansion. Translated into parameters of growth analysis, there should be an effect on both NAR and LAR, and since RGR is their product, a decline in either leaf-area formation or its photosynthetic effectiveness will result in lowered growth.

Leaf expansion seems especially sensitive to decreased water potential (photosynthetic response was less acute), and some data from Boyer's (1970) experiments on sunflower dem-

Table 3-2 Both tomato and bean plants show a growth response to CO_2 enrichment. RGR is enhanced, and NAR is similarly affected despite the *decrease* in LAR at elevated CO_2. The improved growth is also reflected in better root development (Tognoni et al., 1967).

	Tomato		Bean	
	300 ppm CO_2	1,000 ppm CO_2	300 ppm CO_2	1,000 ppm CO_2
RGR, mg/g-day	222	254	122	172
NAR, mg/dm²-day	71	89	46	80
LAR, dm²/g	3.0	2.8	3.2	2.7
Root-to-top ratio	0.19	0.21	0.18	0.25

onstrate this effect (Fig. 3-5). Corn and soybean leaves were completely analogous in this respect. Boyer's observations for maize have been confirmed by Acevedo et al. (1971), who demonstrated a reduction in the rate of laminar extension once leaf-water potential fell to only −2.8 bars. With further increase in tension, growth stopped well before photosynthesis declined noticeably.

The more general consequences of moisture stress on growth are shown in Fig. 3-6. These data illustrate the response of young aspen trees to prolonged moisture stress. Growth indexes provide a convenient form of analysis and demonstrate how the components of RGR differ in their sensitivity to lowered soil water potential.

Nutrition Mineral nutrition impinges on all phases of plant growth, and adequate supplies of essential elements are required to sustain normal growth and development. Nutrient deficiency reduces growth in general and leaf-area expansion in particular. Data in Fig. 3-7 illustrate this point. Other nutrient deficiencies elicit a comparable response. Bouma's (Bouma, 1967; Bouma and Dowling, 1966) work on nitrogen, phosphorus, potassium, sulfur, calcium, and boron yielded different concentration responses but the same overall pattern as those in Fig. 3-7. Under conditions of minimal phosphate (0.25 ppm in solution culture) reduced growth in subterranean clover was a consequence of both lower NAR (0.41 at minimum phosphorus; cf. 1.01 mg/cm²-day at maximum phosphorus), and reduced leaf growth.

The consistency with which subterranean clover reacted to both withdrawal and supply of essential elements enabled Bouma and coworkers to employ growth analysis and subsequently direct measurements of photosynthesis to assess a plant's nutritional status.

Species Comparisons

Once growth analysis became widely used by whole-plant physiologists, sufficient data were assembled to enable Heath and Gregory (1938) to compile values from numerous workers re-

Fig. 3-5 When potted sunflower plants were allowed to dry out, the decline in leaf enlargement occurred at markedly lesser water deficits than the decline in photosynthesis (Boyer, 1970).

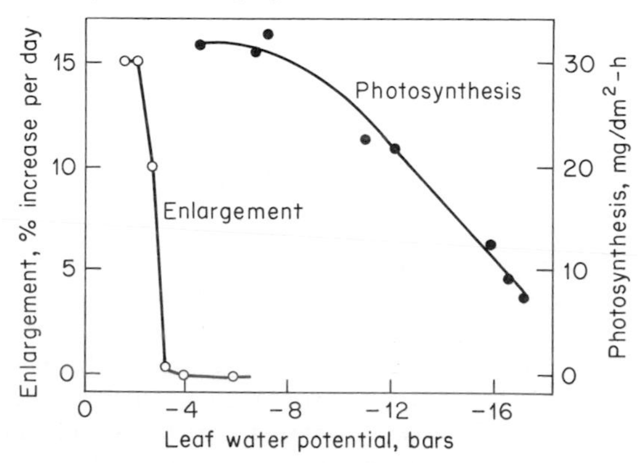

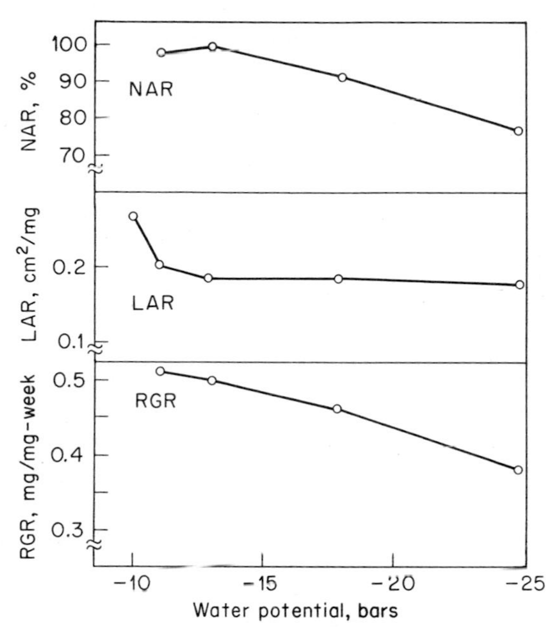

Fig. 3-6 As soil-water potential decreases, young aspen trees show a deterioration of LAR at lesser water deficits than NAR. The RGR declines with the NAR (Jarvis and Jarvis, 1963).

lating to a large number of species. Tabulated data for NAR varied between 0.12 and 0.72 g/dm^2-week, but despite this range, Heath and Gregory (1938) asserted NAR to be reasonably *constant* for all species and environments. Admittedly, the variation occurs within a certain range, but NAR is no longer regarded as constant (Watson, 1947, 1952), and consistent differences between species do exist, as illustrated in Table 3-3. If high NAR is coupled with greater LAR, the consequences for growth are confounded. The performance of sunflowers, compared with two woody perennials, can be analyzed in these terms (Table 3-4). *Helianthus* excels under this high irradiance due to greater values for both LAR and NAR.

Watson (1952) has suggested that LAR, rather than NAR, is more often the prime determinant of growth (further discussed in the section on Growth of a Plant Community, below) although appreciable variation in NAR does exist (Tables 3-3 and 3-4) and small but consistent differences occur even within a species (Thorne and Evans, 1964).

Since NAR is a direct reflection of a plant's photosynthetic capacity, we would expect some correlation between these two parameters. Data for photosynthesis were listed previously (Table 1-1), and we singled out C_4 plants for their high rates, which contrasted with the low values for tropical evergreen trees. A similar distinction exists with respect to NAR (see Table 3-3), where C_4 plants (*Zea* and *Amaranthus*) show substantially higher values than other herbaceous or perennial C_3 plants.

Herbaceous plants do, on the average, have higher rates of photosynthesis than woody perennials, and their values for NAR tend to differ in the same direction. Kozlowski and Keller (1966) summarize these differences as follows. Young woody plants have low values for NAR, 20 to 50 g/m^2-week compared with 70 to 150 g/m^2-week for rapidly growing herbaceous plants. The operative term is "rapidly growing," because photosynthesis does diminish during phases of quiescent growth; i.e., some form of internal regulation must be operating.

Fig. 3-7 Young plants of *Trifolium subterraneum* maintained 14 days at five different phosphate levels show growth responses largely attributable to nutritional status. Deficiency reduced growth in general but leaf area in particular. N, P, K, Ca, S, and B deficiencies all yield similar patterns, although critical levels differ (Bouma and Dowling, 1966).

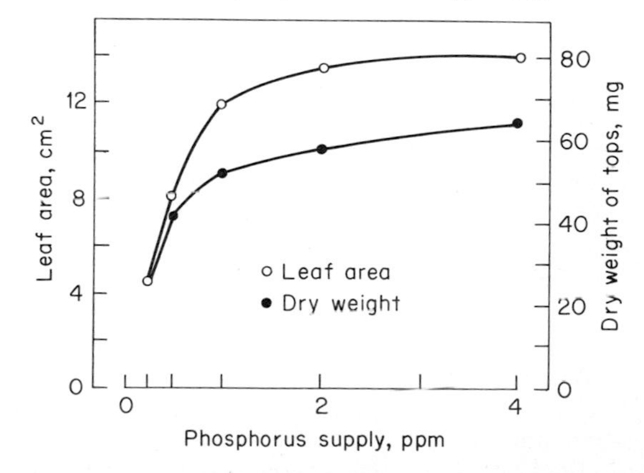

Table 3-3 Herbaceous plants, and especially C_4 species, generally show greater rates for RGR and NAR than woody perennials of either deciduous or evergreen habit (Jarvis and Jarvis, 1964).

Species	RGR, mg/g-week	NAR, g/m²-week
Herbaceous:		
Zea mays	2,310	152
Amaranthus viridis	2,590	147
Hordeum vulgare	920	68
Deciduous:		
Malus sylvestris	—	69
Acer pseudoplatanus	—	33
Fraxinus excelsior	300	28
Evergreen:		
Citrus aurantium	140	18
Theobroma cacao	140	13
Picea abies	58	20

Internal Regulation

Assimilate supply and demand In addition to the broad-scale growth comparisons made previously, intraspecific differences in NAR emphasize the existence of internal controls over assimilation. Two varieties of *Beta vulgaris,* spinach beet and sugar beet, provide a nice example; sugar beet always develops a larger storage root despite its similar-sized top. The question now arises whether one variety grows faster due to naturally greater NAR or whether an inherently stronger demand for assimilates, by a more vigorous storage organ, elicits some photosynthetic response (referred to previously in Chap. 1).

Thorne and Evans (1964) working at Rothamsted, England, were able to answer this question in growth-analysis terms by following the performance of grafted plants. Reciprocal grafts were made between the tops and roots of sugar beet and spinach beet in all four possible combinations. Results are presented in Table 3-5. Irrespective of aerial components, the highest NAR was associated with the sugar beet root system. This difference became accentuated with time and eventually reached 60 percent.

Burt (1964) reports an analogous situation for potato (*Solanum tuberosum*), where the removal of tubers 21 days after their initial formation reduced NAR by 75 percent; Spence and Humphries (1972) have reported enhanced NAR in small sweet potato plants (*Ipomea batatas*) associated with tuber formation.

A wealth of additional evidence (as reviewed by Neales and Incoll, 1968) supports the general contention that NAR has a dependence upon growth as well as a causal involvement.

Leaf age As with photosynthesis (Fig. 1-20), NAR also declines with age. Thorne (1960) has studied this situation in some detail for sugar beet, potato, and barley growing in a controlled environment and reports that NAR fell linearly with time for all species. This decline is particularly evident when NAR is expressed on a leaf-weight rather than area basis (Fig. 3-8). Other indexes, namely, RLGR, LAR, and RGR, also decreased. This general falloff in performance is of course a composite situation because individual leaves are still being produced and are contributing photosynthetically, but their individual contributions are submerged by the general decline in plant vigor.

Mathematical Models

Every plant or animal species has a characteristic form. This immediately implies some control over apportionment of substrates for growth and development. The resultant form can then be translated into mathematical terms.

Growth interrelationships within an organism were first demonstrated by Huxley (1932) for pincer size and body weight in crabs. From com-

Table 3-4 The high relative growth rates characteristic of herbaceous plants such as *Helianthus* growing under strong irradiance (64 to 75 W/m²) is associated with higher NAR and LAR values than representative woody plants (Jarvis and Jarvis, 1964).

Species	NAR g/m²-week	LAR, cm²/g	RGR, mg/g-week
Betula verrucosa	51.1	162	817
Populus tremula	48.1	194	919
Helianthus annuus	74.0	226	1674

Table 3-5 NAR in two varieties of *Beta vulgaris* is altered by the root system. Sugar beet has an inherently larger storage organ than spinach beet and elicits a stronger assimilation rate irrespective of which varietal top is provided (Thorne and Evans, 1964).

NAR, g/m^2-week	Intact plants		Grafted plants			
	Spinach beet	Sugar beet	Sp. beet / Sp. beet	Sp. beet / Sug. beet	Sug. beet / Sp. beet	Sug. beet / Sug. beet
June 20– July 11	44	73	59	69	73	80
July 12– August 10	19	30	22	42	40	60

parisons of final sizes or growth rates of different organs, a relationship emerged which took the form $y = bx^k$, where x and y are parameters of growth, e.g., weight or linear dimensions, while b and k are constants.

In higher plants, a similar relationship exists between relative sizes or internal dimensions of their different organs; e.g., the weight ratio of shoot to root, length to breadth in leaves, diameter to length in cucurbit fruits, and even between the dry weight of thorns and remaining plant parts in citrus (Maggs and Alexander, 1969).

These growth interrelationships can be analyzed as follows. If $y = bx^k$, then $\ln y = k \ln x + \ln b$. A logarithmic plot of the two variables x and y should then yield a straight line (Fig. 3-9). Where this occurs, growth is said to be *allometric* and the slope of such a logarithmic plot (Fig. 3-9) is referred to as the *coefficient of allometric growth*. This coefficient then provides a basis for making quantitative comparisons of growth patterns.

Growth interrelationships can be viewed in a more fundamental way if we now develop the equations behind Fig. 3-9. The constant k represents slope, while b can be derived from the intercept on the ordinate when $\ln x = 0$. If the equation is now differentiated, $\ln y = k \ln x + \ln b$ becomes $dy/y = k\, dx/x$; and with respect to time, $dy/y\, dt = k\, dx/x\, dt$.

These last terms are reminiscent of expressions for exponential growth (see the section on Basic Concepts of Growth Analysis, above) and do in fact represent relative growth rates of the two components (x and y) in our original equation. Consequently, for allometric growth to occur, there must be a constant ratio between the RGRs of the two parameters being examined.

All this mathematical theory does have some relevance to physiology. Hammond (1941), for example, utilized the principles of allometric growth to study genetic control of leaf form in different varieties of *Gossypium arboreum*. Some

Fig. 3-8 Photosynthetic efficiency of cultivated plants, shown here as NAR, generally shows a decline with advancing age, irrespective of environmental conditions (Thorne, 1960).

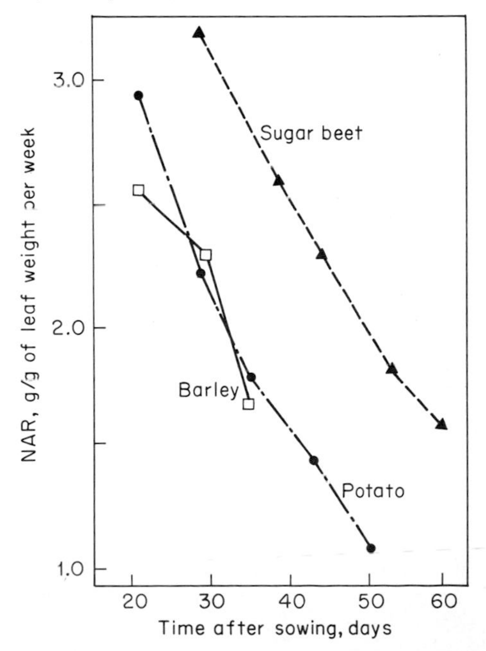

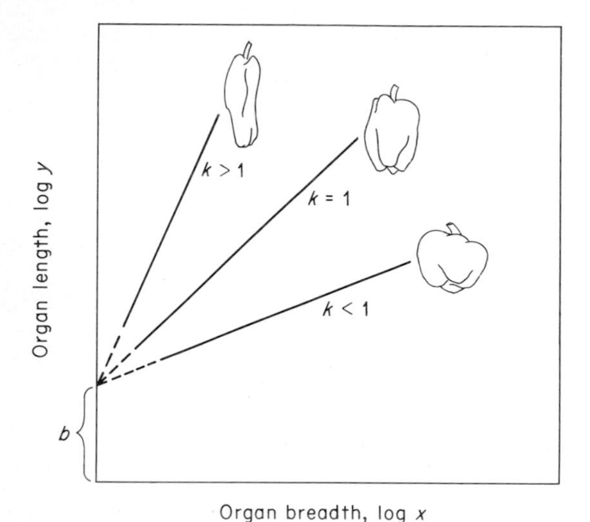

Fig. 3-9 The shape of an organ can be described as a plot of its length against its breath, as illustrated here for fruits of *Capsicum*. If the dimensions describe a linear relationship between length and breadth, allometric growth has occurred and it can be described as the slope of the plot k and a constant b.

of her data on k and b values are shown in Table 3-6. Normal type *Gossypium* is dominant with respect to k value, whereas b is subject to change and shows an intermediate value in the F_1 hybrid. In F_2 progeny, b and k values are transmitted together without separation. This suggests that a single gene is concerned with inheritance and subsequent control of leaf form.

Integration of growth at the whole-plant level is also amenable to mathematical analysis. Thornley (1972*a,b*) has produced a sophisticated model of vegetative growth and has formulated equations which describe steady-state exponential growth. After detailed mathematical treatment, he concludes for such a plant that the total activities of root and shoot are in constant ratio to one another. Such an interrelationship would be a feature in allometric growth, as described previously.

Thornley's (1972*a,b*) model should also help us to describe, in finite terms, how a plant reacts to changes in environmental conditions; members of the Tokyo school have already made some advances in this area. According to Monsi (1960), whole-plant dry weight W can be apportioned into two systems, photosynthetic F and nonphotosynthetic C. The C/F ratio following a period of growth is naturally an outcome of assimilation and subsequent distribution and can be described in the general formulas

$$F = f(W) \qquad \text{and} \qquad C = c(W) = W - f(W)$$

This C/F ratio need not be considered as unrelated to growth concepts developed so far because NAR can be expressed in a form that includes C/F ratio. Iwaki (1959) and Hiroi and Monsi (1963) provide comparable equations that can be summarized as

$$\text{NAR} = (p - r) - \left(\frac{C}{F}\frac{F}{\overline{F}}R\right)$$

p and r are photosynthetic and respiratory rates, respectively, so that $p - r$ represents net CO_2 fixation per unit area of leaf, C is the dry weight of non photosynthetic organs, and R is their respiratory rate. F is leaf dry weight and $\overline{F}$ is leaf area.

The distribution ratio, called δ by Kuroiwa et al. (1964), was then defined as the differential of $f(W)$ (used above) and becomes

$$\frac{dF/dt}{dW/dt}$$

This term can now be used in analyzing a plant's response to environmental conditions. Light intensity provides such an example. Shaded plants often develop greater foliar surface and have higher values for LAR (Fig. 3-2 and Table 3-1). Such data give the superficial impression that more photosynthate is being diverted into leaf growth to compensate, as it were, for reduced photosynthesis. Analysis in terms of distribution ratio δ suggests that the exact opposite has happened (Fig. 3-10). Although LAR is certainly larger under shaded conditions (Table 3-1), the leaves are actually thinner, and in terms of dry-matter distribution, the shaded sunflower plants diverted more substrate to nonphotosynthetic tissues. This tendency became accentuated with age, hence the fall in δ with ad-

Table 3-6 Logarithmic plots of leaf length against a basal dimension (distance from sinus to petiole) yield values for k and b which suggest genetic control over leaf shape in upland cotton. k is the coefficient of allometric growth, and b is a constant as described in Fig. 3-9 (Hammond, 1941).

Constant	Normal type	Okra type	Normal × Okra				
			F_1	F_2			
				Broad	Intermediate	Narrow	Super-okra
k	0.87	0.93	0.87	0.87	0.86	0.93	0.93
b	2.19	4.90	2.95	2.14	2.82	4.47	5.43

vancing size (Fig. 3-10). Kuroiwa et al., (1964) emphasize that δ is basic to the determination of *C*/*F* ratio and therefore to productivity and can also serve as an index of shade tolerance. The sunflower is intolerant of shaded conditions, as indicated by the decrease in δ with shading, especially with advancing age, i.e., increased size. By contrast, the constant δ in *Phaseolus aureus* indicates shade tolerance and implies that assimilate distribution is unaffected by shading.

Impatiens parviflora, known in Europe as a woodland plant, provides a further illustration of shade tolerance. In terms of growth analysis (see Table 3-1), LAR showed a remarkable increase due to shading so that RGR was maintained despite a substantial reduction in light intensity. Even though LAR increased, there was no change in the proportion of whole-plant dry weight to be found in leaves: they were simply thinner.

Such adaptation in *Impatiens,* as in other woodland herbs, confers some competitive advantage at low light intensity. Ecological consequences are discussed in Chap. 15.

Dry-matter distribution can therefore reflect a plant's adaptability to changing conditions, and this capacity is in turn a consequence of genetic makeup; but how is such control exercised? The mechanism is undoubtedly complex, and our understanding is scanty, but mathematical analysis of growth at least offers the prospect of describing a plant's self-regulated development and of predicting its response to environmental changes. This approach should eventually lead to an understanding of key points in control mechanisms.

Fig. 3-10 As *Helianthus annuus* plants increase in weight with increasing age, less of their growth is apportioned to leaves, as indicated by the decline in δ (the ratio of dry matter between photosynthetic and nonphotosynthetic organs). (*b*) As *Phaseolus aureus* plants increase in weight, they maintain a fixed ratio. At lowered light intensities, the sunflower still showed a decline in the ratio, and mung bean still showed a constant ratio (Kuroiwa et al., 1964).

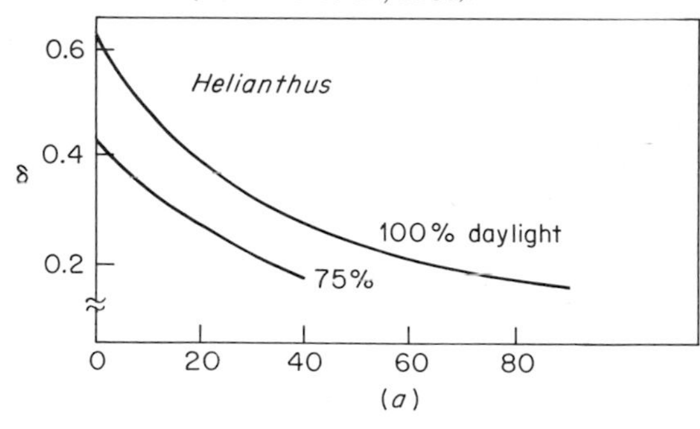

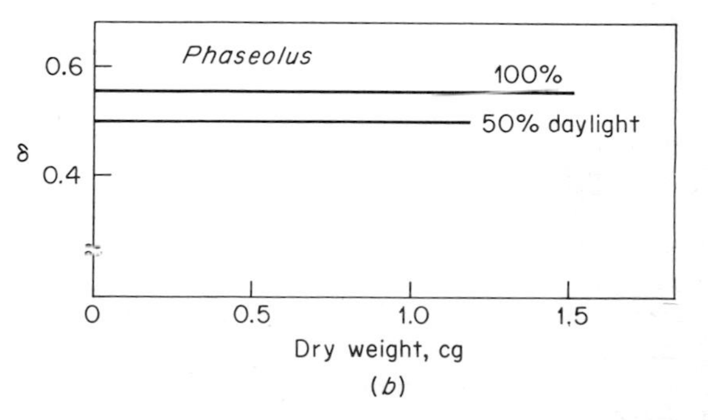

GROWTH OF A PLANT COMMUNITY

Interaction between Plant and Environment

The principal factor distinguishing growth patterns of individual plants from the performance of a community is mutual interference; i.e., individuals must make the best use of shared resources. Such competition is especially acute under natural conditions, where survival of the species within a given location is at stake, but it is no less intense in artificial communities based upon a common species. Within limits set by a plant's inherent capacity for growth, its overall performance is a direct consequence of how successfully it exploits the local environment. Such factors as light, temperature, water supply, and mineral nutrition assume major importance (they are discussed further in Part 4 under Ecological Physiology). Our immediate concern is an analysis of community growth by extending the indexes already described in the preceding section. Growth analysis aimed at a community level can help ascertain how plants effectively utilize their aerial environment.

Analysis of community growth Growth of a single plant can be usefully analyzed in terms of dry-matter increment per unit time and as a function of leaf area, that is, NAR. Community or crop growth, on the other hand, cannot be adequately described in the same terms because factors beyond NAR help determine total dry-matter production. Even at wide spacing, yield is not necessarily correlated with NAR because high photosynthetic activity is easily offset by low LAR. Leaf area does tend to be a more common determinant of plant growth than the photosynthetic capacity of individual leaves (Watson, 1952), so that effective exposure of the leaf surface becomes critical. High density affords greater leaf area per unit land area and therefore a greater potential for yield, but the photosynthetic rate of individual leaves will tend to be reduced due to mutual shading. Canopy density and NAR will therefore interact to determine total productivity, and to analyze such a situation we need parameters which quantitatively describe growth and foliar density at a community level.

Watson provided such an analysis of crop growth on the basis of his field work at Rothamsted, England, in the mid-1940s (Watson 1947, 1952). His indexes can be related to the growth-analysis terms described earlier in this chapter. RGR for an individual plant was

$$\frac{1}{W}\frac{dW}{dt} \qquad \text{where } W = \text{whole-plant dry weight}$$

while NAR was

$$\frac{1}{A}\frac{dW}{dt} \qquad \text{where } A = \text{leaf area}$$

By analogy, crop growth rate (CGR) is then defined as

$$\frac{1}{L}\frac{dW}{dt} \qquad \text{where } L = \text{ground surface}$$

CGR represents total dry-matter productivity of the community per unit land area over a certain time span. Watson had expressed photosynthetic cover in a community by a pure number, the ratio of leaf area (one side) to ground area, and called this ratio *leaf-area index* (LAI). This concept is basic to any analysis of community growth or light interception and especially to the performance of individual leaves (NAR). Since LAI is simply the ratio of A/L (see above), it becomes a fulcrum between NAR and productivity because CGR = NAR × LAI.

The concept of LAI has been applied with good results to dense stands of agronomic plants or established pastures and can be measured experimentally by harvesting all material from a quadrat of fixed dimensions, determining total leaf area, and relating this quantity to ground surface. Although reliable, the method is laborious; fortunately Warren-Wilson (1959) provided an alternative by determining LAI with inclined point quadrats. One traditional method in grassland analysis is to pass long needles (point quadrats) into the sward and record the contacts each needle makes on its way to the ground. Warren-Wilson (1959) modified this approach. He introduced inclined (rather than vertical) needles and operated at two inclinations (13 and 52°). LAI was obtained to within ±2 percent after adding the mean number of contacts at the two angles and making certain adjustments. In summary, the relationship Warren-Wilson (1959) derived was LAI $\approx 0.23L_{13} + 0.78L_{52}$. This formula allows rapid and nondestructive determination of LAI in the field.

Under continuous plant cover, e.g., natural heathland, established pasture, or dense crops, values for LAI generally fall between 1 and 8. Where solar radiation imposes a distinct limitation on growth (substory communities), values close to 1 are common, whereas deciduous forests might develop an LAI approaching 8. Where growth is not limited by other environmental factors, a community of leaves will exploit all available sunlight to the fullest extent, and total leaf surface (relative to ground area) required to achieve this situation will be determined more by leaf arrangement than by optical properties of individual leaves. Light extinction within the canopy is therefore more closely related to LAI than to chlorophyll "concentration" within the stand because variation in leaf chlorophyll content has surprisingly little effect on gross optical properties.

If all of the solar radiation falling on a crop could be intercepted usefully by photosynthetic tissue, maximum yield per unit land surface would be ensured, and since crop growth rate is the product of net assimilation rate and leaf area index, highest productivity should occur when these terms are maximized. CGR shows a close dependence upon LAI, and for a community of plants such as sunflower (Fig. 3-11) a clear optimum exists. As LAI is increased, the degree of mutual shading must intensify so that NAR will at some stage begin to decline. A point is reached where further increases in leaf surface, i.e., greater LAI, no longer offset the reduction in net photosynthesis due to less effective illumination, and a greater proportion of respiratory losses from nonphotosynthetic tissue will prevail. At this stage the optimum LAI will have been exceeded. The community is still gaining weight at high LAI, but the *rate* of increase has diminished.

Light intensity and plant form are both in-

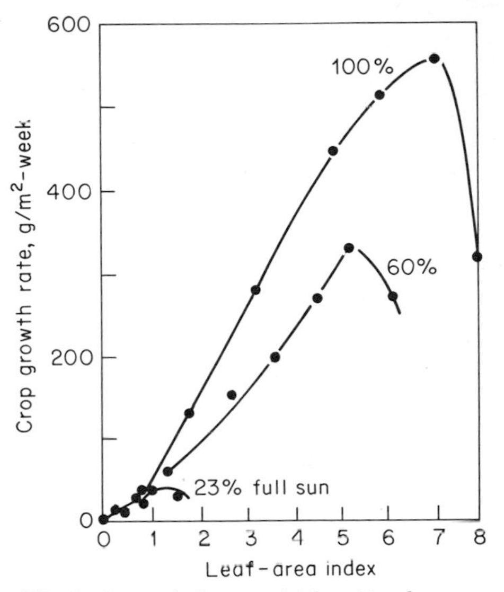

Fig. 3-11 The leaf-area index at which optimal crop growth rate is obtained depends upon the light intensity. Thus in a community of sunflower plants (100 plants per square meter) the optimum LAI is 7 under full sun, 5 under 60 percent full sun, and only 1.5 under 23 percent full sun (Hiroi and Monsi, 1966).

volved in determining the relationship between CGR and LAI. For a given species CGR is maximized at an LAI which itself increases with light intensity (Fig. 3-11). At supraoptimal LAI, respiratory losses from nonphotosynthetic tissues begin to assume greater significance; i.e., the *C*/*F* ratio has increased to the point where productivity is unfavorably affected because crowded plants have much greater sink demand (*C*) than source capacity (*F*).

Net assimilation rate (NAR in Fig. 3-12) must also respond to increased LAI, but the effect is dependent upon plant form. Leaf disposition in kale tends to be horizontal whereas sugar beet has more erect foliage. The decline in NAR is therefore more accentuated in kale than in sugar beet (Fig. 3-12). As a result fewer leaves are adequately illuminated within the canopy of kale compared with sugar beet. The consequence of these combined factors in determining CGR is evident in Fig. 3-13.

Leaf orientation assumes such major importance because a community of plants with ver-

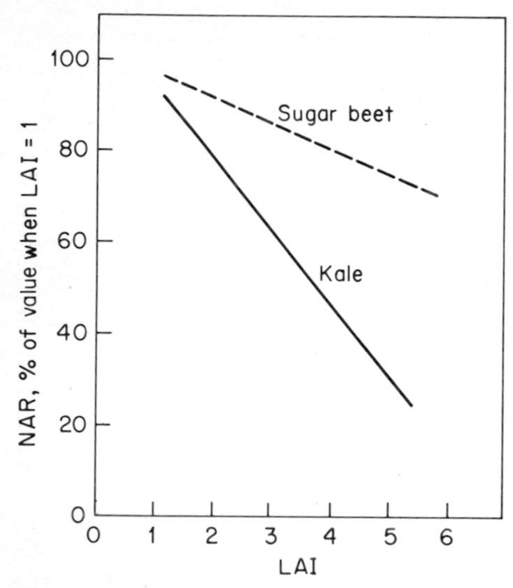

Fig. 3-12 Net assimilation rate falls as leaf area index increases, but the reduction is less severe in sugar beet, where leaves tend to be more erect, than in kale, whose foliage is more or less horizontal (Watson, 1958*b*).

tically oriented foliage makes good use of both diffuse and direct sunlight (Kriedemann and Smart, 1971) and can achieve greater LAI before fully intercepting incident sunlight than a group of individuals with more or less horizontal leaves. Brougham (1958), for example, showed 95 percent interception of available sunlight at an LAI of 7.1 for ryegrass compared with 3.5 for white clover. CGR should therefore be enhanced by erect foliage, especially at high planting density, i.e., high subsequent LAI. Data taken from Watson (1958b) and shown as Fig. 3-13 illustrate this point and also offer an interesting species contrast: sugar beet fails to achieve even a ceiling yield, whereas kale CGR shows an obvious maximum at an LAI between 3 and 4. Data shown in Fig. 3-14 also emphasize this principle by demonstrating how the response of community photosynthesis to LAI varies according to leaf orientation (Pearce et al., 1967).

As a matter of historical interest, as early as 1932, Boysen-Jensen had already stressed the importance of canopy shape and leaf inclination in relation to light utilization. He explained that dry-matter production may differ according to the habit of the *assimilation system* even when there is no difference in either leaf area or photosynthetic efficiency.

An Optimum LAI? Increased plant density results in a greater photosynthetic surface, but photosynthetic efficiency within the community becomes offset by other factors such as accelerated senescence in lower layers of foliage and greater respiratory losses relative to photosynthesis. Irrespective of original spacing, a community of plants will tend to achieve a common LAI whose magnitude depends upon inherent form (leaf orientation) and available light. The sunflower community referred to in Fig. 3-11 achieves an LAI of 7 under full sun compared with 1.5 under reduced illumination. Ceiling LAI does not necessarily imply maximum CGR, and Donald (1961) has developed a model for pasture growth which illustrates the principles involved. In models of this kind, respiratory losses are assumed proportional to biomass and independent of current photosynthetic rate, with the result that net photosynthesis for the community, and hence growth, passes through a point we have referred to as *optimum-leaf-area index* (see Fig. 3-16).

Although this concept was supported by obser-

Fig. 3-13 Stand geometry also influences the relationship between crop growth rate and leaf-area index by its effect on light penetration (Watson, 1958).

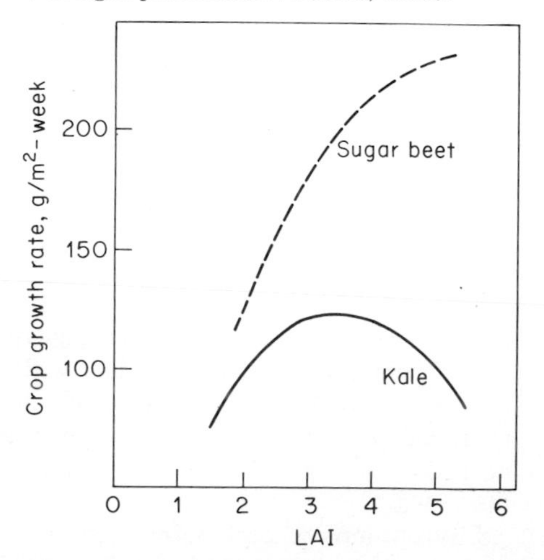

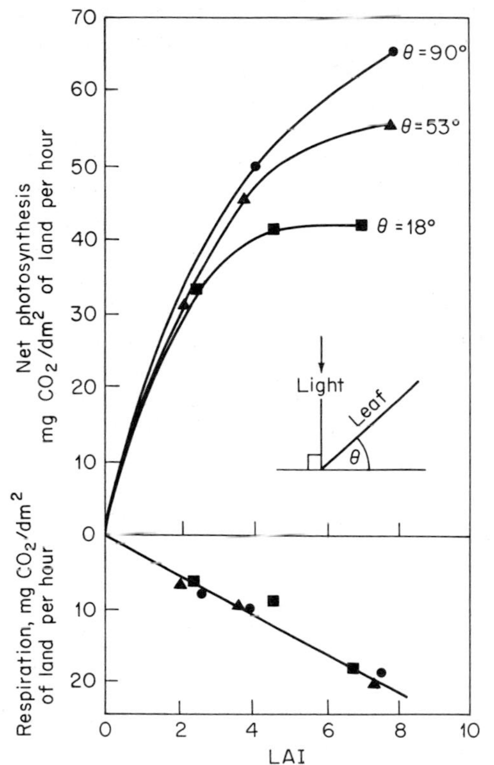

Fig. 3-14 In a stand of barley plants, the greatest net photosynthesis can be achieved at high leaf-area index (a high-density planting), and the maximum achievable is markedly influenced by the angle of leaf orientation (Pearce et al., 1967).

vation (Stern and Donald, 1961) and gained wide acceptance, it can be misleading for at least two general reasons: (1) The optimum can be biased toward the production of new material rather than total biomass because underground parts plus dead and senescent leaves are not always retrieved quantitatively. To provide accurate data for growth of the whole plant, it is necessary to count all material, including the tissue that dies between harvests. (2) Models of this type are not immediately applicable to field conditions because they are based upon light as the sole limiting factor. Additional influences such as temperature, humidity, water supply, wind speed, and nutrient availability can easily predominate in limiting productivity.

Notwithstanding these problems in extrapolating from model to field situations, one further assumption about respiratory losses requires confirmation: Is CO_2 evolution in constant proportion to biomass? Data from several independant sources suggests not. McCree and Troughton (1966*a*), for example, demonstrated that young clover plants show an adaptation in respiration rate according to incident-light intensity. If this behavior applies in a plant community, shaded leaves will not be as "parasitic" because their substrate demand would fall in proportion to local photosynthetic activity. Community photosynthesis should then achieve a ceiling value at high LAI rather than show a decline. McCree and Troughton (1966*b*,) subsequently confirmed this behavior for clover plants under controlled conditions.

By relating community photosynthesis to changing LAI Ludwig et al. (1965) again demon-

Fig. 3-15 The CO_2 assimilation and respiration rates of a stand of cotton plants show that both assimilative and respiratory rates are related to leaf-area index. Respiratory losses assumed the greatest significance at the highest temperature (40°C) (Ludwig et al., 1965).

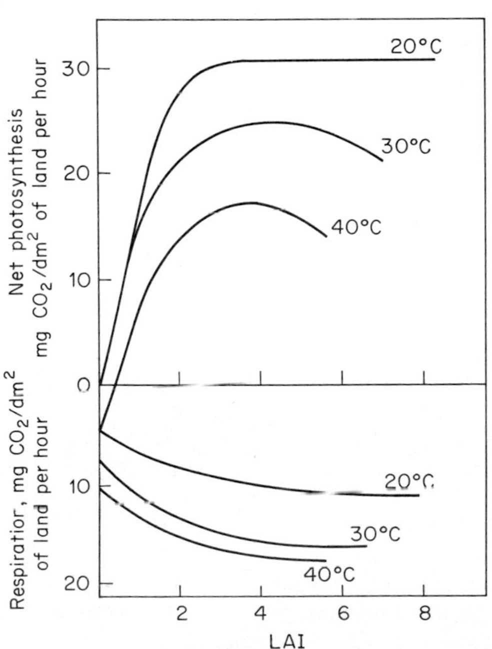

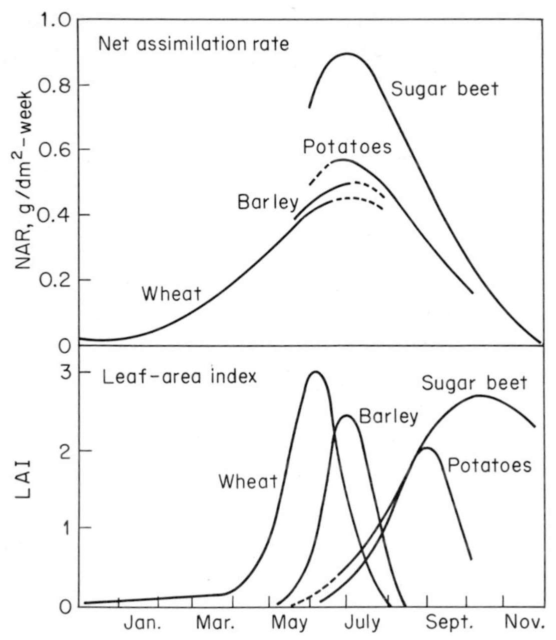

Fig. 3-16 The seasonal maximum in net assimilation rate may coincide with the period of greatest leaf area index in some crops but not in others (Watson, 1947).

strated this adaptation in leaf respiration. These authors started out with an intact stand of cotton in a growth chamber and measured CO_2 exchange by the enclosed community. Plants were then defoliated progressively, from lower layers upward, and gas exchange was plotted against LAI at successive stages of defoliation. Relevant data are shown in Fig. 3-15. Adaptation in respiratory rate by lower leaves is a feature of the established community and has important implications for photosynthesis within a stand: light-compensation point is not a fixed quantity (Cf. Chap. 1) but declines with depth, so that lower layers of foliage can be self-sustaining under a light intensity which would be insufficient to offset respiratory losses in more exposed foliage.

With the artificial community of Fig. 3-15, a clearly defined optimum LAI for net photosynthesis (and hence growth) was not apparent at moderate temperature. Only when respiratory losses were amplified by high temperature (40°C) did CO_2 fixation by the community show a decline as LAI was extended beyond 4.

A dense community of rapidly growing herbaceous plants could also show an optimum LAI for productivity (cf. Fig. 3-11) because respiratory loss would again represent a substantial slice of photosynthetic gain. Conversely, a stand of well-established plants growing under moderate conditions would show a relationship between productivity and LAI which simply approached an asymptote with increasing density. Sugar beet appears to behave in this way (Fig. 3-13), as do growth-cabinet communities of wheat or alfalfa for LAI values as high as 10 (King and Evans, 1967).

Microclimate In dealing with community microclimate, the only assumption to be made with absolute certainty is that environmental conditions will vary unceasingly! Foliage experiences fluctuations in its surroundings and presumably reacts to them; e.g., in the intensity, direction, and quality of illumination; CO_2 concentration profiles; temperature, relative humidity, and wind speed, i.e., conditions governing evaporation and heat transfer.

A plant community could hardly come to equilibrium with such an environment, and even at the level of its individual members, environmental conditions differ enormously from base to apex. Despite such complexities, the interaction of local weather and crop architecture more often than not determines plant performance and therefore must be understood. The present analysis cannot be exhaustive; it simply refers to those features of the aerial environment which clearly relate to community photosynthesis and growth, i.e., light and CO_2 supply. Other factors undoubtedly enter the picture and can sometimes assume even greater significance (Part 4), but one simple feature of higher plants attests to the primary importance of these two basic requirements for photosynthesis: they evolved with flat leaves, which enhance both light absorption and CO_2 transport to photosynthetic sites, a difficult compromise for any organism.

Light A community of plants growing outdoors intercepts direct solar radiation plus a certain amount of diffuse skylight which varies according to cloud cover. Some radiation penetrates directly, e.g., sun flecks, but is supplemented within the canopy by energy transmitted through leaves plus that reflected from plant and soil surfaces. Although heterogeneous, the radiation climate permits two generalizations: intensity and quality change with depth. Leaves cause a strong attenuation of solar radiation, and they also show a preferential absorption of energy in the 400- to 700-nm wavelength range (Fig. 1-9), so that longer-wave radiation (including far red) becomes more predominant at lower levels.

Against this background the term *light* can be seen as a gross oversimplification, but common usage rather than strict accuracy has prevailed. Photosynthetically active radiation (PhAR) is a more acceptable term for energy at wavelengths responsible for photosynthesis.

In a dense canopy, light intensity obviously diminishes with depth, and this attenuation can be described mathematically. On the basis of both experimentation and theoretical calculation, Monsi and Saeki (1953) suggest that light intensity at one height in a homogeneous plant community decreases with increasing leaf area in a manner resembling the Lambert-Beer law, so that $I_i/I_o = e^{-K \cdot \mathrm{LAI}}$, where LAI is the familiar leaf-area index, I_i is the light intensity inside the canopy, I_o is the light intensity outside the canopy, and K is an extinction coefficient characteristic of each canopy.

In arrays of more or less vertical leaves, K ranges between 0.3 and 0.5, whereas communities with nearly horizontal leaves have extinction coefficients between 0.7 and 1.0 (Hiroi and Monsi, 1966).

Although useful for growth analysis, a single parameter such as LAI cannot fully characterize the spacial complexities of leaf arrangement; furthermore, Beer's original proposal applied to the extinction of a monochromatic beam of parallel light rays in a homogeneous solution of particles of molecular dimensions. Diffuse light entering a canopy of leaves hardly fulfils these requirements, but in spite of its simplicity, Monsi and Saeki's relationship holds true for more practical purposes. Their model predicts, for example, that light intensity will show a logarithmic decline with depth so that percent visible radiation plotted as a function of height on semilogorithmic coordinates should yield a straight line. Data plotted in Fig. 3-17 provide one example, among many others, which confirm the model.

In general terms, both light climate and community productivity can be related to leaf-area index because stand geometry offers a common denominator. Since the constant K characterizes a stand in this respect, the form of any relationship between productivity and LAI will depend upon the value for K. Saeki (1960) developed this theme for a stand of *Celosia cristata* and generated the relationships shown in Fig. 3-18. The merit in having dense communities with erect foliage, i.e., low K due to low opacity, is self-evident.

Fig. 3-17 Attenuation of solar radiation within a corn crop approximates an exponential function so that intensity falls logarithmically with depth. The discrepancy between theoretical and measured values close to the ground is related to stand geometry and spectral composition of transmitted energy (Allen et al., 1964).

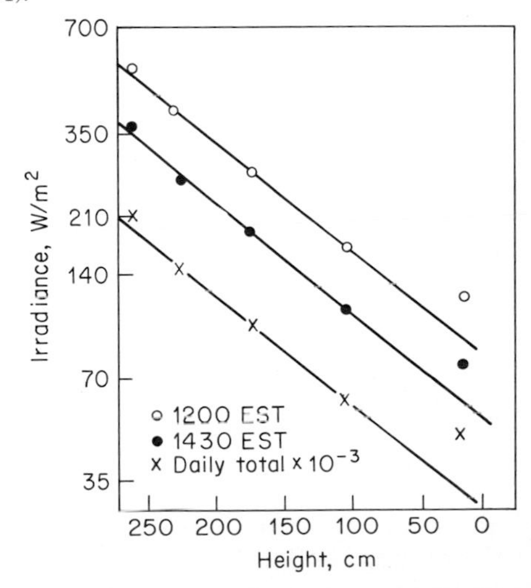

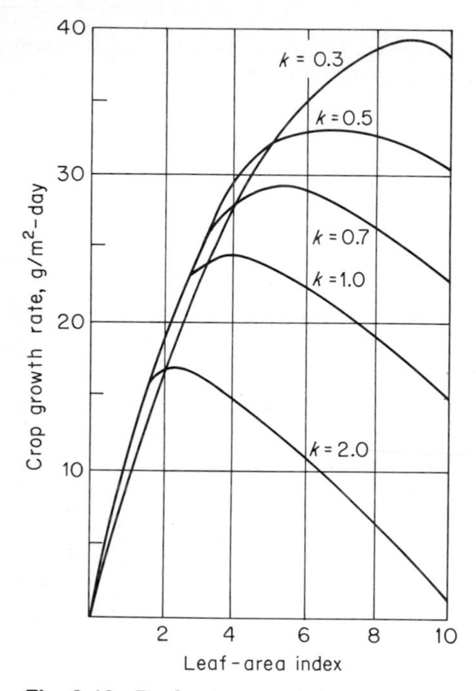

Fig. 3-18 Productivity in *Celosia* under full sunlight can be shown, theoretically, to depend upon foliar density within the stand, that is *K*, and leaf-area index. This prediction is borne out by observation (compare kale and sugar beet in Fig. 3-13, where the crop with horizontally oriented leaves i.e., high *K*, shows a clearly defined optimum LAI, whereas low *K* results in a ceiling value) (Saeki, 1960).

Carbon Dioxide Atmospheric CO_2 is not immediately obvious to the unaided observer, and this feature of the aerial environment generated much less interest than light profiles or water vapor and heat fluxes during early studies on community microclimate. CO_2 supply to leaf surfaces is a function of concentration and turbulence, so that CO_2 effects on photosynthesis and productivity must be considered in association with air movement. Leaf transpiration and evaporation from soil surfaces, as well as heat dissipation, all depend upon wind, so that turbulent transfer of CO_2 and H_2O becomes a major concern of micrometerology (see below).

Wind speed through the canopy will of course diminish with depth due to roughness and frictional drag, so that boundary-layer resistance of individual leaves will increase (Chap. 1). Consequently, CO_2 transfer from atmosphere to leaf surface will be hampered by the combined effects of reduced CO_2 concentration and lower turbulence. A corn field under calm sunny conditions provides a good illustration (Fig. 3-19, solid line). Corn leaves represent a strong sink for CO_2 (typical of C_4 plants) and commonly reduce CO_2 concentration by 100 ppm at the top of the canopy. Immediately adjacent to leaf surfaces, this depletion can exceed 200 ppm (Lemon, 1970). Under calm conditions, thermal convection currents would help improve CO_2 supply, but the situation on a windy day would still be more favorable with respect to CO_2 supply (Fig. 3-19).

Community Photosynthesis

Plant productivity depends upon a protracted sequence of events whose time scale is not of the same order as that of changes in environmental conditions. Growth analysis offers some improvements in this respect and sharpens our understanding of plant response to external conditions and interactions within the community, but this form of analysis gives a time-lapse pic-

Fig. 3-19 CO_2 depletion within the atmosphere at the top of a corn canopy is accentuated by calm sunny conditions. Close to the leaf surfaces, this reduction is even more severe (Uchijima, 1970).

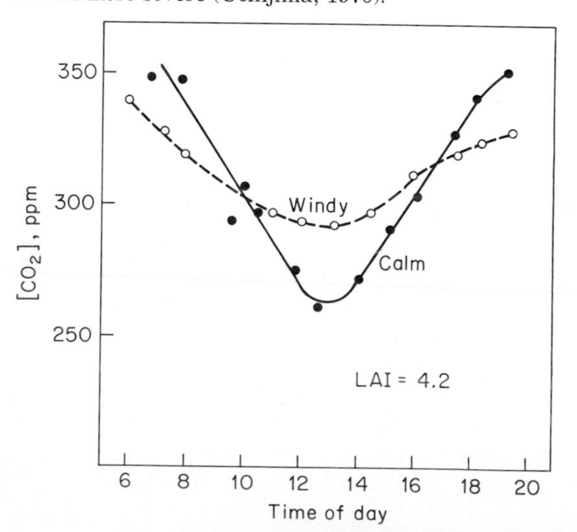

ture of productivity because it relies upon successive harvests spaced over days or weeks. By contrast, environmental fluctuations occur on a time scale of hours, minutes, or even fractions of a second. To appreciate plant interactions and to assess the importance of environmental conditions (including short-term changes) an instantaneous rather than an integrated measure of plant performance is required. Community gas exchange, measured *en masse*, provides an answer.

In dealing with large populations, one outstanding difficulty emerges: How can we measure community gas exchange without disturbing either the community or its environment? Two distinct approaches have been adopted: (1) enclose a large group of plants in a huge assimilation chamber and control the environment or (2) utilize the existing natural environment and gauge community performance from micrometeorological data.

Controlled conditions This approach can be adopted for either natural or artificial communities. Growth chambers, or *climate rooms*, can be hermetically sealed for gas-exchange measurement, and environmental conditions or community geometry can be modified at will. CO_2 concentration within the recirculating atmosphere is generally monitored with an infrared CO_2 analyzer, which drives a servo system designed to maintain atmospheric CO_2 at a predetermined level (say 300 ppm), thereby compensating for assimilation. Photosynthesis measurement then depends on recording the amount of CO_2 injected into the system to maintain a preset level. Alternatively, a large volume of air passes continuously through the chamber and photosynthesis is computed from flow rate and CO_2 depletion.

This method of total enclosure applies equally well to growth cabinets and field situations. Equipment for CO_2 measurement and atmospheric modification can be transferred outdoors and coupled to a transparent tent which may cover 1 m^2 or more of ground, depending upon the species under study. Thomas and Hill (1937) originally devised such a method for stands of wheat and alfalfa growing at Logan, Utah. Their technique has since been modified extensively, but the general approach has been adopted at numerous research centers around the world. Such an installation is shown in Fig. 3-20.

Fig. 3-20 Community gas exchange can be measured under field conditions with these large assimilation chambers. An air conditioner stabilizes temperature and humidity, while CO_2 is metered into each unit with rotameters and timers. A plastic floor is installed to seal off the soil so that net CO_2 uptake represents photosynthesis minus respiration of above-ground parts. Since the soil surface was sealed off, water collecting on the air conditioner's heat exchanger gives a direct measure of transpiration (Baker and Musgrave, 1964). (*Photograph courtesy of Dr. R. B. Musgrave, Cornell University.*)

Fully controlled conditions are more a feature of artificially lit growth cabinets, and community photosynthesis can be studied in relation to any environmental or plant parameter. Figure 3-15, based on the behavior of a cotton community, demonstrates one application where LAI was varied by defoliation; here one aspect of the field situation was tested under controlled conditions to answer a question about optimum LAI. Data from such controlled environments are also utilized in the other direction, i.e., in extrapolating from phytotron to field. Basic information about plant performance is built up by deliberately modifying a community's aerial environment and then monitoring gas exchange. Information generated in this way becomes incorporated into equations for use in computer simulations of community growth (discussed later in this chapter under Community Models).

If gas analysis is coupled with detailed growth analysis, a complete balance sheet of the plant's dry-matter economy can emerge (see, for example, Bate and Canvin, 1971). Physiologists at Rothamsted (Ford and Thorne 1967) have also completed such an undertaking for certain crop species (sugar beet, barley, kale, and corn). Plant populations were exposed to three CO_2 concentrations (300, 1,000 and 3,300 ppm) and grown at light intensities ranging from 3.7 to 11.6 cal/dm^2-min of visible radiation. Environmental effects revealed by classical growth analysis were viewed against data on gas exchange of individuals sampled from the same population. Extra CO_2 was found to increase photosynthesis (gas-exchange measurement) to a greater extent than resultant NAR (dry-weight–leaf-area measurement). The explanation lay in how dry matter became distributed. At elevated CO_2 there was relatively less tissue photosynthesizing and more tissue respiring, i.e., LAR had *decreased* (this confirms the CO_2 effect in Table 3-2). Results from this dual approach show how the benefit of supplementary CO_2 must ultimately depend upon faster photosynthesis, but its final expression in dry matter will depend upon partitioning assimilates between photosynthetic and nonphotosynthetic tissues. Environmental factors affecting this distribution will interact with photosynthetic activity in dictating final productivity.

In a similar vein, McCree and Troughton (1966*b*) subjected white clover plants (*Trifolium repens* L.) to different levels of constant light (holding other environmental factors constant) and compared total dry-matter increase with net CO_2 uptake. Total respiratory losses over 24 h amounted to 38 to 43 percent of gross photosynthesis. When this factor was taken into account, predicted vs. observed growth rates showed close coincidence (Fig. 3-21). Their data revealed an additional principle about respiratory losses which relates to optimum LAR discussed previously; i.e., losses are more closely related to current photosynthetic activity than to plant size; changes in environmental conditions lead to adjustment in both photosynthesis *and* respiration.

Fig. 3-21 Growth rate of white clover can be predicted by measuring net daily uptake of CO_2 and making allowances for respiratory losses. Such close agreement with observed growth helps validate the modeling system (McCree and Troughton, 1966*b*).

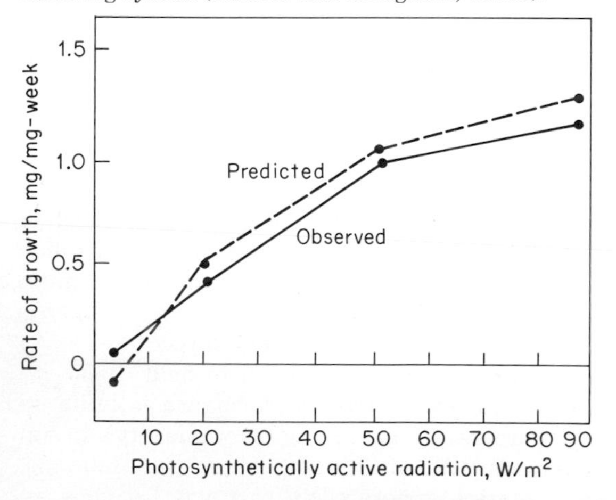

Natural conditions A clear distinction exists between field measurements of community photosynthesis using large air-conditioned enclosures (Fig. 3-20) and photosynthesis of an equivalent undisturbed community. Nevertheless, these chambers provide usable data and offer unique possibilities for precise work. For example, Puckridge (1972) established the contribution made by different plant parts to total photosynthesis of a wheat community by administering $^{14}CO_2$ and then assaying the radioactivity of different organs; such detailed work requires an enclosure.

Evaluation of cultural treatments such as planting density (hence stand geometry and LAI), fertilizer level, soil-water potential, or even varietal comparisons are all made possible with field chambers. Environmental factors such as solar radiation, temperature, humidity, wind speed, and CO_2 concentration are also amenable to study. Predictions of community photosynthesis, based upon growth-chamber experience, can be tested, a clear prerequisite for mathematical modeling.

Field chambers therefore simulate the more important features of the field environment and reveal community responses, but they do not *duplicate* natural conditions. The community has to be enclosed so that photon flux density and spectral composition no longer coincide with natural sunlight; walls and ceiling of the cabinet are often warmer than enclosed foliage, adding to their radiation load; and while air flow is substantial, the scale of turbulence differs from that in the field. Photosynthesis measurement within the undisturbed community clearly necessitates a different approach.

Micrometeorology Although the natural microclimate is characterized by a chaotic variation in its physical parameters, some overall patterns are discernible. Meteorologists working on a global scale are able to relate the exchange of heat and momentum near the earth's surface to rates of turbulent mixing. A micrometeorologist utilizes these same principles because a plant community has to depend upon turbulent transfer processes for exchange of carbon dioxide, oxygen, water vapor, and heat.

When leaves intercept solar radiation, they act as sources of heat to evaporate water (latent-heat exchange) and in turn warm the ambient air (sensible-heat exchange). At night, leaves lose energy by longwave radiation and act as heat sinks. Where net exchange of heat (or some other physical entity) between a plant community and the surrounding atmosphere is governed by turbulent mixing, a simple equation describes its vertical flux.

$$\text{Flux intensity } q = \text{diffusivity coefficient } K \times \text{gradient } \frac{dc}{dz}$$

where dc/dz represents change in concentration with height. If flux density is measured in different horizontal planes above the crop (see solid points in Figs. 3-22 and 3-23), the flux difference between successive layers $q_{z2} - q_{z1}$ represents the source or sink intensity Q of the community, i.e.,

$$Q = q_{z2} - q_{z1} = \left(K_{z2}\frac{dc}{dz}\right)_{z2} - \left(K_{z1}\frac{dc}{dz}\right)_{z1}$$

Transpiration can be defined in this way, in which case the community represents a source of water vapor and the flux is upward (positive); with photosynthesis the plant population is a sink for CO_2 and the flux is downward (negative).

Micrometeorologists develop this principle for practical purposes by measuring physical parameters at a number of finite levels within and above the vegetation and construct vertical profiles (Fig. 3-23); fluxes over any desired height interval are then computed.

The key problem in measuring community gas exchange is the determination of K. As implied by the equation above, K varies with height (air-density effect); its magnitude depends upon the degree of atmospheric turbulance, which in turn is a function of wind speed, surface roughness, and thermal instability. Typically, the magnitude of K is around 10^3 cm^2/s at heights of 1 to 2 m (several orders of magnitude greater than molecular diffusivity referred to in Chap. 2).

K is commonly determined by measuring the fluxes of two climatic entities simultaneously under weather conditions that are reasonably uniform in time and space. If diffusivities of the two entities are the same and the flux of one is known (or can be derived from measurement), the flux of the other can be inferred from profile data.

Two approaches are commonly adopted when measuring community photosynthesis. K can be derived from either *momentum balance* or *heat balance*. Momentum is extracted out of wind as it blows across vegetation. Leaf flutter and branch swaying are manifestations of frictional drag, and since air has mass, a change in velocity in the community means that momentum is being transferred downward. A diffusionlike equation describes this flux of momentum, and to arrive at a balance for K determination, only two field measurements are required, namely, leaf-area distribution as a function of height

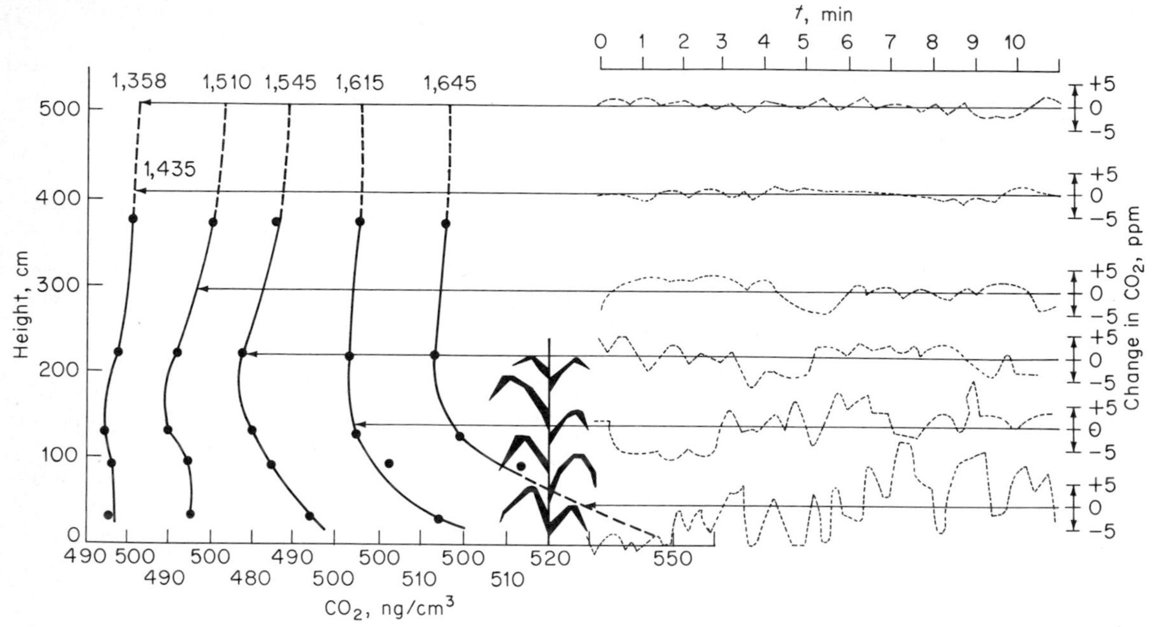

Fig. 3-22 As the wind blows across a plant community, CO_2 concentration shows fluctuations, due to photosynthetic activity, whose amplitude varies according to height and stand geometry. These fluctuations can be used to develop CO_2 concentration profiles for different periods in the day (time indicated above diagram). Changes in profiles are then used to derive net CO_2 flux into or out of the community (Lemon et al., 1969).

and mean-wind profile over discrete time intervals (say 10 to 20 min). Figure 3-23 describes such quantities.

The energy, or heat-balance, approach also yields a value for K and without having to obtain and interpret wind-profile data, but more observations are required. This method relies on the principle of energy conservation to balance storage and dissipation of heat within a community. In simplified terms, the energy balance of a community can be given as

$$R = E + H + G + P$$

where R is net radiation (the difference between incoming and outgoing radiation fluxes), E is evaporative heat flux (energy consumed in evapotranspiration), H is the sensible-heat exchange between surface and air, G is heat flux to the soil, and P is energy used biologically by the community.

E and H, that is, the rates at which water vapor and sensible heat are transferred from plant community to surrounding atmosphere, are proportional to water vapor and temperature gradients, respectively. By measuring net radiation flux plus gradients in temperature and water vapor over the vegetation it is possible to calculate heat exchange and water-vapor exchange, complete an energy budget, and so estimate K. According to Lemon (1968), K is best determined by heat budget under low steady winds, while the momentum-budget method is most successful under moderate steady winds.

Community architecture has a profound influence on plant microclimate by its effect on ventilation and air movement. The turbulent diffusivity coefficient K as it applies to heat, water vapor, carbon dioxide and oxygen is the key parameter. Community photosynthesis P is then obtained from the expression

$$P = -K_c \frac{dc}{dz}$$

As an illustration of the type of data gained by micrometeorological methods, Fig. 3-24 shows the dependence of photosynthesis in a rice community on solar radiation and wind speed.

A primary aim in micrometeorology is to measure community photosynthesis in undisturbed natural situations over periods short enough to detect environmental effects (as shown above). The outstanding advantage of aerodynamic studies over field-chamber methods is their minimal disturbance of the natural environment; nevertheless, there are some drawbacks:

1 A large area of uniform vegetation is required and with adequate fetch (a free run for air movement upwind of the measurement site).

2 The range of environmental conditions under study cannot be generated at will or artificially increased.

3 Multiple regression analysis is often employed to sort out key factors controlling photosynthesis so that extensive mathematical modeling has to be pursued concurrently.

4 It is difficult to estimate contributions from individual leaves without determining the fine structure of both community and atmosphere in great detail.

5 Long-term changes in community performance (over weeks or months) cannot be appreciated within this context of short-term environmental fluctuations.

Notwithstanding these limitations, micrometeorological methods have become established as tools for field research; simulation experiments on analog and digital computers allow extrapolation from cuvette and chamber experiments to the field situation; model predictions then need testing in the natural environment—hence micrometeorology.

Fig. 3-23 Micrometeorologists obtain a detailed picture of plant performance under *natural* conditions by measuring physical parameters above and within the canopy. Profiles shown in this diagram are drawn from two stands of corn which differ in plant density (climate data are integrated 30-min values from 11:45 to 12:15 A.M. (Lemon, 1970).

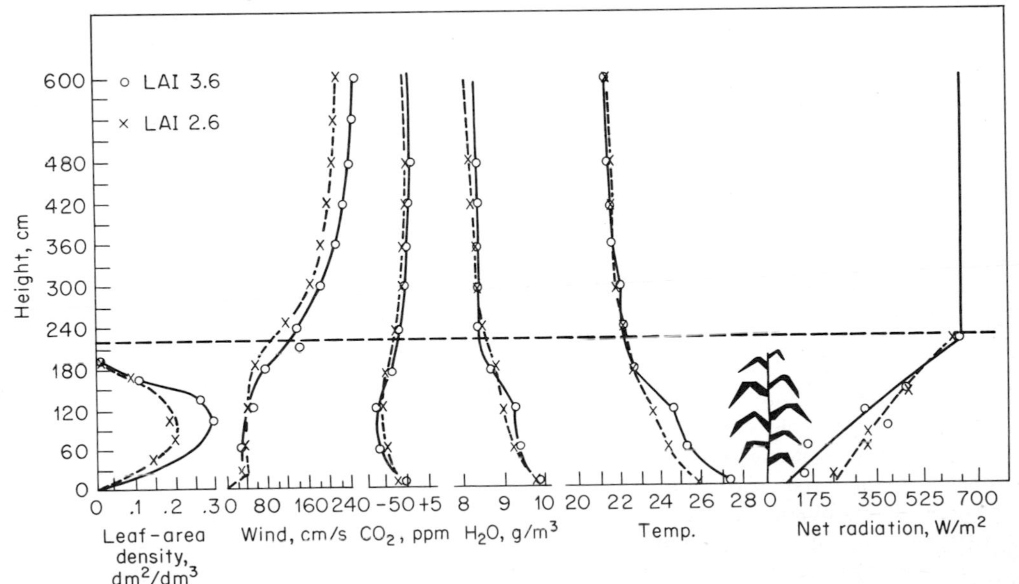

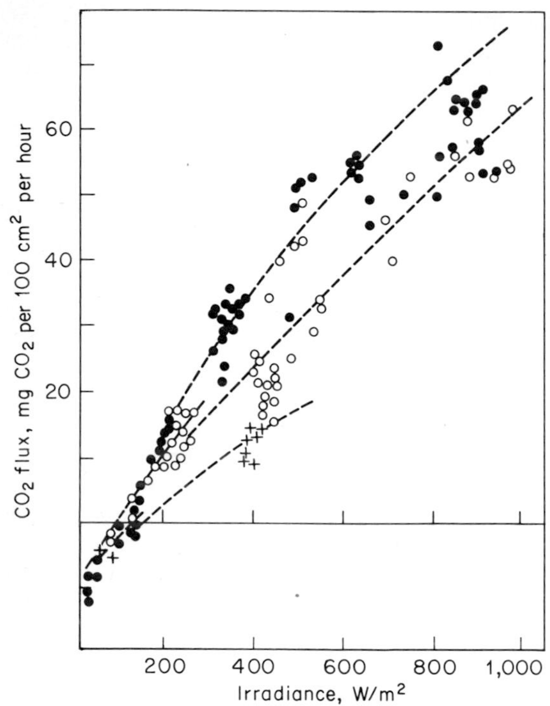

Fig. 3-24 Downward CO_2 flux into a rice field depends upon solar radiation and wind speed. These data were obtained by micrometeorological methods in a rice field. Average height of the community was 70 cm. Wind-speed categories: +, 1.5 to 1.9 m/s; o, 2.0 to 2.9 m/s: •, 3.0 to 3.9 m/s (Yabuki et al., 1971).

Community Models

Models are simply ideas put down on paper, i.e., precise statements of how systems are *thought* to operate. Physical scientists have been using this approach for a long time to predict untested conditions. They know the behavior of electrons well enough to be able to simulate real systems by simply scaling components on an analog computer. Biological models, on the other hand, are of a different character because higher plants can display a wide range of behavior and our understanding of biological principles is in its infancy compared with the physical sciences. Nonetheless, biological models do have a long history. Mendel's experiments on color inheritance in sweet peas, which eventually led to the concept of genes, amounts to a kind of model. Similarly, enzyme action and nucleic acid replication have been usefully represented by model systems.

The rapid emergence of mathematical modeling within environmental spheres of biology has paralleled developments in fast data capture and processing. Literally thousands of pieces of information relating to environmental conditions and plant response need to be collected and processed in any sizable experiment. High-speed computers have made this possible.

In simplified terms, dry-matter yield Y from any crop is a function of time t and effective photosynthesis (net P), i.e.,

$$Y = \int_{t_1}^{t_2} (\text{net } P) \quad dt$$

Net photosynthesis, as the major contributing process, represents the output from all simulation models, including those of Japanese pioneers (e.g., Saeki 1960) and other schools (de Wit, 1965; Monteith, 1965; Duncan, et al., 1967; Lemon et al., 1971). The central theme in all these models has been to produce realistic estimates of community productivity when photosynthesis is unconstrained by limitations in the supply of water and nutrients or by unfavorable temperature; i.e., production is regarded as being determined primarily by solar energy.

Predictive models If the photosynthesis–light-response curves for single leaves measured under laboratory conditions are obtained (cf. Figs. 1-25, 15-5 and 15-7), CO_2 fixation by a community of this same species can be estimated from foliar distribution and radiation climate within the canopy. Saeki (1960), Monteith (1965), and Duncan et al. (1967) have all adopted such an approach.

Light-response curves for photosynthesis in single leaves can be fitted rather closely by several functions including the rectangular hyperbola

$$P = \frac{P_{\max} I}{I + k} - R$$

where $P_{\max}$ is the value for photosynthesis at light saturation, I is light intensity, R is respiratory loss, and K is a constant which equals I at $P_{max}/2$. Monteith (1965) was able to develop an alternative expression couched in terms of photochemical efficiency and diffusive resistances, more applicable to the model of the photosynthesizing leaf discussed previously (Chap. 1). According to Monteith's expression, the rate of photosynthesis p on light intensity I is fitted by $p = (a + b/I)^{-1}$. The term b/I is in effect a photochemical *resistance*, which appears surprisingly consistent over a wide range of plant species at about 30 k cal/g. The other variable, a, is proportional to total diffusive resistance to CO_2 (see Fig. 1-19).

The next component of a predictive model, i.e., radiation climate within the plant community, is more difficult to define with close precision due to fluctuations in energy sources such as diffuse skylight, alterations in solar angle, leaf reflectivity, and intermittent sun flecks. Light attenuation with depth does bear some analogy to the Lambert-Beer law (discussed earlier in this chapter) but is more accurately described in Monteith's (1965) binomial expression, which deals with a canopy in terms of successive layers of leaves. For the nth leaf stratum

$$\frac{I_n}{I_0} = [S + (1 - S)T]_n$$

where S is that fraction of total incident radiation not intercepted and T is the mean transmission coefficient of leaves.

Given these relationships, Monteith (1965) was able to calculate the total photosynthetic activity of successive leaf layers; summation then yielded a value for the whole stand. After some trial calculations and successive approximations of constants, he obtained reasonable agreement between measurement and prediction (Fig. 3-25). These data, originally from Watson's (1958*b*) publication, were discussed (Fig. 3-13) in connection with stand geometry and LAI. Figure 3-25 highlights the difference between kale and sugar beet in succinct mathematical terms rather than the more descriptive terminology of leaf orientation and light interception (compare a values on these two crops).

Further refinement of Monteith's (1965) model (see, for example, Osman, 1971; Patefield and Austin, 1971) has achieved even closer coincidence between projection and observation. Such developments emphasize the comprehensive nature of predictive models that simulate plant growth.

Subsequent to pioneering work in Japan (Saeki, 1960), Duncan et al. (1967) published their model for community productivity in the form of a computer program. In summary, they had to account for leaf illumination over varying areas and angles under an ever-changing light source. Their expression of stand geometry and effective illumination of successive leaf layers (10 cm deep) was derived from point quadrat theory (discussed briefly in connection with LAI). Once leaf illumination had been computed, this value was substituted into an equation relating illuminance to photosynthesis for leaves in that layer. Summation of individual layers yielded a value for community photosynthesis; correction for respiratory losses (taken over 24 h as 40 percent of leaf photosynthesis) then gave an estimate of dry-matter productivity. Predicted vs.

Fig. 3-25 Watson's (1958) observations on dry-matter productivity in kale and sugar beet communities form a basis for Monteith's (1965) test of his predictive mode. Input data were limited to photosynthetic characteristics of leaves and community light climate. a, one of the basic terms used by Monteith (1965) in his mathematical description of a leaf's light-response curve for photosynthesis, is proportional to the sum of external, stomatal, mesophyll, and carboxylation resistances (see Fig. 1-19 for further explanation).

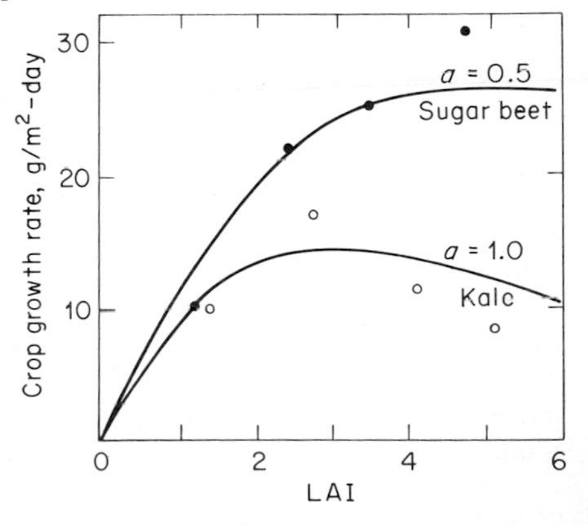

observed yields on corn growing at Davis, California, showed sufficient coincidence to validate their model.

Further applications for modeling Computer simulation of plant-growth processes or of overall community production has been applied successfully to situations where environmental factors dictate current performance. The realm of stress physiology, where internal controls override environmental factors, is less amenable to predictive modeling because underlying biological principles are less well defined. As our understanding develops, models will be applied in a number of ways:

1 To establish components in complex biological systems. A jigsaw puzzle offers a reasonable analogy. We cannot appreciate the significance of individual pieces until the entire picture is complete.

2 In correcting erroneous concepts. The growth of maize at dense planting provides an example. Duncan's model based on fixed respiratory loss, relative to photosynthesis, predicted negative growth at high density. In practice this simply does not occur, and the model led to new concepts about respiratory losses, confirmed by McCree and Troughton (1966*a*). The readjustment in respiratory rate that occurs when environmental conditions change was then described mathematically by McCree (1970), and his equation for white clover respiration will undoubtedly lead to modification in productivity models for other species.

3 Evaluating hypotheses. Descriptive schemes can be made quantitative and tested in laboratory or field. Models of a photosynthesizing leaf (Chap. 1) led to laboratory experiments that improved concepts of gas exchange. Similarly, short-term measurements of community microclimate and photosynthesis provide starting data for computer simulation. Values generated for photosynthesis within the model are then compared with direct measurements. Agreement lends confirmation to the experimenter's theories; complete disparity highlights inadequacies in the model.

Whether a model eventually stands or falls is irrelvant; new concepts will have emerged in either event.

SUMMARY

In simple plants such as the unicellular algae, growth may be considered as a relatively simple expression of photosynthesis, storage of the synthesized products, and their utilization. The evolution of higher plants has demanded an elaboration of this simple pattern into much more complicated expressions of these basic processes which enable individuals to adjust to their ever-changing microclimate.

As plants came to utilize a broader range of growing conditions, their assimilatory system responded with a surprising array of adaptations to environmental limitations, thereby creating a community which is finely tuned to its environment. Some of the most impressive of these adaptations include the development of sun leaves and shade leaves, with strikingly different photosynthetic features, including different efficiencies of light utilization and distinctive morphologies.

In brief, examination of the assimilation and growth of higher plants in quantitative terms has illustrated some of the physiological devices which plants employ in adapting themselves with increasing refinement to the limitations imposed by the structural nature of both individual plants and their communities, as well as by the environmental stresses which they normally experience.

GENERAL REFERENCES

Duncan, W. G. 1967. Model building in photosynthesis. In "Harvesting the Sun" A. San Pietro, F. A. Greer, and T. J. Army, eds. Academic Press, New York, pp. 309–320.

Evans, G. C. 1972. "The Quantitative Analysis of Plant Growth." University of California Press, Berkeley. 734 pp.

Loomis, R. S., W. A. Williams, and A. E. Hall. 1971. Agricultural productivity. Ann. Rev. Pl. Physiol. 22:431–468.

Milthorpe, F. L., and J. Moorby. 1974. "An Introduction to Crop Physiology." Cambridge University Press, London. 202 pp.

Rees, A. R., K. E. Cockshull, D. W. Hand, and R. G. Hurd (eds.). 1972. "Crop Processes in Controlled Environments." Academic Press, London. 391 pp.

PART 2 GROWTH REGULATION

AUXINS

GIBBERELLINS

CYTOKININS

ETHYLENE

INHIBITORS

DIFFERENTIAL GROWTH

In the single-celled organism growth is complicated enough, involving the processes of cell division, enlargement, and organelle production, but with the evolution of the multicellular higher plants, the phenomenon of growth has become enormously complicated. Unlike cell colonies, the large organism faces the problem of growth with more complicated requirements for the localization of cell division, cell enlargement, the differentiation of tissues and organs, and the need to elicit growth responses to environmental stimuli. There are two classes of discernible mechanisms which might be employed by the organism to bring about these types of regulation: (1) systems of chemical messengers, which direct cells to carry out various functions of growth and differentiation, and (2) systems of physical or field forces. The former class includes the plant hormones or chemical regulators, and the latter might include electric gradients, mechanical or structural constraints, and metabolic or gas-concentration gradients. Other types of mechanisms may exist, but we are unaware of them.

Although little is known about physical and field forces which may be involved in the multicellular plant organization and regulation, much information has been accumulated about hormone systems and chemical control mechanisms. Considering the present state of knowledge, therefore, most of our attention will be directed to these hormonal and chemical control systems in developing the biologist's picture of how plant cells are regulated in growth and differentiation.

The plant physiologist knows of five types of chemical growth-regulator systems: the auxins *and* gibberellins, *which stimulate cell elongation; the* cytokinins, *which stimulate cell division;* ethylene *gas, which stimulates the swelling or isodiametric growth of stems and roots; and the* inhibitors. *Unlike most animal hormones, which may have relatively specific types of physiological regulatory functions, plant-growth regulators have widely overlapping functions. While each is distinctive, both in chemical characteristics and in being able to bring about characteristic growth responses, each of the five types of regulator is capable of altering most aspects of growth, including cell division, cell enlargement, differentiation, and differential growth phenomena. The same type of overlap holds for the various stages of plant development, as is illustrated in Table 18-1. Because of these overlapping functions, it becomes difficult to define the five classes of growth regulators without being simply empirical; auxins are substances which generally resemble indoleacetic acid and have the ability to stimulate elongation of coleoptiles; gibberellins are diterpenoids, such as gibberellic acid, which have the ability to stimulate the elongation of the stems of green seedlings, especially certain dwarf and rosette types; cytokinins are usually substituted adenines which resemble zeatin and have the ability to stimulate cytokinesis in cultures of tobacco cells. Ethylene is a gaseous regulator which stimulates isodiametric growth in the apices of dicot seedlings; and inhibitors are regulators of growth which ordinarily depress cell-enlargement activities. Early hopes of assigning to one substance the responsibility for cell elongation, to another the responsibility for cell division, to another leaf growth or root or fruit growth have faded as the growth and differentiation processes appear more and more to be regulated by concerted interactions between the various types of growth regulators.*

Growth regulation in higher plants is like contrapuntal music, each of the various growth regulators playing concertedly on the instruments of growth. In some cells, one specific growth regulator may be the principal limiting factor in growth, as the elongation of the oat coleoptile is limited principally by auxin; yet gibberellins may be contributing to the growth of these same cells. In other cases, cytokinins may be the principal limiter of cell division, as the tobacco callus in tissue culture is limited principally by cytokinin, yet auxins may be contributing to the same growth function. The simultaneous interaction of two or more of the five known types of growth regulators in regulating various aspects of growth and differentiation appears to be a general feature of higher plants. The various classes of plant-growth regulators are therefore more readily distinguished by their chemical relationships and their activities in certain bioassays than by areas of specific physiological responsibility or control.

From 1930 to 1950 plant physiologists had precise knowledge of just one endogenous growth regulator, or hormone, in plants. In the next decade two more were discovered, and in the next decade two more, bringing the list to five. It is probable that still more await detection.

The evolution of higher plants with their complex regulatory systems can be viewed as the manifestation of increasingly elaborate adaptations of chemical regulatory systems to achieve a finer adjustment of internal controls and the tuning of these controls for finer sensing and response of the plant to its environment.

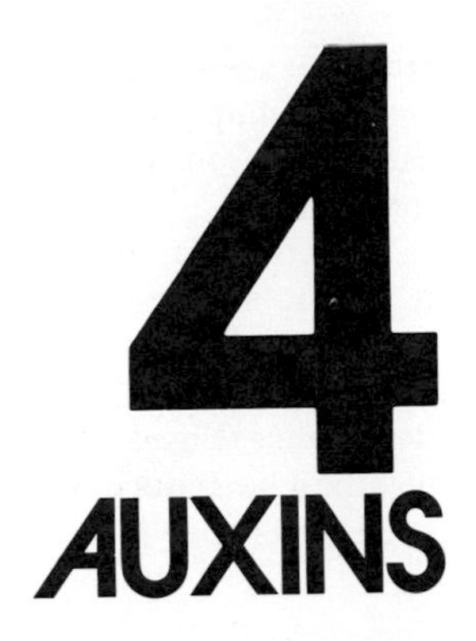

4 AUXINS

The discovery of auxins was the outcome of experiments designed to explain a correlation effect, phototropism. Darwin (1897) found that the tip of a grass coleoptile is essential to the tropistic response of the whole coleoptile; Paál (1919) concluded that a *correlation carrier* is supplied by the tip and that its movement leads to the development of curvature. Went (1928) actually found such a substance in diffusates from coleoptile tips and was able to explain both the correlative nature of the tropistic response and the endogenous control of growth rates on the basis of this substance, auxin.

Kögl et al. (1934) described indoleacetic acid (IAA) as an auxin. This chemical was purified from plant materials by Kögl and Kostermans (1934) and by Thimann (1935). Presuming that this hormone was a universal control of growth, Went (1928) proposed that "Ohne Wuchsstoff, kein Wachstum" (without auxin, no growth). With the later recognition of the gibberellins and cytokinins, the universality of auxin control over growth has become less certain, but the term *growth hormone* became synonymous with auxin, and for about 25 years studies of the hormonal regulation of growth were mainly studies of the effects of auxin on cell enlargement. This hormone, auxin, is certainly intimately involved in the catalysis of cell enlargement in stems and coleoptiles; however, its role as a stimulant of growth in other organs such as roots, leaves, and fruits is still uncertain.

Since the search for auxin was initially a study of tropistic curvature of coleoptiles, it is appropriate that the first good bioassay for it should utilize the stimulation of coleoptile curvature. The data in Fig. 4-1 show the quantitative curvature which develops when auxin is applied to one side of the decapitated *Avena* coleoptile. The precision of the response depends on the polar transport of the auxin down the coleoptile on the treated side and the preferential stimulation of growth there.

When pieces of coleoptile or stem are immersed in an auxin solution, quantitative stimulations of elongation are obtained, as shown in Fig. 4-2. These large stimulations are roughly proportional to the logarithm of the concentration of

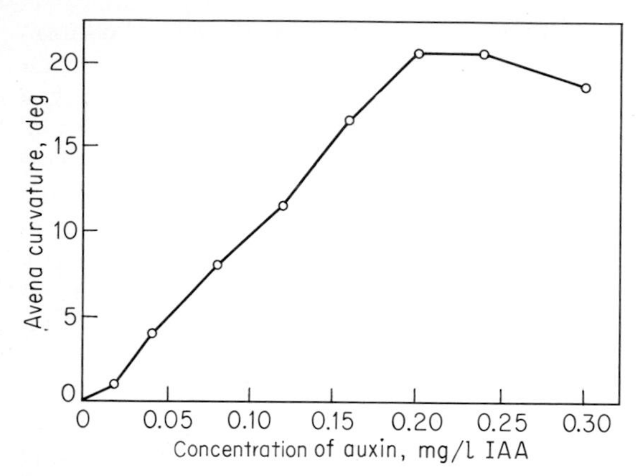

Fig. 4-1 The unilateral application of auxin to decapitated oat-coleoptile tips produces quantitative curvature responses away from the treated side. This technique provides a means of quantitative bioassay for auxin (Went and Thimann, 1937).

auxin, and etiolated stems are more sensitive than green ones.

In addition to bioassays, much work has been done using colorimetric measurements of indoleacetic acid in plant extracts or diffusates (see, for example, Gordon and Weber, 1951); more recently mass spectroscopy has proved useful in auxin identification (Greenwood et al., 1972). A fluorometric assay has proved useful for several synthetic auxins (Hertel et al., 1969), and auxin derivatives can be assayed very precisely using either fluorometric (Knecht and Bruinsma, 1973) or gas-chromatographic methods (Grunwald et al., 1968).

Inhibitory growth responses to auxin are very general. Roots are particularly sensitive to inhibition by auxins; an example of quantitative inhibition is shown in Fig. 4-3.

In addition to their abilities to regulate growth, the polar movement of auxins plays a major role in the regulation of correlation effects in the plant. These are regulatory effects imposed upon one part of a plant by a remote part. The central role of auxins in the tropisms is one such correlation effect; auxins also play an important part in apical dominance (Thimann and Skoog, 1934) and abscission (LaRue, 1936). Other correlation effects in which auxin plays a role include the polar differentiation of organs and tissues, e.g., effects on the induction of root initiation (Thimann and Went, 1934) and on the polar differentiation of xylem (Jacobs, 1954; Wetmore, 1955).

The tropistic stimulation shown in Fig. 4-1 is a classic illustration of a correlation effect of auxin; another is the ability of auxin to inhibit the enlargement of the lateral bud in apical dominance (Fig. 4-4); and the quantitative ability of auxin to induce rooting in the basal ends of cuttings is shown in Fig. 9-6.

NATURAL OCCURRENCE

The stimulation of coleoptile or stem growth by auxin has been the most common basis for measuring the amount of auxin in extracts or diffusates from plant parts. Two principal assay methods have been widely used; the straight-growth tests, in which the simple elongation of coleoptile or stem sections in a test solution is measured (Fig. 4-2), and the curvature test, in which coleoptiles are given a unilateral treatment of auxin in an agar block and the stimulation of growth on the treated side results in a growth curvature (Fig. 4-1). The straight-growth

Fig. 4-2 Sections of pea stem show quantitative stimulations of growth when floated on auxin solutions. Sections from etiolated plants are more responsive than those from green plants and are commonly used as an auxin bioassay (Galston and Hand, 1949; Galston and Baker, 1951).

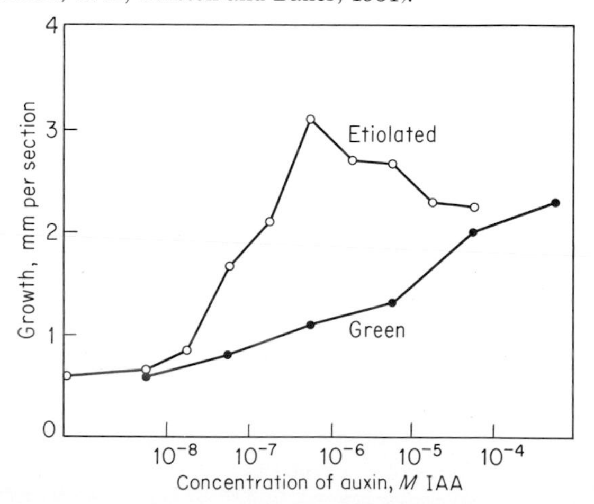

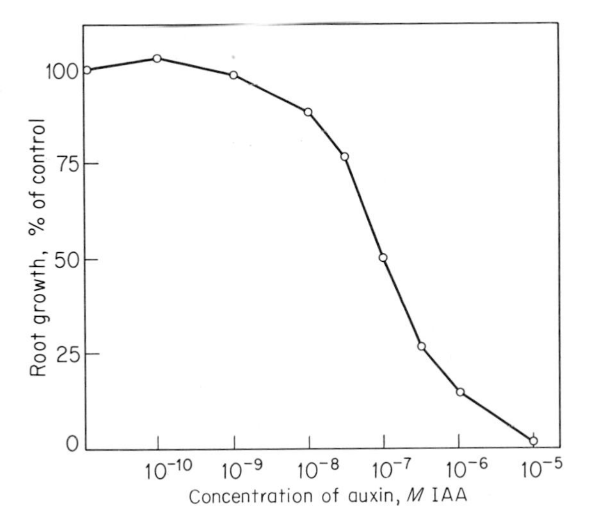

Fig. 4-3 Auxin responses of intact pea roots show quantitative inhibitions of growth (Aberg and Jonsson, 1955).

tests are handicapped by having only a logarithmic sensitivity to concentrations of auxin, and the curvature test is limited by being sensitive only to auxins which are rapidly transported in a polar manner by the coleoptile tissue. Before about 1952, most workers used the curvature test as a bioassay, and since indoleacetic acid is almost unique in this transport characteristic, nearly all isolations from plants led to the identification of IAA. Since the advent of paper chromatography, however, the straight-growth assay has become widely accepted, as it adapts more readily to the bioassay of chromatographs. With this assay technique, evidence for other growth-stimulating compounds in plants has been found, and the concept of plants containing other auxins than IAA has frequently arisen. Over a period of about 20 years, numerous claims of non-indolic auxins in plants have gradually been resolved, but naturally occurring auxins have all turned out to be indole compounds. From the days of the first discovery of auxin, IAA has held a central position as a plant auxin; numerous other indole compounds have been reported in plants, including indoleacetaldehyde, indolepyruvic acid, and indoleacetonitrile, but the auxin activity of each of these can probably be accounted for through conversion into IAA. Some of the conversions involved in the biosynthesis of IAA are shown in Fig. 4-5.

Indoleacetaldehyde (Fig. 4-5) was identified as an auxin-active material in plants by Larsen (1944), and subsequent work (Larsen, 1951; Bentley and Housley, 1952) established that its activity in stimulating growth is due to its conversion into IAA. It is generally considered to be the immediate precursor of IAA in the pathway from tryptophan (Gordon, 1956).

Indolepyruvic acid was reported to occur as an auxin in corn (Yamaki and Nakamura, 1952). While these reports were denied by Bentley et al. (1956), similar findings with other plant materials by four other laboratories serve to confirm this material as a natural plant constituent (Gordon, 1961). This indole compound is considered to be the major intermediate in the conversion of tryptophan to IAA (Schneider et al., 1972).

Tryptamine is another compound which may be an intermediate in IAA biosynthesis, and while it has been detected as a natural plant substance (Udenfriend et al., 1959), it has not been

Fig. 4-4 Decapitation of green seedlings of *Vicia faba* causes rapid growth of the lateral bud; auxin solution applied to the severed stem suppresses the lateral bud. Auxin was an extract of *Rhizopus sinuis* and was applied for 8 days (Thimann and Skoog, 1934).

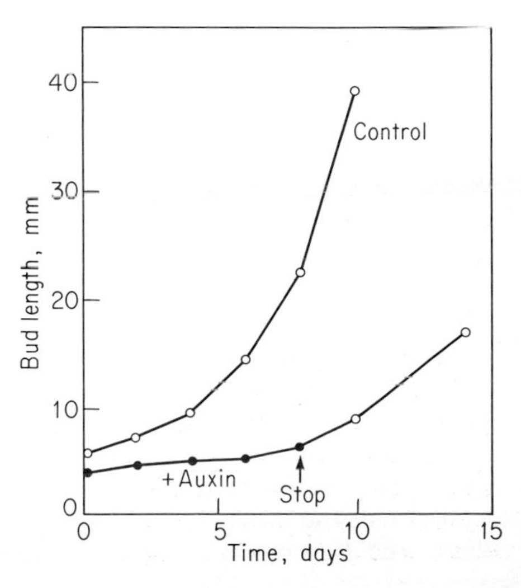

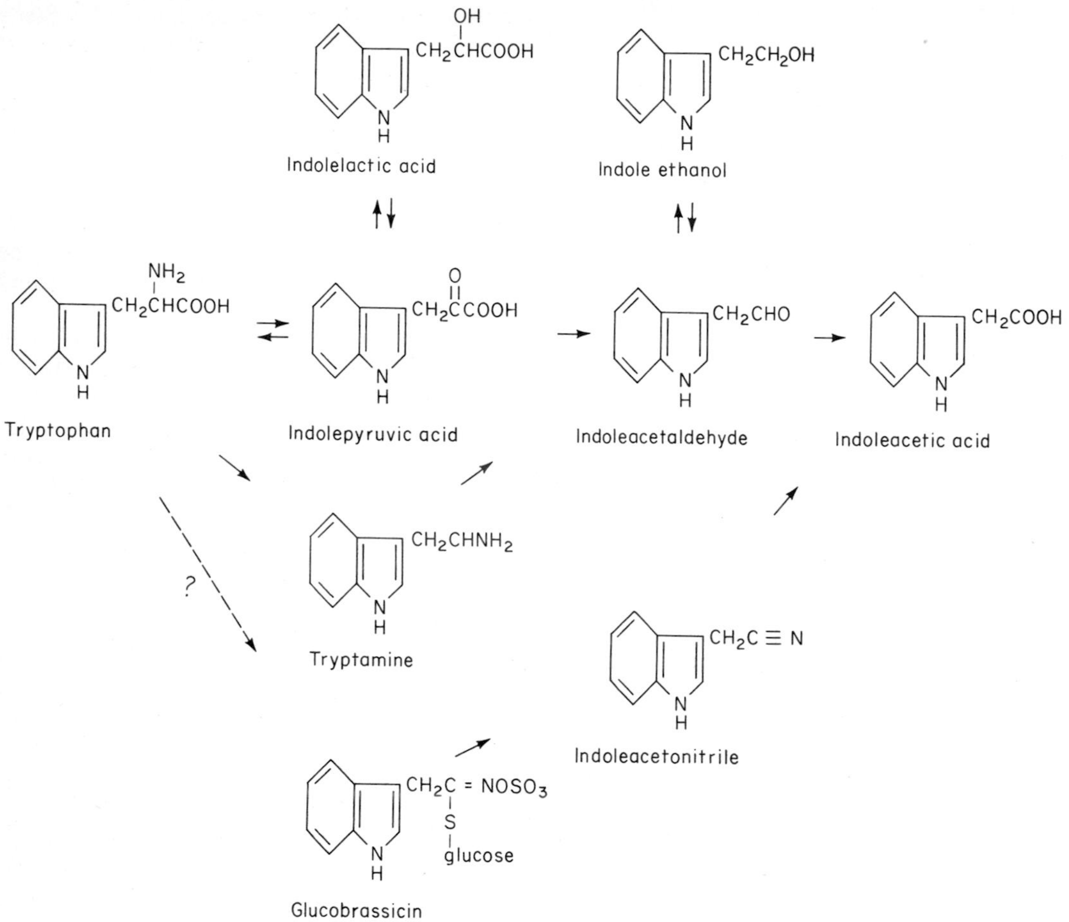

Indolelactic acid

Indole ethanol

Tryptophan

Indolepyruvic acid

Indoleacetaldehyde

Indoleacetic acid

Tryptamine

Indoleacetonitrile

Glucobrassicin

Fig. 4-5 Some pathways in the biosynthesis of indoleacetic acid.

claimed to be an auxin. It is readily converted into IAA in plants (Thimann and Grochowska, 1968) and appears to be a natural intermediate in IAA biosynthesis in some plant species (Muir and Lantican, 1968).

Since the earliest days of auxin research it has been presumed that tryptophan is the major substrate for IAA biosynthesis (Thimann, 1935). The unpleasant possibility has been raised that the formation of IAA from tryptophan may be due to the actions of microorganisms in the plant assay system rather than to the natural synthetic activities of the plant (Libbert et al., 1966; Thimann and Grochowska, 1968); however, in at least some instances the conversion into IAA can occur readily in sterile conditions of plant culture (Muir and Lantican, 1968); hence the sequence shown in Fig. 4-5 appears tenable.

Indoleacetonitrile has been isolated from numerous species of plants, especially from members of the Cruciferae, and has been claimed to be another natural plant auxin (Jones et al., 1952), but this compound, too, depends for its auxin activity on its conversion into IAA. This conversion takes place through the action of a nitrilase (Thimann and Mahadevan, 1958). The indoleacetonitrile arises in plants from the glucoside oil glucobrassicin, a natural component of the Cruciferae, through the action of the enzyme myrosinase (Gmelin and Virtanen, 1961). In fact, the isolations of indoleacetonitrile from plants may generally be due to the transformation of glucobrassicin into the nitrile during the isolation procedure (Kutacek and Prochazka, 1964). Experiments with radioactive tryptophan indicate that tryptophan is somehow incorporated into glucobrassicin and hence is a probable substrate for IAA biosynthesis through this pathway as well (Kutacek et al., 1960).

Indoleethanol has been shown to be a metabolite of tryptophan in bacteria (Larsen et al., 1962), and following its isolation from higher plants (Rayle and Purves, 1967*a*) it may take its place as another intermediate in the biosynthesis of IAA (Rayle and Purves, 1967*b*).

The ethyl ester of IAA has been found in apple and corn extracts (Redemann et al., 1951) and in willow and tobacco (Hinsvark et al., 1954; Schwarz and Bitancourt, 1957). The natural presumption is that the ester can be converted into IAA in the plant.

Not shown in Fig. 4-5 is indoleacetamide, which has been identified chromatographically in various tissue extracts (Andreae and Good, 1957). Experiments by Zenk (1961) have shown that the amide is formed from the IAA glucoside during extraction if ammonium hydroxide is used as a solvent, and so it seems doubtful at present whether indoleacetamide is a natural precursor of auxin even though it is readily converted into IAA.

Discussions of enzyme systems involved in auxin biosynthesis are available in Carr (1972).

AUXIN METABOLISM

As a basic component of any hormone system, there must be some means by which the organism can remove or dispose of the hormone. The plant has four known options for disposal of IAA: it can be (1) bound onto some sites in the cytoplasm, or (2) converted into several types of derivatives, or (3) enzymatically degraded, or (4) excreted.

Bound auxin may be analogous to the storage forms of some animal hormones, e.g., the neurohumors. Thimann and Skoog (1940) first suggested that IAA may be bound to some proteins, following their observation that IAA is released from some plant tissues by proteolytic enzyme action. Reports of the binding of IAA to protein components have been scattered through the subsequent years (Gordon, 1946; Zenk, 1963; Winter and Thimann, 1966), but the importance of this fate of IAA in the living plant has not yet been clarified.

The conversion of IAA into derivatives affords the plant a relatively rapid and effective means of disposing of auxin (Fig. 4-6). The first derivative to be discovered was the peptide indoleacetyl aspartate (Andreae and Good, 1955). In higher plants no other IAA peptide than the aspartate has been found. When auxin has been added to various plant tissues, the IAA aspartate begins to accumulate after about 4 or 5 h (Andreae and van Ysselstein, 1960), and this derivative shows no biological activity; hence the conversion to the peptide is apparently tantamount to removal from the hormonal system. A less permanent disposal system is the formation of the IAA glucoside, first discovered by Zenk (1961, 1962). This derivative is formed in a wide array of higher plant tissues, ordinarily somewhat more rapidly than the peptide (Fig. 4-7). It does show biological activity, presumably as a result of being hydrolyzed to the free IAA again. With synthetic auxins, both the aspartate and the glycoside derivatives are formed (Zenk, 1962, 1968), though in the case of 2,4-dichlorophenoxyacetic acid (2,4-D) rather more slowly.

Two other derivatives have been reported; an IAA arabinose was found in corn (Shantz and Steward, 1957) and IAA-inositol-arabinose glycosides in corn (Labarca et al., 1965).

AUXIN DEGRADATION

An enzyme preparation capable of carrying out the oxidative degradation of IAA was first made from pea seedlings by Tang and Bonner (1947). A large amount of research has been done on this enzyme, termed *indoleacetic oxidase*. Early work on this subject was hindered because only relatively crude preparations could be made, but

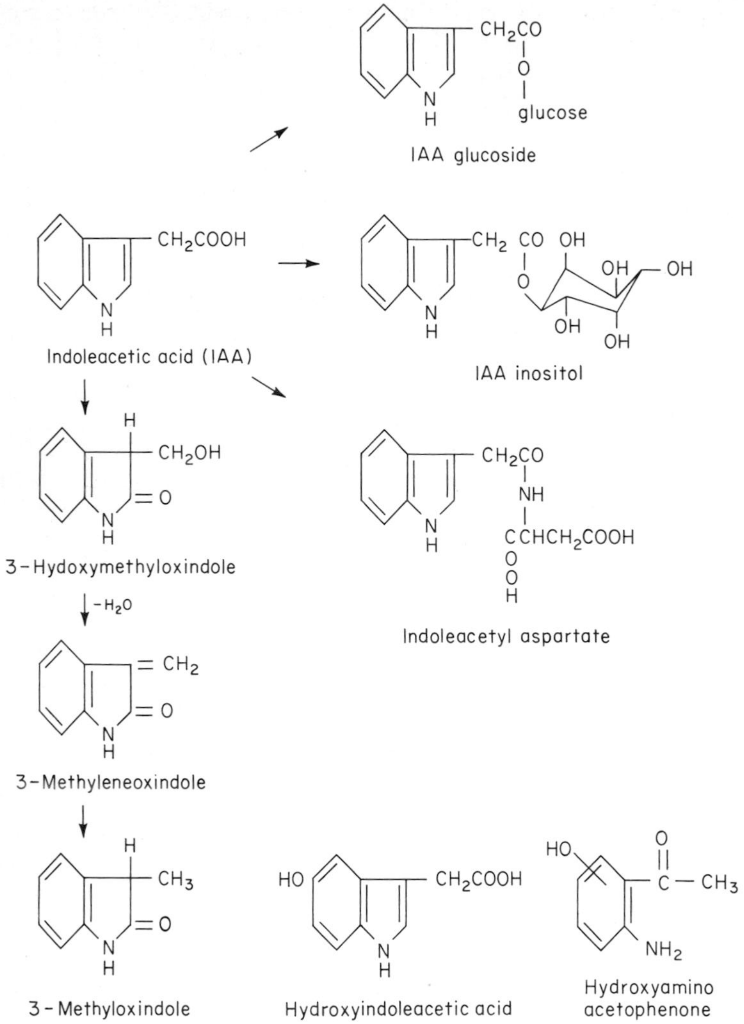

CH_2CO
O
glucose
IAA glucoside
CH_2COOH
Indoleacetic acid (IAA)
CH_2 CO OH
OH
OH
OH
OH
IAA inositol
CH_2OH
=O
3-Hydoxymethyloxindole
CH_2CO
NH
$CCHCH_2COOH$
Indoleacetyl aspartate
$-H_2O$
= CH_2
3-Methyleneoxindole
CH_3
3-Methyloxindole
HO
CH_2COOH
Hydroxyindoleacetic acid
C — CH_3
NH_2
Hydroxyamino acetophenone

Fig. 4-6 Some known products of indoleacetic acid metabolism.

it was early established that the IAA oxidase has the characteristics of a peroxidase (Galston et al., 1953). Probably no other plant enzyme is so famous for having a wide array of isozymes as peroxidase; it is not uncommon for over 10 isozymes to be separated electrophoretically from plant extracts (Siegel and Galston, 1967). A careful separation of the isozymes of peroxidase from tobacco roots was carried out by Sequiera and Mineo (1966) (Fig. 4-8), and they made a comparison of the relative effectiveness of the separate isozymes for oxidizing IAA and a standard peroxidase substrate such as guaiacol. These experiments led to the finding that the peroxidases generally can oxidize IAA but that an enzyme fraction which lacks peroxidase activity is markedly more effective as an IAA oxidase (Fig. 4-9). The affinity of this enzyme for IAA (K_m of about 0.1 mM IAA) was markedly greater than the affinity of peroxidase for IAA. The IAA

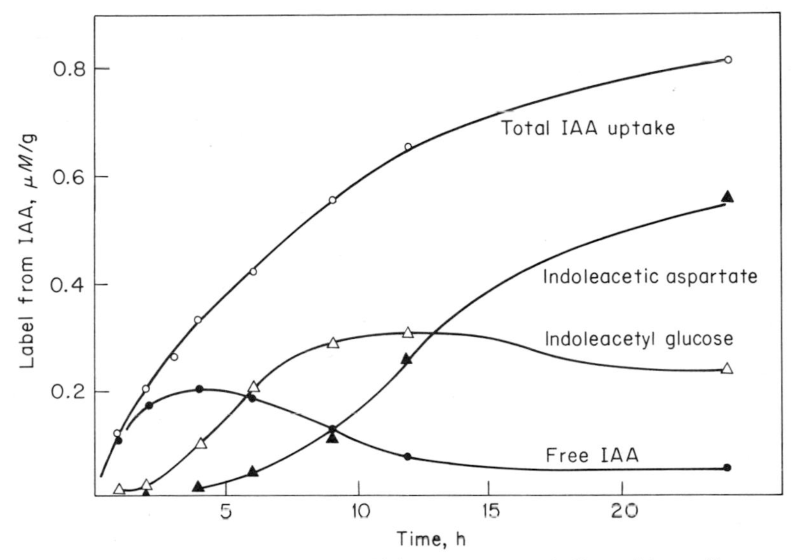

Fig. 4-7 After radioactive indoleacetic acid has been added to leaf disks of *Hypericum hircinum,* there is a rapid formation of IAA glucose, followed later by the formation of IAA aspartate, as indicated by radioactivity on chromatograms (Zenk, 1963).

oxidase was essentially ineffective against the peroxidase substrate guaiacol, and the lag period observed in its action against IAA was not removed by addition of peroxide, as the lag period of peroxidase was. This IAA oxidase is ordinarily present in horseradish peroxidase preparations. It appears, then, that the oxidation of IAA can be catalyzed to some extent by peroxidases but more effectively by a related but not peroxidative enzyme.

The distribution of IAA oxidase is ordinarily related to the growth rate. Stem tips and root tips have generally lesser amounts of the enzyme than older tissues, and roots are often markedly richer in the enzyme than stems (Fig. 4-10). Its inverse relationship to growth suggests that it may contribute to the termination of growth as tissues mature.

The oxidation of IAA is well known to involve the evolution of CO_2 and the consumption of O_2 in approximately equivalent amounts (Wagenknecht and Burris, 1950). Two cofactors are needed for the enzyme function, Mn^{2+} and a phenolic. The Mn^{2+} is considered by some to be the agent which directly oxidizes the IAA, since manganic ions can directly oxidize IAA with the production of CO_2 (Waygood et al., 1956). Alternatively, it has been suggested that Mn^{2+} may serve as an electron carrier between the enzyme and phenolic cofactor (Ray, 1962).

The requirement of a phenolic cofactor for IAA oxidase activity was first recognized by Goldacre et al. (1953). In the plant such phenolics as

Fig. 4-8 When partially purified peroxidase preparation from tobacco roots is partitioned on a Sephadex column, IAA oxidase activity is found to be associated with peroxidase activity but also in a fraction which lacks peroxidative activity (Sequiera and Mineo, 1966).

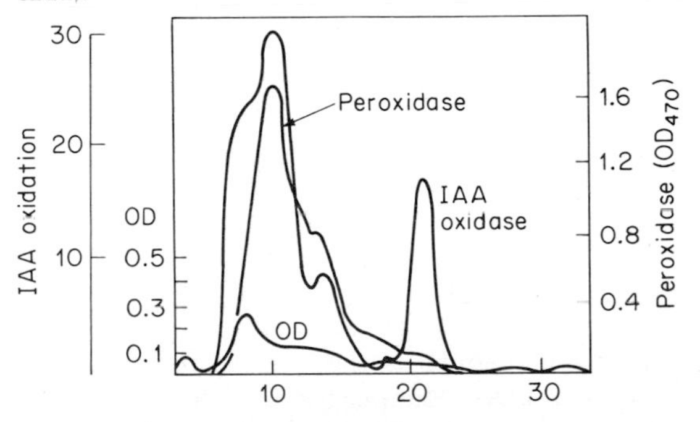

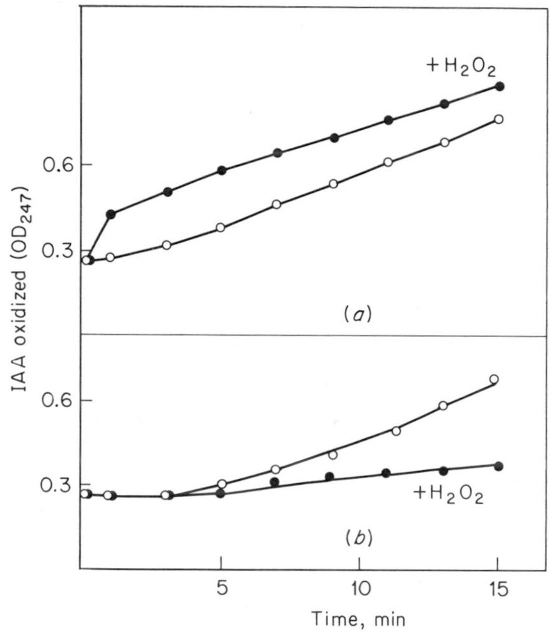

Fig. 4-9 A comparison of the IAA oxidase-peroxidase mixture (*a*) and the isolated IAA oxidase (*b*) (from Fig. 4-8) for the stimulatory effect of hydrogen peroxide on IAA oxidation. Unlike peroxidase, the IAA oxidase is not stimulated by peroxide (Sequiera and Mineo, 1966).

p-coumaric or ferulic acid have been reported to serve as natural cofactors for IAA oxidation (Gortner et al., 1958) or, more commonly, various monophenolics such as 4-hydroxybenzyl alcohol (Mumford et al., 1963) or kaempferol (Furuya et al., 1962) (see Fig. 4-11). These cofactors can enhance the oxidation of IAA and hence bring about inhibitions of growth, as illustrated in Fig. 4-12. Experimental studies of IAA oxidase commonly utilize 2,4-dichlorophenol as the phenolic cofactor (Goldacre et al., 1953).

The distinction between cofactors and inhibitors of IAA oxidase is rather fine. Whereas monophenolics are cofactors, *o*-diphenolics serve as inhibitors of IAA oxidase, as in the case of chlorogenic acid (Gortner and Kent, 1953; Nitsch and Nitsch, 1962). The close analog of kaempferol, quercitin (Fig. 4-11), with an *o*-diphenolic side ring, is identified as the main IAA oxidase inhibitor in peas (Mumford et al., 1961). Phenolics which would inhibit the oxidation of IAA should bring about an enhancement of growth in an auxin system, as illustrated in Fig. 4-12.

Light has some rather complicated effects on IAA oxidation, including stimulatory effects in etiolated plant materials (Galston and Baker, 1951) and inhibitory effects in others (Mumford et al., 1961). One of the interesting light effects operates through the alteration of quercitin content in peas; red light can stimulate the formation of quercitin and thus depress the IAA oxidation, resulting in an enhancement of growth (Mumford et al., 1961). In the cases studied, kaempferol levels are not apparently responsive to light. At least in this instance, then, light may regulate IAA oxidase through the alteration of IAA oxidase inhibitor levels.

The products formed from the oxidation of IAA are multiple and complex. Following the reaction spectrophotometrically, one can detect that numerous products are being interconverted (Ray and Thimann, 1955). Even chromatographic separation of the IAA oxidase action shows the accumulation of several products (Pilet, 1960). In view of the utilization of 1 mol of oxygen and

Fig. 4-10 Indoleacetic oxidase activity is markedly greater in pea roots than in pea stems, and in both cases, activity rises with increasing distance from the meristematic tip (Galston and Dalberg, 1954).

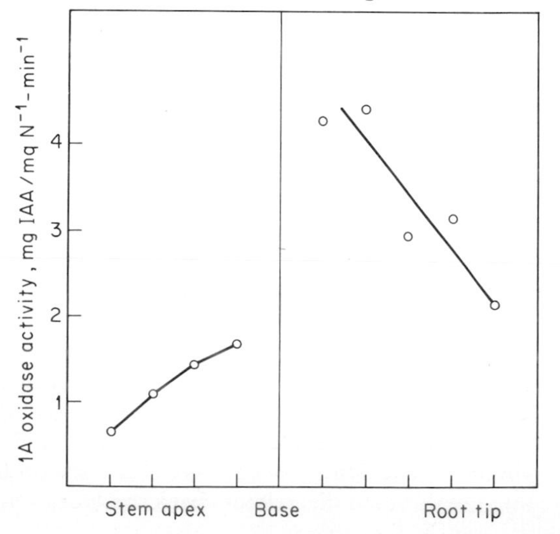

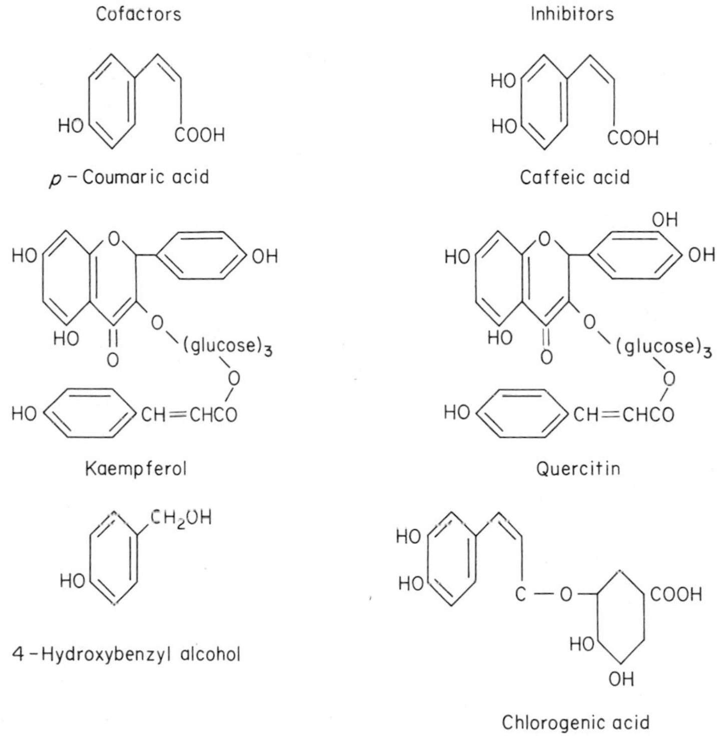

Cofactors

Inhibitors

p-Coumaric acid

Caffeic acid

Kaempferol

Quercitin

4-Hydroxybenzyl alcohol

Chlorogenic acid

Fig. 4-11 Naturally occurring cofactors and inhibitors of IAA oxidase.

the production of 1 mol of CO_2, one would anticipate simple oxidative decarboxylation of the side chain. This reaction would lead to the production of indolealdehyde, but this compound is only occasionally identified as a product of the reaction and then accounts for only about 5 percent of the IAA destroyed (Racusen, 1955). Another product appears to be 3-methyldioxindole (Ray and Thimann, 1955) or 3-hydroxymethyloxindole (Still et al., 1965). Once the latter has been formed, it may spontaneously react with reducing compounds to give 3-methyleneoxindole, which may (Tuli and Moyed, 1969) or may not (Audyse et al., 1972) have growth-stimulating properties and is in turn readily reduced to 3-methyloxindole (Fig. 4-6). The further oxidation of the indole ring probably leads to the formation of hydroxyformamidoacetophenone or hydroxyaminoacetophenone (Galston, 1955). The insertion of a hydroxyl group into the indole ring is known to occur in plant materials (Titus et al., 1956) and may precede the rupture of the indole ring in the IAA degradation reactions. In this sequence, the initial products of the enzymatic degradation of IAA are unstable compounds which can then go through sequential transformations.

While IAA oxidase and other peroxidases are surely capable of oxidizing indoleacetic acid and in some cases the activity of the enzyme can be correlated with growth, e.g., high enzyme activity in regions of least growth or in stages of minimal stem elongation in rosette plants (Konishi, 1956), yet there is reason to be cautious about its role in the intact plant. For example, there are developmental changes involving lowered auxin levels in which peroxidase activities may not rise (see, for example, Gahagan et al., 1968).

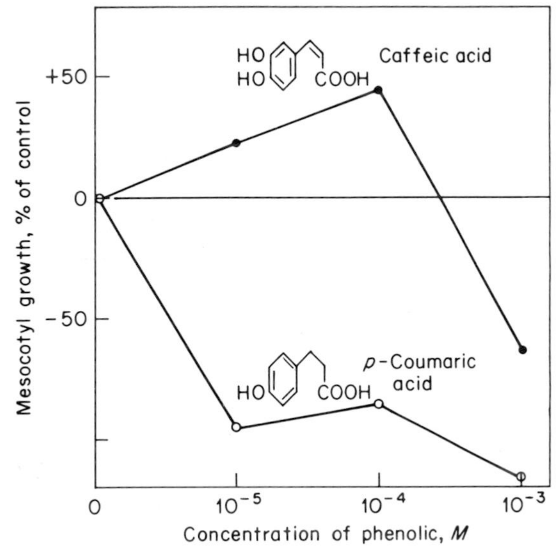

Fig. 4-12 A comparison of the effects of an *o*-diphenol (caffeic acid) and a monophenol (*p*-coumaric acid) on mesocotyl growth in the presence of 10 μM IAA; the *o*-diphenol is an inhibitor of indoleacetic oxidase, and the monophenol is a cofactor for the enzyme (Nitsch and Nitsch, 1962).

In short, the plant has several means at its disposal for the metabolic removal of auxin from the growth-regulating system; it can convert IAA into derivatives like the glucoside, from which the IAA can be regenerated rather easily; it can make more stable derivatives which remove the auxin rather permanently; or it can degrade the IAA into a series of oxidation products.

RELATION TO GROWTH

Correlations between auxin content and growth have been shown in a variety of ways. Some results of Scott and Briggs (1960) (see Fig. 4-13) show that the growth rate in pea stems dwindles from the apex toward the base of the plant, as does the auxin content. The correlation with time is illustrated by some data of Hatcher (1959) (see Fig. 4-14). The auxin content in apple twigs rises in the spring as growth gets under way, and it subsequently declines through the growing season; trailing after it is a decline in the growth rate until autumn. Another type of correlation is found in comparisons of *long shoots* and *short shoots*. Branches of *Ginkgo* which will develop into long shoots form large and continuing supplies of auxin; those which will be short shoots develop a diminutive supply, as illustrated in Fig. 4-15.

Correlations between auxin content of tissues and growth rates have frequently been found, but there are also many instances in which no such correlation was observed. In their data for pea stems (Fig. 4-13), Scott and Briggs (1960) observed a slight decline in diffusible auxin down the stem over the region of declining growth rate, but extractable auxin showed no appreciable change over the whole region from the rapidly growing stem apex to the point where growth had essentially stopped. They deduced that the auxin obtained from peas by diffusion is more relevant to the growth-regulating action than that obtained by extraction. In contrast, Went (1942) reported that in oat coleoptiles the extractable auxin correlated well with growth rate, whereas the diffusible auxin correlated instead with the tropistic reactions. This sufficiently illustrates that the means of sampling the auxin content of a tissue will influence what correlations will be obtained. It is not possible to generalize that diffusible auxins are relevant to growth activi-

Fig. 4-13 A comparison of the distribution of growth activity down a pea stem and the distribution of diffusible auxin (Scott and Briggs, 1960).

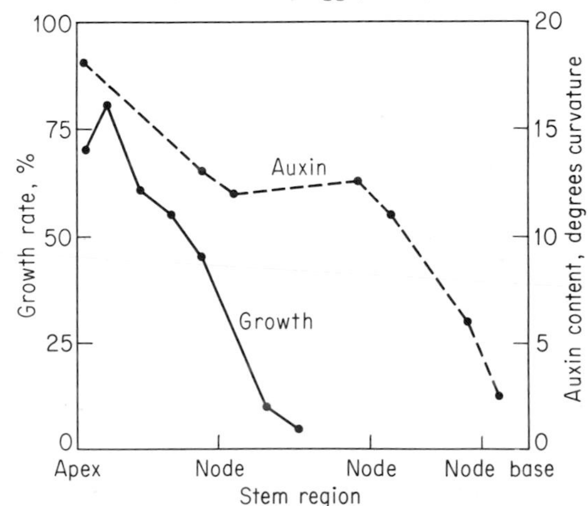

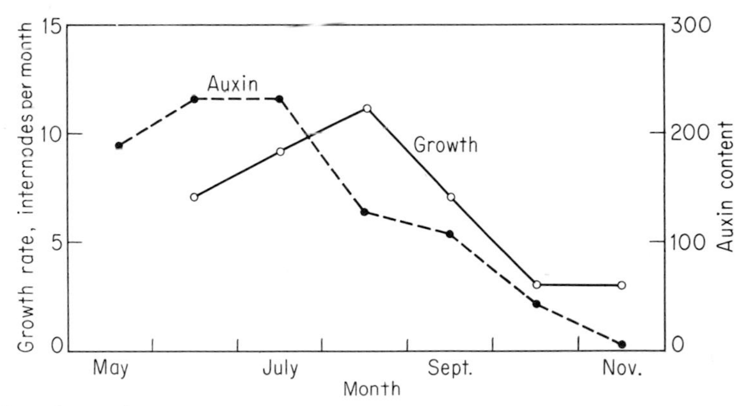

Fig. 4-14 A comparison of the distribution of growth of apple stems through the growing season and the content of diffusible auxin from the first five nodes (Hatcher, 1959).

ties in all plants or that extractable auxins are relevant to some other types of activities in plants.

Auxin often occurs most abundantly in the most actively growing tissue, even in tissues where it apparently does not stimulate growth. In roots, for instance, Pilet (1951a) found that extractable auxin is most abundant about 2 to 8 mm from the tip of the *Lens* root and that the most active growth is about 2 mm from the tip (Fig. 4-16). Furthermore, with increasing age of the plant, the subterminal auxin content continues to rise in time in a manner suggestive of the increasing root-growth rate. Very young *Lens* roots give small positive growth responses to added auxins; older roots are not stimulated. The amount of auxin which is found by extraction of the roots greatly inhibits growth if applied to other roots (Audus, 1972).

Thimann (1937) proposed that roots have a lower threshold of auxin sensitivity and that their natural auxin content is above the optimum for growth. Thus, the removal of the root tip as the source of auxin should increase the root growth, and, in fact, Cholodny (1924) succeeded in showing such an increase. Yet many other investigators have been unable to obtain promotions of root growth following decapitation (Gorter, 1932; Younis, 1954; Vardar and Tözün, 1958). There is some doubt about auxin's role in stimulating root growth, though some authors feel that it does play such a role (Scott, 1972).

The situation in fruits presents another relevant illustration of auxin production which may or may not be associated with growth. As some fruits enter rapid phases of growth, there is a preceding surge in auxin production (Nitsch, 1950; Wright, 1956), but in other species it may not be related to any stage of active fruit growth at all (Stahly and Thompson, 1959; Coombe, 1960).

The evidence on the auxin contents of various tissues as they may relate to growth rates generally supports the concept of auxin as a growth

Fig. 4-15 Markedly more auxin is diffusible from buds of *Ginkgo* which will develop into long shoots than from those which will not elongate, remaining as short shoots (Gunckel and Thimann, 1949).

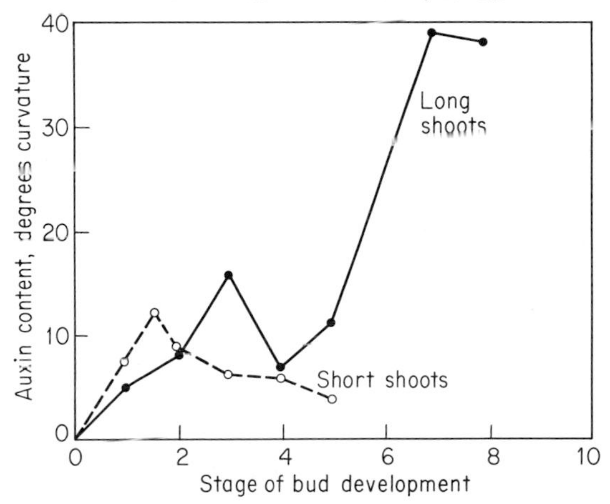

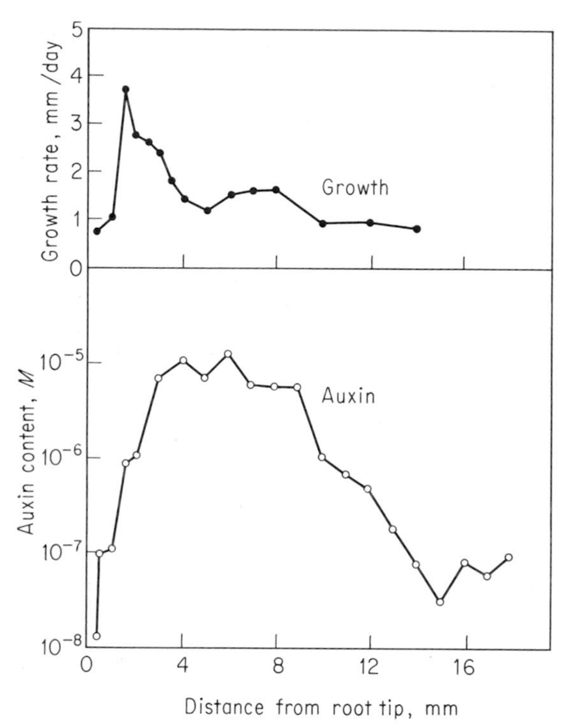

Fig. 4-16 The auxin extractable with chloroform from *Lens* roots is most abundant at the region of most rapid growth or shortly behind it (Pilet and Meylan, 1953).

hormone in stems and coleoptiles, but in roots and fruits the role of auxin is less certain. Auxin occurrence is usually associated with growing tissues, and in some of these it can exert a strong regulatory influence.

AUXIN SOURCES

The ability of auxin (or at least indoleacetic acid) to move through many plant tissues in a polar manner results in its physiological influences over a wider range of locations than those in which it is formed. The usual sites of auxin formation in vascular plants are meristems and enlarging tissues. The classic case of hormone synthesis in the apex of the grass coleoptile, implied by the experiments of Darwin (1897) and specifically suggested by Paál (1919), was experimentally proved by Went (1928) to explain the physiological mechanism of tropistic movements in plants. The apex is the principal site of auxin formation in the coleoptile, and the polar transport of the hormone down the coleoptile provides the stimulus for growth in the regions below the tip. In dicotyledons the meristematic apex is usually the main locus of auxin production, and its removal results in depressed auxin levels and depressed growth in tissues below the apex. If one plots the amount of auxin as it relates to the distance from the coleoptile apex, one usually obtains a declining curve similar to that shown in Fig. 4-13 and roughly parallel to the declining gradient of growth rate. There are some interesting variants of this situation; Mirov (1941) found that in pine shoots auxin occurrence is much greater at the more basal parts of the new growth than at the apical; Gunckel and Thimann (1949) reported that in *Ginkgo* the early stages of stem growth seem to be supplied with auxin from the apical meristem but that as growth progresses, a larger supply is formed in the more basal internodes in a manner similar to that in pine.

Embryos are another type of meristem which often produce large amounts of auxin. Nitsch (1950) and Luckwill (1949) have shown that embryos are a principal site of auxin production in strawberry and apple fruits during fruit growth. In seeds, too, the embryo is a major source of the hormone (Hemberg, 1955).

Expanding tissues are common sources of auxin; the lengthening internodes of *Ginkgo* and expanding cells in the tip of the oat coleoptile are good examples. During the enlargement of leaves of higher plants, there is generally a large production of auxin (see, for example, Wetmore and Jacobs, 1953). There is a similar auxin production during the expansion of fern pinnae (Steeves and Briggs, 1960). The auxin production by leaves is especially interesting since the effects of auxin on growth of the leaf blade are generally small (Miller, 1951). This is another example of the production of auxin in a tissue without a concomitant role in stimulating growth of the tissue. It should be noted, though, that

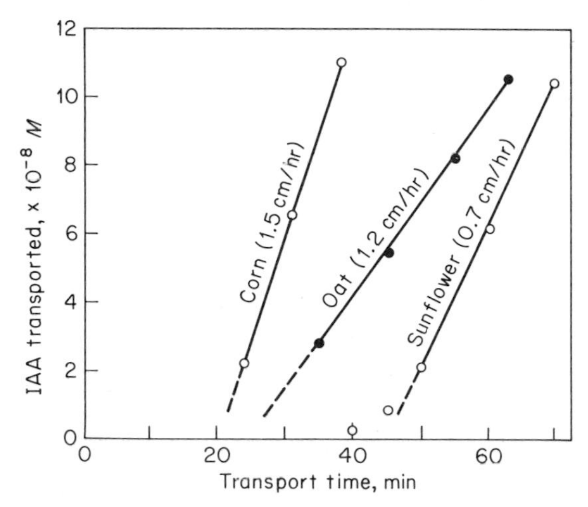

Fig. 4-17 The velocity of the transport of auxin through tissues can be measured as the time of the first arrival of detectable auxin through 5-mm pieces of tissue; here the velocities are compared for corn coleoptile (1.5 cm/h), oat coleoptile (1.2 cm/h), and sunflower stem (0.7 cm/h) (Hertel and Leopold, 1963; van der Weij, 1932).

auxin produced in the leaf may have extensive correlation effects, i.e., hormonal effects on more or less remote tissues.

Auxins are frequently produced by parasitic or symbiotic organisms. Root growth is stunted by auxin-producing mycorrhizae (MacDougal and Dufrenoy, 1944); the formation of nodules on legumes is associated with auxin-forming *Rhizobium* (Kefford et al., 1960); and some plant swellings are associated with invasion by auxin-producing pathogens (Wolf, 1952). Other pathogens apparently reduce the auxin content of host tissues (see, for example, Sequeira and Steeves, 1954; Shaw and Hawkins, 1958).

TRANSPORT

There are hormones in the plant with more dramatic powers of stimulating growth than auxin but none which have such distinctive abilities to carry regulating messages from one part of a plant to another, providing a remote control over growth and form. The transport system for auxin and its directional or polar quality are quite evidently responsible for its systemic regulatory influences.

The *polarity* of auxin movement was recognized at the time of its discovery (Went, 1928), and the major characteristics of this polar transport in oat-coleoptile sections were described in a series of experiments by van der Weij (1932). Among these characteristics were a velocity of about 1 to 1.5 cm/h (Fig. 4-17), a directional feature which moves auxin almost exclusively in the basipetal direction, and an ability to transport IAA against a concentration gradient. The polar transport was most marked in apical sections of the coleoptile and lessened in older, more basal sections. This gradient in polarity has been noted in numerous stems and coleoptiles since that time (Fig. 4-18). The polarity is markedly more evident in some tissues than in others, coleoptiles being usually very polar and some stems and petioles being less so (Jacobs, 1961). In roots, transport of auxin is very much less, showing a velocity of 0.1 or 0.2 cm/h (Kirk and Jacobs, 1968) and a polarity of movement generally toward the root tip. The direction of auxin movement in the root is somewhat unclear, some workers finding movement exclusively toward the tip (Wilkins and Scott, 1968),

Fig. 4-18 The ability of bean stems to transport auxin basipetally declines in a gradient down the stem. Stem sections 5 mm long transported for 3 h from 2 mg/l IAA donor blocks. Only small amounts of auxin moved acropetally (Jacobs, 1950*a*).

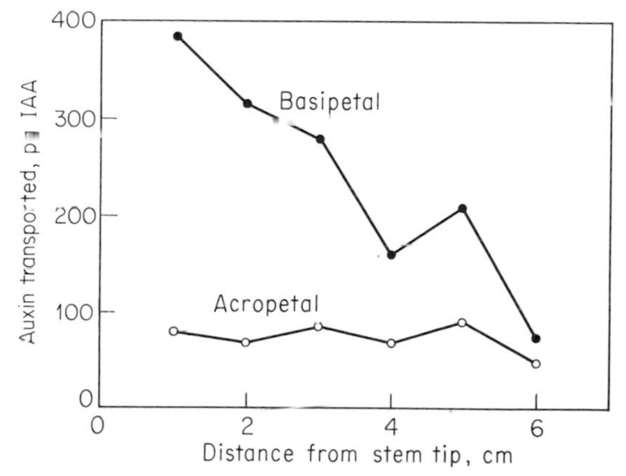

some away from the tip (Nagao and Ohwaki, 1968), and others in both directions (Hillman and Phillips, 1970). In the root, where auxin stimulations of growth are markedly less clear, so also are auxin transport and polarity.

That *active transport* is involved in the movement of auxin can be inferred from the five criteria of an active-transport system (Danielli, 1954):

1 It should have a velocity greater than that of diffusion. The velocities of 1 or 1.5 cm/h for auxin transport in stems and coleoptiles (Fig. 4-17) are about tenfold greater than diffusion.

2 It should be driven by metabolic forces. The susceptibility of IAA transport to inhibitors has been known for many years (du Buy and Olson, 1940; Niedergang-Kamien and Leopold, 1957). Metabolic inhibitors are generally less effective than certain poisons such as 2,3,5-triiodobenzoic acid (Niedergang and Skoog, 1956) and naphthylphthalamic acid (Morgan and Soding, 1958). The dependence of transport on aerobic metabolism is elegantly shown by experiments illustrated in Fig. 4-19.

Fig. 4-19 The metabolic nature of auxin transport in corn coleoptiles is evident from the sensitivity of transport to anaerobiosis. A pulse of IAA-^{14}C was applied at the apical end (*left*) for 15 min, and the subsequent movement of the pulse down the section in air is contrasted with the paucity of movement under nitrogen (Goldsmith, 1967).

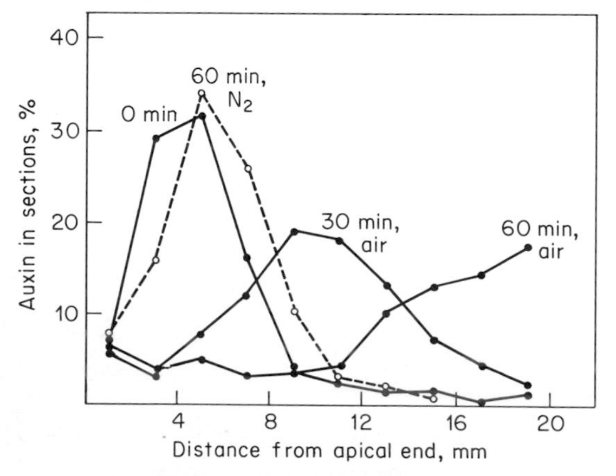

3 It should be able to move a solute against a concentration gradient. This was first inferred for auxin transport by van der Weij (1932), who observed that sections of coleoptile can transport IAA into an agar block at its basal end until the concentration there exceeds that of the donor agar block at the apical end. Similar results can be obtained with radioactive-tagged IAA (Goldsmith and Wilkins, 1964).

4 It should show a specificity for certain substrates. The polar transport system will transport naphthaleneacetic acid almost as well as IAA and with similar velocity (Hertel et al., 1969); 2,4-D can be transported with the same polarity but with very much less velocity (McCready, 1963). Among the analogs of IAA, the transport system can even discriminate between the optical isomers of indole-α-propionic acid, the (+) enantiomorph being more active as an auxin and more effectively transported than the (−) (Hertel et al., 1969). With naphthaleneacetic acid, the α-substituted acetic is active as an auxin and is actively transported, but the β-substituted acetic is inactive and is not transported. Thus, within the range of compounds tested, there is a good correlation between structural requirements for auxin activity and for auxin transport.

5 It should show a saturation effect. Assuming that transport occurs through the attachment of auxin to some transport site, a high enough concentration of the auxin should result in saturation of the transport sites and an inability to transport proportionally more auxin when higher concentrations are provided. Saturation effects of transport are very generally observed (Goldsmith and Thimann, 1962). Hertel et al. (1972) have estimated the transport sites as being between 0.1 and 1 μM per tissue volume, based on experiments on the specific binding of IAA onto membrane preparations from corn coleoptiles.

Collectively these lines of evidence confirm that the polar movement of auxin is an active transport. An active-transport system can pump

Table 4-1 When auxin transport is inhibited by low temperature, anaerobiosis, or triiodobenzoic acid, there is an associated increase in the auxin held in the tissue; this indicates a secretive quality of the auxin transport system (Hertel and Leopold, 1963).

Treatment	Radioactivity from ^{14}C-IAA, cpm	
	In tissue	In receptor block
Control, 24°C	2,280	4,720
Low temperature (6°C)	3,980	2,160
Nitrogen atmosphere	3,230	2,930
Plus TIBA	4,200	2,020

substrates either into or out of cells, and Hertel and Leopold (1963) have found that various chemical and physical means of slowing transport result in increased amounts of IAA accumulating in the tissue (Table 4-1), from which they conclude that auxin transport is an active secretion of auxin from cells. One might infer, therefore, that the site of auxin transport is located on the plasmalemma and that polarity of transport is achieved by preferential activity of transport sites at the basal end of the cell (Leopold and dela Fuente, 1968).

The polar feature of auxin transport is diminished when one measures it in short tissue sections, approaching an equal movement in acropetal and basipetal directions when very short sections are used (dela Fuente and Leopold, 1966). An exponential amplification of transport polarity with increasing cell number along the length of a stem section suggests that each cell has a small preferential secretion of auxin out of its basal end, and repetition of this bias as the auxin moves through successive cells results in tissue polarity which is much greater than that in individual cells (Leopold and Hall, 1966). It is often observed that polar bud and root differentiation are markedly greater in larger plant pieces than in smaller ones (Bonnett and Torrey, 1965; Okazawa et al., 1967).

As a developmental control system, the polar transport of auxin has its greatest impact in stems and coleoptiles, where it provides a regulatory influence down the stem, declining in intensity with increasing distance from the apex. At least in stem and coleoptile tissues, the transport system carries the auxin to regions of cell enlargement, where it can serve to regulate growth. A phototropic or geotropic stimulus can alter the transport system so that a lateral redistribution of auxin takes place through change in the polar feature (Goldsmith and Wilkins, 1964). Again in the petioles of leaves, the polar transport system provides a regulatory influence, in this case principally an auxin suppression of abscission development. The movement of auxin in roots is less clear, and the regulatory role of transport there is still obscure.

AUXIN EFFECTS

The best-known growth stimulations by applied or exogenous auxins are the elongation of stems and coleoptiles. The concentrations of applied auxin which stimulate growth are in the same range as the endogenous auxin concentrations, and so the idea that auxin controls this type of growth is natural and logical. Another type of growth stimulation by auxins is cell enlargement by nonpolar swelling, which occurs in potato and artichoke tubers, in callus cells in tissue cultures, and in some fruit growth. Whether auxin occurs in concentrations needed for stimulating growth of these nonpolar tissues is not yet convincingly established. In many cases there is a lack of correlation between growth and the amount of auxin in these less polar types of tissues.

The stimulatory effects of auxin on growth vary considerably between tissues. In general, the most marked stimulations are obtained in seedling stems and coleoptiles, green stems being often less responsive than etiolated stems. Most other plant tissues are very much less stimulated by auxin than stems and coleoptiles are (Thimann, 1937), and in nearly all growing tissues high concentrations of auxin bring about growth inhibition rather than stimulation.

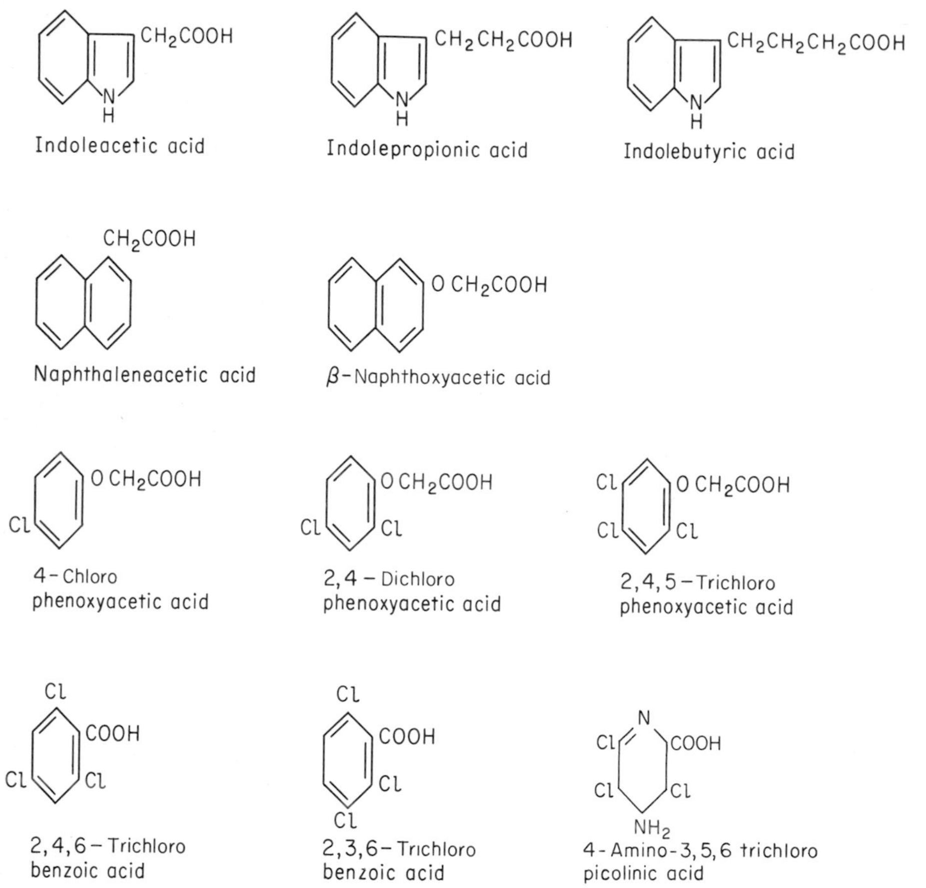

Fig. 4-20 Examples of common auxins, including some indole acids, naphthalene acids, phenoxy acids, and benzoic acids.

In addition to the effects on growth, auxin participates widely in the overall organization of plant processes, including the regulation of differential growth rates, i.e., the tropisms and apical dominance, and the regulation of differentiation phenomena, e.g., the polar differentiation of xylem strands in leaves or stems or the polar differentiation of roots at the base of a cutting or at other barriers to polar transport. In abscission phenomena, auxin may be involved in regulating both the differentiation of a separation layer and the new growth on the proximal side of the separation zone. Since these correlation effects all involve more regulators than auxin alone, they will be discussed in Chap. 9 and are mentioned here only to emphasize the multiplicity of auxin regulatory effects in the growing plant.

SYNTHETIC AUXINS

The wide range of synthetic chemical species which can engender auxin responses is striking; especially impressive is the fact that close analogs of IAA may be inactive as auxins and yet chemicals of such diverse types as the phenoxyacetic, naphthaleneacetic, benzoic, and picolinic acids are highly active as auxins. The structural requirements are generally consistent with the concept of attachment to a site with specific stereo dimensions (Porter and Thimann, 1965; Hertel et al., 1969a).

The ability of various indole compounds to produce growth and formative effects was reported almost simultaneously from several laboratories (Went and Thimann, 1937). Activities of the naphthalene acids were observed by Zimmerman et al. (1936) and of the naphthoxy acids by Irvine (1938). The very strong activity of some phenoxyacetic acids in causing formative effects was realized during the early period of World War II (Zimmerman and Hitchcock, 1942), and for a time the research on this class of auxins was carried on under military secrecy. The strong activity of the benzoic (Bentley, 1950) and picolinic acids (Kefford and Caso, 1966) was not discovered until much later. Among the most active of these four types of common auxins, toxicity and persistence increase generally in the series of indoles, naphthyls, phenoxy, benzoic, and picolinic auxins (Fig. 4-20). Their usefulness in agricultural applications reflects these properties.

The interesting question of the relation of structure to activity of the auxins has received much attention, in the expectation that some light would be shed on the nature of the auxin action. Three general features of the molecule are ordinarily considered to be necessary for auxin activity: an unsaturated ring, an acidic side chain, and some spatial relationship between the two (Koepfli et al., 1938). Some representative examples of these three requirements are illustrated for IAA and 2,4-D analog in Fig. 4-21. The ring requirement is not fulfilled when various positions have been filled there. For example, *N*-acetyl-IAA is inactive, and 2,4,6-trichlorophenoxyacetic acid is ordinarily inactive as an auxin. Saturation of the ring can also remove auxin activity. The requirement for an acid side chain is illustrated by skatole and 2,4-dichloroanisole, neither of which is active. The acid group need not be a carboxyl group, sulfonate and phosphonate groups being able to satisfy this requirement. The length of the side chain has a characteristic effect. Even-numbered carbons are active in these two series, and the even-numbered side chains are metabolized by β oxidations to the acetic side chain, which is active (Fawcett et al., 1952). The requirement for a certain spatial relationship between the ring and the side chain can be illustrated by the isobutyric side chains, as indole isobutyric or 2,4-dichlorophenoxyisobutyric acids; the former is inactive in some plant assays but does have auxin activity in others (Thimann, 1958), and the latter is ordinarily an effective antagonist of auxin. In the optically active α-propionic derivatives, if the methyl group on the side chain is on the side which results in dextrorotatory action, the compound is active as an auxin in either series. If the methyl group results in levorotatory action, the compound is less active (L-indole-α-propionic acid) or inactive (L-2,4-dichlorophenoxy-α-propionic acid). A classic case of spatial interference is that of cinnamic acid: the cis form is active as an auxin, but the trans form is an antiauxin (van Overbeek et al., 1951).

If one accepts the idea of auxin becoming attached to some stereospecific site, one can interpret many of the structural requirements for auxin activity as reflections of the features of such a site of attachment. Thimann (1963) has proposed that in order to fit into an attachment site, the auxin molecule needs a negative charge and a positive charge at a specific distance from one another, the negative charge being that of the ionized carboxyl group, and the positive charge being at the unsubstituted ortho position in 2,4-D or the indolic nitrogen in IAA. A distance of 5.5 Å between the negative and positive charge is consistently obtained in each of the major classes of auxins, including the indoles and phenoxyacetic, benzoic, and picolinic acids (Fig. 4-22). Also shown is the auxin which lacks an unsaturated ring entirely, carboxymethyl dimethyldithiocarbamate. This compound can assume a flat structure resembling a ring (van der Kerk et al., 1955; Veldstra, 1956), and the

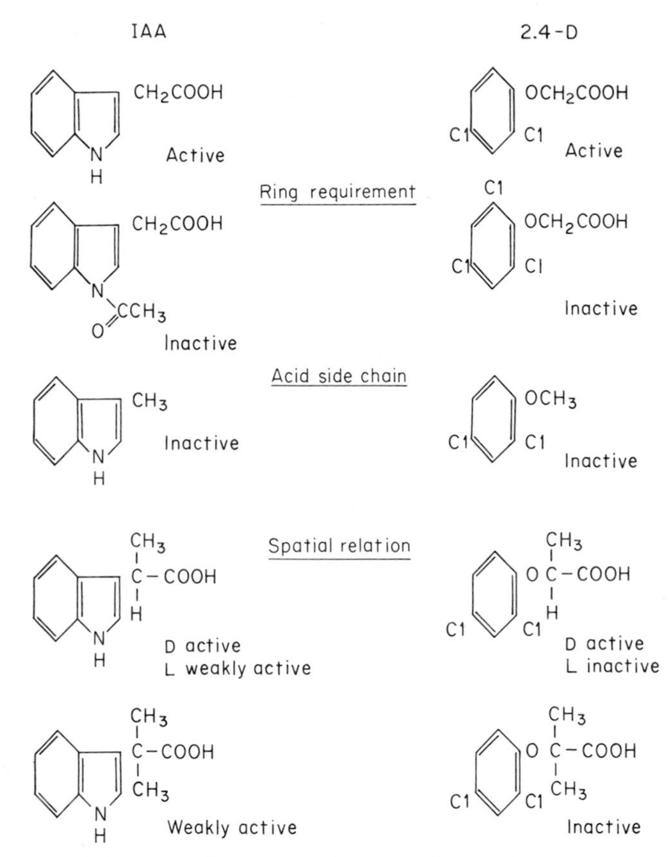

Fig. 4-21 Examples of molecular variations which affect the auxin activity of the IAA series (*left*) and the 2,4-D acid series (*right*), with special reference to the three presumed structural requirements for auxin activity.

nitrogen takes on a slight positive charge at the usual 5.5 Å from the carboxyl (Thimann, 1963).

Collectively, then, the structural requirements for auxin activity seem to relate to requirements for an attachment to some site in the cell, with characteristic charge requirements and steric limitations. The experiments of Hertel et al. (1972) on the actual binding of auxin onto isolated membrane fragments and the stereospecificity of this binding have already been mentioned in connection with auxin transport. The requirements for a presumed catalytic activity at the binding site might relate to some chemical effects or, more likely, to some allosteric effects of the molecule in its attached position (van Overbeek, 1966).

Antiauxins

From the model of auxins as molecules which catalyze some processes by becoming attached to a stereospecific site, it is a natural extrapolation that molecules which cannot entirely satisfy the requirements for attachment or catalytic action might serve to antagonize the stimulations by auxins. The first report of a substance which could limit the action of auxin was that of γ-phenylbutyric acid (Skoog et al., 1942). The antiauxin effects of *trans*-cinnamic acid have already been mentioned (van Overbeek et al., 1951). The antiauxin concept was more specifically defined by McRae and Bonner (1953), who suggested that antiauxins should be expected to be chemicals closely resembling auxins but lacking

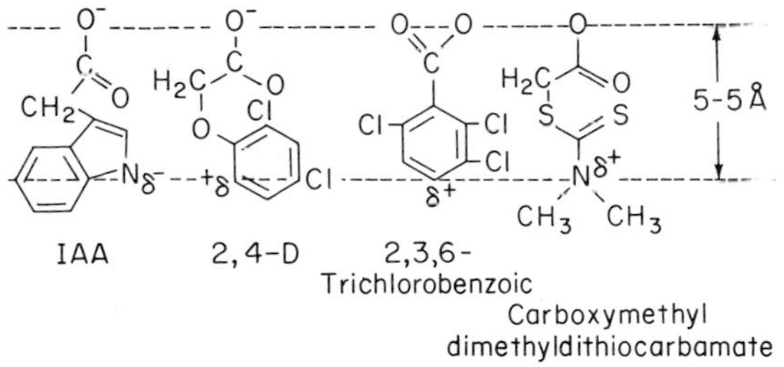

IAA 2,4-D 2,3,6-Trichlorobenzoic Carboxymethyl dimethyldithiocarbamate 5-5 Å

Fig. 4-22 Illustration of the theory that a 5.5-Å spacing is necessary between the carboxyl (−) and the ring (+) charge on the auzin molecule. (*From Porter and Thimann, 1969.*)

at least one requirement for activity. In view of the three structural requirements for auxin activity (Fig. 4-21) they proposed that antiauxin activity might arise from an inadequate unsaturated ring, an inadequate acid side chain, or an inadequate spatial relation between the two. Studies of the analogs of 2,4-D have revealed strong antiauxin activities for 2,4,6-trichlorophenoxyacetic acid and 2,4-dichlorophenoxyisobutyric acid and lesser such activity for 2,4-dichloroanisole (Fig. 4-21). The interpretation that these antiauxins each lack one of the structural requirements for auxin activity is weakened, however, by the absence of parallel activities in the indoleacetic acid analogs; thus *N*-acetylindoleacetic acid, skatole, and indoleisobutyric acid are ineffective as antiauxins (Thimann, 1958) or are antiauxins only in the regulation of root growth (Åberg, 1958). With 2,6-dichlorophenoxyacetic acid, antiauxin activity can be converted into auxin activity by further substitutions in the side chain (Osborne et al., 1955). Thus the concept of antiauxins interfering with the three basic structural requirements for auxin activity regrettably is inadequate.

MECHANISM OF ACTION

Inspection of the problem of how a hormone at very low concentration levels can exert very large physiological effects leads one naturally to the deduction that any mechanism of action must involve a large amplification effect. In bringing about its action the hormone can be expected to exert an influence on some process that alters a large number of other molecules. Three logical functions through which large amplification effects could be obtained would be an alteration of the nucleic acid–directed protein synthesis, the regulation of a pace-setter enzyme, or the regulation of a permeation phenomenon.

Hormonal regulation of nucleic acid synthesis and thus of protein synthesis would be a prime possibility for a hormonal mechanism. Silberger and Skoog (1953) observed that the growth of tobacco pith tissue induced by IAA is preceded by a proportional increase in RNA and that this increase is maximal at the concentration of auxin which will later produce maximal growth. When inhibitors of protein synthesis became available, it was soon found that interference with leucine incorporation into protein is closely correlated with interference with growth, e.g., using chloramphenicol (Fig. 4-23). Using the incorporation of $^{32}PO_4$ into RNA as a measure of

Fig. 4-23 The inhibition of growth with chloramphenicol is closely correlated with an inhibition of leucine-^{14}C incorporation into protein. Oat coleoptiles in 10 ppm IAA (Nooden and Thimann, 1965).

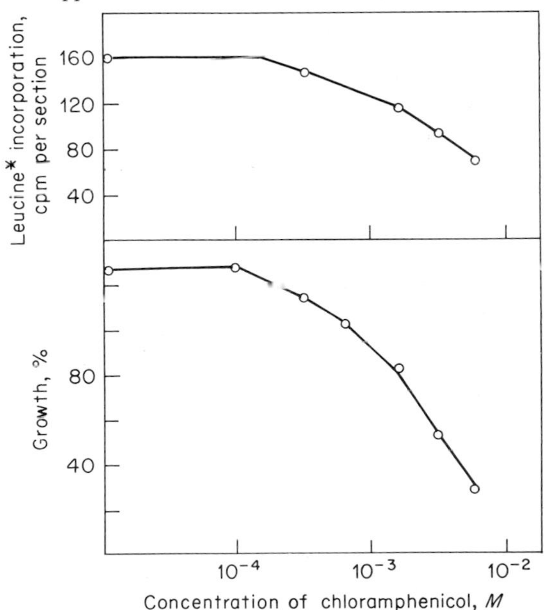

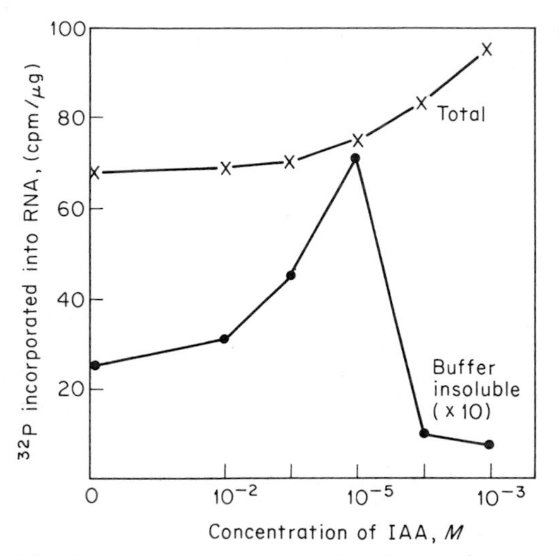

Fig. 4-24 Auxin treatment of pea internode sections causes a stimulation of $^{32}PO_4$ incorporation into RNA, including a two-phase effect on incorporation into an RNA fraction which was insoluble in buffer (Roychoudhury and Sen, 1964).

RNA synthesis, Roychoudhury and Sen (1964) found that the application of IAA to peas results in an enhanced RNA synthesis and that a phenol-extractable (buffer-insoluble) fraction of RNA shows the two-phase curve in response to IAA which is characteristic of growth responses (Fig. 4-24). Similarly, Key and Shannon (1964) found that the incorporation of labeled nucleotides into the nucleic acids is stimulated by either IAA or 2,4-D. Collectively, these experiments imply that the regulation of growth by auxin may involve a regulation of RNA synthesis and hence of RNA-directed protein synthesis (Key, 1964).

Separation of the various fractions of nucleic acids by methylated-albumin-kieselguhr (MAK) columns permits the examination of the incorporation of label (either as $^{32}PO_4$ or labeled nucleotides) into soluble RNA (sRNA), DNA, light and heavy ribosomal RNA, and a fraction which has the characteristics of mRNA. It was found that auxin stimulations involve increases in the incorporation into all these nucleic acid fractions, especially into the ribosomal fraction (Fig. 4-25). From comparative studies of the effects of various types of nucleic acid inhibitors, Key and Ingle (1964) found that 5-fluorouracil (5-FU) can bring about a 60 percent inhibition of RNA synthesis without interfering with auxin-stimulated growth (Fig. 4-26). The 5-fluorouracil inhibited each of the RNA fractions separable on the MAK column except the last, the fraction containing mRNA. The inhibition of growth by nucleic acid inhibitors therefore appears to result principally from the suppression of synthesis of some component of this last elution fraction, most probably mRNA.

The auxin-stimulated synthesis of RNA can be antagonized by an antiauxin in a manner quite similar to the antagonism of growth (Masuda and Tanimoto, 1967).

An auxin regulation of DNA-directed RNA synthesis might take place via an increase in the template activity of DNA or, alternatively, via an increase in the effectiveness of RNA polymerase. In studies of 2,4-D-stimulated RNA synthesis in soybean tissue, O'Brien et al. (1968) found that the addition of RNA polymerase

Fig. 4-25 Treatment of soybean hypocotyls with 10 ppm 2, 4-D results in a marked stimulation of $^{32}PO_4$ incorporation into the nucleic acid fractions separable on a MAK column. The six peaks in optical density (solid line) are I: sRNA, II: sRNA, III: DNA, IV: and V: rRNA, and VI: DNA-like RNA (Key et al., 1967).

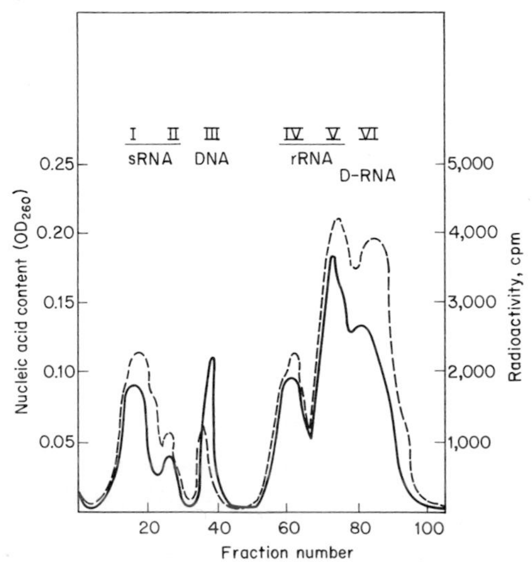

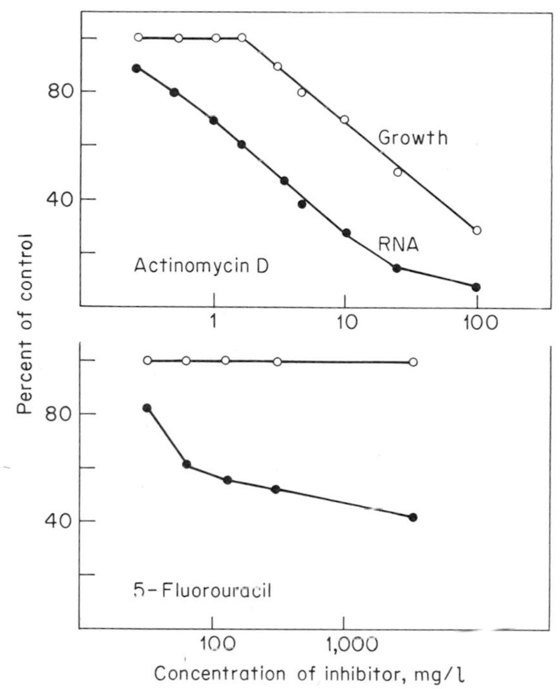

Fig. 4-26 The inhibition of growth by actinomycin D is associated with a strong inhibition of RNA synthesis; 5-fluorouracil, however, can profoundly inhibit RNA synthesis without inhibiting growth. Soybean hypocotyls in 10 ppm 2,4-D; RNA synthesis taken as the incorporation of ^{14}C-ADP (Key et al., 1967).

to chromatin from auxin-treated soybeans yields about the same maximum RNA synthesis as chromatin from untreated controls; this suggests that a major effect of the auxin on RNA synthesis may be through an enhancement of RNA polymerase activity.

The auxin stimulus could involve an enhancement of more RNA, or of a qualitatively different RNA. Thompson and Cleland (1971) compared the competitive characteristics of RNA for binding to DNA, and found that the IAA-stimulated RNA was not measureably different from RNA from control tissues.

Hormonal regulation of an enzyme could result either from an alteration of nucleic acid–directed protein synthesis or, more directly, through a release or activation effect on the enzyme. Without attempting to distinguish between these two types of enzyme effects, let us first ask what types of enzymes might serve as critical or pace-setting enzymes for growth. Of course the possibility of respiratory enzymes comes up for consideration; but there seems to be an impressive difference between the quantitative effects of auxins on growth and on respiratory systems (Thimann, 1952), and the possibility of an auxin action through the respiratory system has been set aside (Cleland, 1961). The alternative possibility seems to be an action through the enzymes which alter the characteristics of the cell wall. The stimulation of growth by auxin was early shown to be attributable to a softening of the cell walls (Heyn, 1931). When an experimental technique which could distinguish between plastic and elastic changes in the cell wall was used (Fig. 4-27), the auxin stimulation of growth was found to be closely associated with changes in plasticity (Fig. 4-28).

Fig. 4-27 As a means of measuring the plasticity and elasticity of oat coleoptiles, the tissue can be held horizontally and deflected by hanging a weight on it; the degree of restoration following removal of the weight can be taken as a measure of elasticity and the unrecovered deflection as a measure of plasticity (Tagawa and Bonner, 1957).

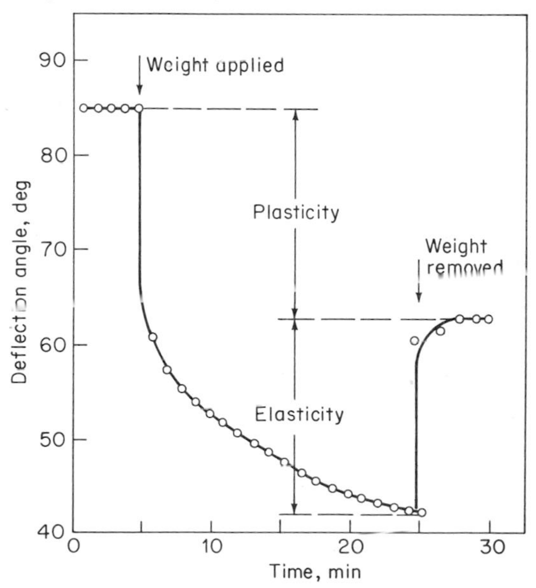

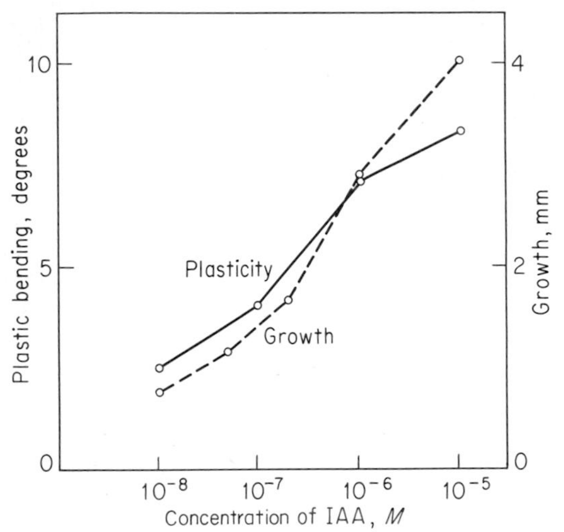

Fig. 4-28 With the measure of plasticity shown in Fig. 4-27 the oat coleoptile is seen to respond to auxin treatments with an increase in plasticity which is similar in magnitude to the increase in growth (Bonner, 1960).

Since the cementing substance of young walls involves pectins and hemicelluloses, the possibility was raised that auxins serve to remove the calcium ions which cross-link the carboxyls of these cementing polymers; however, experiments with radioactive calcium have failed to show a release of calcium sufficient to account for the auxin stimulations of growth (Cleland, 1960a). Enzymatic alteration of the extent of methylation of the pectic acid groups has been considered as another possibility. Schrank (1956) found that ethionine, which is a strong inhibitor of methylations, is a very effective inhibitor of oat-coleoptile growth; however, the auxinstimulated growth of corn mesocotyls does not show a methylation requirement (Cleland, 1960b).

Instead of attacking the cementing properties of walls enzymatically, perhaps auxin alters the cellulose fibrils themselves. Fan and MacLachlan (1966) found an impressive correlation between the auxin stimulation of growth and the associated increase in cellulase in pea stems (Fig. 4-29). They also found that like the auxin stimulation of growth, cellulase increases are prevented by actinomycin D and puromycin but are insensitive to the DNA synthesis inhibitor 5-fluorodeoxyuridine (FUDR) (Fan and MacLachlan, 1967). Another hydrolytic enzyme which has been implicated is β-1,3-glucanase. Tanimoto and Masuda (1968) have found that this enzyme increases in activity during the auxin stimulation of growth, and further that applications of the enzyme directly to the tissue can induce some initial growth responses (Wada et al., 1968), principally altering the elasticity of cell walls (Masuda, 1968). Ruesink (1969) has cast some doubt on the cell-wall-hydrolyzing enzymes as being central to auxin action, since he cannot obtain an enhancement of growth with the application of cellulase to oat coleoptiles, even though physical measurements of the wall indicate that the extensibility has been changed by the enzyme.

A second part of the growth reaction, distinguishable from cell-wall plasticization, is a synthesis of new cell-wall materials. Several reports indicate that auxin treatments increase cell-wall components, including cellulose and hemicellulose in pea stems (Christiansen and Thimann, 1950) and hemicelluloses and pectins in oat coleoptiles (Bayley and Setterfield, 1957). Albersheim and Bonner (1959) reported large increases in the synthesis of both cold and hot-

Fig. 4-29 The stimulation of growth of peas by auxin is correlated with a rising cellulase level in the tissue; 500 ppm IAA in lanolin (Fan and Maclachlan, 1966).

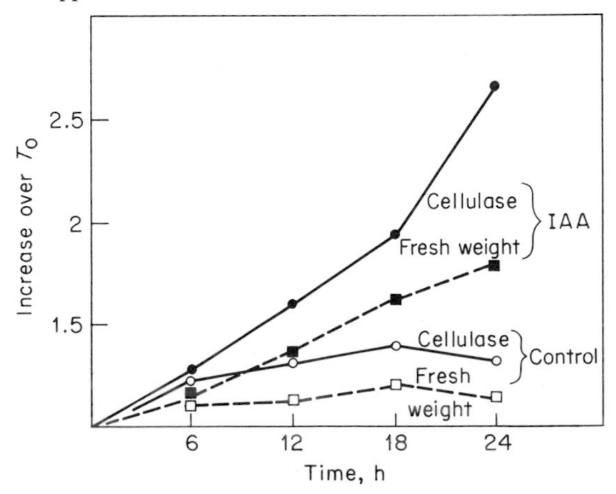

water-soluble pectic substances with auxin treatment, along with increases in the incorporation of radioactivity from glucose substrate (Morré and Olson, 1965). Whether cell-wall synthesis is a part of the basic growth stimulus by auxin is unclear. Numerous investigators have failed to detect cell-wall synthesis associated with growth stimulations of coleoptiles (Bennet-Clark, 1956; Ordin and Bonner, 1957).

A third component of the growth reaction is the osmotic uptake of water, which maintains a swelling force against the softening cell wall. Early reports indicated that this might involve a metabolic uptake of water (Bonner et al., 1953), but more recent evidence supports the view that water-uptake activities are osmotic and not driven by metabolic activity. Two lines of evidence favor this osmotic mechanism: increases in the osmotic values of the ambient solution result in exponential depressions of cell enlargement (Ordin et al., 1955), and as growth proceeds there may be an actual dilution of the osmotic components associated with water uptake (Hackett, 1952). Such a dilution would be expected if growth were initiated by a softening of the wall followed by osmotic entry.

The possibility that auxin exerts its hormonal action through some permeation effect is largely speculative since there is little experimental information on the subject. Commoner et al. (1943) suggested that the auxin stimulation of growth involves a stimulation of cation uptake, but the rapid effects of auxin on cell walls has stifled interest in this possibility. Gregory and Cocking (1966) have made the striking observation that isolated protoplasts taken from tomato fruits and entirely without any cell wall respond to auxin by the uptake of water, probably resulting from an uptake of solutes. Weigl (1969) has suggested that auxin molecules could become attached to lipid components of membranes and there greatly alter the ion permeability of the membrane. He has set up model membranes and shown that auxins do bind to his models, with a good specificity for active auxins; he attributes the facility of binding to the spacing of positive and negative charges on the auxin matching

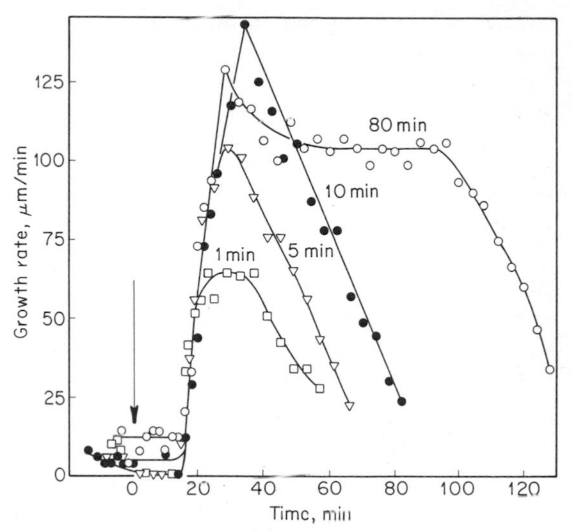

Fig. 4-30 Kinetics of changes in growth rate of corn-coleoptile sections after the introduction of 10^{-5} *M* IAA for various time periods beginning at the arrow marker. In each case there was a lag time of about 15 min before growth rate increased; and growth rate decayed after removal of the auxin (dela Fuente and Leopold, 1970).

opposite charges on the lecithin component of his model membranes.

Some important evidence that pertains to the type of mechanism of auxin action can be obtained from kinetic experiments, in which the timing of the growth response is measured. The subject was opened by Ray and Ruesink (1962), who developed an optical system for following the growth of coleoptile sections continuously. The first parameter that can be obtained from this type of experiment is the lag time between auxin application and the first detectable increase in growth rate; Ray and Ruesink found lag times ranging from 10 to 15 min, and similar values have subsequently been reported for numerous other types of tissue. The lag time for corn coleoptiles is illustrated by the growth rates shown in Fig. 4-30. By applying higher auxin concentrations and by doing the experiment at elevated temperatures Nissl and Zenk (1969) shortened the lag time to essentially zero, indicating that the auxin was able to bring about growth responses almost at once upon entry into the cell. Another parameter of interest is the aftereffect, or the time following the withdrawal of the auxin supply until the stimulated growth rate begins to revert to the lower endogenous rate. Ray and Ruesink estimated the after effect at about 40 min. Similar values can be run for the after effect in the data shown in Fig. 4-30. Evans and Hokanson (1969) obtained shorter aftereffect times by employing lower auxin concentrations, indicating that much of the aftereffect was the time required for the auxin to be transported out of the sections. Dela Fuente and Leopold (1970) were able to correct for the transport time and estimated that about 30 min was needed for the aftereffect plus 50 percent decline in growth rate. Collectively these kinetic experiments indicate that auxin has a stimulating effect on growth very rapidly upon entry and that the stimulation decays very rapidly upon auxin exit. Since the nucleic acid–directed synthesis of new enzymes should take longer than a few minutes, and since the expected half-life of the enzymes once formed should be at least several hours, it seems very unlikely that these short-time effects of auxin can be attributed to a hormone regulation of enzyme synthesis. Nissl and Zenk have suggested the alternative of an auxin effect on some preformed system which regulates wall synthesis. The longer-term effects of auxin on growth may well involve nucleic acid–directed enzyme synthesis, but the short-term effects imply an action of auxin in releasing growth-limiting substances already present in the cell which have fairly direct actions in softening the cell wall and thus leading to growth.

Fig. 4-31 The time curves for cell lengthening in wheat roots after introduction of three concentrations of the auxin naphthaleneacetic acid. The auxin stimulated elongation but terminated growth at a lesser final size (Burström, 1969).

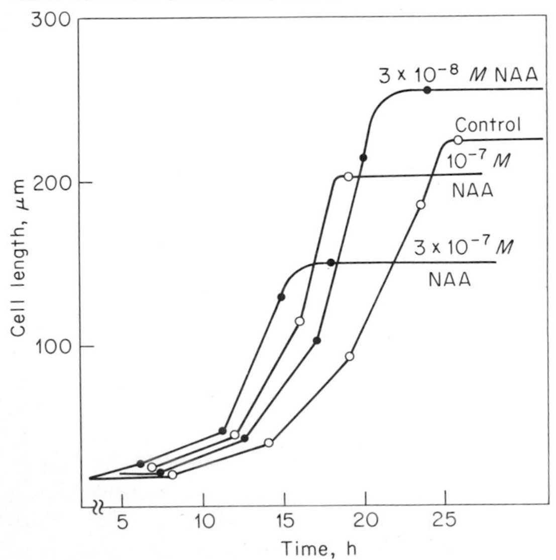

What may the nature of the rapid response to auxin be? Rayle and Cleland (1970) have found that the immersion of coleoptiles in a solution with a low pH such as 3.0 can cause rapid softening of the wall and immediate growth stimulation. Evans et al. (1971) have also obtained rapid growth bursts upon exposure of coleoptile to solutions saturated with CO_2. They suggest that the early responses to auxin may be through the auxin activation of a proton pump on the plasmalemma and that the consequent decrease in pH in the cell-wall region may be responsible for the rapid responses. Hager et al. (1971) extended the proton-pump theory, with evidence that the metabolic role in the early auxin responses is providing the energy for a proton pump and the resultant lowering of the pH in the wall enhances the activity of wall-loosening enzymes there. Whether the lower pH serves to loosen certain bonds in the wall directly or to enhance the activity of wall-loosening enzymes, Barkley and Leopold (1973) believe that the rapid auxin effect is not due to the lowered pH because some tissues that respond well to auxin do not respond to the lowered pH.

The drop in pH has been shown to be correlated with growth in a variety of ways: Cleland (1973) reported that blocking auxin-stimulated growth with inhibitors such as CCCP also blocked the pH drop. Marre et al. (1973) reported that the extent of the pH drop was proportional to the growth stimulatory activity of 4 different auxins and the fungal inhibitor fusicoccin. This fungal agent has very dramatic short-term stimulatory effects on growth of a wide array of plant tissues, and always causes a simultaneous drop in pH (Marre et al., 1974).

A different approach to the rapid responses to auxin was proposed by Evans and Ray (1969), who suggested that some growth-limiting protein (or other limiting substrate) may be used up in the rapid growth reaction. Penny (1971) did some clever experiments which led her to conclude that there is indeed such a growth-limiting protein; when she blocked its synthesis with cycloheximide, she estimated that this pool has a half-life of 12 to 17 min; she asserts that once this pool is exhausted, new protein synthesis is necessary before the rapid auxin response can be obtained again.

It has been reported that cyclic AMP can markedly enhance auxin stimulations of growth (Kamisaka and Masuda, 1970), and one might infer that this mediating factor for the actions of many animal hormones may be involved likewise in auxin action. The effects of added cyclic AMP are certainly small (Hartung, 1972), but, on the other hand, some plants are very rich in the enzymes which split the cyclic phosphate ring (Lin and Varner, 1972), and this could make in vivo tests rather insensitive to added cyclic AMP. Another candidate for a cofactor of hormone action is a stimulator of RNA polymerase, which can be displaced off membrane preparations by auxin (Hardin et al., 1972). Such a cofactor could well provide a link between a plasmalemma site of auxin attachment and the auxin regulation of nucleic acid–directed protein synthesis.

AUXIN INHIBITIONS

Since inhibitory actions of auxins may be even more widespread than promotive actions on growth, it is unfortunate that so little information is available on their mechanisms. Three suggestions have been made to explain auxin inhibitions. Burström (1955) has concluded from work with wheat-root inhibitions that auxin may catalyze the hardening of the cell wall, thus leading to a shorter duration of growth and a shorter final cell length. The enhanced hardening action is deduced from data like those in Fig. 4-31. A second suggested explanation of auxin inhibitions is that the hormone exerts a self-inhibition by the attachment of separate molecules to each of the two presumed sites of auxin attachment (Foster et al., 1952). This type of action would be similar to the substrate inhibition of some enzymes, in which there is more than one attachment site for a substrate. Each of the two molecules would serve to prevent the completion of two-point attachment by the other molecule. A third suggested explanation of auxin inhibition is that ethylene production is stimulated by auxin and the ethylene is responsible for the inhibitory effect (Burg and Burg, 1966b). In this case, the inhibition would be due to an action quite separate from the mechanism of growth stimulation. The ethylene explanation rests principally on the finding that inhibitions occur at concentrations of auxin which do stimulate ethylene formation, as shown in Fig. 4-32. It may serve to explain auxin inhibitions of pea-stem growth and inhibitions of bud growth but apparently not inhibitions of coleoptile growth since these tissues produce little ethylene and are fairly insensitive to it (Burg and Burg, 1968a).

It is appealing to think that the characteristic two-phase curve of auxin responses, like those in Fig. 4-2 and 4-3, might reflect a single auxin action resulting in diverse growth responses; however, since many tissues show very strong auxin inhibitions and little or no auxin promotive responses, e.g., in roots, the idea of self-inhibition or inhibition by self-competition reactions becomes less attractive and the idea of inhibition by mechanisms quite separate from promotive actions seems a more reasonable alternative. Thus the auxin inhibition of wheat roots may be through a stimulation of cell-well hardening; the inhibition of pea stems may be through the stimulation of ethylene production;

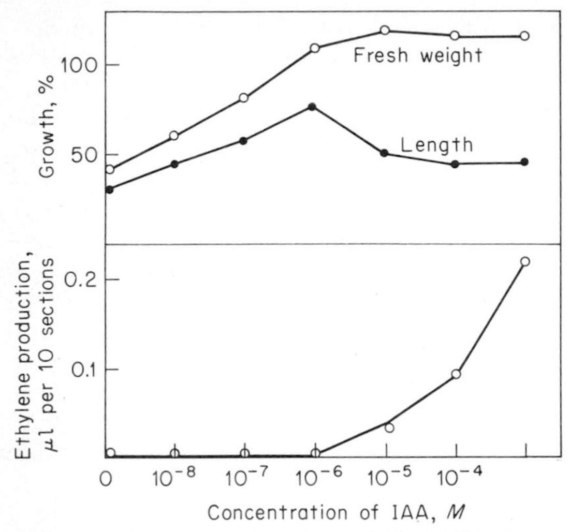

Fig. 4-32 Auxin concentrations which inhibit the growth of pea-stem sections, as indicated by the decrease in length achieved, are correlated with those concentrations which bring about ethylene biosynthesis (Burg and Burg, 1966).

the inhibition of bud growth may be a result of ethylene production and the consequent inhibition of cell division (Burg and Burg, 1968b); and the inhibition of abscission may be through a suppression of senescence (Abeles et al., 1967).

SUMMARY

Three major features of auxins stand out strikingly: (1) the diversity of auxin effects, (2) the diversity of the other chemical controls which may be interwoven with the auxin effects, and (3) the systemic patterns of the auxin effects.

The diverse effects of auxins are readily apparent in the large and expanding list of growth and differentiation activities which are influenced by endogenous auxins or altered by exogenous auxins (cf. Table 18-1). In addition to cell-elongation, apical dominance, and abscission processes, there are effects on flower initiation and development, pollen-tube growth, fruit-set and fruit growth, the formation of compression wood in conifers, tuber and bulb formation, and seed germination. Almost every dynamic part of plant growth and development seems to be affected by auxin. In the individual cell there are effects on the plasticity and elasticity of the wall, on cytoplasmic viscosity, on protoplasmic streaming, on respiration rates, on metabolic pathways, on changes in oxidative states, on the contents of nucleic acids, and on the activities of many enzymes.

Allied with the diversity of physiological effects of auxins is a diversity of molecular species which can bring about the responses. The fact that IAA can share its biological effectiveness with such diverse synthetic chemicals as naphthaleneacetic acid, 2,4-dichlorophenoxyacetic acid, 2,3,6-trichlorobenzoic acid, and a host of others attests to the wide range of compounds which can act as auxins in plants. Most auxins are able not only to stimulate growth of stems and coleoptiles but also to alter differentiation, abscission, and other developmental effects. The systemic correlative effects are not shared by all auxins, for the limited ability of many to move in the auxin-transport system restricts their participation in normal correlation effects. They are readily swept along in the stream of phloem translocation, but such movement does not conform to the polar qualities of the hormone-transport system.

The diversity of agents which can participate in growth regulation is a strikingly repetitive theme. Cell enlargement may be directed by gibberellins, auxins, kinins, ethylene, and various inhibitors; there are strong growth-promoting effects of chelating agents, fatty acids, and even organic acids. When auxin was first discovered, Went (1928) said, "Ohne Wuchsstoff, kein Wachstum." Now there is evidence that such a concise, all-or-none dependence on auxin is not a general characteristic of cell enlargement (cf. Kefford and Goldacre, 1961). Many cells do not appear to need auxin for growth. The auxin influences on physiological and developmental processes generally require other substances as cofactors, including the factors for cell division, cell enlargement, xylem differentiation, root initiation, and apical dominance.

The chemicals regulating growth and development are manifold.

The systemic patterns of auxin effects in the plant, representing the correlation effects on growth and differentiation, reveal auxin as a chemical messenger influencing many patterns of plant development. A system of chemical messengers is a principal ingredient for the creation of a multicellular organism out of what would otherwise be only a multicellular colony, and auxin is the outstanding known participant in such a control system.

GENERAL REFERENCES

Audus, L. J. 1972. Plant Growth Substances. Leonard Hill Books, London. 533 pp.

Evans, M. L. 1973. Rapid stimulation of plant cell elongation by hormonal and non-hormonal factors. *Bioscience*, 23:711–718.

Thimann, K. V. 1972. The natural plant hormones. In F. C. Steward, *ed*.. "Plant Physiology, a Treatise." Academic Press, New York, pp. 3–332.

5
GIBBERELLINS

OCCURRENCE

BIOSYNTHESIS AND METABOLISM

TRANSLOCATION

GIBBERELLIN ANTAGONISTS

REGULATORY ROLES

MECHANISM OF ACTION

SUMMARY

At about the same time auxins were first recognized as chemical constituents of plants, work in Japan led to the discovery of another group of growth substances, the gibberellins. Kurosawa (1926) was studying a disease of rice plants caused by the fungus *Gibberella fujikuroi*, which caused a characteristic excessive growth of the rice plant. He found that extracts of the fungus applied to an uninfected plant could bring about the growth stimulations associated with the disease. The separation of this growth regulator closely preceded the first separation of auxin from plants in 1928. The heat-stable substance eluded purification for several years, partly because of the presence in the extract of an inhibiting substance, fusaric acid (cf. Chap. 8); but in 1935, Yabuta crystallized the compound and named it *gibberellin*. While the auxins were widely heralded through the next two decades, the research on gibberellins proceeded in Japan almost unnoticed by the Western world.

In the early 1950s there was a sudden widespread interest in the gibberellins. In addition to the Japanese work, headed by Hayashi and Sumiki, active centers of interest sprang up in England, under Brian, and in the United States, under Stodola and others. The impressive ability of gibberellin (GA) to stimulate plant growth (Fig. 5-1) had finally attracted the attention of Western physiologists, and soon the molecular structures of several gibberellins were clarified (Cross, 1954). More than 30 separate gibberellins have been isolated from fungus or higher plants. In its ability to increase growth in intact plants, gibberellin is far more impressive than auxin. Applications of gibberellin to grasses such as rice cause both leaf and stem elongation. In the broad-leaved plants, the stimulations of stem growth can be very dramatic. Following GA applications a cabbage may elongate to a height of 6 ft, and a bush bean may become a climbing pole bean (Fig. 5-2). The stimulative effects on stems are more pronounced with intact plants than with excised pieces.

Particularly responsive to GA applications are dwarf or short-stemmed plants such as rosette species, and often the natural tendency toward extensive growth is associated with high natural

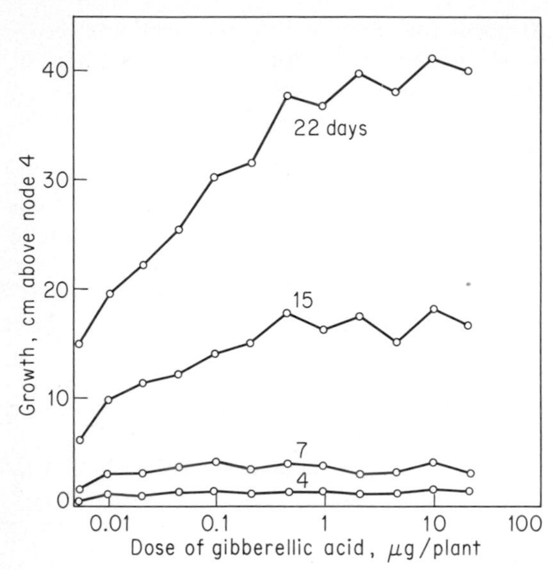

Fig. 5-1 Applications of gibberellin to Meteor peas result in large stimulations of the growth of the plant; as little as 0.1 µg per plant causes doubling of plant height in 22 days (Brian and Hemming, 1955).

GA levels whereas the dwarfing or rosette types of growth are associated with low GA levels.

In addition to the powerful effects of GA on stem growth, it has many regulatory effects on plant development (Table 18-1). Frequently dormancy of buds and seeds can be relieved by GA, and some germination processes, e.g., the synthesis of hydrolytic enzymes for digestion of the endosperm reserves, are driven by GA. Flowering can be induced by applications of GA to some species or by environmental experiences which lead to natural elevation of GA content. The sex expression of numerous species is subject to control by GA, and the fruit-set and growth rates of some fruits as well as maturation and ripening phenomena are subject to GA control. Finally, the senescence of some plant parts, especially of leaves, is subject to GA regulation. In short, this is a hormone of many and powerful regulatory functions.

OCCURRENCE

Gibberellins are diterpenoids; diterpenes are composed of four isoprene units, usually arranged to form three rings, and are common constituents of the resins of conifers; diterpenoids, which usually have four rings, are common plant constituents, especially as glycosides.

The gibberellin diterpenoids have an additional lactone bridge, as shown in Fig. 5-3. The number of known gibberellins from fungus and plant sources continues to increase (it was 34 in 1974); variations are mostly with respect to the presence or absence of an unsaturated bond in the A ring, the number and location of hydroxyl substitutions, and the number of carboxyl groups. GA_4, GA_1, and GA_3 (gibberellic acid) are some of the most widespread gibberellins in plants, and the probable manner of interconversion of these and some related gibberellins is indicated in Fig. 5-3. Apparently the interconversions are not always easily carried out in plants, however, judging from the specificity of responsiveness to the different gibberellins from one species to another. Illustrating the selectivity, Table 5-1 indicates the responsiveness or lack of responsiveness of six bioassays to six different gibberellins.

Besides the diterpenoid lactone gibberellins, activity in the form of growth stimulation has been observed for several other types of natural plant constituents. Kaurenoic acid has been reported to be a naturally occurring compound with gibberellin activity in the rice assay (Katsumi et al., 1964) and so also has steviol (Ruddat et al., 1963). It is highly probable that these compounds show activity as a result of conversion within the plant to the gibberellin type of compound. However, two other compounds surely fall outside the range of possible gibberellin precursors and still show the gibberellin type of activity: helminthosporic acid (Tamura and Sakurai, 1967) (Fig. 5-4), a product of *Helminthosporium sativum*, and phaseolic acid (Redemann et al., 1968), isolated from bean seeds. These, plus the possible gibberellin activity of some condensation products of acetone and unidentified fatty acids (Mitchell et al., 1969), suggest that the types of growth regulation associated with the gibberellins may be shared, at least in part, by some chemicals of very different structure.

Fig. 5-2 Application of 20 μg of gibberellin to a Contender bean plant results in large stimulation of growth and a change from bush to vine habit (Wittwer and Bukovac, 1957).

Gibberellin assays are ordinarily based on the responses of plants which have a very low endogenous GA content. Dwarf lines are frequently a consequence of genetic limitations of GA biosynthesis, and several lines of dwarf corn having such genetic limitations (Phinney, 1956) are highly responsive to added GA. An assay using this dwarf material was developed by Phinney et al. (1957); a GA solution is applied to the ligule of the first leaf of dwarf corn seedlings, which causes a marked stimulation of the next node or leaf sheath, as shown in Fig. 5-5. Dwarf peas are also effective materials for GA bioassay. In this case a solution is applied to the seedling,

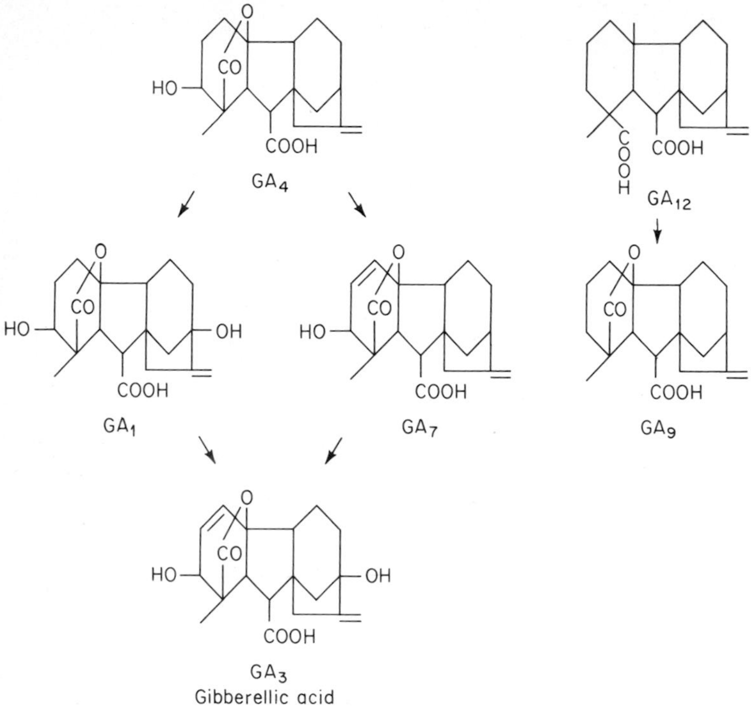

GA₄ → GA₁, GA₇; GA₁₂ → GA₉; GA₁, GA₇ → GA₃

GA₃
Gibberellic acid

Fig. 5-3 The structures of some plant gibberellins and suggested interconversions between them.

and the resulting stimulation of stem elongation is measured, as shown in Fig. 5-1. Another type of sensitive material is the lettuce seedling, which has a low GA level when it is grown in the light and is therefore quite sensitive to added GA. The entire seedling is placed in the GA solution, and elongation of the hypocotyl is taken as the measure of GA response, as shown in Fig. 5-6. In some instances, it is preferable that the bioassay for GA not involve growth responses. In the germinating barley seed, the aleurone layer responds to GA produced in the embryo by forming hydrolytic enzymes such as α-amylase (Hayashi, 1940; Paleg, 1960). Isolated

Table 5-1 A comparison of the effectiveness of six different gibberellins in a series of bioassays, indicating striking differences in the ability of plants to respond to the various molecular species of gibberellins (modified from Brian, 1964).

	Responsiveness to various gibberellins					
Bioassay	GA_1	GA_3	GA_4	GA_5	GA_7	GA_8
Dwarf corn $d-1$	+	+	+	+	+	0
Bean-stem elongation		+	0	+	+	0
Dwarf-pea stem elongation	+	+		0		
Cucumber-hypocotyl elongation	+	+	+		+	0
Myosotis flowering	+	0	0	0	+	0
Silene flowering	0	0	0	0	+	

seed halves can be very sensitive assay material for GA without involving growth, as shown in Fig. 5-7. A very simple and convenient bioassay with isolated tissue pieces involves the ability of GA to defer the onset of senescence in leaf disks of *Rumex*. The isolated disk loses its chlorophyll content unless GA is supplied, and this chlorophyll retention is a quantitative GA response (Fig. 5-8).

Gibberellin can be detected in a wide range of plant parts and in various stages of growth. In the early stages of seed germination, GA may be present (Wheeler, 1960), and this hormone is certainly involved in the regulation of some growth processes in the young seedling (Sebanek, 1965). In stems, GA is of general occurrence, and its important role in the regulation of growth is well known. Substantial amounts of GA may be found in roots (Carr et al., 1964) and leaves, but a regulatory role over the growth of roots or leaves seems doubtful (Cathey, 1964). Several types of fruits are known to have generous amounts of GA, and at least in some instances the GA content correlates well with growth rates, from which it may be inferred that the hormone serves to regulate growth there (Murakami, 1961). In seeds, too, substantial amounts of GA can be found during seed enlargement (Baldev et al., 1965) and maturation (Corcoran and Phinney, 1962).

Fig. 5-4 Examples of compounds which show gibberellinlike activities in stimulating growth but which are not diterpenoid lactones.

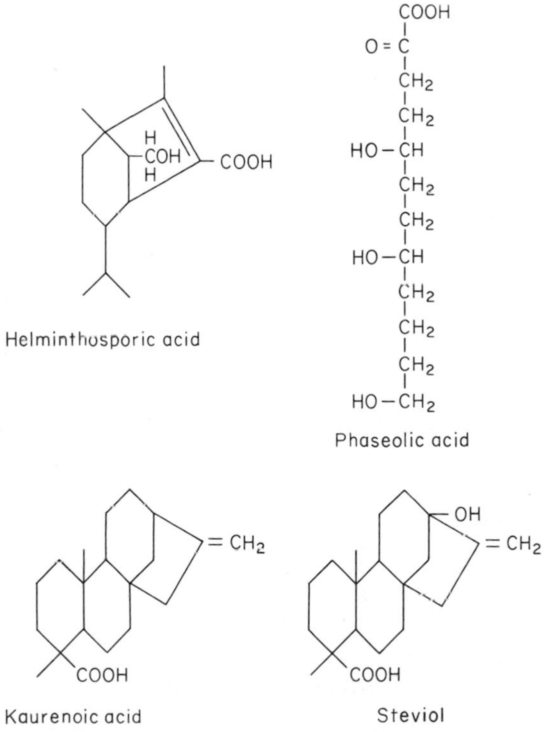

Helminthosporic acid

Phaseolic acid

Kaurenoic acid

Steviol

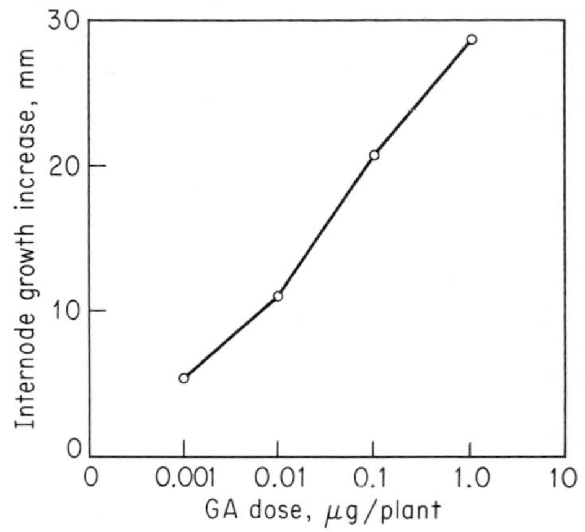

Fig. 5-5 The quantitative response of dwarf corn $d-5$ to GA applied to the first leaf ligule (Smith and Rappaport, 1961).

BIOSYNTHESIS AND METABOLISM

The biosynthesis of the gibberellins springs from the isoprenoid pathway of acetate metabolism. Birch et al. (1959) first showed that labeled acetate can be incorporated into GA through mevalonic acid. Details of this pathway are described by Cross (1968), Lang (1970), and West and Fall (1972); the general outline is represented in Fig. 5-9. It may be considered to involve three groups of reactions:

1 The stringing together of the isopentenyl groups, four of which combine to form the geranylgeranyl pyrophosphate intermediate

2 The cyclization to form kaurene

3 A series of oxidation steps, including a con-

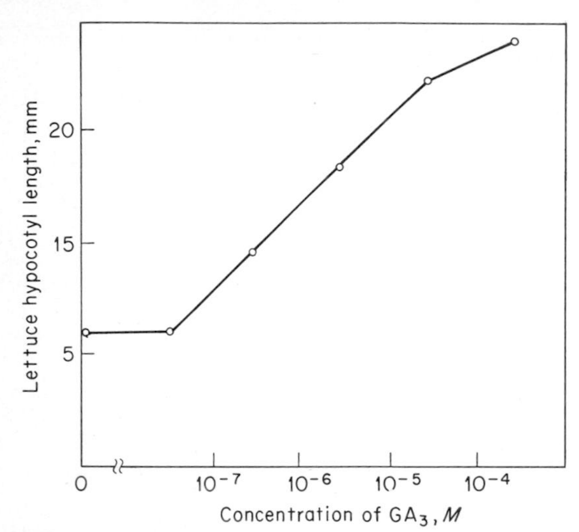

Fig. 5-6 The effect of GA in stimulating lettuce-hypocotyl growth is utilized as a bioassay for GA (Frankland and Wareing, 1960).

traction of the B ring to form the presumed first synthesized 20-carbon gibberellin, GA_{12}, and the 19-carbon GA_4

Some suggested final interconversions are indicated in Fig. 5-3. Elucidation of the steps from mevalonate to kaurene has been greatly facilitated by the discovery that soluble enzyme preparations from higher plants can carry out these synthetic steps (Graebe et al., 1965). The steps following kaurene are less well understood, and most of the experimental work in this sector has been done with fungal cultures (Geissman et al., 1966; Lang, 1970), where the enzyme systems are located on microsomes.

Numerous growth retardants are known which can bring about dwarflike growth in plants, and these inhibitory effects are generally overcome by the addition of GA. Examination of the possible effects of growth retardants on GA biosynthesis has been very productive, for in several instances these compounds have proved to be effective inhibitors of GA biosynthesis, both in cultures of the fungus (Kende et al., 1963) and in higher-plant tissues (Baldev et al., 1965). Examples of growth retardants which serve to inhibit GA biosynthesis are AMO-1618 and CCC (molecular structures are given in Fig. 18-18), which interfere with the initial condensation of geranylgeranyl pyrophosphate into a ring structure as copalyl pyrophosphate (Dennis et al., 1965). Phosfon, another retardant, inhibits both this step and some subsequent step in the formation of kaurene (Lang, 1970). While these inhibitory effects on GA biosynthesis are well established, it is rather puzzling that the addition of kaurene does not restore GA biosynthesis (Harada and Lang, 1965). The ability of AMO-1618 virtually to eliminate GA biosynthesis in pea fruits is illustrated in Fig. 5-10.

Where in the plant does GA biosynthesis occur? Examination of the amounts of GA diffusible from sunflower-stem sections at intervals along the stem indicates that the amount of GA present increases with greater proximity to the stem apex (Fig. 5-11). From this type of analysis, combined with examination of extractable GA, Jones and Phillips (1966) inferred that the GA is mainly supplied by the apex rather than synthesized in the stem. Earlier experiments by Lockhart (1957) had shown that cutting off the apex of pea plants results in large inhibitions of

Fig. 5-7 Quantitative effects of GA on the production of α-amylase in barley-endosperm pieces. Three seed halves are incubated per milliliter for 48 h, and amylase activity is taken as the amount of reducing sugar formed (Coombe et al., 1967).

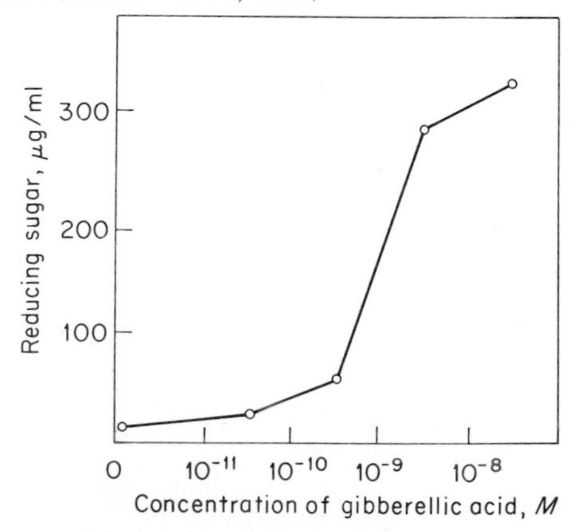

stem growth, and application of GA to the severed stem restored growth; from these experiments he deduced that the apex is a major source of GA to the stem. More precise examination of the apex by Jones and Phillips (1966) revealed that the leaf primordia in the apex are the main source of GA rather than the meristem itself. The leaf primordium appears to continue to produce GA through the period of cell division (Humphries and Wheeler, 1964). The exudate from severed roots of several species of plants provides enough GA to account for the entire plant supply, indicating that the root system is a major source of GA (Carr et al., 1964). GA biosynthesis may be inferred to occur in embryos (Paleg, 1960), cotyledons (Sebanek, 1965), fruits (Baldev et al., 1965), and seeds (Weaver and Pool, 1965). Wounding can stimulate GA biosynthesis in tubers (Rappaport and Sachs, 1967).

The role of the roots in providing gibberellins to the tops of the plant was dramatically illustrated by Crozier and Reid (1971), who report that repeated removal of the roots of bean plants leads to increased dwarfing growth characteristics. In the absence of the roots there is a marked depletion of GA_1 content and a marked rise in a related compound GA_{19}, which they believe to be a precursor of GA_1. They suggest that GA_{19} is ordinarily translocated to the roots and there converted into GA_1 in the intact plant.

The fate of GA in the plant was opened up as an experimental subject by Nitsch and Nitsch (1963), who applied radioactive GA_3 to plants and observed the loss of about 80 percent of its biological activity in 3 days' time, with the simultaneous appearance of new radioactive products on chromatographic separations. Two types of products are known, one the glucoside of GA (Fig. 5-12) and the other some bound form. The discovery of the GA glucoside was made by Tamura et al. (1968); since the initial work, they have found glucosides of several types of GA in *Pharbitis* (Yokota et al., 1969). It is attractive to think that the glucoside may be a temporary storage form from which the GA may

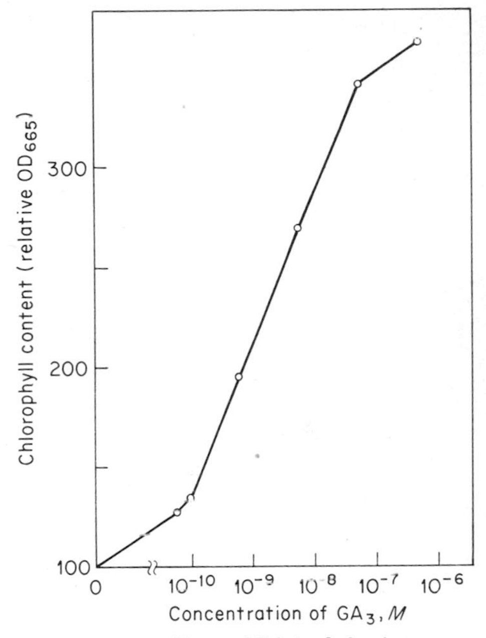

Fig. 5-8 Quantitative effects of GA in deferring senescence of disks of *Rumex* leaves. The progress of senescence is measured as the amount of chlorophyll disappearance from the leaf disks (Whyte and Luckwill, 1966).

be regenerated and in which GA may be translocated, since it has been found in the bleeding sap of maple and elm (Sembdner and Wieland, 1968; Lang, 1970). The nature of the bound form of GA is obscure, and some workers have regenerated GA from a protein fraction by hydrolytic or protease actions (McComb, 1961; Hashimoto and Rappaport, 1966), whereas other workers have been unable to find such a binding of GA (Kende, 1967; Jones and Lang, 1968).

Two cases have been reported of the internal increase of natural GA levels in the plant through the release of the GA from a bound form. Aung et al. (1969) have reported that low-temperature experiences will bring about the release of GA from some bound form, thus increasing the GA level in tulip plants. Loveys and Wareing (1971*b*) have found that the leaf-unrolling response to red light in small grain

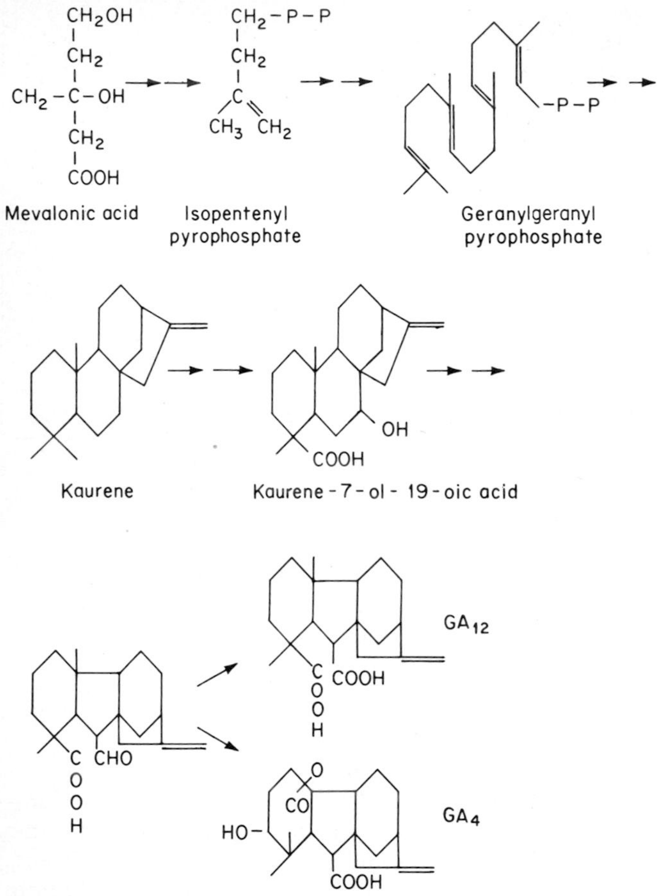

Fig. 5-9 Diagrammatic representation of the pathway of gibberellin biosynthesis from mevalonic acid. (*Modified from West and Fall, 1972.*)

seedlings is driven by an increase in GA levels in the leaf and that this, too, is obtained through the release of GA from some bound form.

To illustrate the metabolism of GA in the plant, some data of Barendse et al. (1968) are shown in Fig. 5-13. Tritiated GA_1 was applied to developing pea pods in tissue culture, and the conversion of the GA into a form not soluble in ethyl acetate and a form associated with the tissue residue can be observed over a period of 12 days. When the peas from this experiment were germinated, there was a subsequent release of an ethyl acetate-soluble GA from the aqueous and tissue fractions. The later finding of gibberellins with enough hydroxyl groups to make the molecule really water-soluble (Coombe, 1971) makes this kind of solvent separation of free gibberellins subject to some question. In another time-course study, Stoddart (1966) found that tritiated GA_3 had largely disappeared from the treated red clover plants in 10 days' time and that the disappearance was faster in short photoperiods than in long ones, associated with slower growth rates. Conversely, van den Ende and Zeevaart (1971) found a more rapid GA_3 metabolism in *Silene* in long photoperiods than in short ones. Each of these studies illustrates the removal of GA from the free and active form in what appears to be a hormone-disposal system.

TRANSLOCATION

In contrast to auxins, it seems uniformly true that gibberellins applied to one part of the plant can have regulatory effects on all other parts. There does not seem to be a directional quality to its regulations. Early experiments on stem sections by Kato (1958) showed that GA moves with equal facility in the basipetal and acropetal direction. Experiments with radioactive GA have provided more convincing evidence that the movement is without polarity (Chlor, 1967; Hertel et al., 1969; but cf. Jacobs and Kaldewey, 1970).

Translocation of this hormone appears to take place readily in either the phloem or the xylem translocation system. Hoad and Bowen (1968) have found GA to be present in authentic phloem sap; McComb (1964) reported that GA moves in the same pattern as the carbohydrate-translocation system and with a similar velocity (5 cm/h). Mention has already been made of the presence of GA in the xylem sap collected from severed roots (Carr et al., 1964). While GA can move freely along the stem in either the acropetal or basipetal direction, Lang (1970) has pointed out that neither the phloem nor the xylem system will account for its movement from the apical meristem into the stem below. Since the apex is probably a major site of GA biosynthesis, this remains perplexing.

The facile translocation of GA around most parts of the plant makes this hormone a fairly systemic regulator, in contrast to auxin, which carries a directional or polar set of regulatory influences.

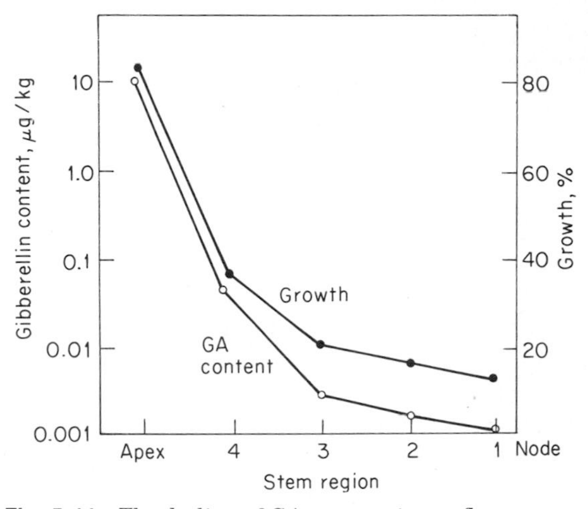

Fig. 5-11 The decline of GA content in sunflower stems with increasing distance from the apex is roughly correlated with the decline in growth rate. GA obtained by diffusion from stem sections (Jones and Phillips, 1966).

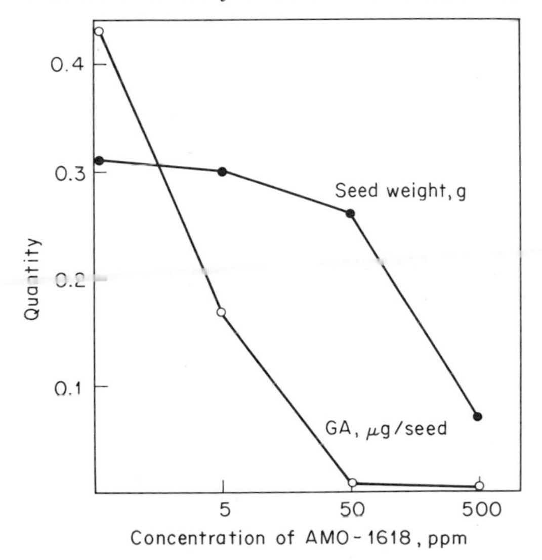

Fig. 5-10 The effects of AMO-1618 on the biosynthesis of GA and the growth of pea fruits in tissue culture, showing its particular effectiveness as an inhibitor of GA biosynthesis (Baldev et al., 1965).

GIBBERELLIN ANTAGONISTS

It is difficult to assign a growth inhibitor to an antagonism of a specific hormone, but the possibility of the plant's utilizing hormone antagonists in the regulation of growth makes the inhibitor question especially interesting. In developing the concept of antiauxins, McRae et al.

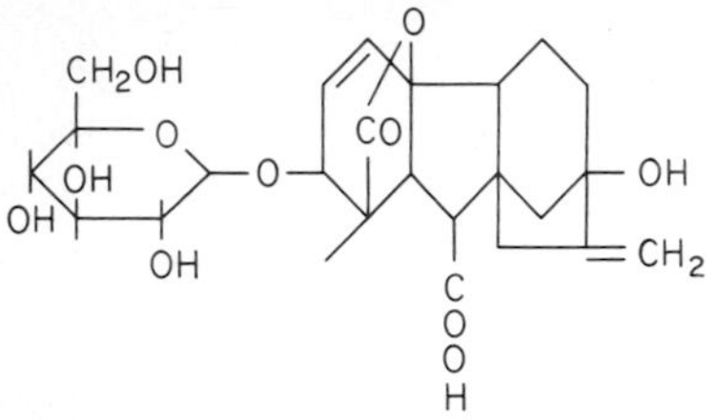

Fig. 5-12 The glucoside of gibberellic acid.

(1953) suggested that an antiauxin should bear a molecular resemblance to the hormone so that it can attach to the regulatory site; once there, it would inhibit either because of incomplete attachment or because attachment prevents catalytic action. Lona (1964) has found some growth-inhibitory effects for two diterpenoids, epiallogibberellic acid and atractyligenin, but there is little evidence to imply a competitive antagonism of gibberellin by these materials. Kato and Katsumi (1967) found slight inhibitory action for pseudogibberellic acid, an analog of GA_3 which has the β hydroxyl in the wrong stereo position; but the apparent antagonism applied only in the rice GA bioassay and did not show up in the regulation of cucumber-seedling growth. A synthetic diterpene, chlorofluorenol (see Fig. 18-18) was reported by Ziegler et al. (1966) as a possible GA antagonist, but Krelle (1967) established that it is not.

Several unidentified natural inhibitors have been suggested as GA antagonists, e.g., inhibitors extracted from *Ceratonia* (Corcoran et al., 1961), lima bean (Köhler, 1964), and conifer seeds (Michniewicz and Kopcewicz, 1966).

Abscisic acid exerts a powerful inhibition against GA actions. It has been found to depress the GA stimulation of growth of dwarf corn and dwarf peas (Thomas et al., 1965) and to depress the GA stimulation of α-amylase synthesis in the barley endosperm (Chrispeels and Varner, 1966). It seems unlikely, however, that abscisic acid could be considered a GA antagonist in view of the fact that its inhibitory effects apply equally well to growth stimulations by auxin, gibberellin, or cytokinin (Leopold, 1971).

Ethylene, too, exerts an inhibitory action against some gibberellin actions. Scott and Leopold (1967) reported that ethylene can interfere with the gibberellin stimulations of growth, delay of senescence, and a part of the stimulation of α-amylase production in barley (but see Jones, 1968). As with abscisic acid, however, it seems unlikely that the ethylene actions could be construed as being genuinely antagonistic to ethylene at the site of GA action. In fact there are a few instances in which ethylene and GA have roughly similar rather than antagonistic actions, e.g., breaking seed dormancy (Burdett and Vidaver, 1971) and stimulating rice-seedling growth (Suge, 1971).

The ability of some synthetic growth retardants to inhibit GA biosynthesis is not to be construed as a gibberellin antagonism, even though their inhibition may be relieved by added GA (Cathey, 1964).

The possibility of the regulation of plant growth by natural or synthetic antagonists of GA remains only an interesting possibility to date.

Fig. 5-13 After the application of tritiated GA_1 to excised pea fruits, there is a gradual conversion of the GA into a water-soluble form and a form associated with the insoluble tissue fraction as evidenced by the distribution of radioactivity (Barendse et al., 1968).

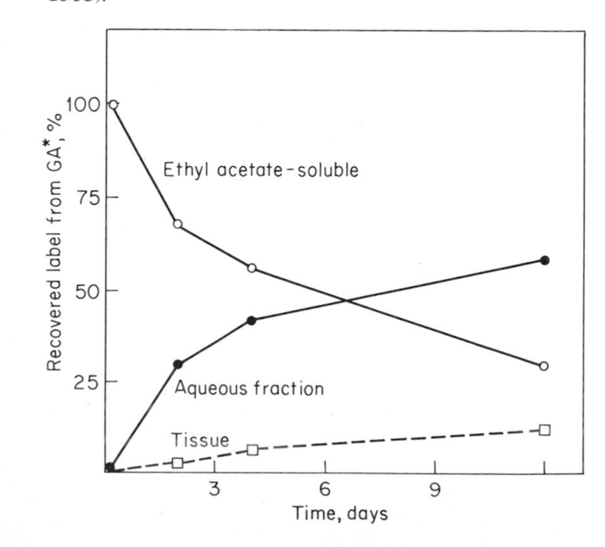

Fig. 5-14 Gibberellin applications to dwarf corn d–1 cause it to grow like standard corn. *Left to right:* standard control, standard with gibberellin applied (60 μg per plant), dwarf control, dwarf with gibberellin applied (Phinney, 1956).

REGULATORY ROLES

The regulation of growth by GA relates almost exclusively to stem elongation, since roots and leaves show only weak responses to GA. Some of the most dramatic effects of GA are on the growth of dwarf mutants. Phinney (1956) reported that GA applications restored some mutant lines of dwarf corn to standard or normal size (Fig. 5-14). The dwarfing habit was found to be associated with a very low endogenous level of GA, and the dwarfing habit in these particular lines can be attributed to a shortage of GA. Similar studies have been made of *Pharbitis* dwarfs; there, too, the dwarf plants have very low GA levels which appear to account for the dwarfing habit. A curious anomaly is some dwarf lines of pea which are highly responsive to added GA but apparently do not have naturally lower GA levels than standard lines (Jones and Lang, 1968). Kuraishi and Muir (1964*b*) have suggested that the GA role in stimulating growth of dwarf peas is through an enhancement of auxin levels. It should be noted that some dwarfing genes do not involve GA (McComb and McComb, 1970).

The other extreme of stem elongation is climbing or twining stems; since it is known that GA applications to bush beans can convert them to the climbing habit (Fig. 5-2), the climbing varieties may be inferred to contain markedly higher GA levels than bush varieties. The same situation is found for rosette habits; bolting can be induced with GA (Sachs et al., 1959), and natural bolting is associated with a rise in GA content (Lang, 1960). The growth stimulations achieved by GA appear to be attributable to stimulations of both cell division (Sachs et al., 1959) and cell enlargement (Haber and Luippold, 1960).

The regulatory roles of GA in plant development include nearly a complete listing of developmental functions in plants. Kahn et al. (1957) found that GA treatments can overcome the dormancy of some seeds, and subsequent workers have found that the emergence from dormancy correlates with an increase in natural GA levels (Smith and Rappaport, 1961). In rare instances GA applications can actually induce dormancy (Nagao and Mitsui, 1959). Paleg (1960) established that GA originating from the embryo is responsible for the hydrolysis of starch reserves in the endosperm during germination of grains; Loveys (1970) found another GA role in the germinating seedling, where it is involved in the unrolling of young leaves in response to light.

GA treatments can induce flowering of some photoperiod-sensitive and cold-requiring species (Lang et al., 1957), discussed in Chap. 12. Lang and Reinhard (1961) have subsequently shown that the induction of flowering can be correlated with a natural rise in GA content in some plants. In the LDP spinach there may be no rise in GA (Zeevaart, 1971). Inhibition of GA biosynthesis with growth retardants was found by Zeevaart and Lang (1962) to prevent flower induction under long photoperiods. A similar blockage of vernalization by growth retardants has been reported by Suge and Osada (1966). The regulation of the sex of flowers is also subject to GA control, with GA treatment ordinarily inducing male flowers (Mitchell and Wittwer, 1962) but in a few cases inducing female flowers instead (Shifriss, 1961).

The regulatory actions of GA in fruiting start with its ability to stimulate fruit-set in some species (Weaver, 1958). Gibberellins have been extracted from newly set fruits and from pollen (Coombe, 1960). Jackson and Coombe (1966) have found that the rate of fruit growth is in some instances under the control of endogenous GA levels. Fruit ripening, too, can be altered by GA applications (Coggins et al., 1960).

Another developmental function sometimes subject to GA control is leaf senescence. Fletcher and Osborne (1965) were the first to report deferral of leaf senescence with GA applications. In some instances, GA can have the opposite effect and actually enhance the development of leaf senescence (Halevy and Wittwer, 1965).

In short, GA can be envisioned as a hormone which can be involved in the regulation of growth and form of the plant and the entire range of developmental stages from dormancy to reproduction and senescence.

MECHANISM OF ACTION

In the light of the great diversity of GA effects on plants, it becomes a matter of some selection when one decides what function of GA should be used in a study of mechanism of action. The GA regulation of growth itself is involved with both cell-division effects (Sachs et al., 1958) and cell enlargement without cell division (Haber and Luippold, 1960; Haber et al., 1969).

When the Western world realized that GA is a major regulator of plant growth, the concept of auxin as the growth hormone was entrenched in physiological thought, and so close attention was given to the possibility that the GA effects are the result of some interaction with auxin. Mention has already been made of the finding that in dwarf peas, GA stimulations of growth are associated with sizable increases in auxin content of the tissues (Kuraishi and Muir, 1964*b*). The idea that GA might serve as a growth stimulant specifically through a stimulation of auxin content seems unlikely since so many GA effects are not obtainable with auxin at all. A specific example is the stimulation of α-amylase formation by barley endosperm, which can be inhibited by the antiauxin *p*-chlorophenoxyisobutyric acid but cannot be stimulated by any auxin tested, either in the presence of the antiauxin or in its absence (Cleland and McCombs, 1965).

Pilet (1957) reported that gibberellin suppresses the activity of IAA oxidase, but Kato and Katsumi (1959) noted no such suppressive effects. In a clever experiment, Kefford (1962) prevented IAA oxidase from acting and found that the growth response to gibberellin is not suppressed under such conditions. And again gibberellin can still promote growth when supraoptimal auxin concentrations are supplied (Hillman and Purves, 1961).

If auxin and GA systems are both operating

to regulate growth, surely they must exert their actions through alteration of somewhat different components of the growth system and one would expect that their effects would not always be equally limiting. Wright (1961) has made careful measurements of the growth responses of wheat coleoptiles to added GA, auxin, and cytokinins and has defined distinctive times of responsiveness to the three hormones. GA responsiveness appears first in the ontogeny of the coleoptile, followed by cytokinin responsiveness and then, during the period of maximal cell enlargement, auxin responsiveness. He suggests that the separate hormones regulate separate aspects of growth and that they are not limiting at the same time during the growth of the coleoptile.

The mechanism of GA action can be more precisely examined in simpler systems than those involving cell division and cell enlargement. The regulatory actions of GA on enzyme formation are attractive possibilities for mechanism studies since in such systems many fewer reactions must be involved. The conceptual model of a hormonal regulation of genetic information, determining whether a given set of enzymes will be synthesized or not synthesized, is brought to the forefront by the findings of GA regulations of enzyme formation. In barley endosperm, GA can stimulate the formation of α- and β-amylase, protease, and ribonuclease (Paleg, 1961; Srivastava and Meredith, 1962; Chrispeels and Varner, 1966); in leaves GA can stimulate the formation of RuDP carboxylase (Treharne and Stoddart, 1968) and nitrate reductase (Lips and Roth-Bejerano, 1969). In other tissues, GA can suppress the formation of specific enzymes; for example, GA can inhibit the formation of invertase or peroxidase in cut sugarcane-stem pieces (Glasziou et al., 1968), although the same enzymes may be stimulated by GA in cut tuber pieces (Edelman and Hall, 1964).

The ability of GA to turn some enzymes on and others off raises the question whether the regulation is an alteration of RNA-directed protein synthesis. Using isolated barley-aleurone cells, the metabolically active cells surrounding the endosperm, Varner and Chandra (1964) showed that the GA stimulation of α-amylase activity is readily inhibited by actinomycin D or p-fluorophenylalanine (Table 5-2). If the hormone serves to regulate the RNA synthesis, one would expect the greatest susceptibility to RNA-synthesis inhibitors like actinomycin D to occur in the first few hours after the hormone is added; the data in Table 5-2 show that after 7 h in GA, actinomycin D inhibition of amylase formation had almost passed whereas the protein-synthesis inhibitor was still highly effective. Even without added GA, the aleurone cells are competent to synthesize protein (Varner et al., 1965), but in the presence of GA large amounts of a new protein are synthesized (as indicated by incorporation of radioactive phenylalanine, Fig. 5-15) with the same column-elution characteristics as α-amylase. That the α-amylase is newly synthesized was conclusively ascertained (Filner and Varner, 1967) by supplying

Table 5-2 The stimulation of α-amylase formation in barley aleurone cells treated with GA is sensitive to inhibitors of RNA synthesis (actinomycin D) and inhibitors of protein synthesis (p-fluorophenylalanine); after 7 h in GA, the RNA-synthesis inhibitor no longer depresses amylase formation, but the protein-synthesis inhibitor still does (Varner and Chandra, 1964).

	Treatment	α-Amylase, μg
1	Control	13
2	1 μM GA	66
3	1 μM GA + 100 ppm actinomycin D	24
4	1 μM GA + 1 mM p-fluorophenylalanine	12
5	Treatment 3 after 7 h	55
6	Treatment 4 after 7 h	12

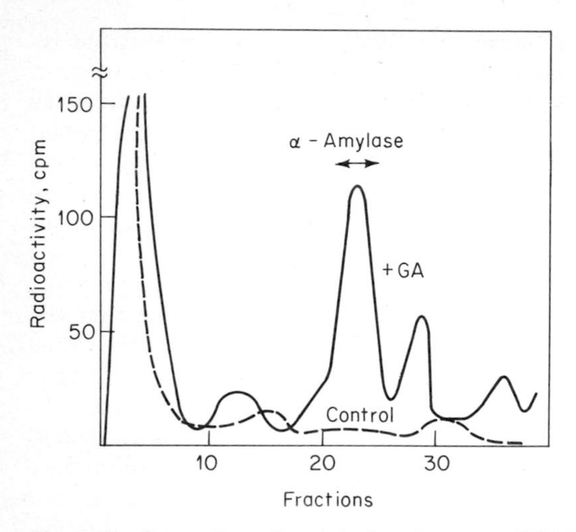

Fig. 5-15 Separation of protein fractions on a DEAE column of ^{14}C-phenylalanine–labeled barley aleurone, showing a contrast between control (*dashed line*) and GA treatment (*solid line*, with μM GA). The GA treatment has stimulated the synthesis of new protein peaks, the largest of which occurs in the same fraction as α-amylase (Varner, 1964).

amino acids in which the carboxyl group had been labeled with heavy oxygen, ^{18}O, during the GA stimulation and showing that the amylase formed in response to the GA has a greater density, as expected if it is assembled out of the heavy-isotope-labeled substrates (Fig. 5-16).

The GA stimulus, then, may involve the formation of a new RNA (a new mRNA?), and definitely involves the *de novo* synthesis of the enzyme in question. In addition to this level of hormonal control, there was evidence from early stages in this work that the hormone had yet another regulatory influence, regulation of the release of enzymes from the aleurone cells. This release is illustrated by the data in Fig. 5-17, where the first appearance of α-amylase occurs on the aleurone particles, and later large amounts of enzyme appear in the medium itself. Similar data can be seen for ribonuclease in the barley aleurone (Chrispeels and Varner, 1967). An analogous hormonal regulation of enzyme release has been reported for the auxin stimulation of cellulase in bean petioles (Abeles and Leather, 1971). The release may involve a loss of binding of the enzyme from some surface or, more probably, a secretion of the enzyme out of the cells.

With the conceptual model of GA serving to regulate RNA-directed protein synthesis, two questions about the nature of this control arise: Does it involve an alteration of the DNA? Is the hormonal regulation exerted in the nucleus, where much of the DNA is synthesized? Nitsan and Lang (1965) report that inhibitors of DNA synthesis such as 5-fluorodeoxyuridine can inhibit GA stimulations of growth, and in lentil epicotyls they found evidence of GA enhancement of DNA synthesis. But Haber et al. (1969) prevented cell divisions in lettuce by treating the seeds with heavy radiation doses, and in the seedlings from this material they obtained GA stimulations of growth without cell division, without DNA doubling in existing cells, and

Fig. 5-16 The GA stimulation of α-amylase synthesis in barley-aleurone cells, carried out in the presence of amino acids labeled with ^{18}O, results in enzyme with slightly greater weight, as shown by distribution on a CsCl density gradient. Carrier or normal α-amylase was marked with tritium for easy identification. The incorporation of the heavy amino acids into the enzyme conclusively establishes that the hormone brought about the *de novo* synthesis of this enzyme (Filner and Varner, 1967).

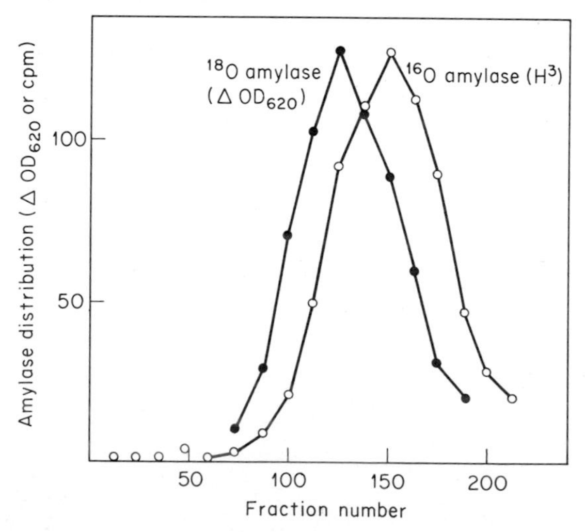

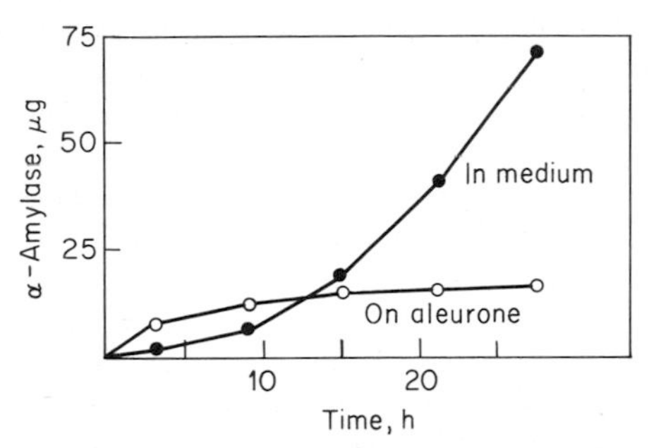

Fig. 5-17 Following the addition of GA to barley aleurone cells, α-amylase appears first associated with the aleurone cells; after several hours, the enzyme begins to accumulate in a soluble form in the medium (Varner et al., 1965).

without detectable DNA synthesis. Thus, with respect to the first question, it seems doubtful that DNA synthesis is ordinarily involved in GA regulations. The possible action of GA in isolated nuclei was examined by Johri and Varner (1968), using the nuclei of dwarf peas; they detected marked increases in RNA synthesis in the nuclei when GA was present during the isolation procedure (Fig. 5-18). Experiments designed to examine the component of RNA synthesis that might be regulated by GA produced data illustrated in Fig. 5-19, where a time sequence seems evident. GA application to hazel seeds led first to an increase in template activity of the DNA, followed some 15 or 20 h later by an increase in RNA polymerase activity, and then an actual rise in RNA synthesis. These data imply that the hormone may unveil some sectors of the chromatin DNA, making it available for RNA synthesis, and it may also increase the activity of RNA polymerase. These effects should lead to the formation of new enzymes (cf. Figs. 5-15 and 5-16) and enhanced RNA synthesis (Fig. 5-18).

The concept of GA altering the readout from the DNA, hence new RNA and new enzymes, finds some encouragement from the experiments of Grieshaber and Fellenberg (1972), who detected differences in histone binding patterns onto DNA after GA treatment. But if new species of RNA were being assembled, one might expect to see differences in the binding of the newly formed RNA onto chromatin in competition experiments with the RNA of non-GA-treated material; Thompson and Cleland (1972), however, could not find evidence of such an alteration in the RNA of peas.

The time sequence involved in GA regulation of enzyme formation is a challenging one, as it was in auxin stimulations. Some careful timing measurements by Pollard and Singh (1968) indicate that the barley-aleurone cell responds to GA by an increased amylase activity in 4 h and an increased RNA synthesis in 6 h. But if the GA effect were initially on RNA synthesis, one would certainly have expected that parameter to show the first rise, followed by protein synthesis and increased amylase. The GA stimulation of growth in peas can be measured 25 min after GA application (Warner and Leopold, 1969). It would seem possible that the earliest aleurone responses to GA might be due to the release of preformed amylase and that the response after 6 h may be due to the newly synthesized enzyme. The rapid action of GA on pea growth may be

Fig. 5-18 The application of GA during the isolation of nuclei from pea seedlings results in an increased RNA synthesis in the nuclei, as indicated by the incorporation of ^{14}C-cytidine triphosphate (CTP) (Johri and Varner, 1968).

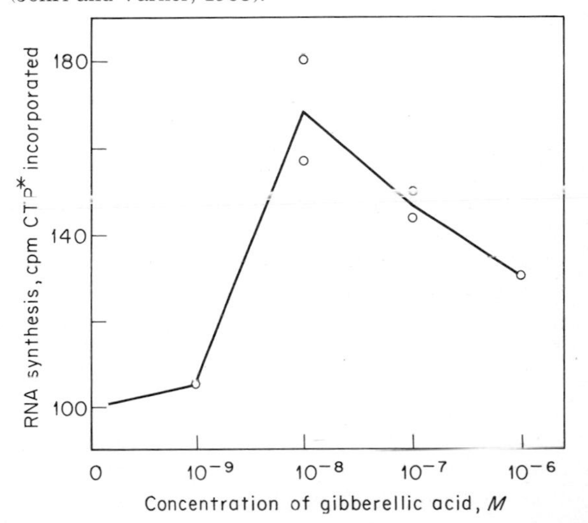

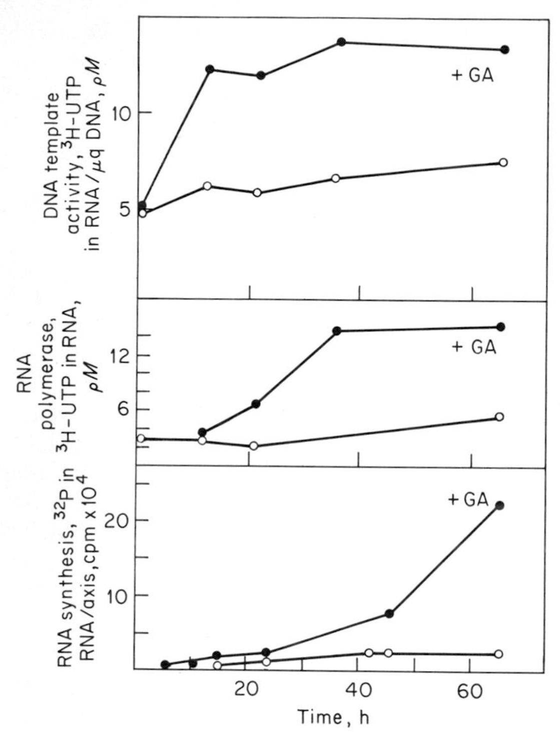

Fig. 5-19 The application of GA to hazel seeds during germination results in an early increase in DNA template activity (*above*) and somewhat later in a rise in RNA polymerase activity (*middle*) and actual RNA synthesis (*below*). Template activity was measured as ^{3}H-uridine triphosphate incorporation into RNA in the presence of chromatin and an excess of RNA polymerase. RNA synthesis was measured as $^{32}PO_4$ incorporation (Jarvis et al., 1968).

too rapid for a nucleic acid regulatory action (cf. Evans and Ray, 1969).

The several lines of evidence about timing of the various components of GA response make it attractive to suppose that the first actions of GA may relate to preformed systems rather than to the formation of new species of RNA and hence new enzymes. In the barley-aleurone response to GA, Jones (1969) found that an increase in endoplasmic reticulum is evident sooner than the appearance of the α-amylase. Evins (1971) detected membrane changes in the aleurone cells within 1 or 2 h after GA treatment. The interest in the possible involvement of membranes in the first GA responses has increased further since several components of membrane structures, e.g., phospholipids and lecithins, show an enhanced rate of synthesis considerably before the first amylase is formed (Johnson and Kende, 1971; Evins and Varner, 1971). The alteration of membrane-permeability functions by animal steroid hormones was an attractive analogy to Wood and Paleg (1972), who constructed synthetic membranes with some phospholipids, lecithins, etc., and observed some striking increases in permeability of their model membranes upon addition of GA.

As with auxin, respiratory metabolism seems to be essential for the GA hormonal action, but there is no evidence of an important alteration of metabolic activity which can be interpreted as evidence that the hormone action is via an effect upon a metabolic-system component. In the barley system, aerobic respiration is essential for the RNA and protein synthesis (Varner, 1964), but no respiratory alteration is evident from the hormonal treatment (see also Palmer, 1966).

Using another analogy with animal hormone systems as a basis for experimental design, Galsky and Lippincott (1969) discovered that cyclic AMP can stimulate α-amylase formation in the barley-aleurone test in a manner similar to GA, though at markedly higher concentrations. Other adenosine compounds did not have activity. The lettuce-hypocotyl test also shows some activity for cyclic AMP (Kamisake et al., 1972). The effects of cyclic AMP are inhibited by ABA, cycloheximide, and 6-methylpurine, as the GA effects are (Gilbert and Galsky, 1972). The difficulties of assuming a cyclic AMP hormonal effector were mentioned at the end of Chap. 4.

SUMMARY

In retrospect, the mechanisms of GA action, even in such relatively simple systems as the barley-aleurone responses, seem to be too complex to unravel at present. It is clear that the hormone provokes the *de novo* synthesis of new enzymes and that this is associated with stimulations of RNA synthesis; but there are earlier effects of the hormone, which appear to involve alterations in the formation of some membranes and membrane components in the cell. The hormone must be present continuously for the regulatory action to proceed (Chrispeels and Varner, 1967), and this implies that the hormone may become attached to some site of action through a relatively facile attachment-detachment mechanism (such as adsorption, rather than a covalent bonding). As with auxin, GA appears to involve an alteration of nucleic acid-directed protein synthesis in some longer-term regulatory actions but also to involve some other type of activation phenomenon in short-term regulatory actions.

GENERAL REFERENCES

Jones, R. L. 1973. Gibberellins: Their physiological role. *Annu. Rev. Plant Physiol.*, 24:571–598.

Lang, A. 1970. Gibberellins: Structure and metabolism. *Annu. Rev. Plant Physiol.*, 21:537–570.

Paleg, L. G. and G. A. West. 1972. The gibberellins. In F. C. Steward (*ed.*), Plant Physiology, a Treatise." Academic Press, New York, 6B:146–180.

West, C. A. 1973. Biosynthesis of gibberellins. In B. V. Milborrow (*ed.*), "Biosynthesis and its Control in Plants" Academic Press, London, pp. 143–177.

6 CYTOKINİNS

In the discussion of the auxin and gibberellin growth substances, cell enlargement as the basic unit of growth was emphasized. The logical need for a control system for cell division led Wiesner (1892) to suggest long ago that there might be such a chemical regulator in plants. In a startling projection into experimental approaches which were to follow 40 years later, Haberlandt (1913) found that diffusates from phloem tissues could induce cell division in potato parenchyma. Crushed cells yielded a material which provoked cell division around a wound, and rinsing the region of the crushed cells removed the stimulus (Haberlandt, 1921). Using modern tissue-culture techniques, Jablonski and Skoog (1954) essentially repeated Haberlandt's experiment. They added to the evidence for a substance regulating division by showing that a piece of vascular tissue cultured on top of pith tissue can bring about division in the pith cells, which otherwise do not divide. In 1955, Miller et al. separated from herring sperm DNA, the first known stimulant of cell division, which they named *kinetin* and identified as 6-furfurylaminopurine. Later work revealed that kinetin is probably not a natural component of plants. The first cytokinin hormone in plants was identified by Letham et al. (1964) from corn seeds; it is named *zeatin*, and has the structure shown in Fig. 6-1. Several other purine cytokinins subsequently isolated from plant materials are shown in the figure. The ability of representative cytokinins to stimulate cell division is illustrated in Fig. 6-2.

The generic name *cytokinin* is used for chemical substances which can stimulate cell division, or cytokinesis, in the manner of zeatin. Earlier use of the term *kinin* led to some confusion, as the same term is employed by animal physiologists to refer to a class of polypeptides in stings and venoms which irritate smooth muscles and nerve endings (cf. Collier, 1962).

The cytokinins began to receive attention in the physiological literature in the 1940s. Van Overbeek et al. (1942) found that coconut milk contains some material which greatly promotes the growth of embryos in tissue cultures. In Steward's laboratory a refined method for studying the requirements for growth substances in

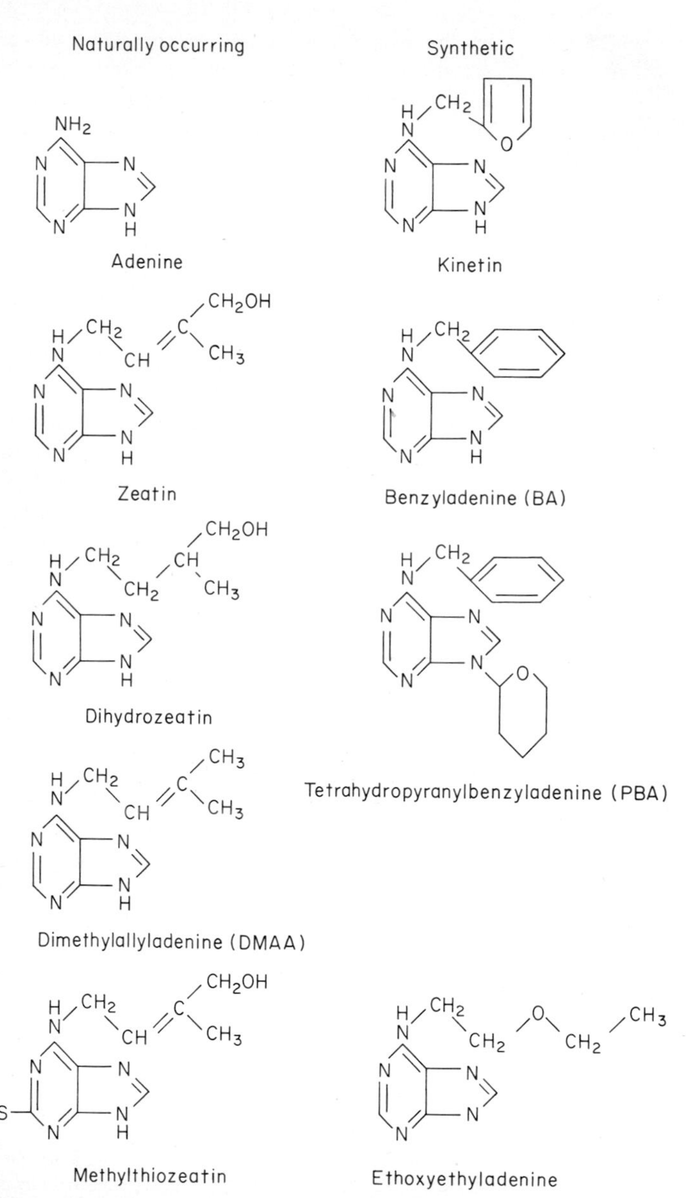

Fig. 6-1 Structures of some naturally occurring and synthetic purine cytokinins.

cultures was developed using carrot-root tissue (Steward and Caplin, 1952) and made possible the analysis of the components of coconut milk and other sources of cytokinin.

While cytokinins are considered as hormones which regulate cell-division activities, their role in the plant has gradually been recognized as being so widespread that it includes some aspects of every part of growth and development. Far from being cell-division hormones in the strict sense, cytokinins participate in cell enlargement, tissue differentiation, correlation effects, dormancy, several phases of flowering and fruiting, and the regulation of senescence (cf. Table 18-1). Thus the theme is repeated that a substance isolated from plant cells because of its effectiveness in a relatively precise growth phenomenon is found to be involved in nearly all aspects of plant growth and development.

OCCURRENCE

The earliest assays for cytokinins utilized the hormonal stimulation of cell division as their basis; the most widely used assays of this sort are the stimulation of growth in tobacco pith-tissue culture (Skoog and Miller, 1957), shown in Fig. 6-2, and the somewhat simpler soybean-cotyledon assay (Miller, 1963). In each case, the increase in fresh weight of the tissue in culture is taken as a measure of cell divisions (Miller, 1961). Many other biological activities of cytokinins can be utilized in bioassays; these especially include the cytokinin retardation of leaf senescence and stimulation of cotyledon expansion. After Richmond and Lang (1957)

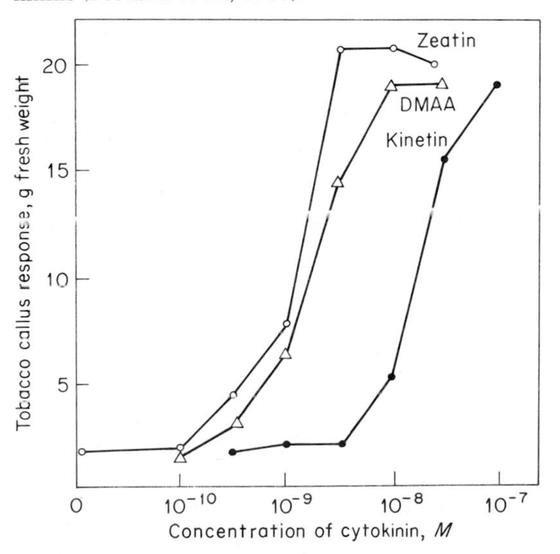

Fig. 6-2 The stimulation of tobacco callus-tissue growth (cell division) by three representative cytokinins (Leonard et al., 1968).

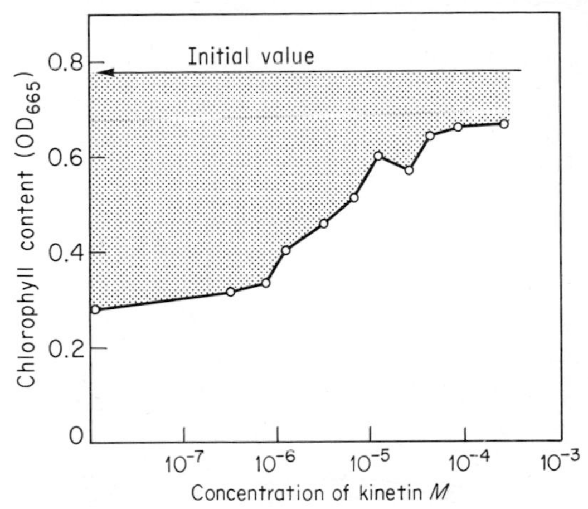

Fig. 6-3 The ability of cytokinins to suppress the loss of chlorophyll from darkened disks of mature green leaves has been used as an assay. *Xanthium* leaf disks are floated on kinetin solutions for 48 h, and then the chlorophyll is extracted with 80% ethanol and measured at 665 nm. The initial chlorophyll content before the disks were placed in darkness is indicated by the arrow (Osborne and McCalla, 1961).

found that cytokinins applied to *Xanthium* leaves are very effective in deferring senescence, Osborne and McCalla (1961) devised a procedure for quantitatively following the effect as a bioassay for cytokinins. Representative responses of this test, as measurements of chlorophyll remaining in the leaf after a period of 48 h in the dark, are shown in Fig. 6-3. The striking ability of cytokinins to stimulate cotyledon enlargement, first discovered by Ikuma and Thimann (1963), was later utilized as a basis for assay using radish cotyledons (Letham, 1968) or *Xanthium* cotyledons (Esashi and Leopold, 1969a). Representative results from such an assay are shown in Fig. 6-4; an optimum cytokinin level can produce a fourfold increase in fresh weight over the control.

With these various bioassays, rather extensive arrays of cytokinins have been identified in plants, some of which are shown in Fig. 6-1. The natural cytokinins in plants are ordinarily amino purines. Adenine, the parent compound, was first shown to be active in the stimulation of bud differentiation in tissue culture (Skoog and Tsui,

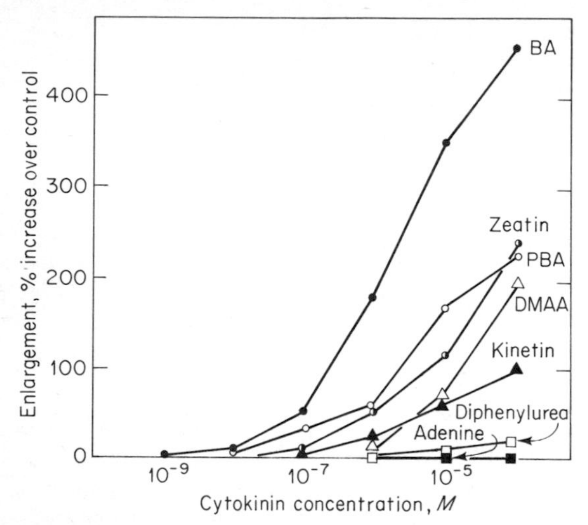

Fig. 6-4 A comparison of stimulations of *Xanthium* cotyledon enlargement by seven purine cytokinins, including benzyladenine (BA), zeatin, tetrahydropyranylbenzyladenine (PBA), α, α' -dimethylallyadenine (DMAA), kinetin, diphenylurea, and adenine (Esashi and Leopold, 1969*a*).

1948); its activity as a cytokinin is ordinarily small or even absent in some tests (Fig. 6-4). However, it is a natural component in plants and does have some activity in the tobacco and soybean tissue-culture tests (Miller, 1968). Zeatin, with the isoprene side chain (Fig. 6-1), is the most active naturally occurring cytokinin known to the present. After it was identified in extracts from corn (Letham et al., 1964), it was found in numerous higher plants, e.g., *Begonia, Bryophyllum,* chicory, spinach, pea, and sunflower (cf. Fox, 1969). The readiness of purines to form the riboside and ribotide derivatives would lead one to expect these derivatives of zeatin to occur in plants, and they, too, are apparently common there (Miller, 1965).

Most of the cytokinins that have been isolated from higher plants have been obtained by hydrolysis of sRNA extracts, the fractions principally composed of tRNA. The existence of cytokinins as structural components of serine tRNA and tyrosine tRNA was reported by Hall et al. (1966) and Biemann et al. (1966) from work on the tRNA of yeast. γ, γ-Dimethylallyladenine (DMAA), an analog of zeatin lacking the hydroxyl group (Fig. 6-1), was found in yeast tRNA (Biemann et al., 1966) and then also in higher-plant sRNA (Hall et al., 1967). Other analogs of zeatin isolated from higher-plant sRNA include dihydrozeatin, isolated from lupine seeds (Kushimitzu et al., 1967); the *cis* isomer of zeatin, isolated from corn sRNA (Hall, 1967); and the methylthio derivative (Fig. 6-1), found in the tRNA of wheat (Hecht et al., 1969).

During investigations of natural cytokinins, it has often been observed that as greater purity is achieved, activity is lowered or lost. In several instances this appears to be due to the interactions of cytokinins with other limiting compounds, and without the latter, cytokinin activity may sometimes not occur. The most frequent cofactor, if the term may be used in this fashion, is *myo*-inositol, which occurs in the cytokinin system of coconut milk (Steward and Shantz, 1956) and corn (Letham, 1963*b*) and can be a necessary component in some cytokinin functions, e.g., the regulation of root thickening in radish (Loomis and Torrey, 1964).

DISTRIBUTION

Information on the amounts of cytokinins that normally occur in plants and how the levels change during growth and development is surprisingly scarce. Early work suggested that the levels of cell-division factors are high in young developing fruits and seeds; the isolation of zeatin was done from extracts of developing fruits of plum and later of corn. Some data of Miller (1967) suggest that cytokinin levels rise during the early development of corn kernels and then drop again as the fruits mature. Embryos and young fruits have been the commonest sources of cytokinins in plant extracts (Steward and Shantz, 1956). Roots (Weiss and Vaadia, 1965) and especially root exudates (Kende, 1965) are fairly rich in cytokinins. In pea roots there is a striking concentration of cytokinins in the first mm of the root tip (Short and Torrey, 1972).

If the xylem sap contains substantial amounts

of this hormone, one would expect that roots might be a major source of cytokinins for the plant and that variations in the synthesis or supply of the hormone from roots might have developmental significance. Evidence that in fact the supply from the roots can respond to developmental changes in the plant was produced by Sitton et al. (1967). If cytokinins defer senescence, the normal development of senescence might be expected to be correlated with a depression of cytokinin supplied by the roots. Sitton et al. found that this is in fact the case with sunflower plants, and they noted that the cytokinin produced in the fruit do not have such strong systemic effects on the plant as that supplied by the root. Environmental stress can likewise be reflected in a depression in the cytokinin levels in xylem sap (Itai and Vaadia, 1971). Leaves, too, yield cytokinins upon extraction (Heide and Skoog, 1967).

In general, it appears that cytokinins are widespread in plants, not only as components of tRNA but as free hormonal substances. They are perhaps most abundant in embryos and developing fruits. In terms of systemic influences in the plant, it seems probable that the supply coming to the shoot from the root has the greatest promise of regulating plant performance in terms of stages of plant development and the responses of the plant to changes in the environment.

TRANSLOCATION

Our knowledge of cytokinin movement in the plant is full of conflicts. We know, first, that applications of cytokinins to leaves or stem pieces have very localized effects on the release of buds from apical dominance (Thimann and Wickson, 1958) or on the directed translocation of solutes (Mothes, 1961). Such experiments with applied materials imply that the hormone may be very immobile in the plant. Second, the occurrence of cytokinins in the xylem sap, especially in the exudate from severed root systems (Kende, 1965), implies that the hormone may move freely with the xylem sap. Third, the movement of benzyladenine through petiole sections has been reported to be polar, like auxin transport (Osborne and Black, 1964). How can such diverse findings be reconciled?

The immobility of cytokinins in leaves is consistent in many studies of the subject. Likewise, the cytokinins produced in fruits show no apparent mobility; so they, too, may be presumed to have localized effects (Sitton et al., 1967). The dynamic actions of roots in regulating the growth of the above-ground parts of plants can be generally attributed to the gibberellins and the cytokinins in the xylem sap (Carr and Reid, 1968). The ability of cytokinins to move upward in the sap has been confirmed with experiments involving injections of cytokinins into stems and measuring the resultant systemic effects (Pieniazek and Jankiewicz, 1965; Guern et al., 1968). The evidence for a polar transport of cytokinins is not obtained by all workers (Fox and Weis, 1965), and it seems likely that the small amounts of labeled benzyladenine (or its derivatives) observed to move in a polar manner through petiole sections may have been mobilized there by cell-division activities at the base of the section, an activity observed in the same type of plant material by Zaerr and Mitchell (1967).

Seth et al. (1966) have reported an enhancement of kinetin movement in bean stems by the addition of auxin.

BIOSYNTHESIS

The biosynthesis of cytokinins of the purine type may be assumed to occur via the substitution of the side chain onto the common plant constituent adenine. The five-carbon side chain of most cytokinins suggests, of course, that it may come from an isoprene source, and this suggestion was supported by Peterkovsky (1968), who was able to show that radioactivity from mevalonate is incorporated into the tRNA of *Lactobacillus* and that the label is probably in the dimethylallyl side chain. Chen and Hall (1969) have worked on the same question using tobacco-tissue cultures and found excellent incorporation of mevalonate into DMAA. Further, they found an enzyme preparation which would actually introduce the dimethylallyl side chain into the tRNA. It should be noted that the formation of this cytokinin within the tRNA may or may not relate to the biosynthesis of the free cytokinin hormone.

Using *Rhizopus* cultures, Miura and Miller (1969) have found that DMAA can be converted in that organism into zeatin; they infer then that zeatin may be a product of the oxidation of DMAA.

METABOLISM

We have very little information about a cytokinin disposal system. Some early evidence of Loeffler and van Overbeek (1964) indicates that there may be a glycoside formed in the plant, or some other product which would release a cytokinin upon acid hydrolysis. The glycoside of benzyladenine (BA) has subsequently been identified (Deleuze et al., 1972) and so also the 7-glucoside of zeatin (Parker and Letham, 1973). The oxidation of purine rings by xanthine oxidase can apply to the 6-substituted adenines (Henderson et al., 1962), and the products of the oxidation may serve as inhibitors of xanthine oxidase. The riboside seems to be more susceptible to xanthine oxidase action than the free purine or those variously substituted in the 9 position (Guern et al., 1968). Radioactive BA is broken down into uric acid and allantoin or allantoic acid (McCalla et al., 1962). A small amount of the label from cytokinins such as BA is fixed into the RNA (Wollgiehn, 1965; Fox, 1966).

Perhaps the most labile part of the purine cytokinin molecule is the 6-amino bond to the side chain. The side chain is easily removed by photooxidation to yield adenine and the side-chain fragment (Rothwell and Wright, 1967). A similar cleavage of the molecule can occur in the presence of acid (Leonard et al., 1968).

After the side chain of a cytokinin is removed, further metabolism might be expected to proceed in either of two directions: the purine ring may be successively oxidized by xanthine oxidase to produce uric acid and then urea, as shown in Fig. 6-5, or carbon atom 8 may be oxidized away to leave a skeleton like triaminopyrimidine. Schlee et al. (1966) expect this carbon to be added to glycine to form serine, which may be utilized again in the biosynthesis of adenine. Mention has been made of the inhibition of xanthine oxidase by the products of cytokinin oxidation (Henderson et al., 1962). Schlee et al. (1966) further show that cytokinins can inhibit the initial deamination of adenine, as illustrated in Fig. 6-6. Altogether, this evidence implies that cytokinins may exert a braking action on their own oxidation. The contrast of this situation with the self-stimulated oxidation of IAA acid seems to suggest that the metabolic removal of cytokinins may not be a very critical part of the hormonal control mechanism.

SYNTHETIC CYTOKININS

Since all the naturally occurring purine cytokinins have side chains of five carbons, one might expect an aliphatic side chain of five carbons to be optimal for cytokinin activity. Tests of the types of substitutions onto purines have revealed that the five-carbon side chain is optimal, but nevertheless there is a wide range of side-chain possibilities. For example, the length of an aliphatic side chain may be increased from the optimal five-carbon length to as much as ten carbons and still show activity, though declining with increasing numbers of carbons (Skoog et al., 1967). Introduction of an oxygen in the side chain, as in ethoxyethyladenine (Fig. 6-1), does not lessen cytokinin activity (Rothwell and Wright, 1967). An unsaturated bond in the side chain also enhances activity (Skoog and Leonard, 1968). Among the synthetic purine cytokinins, those substituted with a benzyl ring can be more active in some assays than even the most active natural cytokinin, zeatin; BA is the most outstanding of these (see Fig. 6-4).

Alterations of the purine ring lower the cytokinin activity. For example, the riboside and ribotide derivatives (Fig. 6-5) are invariably less active than the adenine derivatives (Leonard et al., 1968). Methyl or amino groups in the purine ring lower activity (Leonard et al., 1966), and hydroxy substitutions eliminate activity (van Eyk and Veldstra, 1966). A naturally occurring

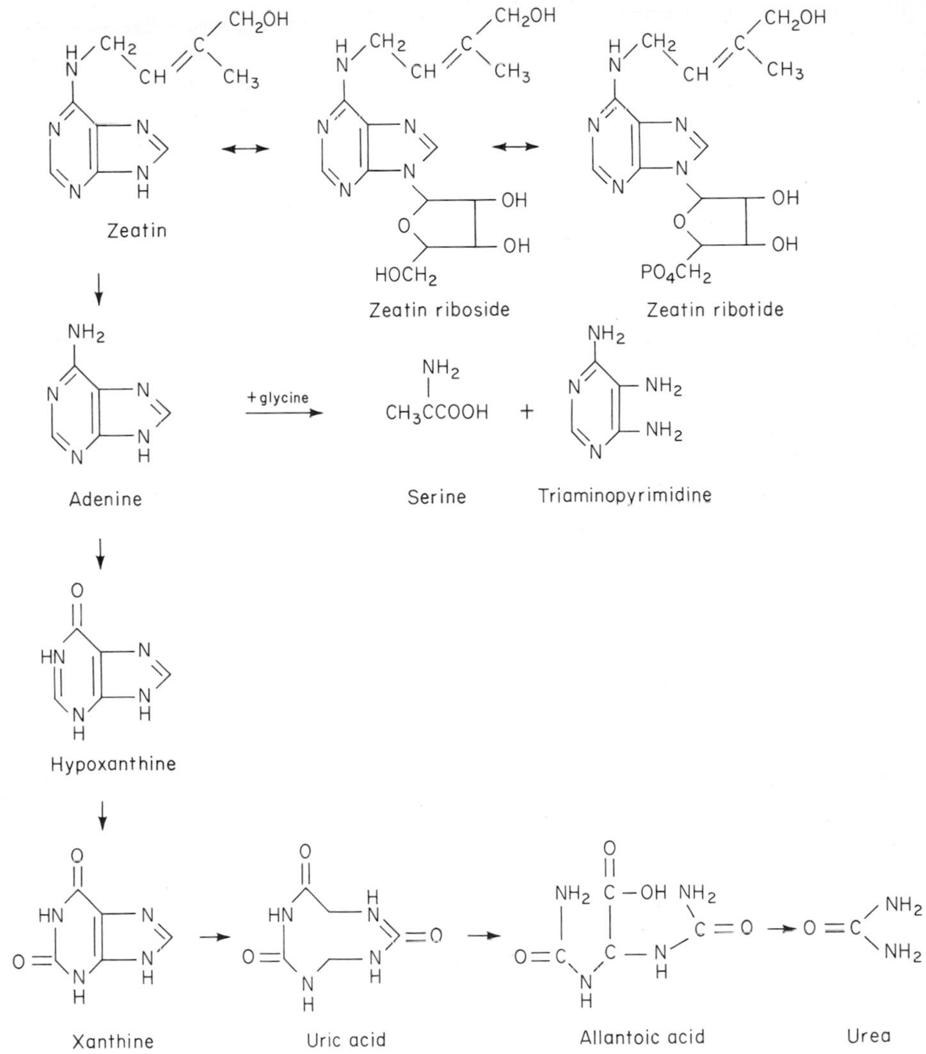

Zeatin ↔ Zeatin riboside ↔ Zeatin ribotide

↓

Adenine $\xrightarrow{+glycine}$ Serine + Triaminopyrimidine

↓

Hypoxanthine

↓

Xanthine → Uric acid → Allantoic acid → Urea

Fig. 6-5 Some reactions in the metabolism of a cytokinin. (*Adapted from Schlee et al., 1966.*)

alkaloid in which a side chain similar to that of DMAA is located in the 3 instead of the 6 position of the purine has been reported to have cytokinin activity (Hamzi and Skoog, 1964).

The structure of the purine ring can be altered by replacement of two nitrogens with carbons and still have activity in several assay systems, as in benzylaminobenzimidazole (Kuraishi and Yamaki, 1967) (Fig. 6-7). Another departure from the basic cytokinin structure is benzthiazolyloxyacetic acid, an active cytokinin. A more drastic alteration of the purine structure to 3-methylpyrazolpyrimidine yields compounds which are impressive cytokinin antagonists (Skoog et al., 1973). A further departure from the purine cytokinins are the phenyl ureas and related biurets (Fig. 6-7). Shantz and Steward (1955) first reported diphenylurea as a cytokinin,

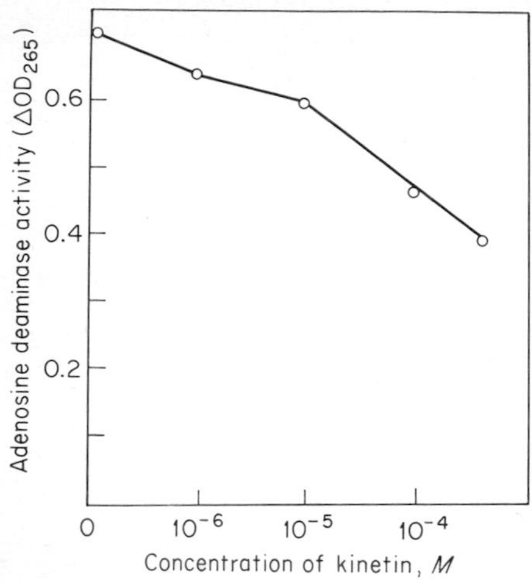

Fig. 6-6 Kinetin can inhibit the activity of adenosine deaminase, the first step in the metabolism of a cytokinin as represented in Fig. 6-5 (Schlee et al., 1966). Adenosine substrate at 4 m*M; Pelargonium* leaves.

but their finding was not pursued further until much later, when Bruce and Zwar (1966) found that out of about 500 derivatives of this compound, nearly half showed some cytokinin activity. Chlorophenyl-phenylurea was one of their most active compounds. Extension of the urea moiety to a biuret, as in fluorophenylbiuret, again gives high activity (Mitchell, 1971).

The astonishing range of chemical substances which can show cytokinin activity might make one wary about deciding whether the actions of the various chemicals are really pertinent to the biological activity which we associate with the purine cytokinins (Fox, 1969). The wide range of cytokinin structures may be attributable in part to the fact that diverse bioassays have often been used to establish the degree of activity. Perhaps the most definitive evidence to indicate that such diverse compounds as the phenyl ureas are really acting on the same biological site as the purines is the work of Kefford et al. (1968), who examined the antagonism of phenyl ureas by benzyl ureas. The extra carbon between the phenyl ring and the urea moiety converts the compound into an effective antagonist of cytokinins in the tobacco assay, the carrot assay, the soybean assay, leaf-senescence tests, and bud-growth tests. Most relevant here is the observation by Kefford et al. that the benzylurea antagonism applies equally well to the purine cytokinins and the phenylurea cytokinins. As with the auxins, we must deduce that remarkably different types of molecules can serve to regulate the quite specific hormonal functions of cytokinins in plants.

REGULATORY EFFECTS

Like the other growth substances in plants, cytokinins have a breathtaking range of regulatory effects, including effects on growth, differentiation, and the various stages of plant development (Table 18-1).

Turning first to the effects on growth, we see that cytokinins can serve to stimulate the doubling of nuclear DNA, a property it shares with some other regulators (Patau et al., 1957), as well as mitosis and cytokinesis (Das et al., 1956). Cytokinin effects on cell division are illustrated in Fig. 6-2. While the stimulations of DNA doubling and mitosis are shared with auxin and gibberellin, the ability to stimulate cytokinesis is thought to be exclusive with cytokinins (Torrey, 1961). The cytokinin effects on growth are quite distinctive, for in contrast to the promotions of stem growth by auxins and gibberellins, cytokinins ordinarily inhibit elongation of stem sections (Fox, 1964) and stimulate leaf enlargement (Miller, 1956). Cytokinins may also stimulate growth by swelling rather than elongation (Katsumi, 1962), an effect which may participate in the formation of such natural swellings as the radish root (Loomis and Torrey, 1964). Root growth is ordinarily inhibited by cytokinins (Gaspar and Xhaufflaire, 1967).

The differentiation of buds was one of the first properties of cytokinins discovered (Skoog and Tsui, 1948; Skoog and Miller, 1957). Buds can be induced in callus tissue, leaves, roots, cotyledons, or stem pieces (Miller, 1961), and the tendency for auxin to inhibit this cytokinin action ordinarily imposes a requirement for a balance between the auxin and cytokinin levels to achieve the differentiation of buds (Skoog and Tsui,

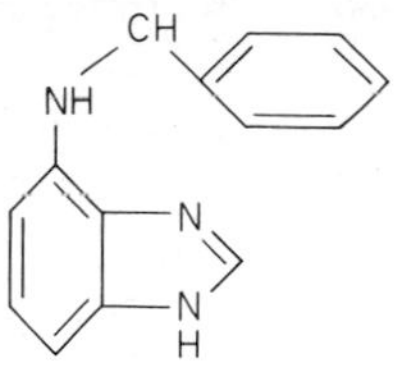

Benzylaminobenzimidazole

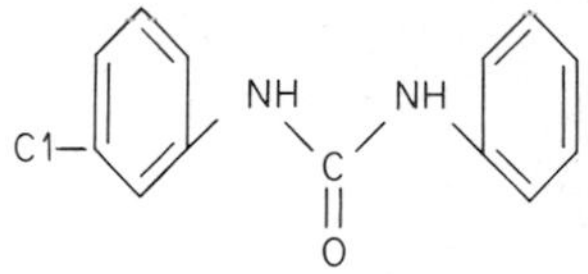

Chlorophenylphenylurea

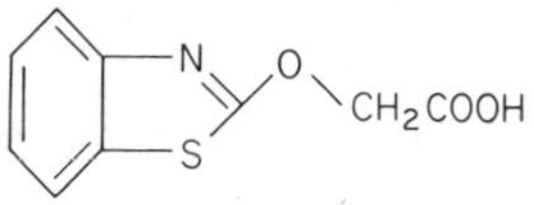

Benzthiazolyloxyacetic acid

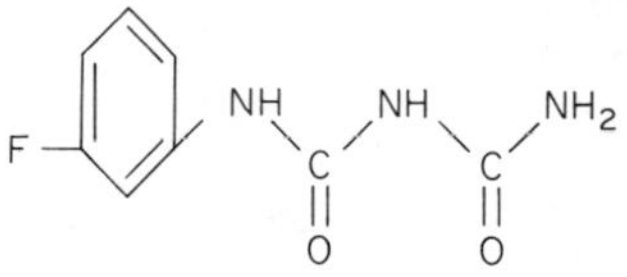

Fluorophenylbiuret

Fig. 6-7 Some synthetic compounds with good cytokinin activity lack the purine structure.

1948; Fox, 1964). The sensitivity of bud differentiation in tissue cultures is illustrated in Fig. 6-8.

Another control of differentiation exerted by cytokinins is an effect on chloroplast development. Cultures of tobacco callus become green in the light if cytokinin is supplied, and Stetler and Laetsch (1965) have found that the conversion of proplastids into chloroplasts with grana is specifically stimulated by cytokinin.

Effects of cytokinins on the various stages of plant development are manifold. Applications of this hormone can stimulate germination (Miller, 1956) and even break seed dormancy (Khan, 1966). At least a part of this effect is probably the stimulation of cotyledon growth illustrated in Fig. 6-4. Flowering, too, can be enhanced in some species by cytokinin (Nakayama et al., 1962) or even induced in photoperiod-requiring species (Michniewicz and Kamienska, 1965; Maheshwari and Venkataraman, 1966). The subsequent stages of sex expression (Negi and Olmo, 1966), fruit-set (Crane and van Overbeek, 1965), and fruit growth (Weaver and van Overbeek, 1963) are all susceptible to alteration in at least some instances by cytokinins.

One of the most studied regulatory actions of cytokinins is the delay of leaf senescence, discovered by Richmond and Lang (1957). The data in Fig. 6-3 illustrate this activity. The discovery of this action was soon followed by the finding (Mothes, 1961) that localized applications of cytokinins to leaves can lead to the mobilization of various nutrients. The mobilization effect, illustrated in Fig. 6-9, probably is a major function not only in the regulation of leaf senescence

Fig. 6-8 The stimulatory action of a cytokinin on bud differentiation in root cultures of *Isatis tinctoria* (Danckwardt-Lilliestrom, 1957).

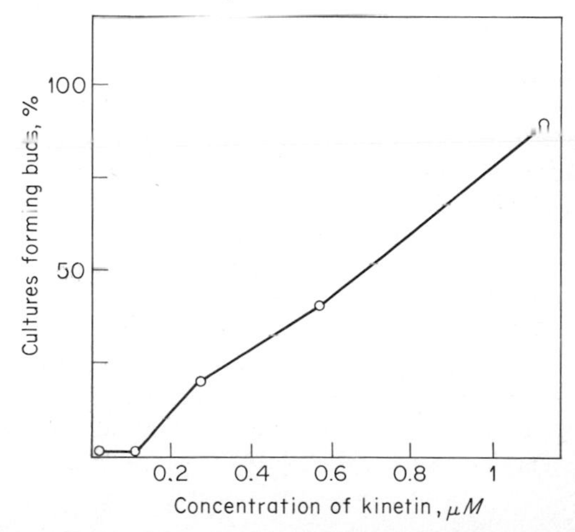

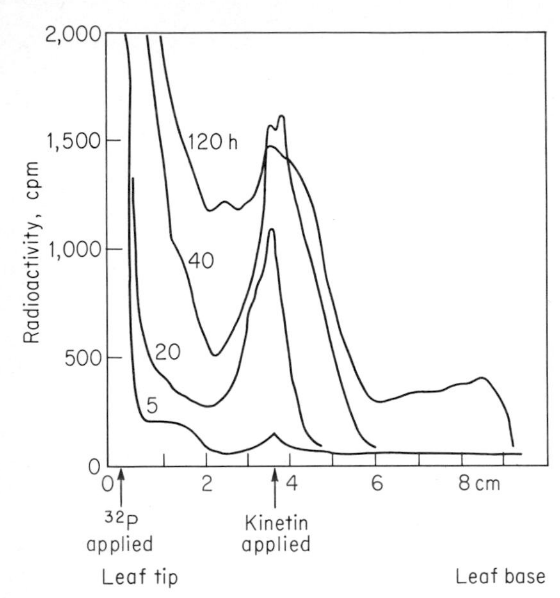

Fig. 6-9 A localized application of kinetin (30 mg/l) to corn leaves results in a mobilization of radioactivity from $^{32}PO_4$ applied near the leaf tip. Scanning the same leaf at intervals reveals the accumulation at the position of kinetin application (Müller and Leopold, 1966).

but also in the cytokinin regulation of bud growth (Thimann and Wickson, 1958) and in the regulation of apical dominance and the growth of fruits (Weaver and van Overbeek, 1963).

CYTOKININS IN tRNA

Among the hormones, cytokinins have a unique characteristic in that they can be a structural component of RNA. Finding adenine nucleotides in the serine and tyrosine tRNA of yeast, with side chains exactly like those of the most active cytokinins, aroused great interest (Biemann et al., 1966; Hall et al., 1966). In each case, the substituted adenine was located immediately adjacent to the anticodon, as shown in Fig. 6-10, suggesting that the cytokinin in the tRNA may serve some regulatory function in fitting the charged tRNA onto the messenger in protein synthesis. An intensive examination of tRNA fractions in higher plants was touched off, and hydrolyzed RNA fractions yielded cytokinin activity specifically from the tRNA fractions, as illustrated in Fig. 6-11. Although the first cytokinin found in tRNA was DMAA, it soon became evident that many higher plants contain zeatin (Hall et al., 1967) or the methylthio derivative of zeatin shown in Fig. 6-1 in their tRNA (Hecht et al., 1969). These cytokinin components appear to be distinctive to serine and tyrosine tRNA.

Theoretically there might be as many as six different codons in mRNA for any given amino acid, and so, as a corollary, there might be six tRNA species for any amino acid. Cherry (1967) has outlined evidence for the concept of various species of tRNA serving as a regulatory mechanism over protein synthesis, and later work in his laboratory (Anderson and Cherry, 1969) reported the actual separation of six different tRNA fractions which could be charged with leucine. Since these profiles were distinctively different for different plant tissues, the possibility that enzyme synthesis is controlled by the tRNA species is attractive and the possibility that cytokinins are involved in anticodon effectiveness of at least two amino acid tRNAs is very suggestive.

Fox and Chen (1967) used radioactive BA (see Fig. 6-1) to study the possible incorporation of the cytokinin into various RNA fractions, and they found the radioactivity from BA in the sRNA fraction which contained tRNA (Fig. 6-12). Their experiments were repeated by Kende and Tavares (1968), who confirmed that BA is incorporated into the sRNA. These later experiments were extended to include 9-methyl-BA, also active in growth, which was not incorporated into the sRNA. Kende and Tavares reasoned that if the substituted adenine in tRNA served as a regulator, one should expect it to be possible to induce organisms with such components to form mutants which require added cytokinin for growth. They looked for such mutants in *Escherichia coli* without success and concluded that the cytokinin presence in tRNA probably does not act as a regulator in the sense of the cytokinin hormonal action.

Bick et al. (1970) studied the effects of BA on soybean-cotyledon tRNA and found that while two leucyl tRNA species are increased by BA treatment, tyrosyl tRNA profiles are unaffected. The latter would be the ones expected to have cytokinin nucleotides as structural components.

Possible hormonal regulation through tRNA remains a point of argument. The presence of the cytokinin adjacent to the anticodon is essential for tRNA functioning in protein synthesis (Fittler and Hall, 1966). It seems less than likely that this is the manner of cytokinin regulatory actions.

MECHANISM OF ACTION

Since the report by Guttman (1956) that cytokinin treatment of roots results in a large increase in the nucleic acid levels detected by staining techniques and that the increase is in fact distinctive to the RNA, many experimenters have looked to the nucleic acid alterations for the possible mode of action of cytokinins. In leaves of tobacco, kinetin causes marked increases not only in RNA but DNA and protein synthesis as well (Parthier and Wollgiehn, 1961), which implies that in the deferral of senescence, kinetin acts on the overall synthetic activities of the leaf. This cytokinin action is prevented effectively by inhibitors of RNA and protein synthesis (Wollgiehn and Parthier, 1964). In some experiments with isolated nuclei from coconut milk, Roychoudhury et al. (1965) found that, like auxin and GA, kinetin causes an increase in nuclear RNA synthesis and may regulate the release of RNA into the cytoplasm. In peanut cotyledons Carpenter and Cherry (1966) observed an increase in all five fractions of nucleic acids after BA treatment, but in soybean hypocotyls Vanderhoef and Key (1968) report an inhibition of

Fig. 6-10 Structure of yeast tRNA for tyrosine (left) and for serine (right), showing location of the DMAA adjacent to the anticodon of each molecule (Hall, 1968).

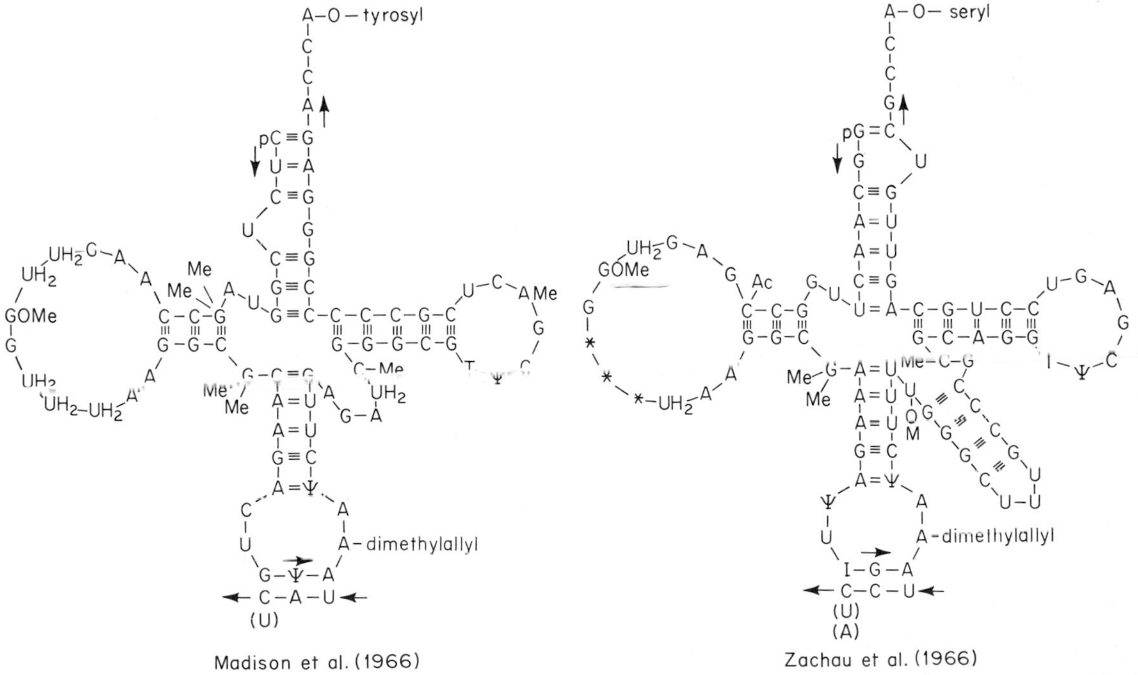

Madison et al. (1966)

Zachau et al. (1966)

rRNA synthesis by kinetin. One cannot readily see a general expression of cytokinin regulation of any particular component of the nucleic acid–protein-synthesis system, and yet each cytokinin action appears to be associated with marked stimulations of some of these systems.

Cotyledon expansion is a particularly appealing system in which to study cytokinin actions since (at least in some cases) growth is here sensitive only to this one class of plant growth regulator (Rijven and Parkash, 1970). The kinetin stimulation of growth is associated with a large increase in rRNA synthesis (Rijven and Parkash, 1971) followed by enhanced protein synthesis. In experiments using selective inhibitors of RNA synthesis like those used by Key et al. (1967) to examine auxin actions (Fig. 4-26), Rijven and Parkash found that 8-azaguanine inhibits growth and RNA synthesis in tandem; however 5-FU and FUDR, which preferentially inhibit rRNA and tRNA synthesis, did not inhibit growth (Fig. 6-13). These ex-

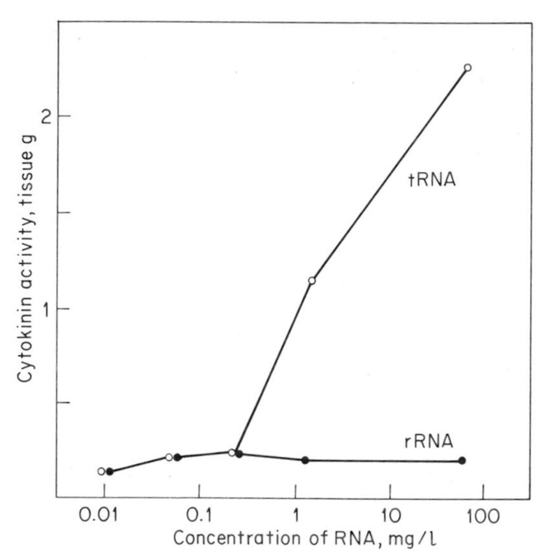

Fig. 6-11 Hydrolysates of RNA fractions from yeast show good cytokinin activity in the tRNA hydrolysate but not in rRNA hydrolysate. Assayed by the tobacco-callus bioassay (Skoog et al., 1966).

Fig. 6-12 After ^{14}C-benzyladenine is introduced into soybean seedlings, radioactivity from the cytokinin can be detected in the first two sRNA peaks, i.e., in the fractions containing tRNA (Fox and Chen, 1967). The numerals identifying the nucleic acid peaks from a MAK column are as in Fig. 4-25.

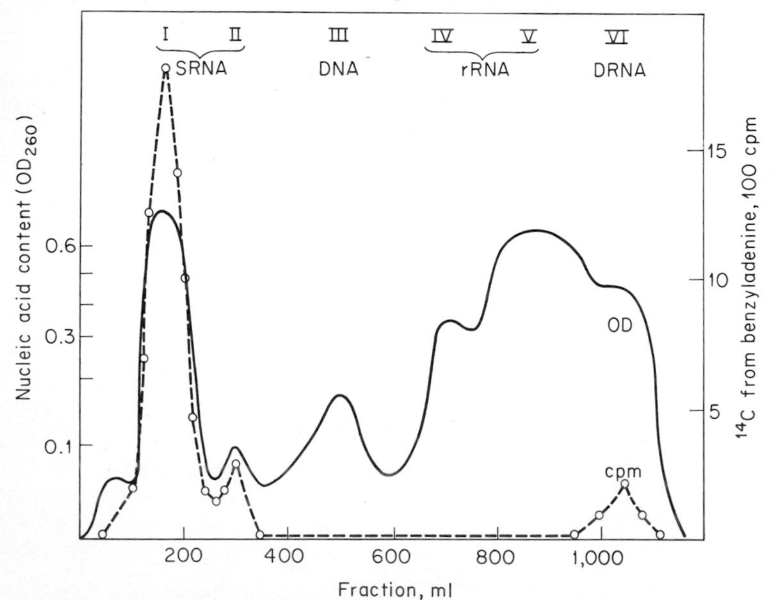

periments imply that the growth stimulation does not depend upon rRNA or tRNA synthesis. Addition of exogenous tRNA in their in vitro system did not enhance protein synthesis but added mRNA did. Like Bick et al. (1970), they found that cytokinin application had no preferential effect on the tyrosine tRNA and concluded that cytokinin effects might be obtained through mRNA synthesis but probably not through tRNA synthesis.

From tobacco and soybean nuclei, Matthysse and Abrams (1970) have isolated a protein which mediates the kinetin stimulation of RNA synthesis in the nuclei. One is tempted to believe that such a protein serves as the site of cytokinin action and that the ensuing nuclear responses are due to the action of the hormone-protein combination.

In several instances, the action of cytokinin has been identified with the formation or enhanced formation of specific enzymes, among them such diverse entries as tyramine methylpherase (Steinhart et al., 1964), proteinase (Penner and Ashton, 1967), some enzymes in the synthesis of thiamine (Dravnieks et al., 1969), and nucleases (Srivastava, 1968).

Bezemer-Sybrandy and Veldstra (1971) have repeated the studies of the radioactivity from BA incorporated into RNA, using *Lemna minor*, and they conclude that the small radioactivity which appeared in the sRNA fraction does not represent incorporation of BA at all. Even if one assumes that the cytokinin does not act through an incorporation into nucleic acids or polynucleotides, one might conceive of a modulation effect of the free cytokinin on the tRNA. For example, Cherry and Anderson (1972) have suggested that the free cytokinin might serve to protect the tRNA through an inhibition of some ribonucleases which were specific for the zeatin type of nucleotide.

With any hormone one would expect some specific site of attachment where the hormone can serve to regulate cellular functions; the studies of relationships between cytokinin structure and activity do not give as clear a picture of stereospecificity as for auxin structural requirements. Nevertheless, for zeatin there is a stereo requirement for the trans configuration of the isopentenyl side chain, the *cis* form being much less active than the *trans* form in bioassays. Berridge et al. (1972) searched for possible sites of cytokinin attachment in chinese cabbage tissues and found a reversible binding to the ribosomes. They were not able to detect any alteration in the effectiveness of the ribosomes with the attached cytokinin in terms of protein synthesis.

SUMMARY

While there is a possibility that this hormone may act through a direct participation in nucleic acid functioning, aside from the occurrence of cytokinin components in tRNAs we have little evidence upon which to base an explanation of its function through such a structural involvement. As with auxin and gibberellin, we see

Fig. 6-13 Effects of some inhibitors on growth and RNA synthesis in the fenugreek cotyledons. 8-Azaguanine inhibits both growth and RNA synthesis, whereas FUDR and 5-FU inhibit RNA synthesis without apparent effects on growth (Rijven and Parkash, 1971).

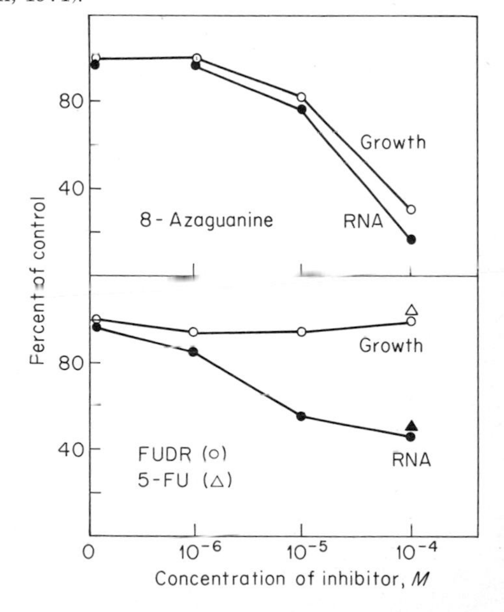

large changes in the nucleic acids and in protein synthesis and degradation associated with the hormone action, but again it becomes very risky to attribute the changes in these synthetic systems to the direct action of the hormone. Here too we are uncertain where the actual site of hormone action might be in terms of intracellular structures. Besides the multiple effects of cytokinins on nucleic acids and protein synthesis, we see numerous instances where this hormone brings about dramatic changes in the biosynthesis or the content of other hormones; e.g., cytokinins may bring about increases in auxin content (Skoog and Armstrong, 1970), gibberellin content (Loveys and Wareing, 1971*a*), or ethylene content (Fuchs and Lieberman, 1968). In at least some instances, then, cytokinin regulatory effects may involve alterations of the spectrum of interacting hormones in the plant.

GENERAL REFERENCES

Hall, R. H. 1973. Cytokinins as a probe of developmental processes. *Annu. Rev. Plant Physiol.,* 24:415–444.

Skoog, F. and R. Y. Schmitz. 1972. Cytokinins. In F. C. Steward (*ed.*), "Plant Physiology, a Treatise." Academic Press, New York, 6B:181–212.

Skoog, F. and D. J. Armstrong. 1970. Cytokinins. *Annu. Rev. Plant Physiol.,* 21:359–384.

7 ETHYLENE

OCCURRENCE

MOVEMENT

BIOSYNTHESIS

STRUCTURE AND ACTIVITY

REGULATORY ACTIONS

MECHANISM OF ACTION

SUMMARY

The discovery of ethylene as a plant-growth regulator might be said to have occurred as the inverse of the discoveries leading to the other plant hormones. Instead of starting with a plant function like tropism, cell enlargement, or cell division and looking for a regulator in the plant that might be responsible, experimenters first knew ethylene as an exogenous chemical with strong physiological effects, and only gradually over a series of decades did it become evident that this compound is in fact a natural plant hormone.

The story began with the observations of Neljubow (1901) that ethylene gas can markedly alter the tropistic responses of roots. The synthetic gas was observed by Rosa (1925) to be fairly effective in breaking potato dormancy, by Denny (1924) to be highly effective in inducing fruit ripening, by Zimmerman et al. (1931) in inducing leaf abscission, and by Rodriguez (1932) in inducing flowering in pineapple plants. These empirical experiments suddenly took on a different quality when Gane (1934) established that ethylene is actually a natural product of ripening fruits and its presence might therefore account for their stimulatory mutual ripening effect (Gane, 1935). Denny and Miller (1935) exploited this idea and found evidence of ethylene production not only by ripening fruits but also by flowers, seeds, leaves, and even roots; ethylene was also found to be a natural product in self-blanching celery (Nelson and Harvey, 1935). The natural extrapolation from these facts—that ethylene has such profound regulatory activity and that ethylene is formed naturally in plants in amounts sufficient to bring about regulatory effects—was that ethylene might be considered a plant hormone. Crocker et al. made such a proposal in 1935 but drew some strong criticism for it (Went and Thimann, 1937). Since auxin was at that time the central plant hormone, it seemed much more reasonable to attribute the effects of ethylene to alterations of auxin content, Michener (1938) showed a drop in auxin levels in some tissues after ethylene exposure and also a synergism between auxin and ethylene in the swelling and rooting responses. That auxin and ethylene have opposite effects in several plant functions did not seem to interfere with the synergistic-interaction idea. At this early stage, however, Zimmerman and Wilcoxon (1935) did observe that auxin treatments sometimes stimulate ethylene production by the treated tissues and suggested that some of the similar effects of auxins and ethylene might be attributable to this fact.

After this early spurt of attention, ethylene was relegated to the category of an exogenous chemical which could do some interesting things to plants—except by the fruit physiologists, for whom it remained as an important self-regulator of ripening (Biale, 1950). The reemergence of ethylene as an important hormonal regulator of plant physiological processes had to wait for the development of gas chromatography; once this technique could be applied to the analysis of naturally occurring levels of ethylene in various plant parts (Burg and Thimann, 1960), there followed an avalanche of experimental work on this compound and ethylene finally emerged as an accepted plant hormone (Pratt and Goeschl, 1969).

Accepting ethylene as a plant hormone requires a certain stretching of the concept. Animal physiologists originally defined a hormone as a substance which is formed in ductless glands and moves in the bloodstream to carry a physiological control to another organ, and the definition was somewhat altered to accommodate the situation in plants, where ductless glands and bloodstream are both absent. Went and Thimann (1937) defined a hormone as "a substance which, being produced in any one part of the organism, is transferred to another part and there influences a specific physiological process." The concept of a hormone being transferred would make physiologists think first of the auxin transport system or the translocation of gibberellins and cytokinins in the vascular system; the idea of transfer by gaseous diffusion is a novel variation.

OCCURRENCE

After the first proof that ethylene is an emanation from ripening fruits, occasional researchers exercised the self-discipline to carry out the exacting chemical techniques necessary to establish the identity of ethylene from plants (e.g., Pratt, 1954), but for the most part the quantitative measurement of ethylene awaited the advent of gas chromatography. This technique is so precise that the bioassay of ethylene is ordinarily not necessary. However, simplified methods of bioassay utilize the blockage of curvature development in sections of etiolated pea stems (Burg and Burg, 1967c) or the swelling reaction of etiolated pea seedlings (Warner and Leopold, 1967).

The general occurrence of ethylene as a plant emanation was recognized from the time it was first positively identified (Denny and Miller, 1935). One of the earliest observations was that auxin applications can greatly stimulate ethylene production by plants (Zimmerman and Wilcoxon, 1935), an effect rediscovered some decades later (Morgan and Hall, 1962).

Representative of the time course of ethylene production are some data for ethylene from germinating seeds (Fig. 7-1) and from ripening fruits (Fig. 7-2). Ethylene production commonly ranges

Fig. 7-1 Seed germination is sometimes associated with the production of ethylene, as illustrated here for germinating seed of oat (Meheriuk and Spencer, 1964), *Xanthium*, and lettuce (Leopold, unpublished).

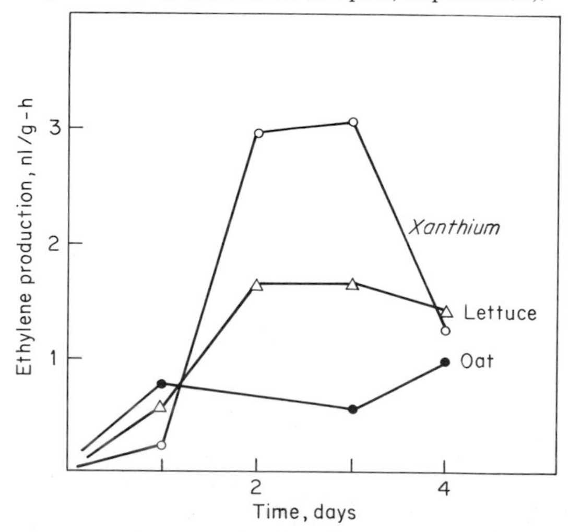

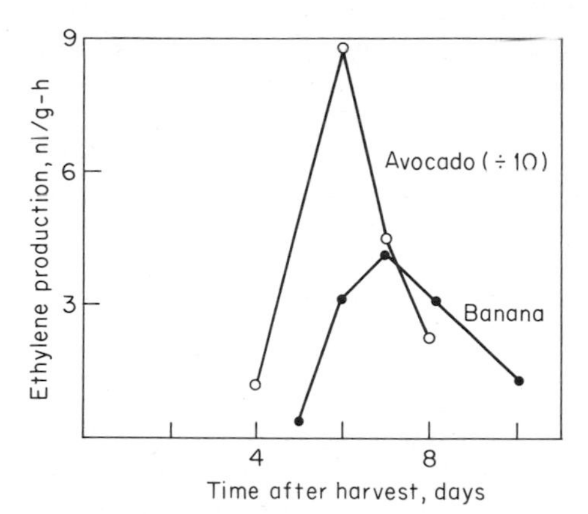

Fig. 7-2 Ripening often follows after fruits are picked. In some fruits there is a large biosynthesis of ethylene during the ripening stage, as illustrated here for avocado and banana fruits. The ethylene values for avocado are one-tenth the actual amounts (Biale et al., 1954).

from 0.5 to 5 nl/g-h; but in some fruits, for example, the rates are around 100 nl/g-h, as in Fig. 7-2. Auxin applications can increase ethylene production by factors of from 2 to 10 (Abeles and Rubinstein, 1964) (Fig. 7-3). Possibly the highest ethylene production recorded is in the fading blossoms of *Vanda* orchids, which produce 3,400 nl/g-h ethylene (Akamine, 1963).

Ethylene production rates are very susceptible to environmental factors. Burg and Thimann (1959) described some effects of temperature and oxygen on ethylene production by apple fruits. As shown in Fig. 7-4, either low or high temperatures can greatly depress the ethylene-production rate. Oxygen is essential, and at oxygen levels below about 2 percent, ethylene production is essentially terminated. Both the low-temperature and the low-oxygen effects are usefully exploited in fruit-storage techniques, where limitations of ethylene production can greatly improve the storage life of fruits.

Light has a great influence on ethylene production by young etiolated seedlings. Kang et al. (1967) found that the epicotyl-hook formation of etiolated pea seedlings is a morphological response to the ethylene produced by the seedling apex; and red light, which is known to convert the seedling from a hooked- to a straight-stem form, causes a sharp depression in the amount of ethylene formed. Goeschl et al. (1967) reported that this light control is a phytochrome-mediated system, since the depression of ethylene production is caused by red light and reversed by far-red light. Irradiation of some fruits with gamma rays can greatly increase ethylene production (Young, 1965).

Another influence over ethylene production is stress or injury. In many tissues, a burst of ethylene production is evident after bruising or cutting. In some tissues this burst may last for 2 to 10 h and even up to 24 h (Abeles and Rubinstein, 1964; McGlasson and Pratt, 1964). In fruits, the stimulation of ethylene production by cuts or bruises (Meigh et al., 1960) may be very large, e.g., the tomato, and bear considerably on the effectiveness of storage of the fruits. In other fruits, e.g., the apple, there may be no stimulation by wounding (Robitaille, 1973). An impressive effect of stress on ethylene production occurs in pea seedlings, which give rise to a large burst of ethylene production as they meet inter-

Fig. 7-3 Ethylene production by sections of bean petioles is greatly stimulated by indoleacetic acid. Data for 6-h time period (Abeles and Rubenstein, 1964).

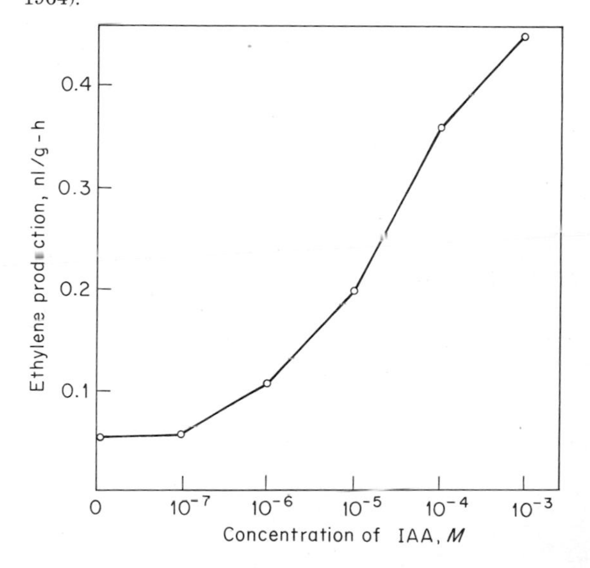

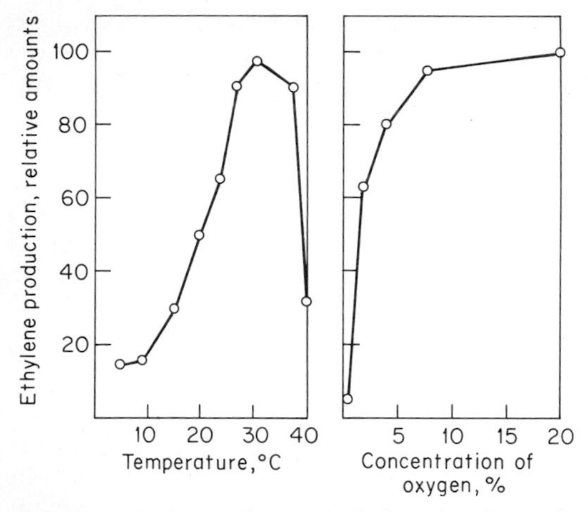

Fig. 7-4 The biosynthesis of ethylene by pieces of apple fruit is very sensitive to temperature and oxygen concentrations. Either high or low temperatures greatly reduce the amounts of ethylene formed, and low oxygen tensions almost eliminate it (Burg and Thimann, 1959).

ference with their growth through the soil. This ethylene may serve to restrict the stem tip into a tighter hook and cause a swelling or widening of the stem which may strengthen the seedling's ability to emerge from the soil (Goeschl et al., 1966). In trees, the bending of branches may likewise cause a marked increase in ethylene levels in the wood (Leopold et al., 1972).

In tissues which can produce ethylene, the experience of exposure to some ethylene may cause an avalanche of ethylene production like an autocatalytic response. This multiplication effect, first described by Hall (1951), has been restated by Burg and Dijkman (1967). In their study, ethylene production by *Vanda* orchid flowers was examined; not only was ethylene production stimulated by ethylene exposure, but a very localized auxin application, which caused a localized ethylene production, could then lead to the spreading of ethylene production over the entire flower, causing the bloom to fade. The autocatalytic effect of ethylene on its own production is evident in the ethylene triggering of climacteric fruit ripening (Burg and Burg, 1962) and the ethylene stimulation of leaf abscission (dela Fuente and Leopold, 1968).

Most hormonal responses of plants can be quantitatively studied by supplying known molarities of the hormone to plant pieces and measuring their quantitative response. For ethylene the concentration effects may be less clear since tissues which are responsive to this hormone may also be expected to respond to ethylene applications by generating more ethylene, making interpretations of plant responses on a molar basis much more difficult. Most ethylene responses can be obtained with the application of about 0.1 or 0.01 ppm ethylene, i.e., about 3 or 0.3 μM concentrations. The levels of ethylene achieved inside the tissues with these treatments, however, may be an order of magnitude different from the applied levels.

MOVEMENT

Ethylene is a small enough molecule to move readily through plant tissues by diffusive processes. Not only does its small size facilitate its movement, but its solubility in water and even greater solubility in lipophilic systems permits easy movement through plants. Zimmerman et al. (1931) applied ethylene to various plant parts, e.g., tip, individual leaves, and petioles, and observed that the plant responds systemically. Ethylene can move across a stem region killed by electric treatment, suggesting a passive movement through the plant. These studies used very high ethylene concentrations, but studies of ethylene movement through tissues at more physiological concentrations (Burg and Burg, 1965b) show that ethylene moves much like CO_2. Both gases follow Fick's law of diffusion, indicating that within fruit tissues, resistance to movement is systemically uniform.

Movement of ethylene through tissues may be through air spaces and hence related to tissue porosity (cf. Soudain, 1965), but its lipophilic quality should also permit ready movement through membranes. Burg (1968) points out that cuticular barriers probably keep the internal concentrations substantially higher than the outside concentrations; as a consequence, ethylene levels may generally be underestimated when ambient concentrations are used as the measure.

BIOSYNTHESIS

More research has been published on ethylene biosynthesis than any other sector of ethylene physiology. Yet there is much that is not clear and considerable disagreement obtains between research workers about the natural source of ethylene. The most prominent candidate for a precursor of ethylene in plants is methionine. Lieberman et al. (1966) showed that this amino acid is effectively converted into ethylene by slices of ripening apple. Its conversion into ethylene requires oxygen and is susceptible to inhibitors, as is the natural biosynthetic system. Labeling the molecule in various positions has established that carbon atoms 3 and 4 are exclusively converted into the ethylene molecule. Since this first report, the substrate effectiveness of methionine has been shown to hold for several types of plant tissues which are normally capable of ethylene biosynthesis. Baur et al. (1971) have pointed out that tissue from immature fruits (which biosynthesize almost no ethylene) does not convert methionine into ethylene, whereas ripening fruit tissue does; they deduce that the metabolic controls over ethylene biosynthesis must act between methionine and ethylene. Methionine levels in plants are ordinarily quite low (Spencer, 1969), and Baur et al. showed that the natural methionine content of avocado fruits during ripening would be enough to support ethylene biosynthesis for only 3 h. Applications of methionine are known to stimulate physiological processes in the manner of ethylene, e.g., abscission (Rubinstein and Leopold, 1962) and flower initiation (Nitsch, 1968).

A limitation to acceptance of methionine as an ethylene precursor is that many tissues and model ethylene-generating systems convert it to ethylene rather poorly, especially at physiological pHs. Ku and Leopold (1970) have shown that peptides with methionine end groups are more effective precursors, and Demorest and Stahmann (1971) have brought forth evidence that protein hydrolysis in senescing or ripening tissues may provide substrates for ethylene biosynthesis as methionine peptides.

Another suggested substrate for ethylene biosynthesis is fatty acids. Lieberman and Mapson (1962) suggested that in apple fruits linolenic acid may serve as a precursor of ethylene. Spencer and Olson (1965) showed a correlation between ethylene formation and fatty acid metabolism in germinating castor beans. Lipoxidase, too, has been correlated with ethylene formation in apples (Wooltorton et al., 1965), and its activity may provide not only for the metabolism of fatty substances but also peroxides utilized in the conversion of methionine into ethylene (Meigh et al., 1967).

β-Alanine has been reported to increase ethylene biosynthesis in the presence of some tissue extracts (Meheriuk and Spencer, 1967). Ethionine has been similarly reported in some tissues (Shimokawa and Kasai, 1967), but that amino acid inhibits ethylene formation in others (Burg and Clagett, 1967).

Several nonenzymatic systems have been used to study ethylene formation following the finding of Abeles and Rubinstein (1964) that boiled plant tissues can generate ethylene. In one of the most useful of these, the model system of Yang et al. (1966), methionine is converted into ethylene in the presence of a pigment (flavin mononucleotide, FMN) and light. This system generates ethylene more promptly after the introduction of methional than methionine (see Fig. 7-5), which has led to the proposal that methional serves as an intermediate in the conversion of methionine into ethylene (Yang et al., 1967). Another suggested intermediate is α-keto-γ-mercaptobutyrate (KMB) (Fig. 7-5) (Yang, 1968). Several papers have argued this possibility, and Baur et al. (1971) may have closed the argument by showing that radioactive-

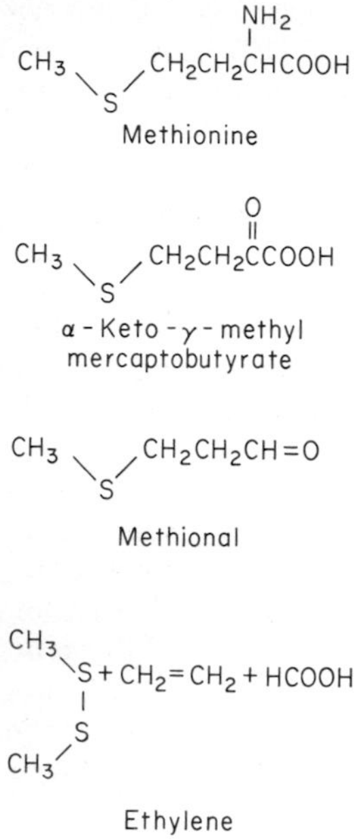

$CH_3{-}S{-}CH_2CH_2CH(NH_2)COOH$

Methionine

$CH_3{-}S{-}CH_2CH_2C({=}O)COOH$

α-Keto-γ-methyl mercaptobutyrate

$CH_3{-}S{-}CH_2CH_2CH{=}O$

Methional

$CH_3{-}S{-}S{-}CH_3 + CH_2{=}CH_2 + HCOOH$

Ethylene

Fig. 7-5 Some suspected intermediates in the biosynthesis of ethylene.

tagged KMB is converted into ethylene with a longer lag period than methionine; so at least in the apple tissue it appears that the KMB may not be an intermediate between methionine and ethylene.

The cellular location of ethylene biosynthesis has also been a subject of argument. Lieberman and Craft (1961) showed that ethylene can be synthesized by preparations rich in mitochondria from tomato fruits, but Ku and Pratt (1968) have shown that carefully cleaned mitochondria do not synthesize ethylene and that the synthesizing system is in the soluble components of the cytoplasm.

The question of what enzymes are involved in the biosynthesis of ethylene will of course presuppose knowledge of its precursor or precursors. There is still considerable disagreement on this point: in addition to methionine or peptides containing methionine, suggested candidates for precursors include fatty acids, β-alanine, glucose, fumarate (Burg and Burg, 1964), propanal (Baur and Yang, 1969) and pyruvate (Shimokawa and Kasai, 1967). Several of these possible precursors may be converted into ethylene by the action of peroxidase; peroxidase has been used in a model system for the conversion of methional (Yang, 1967) or KMB (Ku et al., 1969) or peptides of methionine into ethylene (Ku and Leopold, 1970). Other systems containing peroxides are also capable of generating ethylene from various substrates (Kumamoto et al., 1969; Takeo and Lieberman, 1969). It is puzzling, then, that Kang et al. (1971) found a singular lack of correlation between the peroxidase levels and ethylene production in pea tissues. Their studies suggested the conclusion that peroxidase activity probably does not limit ethylene biosynthesis and that if the enzyme is involved, it is in some nonlimiting step. Mapson and Wardale (1971) have proposed a sequence of enzymes for the conversion of methionine into ethylene in tomato fruit: first, transaminase removes the amino group, and, second, an oxidase generates peroxides which permit the third enzyme, peroxidase, to carry out the final oxidations to produce ethylene.

As a speculative opinion, one might suppose that the regulatory controls of ethylene biosynthesis, and even perhaps the substrates for biosynthesis, are different for different physiological conditions. For example, endogenous ethylene formation in undisturbed tissues has characteristics different from that in bruised or wounded tissue and possibly auxin-stimulated tissue and climacteric fruit tissue. Injury may increase the activity of peroxidases, whereas ripening or senescence may increase the available amounts of methionine or peptides of methionine, for example. The requirement for oxygen (Gane, 1934; Burg and Thimann, 1959) and the inhibition by cyanide (Kang et al., 1971) are certainly consistent with a role of peroxidase. The strong inhibition of ethylene biosynthesis in peas by light (Burg and Burg, 1968a) implies that there is an effective control of the enzyme system in the plant.

Ethylene formation can be stimulated by a wide spectrum of chemical treatments. The effects of auxin as an ethylene stimulator have been known for many years (Zimmerman and Wilcoxon, 1935; Morgan and Hall, 1962). Various growth regulators which serve to stimulate growth can likewise stimulate ethylene formation (Abeles and Rubinstein, 1964; Fuchs and Lieberman, 1968). Several chemicals have been found to stimulate ethylene production and thus stimulate abscission (Cooper et al., 1968), e.g., ascorbic acid, cycloheximide, and iodoacetate; these effects might be interpreted as disruptions to the tissue not unlike wounding. Another group of chemicals can be applied to plants to stimulate ethylene production through their own breakdown into ethylene. The most prominent are β-hydroxyethylhydrazine (Palmer et al., 1967) and chloroethylphosphonic acid (Warner and Leopold, 1967; Cooke and Randall, 1968).

The possibility that plants may have components that can serve as depressors of ethylene biosynthesis has been opened by the reports of Sokai and Imaseki (1973, a, b) of the existence of a protein in mungbean hypocotyls which suppresses ethylene production. Present most abundantly in the epidermal cells, this protein appears not to be a simple peroxidase inhibitor, and it can depress ethylene formation in several different types of tissues. Certain exogenous chemicals have also been found to suppress ethylene synthesis (Patil and Tang, 1974; Liebermann, 1974).

STRUCTURE AND ACTIVITY

The relative effectiveness of various unsaturated hydrocarbons in the stimulation of epinasty in tomato plants was measured by Crocker et al. (1935). Using more refined bioassays, Burg and Burg (1967c) and Abeles and Gahagan (1968) have made such comparisons for activity in suppressing pea curvature and in the stimulation of abscission, respectively. There is fairly good agreement between these various comparisons, ethylene being roughly two orders of magnitude more effective than its nearest analog, propylene. Some comparative values for six gases are shown in Table 7-1. The exponential falloff of biological activity with increasing molecule size has been interpreted by Burg and Burg as reflecting steric hindrance; the partitioning coefficient of gases between water and lipids is known to fall off very rapidly with increasing molecular size (Nelson and Hoff, 1968), and it is possible that the lesser water-lipid partitioning may be contributing to the lesser activity for the larger gases in Table 7-1. Another molecular requirement for ethylenelike activity seems to be an unsaturated bond next to a terminal carbon. The freedom of resonance of the terminal double-bond carbon seems to be critical, for a halide substitution markedly lowers the biological activity, e.g., vinyl chloride. The effectiveness

Table 7-1 Comparative effectiveness of ethylene and some other gases in the pea-curvature test, showing the far greater activity of ethylene than any other known compound in this system (Burg and Burg, 1967c).

	Relative activity	
Compound	In gas phase for half-maximum activity, ppm	Relative moles per unit effectiveness K_A'
Ethylene	0.1	1
Propylene	10	130
Vinyl chloride	140	2,370
Carbon monoxide	270	2,900
Acetylene	280	12,500
1-Butene	27,000	140,000

of carbon monoxide in ethylene systems was noted long ago by Zimmerman (1935).

The relations of molecular structure to biological activity are interpreted by Burg and Burg as reflecting the ability of the gas to bind to a metal. They suggest that the ethylene binds to a metal-containing receptor site and point to the parallel affinities of unsaturated gases and carbon monoxide to metals such as silver or mercury as being consistent with this interpretation of structural requirements.

Ethylene is the most abundant of the unsaturated hydrocarbons formed by plants (Meigh, 1959; Jansen, 1965). The minute quantities and low activities of other gases such as propylene, ethane, and acetylene (Meigh, 1959) make it unlikely that other hydrocarbon gases perform regulatory functions in plants.

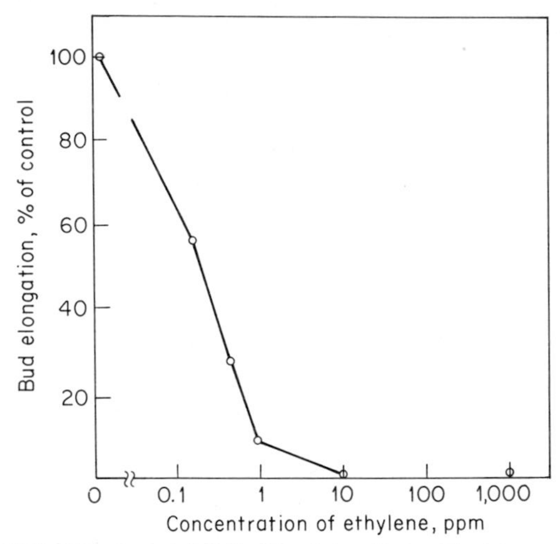

Fig. 7-7 Ethylene inhibits the elongation of buds on cuttings of etiolated pea, reaching half saturation at less than 1 ppm ethylene. Cuttings held in the light in 1% sucrose for 7 days (Burg and Burg, 1968*b*).

REGULATORY ACTIONS

As with the other plant hormones, a list of regulatory roles for ethylene looks like a complete catalog of growth and developmental functions (Table 18-1). Foremost among functions which may be regulated by ethylene are the inhibitory effects on growth. The inhibition response of pea-stem sections is illustrated in Fig. 7-6; associated with the lessened growth in length is a radial enlargement of the tissue (swelling response) in pea seedlings, which may be helpful in the emergence of the seedling through the resistance of the soil (Goeschl et al., 1966). Likewise the lateral enlargement of some roots such as radish may be a reflection of the same ethylene stimulation of lateral growth (Radin and Loomis, 1969). It is proposed that ethylene inhibition of elongation growth is responsible for the geotropic bending of roots, through the differential formation of ethylene on the geotropically lower side in pea roots (Chadwick and Burg, 1967; cf., however, Andreae et al., 1968, and the reply of Chadwick and Burg, 1970). A striking contrast is the *diageotropica* mutant of tomato, in which ethylene biosynthesis is apparently blocked, and ethylene must be supplied exogenously before normal geotropic behavior of the stems will occur (Zobel, 1973). Another growth-inhibition effect of ethylene is the suppression of bud growth in the pea seedling (Fig. 7-7). Such an inhibition of the growth of lateral buds has been considered for years to be a function of auxin, and now evi-

Fig. 7-6 Ethylene inhibition of the elongation growth of pea-stem sections is about half saturated at less than 1 ppm. Sections from etiolated seedlings, supplied with 0.2 μM indoleacetic acid (Burg and Burg, 1968).

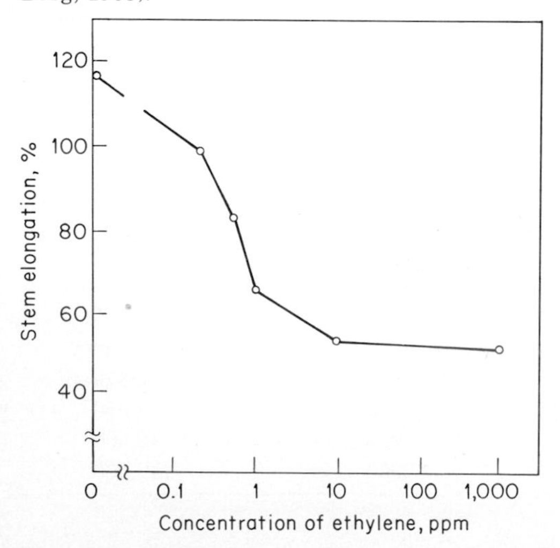

dence is at hand that in peas the auxin functions in part through the biosynthesis of ethylene and consequent bud inhibition.

In addition to the inhibitory effects of ethylene, small but definite promotions of growth by this hormone have been reported for several species and under certain conditions. The cases reported are principally for species which grow well under water, e.g., rice (Ku et al., 1970) and water plants (Musgrave et al., 1972). The growth of fruits is dramatically stimulated by ethylene in some species (Maxie and Crane, 1968).

The participation of ethylene in the endogenous control of growth rates appears to be widespread in higher plants and includes not only the inhibitions of elongation growth but also in certain instances the stimulation of lateral growth and occasionally the stimulation of elongation growth as well.

Several types of differentiation are known to be within the regulatory domain of ethylene. Foremost among them is the formation of a *separation layer* or *abscission zone* (Fig. 9-27). In the early days of ethylene research it was noted that leaf abscission can be stimulated by this gas (Zimmerman et al., 1931). Studies of natural and induced abscission led Hall and Lane (1962) to suggest that abscission is regulated by a balance between auxin (as a retardant of abscission) and ethylene (a stimulant). Still another type of differentiation under ethylene control is flowering. In pineapple, Rodriguez (1932) observed that ethylene applications can induce flowering, and Clark and Kerns (1942) described a similar control in that species with auxin. Later experiments have revealed that the auxin stimulation of pineapple flowering is a consequence of the stimulation of ethylene biosynthesis following auxin treatment (Burg and Burg, 1966c). Several other species are forced to flower by ethylene (Stuart et al., 1966; Nitsch, 1968). Another type of differentiation which may be at least influenced by ethylene is root initiation. Zimmerman and Wilcoxon (1935) were able to stimulate rooting with ethylene, and it is possible that the auxin stimulations of rooting may be due in part to the ethylene formed in response to the applied auxin.

Dormancy of buds and seeds is sometimes relieved by ethylene applications, first reported by Rosa (1925) and Vacha and Harvey (1927). It is curious that in many of the early surveys of chemicals which might break dormancy, ethylene ordinarily was reported as being ineffective (see, for example, Denny, 1926). Modern experiments have shown that not only ethylene can break dormancy under some conditions (Toole et al., 1964) (Fig. 7-8) but endogenous formation of ethylene may be responsible for breaking dormancy (Ketring and Morgan, 1969; Esashi and Leopold, 1969*b*).

The effectiveness of ethylene in stimulating fruit ripening first drew the attention of physiologists and horticulturists (Denny, 1924). Its stimulations of ripening are markedly more impressive in climacteric than in nonclimacteric fruits (Biale, 1964). That the climacteric ripening phase is triggered by ethylene has been indicated by the work of Burg and Burg (1962) (see Fig. 13-31).

MECHANISM OF ACTION

The important and intriguing question of how hormones exert their regulatory effects can be

Fig. 7-8 Ethylene can stimulate the germination of otherwise dormant seeds of subterranean Dinninup clover (Esashi and Leopold, 1969*b*).

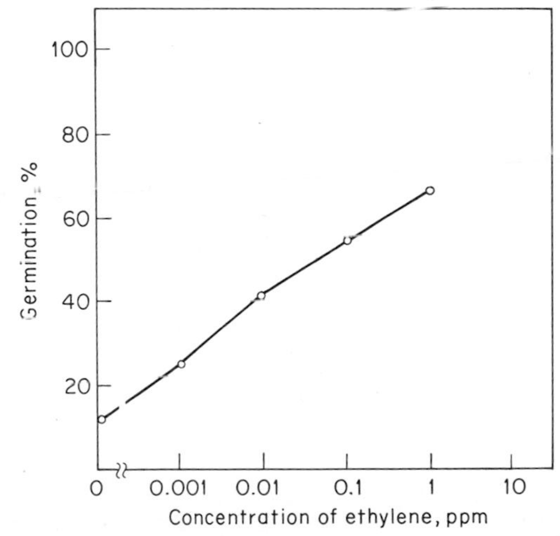

answered no more clearly for ethylene than for the other known plant hormones. Three theoretical lines have been suggested:

1 Ethylene becomes attached to some metalloprotein site in the cell which can serve in some regulatory manner.

2 Ethylene becomes attached to membrane layers, altering their function in some way.

3 Ethylene serves to regulate plant processes through an alteration of RNA and resultant alteration of RNA-directed protein synthesis and thus of enzyme patterns.

The metal-adsorption theory of Burg and Burg (1967c) is based on two principal lines of evidence: (1) the changes in biological activity with molecular structure like those represented in Table 7-1 are similar to the changes in adsorption of the gas onto a heavy metal such as silver; (2) the attachment of ethylene to a heavy metal

Fig. 7-9 Carbon dioxide apparently acts as a competitive inhibitor of ethylene in the pea-seedling curvature test. When the inverse of the ethylene concentration is plotted against the inverse of the curvature inhibition, the plots with and without CO_2 show the same intercept on the ordinate, which is consistent with the competitive-inhibitor interpretation (Burg and Burg, 1967).

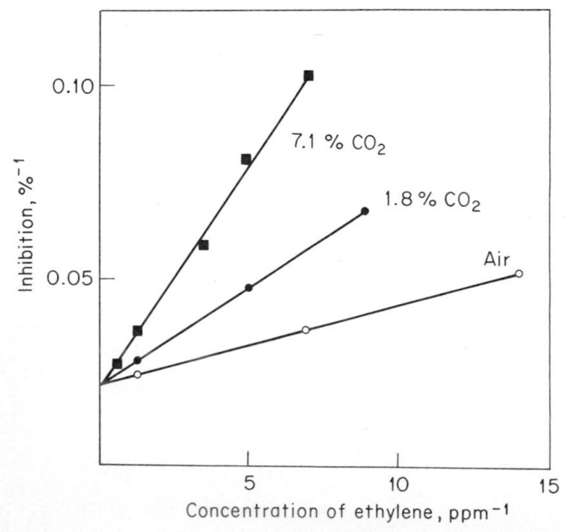

is inhibited by carbon dioxide in a manner similar to the inhibition of many biological responses to ethylene. The competitive nature of CO_2 interference with ethylene action is illustrated in Fig. 7-9. Such CO_2 interference with ethylene responses has been recorded for many ethylene systems, including the stimulation of abscission (Abeles and Gahagan, 1968), epicotyl-hook formation in etiolated seedlings (Kang et al., 1967), inhibition of root growth (Chadwick and Burg, 1967), and the stimulation of fruit ripening (Burg and Burg, 1965a, c). CO_2 reversal has become a standard test for ethylene-stimulated actions.

In their experiments with ethylene inhibition of pea-stem elongation, the Burgs were also able to show an interaction of ethylene with oxygen (Fig. 7-10). The requirement for oxygen in ethylene responses had been known since the early experiments of Denny (1924); the Burgs established that depressed oxygen levels inhibit ethylene action, the effects having the characteristics of a competitive inhibition. (Double reciprocal plots of ethylene and oxygen-deficient ethylene conditions give a common intercept in the ordinate, as shown in Fig. 7-10.) They suggest that the oxygen is necessary for oxidation of the metal-receptor site to which ethylene becomes attached. Abeles and Gahagan (1968) found that the depressed ethylene response at low oxygen levels can be accounted for on the basis of a simple depression of respiration and disagreed with the concept of oxidation of a site of attachment. The oxygen requirement is consistent with the great prolongation of fruit storage life in controlled atmospheres with low oxygen and high CO_2 levels.

The nature of the binding of ethylene onto sites in the plant has been examined by supplying deuterated ethylene and examining it after exposure to the plant for a deuterium displacement by hydrogen atoms (expected for a covalent attachment) or an alteration of the arrangement of the specific deuterium atoms in the molecule (expected for a hydrogen-bonding attachment). Experiments by Abeles et al. (1972) and Beyer (1972b) agree that the recovered ethylene is unaltered, which is consistent with an attachment neither by covalent nor by hydrogen bonding.

The presence of deuterium would be expected to result in enhanced affinity of the ethylene for a heavy metal such as silver, and the lack of a difference in biological activity between the deuterated and the hydrogen-containing ethylene does not provide support for the metalloprotein attachment.

Burg and Burg (1967) suggested zinc as an attractive possibility for the metal site for ethylene attachment, and Ku and Leopold (see Leopold, 1972) have found that the zinc-containing enzyme carbonic anhydrase does in fact adsorb ethylene but that the measured effectiveness of the enzyme is not affected by such adsorption.

A second theory holds that ethylene may find its active site on a membrane surface, where it may become attached to a lipid layer. Lyons and Pratt (1964) tested this possibility using the swelling of isolated mitochondria as a measure of increased membrane permeability; they observed a more rapid swelling in the presence of ethylene and concluded that the gas was altering permeability. Later experiments (Mehard and Lyons, 1970) extending the measurements on permeability to other unsaturated gases forced the conclusion that the permeation effects were not specific to ethylene and hence probably not responsible for ethylene actions.

Mehard et al. (1970) tested the membrane-alteration idea in another manner, using artificial membranes and measuring the effects of a series of alkenes and alkanes on surface tension; they found that ethylene has very little effect on the surface tension of the membrane, indicating that such physical actions on membrane structure are not likely to account for biological actions of ethylene.

The third type of ethylene mechanism might be the alteration of enzyme patterns in the plant through an alteration of RNA-directed protein synthesis. Ethylene is known to bring about alterations in the amounts or effectiveness of numerous enzymes. Probably the first enzyme known to be turned on by ethylene is peroxidase. The response of sweet potato tissues to infection by the black rot fungus involves a dramatic increase in peroxidase levels; Stahmann et al. (1966) suspected that the production of ethylene

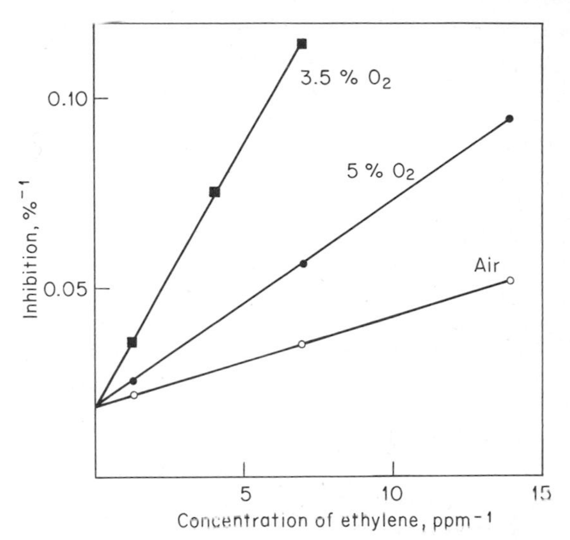

Fig. 7-10 When oxygen levels are less than the 20 percent in air, there is a suppression of ethylene action on the pea-curvature test. Plotted as in Fig. 7-9; the common ordinate intercept is interpreted as indicating that oxygen must act on the site of ethylene action before the ethylene can bind to the action site (Burg and Burg, 1967).

by the infection system might be involved in the enzymic changes. They were in fact able to induce the peroxidase increase by supplying ethylene and no fungus. Gahagan et al. (1968) extended these studies by showing that CO_2 can prevent the ethylene enhancement of peroxidase and that inhibitors of RNA protein synthesis (actinomycin D and cycloheximide) can prevent the peroxidase increase.

Enzymes involved in phenolic metabolism are also altered by ethylene. Chalutz et al (1969) report that ethylene exposure of carrot roots leads to an enhancement of isocoumarin biosynthesis and that this change is prevented by elevated CO_2 levels. In citrus peels, ethylene induces an increase in phenylalanine ammonia-lyase, which is again prevented by CO_2 or by cycloheximide (Riov et al., 1969); this enzyme is often a limiting step in the biosynthesis of phenolics in plants. Other enzymes known to increase in at least some tissues following ethylene exposure include phosphatases (Olsen and Spencer, 1968) and numerous enzymes involved in fruit ripening (e.g., organic acid–metabolizing enzymes; Rhodes et al., 1968).

The ethylene effects on cellulase deserve special mention. The stimulation of abscission in bean leaves by ethylene is known to be associated with an increase in some cellulase enzymes in the region of separation (Abeles, 1969; Ridge and Osborne, 1969). At least one of these cellulases appears through *de novo* synthesis (Lewis and Varner, 1970). The rapidity of ethylene actions on abscission development (dela Fuente and Leopold, 1969) was not consonant with the enzyme-synthesis kinetics (Abeles and Leather, 1971), and when the distribution of the enzymes was reexamined, Abeles found that the rapid response to ethylene correlated with a rapid release of cellulase from a bound form into a soluble form in the intercellular spaces of the petiole. An ethylene-stimulated release of enzymes from a bound form has also been reported by Jones (1968); using the barley-aleurone layer, which forms α-amylase in response to gibberellin, Jones found that introduction of ethylene enhances the release of amylase into the liquid medium. From these results, it is evident that enzymatic changes induced by ethylene may not necessarily involve an alteration in enzyme biosynthesis.

Numerous ethylene actions have been effectively inhibited by nucleic acid or protein-synthesis inhibitors, as already mentioned, and this evidence seems to imply that ethylene serves to regulate enzyme synthesis through RNA-directed synthesis. Holm and Abeles (1967) observed a small increase in $^{32}PO_4$ incorporation into RNA of bean abscission zones in the presence of ethylene; this fact plus the inhibition of abscission by actinomycin D and certain other inhibitors led them to suggest that ethylene acts through a synthesis of some new RNA, possibly a mRNA. Holm et al. (1970) went on to show that in soybean hypocotyls ethylene exposure leads to a marked increase in the DNA-directed RNA synthesis and brings about the formation of RNA with an altered base composition. In Japan, Shimokawa and Kasai (1968) applied radioactive ethylene to morning glory seeds and recovered some of the label in a light RNA fraction, perhaps a tRNA.

The numerous lines of evidence that ethylene may regulate growth functions through RNA and RNA-directed protein synthesis may offer too simple an explanation, however. Three types of evidence suggest that ethylene regulatory actions may not occur entirely through changes in protein synthesis:

1 As already mentioned, some rapid ethylene actions seem to involve alterations of enzyme secretion rather than synthesis (Abeles and Leather, 1971; Jones, 1968).

2 Timing measurements of ethylene inhibitions of growth show that altered growth rates set in within about 5 min after ethylene exposure, and this is probably too rapid for an effect via protein synthesis (Warner and Leopold, 1971).

3 At least some ethylene responses are not inhibited by cycloheximide or actinomycin D (Burg and Burg, 1967a).

Many of the ethylene effects on growth involve an apparent alteration in the polarity of growth, the familiar swelling response of etiolated peas and the inhibition of tropistic responses being good illustrations of this polar alteration. Borgström (1939) first expressed the idea that ethylene acts through a disturbance of polarity and supported this suggestion with experiments which showed a stimulation of lateral cell enlargement in pea stems by ethylene. The possibility that ethylene effects might be connected with an alteration of polar auxin transport was opened by Morgan and Gausman (1966), who showed that in plants whose growth had been inhibited by previous ethylene exposure there was a marked decrease in polar auxin transport. Later experiments from the same laboratory (Beyer and Morgan, 1969) supported the idea of ethylene inhibition of auxin transport and eliminated the possibility that auxin had been removed from the polar system through increased decarboxylation. Burg and Burg (1967b) examined the possible inhibition of lateral auxin transport in geotropically stimulated pea seedlings and showed that lateral transport is less susceptible to ethylene inhibition than longitudinal or polar transport; they utilized this difference to develop a model of tissue swelling as a preferential interference with longitudinal auxin transport. Burg et al. (1971) have expanded the idea of polar disorganization by ethylene by showing that both mitotic activity and the orientation of cellulose microfibrils in the cell walls of pea seedlings are disrupted by ethylene treatment, and they indicate that each of these events could result from disorientation of microtubules which may be involved in mitotic spindle mechanisms as well as cell-wall deposition patterns.

It would seem unlikely, however that such effects on mitotic apparatus and on cell wall deposition patterns could be involved in the growth stimulatory responses to ethylene.

SUMMARY

The dynamic part taken by ethylene in the regulation of growth and of many diverse developmental phenomena has forced the realization that this tiny molecule must be accepted as a hormonal agent in plants. The existence of an ethylene-requiring mutant puts the final touch on the case.

The mechanism of action of this hormone is no clearer than that of the other plant hormones. Despite strong evidence that the nucleic acid system is intimately involved in ethylene actions, some fast responses to ethylene seem to be independent of nucleic acid and protein synthesis.

GENERAL REFERENCES

Abeles, F. B. 1973. "Ethylene in Plant Biology." Academic Press, New York. 302 pp.

Abeles, F. B. 1972. Biosynthesis and mechanism of action of ethylene. *Ann. Rev. Plant Physiol.,* 23:259–292.

Osborne, D. J. 1973. Ethylene and protein synthesis. In B. V. Milborrow, (*ed.*), "Biosynthesis and its Control in Plants." Academic Press, London, pp. 127–142.

8 INHIBITORS

The design of regulatory systems might reasonably include not only means by which growth can be stimulated but also the means by which growth can be restrained. Hormonal systems which suppress growth include auxin at high concentrations (see Fig. 4-3) and ethylene (see Fig. 7-6); there are instances, too, in which gibberellins and cytokinins may serve as suppressors of growth, though they are less common. A hormone system which appears to be much more widespread as a suppressor of growth is abscisic acid (ABA), an isoprenoid compound which shares the mevalonic acid–synthesis pathway with gibberellins (Fig. 5-9) and cytokinins.

Besides hormonal suppressions of growth, the plant has at hand a wide range of secondary plant substances, which accumulate in the plant and have no apparent role in a metabolic sequence. In a wide range of growth phenomena, these secondary plant chemicals serve as inhibitors in the plant. They may also be toxic to insects or browsing animals and thus serve to limit attacks on the plant or in other instances to act as attractants. Among the secondary plant chemicals which act as inhibitors of the plant that bears them, phenolic acids, phenolic lactones, and the related flaviniums are the most common.

A central theme in the area of the plant-growth regulators is their involvement in sensing environmental cues by the plant. The inhibitor systems, like the other regulatory systems, show some dramatic alterations in response to such environmental events as changes in light, photoperiod, and temperature and water stress.

The participation of inhibitors in regulation of growth has been recognized since the late 1940s, and yet the physiological understanding of inhibitors and their roles has developed very slowly, especially in comparison with the understanding of auxins, gibberellins, and cytokinins. Three types of difficulties have retarded the experimental development of the inhibitor field:

1 When one extracts secondary chemicals from a plant, the material one obtains in a test tube or on a chromatogram may differ considerably qualitatively and quantitatively from that inside the plant; this is often a consequence of the ready conversion of plant chemicals during extraction and purification procedures and the actual loss of some principal ingredients (Guern, 1964).

2 The inhibitors may be difficult to assay. Many of the most common bioassays for growth regulators are rather insensitive to secondary plant substances such as the phenolics, and furthermore assays have frequently been employed which are remote from the physiological phenomenon which might be regulated by the inhibitor, leaving room for some serious misconceptions.

3 Reintroduction of inhibitors into the plant presents some problems of its own, and a suspected inhibitor will often show very poor activity in a bioassay, perhaps because it cannot readily reenter the plant.

Illustrating some of the perplexing results that can be obtained, von Guttenberg and Leike (1958) found that the inhibitor content in extracts of lilac became very high during the period of most rapid growth in the spring, and no decline was evident as the buds emerged from dormancy. Buch and Smith (1959) found that the inhibitor extractable from dormant potato buds was evident only when *Avena* coleoptile sections were used for assay. No inhibition was detected when potato buds were used for the assay, even though they could confirm that the inhibitor had really entered the tissue. Guern (1964) has shown that ordinary extraction and isolation procedures may involve the entire loss of some important inhibitors, e.g., chelidonic acid, and that others, e.g., caffeic and *p*-coumaric acids, are formed during the extraction procedure. In short, although work with plant-growth inhibitors can provide some promising correlations with inhibited growth states, there are many instances where correlation is lacking.

Fig. 8-1 The inhibitory effects of abscisic acid (ABA) and a lactone such as coumarin are compared in the germination of lettuce seeds: note the much greater sensitivity to ABA (Esashi and Leopold, unpublished). Grand Rapids lettuce germinated at 23°C in dark in 20 h.

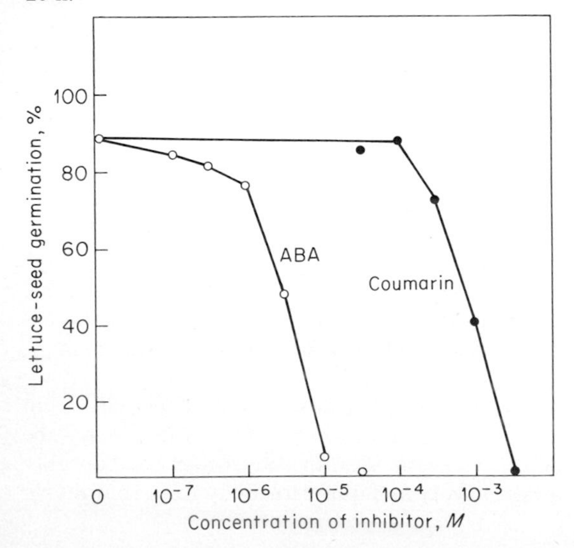

CHEMICAL NATURE OF INHIBITORS

The central position among the growth inhibitors must go to the hormone abscisic acid (ABA), a sesquiterpene isolated and identified almost simultaneously in two different laboratories (Ohkuma et al., 1965; Cornforth et al., 1965a). For the preceeding 10 years, an unidentified inhibitor, termed inhibitor β by Bennet-Clark and Kefford (1953), was known to be a widespread and powerful growth inhibitor in plants. The principal component of inhibitor β is now known to be abscisic acid (Milborrow, 1968). It is noteworthy not only for its widespread distribution in plants but also for its level of effectiveness; in a lettuce-seed germination test, ABA is 100 times more effective than a lactone inhibitor like coumarin (Fig. 8-1). On a molar basis, ABA is active in roughly the same range of concentrations as the auxins, gibberellins, cytokinins, and ethylene.

A second category of inhibitors is the phenolics, including phenolic acids of the benzoic acid and the cinnamic acid families, the lactones of the coumarin family, and flavonoids (see Figs. 8-2 and 8-3 and cf. Figs. 4-11 and 4-12). Unlike the hormonal regulators, these phenolic inhibitors may occur in very high concentrations in plants, either as the free phenolic or as a derivative such as the glycoside. Pridham (1965)

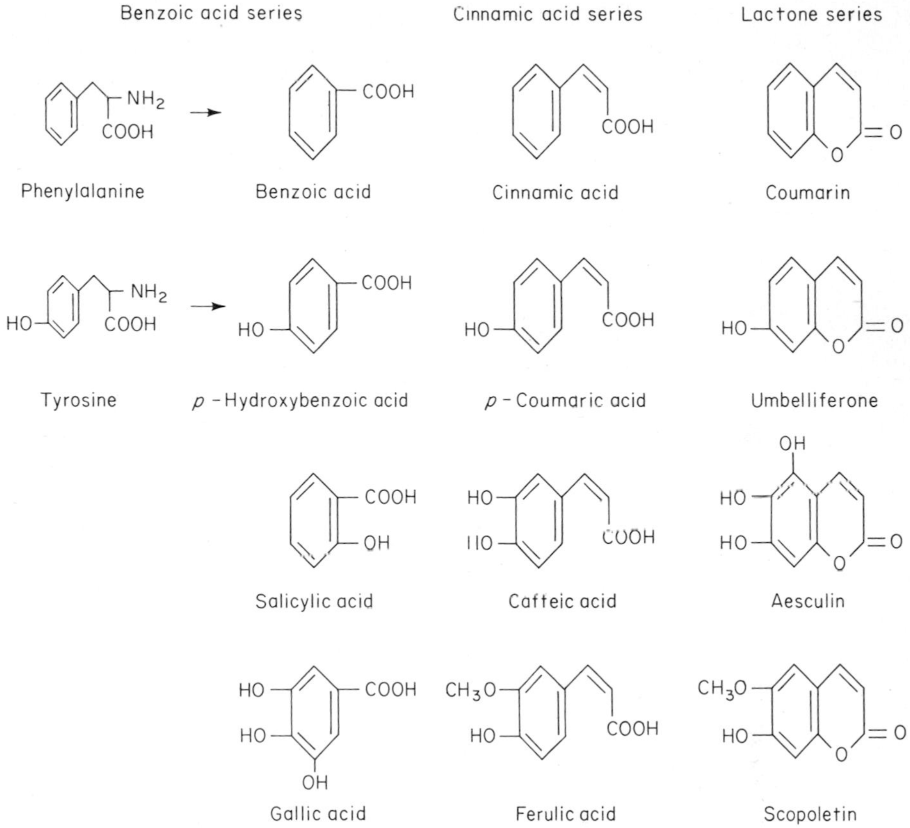

Benzoic acid series | Cinnamic acid series | Lactone series

Phenylalanine → Benzoic acid | Cinnamic acid | Coumarin

Tyrosine → *p*-Hydroxybenzoic acid | *p*-Coumaric acid | Umbelliferone

Salicylic acid | Caffeic acid | Aesculin

Gallic acid | Ferulic acid | Scopoletin

Fig. 8-2 Some representative phenolic inhibitors. The origin from phenylalanine and tyrosine is indicated at the left.

has asserted that the phenolics are second in abundance in plants only to the carbohydrates. Although they commonly show inhibitory effects on growth only at fairly high concentrations (see, for example, Zinsmeister and Hollmuller, 1964), they often do occur in just such concentrations. Goren and Monselise (1964) estimate the flavonoid hesperidin to make up over 30 percent of the dry weight in some citrus tissues. The phenolics arise from the shikimic acid pathway and are generated mostly out of phenylalanine or tyrosine, as sketched in Fig. 8-2. There are other pathways of synthesis, however (Pridham, 1965).

The remaining inhibitors among the secondary plant chemicals are divided among relatively small categories. Among the quinones, juglone (Fig. 8-4) is a well-known representative. Related to the phenolic acids, quinic acid is of common occurrence, and is a component of the depside chlorogenic acid. Intermediate between the lactones and the quinones is chelidonic acid. The cyanogenic compounds, including the mustard oils and mandelonitrile derivatives, occasionally occur in high concentrations in some plants. Among the terpenes are many aromatic or volatile compounds which serve as effective growth inhibitors (Bode, 1939; Muller et al., 1964). Fatty acids include many natural growth inhibitors, which have received surprisingly little attention (Poidevic, 1965). Amino acids may serve as growth inhibitors, e.g., hydroxy-

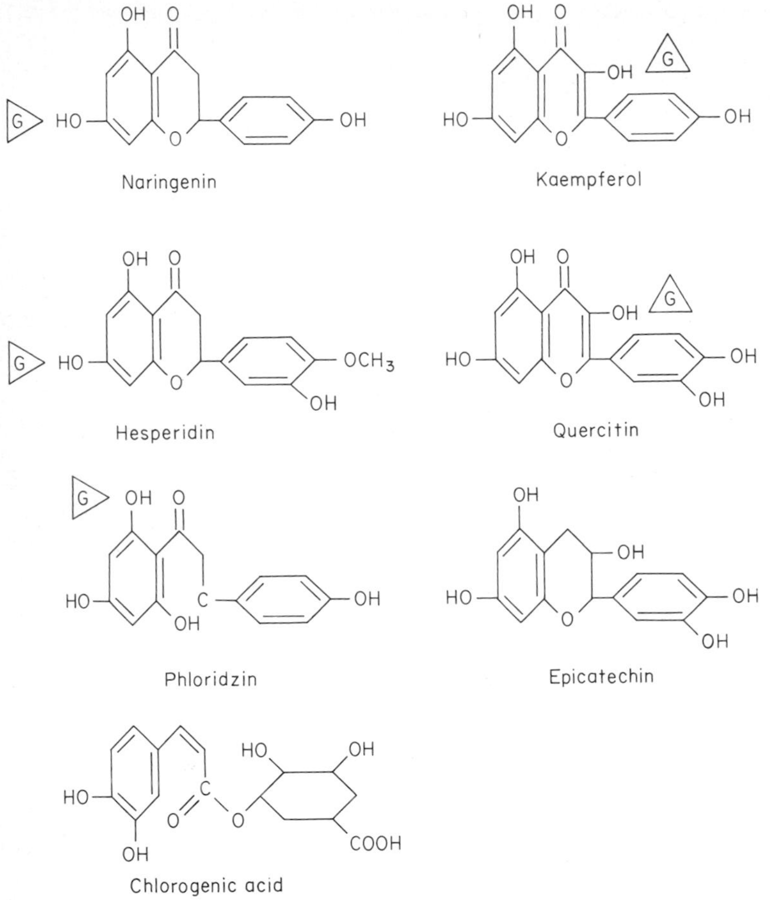

Naringenin

Kaempferol

Hesperidin

Quercitin

Phloridzin

Epicatechin

Chlorogenic acid

Fig. 8-3 Some representative phenolic inhibitors of the flavinium, chalcone, and depside categories. Common location of glucoside attachments are indicated by the G pointers.

proline (Cleland, 1963) and naturally occurring analogs of amino acids such as mimosine (Suda, 1960; Smith and Fowden, 1966). Proteins or polypeptides are occasionally reported as growth inhibitors (Elliott and Leopold, 1953). The range of naturally occurring growth inhibitors is very great, and many of them are scarcely understood or even recognized.

ABSCISIC ACID

This inhibitor can be obtained readily by alcoholic extraction of many plant materials (Milborrow, 1968). Bioassays for ABA include the coleoptile section test ordinarily utilized for auxins (Fig. 8-5), the inhibition of seed germination (Addicott and Lyon, 1969), and the inhibition of growth responses to cytokinins and gibberellins (Cornforth et al., 1966b). An extremely precise means of measuring ABA levels is through its strong *Cotton effect,* an interaction between the rotatory effect on light and optical absorbency in the ultraviolet wavelengths. ABA is unique among hormones in having an asymmetric carbon (the ring carbon to which the side chain is attached; see Fig. 8-6); this structural feature gives (+)-ABA a very strong optical rotatory power, among the highest known for any compound. The (+)-ABA can be measured either through the optical-rotatory dispersion or by mixing in a known amount of racemic mixture (±)-ABA and calculating the amount of (+)-ABA originally in the plant extract through the ratio of the optical density at 260 nm to the optical-rotatory dispersion at 246 nm (Milborrow, 1968). These optical systems of detection show that (s)(+)-ABA is common in plant materials, ranging in concentration between 0.03 to 4.0 mg/kg fresh weight.

Two other detection methods should be mentioned here, the gas-chromatographic method (after formation of the methyl ester, Lenton et al., 1968) and a bioassay involving the closure response of stomata to ABA (Tucker and Mansfield, 1971). This second method utilizes leaf pieces of *Commelina* leaves floated on ABA solution or treated with droplets of the solution in a CO_2-free atmosphere; the extent of stomatal

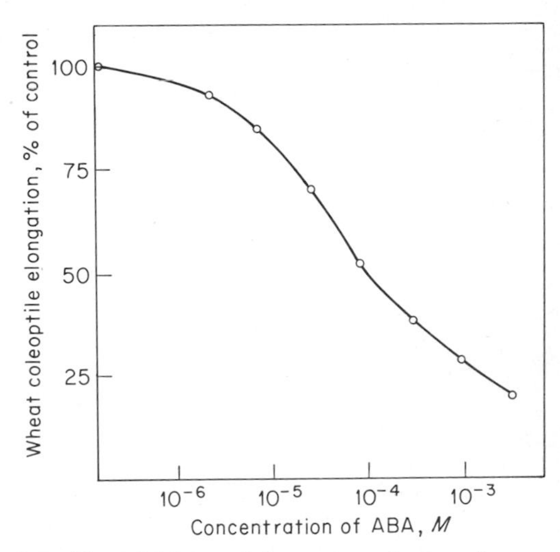

Fig. 8-5 The inhibition of elongation of coleoptile sections is an effective means of bioassay for abscisic acid (Wright and Hiron, 1972).

Fig. 8-4 Some other types of growth inhibitors.

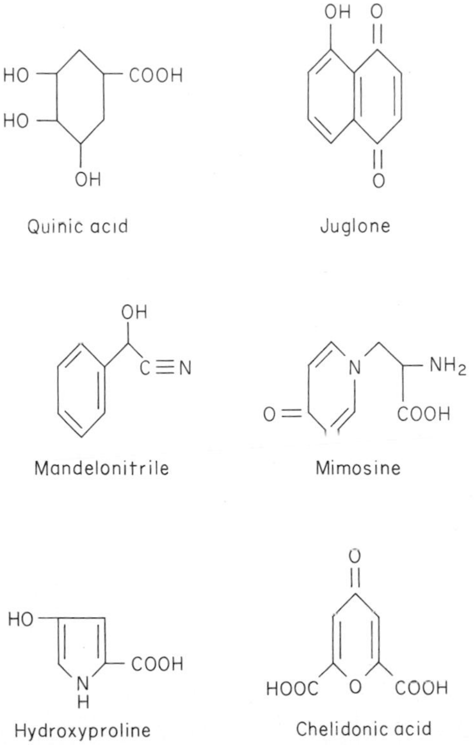

Quinic acid

Juglone

Mandelonitrile

Mimosine

Hydroxyproline

Chelidonic acid

closure is followed optically, as illustrated in Fig. 8-7.

Naturally occurring ABA is entirely of the dextrorotatory (+) type, and the synthetic is the racemic (±) mixture (Cornforth et al., 1966a). Milborrow (1968) found that both the racemic forms are biologically active.

Another type of isomerism exists in the ABA side chain: each of the two double bonds in the side chain has the alternatives of *cis* and *trans* configuration. Naturally occurring ABA has the *cis* configuration in the double bond nearest to the carboxyl and a *trans* configuration in the one nearest to the ring (Fig. 8-8). The 2-*trans*-ABA has little or no biological activity as a growth inhibitor in tests carried out in darkness (Cornforth et al., 1965a). Light causes an interconversion between the *cis-trans*-ABA and the 2-*trans*-ABA (Mousseron-Canet et al., 1966), and so the reports of small amounts of activity of the 2-*trans*-ABA in biological tests carried out in the light are undoubtedly reflections of the photoracemation (Milborrow, 1970). In plant materials generally, ABA exists mainly as the *cis-trans* form, but stereolabeling experiments with mevalonic acid substrates indicate that the initial product of ABA biosynthesis is the 2-

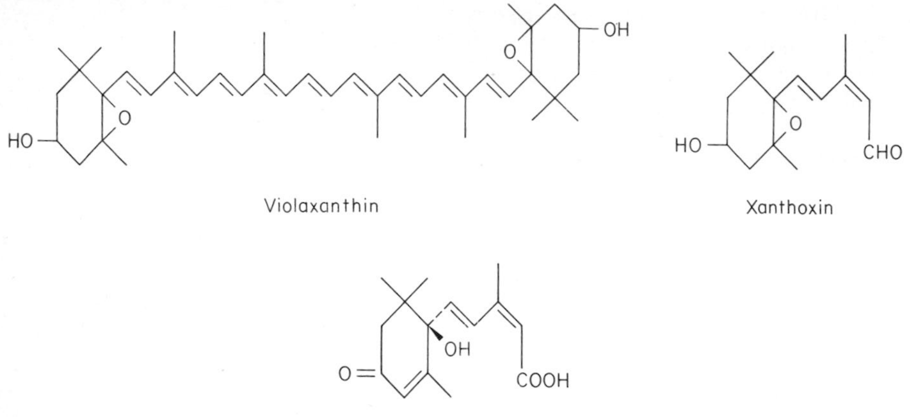

Violaxanthin

Xanthoxin

Abscisic acid

Fig. 8-6 The formation of abscisic acid may occur in part through the oxidation of some xanthophylls such as violaxanthin; xanthoxin, a keto derivative, which may be an intermediate, has biological activity (Taylor and Burden, 1970).

trans-ABA (Robinson and Ryback, 1969); it is probable that the final conversion into the *cis-trans* configuration in the plant occurs as a response to light (Milborrow, 1970). Little more can be said of the relation of structure to activity of ABA. Many analoges have been synthesized, but all changes that have been examined, either to the ring or to the sidechain, result in marked depression of biological activity (Milborrow, 1974).

Two attractive pathways for the biosynthesis of ABA would be directly through the isoprenoid pathway from mevalonic acid or, alternatively, through the oxidation of xanthophylls with similar basic structures, e.g., violaxanthin (Addicott et al., 1966). Taylor and Burden (1970) have found that either the photooxidation or biological oxidation of violaxanthin can give rise to xanthoxin (Fig. 8-6), a keto epoxide derivative of ABA which shows growth-inhibitory action in bioassays. The epoxide derivatives of ABA were synthesized by Tamura and Nagao (1969), and subsequent experiments by Milborrow and Noddle (1970) showed that these can be converted into ABA by various plant materials. Wright and Hiron (1969) found that the biosynthesis of ABA in wheat leaves is enormously stimulated by a period of water stress, and Milborrow and Noddle (1970) have utilized this stimulatory effect to examine the pathway of ABA synthesis. Their work and other labeling experiments (Robinson and Ryback, 1969; Milborrow, 1972) indicate that ABA biosynthesis normally occurs directly from mevalonate, rather than from the more indirect oxidation of xanthophylls. Milborrow (1972) has found that excess ABA in a

Fig. 8-7 The effectiveness of abscisic acid in causing stomatal closure can be utilized as a bioassay. Leaf pieces from *Commelina communis* floated on ABA solutions in the light were measured for stomatal aperture after 3 h (Tucker and Mansfield, 1971).

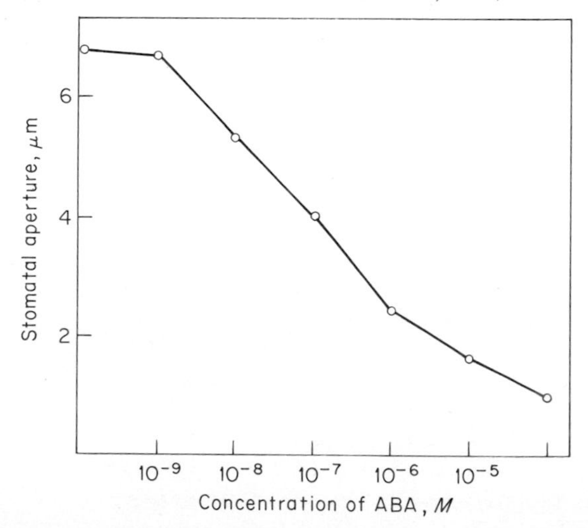

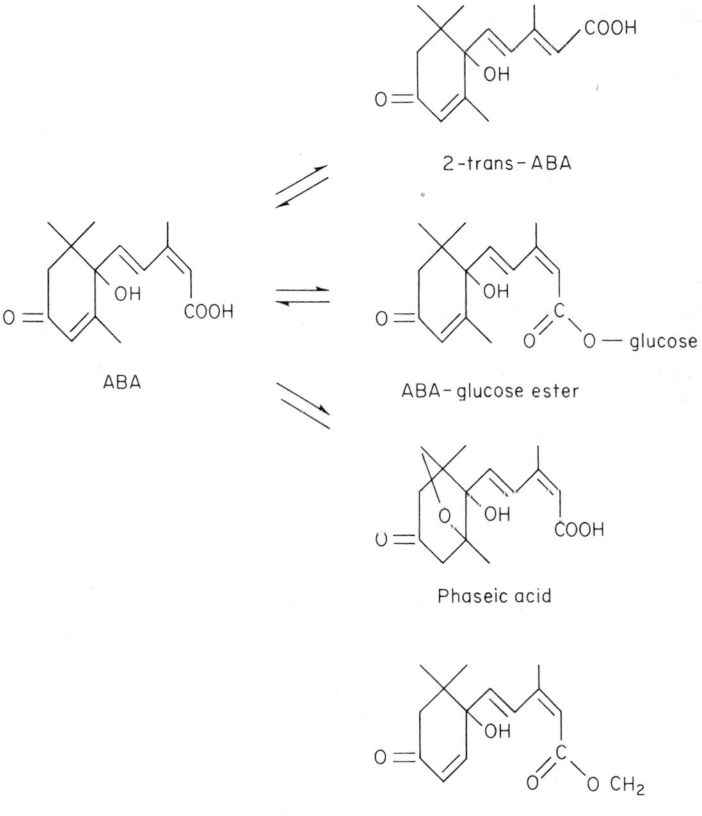

Fig. 8-8 The metabolism of abscisic acid in the plant is known to give rise to the 2-*trans*-ABA, the glucose ester, and possibly to phaseic acid. Methyl-ABA has not been detected as a natural product.

tissue causes a feedback inhibition of ABA biosynthesis. An important site for ABA biosynthesis is the plastids, especially the chloroplasts (Milborrow, 1974).

The distribution of ABA is very widespread indeed. It is of general occurrence in monocots and dicots and also in ferns. Using the racemate-dilution method, Milborrow (1968) found ABA in higher concentrations than had been estimated using the optical-rotation method of measurement. Fruits contained the highest ABA concentrations, rose fruits having the highest level of the various tissues examined.

The translocation of ABA through the plant has been examined in experiments similar to those concerned with auxin transport through stem or petiole sections, and ABA moves with a greater velocity than auxin, in the range of 20 mm/h (Ingersoll and Smith, 1971). The movement of this hormone appears to lack consistent directional or polar qualities.

The metabolism of ABA has been studied using radioactive ABA, and some evidence for three different metabolites has been found: (1) The ready conversion to the 2-*trans*-ABA has already been mentioned. (2) Extracts contain an oxidation product of ABA, phaseic acid (Fig. 8-8). Both these products are apparently without biological activity. (3) The glucose ester, first isolated from plants by Koshimizu et al. (1968), is formed much more readily from the 2-*trans*-ABA than from the *cis*-*trans*-ABA, and while it is not hydrolyzed readily by β-glucosidase, it is readily deesterified by cell sap to release ABA (Milborrow, 1970). The methyl ester of ABA has been reported to have some biological activity as an inhibitor in growth responses (Koshimizu et al., 1966), but short-term stomatal closure is not responsive to the ester (Kriedemann et al., 1972), suggesting that the growth responses may occur only after the hydrolysis of the ester. The ethyl ester has no biological activity in bean-stem sections (Walton and Sondheimer, 1972). At present, neither the methyl nor the ethyl ester has been found as a natural metabolite of ABA in plants.

There are surprising differences in the readiness with which various enantiomorphs of ABA are metabolized. The readier formation of the glucose ester from the 2-*trans*-ABA has been already mentioned; Sondheimer et al. (1971) have also found that the natural (S) isomer of ABA is metabolized almost tenfold faster than the unnatural (R) or (−) isomer.

The regulatory role for ABA in plants is best documented for responses to stress, especially water deficiency. The impressive increases in ABA synthesis in wheat leaves following water stress discovered by Wright and Hiron (1969) have been mentioned. The magnitude of the ABA increase is illustrated by the data in Fig. 8-9,

Fig. 8-9 The wilting of brussels sprouts plants results in a large increase in ABA content of the leaves, which gradually declines after the plants have been rewatered. Plants brought to 40 percent wilting by a 72-h water stress (Wright and Hiron, 1972).

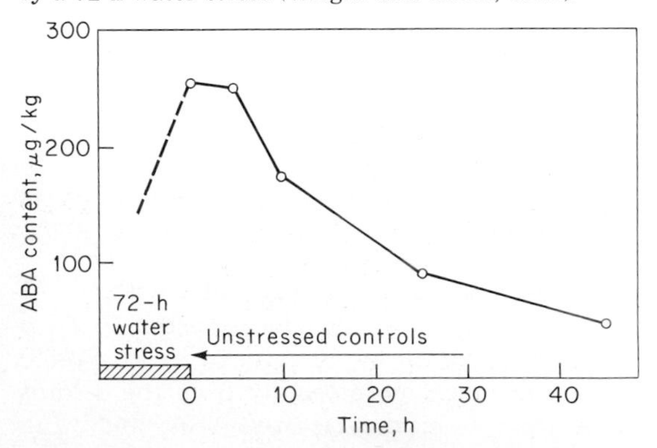

and the same type of increases with water stress have subsequently been reported for quite a wide range of plant species (see, for example, Wright and Hiron, 1972; Milborrow, 1972). As indicated in Fig. 8-9, return of ABA to normal levels may be expected after 1 or 2 days following restoration of normal water supply. Similar increases in ABA are brought about by osmotic stress (Mizrahi et al., 1970) or waterlogging the roots (Wright and Hiron, 1972) or by the stress resulting from mineral deficiencies (Mizrahi and Richmond, 1972). The rise in ABA level following stress serves to bring about a rapid closure of stomata (Mittelheuser and van Steveninck, 1969), a phenomenon illustrated in Fig. 8-7. It seems reasonable to construe the stomatal control as a protective system against transpirational water loss under conditions of stress. Cytokinins are known to stimulate the opening of stomata (Meidner, 1967), and there is some evidence for a balancing between cytokinins and ABA in the regulation of stomatal behavior (Mizrahi et al., 1970).

A mutant tomato has been described (Tal et al., 1970) which suffers from chronic wilting of leaves; the mutant appears to lack a normal ability to close the stomata (cf. Table 17-3). Since it was suspected that the difficulty might be due either to excessive cytokinin levels or deficient ABA levels, the content of these hormones was measured, and the *flacca* condition was found to be associated with a deficient ABA content. Application of ABA readily restores the turgid condition. Graftage of normal tomatoes onto the *flacca* rootstock induced the wilty condition, suggesting that the root system failed to provide some component of the stomatal regulatory system.

The regulatory role of ABA in the detection of seasonal cues is less clear. Numerous reports in the literature describe the accumulation of inhibitors under photoperiods which induce dormancy. The inhibitor is readily shown to have the chromatographic characteristics of inhibitor β (Fig. 8-10). The accumulation of inhibitors under short photoperiods provided the material which led to the discovery of ABA in extracts of sycamore (Eagles and Wareing, 1963). In that species,

ABA accumulates under short photoperiods and then decreases as dormancy is passed, and the application of ABA actually induces dormancy in sycamore, including the induction of the normal morphological features of dormancy (El-Antably et al., 1967); the dormancy induced by ABA is relieved by applications of gibberellin. In some species, however, ABA levels are increased under long-day conditions instead of short days (Zeevaart, 1971; Lenton et al., 1972), and even in the short-day induction of dormancy in birch seedlings, determinations of ABA levels have not shown ABA to increase under the short photoperiods (Loveys et al., 1974). Nevertheless, there is a wide range of short-day regulatory effects which seem to involve increases in ABA, including the regulation of flowering, senescence, tuber formation (El-Antably et al., 1967), and the induction of coldhardiness (Irving, 1969). In the last instance, ABA levels are apparently increased by low-temperature experiences as well as by short photoperiods. Burden et al. (1971) have suggested that light itself can stimulate ABA formation through the stimulation of the xanthoxin levels.

During the ontogeny of cotton fruits, ABA content shows some dramatic changes, reaching the highest levels during the periods of abscission and dehiscence of the fruits (see Fig. 13-25). Its occurrence during the period when developing fruits are liable to be abscissed from the plant led to the suggestion that it serves as the endogenous stimulator of abscission (Addicott and Lyon, 1969), and it was isolated from this material and named for its stimulatory effects on abscission. The abscission-speeding effect of ABA is not universal by any means, and part of its effect may be a consequence of its stimulation of ethylene production (Cracker and Abeles, 1969).

It should be noted that in some plants at least ABA can cause an increase in growth (Takahashi, 1972). In special cases, applications of ABA can bring about increases in GA levels (Railton and Wareing, 1973).

When we turn to the mechanism of ABA action, it is not at all surprising to find many instances in which this hormone is known to cause

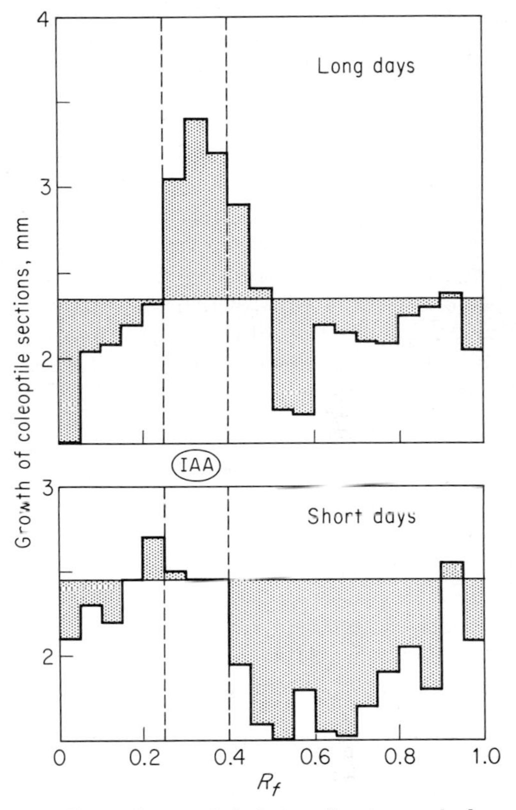

Fig. 8-10 Short photoperiods bring about a marked increase in growth inhibitors in *Rhus typhina*. After 14 long or short days, shoot tips were extracted with methanol and chromatographed in isopropanol-ammonia-water, and the paper was assayed with the oat-coletoptile section test. The dotted lines indicate the expected R_f of indoleacetic acid; the inhibitor seen at R_f 0.5 to 0.7, called inhibitor β, is probably mainly ABA (Nitsch, 1957).

dramatic alterations in the activity of certain enzymes. It can suppress the formation of α-amylase in the barley endosperm, thus acting in a general way against the actions of gibberellin (Chrispeels and Varner, 1966). Some other enzymes recorded as being restricted by ABA include phenylalanine ammonia-lyase (Walton and Sondheimer, 1968) and invertase (Cherry, 1968).

The mechanism of ABA action has been inferred by several authors to involve a modification of the nucleic acid and protein-synthesis systems. Van Overbeek et al. (1967) suggested that ABA might interfere with DNA synthesis, but Haber et al. (1969) showed that lettuce seeds irradiated so that they were incapable of synthesizing DNA could still show the ABA inhibition of germination. Pearson and Wareing (1969) found that ABA depresses RNA synthesis in radish hypocotyls, possibly by a depression of chromatin activity. Walton et al. (1970) have likewise observed a depression of RNA synthesis in bean pieces by ABA, which they suggested was due to an inhibition of translation. ABA can bring about an actual depression in DNA and RNA levels (Belhanafi and Collet, 1970), and it has been suggested that this may result from an increase in nucleases (Leshem, 1971). Leshem and Schwarz (1972) report a preferential decrease in ribosomal RNA following ABA treatment of tobacco.

As with the other hormones, the rapidity of plant responses to ABA raises some questions about whether the mechanism of hormone action depends specifically on an alteration of the nucleic acid–protein-synthesis system. Kriedemann et al. (1972a) have measured stimulation of stomatal closure within a minute of ABA application. The rapidity of response implies that some ABA effects are independent of nucleic acid–directed protein synthesis.

Fig. 8-11 In buckwheat hypocotyls, light causes a marked increase in phenylalanine ammonia-lyase activity; and the activity declines again after removal of the seedlings to darkness. Hypocotyls were extracted and acetone powders placed in buffer in the presence of phenylalanine; the formation of *p*-coumaric acid was followed as the optical density at 280 nm (Scherf and Zenk, 1967).

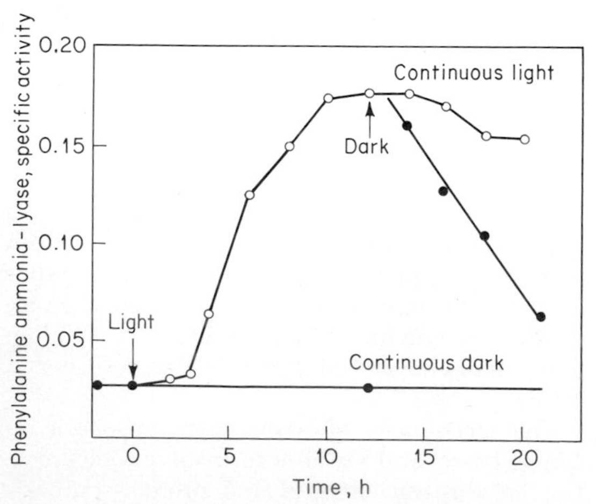

PHENOLIC INHIBITORS

The phenolic acids form the best-known group of inhibitors among the secondary plant chemicals. Most of them are synthesized via the shikimic acid pathway, through the deamination of phenylalanine or tyrosine, although a few are synthesized from acetic acid units. Members of the cinnamic acid series can be converted into the benzoic acid counterparts by β-oxidation (Zenk and Müller, 1964). Elaboration of more complicated hydroxylation patterns can follow the formation of the simple phenolic acids (Nair and Vining, 1965). The lactones of the coumarin series are formed by the ready lactonization of the phenylpropene equivalents, especially after the formation of the *o*-glucoside (Neish, 1965) and the conversion of the *trans*-cinnamic acid to the *cis* configuration by the action of light (Haskins and Gorz, 1961). The phenolic lactones ordinarily are formed via hydrolysis of the glycosides. The more elaborate flaviniums and chalcones (Fig. 8-3) are thought to be derived from the cinnamic acid series with the addition of acetate blocks to constitute the A ring.

The phenolics ordinarily occur in plants as glycosides, usually involving glucose but sometimes several sugars in series. Rupture of plant tissues ordinarily leads to the hydrolysis of phenolic glucosides to the aglycones, which are markedly more reactive (Pridham, 1965).

The inhibitory effects of phenolics on growth are commonly attributed to the enhancement of indoleacetic oxidase, but it is very probable that other actions such as an interference with oxidative phosphorylation are involved (Pridham, 1965). The effectiveness of phenolics as cofactors for IAA oxidase is fairly specific for monophenols such as *p*-coumaric acid. *o*-Diphenols such as caffeic acid can serve as inhibitors of IAA oxidase, hence acting as stimulants of growth under some conditions. A comparison of the effects of these two phenolics is shown in Fig. 4-12. Homologous flavinium derivatives such as kaempferol and quercitin (monophenolics and *o*-diphenolics, respectively) are likewise respectively stimulants and inhibitors of IAA oxidase. The polyhydroxy phenolics are in many instances stimulators of growth at low concentrations and become inhibitory at higher concentrations (Nitsch and Nitsch, 1962).

The ability of phenolic acids to alter the oxidation of IAA is well established; for example, the effects of monophenols as stimulants of IAA decarboxylation are illustrated in Table 8-1, along with examples of the retardation of decarboxylation by *o*-diphenols such as caffeic and chlorogenic acids. However, these effects are often not reflected in alterations in growth rates (Tomaszewski and Thimann, 1966). Aberg and Johansson (1969) have reported that monophenols inhibit both stem and root growth and that *o*-diphenols promote both; this is surprising since we have no reason to believe that auxin is ordinarily a rate-limiting factor in root growth. Furthermore, additions of IAA or 2,4-D did not restore normal root growth. These workers concluded that the regulatory effect of phenols is not primarily through the effects of IAA oxidation. Kefeli and Kadyrov (1971) suggest that the effects of phenolics are primarily on metabolic systems rather than on hormonal systems.

Table 8-1 Effects of various phenolic acids on the decarboxylation of 1-^{14}C-indoleacetic acid by oat coleoptiles (Zenk and Müller, 1963).

Phenolic acid supplied except where noted 0.1 m*M*	Enzymatic destruction of IAA, % of control
IAA (0.01 m*M*) alone	100
p-Coumaric acid	181
Phloretic acid	188
p-Hydroxybenzoic acid	113
Ferulic acid	126
Caffeic acid	79
Chlorogenic acid	73

Regulation of growth and development by phenolics and other secondary plant chemicals is markedly less clear than with ABA. Certainly the best-documented regulatory effect relates to the responses of seedlings to light. Furuya and Thomas (1964) observed a large increase in kaempferol glucoside coumarate in pea seedlings after a brief illumination with red light; this was associated with a rise in the fresh weight of the plumular-hook region of the seedling, but no alteration of the epicotyl growth was reported. The kaempferol increase continued to expand over a 15-h period following the light treatment. Engelsma and Meijer (1965) noted a similar increase in *p*-coumaric acid and other phenolics in gherkin seedlings after illumination with red or blue light, and there was an associated drop in seedling growth. The increase in phenolics might be partly a consequence of an increased conversion of phenylalanine into members of the cinnamic acid series; by assaying for phenylalanine ammonia-lyase activity after supplying phenylalanine to tissue disks and measuring the production of *trans*-cinnamic acid it has been observed that light causes a tenfold increase in activity in potato-tuber disks (Zucker, 1965) and buckwheat hypocotyls (Scherf and Zenk, 1967) (Fig. 8-11). The light enhancement involves the synthesis of new enzymes, as indicated by experiments with protein-synthesis inhibitors and leucine incorporation (Zucker, 1969). The light effect involves photosynthesis in part but is saturated at much lower light intensities than photosynthesis is. Umemoto (1971) has reported a photoperiodic regulation of chlorogenic acid formation in *Lemna,* with much greater synthesis in days longer than 12 h, but the phenolic formed does not seem to participate in growth or flower-induction processes.

OTHER TYPES OF INHIBITORS

In addition to the evidence for phenolics acting against auxin, there are occasional reports of inhibitors which act against the other hormone systems, including the gibberellins and cytokinins. In several instances, plant extracts are reported to contain substances which can inhibit the gibberellin-stimulated growth (Köhler and Lang, 1963; Michniewicz, 1967) or inhibit cytokinin-stimulated growth (Letham,

1963a; Kefford et al., 1968). Since abscisic acid is now known to be capable of inhibiting responses to all the growth hormones (Cornforth et al., 1966b), it is possible that such extracts simply contained ABA. Growth inhibitions by pseudogibberellin (Kato and Katsumi, 1967) and helminthosporal (Tamura et al., 1965) suggest the possibility of more direct gibberellin antagonisms.

SUMMARY

In assessing the roles of growth inhibitors in the regulation of growth and development, there are instances in which it seems clear that either ABA or other types of inhibitors may serve to regulate the growth rates in plants. In the same breath one should say, however, that there are also cases where the content of inhibitors does not seem to correlate with growth rates; this applies to ABA (Zeevaart, 1971), phenolic inhibitors (Goren and Monselise, 1964,) and others such as the cyanogens (Jones and Enzie, 1961). In some instances inhibitory substances appear to be involved in responses of plants to light, including also response to photoperiod. ABA can also participate in the responses of some plants to temperature and drought-stress conditions. Thus, in the growth inhibitors we again find chemical devices through which some plants can translate environmental cues such as light, photoperiod, temperature, and water stress into internal regulatory controls over growth and development processes.

GENERAL REFERENCES

Kafeli, V. I. and C. S. Kadyrov. 1971. Natural growth inhibitors, their chemical and physiological properties. *Annu. Rev. Plant Physiol.,* 22:185–196.

Milborrow, B. V. 1972. The biosynthesis and degradation of abscisic acid. In D. J. Carr (*ed.*), "Plant Growth Substances, 1970." Springer-Verlag, Berlin. pp. 281–290.

Swain, T. (*ed.*). 1966. Comparative Phytochemistry. Academic Press, London. 360 pp.

9 DIFFERENTIAL GROWTH

In the multicellular organism there must be regulatory systems through which cells mutually determine the location and type of growth. The existence of five different types of hormones provides an array of regulatory controls through which differential growth can be achieved. Some of these hormones may contribute polar regulatory signals, e.g., auxin especially, and others may be quite without polar functions, e.g., gibberellin and ABA especially.

In a discussion of the phenomena of differential growth in plants, the theme of regulation through hormonal interactions will recur like a refrain. As our understanding of differential-growth phenomena expands, we increasingly see that nearly every type of regulation in the plant can be influenced by the interactions of several plant hormones. Like notes in a musical scale, hormones can be arranged in numerous combinations and sequences to give a wide range of effects.

POLARITY

We can infer that the emergence of plants from the aqueous to the aerial environment in early evolutionary time increased competition and placed greater survival value on organized aerial or elongation growth. Almost every higher plant is an axially organized organism, and the arrangement of cell division, elongation, and differentiation is organized along the plant axis. Before differentiation can be achieved, the cells in the organism must evolve a polarity. Sinnott (1960) states that polarity is the first step in differentiation.

Polarity appears even in primitive plant forms. Unicellular plants exhibit distinct polarity, developing head ends with flagellae or an eyespot. The multicellular strands of green algae and fungal hyphae characteristically exhibit a continuous polarity over the complete set of cells, all with morphological orientation in the same direction. It appears that a polar orientation can be transmitted from cell to cell, resulting in the long chains of similarly oriented cells such as *Cladophora* or the cladified colonies such as *Ulva*.

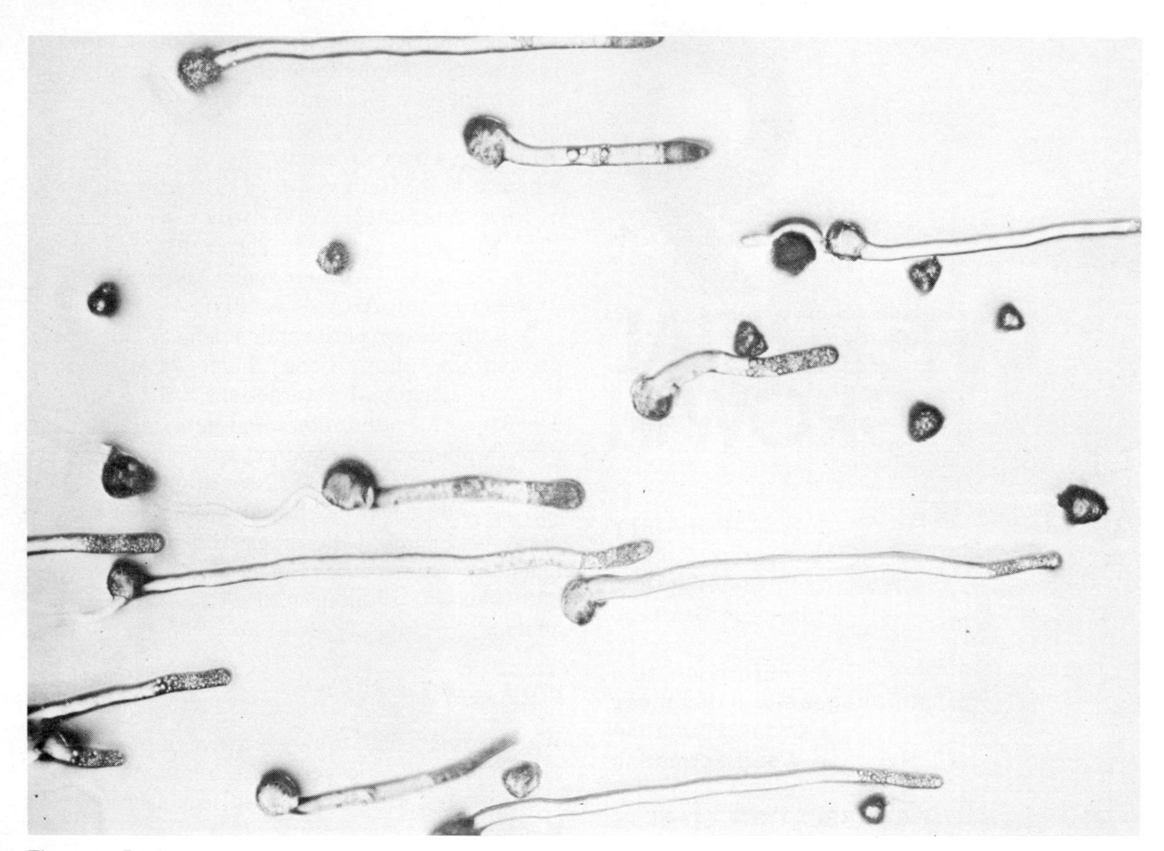

Fig. 9-1 Light can impose a polarity on the germination of fern spores (*Pteridium aquilinum*). Light from the right side of the field has oriented the position of emergence of the protonema from the spores.

Many types of cells have no polarity until they receive some external stimulus. Egg cells or zygotes may be released into the environment without any polar orientation, and then, as growth begins, polarity of the new organism appears. These free nonpolar cells provide excellent experimental material for studying the initiation of polarity. Animal physiologists use frog eggs and sea urchin eggs for such studies, and plant physiologists use the spores of algae, fungi, mosses, horsetails, and ferns.

In the plant, polarity is manifested morphologically by a preferential direction of cell division, cell enlargement, and cell differentiation and by the orientation of the tissues formed from these cellular activities. These preferential-direction qualities are dramatically illustrated by experiments on the induction of polarity and by regeneration and grafting characteristics.

Induction of Polarity

In the lower plants polarity may be initially absent in the new spore and can be established by environmental factors. As spores are shed, the side settling on the ground ordinarily becomes the root end. Light is clearly involved in polar induction; unilateral light induces the polar establishment of the apical end in spores of ferns, horsetails, and fungi and in fern protonemata. Figure 9-1 illustrates the ability of light to orient the polarity of fern-spore germination. Haupt (1957) found that blue light is most effective in invoking polar induction. Several chemicals erase the light-polarizing effect, including chloroform (Haupt, 1957), auxin, and thiamin (von Wettstein, 1953). For a brief period after initial lighting, the polarity of *Equisetum* spores can be reversed by a second light exposure from the opposite direction (Haupt, 1957).

Physical force may induce polarity. Centrifuging individual cells of filamentous algae or pieces of liverwort in such a way that the heavy components move to the apical end can sometimes alter the regeneration pattern so that new rhizoids sprout from the former apical end and new apical shoots from the former basal end. Electric fields have also altered the polarity of regeneration of fungal spores and fern prothallia, rhizoids developing on the electrically positive side and apical shoots developing on the electrically negative side (cf. review by Bünning, 1957).

Pollen grains seem to have little or no polarity, and from which side the pollen tube will emerge can be controlled by centrifugation (Beams and King, 1944*b*). When pollen grains germinate close to one another, there is a tendency to orient emergence so that the pollen tubes are directed away from one another (Beams and King, 1944*a*). This suggests that diffusible substances may participate in the determination of germination polarity.

In most higher plants polarity of a new embryo in the ovule is oriented in a uniform direction, the radicle developing at the end nearest the micropyle (Sinnott, 1960). The polarity of each new cell in a tissue is the same as that of its neighbor, suggesting a possible orienting effect by contact (Bünning, 1957).

In experiments with the culture of individual cells from carrot roots, Steward et al. (1958) found that cell division occurs in a given clone while the culture solutions are rotated continuously but that differentiation proceeds best when the callus is transplanted onto a stable medium. Thus, when the culture is shifted continuously without a positional orientation, differentiation is arrested. It is not clear whether a gravitational force field is involved in establishing polarity in this case.

The induction of polarity in *Fucus* zygotes has been shown by Jaffe (1968) to involve the generation of a bioelectric field. The emergence of the rhizoid involves the synthesis of new enzymes, of course, and Quatrano (1968) has shown that inhibitors of RNA synthesis are effective in preventing the rhizoid emergence only if added in the first 8 h after fertilization. This implies that the RNA which will regulate the rhizoid differentiation is synthesized in the first 8 h; the rhizoid emerges at about 14 h.

Once polarity has been established in a clone or individual cell, it seems to be stable, and new cells added to the organism assume the same polarity.

Polarity in Regeneration

Experiments on regeneration in plants produced the first basic evidences of polarity. Vöchting (1878) established that cuttings of willow stems form roots at their physiologically basal ends and buds at their apical ends regardless of their orientation during the rooting period (Fig. 9-2). He noted that ringing the bark of the cutting acts much like cutting entirely through the stem in regeneration patterns. Whatever size of cutting he tried, there was always the polar orientation of root regeneration at the physiological base; he concluded that polarity is fixed in the plant cell and is irreversible.

Regeneration experiments with some lower plants show weaker polarity patterns. Cuttings from liverwort thalli may show only a weak polarity of rhizoid and protonema formation, and this may be modified or even reversed by environmental influences such as light or gravity (Fitting, 1938).

The polarity of regeneration of stems is the most inflexible. In root cuttings there is sometimes a less pronounced or a more readily modified polarity, and in leaf cuttings roots and buds often form together at the basal end of the leaf piece without regard to apical position on the piece, indicating a relatively weak inherent polarity. The pattern of regeneration of root cuttings is illustrated by experiments on *Taraxacum* (Warmke and Warmke, 1950), in which any cutting regenerates buds at the physiologically

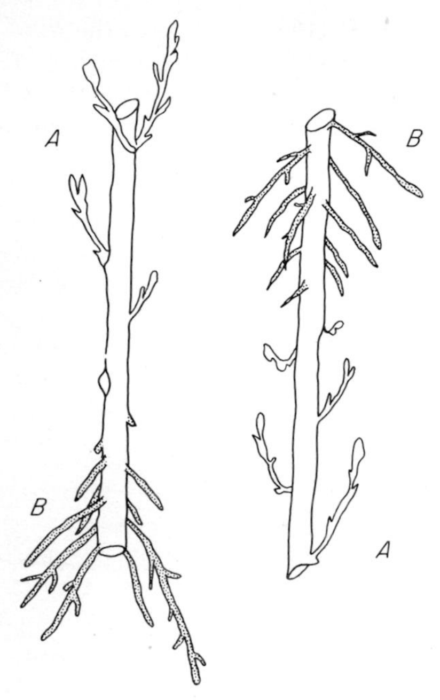

Fig. 9-2 Regeneration of roots in a stem cutting of willow occurs at the physiological base (*B*) of the piece, regardless of the position during regeneration (Pfeffer, 1906; after Vöchting, 1878).

proximal end and roots at the distal end (nearest the root tips), regardless of the orientation of the piece during regeneration (Fig. 9-3).

Regeneration of roots is regulated at least in part by auxin, and the localization of the differentiation of roots at the basal end of cuttings is due to the polar movement of auxin toward the physiologically lower end. This polar movement holds true in both stems (Thimann and Went, 1934) and roots (Wilkins and Scott, 1968). While the polarity of root formation in cuttings may be a reflection of auxin polarity, the case for buds is less clear. Buds generally form at the apical end of cuttings of either stems or roots (Figs. 9-2 and 9-3), and while there is some evidence of a bud-stimulating agent moving to the apex (Lindner, 1940), such a polar movement is not well established.

Cytokinins are the most probable factor in bud regeneration (Dankwardt-Lilliestrom, 1957), and the bud-forming tissues are richest in cytokinins (Heide and Skoog, 1967); however, the polarity of bud formation is not known to be related to any acropetal movement of cytokinin. It is noteworthy that the ability of cuttings to generate buds is often correlated with the size of the piece of tissue (Bonnett and Torrey, 1965; Okazawa et al., 1967).

Polarity in Graftage

Evidence of the polarity of plant cells and tissues is shown in the graftage experiments of Vöchting (1878) with kohlrabi stems and beet roots. Pieces could be readily grafted if they were in the same physiological orientation, but if a stem piece was grafted into another stem in an inverted position, the graft did not take. Instead the scion sometimes invaded the stock with roots, as if it were a soil medium. The explicit polarity for graftage is much less pronounced in fruits. Bloch (1952) found that pieces from *Lagenaria* fruits could be reinserted into the fruit in any position and the graft would be successful. As with regeneration, the polarity of grafts seems to be most pronounced in stem pieces.

Horticulturists have recognized that fruit-tree buds must be grafted into a stock in the proper orientation for good take. Inverted buds may sometimes knit well with cambial tissue, but the vascular connections can be made only by twisting cells so that adjoining ones meet with the same polarity. Apparently cambial cells are somewhat more flexible in their polarity than

Fig. 9-3 Regeneration of roots and buds on root cuttings of *Taraxacum,* showing the formation of roots at the end nearest the root tip (*B*) and buds at the end nearest the stem (*A*) regardless of the orientation of the cutting during regeneration (Warmke and Warmke, 1950).

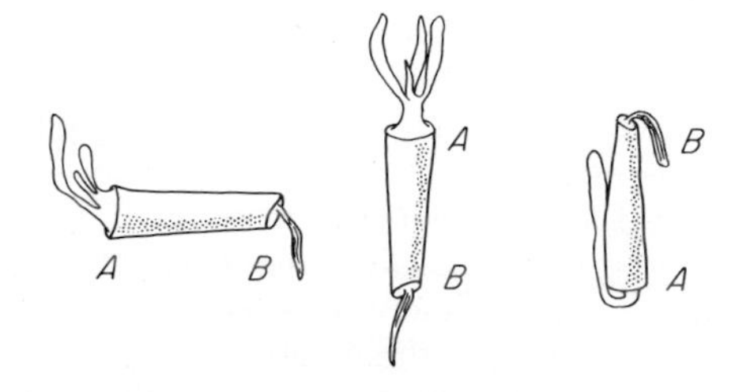

the more completely differentiated tissues (Sinnott, 1960).

DIFFERENTIATION

After polarity has been established in a cell, organized patterns of cell division, enlargement, and tissue differentiation can follow. The beginning of such steps is seen in a callus tissue in culture as it develops local meristem areas with organized cell divisions, cell enlargement, and tissue differentiation.

The meristem not only provides a source of cells for the production of tissues but appears also to provide organizers, or stimuli of differentiation, of the cells into tissues. Decapitation of a stem or root is often followed by a period of partially suspended differentiation.

The requirements for the completion of differentiation may vary phylogenetically. Wetmore (1954) showed that cell clusters from gymnosperms cultured on a simple sugar medium differentiate into complete new plants but explants from angiosperm tissues differentiate new individuals only if the culture medium contains more elaborate nutrients, including auxin, B vitamins, and coconut milk.

Morphogenetic Substances

The ability of some tissues to impose differentiation on others is illustrated with tissue cultures of pith or callus which are maintained in contact with differentiated or meristematic pieces. Jablonski and Skoog (1954) found that pith tissue from tobacco stems cultured in contact with another piece containing vascular strands undergoes cell divisions. They suspected that the nurse piece provides substances which evoke the cell-division response, and they obtained preliminary evidence that extracts from vascular tissue can stimulate cell divisions. Substitution of other substances such as coconut-meat extracts can cause cell division, differentiation of vascular tissues, and formation of buds on the pith piece.

In another nurse-tissue experiment, Wetmore and Sorokin (1955) induced the differentiation of vascular tissue in a block of callus tissue. Simulating the situation in an intact stem, they inserted a small piece of stem tip into lilac-callus tissue in sterile culture. Below the point of contact, scattered rows of xylem tracheids differentiated in the callus (Fig. 9-4). They also found that some xylem differentiation in the callus can be induced by supplying a mixture of auxin and coconut-milk extract in the notch where the meristem would otherwise be introduced. These experiments suggest that substances in the meristem act as the organizers of morphogenesis in adjoining cell regions.

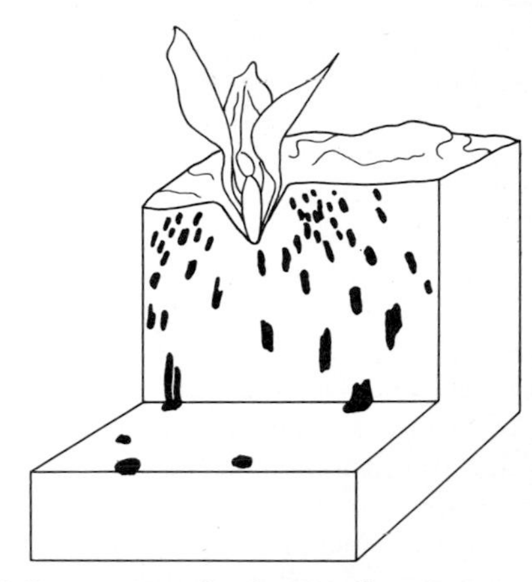

Fig. 9-4 The differentiation of xylem can be stimulated in lilac pith-tissue cultures by implantation of an apical meristem on one side of the cultured piece. This 10-mm^3 piece of callus (*cutaway representation*) was implanted with a lilac meristem, and the xylem cells (*shaded*) were recorded after 54 days (Wetmore and Sorokin, 1955).

The role of auxin in xylem differentiation is suggested by experiments on the regeneration of xylem strands around a cut in the stem of *Coleus*. A cut through a vascular strand resulted in the formation of a new chain of xylem tracheids to rejoin the strand, and the new cells were differentiated principally in a basipetal direction (Sinnott and Bloch, 1944). Jacobs (1952) found that removal of leaves above the cut reduced the xylem differentiation and that addition of auxin paste to the cut petioles restored the differentiation. Later he correlated the rate of xylem formation in expanding leaves with the diffusible auxin content, as shown in Fig. 9-5.

If auxin does stimulate xylem formation, why is xylem not randomly differentiated in tissues grown on a medium containing auxin? Clutter (1960) inserted a vial of auxin solution into the top of a callus piece in culture and obtained a localized formation of xylem near the solution application site. This experiment and that of Wetmore and Sorokin (1955) imply that a localized or oriented occurrence of auxin is critical in the differentiation of xylem.

Experiments with roots have added some interesting facts. To study whether the root apex imposes the pattern of differentiation on the root, Torrey (1955) excised and decapitated root-tip pieces. During regeneration of the tips he found an initial variance of the stelar configuration from the typical triarch to the hexarch stelar pattern, but after some time the stelar pattern returned to triarch. Torrey observed that thicker root tips were more liable to show increases in

Fig. 9-5 The differentiation of new xylem cells in the expanding *Coleus* petiole is correlated with the amount of diffusible auxin obtainable from the base of the petiole. Auxin assayed by *Avena* curvature test (Jacobs and Morrow, 1957).

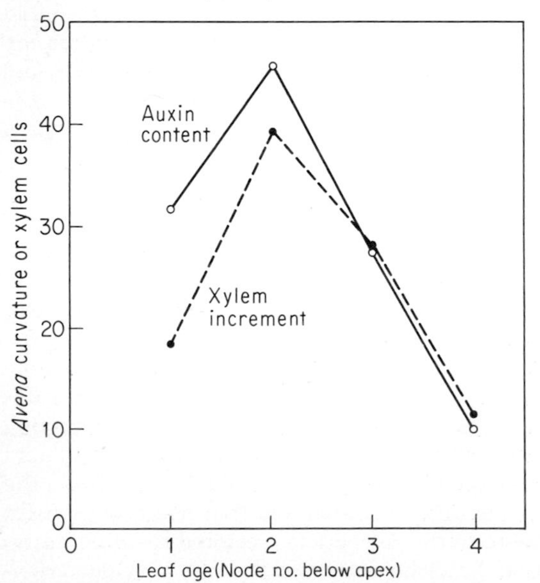

stelar rays. Later (Torrey, 1957) he added auxins to the nutritive medium on which the tips were regenerating, and the resulting thickening of the excised apices was again associated with marked increases in the stelar rays.

Besides the information on morphogenetic substances from the apical meristem influencing xylem formation, there is little evidence about other transmitted influences. There may be some influences from the proximal parts; e.g., phloem differentiation seems to be an invasive action from the older parts of the stem and leaf, maintaining its terminal limits essentially at a constant position with respect to the distance from the developing stem or leaf tip (Jacobs and Morrow, 1958). Wetmore (1959) states that high sugar concentrations in a culture medium encourage the differentiation of phloem cells, and perhaps the sugar availability regulates the projection of the phloem strands.

Tumorous or callus-type growths may be regarded as cells in which the capacity for polar orientation or differentiation is impaired. Numerous chemical materials can induce tumorous growth, including the synthetic auxins. De Ropp (1947) induced tumors on leaves by application of *p*-chlorophenoxyacetic acid plus a yeast extract. In tumorous growth, where polar features are apparently lost, there may be a singularly high effectiveness in the biosynthesis of cytokinins (Wood and Braun, 1961) or of both cytokinins and auxin (Schaeffer and Smith, 1963).

Organ Formation

The differentiation of roots and of buds on cuttings offers some of the most impressive examples of hormonal regulation. And here, as in so many other regulatory functions, we see the interactions of the various growth-regulating substances determining the types and quantities of organs differentiated.

The differentiation of roots was first reported to be a consequence of auxin stimulation (Thimann and Went, 1934), and the location of roots at the base of the cutting is presumed to be a result of the polar transport of auxin to the base of the isolated piece. It has frequently been observed that the number of roots initiated is related to the size of the cutting, and Heide (1968) has shown a general relationship between the auxin levels in *Begonia* leaf cuttings and their ability to form roots. The existence of other natural factors which regulate rooting as an interacting set of influences was described by Hess (1964), and included among these factors was a strongly promotive action by certain phenolics. Biran and Halevy (1973) have further shown that in addition to stimulatory factors, there can be endogenous inhibitors of rooting. Bouillenne (1964), Tomaszewski (1964), and Challenger et al. (1964) all found that striking promotions of rooting can be obtained with the provision of *o*-diphenols in addition to auxin. Catechol, caffeic acid, and chlorogenic acid are among the most effective promoters; the effects appear to be separate from the phenolic sparing action on IAA oxidase (Hess, 1964).

Each of the hormones has effects on root initiation; besides the promotion effect of auxin, there is an inhibition effect by gibberellin (Brian et al., 1955), an even greater inhibition effect by cytokinins (Humphries, 1960), and a promotive effect by ethylene (Zimmerman and Hitchcock, 1933; Kawase, 1971). Illustrating the effects of three types of regulators on rooting of mung bean cuttings are some experiments in Fig. 9-6.

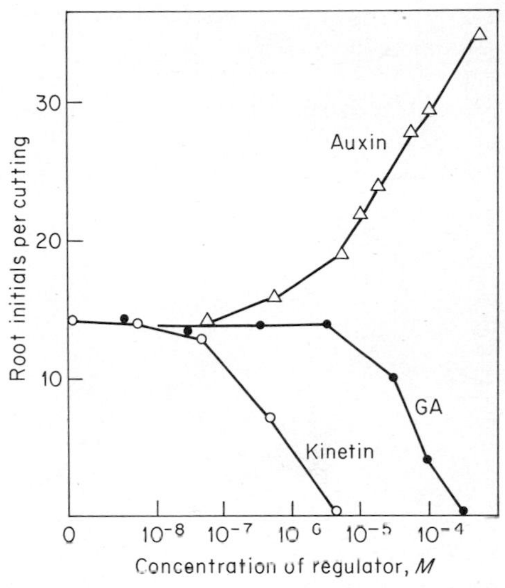

Fig. 9-6 Root initiation in mung bean cuttings can be markedly stimulated by auxin (indolebutyric acid) and effectively inhibited by gibberellic acid or kinetin (Fernqvist, 1966).

The differentiation of buds was first studied as a regulatory problem using tobacco-callus tissues in tissue culture; Skoog and Tsui (1948) were able to obtain good stimulations of bud differentiation by supplying adenine or other purines. They found that optimal bud formation occurs when the tissue is supplied with a balanced combination of adenine and auxin; this led them to propose that differentiation of buds depends upon the balanced interaction of these two materials. Subsequent work by Skoog and Miller (1957) established that the bud-differentiation effect is more properly described as a cytokinin effect. The cytokinin stimulation or enhancement of bud formation on root cuttings was described by Torrey (1958) and on leaf cuttings by Plummer and Leopold (1957). The enhancement of bud formation in tobacco callus is illustrated in Fig. 9-7.

The formation of buds seems to generally require the presence of actively growing roots. Miedema (1973) has found that the response of potato slices to applied cytokinins is quite poor unless there are actively growing roots on the cutting. He was not able to identify the factors being supplied by the roots. Other hormones which can alter the differentiation of buds include stimulatory actions for abscisic acid (Heide, 1968) and inhibitory actions for gibberellin (Heide, 1972).

The concept of differentiation being controlled by balances between common substances in the plant is especially interesting, since otherwise it might be necessary to assume that distinctive substances in the plant control the differentiation of each tissue and each organ. This could become absurd, with bud- and root-controlling substances, flower-controlling substances, and then perhaps stamen-, pistil-, petal-, and sepal-inducing substances. Control of differentiation

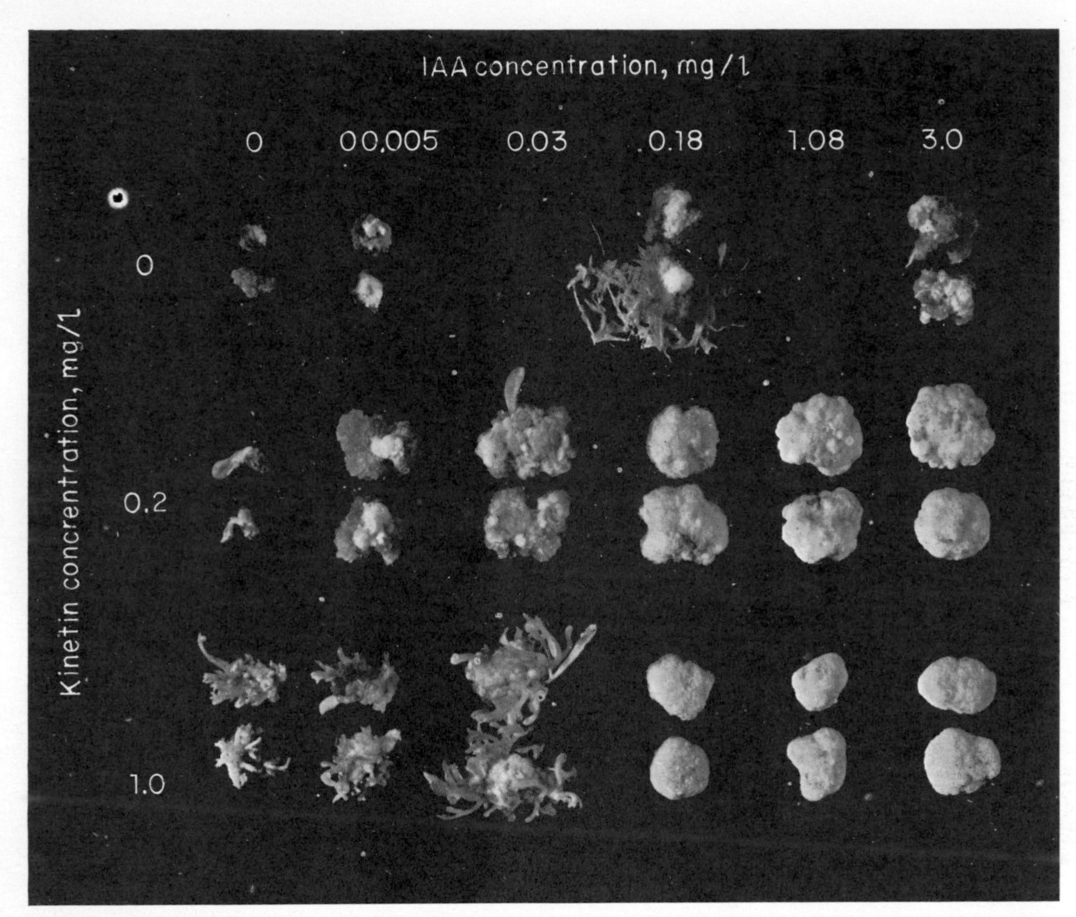

Fig. 9-7 The differentiation of buds on tobacco callus is stimulated by kinetin; the presence of intermediate levels of auxin enhances the bud formation, but high auxin levels prevent it (Skoog and Miller, 1957).

by balances of common regulating substances in the plant seems much more logical.

An illustration of the morphogenetic substances moving about in the plant can be taken from the girdled tree stem. Ringing a stem causes the bark to swell in the region above the ring, with the formation first of a callus and then of adventitious roots. This would be the expected location of accumulation of the polar-transported growth hormone, which would favor the differentiation of the roots. Below the ring there would be a tendency to form adventitious buds. Here one would expect a dearth of auxins and a large accumulation of the organic nitrogenous materials synthesized by the root system. The apical ends of cuttings are found to be high in cytokinins and low in auxin content (Vardjan and Nitsch, 1961).

Reaction-Wood Formation

Another type of tissue differentiation which may be under the control of morphogenetic substances is the formation of reaction wood. This is a wood with a modified structure: in conifers it consists of a thickening and shortening of the tracheids on the *lower* side of a branch or leaning stem and exaggeration of the tracheid wall thickening;

in angiosperms it consists of an exaggerated thickening of the walls of the fibers on the *upper* side of a branch or leaning stem and reduction in number and size of vessel elements (Scurfield, 1973). The formation of reaction wood is usually, though not always, a consequence of a geotropic stimulus of stems or branches in other than vertical orientation, and it serves to help reorient stems into a more vertical position or add support to branches in a lateral position. Reaction wood is undesirable from a lumberman's point of view because of the difference in structural quality of the wood in juxtaposition to the normal wood.

Three types of evidence indicate that morphogenetic substances are involved in the formation of reaction wood:

1 It is generally a georesponse by the plant. Bending a stem or placing it under tension does not itself lead to the formation of reaction wood, but sectors of the stem which are oriented other than vertically will form reaction wood. Thus, when a stem of pine is bent into a loop, reaction wood forms on the lower sides of the horizontal regions, as in Fig. 9-8*d*. Application of tension to some woods will also cause reaction-wood formation as indicated in Fig. 9-8*b* and *c*, especially in regions which are out of vertical orientation.

2 There is a requirement for some stimulatory signal from the growing point or stem apex; removal of the stem tip can prevent the formation of reaction wood with subsequent geostimulation; Wardrop (1956) measured the rate of movement of the signal from the growing apex of *Eucalyptus* and arrived at the value of about 9 mm/h for its movement down the stem.

3 There may be some involvement of auxin; of course the need for some stimulatory signal from the stem tip suggests the involvement of auxin, and the velocity of its movement down the stem is what would be expected for the polar transport of auxin.

Wershing and Bailey (1942) stimulated reaction-wood formation in conifers by the application of auxin to one side of the stem, and Necessany (1958) and some subsequent workers have induced reaction-wood formation in angiosperms with the unilateral application of auxin. In the latter case, reaction wood forms on the side with less auxin (Casperson, 1965); applications of auxin-transport inhibitor 2,3,5-triiodobenzoic acid have also stimulated reaction-wood formation (Morey and Cronshaw, 1968). The case for believing that auxin may be involved in reaction-wood formation is weakened, however, by the fact that Smith and Wareing (1964) find very little difference in the auxin levels along the upper and lower sides of horizontal willow stems; Leach and Wareing (1967) have suggested that another regulatory substance, perhaps a growth inhibitor, might be responsible for the gravity responses. In studies of reaction wood in aerial roots of *Ficus* trees, Zimmerman et al. (1968) were unable to find any stimulation of reaction wood by applications of auxins.

The formation of reaction wood illustrates the existence of substantive influences of tissue differentiation in the plant—morphogenetic substances which can move in the plant and carry the signal for differentiation from one part to another. It illustrates an apparent substantive

Fig. 9-8 The formation of reaction wood (*shaded areas*) in branches of white pine modified by such treatments as tying a branch into an upright position (*b*), decapitating the main stem (*c*), and tying a branch into a circle (*d*). (*Redrawn from Sinnott, 1952.*)

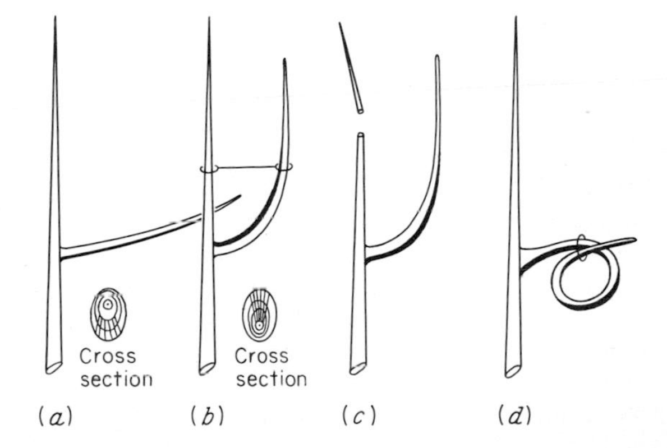

influence on differentiation, an involvement of polarity in the control, and a so-called correlation effect.

CORRELATION EFFECTS

The regulatory effects exerted by one part of the plant on the growth or development in another part have been called *correlation effects*. They represent some of the most intricate involvements of chemical messengers in multicellular plants. The most common correlation effects include the controls of tissue differentiation (xylem, cambium, reaction wood), organ differentiation (buds, roots), and the phenomena of apical dominance, tropisms, and abscission, each of which is regulated in part by influences stemming from the apices or leaves.

Correlative Differentiation

Differentiation of xylem can be induced by an influence from a nearby meristem, and this can be attributed in part to the presence of auxin (Jacobs, 1952). In fact, if one cuts through a vascular strand in a stem, new xylem will be differentiated around the wound mostly in an apex-to-base direction, and the polarity of this differentiation can be correlated with the polarity of auxin transport (Jacobs, 1954; Roberts, 1960).

The cambium is another tissue which is subject to pronounced correlative influences. Jost (1891) noted that the activation of cambium frequently occurs first below leaves, and Coster (1927) described it as a downward flow of activation from leaves and apices. Snow (1932) stimulated cambial activity in cuttings by pressing them up against stems which were already activated, and he deduced that a transmissible substance was responsible. Later (1935) he was able to stimulate cambial activation with the application of auxin. Since that time other growth substances have been implicated in this correlation effect, including gibberellins (Wareing, 1958a), products of wounding (Gouwentak and Hellinga, 1935), and perhaps natural inhibitors (Wareing and Roberts, 1956).

Root differentiation occurs with a strong polarity. Stimulating effects of buds on the rooting of cuttings were noted (van der Lek, 1934) shortly before the stimulatory action of auxin on rooting was discovered by Thimann and Went (1934). The polarity of this correlation effect can be attributed to the polarity of auxin transport and the resultant accumulation of natural auxins at the base of the cutting (Warmke and Warmke, 1950). The presence of leaves, cotyledons, and flower buds on the cutting can greatly influence the rooting effectiveness, indicating that auxin is not exclusively responsible for root differentiation. There are clearly nutritive involvements (Jones and Beaumont, 1937; Pearce, 1943), which may account in part for the promotive effects of leaves on rooting (van Overbeek et al., 1946). Auxins may also be supplied by leaves (Ruge, 1957).

Apical Dominance

It has been known for some time that the shoot apex regulates the growth and development of the lateral buds and branches, a phenomenon termed *apical dominance*. The inhibitory action on branches is widespread throughout the plant kingdom, being evident in bryophytes and pteridophytes (see, for example, LaRue and Narayanswami, 1957; von Maltzahn, 1959) as well as in seed plants. It is common to many different types of organs including stems and tubers (Michener, 1942). The root apex may also inhibit branching in roots (Thimann, 1936). Apical-dominance phenomena may be grouped into three classes: inhibitions of branching, regulation of which branches or parts will grow more rapidly than others, and control of branch angles. In each of these classes the stem (or root) tip contributes regulatory influences over the growth patterns of more or less remote parts of the plant.

The involvement of auxin has been indicated for some cases in each of the three classes of apical dominance. In stems, Laibach (1933) and Thimann and Skoog (1934) showed that auxin applications can replace the stem apex in inhibiting the growth of lateral buds (Fig. 4-4). In roots, on the other hand, Thimann (1936) noted that the apex inhibits branching but that auxin application does not substitute for the apex. The regulation of the growth activities of various branches can be illustrated by the inhibition of normal elongation in the short shoots of *Ginkgo* and *Cercidiphyllum* by the apical bud, an effect which can be replaced by applied auxin (Gunckel and Thimann, 1949). Another illustration of the regulation of relative growth is the phenomenon of compensatory growth, in which the removal of leaves and stems can stimulate the growth of other plant parts. In this case auxin applications do not effectively substitute for the removed parts, but evidently auxin formation by the inhibiting organs does play some role (Jacobs and Bullwinkel, 1953). The third class of apical dominance effects, the control of branching angles, also involves an auxin influence. The growing tip of a young fruit tree causes the lateral branches which grow out to assume positions away from the vertical; decapitation of the apex erases the branch-angle effect and results in vertical branches, but auxin applications restore the effect of the removed apex (Jankiewicz, 1956). The effects of the apex both on the relative growth by laterals and on the branch angle are illustrated in Fig. 9-9. The same type of correlation effect is sometimes imposed on leaf angles by the stem apex, and again it is replaceable with auxin (Dostal, 1962).

The inhibition of growth of lateral buds has received the most attention of the apical-dominance phenomena, and extensive disagreement over the hormonal control has developed. The first hormonal theory of apical dominance was that of Thimann and Skoog (1934), who noted that the inhibitory action of the apex is exerted in a polar manner down the stem; suspecting auxin as the correlation carrier, they removed the apex and found that the lateral buds grew out and that application of auxin in a lanolin paste to the decapitated stump replaced the apex in preventing growth of the lateral buds. Several subsequent studies raised difficult questions about the possible direct action of auxin in apical dominance; e.g., Jacobs et al. (1959) replaced the apex of *Coleus* plants with auxin paste which precisely duplicated the auxin supply of the apex and found no suppression of lateral buds. Experiments with green plants have often provided results which do not support the auxin theory, and some workers (Wickson and Thimann, 1958) have reported that the auxin suppression effect is weakened in proportion to the intensity of light.

Since cytokinins stimulate bud formation and

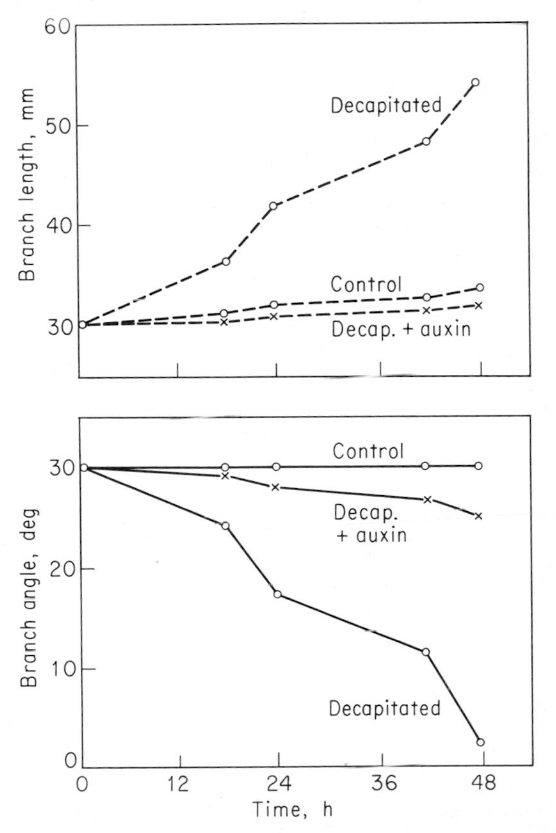

Fig. 9-9 When an apex of *Mirabilis jalapa* is decapitated, growth of the lateral branch below it increases and the angle which the branch makes from the vertical declines. Application of an auxin paste (500 mg/l IAA) to replace the severed apex prevents both the growth increase and the angle change (Vardar, 1955).

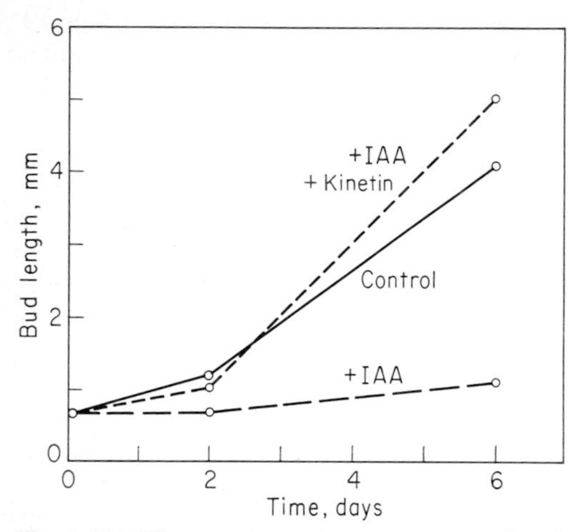

Fig. 9-10 When sections of pea stems are placed in a dish of water, the lateral bud proceeds to grow out. Indoleacetic acid in the solution (1 mg/l) prevents the bud elongation, but the inclusion of kinetin (4 mg/l) with the auxin erases the auxin inhibition (Wickson and Thimann, 1958).

growth in tissue cultures (Fig. 9-7), the possibility of cytokinin serving to abrogate the auxin suppression of lateral buds was examined by Wickson and Thimann (1958). They found that kinetin applications can remove the auxin inhibition, as shown in Fig. 9-10. They suggested that the auxin regulation of lateral buds is subject to relief by cytokinin. Conditions which depress apical dominance, such as high light intensities and rich nitrogen nutrition, may serve to provide higher cytokinin concentrations and hence relieve the auxin inhibitions. The concept of regulation by balance between these two hormones was further developed by Sachs and Thimann (1964), who found that the application of cytokinin to the buds of intact plants forces them to grow out.

Each of the other plant hormones has been suggested as playing a role in apical dominance in various plant materials. For example, gibberellins enhance apical dominance in some species (Scott et al., 1967), although they can stimulate the growth of laterals in others (Kato, 1953). Inhibitor β has been assayed in inhibited buds and found to correlate with the suppressed state; and in fact the application of this inhibitor to lateral buds can enhance the apical-dominance effect (Dörffling, 1966). Ethylene, too, has been invoked to explain apical dominance, and the levels of auxin that serve to suppress lateral growth in peas have been shown to bring about ethylene levels sufficient to inhibit the bud growth; likewise, cytokinin concentrations which relieve apical dominance prevent the inhibitory action of ethylene on the bud growth (Burg and Burg, 1968b).

An effort to bring together the concepts of hormonal action and nutritive action on apical dominance has led to the concept of mobilization, or directed translocation, as a control of lateral-bud growth. Gregory and Veale (1957) established that high nitrogen fertilization can eliminate auxin control of apical dominance and suggested that under low nutritional levels, auxin effects are simply the consequences of deprivation of the lateral buds by mobilization of nutrients to the apex. Booth et al. (1962) and later Seth and Wareing (1964) studied the mobilization of radioactive sucrose and phosphate to the apex of decapitated plants and found that auxin applied to the decapitated stem strongly mobilizes nutrients to the stem tip; the further addition of cytokinin or gibberellin can further enhance the mobilization effect. An indirect effect of hormones on apical dominance might be inferred, through which growth in the stem tip causes a directed translocation to the apex region, depriving the lateral bud of substrates for growth. However, several workers have restricted the auxin to the stem-tip region by application of a ring of triiodobenzoic acid below the growing region, and while this treatment does not depress the growth (and presumably the mobilizing function) of the stem tip, it does inhibit the suppression of the lateral buds (Sebanek, 1966). Sorokin and Thimann (1964) have suggested that a part of the cytokinin effect in relieving apical dominance when applied to the bud may be the stimulation of vascular development connecting the lateral with the main vascular system, thus serving to relieve the nutrient shortage in the lateral bud.

Movements

The immobility of plants puts a premium on their ability to exercise bending movements in response to such stimuli as light and gravity. Geotropism is elementary to the ability of the plant to grow upright and compete for sunlight. Phototropism is scarcely less important in the plant's ability to orient toward the light. Shoot and root bending in response to gravity or light is due to a differential growth on one side; this is usually a result of a differential distribution of auxin, but there are many gaps in our understanding of this type of phenomenon. The agencies of perception (probably a particle for geotropism, a pigment for phototropism) are not clearly understood; neither do we know how the lateral distribution of auxin is achieved or what determines the sign of the response.

Tropisms. From the early work of Darwin (1897) it has been recognized that the apex of the stem or coleoptile is involved in the tropistic responses. Removal of the apex can erase the sensitivity (Fig. 9-11). The function of the apex as a source of auxin was the apparent explanation (Went, 1928). An interesting variant of this pattern, however, is provided by green plants, in which the leaves are important since their removal can greatly depress phototropistic responses (Fig. 9-12), in part because of the functioning of leaves as auxin sources (Lam and Leopold, 1966).

A time curve for the development of curvature is shown in Fig. 9-13, in which there is latent time of about 8 min before the appearance of first curvature. Various organs show geotropic latent times ranging from 7 to 60 min (Larsen, 1962) and as short as 3.3 min for phototropism (Pickard, 1969). Within this time, the sensing of the gravity or light stimulus must occur, followed by the induction of a physiological difference between the two sides of the organ, the lateral redistribution of auxin, and then a sufficient induction of growth for curvature to become evident. Where auxin redistribution drives the motor response to a tropistic stimulus, the auxin redistribution system and the mechanism of auxin action must act within this brief time (cf. Fig. 4-30).

The timing of the tropistic stimulus was studied by Rutten-Peckelharing (1909), who found that the stimulus for geotropism can be quantitatively induced by various intensities of centrifugal force. The minimum stimulation time needed to induce geotropic curvature, called the *presentation time,* varies inversely with the intensity of the tropistic stimulus (Rutten-Peckelhering, 1909). The minimum force for geotropism, i.e., the force value below which no tropistic response is produced, can be estimated at about 1 mg or less. More precise determinations estimate the minimum force at about 0.5 mg when applied for 8 h (Larsen, 1962). The threshold dosage for geotropic responses was estimated by Rutten-

Fig. 9-11 A comparison of the geotropic bending of corn coleoptiles when held continuously in a horizontal position, horozontal for just 1 h, and continuously horizontal but decapitated. The apex plays a critical role in the tropistic response (Hahne, 1961).

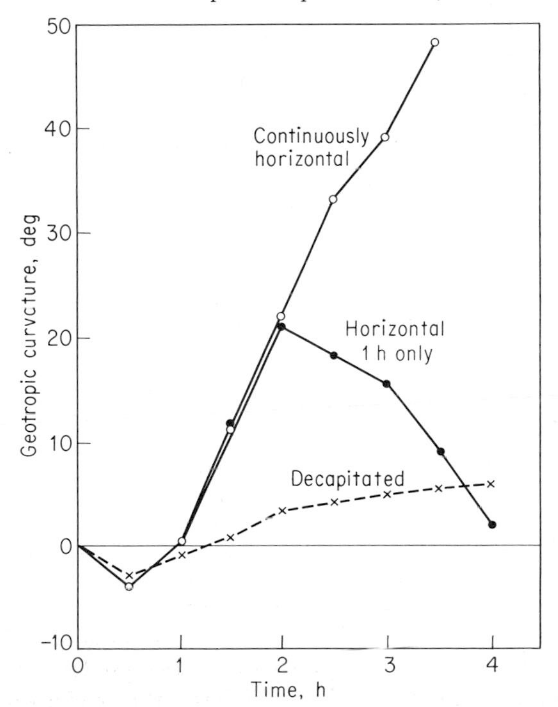

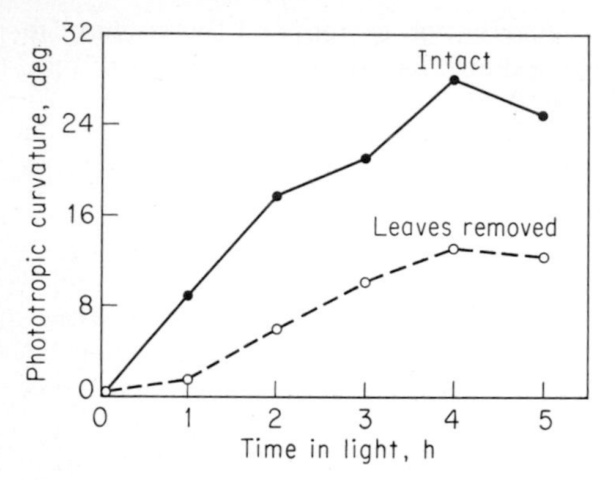

Fig. 9-12 The phototropic curvature of green seedlings of *Helianthus* toward lateral illumination (450 fc) is markedly less if the leaves have been removed (Shibaoka and Yamaki, 1959).

Peckelharing (1909) at about 360 g-s; more refined experiments have shown this to be 240 to 300 g-s (Johnsson, 1965), and for coleoptiles which have been previously gravity-compensated on a clinostat the value is even lower: 100 g-s (Shen-Miller, 1970). For the geotropic stimulus reciprocity holds; i.e., the presentation time is proportional to the inverse of the force times the duration of its application. Johnsson (1965) has illustrated this by plotting the geotropic response resulting from several different centrifugal forces (simulating gravity) for different periods of time, and the intercept of these response curves for each gravity force provides a precise method of determining the presentation times. Reciprocity is seen as the linear change in presentation time with changes in centrifugal force in Fig. 9-14.

The quantitative nature of phototropism is much more complex, first, because the phototropic response is not linear over a wide range of light doses. As shown in Fig. 9-15, curvature increases with light dosage up to about 1,000 meter-candle-seconds (mc-s), and at higher light doses the curvature declines; in some plant materials high intensities actually cause curvature away from the light source. Then at still higher light levels, above 10,000 mc-s, curvature toward the light is again proportional to light dose (Fig. 9-16). Briggs (1960) has shown that reciprocity holds for phototropism for the range of the first curvature response (less than 1,000 mc-s). Intermediate and high light doses do not show reciprocity, suggesting that in this dosage range more than one rate-limiting reaction is involved (Everett and Thimann, 1968; Curry, 1969).

The *perception* of the phototropic stimulus has been studied principally by determinations of the action spectrum. Such a spectrum (Fig. 15-18) forms the basis for an argument between those who attribute the sensing to β-carotene and those who attribute it to riboflavin (e.g., Wald, 1960). The absorption shoulder at 470 nm is distinctive to carotenes such as β-carotene, and the shoulder at 370 nm is distinctive to the flavins. Attempts to match the action spectrum with the absorption spectrum of a particular pigment in a solvent such as methanol or ethanol may be causing more disagreement than is necessary, since Hager (1970) has found that adding water to alcoholic solutions of some carotenoids such as lutein gives marked absorption peaks at 370 nm. Consequently, he has suggested that carotenoids in the cells may lie between aqueous and fatty layers and there can exhibit absorption spectra which would quite

Fig. 9-13 The development of phototropic curvature of oat coleoptiles exposed continuously to unilateral light, showing the latent time of several minutes and the nonlinear development of curvature (Pickard, 1969).

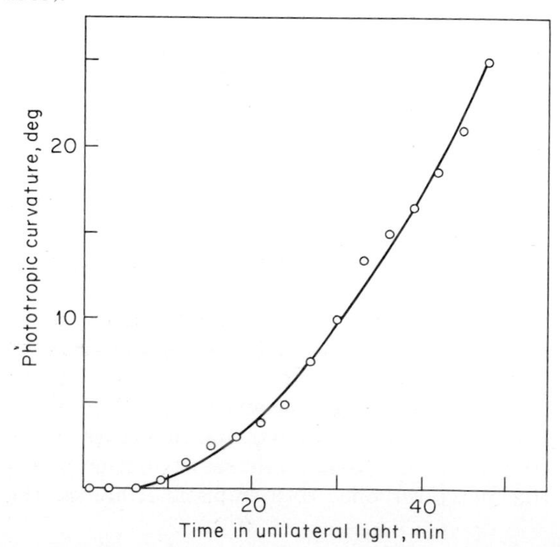

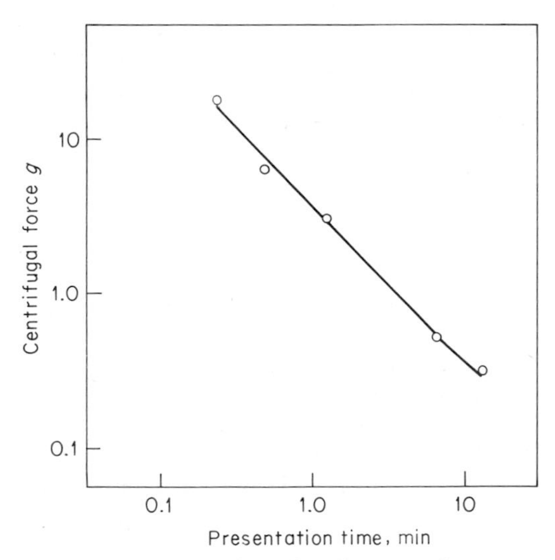

Fig. 9-14 For oat coleoptiles, there is a linear relationship between the centrifugal force applied and the length of the presentation time for initial georesponse. This indicates that reciprocity holds for this geotropic system (Johnsson, 1965).

precisely match the action spectrum for phototropism.

The perception of geotropic stimulus has been attributed to the action of gravity on free-falling bodies in the cytoplasm, the so-called *statoliths* (Nemec, 1900; Haberlandt, 1913). The timing of lateral displacement of starch grains in the cells of various plant parts correlates well with the geotropic responses (Hawker, 1932). Coleoptiles of corn varieties with smaller starch grains have appropriately lessened georesponses (Hertel et al., 1969b). Pickard and Thimann (1966) used gibberellin and kinetin treatments to remove starch grains from wheat coleoptiles and found the geotropic responses much slower but still obtainable; but in similar experiments with cress Iversen (1969) found that geotropic responses were absent until the starch grains were regenerated. In roots the georesponding starch grains are located in the root cap; Juniper et al. (1966) devised a clever trick for removal of the root cap; they were able to show that geotropism was lost until regeneration of a new root cap with starch grains. Especially interesting is that in this kind of experiment the growth rate of the root is not altered; by removal of the root cap only the geotropic response is lost (Könings, 1968).

Audus (1962) has asked whether other bodies in the cell might act as statoliths; the kinetics of geoperception match most nearly the kinetics of displacement of starch grains, but Audus calculates that mitochondria could also serve in part as statoliths. Slower sensing of gravity could involve other organelles; an increase of Golgi bodies on the lower side of geostimulated cells has been reported by Shen-Miller and Miller (1972).

The *induction* of a physiological difference between the sides of a tropistically stimulated coleoptile does not seem to require auxin (Brauner and Hager, 1958). This has been established by depleting the auxin supply by decapitating the seedlings, giving the geotropic stimulus, and then supplying auxin to permit curvature to develop, as shown in Fig. 9-17.

Brauner and Hager (1958) established that the induction of a physiological difference requires metabolic action during the perception stage. For example, holding such depleted plants in a horizontal position in a nitrogen atmosphere

Fig. 9-15 The phototropic curvature of etiolated corn coleoptiles increases with increasing light dosage but not in a simple quantitative way. If light intensities are increased with dosage in order to keep all exposures within 5 min, the curve declines to near zero at about 10,000 mc-s. (*A*). If the intensities are all kept low enough to maintain reciprocity more nearly (*B*), the response curve is less erratic (Briggs, 1960).

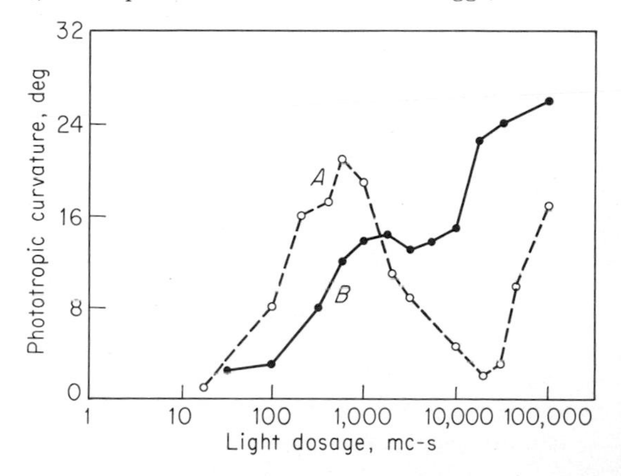

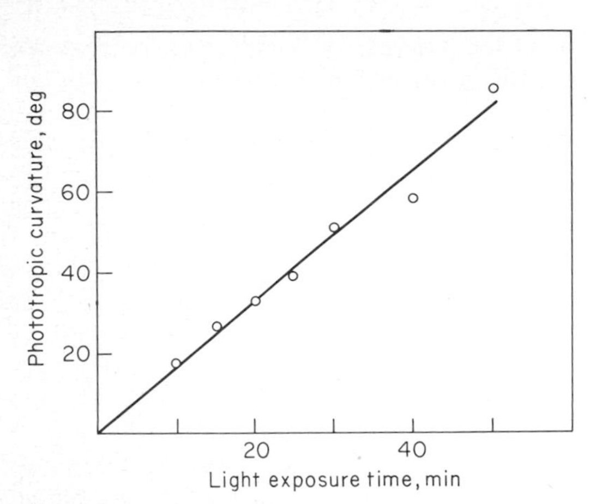

Fig. 9-16 The phototropic curvature of oat coleoptiles given unilateral blue light (500 ergs/cm^2-s) increases in a linear fashion with increases in exposure time (Pickard, 1969).

does not permit induction, and subsequent applications of auxin do not produce the tropistic response (Fig. 9-18). Likewise, the suppression of metabolism with cold during the tropistic stimulation prevents induction. Finally, the persistence of the induced state has been quantitatively measured by Diemer (1961) by interpolating periods of time between the tropistic stimulation of depleted stems and the applications of auxin to permit the actual growth responses. She deduced a half-life of the induced state in the plants she studied at about $4\frac{1}{2}$ h (Fig. 9-19). If depleted, geostimulated *Avena* coleoptiles are held in the cold (2°C), the induced state can be retained unchanged for 7 h (Pickard, 1969).

Response of the plant to tropistic stimuli appears to depend specifically upon the presence of auxin. After the induction of auxin-poor coleoptiles or stems, bending responses do not develop unless some auxin is added exogenously (Fig. 9-17) or supplied naturally by the regeneration of the apex (Brauner and Hager, 1957; Hahne, 1961). In coleoptiles, the response is clearly a consequence of a lateral redistribution of auxin, both for phototropism (Went, 1928) and for geotropism (Dolk, 1929). In roots, the lateral redistribution of auxin occurs following geotropic stimulation, similar to that in shoots (Hawker, 1932). The removal of geotropic sensitivity associated with the removal of the root cap is nicely associated with a loss in the lateral redistribution of auxin (Pilet, 1972).

In coleoptiles numerous precise measurements of the redistribution of auxin following tropistic stimuli are possible. Some representative data for the phototropism of corn coleoptiles are given in Fig. 9-20. These data show that although light does not alter the total amount of auxin obtained by diffusion from the coleoptiles (*a* vs. *b*), it does result in an increase on the shaded side and a corresponding decrease on the lighted side provided the coleoptile is not slit to prevent lateral movement of the auxin (*c*). Goldsmith and Wilkins (1964) followed the extent and timing of such redistribution in the geotropism of coleoptiles with radioactive auxin.

How is the lateral redistribution of auxin brought about? Following minimal stimuli with light or gravity, there is no decrease in total diffusible auxin present, and so a preferential destruction of auxin on one side cannot be the explanation at threshold stimulus levels. Two types of effect can apparently lead to the lateral dissymmetry, a preferential synthesis of auxin on one side and a lateral movement of auxin.

Fig. 9-17 The induction for geotropism can be achieved in decapitated auxindepleted sunflower seedlings. Holding them horizontal for 3 h and then applying auxin solution to the tip (10 mg/l) at the time of restoring the seedling to the vertical position still permits a development of tropistic curvature (Brauner and Hager, 1958).

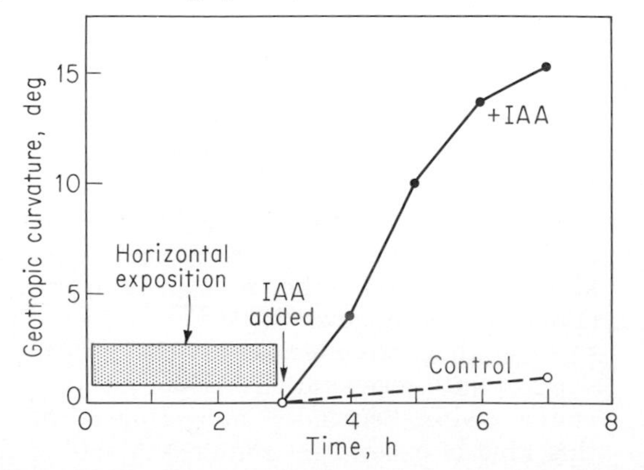

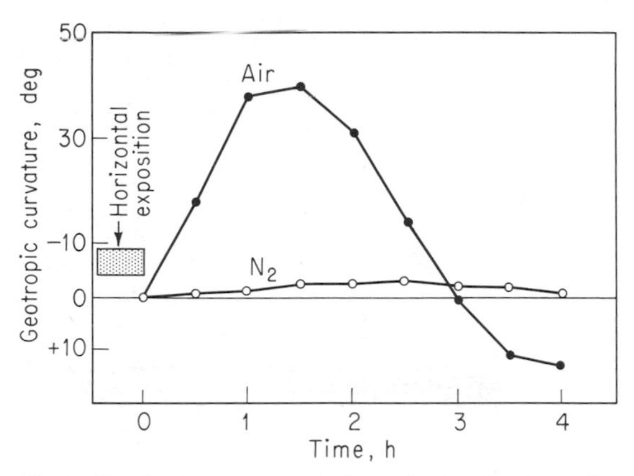

Fig. 9-18 A nitrogen atmosphere during horizontal exposure of *Helianthus* seedlings (30 min) before restoring to air and the vertical position prevents geotropic induction. Time scale begins upon restoring to vertical (Brauner and Hager, 1958).

A synthesis of auxin after geostimulation of roots has been indicated in some timing experiments by Audus and Lahiri (1961). After placing broad bean roots in a horizontal position, they found a large increase in an extractable auxin, separable by paper chromatography, and this increase occurred just before curvature began to be expressed. The increase is suggestive, but its relation to the tropistic response seems to depend on a lateral asymmetry of synthesis, which has not yet been shown. A synthesis of auxin in response to geostimulation has been reported for other roots (Rufelt, 1957) and stems (van Overbeek et al., 1945). In addition, an asymmetric synthesis in response to photostimulus has been reported for green stems (Schmitz, 1933; Shibaoka and Yamaki, 1959) and perhaps for flower pedicels (Zinsmeister, 1960).

The lateral transport of auxin has been directly measured in several studies (Brauner and Appel, 1960; Gillespie and Thimann, 1961). The lateral transport of auxin associated with geotropism appears to be an active transport (Hertel and Leopold, 1963; Goldsmith and Wilkins, 1964). In view of the evidence that transport may be a secretion process (see Table 4-1) and that the secretion activity may shift from the base to the side of the cell upon tropistic stimulation, the induction effect of gravity might be the movement of an energy-producing particle from the base to the side. Ziegler (1953) found that some smaller organelles which act as energy-producing centers are associated with the statoliths. There is a possibility, then, that statolith movement may alter the pattern of availability of metabolic energy for the secretive part of auxin transport (Hertel and Leopold, 1962).

One difficulty with the statolith theory has been that the starch grains often do not fall to the point of contacting the plasmalemma (Larsen, 1971); Sievers and Volkmann (1972) have observed that the falling statoliths in *Lepidium* roots compact the endoplasmic reticulum closer to the plasmalemma on the lower side, and they suggest that this may alter active processes preferentially on the lower end of the cell. Another possible means of activation of the lower side of the cell has been suggested by Shen-Miller and Miller (1972). They observed both an increase of Golgi bodies at the lower side of the cell and an increase in activity in the production of vesicles.

Fig. 9-19 The inducted geotropic state in laterally illuminated *Helianthus* seedlings has a half-life of about $4\frac{1}{2}$ h. This experiment was done with seedlings decapitated for 2 days and was made by giving lateral light (700 fc) for 16 h and then waiting various times before supplying auxin (10 mg/l IAA) to permit the development of curvature response. Curvature recorded 3 h after auxin application (Diemer, 1961).

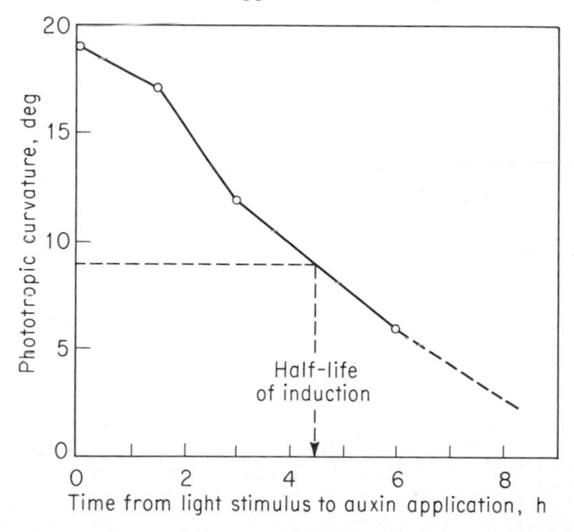

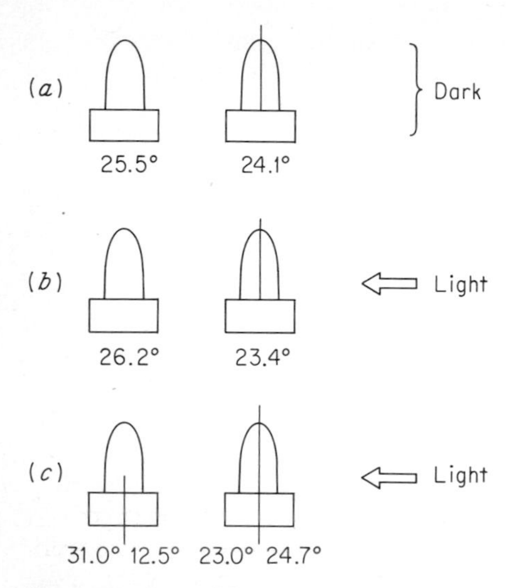

Fig. 9-20 Lateral light causes the lateral redistribution of diffusible auxin coming from corn-coleoptile tips. The amounts of auxin diffused from intact vs. split tips are compared in the dark and in lateral light. Diffusible auxin (3 h) taken from three intact tips (*a* and *b*) or from six halved tips (*c*) for comparison in *Avena* curvature test. Note the increase in auxin on the shaded side without change in total auxin diffusible from tips (*c*) (Briggs et al., 1957).

A difficult question arises: What determines the sign of a tropistic response? Why do stems show negative geotropism and roots positive geotropism? The answer is complicated by the fact that the initial response of stems and coleoptiles to gravity is slightly positive and of roots slightly negative (Brauner and Zipperer, 1961; Rufelt, 1957). Furthermore, the sign of the response can sometimes change with age (Pilet, 1951a), nutrition (Rufelt, 1957), or applied chemicals (van der Laan, 1934).

A suggested mechanism for the positive geotropism of roots has been made by Chadwick and Burg (1967). In experiments with pea roots they noted that the same levels of auxin which cause root-growth inhibition also cause ethylene production. They proposed that the lateral redistribution of auxin in roots of peas may stimulate ethylene formation on the lower side of the root and this may preferentially inhibit growth on the lower side. In support of their argument they found that carbon dioxide can suppress geotropic responses of pea roots; in shoots, where ethylene is not presumed to participate, carbon dioxide does not suppress geotropism. This suggestion of ethylene involvement has been challenged by Andreae et al. (1968), who reported differences in the kinetics of auxin and ethylene inhibitions and concluded that ethylene cannot account for the auxin inhibitory effects. The kinetics were reexamined by Chadwick and Burg (1970), who asserted that the kinetics are consistent with the ethylene theory of geotropic curvature.

Leaf and flower movements. Leaves ordinarily show movements, either through the growth features of their petioles or through the turgor changes in the pulvini. Petiole growth movements are functions of the differential growth of the shaded and lighted sides of the leaf, or upper and lower sides with respect to gravity. This relatively slow movement is apparently

Fig. 9-21 When *Coleus* petioles are placed in a vertical position, they orient themselves to about 25° if the leaf blade is present. Removal of the leaf blade causes a loss in this orientation movement, but orientation can be restored by application of auxin to the petiole stump. The auxin was 10μM IAA (Vardar, 1953).

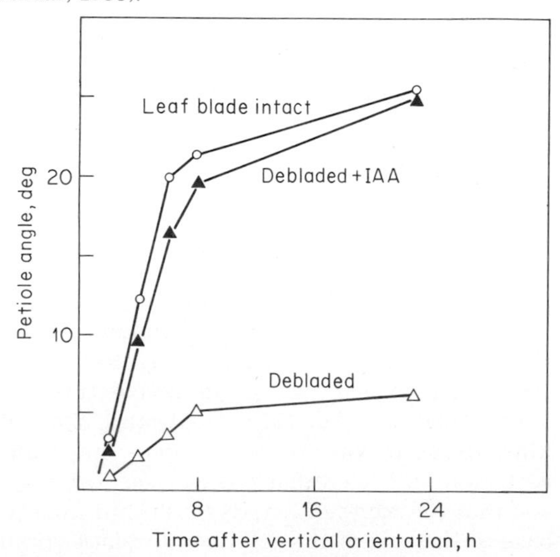

regulated by the supply of auxin coming from the leaf blade, for removal of the blade can almost prevent the leaf movement (Fig. 9-21) and auxin supplied to the severed leaf base can restore the leaf movement. Other leaf movements involve turgor changes in pulvini, the sections of petiole in which the vascular strands are collected into a central cylinder surrounded by parenchyma cells capable of expanding or contracting in a manner which can move the leaf up and down. The leaf movements driven by pulvini include the diurnal *sleep movements* and in some instances the rapid responses to touch, e.g., sensitive plant (*Mimosa*).

Pulvinar leaf movements are very responsive to light and its orientation. While auxin can alter this type of movement, it does not seem to be the natural regulator (Bünning, 1959). The pulvinar movements involve turgor shifts of cells on the upper and lower sides (Fig. 9-22). The rapid seismonastic movements of the sensitive plant and other touch-sensitive leaves are also due to alterations in the turgor of the pulvini. These movements are very rapid, taking less than 2 min. They are preceded by the generation of electric-potential charges, and the movement itself involves the rapid loss of water from cells on the contracting side of the pulvinus (Sibaoka, 1969). In addition to turgor changes there are also drastic alterations in vacuolar components (Weintraub, 1952; Toriyama, 1967). In fact some cell components actually leak out of the pulvinar cells (Jaffe and Galston, 1967a). Satter and Galston (1971) have found potassium efflux during the drop of the leaf, similar to the potassium efflux in the closing of stomata.

Many flowers open their corollas in the daytime and close them at night, e.g., most tulips and crocuses and members of the Compositae and some other families. Occasionally flowers open at night and close during the day, e.g., the night-blooming jasmine (see Fig. 12-29). In some instances these flower movements are driven by light and in others by temperature. For example, tulip and crocus flowers can be made to open by a sudden rise in temperature, or at a constant temperature they can be made to open by exposure to light. Corolla movements are due to differential elongation of the cells in the corolla itself, and sometimes meristematic activity contributes to the differential growth (Sibaoka, 1969).

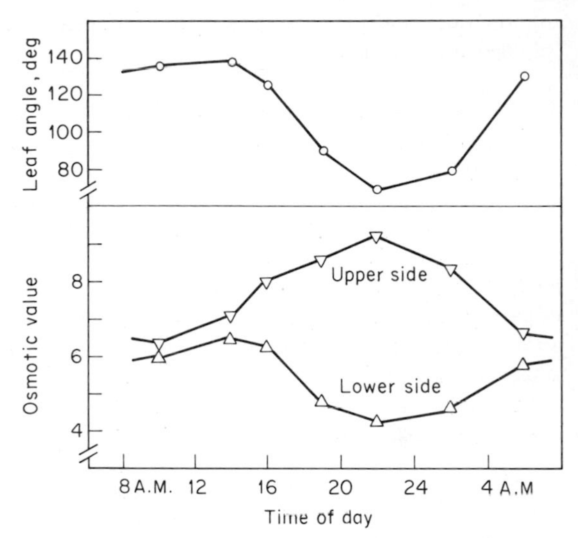

Fig. 9-22 The sleep movements of leaves are driven by osmotic changes in the cells in the upper and lower sides of the pulvinus, as illustrated here for leaves of *Phaseolus multiflorus.* Leaf angles (*above*) given as the angle formed between the petiole and the midrib of the leaf (Bünning, 1959, from data of Zimmerman, 1929).

Some flower parts are able to carry out movements in response to touch; e.g., movements of stamens in *Portulaca, Berberis,* and *Mahonia* occur in response to touch, thus enhancing the placement of pollen on visiting insects; movement of the stigmata of *Mimulus* or *Strobilanthes* or of the style itself in *Stylidium* in response to touch can improve the chance of fertilization by visiting insects. These rapid movements resemble the rapid movements of leaves in that there is characteristically a generation of an electric potential by the stimulated flower part, and the movement of the stamen or stigma involves a rapid loss of water from cells on the contracting side (Sibaoka, 1969).

Movements of flowers themselves, e.g., the heliotropic movements of sunflowers, are a phototropic action of the pedicel similar to the phototropism of the stem.

Contractile roots. Many plants establish the depth of the junction between stem and root through the contraction of roots, which pulls the seedling, corm, bulb, or plant into the soil. For example, seeds of lilies germinating at the soil surface ultimately produce bulbs with basal positions 15 to 10 cm under the surface. Almost all bulb and corm species exercise this type of movement, and smaller movements occur in such species as alfalfa. Contraction movements are also recorded for the aerial roots of the banyan tree (Zimmermann et al., 1968), serving to straighten the root after contact with the soil has been made. Vertical root contractions can cause the movement of the plant base into the soil, and contraction of horizontal roots can move plants along the ground. For example, in certain species of *Oxalis,* bulblets can move as much as 40 to 47 cm due to the contraction of a single root (Galil, 1968).

Morphologically, root contraction occurs as a compaction of some sectors of cortical cells and extensive lateral swelling of others; the

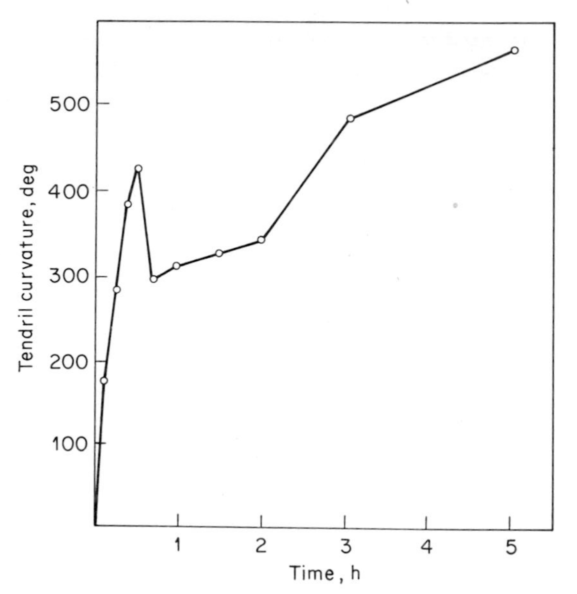

Fig. 9-24 The movement of pea tendrils can be recorded as the increases in tendril curvature after physical stimulation. A rapid curvature occurs for about 30 min followed by a slower (Jaffe and Galston, 1966).

Fig. 9-23 The movement of bulbs and corns into the soil is achieved by contractile movements of roots; *Oxalis incarnata* roots approximately 30 cm long recorded here after planting at the surface of the ground. Contraction averaged almost 2.2 mm/day (Thoday and Davey, 1932).

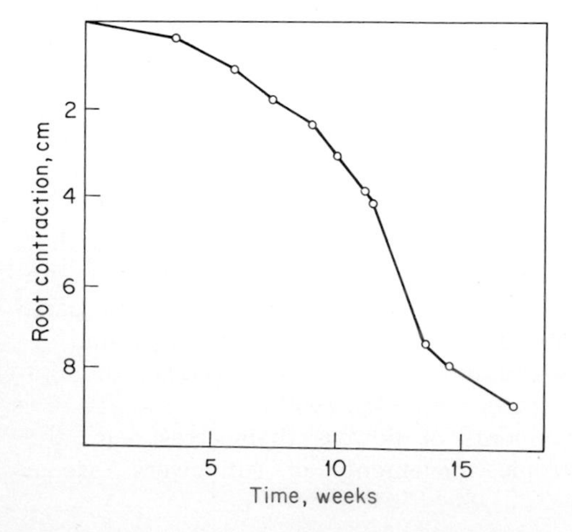

accordionlike folding of the cortical cells is associated with a helical pattern of stelar contraction (Smith, 1930).

Illustrating the root contraction are data for *Oxalis* shown in Fig. 9-23; in this instance the root shortening amounted to an average of 2.2 mm/day over a period of 4 months.

Jacoby and Halevy (1970) have studied how root contraction is stimulated and conclude that in gladiolus the diurnal fluctuation in temperatures near the surface of the soil is the stimulus. When the corm has sunk low enough to experience subthreshold fluctuations in temperature, contraction stops.

Tendril movement. Tendrils are modified stems, e.g., the grape, or modified leaves, e.g., the pea, in which the thigmotropic responses are very highly developed. They serve to hold the plant to supporting structures as a consequence of their coiling movements. In studies of pea-tendril coiling, Jaffe and Galston (1966) have shown that the response to a touch stimulus along the ventral side results in a rapid curvature developing in about 30 min, followed by a slower steady coiling action (Fig. 9-24). The initial, rapid curvature results from an elongation of cells on the dorsal side and shrinkage of cells on the ventral side, and the slower curvature results from the differential growth of the cells on the two sides (Jaffe and Galston, 1966). The rapid reaction appears to be an osmotic process, enhanced by light or ATP and inhibited by anaerobiosis or cold (Jaffe and Galston, 1967b), and involves in part the leakage of solutes out of the cells on the ventral side (Jaffe and Galston, 1968a). Pea tendrils are stimulated to coil by ethylene (Jaffe, 1970), and wild cucumber tendrils are stimulated by auxin (Reinhold, 1967).

In retrospect, plant movements generally involve either the differential growth on the two sides of a plant organ or the preferential expansion or contraction of cells. The latter is usually manifested through shifts in osmotic components and possibly some leakage of solutes out of the shrinking cells; this type of movement includes, of course, the pulvinar movements, the rapid part of tendril movements, and the movement of some floral parts. It seems reasonable for auxin to regulate the first type of differential growth, and there is a possibility that ethylene may dominate the second type.

Abscission

In the angiosperms, there is a widespread occurrence of dehiscence, or the organized separation of cells resulting in tissue separations. Cell separation occurs in the abscission of leaves, flowers, fruits, and stems and also as the opening of some types of anthers upon anthesis and as the opening of many types of fruits, e.g., cotton and the legumes generally.

The phenomenon of abscission has received much attention and stands out as a clear case of a tissue differentiation which is regulated to a great extent by remote tissues, as in the regulation by the leaf of the abscission development in the petiole or the regulation by the fruit of abscission development in the pedicel.

The abscission process is clearly active and not merely a passive falling of the aging organs. Abscission is suppressed by a deficiency of oxygen (Carns et al., 1951) or of carbohydrate (Biggs and Leopold, 1957). It seems reasonable to consider the hormonal influences on abscission to be effects on some dynamic cell processes.

That auxin can inhibit leaf abscission was first observed by Laibach (1933), using the auxin from orchid pollinia. LaRue (1936) found similar inhibitions for leaf diffusates, pollen extracts, and auxin pastes (Table 9-1). The inhibiting effect of the leaf blade on petiole abscission has been attributed to the auxin content of the blade (Mai, 1934; Myers, 1940). The advent of abscission is associated with a drop in auxin content not only in leaves but also in fruits (Luckwill, 1948) and stems (Garrison and Wetmore, 1961).

The ability of auxin to promote abscission was noted by Laibach in 1933 but was not generally recognized until 1951, when Addicott and Lynch reported that auxin pastes applied to tissue on the proximal side of the abscission zone stimulate abscission, in contrast to the inhibitory effect of auxin applied to the distal tissues (Fig. 9-25). Later studies of the effects of various auxin concentrations applied to the two sides showed that low concentrations can bring about promotive effects on either side (Biggs and Leopold, 1958), which implied that the more inhibitory effect of distally applied auxin was partly due to readier movement in the polar direction to the abscission zone. Varying the time of application

Table 9-1 Abscission of debladed *Coleus* petioles is inhibited by leaf diffusates, by pollen, and by auxin pastes. Paired opposite leaves were given either plain lanolin or lanolin containing the tested materials (LaRue, 1936).

Treatment	Average time to abscission: Controls	Treated
Coleus leaf diffusate	43	58
Vallota leaf diffusate	75	123
Populus pollen diffusate	78	124
Auxin paste (5 mg/l):		
In dark	54	131
In light	160	182

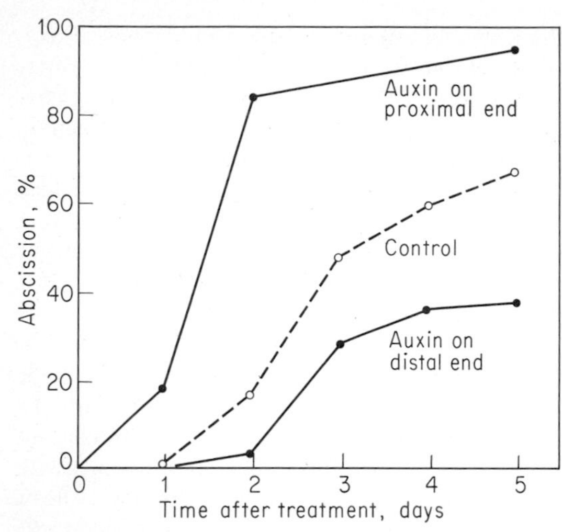

Fig. 9-25 The abscission of bean-petiole explants can be either promoted or inhibited by auxin application, depending on the site of application. Indoleacetic acid (105 mg/l) applied as solution to either end; water to the controls (Addicott and Lynch, 1951).

of one concentration of auxin to the distal side showed that the inhibitory action of auxin is restricted to the first few hours after cutting and that after this aging period auxin applications are only promotive (Fig. 9-26). Therefore, the bean abscission zone passes through two periods of development, an initial period (stage I), which is inhibited by auxin, and a later period (stage II), which is promoted by auxin (Rubinstein and Leopold, 1963). Comparative experiments with leaves of different ages showed that young leaves have a long period of auxin inhibition potential (stage I) whereas with increasing age this inhibitory capacity is gradually lost (Chatterjee and Leopold, 1964). The ability of auxin to inhibit abscission is therefore, at least in part, a function of the aging processes in the leaf. In the bean leaf aging processes which open the way for abscission are known to occur specifically in the pulvinus; the superior inhibiting action of auxin when applied distally (Fig. 9-25) is in part due to its readier deferral of pulvinar senescence (Abeles et al., 1967).

It is curious that ethylene received very little attention as a regulator of abscission before 1964. Its effectiveness in stimulating abscission was reported by Zimmerman et al. in 1931. Hall (1952) considered ethylene as a natural regulator of abscission, working in a balance against auxin. Then in 1964, Abeles and Rubinstein showed that the promotive effect reported for auxin can be accounted for as the result of ethylene production stimulated by the auxin.

The stimulatory role of ethylene applies also to the development of dehiscence of fruits. Lipe and Morgan (1972) have shown that ethylene is formed during the development of dehiscence in cotton and pecan fruits, and the separations of the fruit shells are stimulated by similar amounts of ethylene (see Fig. 13-25).

When freshly cut bean-petiole sections are exposed to ethylene, they lack a responsiveness to added ethylene (Abeles and Rubinstein, 1964b); this period of ethylene unresponsiveness is presumedly the same as the period of auxin inhibitory effects (stage I). Conditions under which abscission develops can also be shown to be periods of natural ethylene biosynthesis (Abeles, 1967).

The other plant hormones, gibberellin, cy-

Fig. 9-26 The time of auxin application after leaf deblading can determine whether abscission is inhibited or promoted. It is suggested that a first stage of abscission is inhibited by auxin and that a subsequent stage is promoted. Debladed petioles of bean treated with lanolin paste of naphthalenacetic acid (5.4 m*M*) on distal side (Rubinstein and Leopold, 1963).

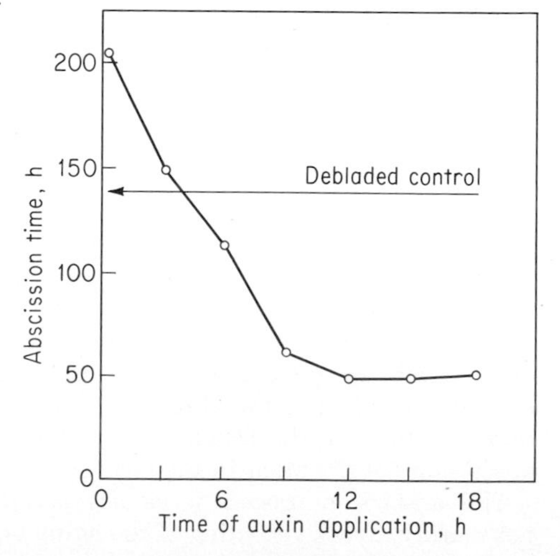

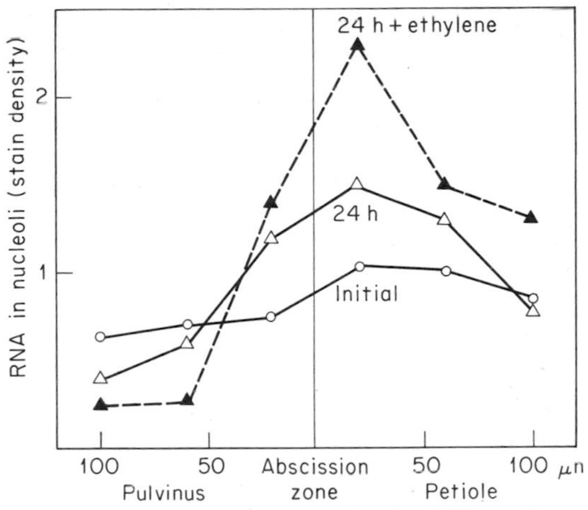

Fig. 9-27 Histochemical staining for RNA in the cells of the region of the bean-leaf abscission zone show a rise in RNA in the nucleoli during a 24-h aging period after cutting and a further rise with 4 h exposure to ethylene (10 ppm). The RNA staining was scored as 0 for none, 1 for pink, 2 for red, and 3 for intense red stain with pyronin Y (Webster, 1968).

tokinin, and abscisic acid, can each promote abscission under some circumstances, but the major hormonal regulators of abscission seem to be auxin as an inhibitor and ethylene as a promoter.

Histochemical studies of the abscission zone during the progressive stages of abscission have shown marked changes in the amounts of RNA in the nuclei of cells in the region of the abscission zone. Webster (1968) has reported that with the advent of stage II, marked increases in RNA are observed in the cells of the bean abscission zone and cells just proximal to it; exposure of the stage II explant to ethylene markedly expands the RNA increase, as shown in Fig. 9-27. The development of the separation process might be envisaged as involving a stimulation of RNA synthesis and a consequent stimulation of protein synthesis as the enzymes for separation are produced. Abeles and Holm (1967) had previously shown that inhibitors of RNA synthesis can inhibit abscission and also that labeling experiments reveal stimulations of RNA and protein synthesis. Webster's experiments placed both the RNA and the protein effects in the histological location of the abscission zone.

In the bean petiole, cell divisions may relate either to the separation activity or the formation of a healing layer over the stump which will be left by the abscission development; thus the protein experiments on incorporation of label into RNA and protein might be clouded by the two different processes. But Stosser (1971) has carried out similar experiments with the abscission zone of cherry fruits, where there is no cell division during the abscission process, and there, too, labeling experiments show enhancement of RNA and protein synthesis during abscission development.

The product of these synthetic processes should be enzymes capable of bringing about separation of the cells of the abscission zone. Pectin methylesterase was the first suggested enzyme, and Osborne (1958) showed that in bean petioles this enzyme declines with age and with abscission development and auxin keeps the levels high. IAA oxidase has been suggested by Schwertner and Morgan (1966) on the basis of enhanced IAA oxidation during the progress of abscission and the parallel effects of phenolic compounds on IAA oxidation and on abscission. Pectinase has been invoked by Morré (1968), who has observed marked increases in this enzyme during abscission development in bean petioles, as illustrated in Fig. 9-28. Cellulase has received the most attention; Horton and Osborne (1967) showed a correlation of cellulase activity with abscission development, and Lewis and Varner (1970) confirmed that this enzyme increases during abscission development (Fig. 9-29). They were also able to show (with labeling experiments similar to those in Fig. 5-16 that the cellulase increase invokes a *de novo* synthesis of enzyme. Kinetic studies show that ethylene stimulations of abscission begin within 1 h of ethylene exposure to stage II petioles (dela Fuente and Leopold, 1969), and this correlates with an ethylene stimulation of the release of cellulase into the apparent free space of the tissue (Abeles and Leather, 1971). Thus the ethylene action may involve both enzyme synthesis and the release of the enzyme from some bound form.

In brief, abscission appears to be a correlative

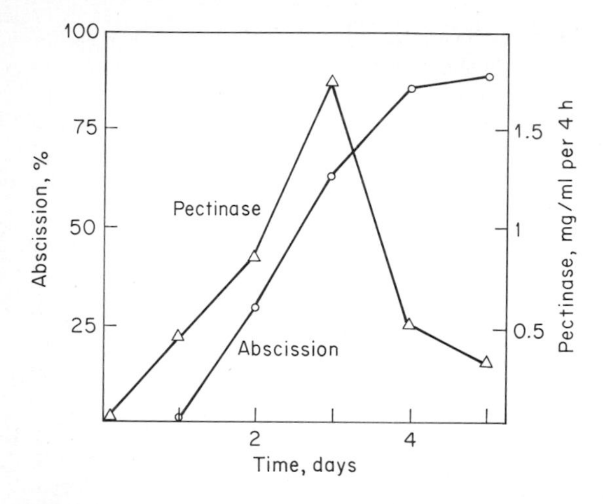

Fig. 9-28 The development of abscission in the bean leaf is associated with a rise in pectinase activity. Abscission is measured as percent of bean explants showing visible separation, and pectinase is assayed as the ability of enzyme preparations to bring about separation of cells in cubes of cucumber-fruit tissue (Morré, 1968).

effect in which the leaf blade (or the flower or fruit) provides the main source of auxin, which suppresses abscission. With increasing age of the leaf or flower or fruit, the auxin supply may dwindle, and the abscission zone which could be inhibited by auxin is now capable of being stimulated by ethylene. The separation of the leaf or other organ is apparently due to the synthesis and release of hydrolytic enzymes in the separation area.

SUMMARY

The most evident means by which growth is regulated in the plant is through the production and transmission of chemical messengers between cells. One of the transcendent features of the chemical regulatory system which has become evident since the 1950s is the extent to which the five known hormone systems interact in the achievement of regulatory control. Each hormone has interactions on the formation, removal, translocation, or action of at least some of the other hormones, and when one attempts to represent such interactions in a rough diagrammatic form, as in Fig. 9-30, their multiplicity becomes striking. While each hormone may have distinctive actions and may dominate in the regulation of some aspects of growth, the amount available in the plant and its distribution and action are subject to the influences of at least some of the other hormones. [See also Jankiewicz (1972) for a cybernetic model of hormone interactions.]

A major contribution to the correlation effects in plants is made by auxin through its characteristic polarity of transport. Primarily synthesized in the apical meristem and moved basipetally from there, it becomes a central carrier of apex-to-base regulatory signals. The meristem is also a principal site of formation of gibberellin, ethylene, and possibly other hormones, and so the contribution of the apical meristem to the growth characteristics of the tissues below is not exclusively through auxin.

In the first quarter of the twentieth century, much attention was given to root development as an important vector in the overall development of the plant; top-to-root ratios were studied assiduously. Now, a half century later, there is a renewed interest in the regulatory functions of the root system in the total plant, and it is increasingly recognized as a central source of gibberellins and cytokinins. There is also a growing body of evidence suggesting that inorganic ions may participate in the effectiveness of hormones (Leopold et al., 1974), and this may

Fig. 9-29 The development of abscission in bean-petiole explants is associated with the development of cellulase (Lewis and Varner, 1970).

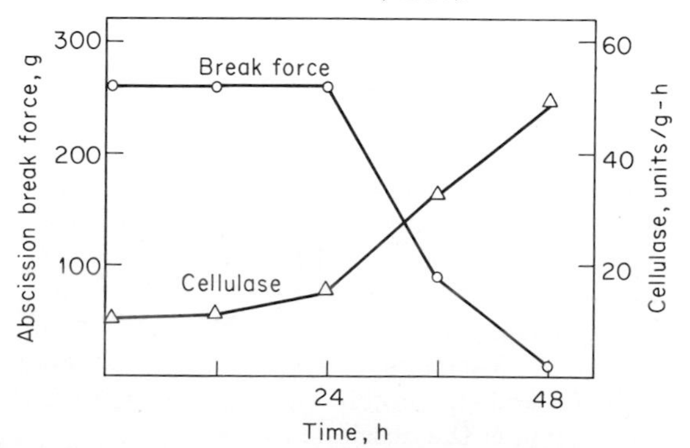

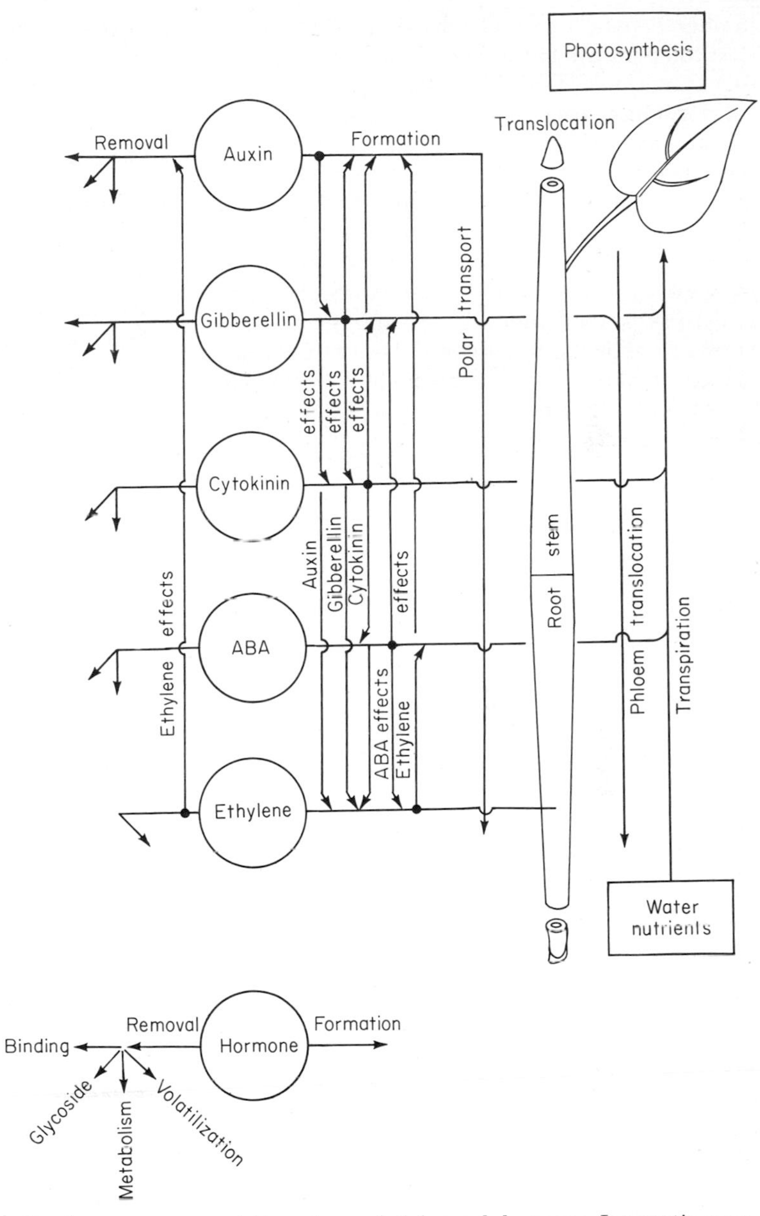

Fig. 9-30 A schematic diagram of the five known hormone systems in plants, showing some of their interactions in terms of removal of hormones (*left*), formation of hormones (*center*), and respective manner of translocation (*right*). The four types of hormone removal are indicated by directional arrows at the extreme left for each hormone. Interactions on hormone formation and removal are indicated by vertical lines, forward arrows indicating promotions and backward arrows inhibitions. Thus auxin promotes gibberellin formation, and ethylene inhibits auxin formation.

further expand our understanding of the roots as a contributors to the homeostasis and regulatory condition of the whole plant.

Efforts have been made to set up ground rules for proving that some particular hormone may be the regulator of a particular plant function (Jacobs et al., 1959) in a manner suggestive of Koch's postulates to prove that a particular microorganism is responsible for a disease. But in view of the multiple interactions between growth substances, and the web of interrelations between particular functions and the various hormones, it hardly seems a reachable objective to try to prove that this hormone is the regulator of that function. In plant physiology it seems that regulation by multiple interacting hormone systems is so widespread that it may be close to the general rule.

GENERAL REFERENCES

Kozlowski, T. T. (*ed.*). 1973. "Shedding of Plant Parts." Academic Press, New York, 560 pp.

Satter, R. L. and A. W. Galston. 1973. Leaf movements: Rosetta stone of plant behavior? *BioScience* 23:407–416.

Wilkins, M. B. (Ed.). 1969. "The Physiology of Plant Growth and Development." McGraw-Hill, Maidenhead. Especially Chap. 5 to 8.

PART 3 DEVELOPMENT

GERMINATION AND DORMANCY

JUVENILITY, MATURITY, AND SENESCENCE

FLOWERING

FRUITING

TUBER AND BULB FORMATION

The assembly of cells, tissues, and organs into an integrated multicellular organism yields not only the characteristics of the separate parts and processes but also some new characteristics which would not have been predictable on the basis of examination of the separate parts. Went (1962) has stated that "We cannot understand an organism by studying it only on the molecular level." Nor can we understand it from exclusive study of the cellular or tissue level. One could never predict the characteristics of a bee colony on the basis of examination of individual bees.

That the whole is more than the sum of its parts becomes an elaborate theme as we observe the complicated changes which the whole plant can experience during the repeating cycle of development. This cycle may start with the germination of a seed and continue with the passage of a juvenile phase of growth and the graduation into maturity, followed by progress into a state of senescence. With maturity the organism is capable of shifting from vegetative to reproductive growth, leading to the initiation and development of flowers, the development of fruits, and the production of a new generation of seed.

The controls by which the plant can regulate development may fill two types of requirements: (1) they are a necessary part of the adjustment of the plant life cycle to some environmental limitations, and (2) they play a major role in achieving outcrossing, i.e., genetic recombination between individual organisms. Some obvious examples of developmental regulation as a means of environmental adaptation are the timing of the onset and emergence from dormancy as an essential part of the seasonal adaptation of plants in temperate and arctic climates. Examples of developmental regulation in the achievement of outcrossing span nearly the entire spectrum of development. If we assume that the main usefulness of flowering is the achievement of outcrossing, the synchronization of flowering activities between individuals of a species becomes crucial. Synchronization can be achieved through the utilization of external cues for flower initiation (photoperiodism, vernalization) or for the anthesis of flowers, e.g., temperature-regulated flower bursts in coffee. Synchronization can also be achieved through internal timing mechanisms, such as the earliness-to-flower feature of many *species; in this case, the timing of germination linked to an internally timed onset of flowering can contribute to the synchronization of flowering between individuals. Outcrossing can also be abetted by the regulation of sexuality of flowers and by the self-incompatibility system. In short, a wide range of the developmental regulations evolved by higher plants can be perceived as devices which adapt the plant to seasonal limitations and facilitate outcrossing.*

A feature common to many developmental controls in plants is the fail-safe mechanism. If the plant has a programmed control restricting a developmental step, e.g., germination, flower initiation, or fruit-set, the failure of the control to be satisfied by an internal or external signal might lead to complete loss of the plant. In discussing regulatory controls of plant development we shall cite numerous instances of fail-safe devices, by which the restrictions of the developmental step can be by-passed if the necessary signal for its release is not perceived. As an example of fail-safe mechanisms, one can cite the gradual emergence of seeds from dormancy even when the normal cue for breaking dormancy has not been experienced (see Fig. 10-27). Another example is the gradually increasing tendency with age for plants to become reproductive even though they have not experienced the photoperiodic or vernalization signal ordinarily needed to induce flowering (Fig. 12-3). Yet another example is the gradual loss of the self-incompatibility restriction; as a flower grows old, the restrictions against selfing may be gradually lowered (Fig. 13-7), and even as the corolla and anthers are abscissed, successful self-pollination may occur.

The physiological nature of the controls of plant development is an area of lively interest and relatively poor understanding. Where external cues are involved in the control of development there must be receptor systems in the plant; e.g., phytochrome is a receptor for photoperiodism cues, or meristematic cells may be receptors for vernalization cues. But even where such cues and receptors are known, how their perception can be translated into an altered development state is not known. As in growth processes, the five categories of growth regulators may play important roles in developmental controls. Interactions of auxins, gibberellins, cytokinins, ethylene, and inhibitors are closely involved in the control of germination, dormancy, juvenility, flowering, and senescence. As in growth regulation, the developmental processes appear to be ordinarily controlled by the interactions of several regulatory agents. The possibility that regulatory agents affect development through an alteration of nucleic acids and their control over protein synthesis is the most attractive current biological model, but to date the extent of its applicability has not been established for any developmental control system in higher plants.

As in growth regulation, the elaborate controls of development in higher plants show an amazingly precise correspondence with the environmental events around the plant. The precision with which the plant-response systems match such environmental cues as the normal changes in light, photoperiod, temperature, or water is impressive evidence of the finesse with which biological mechanisms have been molded by the environment in evolutionary progress.

10 GERMINATION AND DORMANCY

Periods of suspended growth are common to most types of plants, being associated with the seed or with the formation of a resting bud, or both. The seed is a convenient unit in which to suspend growth, as it is so easily transported and dispersed. The suspension of growth of stems is generally associated with the depression of axis elongation and the development of a compact array of nodes enclosed in bud scales. Prolonged dormancy of buds may have evolved from the tendency of many types of perennials to show cyclic growth; e.g., the tropical cacao tree alternates periods of suspended growth ranging from 2 to 8 weeks per cycle (Greathouse et al., 1971). Numerous other perennials show such short cycles of growth, including sycamore, *Syringa,* and *Euonymus alata* (Nitsch, 1957). In most perennials growing in temperate to arctic climates, the cyclic period has been extended to an annual one, the suspended growth period extending through the annual seasons of stress—drought, heat, or cold.

Not much is known about the physiology of seeds and buds in the state of suspended growth except that they become partly dehydrated (see, for example, Abrol and McIlrath, 1966), and respiration falls to a very low level (Spragg and Yemm, 1959). What respiration does occur apparently resembles anaerobic respiration in that it has a high respiratory quotient; i.e., the amount of CO_2 produced per unit of oxygen utilized is high. Pollock (1953) reports that respiratory quotients for dormant buds of maple are commonly 1.6 and may be as high as 2.0 or even 3.0 under conditions of low oxygen tension. Emergence out of dormancy is associated with a marked rise in oxygen utilization for both buds and seeds (Pollock, 1960).

We can expect that the onset of the suspended state, whether it is a true dormancy or a quiescent state, involves a closing down of the synthetic capabilities of the plant. The recommencement of growth may involve expanding the capability for formation of enzyme systems essential for respiration and growth. Also, as a basic component of the dormancy system, we should expect the plant to have evolved cueing systems by which it can time the onset and cessation of growth with changes in the environment.

SEED GERMINATION

Arousing a dry seed to start growth into a new plant involves four groups of processes; the imbibition of water, the formation of enzyme systems, the commencement of growth and radicle emergence, and finally the growth of the seedling with the characteristic features associated with the subterranean plant up to the time of emergence from the soil.

The imbibition of water was described in the classic studies of Shull (1916) on cocklebur seeds, in which a ready comparison can be made between the dormant seed from the lower side of each bud or fruit and the nondormant seed from the upper side. He followed the water uptake as an increase in fresh weight, as illustrated in Fig. 10-1, and recognized an initial period of water uptake, then a plateau, followed by a new increase in fresh weight in the nondormant seeds beginning when the radicle emerges. The initial period of fresh-weight increase, one of hydration of the seed tissues, was completed in the cocklebur experiments in about 6 h; in other types of seeds it may last several days. Shull used salt solutions as osmotic barriers to water imbibition in this first period, and while the entry of salts into his seeds may have introduced some error, he estimated that the imbibition of water takes place with an osmotic force of up to 1,000 atm. He found the temperature coefficient, Q_{10}, to be rather low, 1.5 to 1.8, suggesting that physical rather than metabolic processes dominate this first stage in germination.

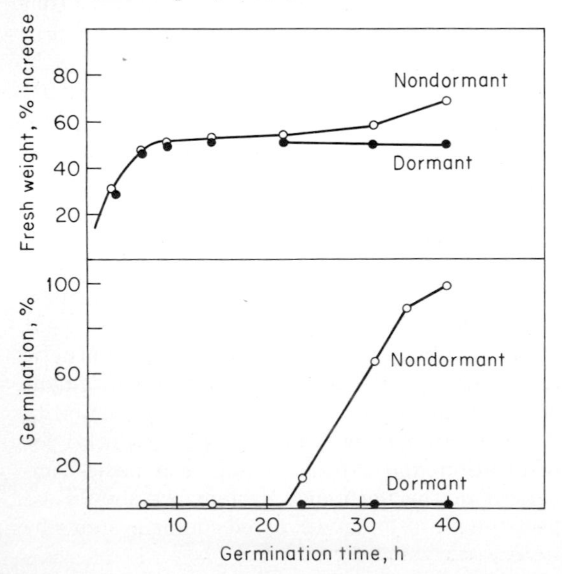

Fig. 10-1 The germination of *Xanthium* seeds involves an initial inhibition period of about 6 h, a period of suspended weight increase, and then a new increase in fresh weight when the radicle emerges. In dormant seeds this suspended period is extended (Esashi and Leopold, 1968*a*).

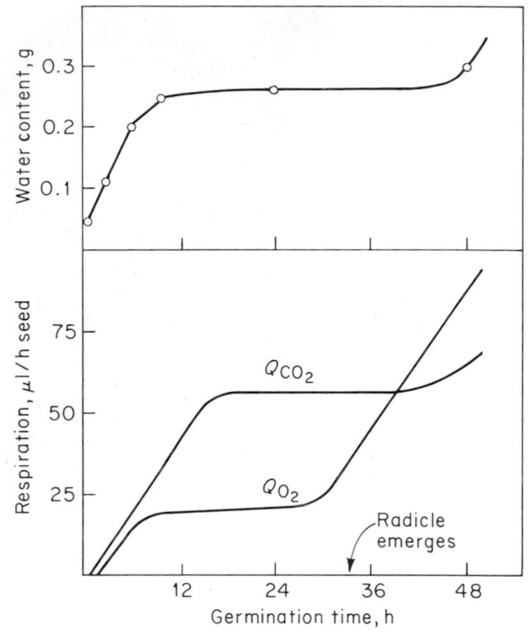

Fig. 10-2 During the germination of peas, the period of suspended weight increase is a period of high respiratory quotient; as the radicle emerges from the seed, there is a marked increase in oxygen utilization (Spragg and Yemm, 1959).

During the plateau period after imbibition, the seed develops the metabolic systems necessary for growth and the enzymic components of these systems. The respiratory behavior of the seed during this period has been examined by Spragg and Yemm (1959), who showed that during the plateau period in pea CO_2 production considerably exceeds oxygen utilization. Their data in Fig. 10-2 illustrate this characteristic, and one can see that at the time of radicle emergence, when the fresh weight begins its second rise, there is a changeover to a greater rate of oxygen utilization. These two types of metabolism may show differences in cyanide sensitivity; the initial respiration with a low Q_{O_2} has a rather large component of cyanide-insensitive respiration (Major and Roberts, 1968) and is thought to involve a different pathway of metabolism (Roberts, 1969). In rice seedlings, the respiration in the first 6 h is reported to be as much as 92 percent cyanide-insensitive. With dormant seeds, respiratory rates may decline over a period of months or even years after the first respiratory period is passed (Ota, 1925).

Achieving an improved capacity for metabolism and growth must depend upon forming the enzyme components of such systems. We come then to the basic problem of the origin of enzymes for germination and growth. In the germinating seed, enzymes may arise from two sources: they may be released or activated from existing proteins, or they may be synthesized *de novo* through the nucleic acid–directed protein synthesis. In comparisons of the two types of sources in the germinating pea (Fig. 10-3) some enzymes appear promptly after imbibition of the seed, as illustrated here by amylopectin glucosidase, and other enzymes appear much later, as illustrated by α-amylase. That the amylopectin glucosidase is released from a preexisting form is evident because its rise in activity has already begun within 3 h of the start of imbibition, is increased by treatment of the seed with a protease such as trypsin, and is not inhibited by poisons of RNA or protein synthesis such as actinomycin D or chloramphenicol (Shain and Mayer, 1968). Such a release from a latent form is also known to occur for β-amylase (Rowsell and Goad, 1962) and phosphatase (Presley and Fowden, 1965). That α-amylase is formed *de novo* during germination is indicated by the fact that its rise is not enhanced by protease treatment and is very effectively prevented by inhibitors of RNA or protein synthesis, e.g., actinomycin D, ethionine, or *p*-fluorophenylalanine (Swain and Dekker, 1969). Filner and Varner (1967) have proved that α-amylase is formed *de novo* in the germinating barley-aleurone cells (Fig. 5-16), and a similar proof of the *de novo* synthesis of isocitritase has been made by Gientka-Rychter and Cherry (1968). Many other enzymes must belong in this category of *de novo* synthesis, including probably phenylalanine ammonia-lyase (Walton, 1968), protease (Penner and Ashton, 1967), lipase (Jones et al., 1967), and nitrate reductase (Rijven and Parkash, 1971).

After the completion of the water-imbibition stage, some seeds begin to synthesize protein, and (Fig. 10-4) protein synthesis accelerates during the plateau period. This increasing momentum of protein synthesis is a common feature of nondormant seeds, but the case for dormant seeds is markedly different.

Tuan and Bonner (1964) first suggested that the state of dormancy represents a state of suppressed chromatin activity. Their experiments on the synthesis of RNA by the chromatin extracted from potato buds are illustrated in Table 10-1. They added the chromatin from dormant

Fig. 10-3 In the germination of pea, some enzymes, e.g., amylopectin glucosidase, increase in activity during the earliest stages of germination, whereas others, e.g., α-amylase, arise relatively late in the germination process (Shain and Mayer, 1968).

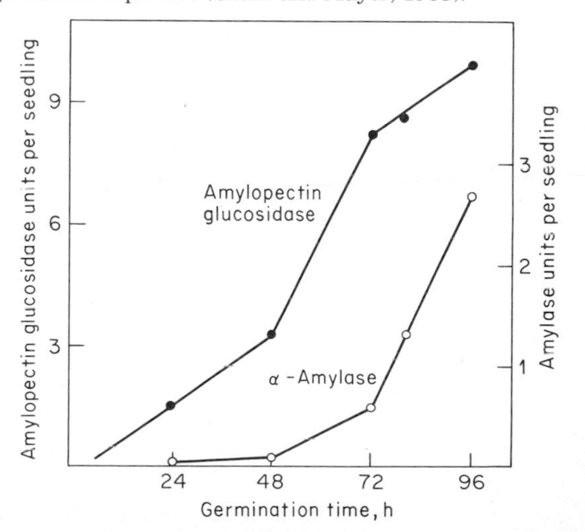

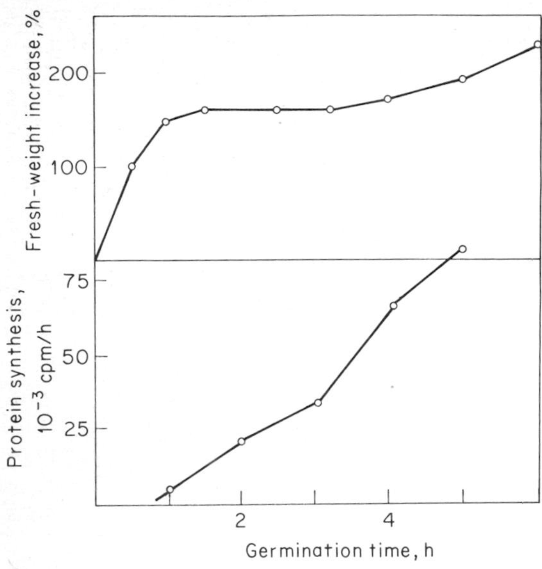

Fig. 10-4 After the completion of the imbibition period in bean seeds, a capacity for the synthesis of protein emerges. This capacity increases during the period of suspended growth. Protein synthesis was measured as incorporation of labeled phenylalanine (Walton and Soofi, 1969).

and nondormant (ethylene chlorohydrin–treated) potatoes to a system in which radioactive AMP was provided as substrate for incorporation into RNA, along with excess RNA polymerase to assure that the template activity of the chromatin would be the limiting factor in RNA synthesis. As their data show, the chromatin from dormant buds was singularly ineffective in supporting RNA synthesis as contrasted with the chromatin from nondormant ones. The repression of template activity is likewise responsible for the ineffectiveness of dormant seeds in synthesizing RNA (Jarvis et al., 1968*a,b*).

The onset of RNA synthesis, like the onset of protein synthesis, occurs during the plateau of seed germination; in dormant seeds, it is conspicuously lacking. The contrast between dormant and nondormant seeds is illustrated for hazel in Fig. 10-5, where gibberellic acid has been applied to some of the seeds to break dormancy. The plateau of fresh-weight gain by the embryo lasted for about 4 days, and the radicle emergence was noted at about 5 days; RNA synthesis was noted as incorporation of radioactive phosphate into RNA beginning on the second day of germination. Essentially no RNA synthesis was observed in the dormant seeds.

The limiting factor in the onset of protein synthesis in wheat and peanut seeds is mRNA, according to Marcus and Feeley (1964). They found that the incorporation of phenylalanine into protein cannot be obtained with the RNA fractions of dry seeds but that if a synthetic messenger such as polyuridylic acid is supplied, protein synthesis proceeds. The same workers showed that by the end of the water-imbibition period of 30 min in wheat, the seeds already contain ribosomes which can effect protein synthesis if mRNA is supplied. The plateau period in that seed lasts only 6 h. In peanut, polyribosomes are not detectable in preparations from dry seeds but can be separated by centrifugation 48 h after moistening the seeds (Jachymczyk and Cherry, 1968). In lettuce seeds, polyribosomes are not detectable as long as the seed remains dormant and first appear at about 8 h after exposure to a light treatment to break dormancy (Mitchell and Villiers, 1972).

The plateau stage of seed germination has been described by Fujisawa (1966) as time during which the seed generates the capacity to synthesize protein. He extended the plateau stage in radish seeds by adding inhibitors of RNA or of protein synthesis such as actinomycin D or puromycin; the extent of prolongation of the plateau was similar to the duration of exposure to the

Table 10-1 The chromatin of dormant potato buds is ineffective in supporting RNA synthesis, whereas buds treated with ethylene chlorohydrin to break dormancy yield chromatin, which is very effective in RNA synthesis. In each treatment, 50 μg DNA was provided with excess RNA polymerase, and AMP* incorporation into RNA was measured (Tuan and Bonner, 1964).

Chromatin source	RNA synthesized, pmol AMP per 10 min
Deproteinized DNA	3,370
Chromatin of dormant buds	122
Chromatin of nondormant buds	1,538

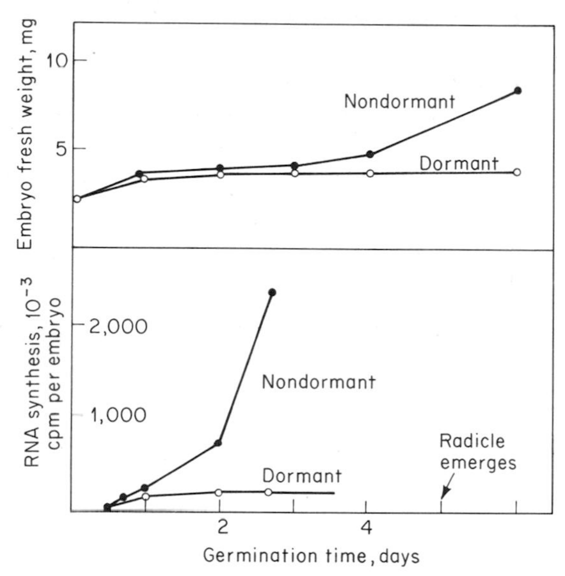

Fig. 10-5 RNA synthesis begins in hazel seeds during the period of suspended weight increase; dormant seeds are essentially unable to carry on RNA synthesis. Dormancy was broken by treatment with 10 ppm GA; RNA synthesis was measured as $^{32}PO_4$ incorporated (Jarvis et al., 1968*a,b*).

inhibitors. In bean seeds, which are not dormant, the plateau stage lasts only about 4 h; RNA synthesis is detectable after 3 h, and protein synthesis is detectable after 1 h (Walton and Soofi, 1969).

The stages of seed germination can be portrayed as an initial water imbibition, leading to the release of preformed enzymes, the subsequent development of ability to synthesize protein and RNA, after which elongation growth may proceed. It is inferred that dormant seeds are unable to develop the ability to synthesize protein and RNA.

DNA synthesis and mitotic activity ordinarily begin after the emergence of the radicle (Davidson, 1966). DNA synthesis is not a component of seed germination, for one can obtain germination (including radicle emergence and geotropic responsiveness) with seeds in which DNA synthesis (and mitotic activity) have been blocked by irradiation (Haber and Luippold, 1960) or by application of inhibitors of DNA synthesis (Tepper et al., 1967).

SEEDLING GROWTH

From the emergence of the radicle from the seed until the emergence of the seedling apex from the soil, the seedling grows as a subterranean organism. In the absence of light, growth is a process of exaggerated elongation, with the stem modified in several ways which facilitate progress through the soil. In grasses, these modifications include the enclosure of the growing point in a closed leaf cylinder, the coleoptile; stem elongation occurs principally in the mesocotyl, or internode between the seed and the coleoptile (see Fig. 10-6). The localization of growth principally in the mesocotyl until the apex emerges from the soil assures the location of the root-shoot junction at a standard depth of soil. Dicotyledonous plants also have several modifications for subterranean growth, including compaction of the stem apex into a hook and lack of leaf expansion. As the seedling emerges from the soil, light is experienced; the stem apex becomes unhooked, and leaf enlargement begins. The cotyledons of some species rise out of the soil and become leaves, e.g., bean and lettuce (Fig. 10-6); in others the cotyledons remain underground, e.g., pea and some lima beans. After the seedling has experienced light, leaf expansion begins, including overall enlargement in dicots and the specialized leaf unrolling in grasses. These events will be discussed in Chap. 15.

A special morphological modification involved in the germination of Cucurbitaceae is the formation of a foot, a lateral projection of the hypocotyl which holds the seed coat down while the cotyledons are lifted out (Fig. 10-7). The location of this remarkable disparity from the usual pattern of stem growth is established by a geotropic sensing of the lower side of the axis during early stages of seed germination.

Most seeds have some form of storage tissue, from which the substrates for growth are derived during the period of subterranean growth; in grasses, this storage tissue is the endosperm, and in most dicots, the storage function is served by the cotyledons. During the growth of the subterranean seedling, hormonal signals are produced in the axis and move into the endosperm or the

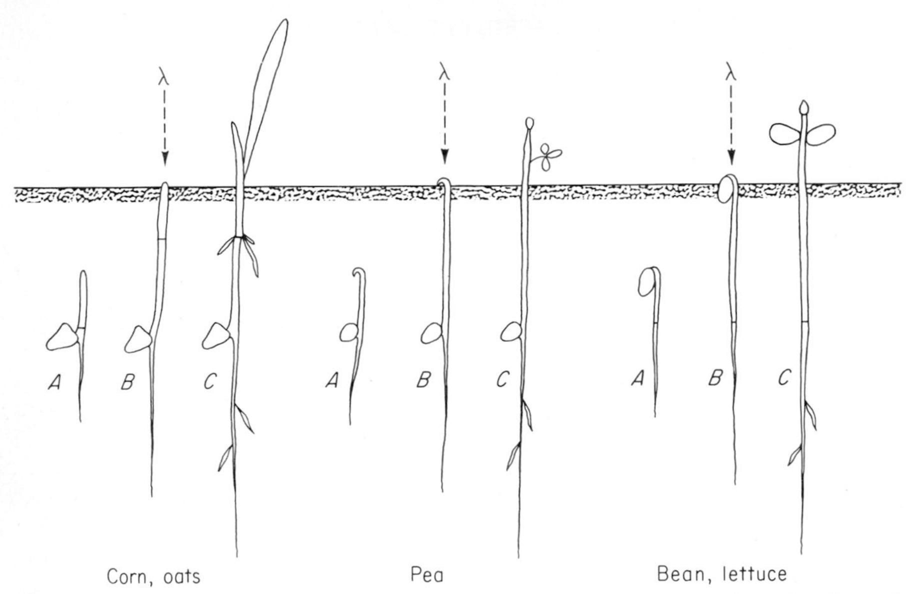

Fig. 10-6 The seedling germinated in the dark penetrates the soil by etiolated growth until its tip experiences light. In grasses the etiolated growth involves elongation of the mesocotyl (the internode between seed and first node), and in dicotyledonous plants it involves the elongation of the epicotyl (pea) or hypocotyl (bean, lettuce). Light suppresses this type of elongation, removes the hypocotyl hook, and favors leaf and internode growth.

cotyledon, where they direct the formation of hydrolytic activities and the subsequent release of simple sugars and amino acids. This type of hormonal control in the seedling is best known for the small grains, in which the embryonic axis is known to bring about the hydrolysis of carbohydrates in the endosperm. Paleg et al. (1962) observed that excission of the embryo prevents the hydrolysis of carbohydrates in the endosperm, as shown in Fig. 10-8, and that application of gibberellin can supplant the axis in bringing about this hydrolysis. The endosperm-mobilizing action of the axis is in fact due to the gibberellin synthesized there; the hormone moves to the endosperm and stimulates synthesis of hydrolytic enzymes in the aleurone cells surrounding the endosperm (see Chap. 5). Among the hydrolytic enzymes participating in endosperm degradation are α-amylase, β-amylase, and ribonuclease (Paleg, 1960; Chrispeels and Varner, 1967). A parallel hormonal function has been reported also for the dicot cucumber; in this seedling, the cotyledons provide the nitrogenous substrates for the axis. Penner and Ashton (1966) found that removal of the axis prevents hydrolytic activities in the cotyledon, especially of protease. At this early stage of germination, protein synthesis in the axis tissue appears to be at the expense of protein components in the cotyledons (Fig. 10-9). A diffusible substance from the axis induces the formation of protease in the cotyledon, and two cytokinins can take the place of the axis in this function. It is possible, also, that conversely the cotyledons may provide some hormonal regulator to the axis (Katsumi et al., 1965). The axis, however, appears to be the principal site of synthesis of hormones involved in the germination and early growth of the seedling (see, for example, Ross and Bradbeer, 1968).

In the course of studying ethylene responses of pea seedlings Goeschl et al. (1966) found that when etiolated peas meet some resistance to their growth progress, the resultant stress leads to ethylene formation by the seedling. Large amounts of ethylene may be produced in this stress response, and ethylene of course causes a thickening and epinastic bending of the pea stem. Both responses may facilitate passage of the seedling around the obstruction. The role of ethylene in holding the apex in the hooked condition is well known (Kang et al., 1967; see also Chap. 7).

The vigor of growth of the new seedling is a matter of great concern to the seed technologist. Lowered vigor is observable as a lessened growth rate of hypocotyl or epicotyl or root, a lowering of P/O ratio of metabolism and an increased tendency of the seed or seedling to leak nutrients out into the ambient solution (Roberts, 1972). Reduced vigor can be observed to result from a temperature stress, mechanical damage to the seed, or aging of the seed (Byrd and Delouche, 1971).

Fig. 10-7 During germination, the cucumber develops a foot, or projection, on the hypocotyl which serves as a lever against which the seed coat is removed. The foot is observable from stage *c* onward, and in stage *e* it has levered the seed coat part way off. In each case a portion of the seed coat has been cut away. (*Photograph courtesy of R. K. dela Fuente.*)

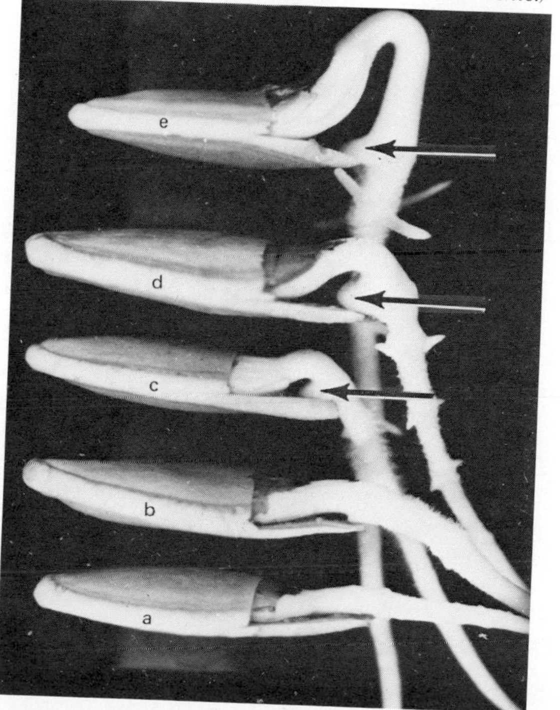

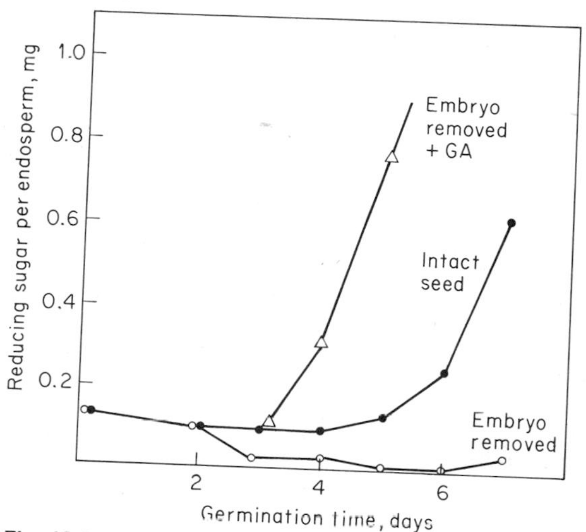

Fig. 10-8 Hydrolysis of the starches in the endosperm of barley seeds depends upon an action of the embryo, since removal of the embryo prevents hydrolytic development; gibberellic acid application replaces the embryo (2 ppm GA for 48 h) (Paleg et al., 1962).

FUNCTIONS OF DORMANCY

Dormancy may constructively limit the time at which seeds will germinate, e.g., desert seeds, which will germinate only at the beginning of the rainy season, or the seeds of apple, which must pass through a cold period (ordinarily the winter) before germination will take place. Another constructive function is the determination of the location for germination, e.g., cypress seeds, which normally require standing water for germination (a desirable site for cypress tree growth), or the spores of corn smut, which are stimulated to germinate by the proximity of corn plants (von Guttenberg and Strutz, 1952). Another function of dormancy is to adapt a species to the seasonal characteristics of the environment, either to the range or to the seasonal limits of ecological variables. As examples, one might cite the requirement of peach buds for cold, which adapts the species to the more northerly climates of the United States; the photoperiodic imposition of bud dormancy on deciduous trees, which adapts them to the winter or the dry seasons of the year; and the requirement of weed seeds for mechanical scarification, which makes them ger-

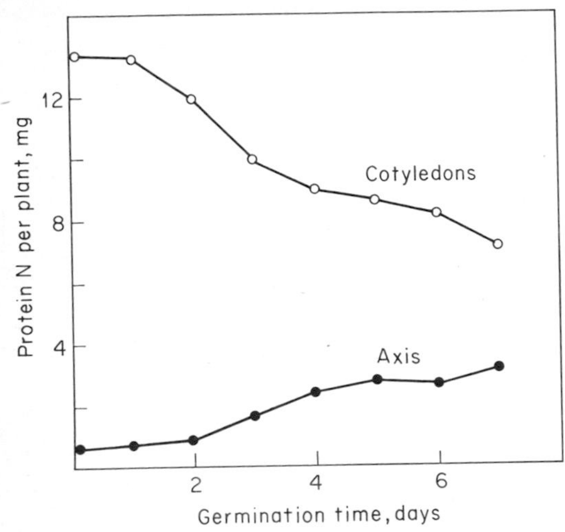

Fig. 10-9 In germinating squash seedlings, the accumulation of protein in the axis appears to occur at the expense of protein in the cotyledons (Wiley and Ashton, 1967).

minate in soil which has been disturbed, thus assuring competitive conditions which will be favorable for their development.

Physiological controls of the dormancy of buds or seeds are often beautifully adapted to meet the functional benefits. The cues utilized by the plant to terminate dormancy may closely match the environmental feature which is being functionally evaded by the interposition of dormancy.

Seed dormancy can also be a component of a seed-dispersal function; the seeds of some plants require passage through the digestive tract of animals before they germinate. Rick and Bowman (1961) were quite unable to germinate the seeds of wild tomato collected on the Galápagos Islands, but seeds that had passed through the digestive tract of a tortoise germinated readily. Such a dormancy-breaking requirement facilitates dispersal of the seed. The extent to which some seeds depend upon passage through the gut of birds, bats, and other animals has been reviewed by van der Pijl (1969).

One must discriminate between a temporary shift in cues that are utilized for breaking dormancy and the genuine changes that are associated with the climatic season favorable for germination. If a desert annual required rain to break dormancy, an aberrant rainstorm could bring it into germination even though the rainy season has not arrived. Two strategies seem to be involved in discrimination of aberrant cues: some plants are able to sum the stimulations given by the cue which regulates their dormancy, and others utilize multiple-cue systems. An example of the former is the requirement of apple seeds for a certain duration of cold exposure to break dormancy; an example of multiple cueing is the requirement for both light and temperature experiences for *Lepidium* seeds (Toole et al., 1955) and for both scarification and a thermoperiod for some weeds (Tager and Clark, 1961).

Some seeds show the interesting feature of 2-year dormancy; i.e., two winters must pass before they germinate. This feature employs multiple cueing; e.g., the seeds of *Trillium* require the passage of one cold period for the afterripening of the radicle end of the embryo; during the next summer, the radicle emerges and becomes established, but the epicotyl remains dormant until the passage of another cold period. The wide range of species which show this 2-year dormancy is discussed by Crocker and Barton (1957).

The development of dormancy is not always systemic; i.e., the dormant condition may be restricted to some parts of the plant. In coffee plants, only the flower buds may show dormancy; in some woody plants dormancy is achieved progressively by individual buds from the base of the branch to the tip, and in some cases cutting a ring through the bark of a twig permits the buds above the ring to continue growth when the rest of the plant is dormant.

We can see dormancy as a developmental intermission which may occur in diverse parts of plants, seeds and buds especially, and which may have many different cues or requirements for being recalled. We can anticipate, then, even before examining the internal physiology of dormancy, that a wide diversity of controls may be possible inside the plant to match the diversity of means of breaking it.

SEED DORMANCY

The seed can be described as a small meristematic axis, often associated with a storage tissue, and enclosed by an enveloping series of membranes and sometimes stony shells which collectively make up the seed coat. These latter structures are frequently critical to the dormancy of seeds, limiting the entry of water and oxygen, mechanically limiting the enlargement of the embryo, or sometimes altering the growth-substance relationships of the enclosed tissues. Many seeds are improved in germination by removal or rupture of the seed-coat layers. This scarification of course alters each type of limitation which the seed coats can impose on germination.

A remarkable mechanism is involved in the limitation of water imbibition by hard seeds, involving a hygroscopically activated valve in the hilum. Its structure has been examined by Hyde (1954), who described the operation of this aperture in the stony coat of some leguminous seeds (Fig. 10-10). Germination is not obtained even in moist conditions because of a limitation of moisture entry by closure of the hilum, brought about through the differential moisture content of internal and external seed-coat layers. If there is more moisture on the outside of the seed coat, the valve closes and water entry is prevented. If there is less moisture on the outside, the valve opens and the seed dries further. Experimental data are given in Fig. 10-11, following the moisture content of normal hard seeds and scarified seeds as they are moved from one humidity level to another. It is evident that the hard seeds lose moisture in the low humidities because of the operations of the hilum valve. Hyde found that the lower the moisture level of the seed, the more pronounced the impermeability of the seed coat and thus the deeper the dormancy.

The seed coat may also limit the movement of gases into and out of the seed. Having noted that oxygen exposures would increase the germination of many seeds, Weisner and Molisch (1890) examined the gas movement through seed coats and found limited permeability to oxygen and (to a lesser extent) CO_2. Some precise measurements have been made by Brown (1940), who found that the membrane around cucumber seeds passes only 4 cm^3/cm^2-h of oxygen, in comparison with 15 cm^3/cm^2-h of CO_2. This limitation of gas exchange is brought about by a living system in the seed coat, since brief exposure of the seed to extreme heat (40°C) or a strong solvent like chloroform relieves the permeability limitation. This observation may help account for the dormancy-breaking effects of high temperature and many chemical treatments of seeds. Brown also found that the water relations of the cucumber seed coat markedly alter its gas permeability. Once the coat has dried thoroughly, the permeability is improved, probably because air spaces are introduced into the structure through which gases can diffuse. Washing the seed also markedly improves the permeability to gases. (Many seeds can be brought out of dormancy by washing.)

In some seeds, germination may be limited through mechanical restriction by the seed coat. Crocker (1906) carried out the first experiments suggesting such a role of the seed coat, and his experiments with hard seeds led him to suggest that the breaking of dormancy by acid treatments or by passage through the digestive system of some birds and animals involves deteriorative effects on the seed coat (Crocker and Davis, 1914). Using the dramatically hard seeds of walnut and hickory, he was able to apply hydraulic forces inside the shell to measure the force necessary to rupture the hard coat. For freshly harvested walnuts, 600 lb/in^2 of pressure was required to rupture the seed coat; after 2 or 3 months in the soil, the shells were markedly

Fig. 10-10 The hard seeds of *Lupinus arboreus* have a moisture valve in the hilum operated by a counterpalisade tissue. When there is more moisture outside than inside, the counterpalisade swells and closes the hilum (Hyde, 1954).

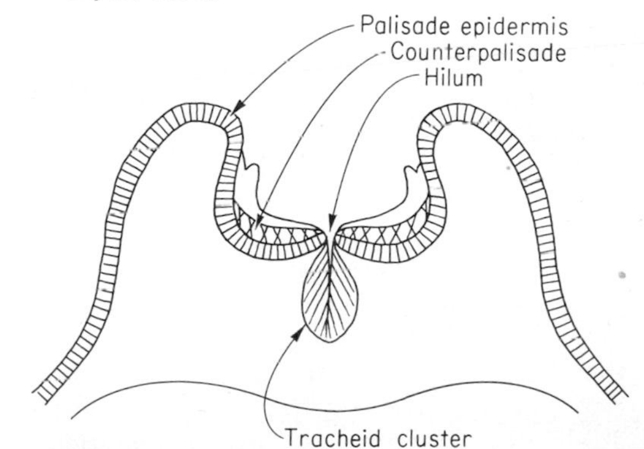

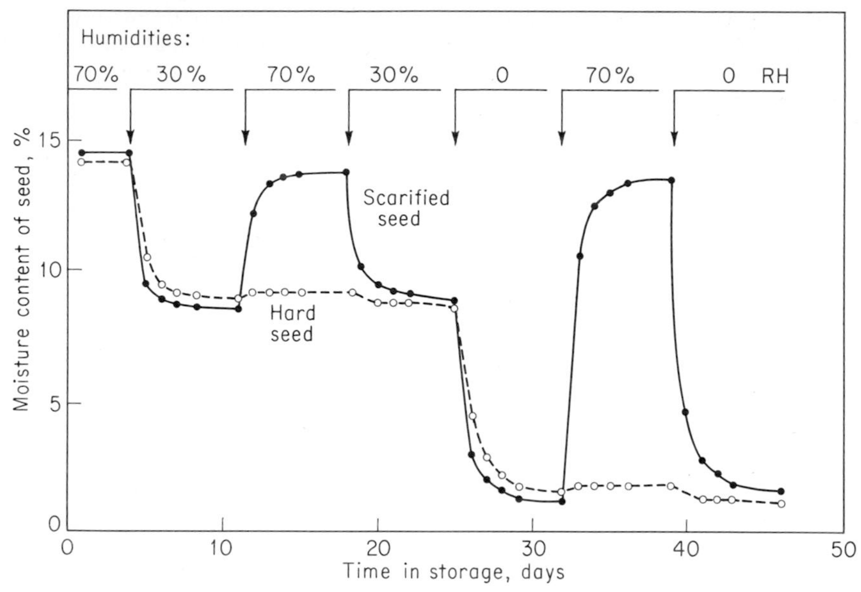

Fig. 10-11 When hard seeds of white clover are placed in alternating high and low relative humidities, the seeds lose moisture in lowered humidities but do not gain it back in elevated humidities because of the operation of the hilum valve. Scarified seeds gain moisture readily (Hyde, 1954).

weakened, and less than a third as much pressure was needed. The decay in breaking force was slower in soils at low temperatures and did not occur in sterilized soil; thus the removal of the restricting force of the seed coat was a consequence of microorganismic attack on the seed coat (Crocker et al., 1946). Even the thin seed coat on lettuce offers mechanical resistance to germination (Böhmer, 1928), and recent experiments by Ikuma and Thimann (1963) indicate that a part of the light regulation of lettuce germination may be the stimulation of formation of enzymes such as pectinase and cellulase which degrade the seed coat and thus allow the radicle to emerge.

The restrictive forces of the seed coat of *Xanthium* seeds have been estimated by Esashi and Leopold (1968), as well as the thrust forces generated by the embryo axis and the cotyledons. By placing whole seeds or isolated embryo axes under a small piston attached to a manometer, they recorded the thrust generated by imbibition and by active growth processes, as shown in Table 10-2. The greatest force contribution was by the axis of nondormant seeds; total forces generated by dormant seeds were insufficient to break through the testa.

Growth Substances and Seed Dormancy

The idea that growth substances serve to regulate the germination and dormancy of seeds springs from the early observations of Köckemann (1934) that inhibitors of germination exist in numerous fruits and apparently serve to prevent seed germination within the fruits. Nutile (1945) made the surprising discovery that a lactone inhibitor such as coumarin can be applied to lettuce seeds and cause them to be dormant and that light can overcome that dormancy; lactone-induced dormancy has been confirmed many times, but the nature of the inhibition and the light reversal are not known. It soon became evident that seeds are rich sources of inhibitors (Evenari, 1949) and that in many seeds they are most abundant in the seed coat; this is without doubt an important part of the effectiveness that removing the seed coat has in releasing a seed from dormancy. Much earlier, Laibach (1929) showed that isolated embryos from dormant iris seeds grow well on a nutrient agar and are not inherently dormant; this is consistent with the evidence that the seed coat is a major source of inhibiting substance in the seed.

Quantitative studies of growth inhibitors as they may relate to dormancy began with the work of Hemberg (1949*b*), who measured large amounts of inhibitors in dormant buds of potato and much lesser amounts after dormancy was broken. In detailed studies of dormancy in *Xanthium* and lettuce seeds, Wareing and Roda (1957) found that dormant seeds contain large amounts of growth inhibitors and that the amounts declined sharply upon emergence from dormancy. They described the testa, or seed coat, in *Xanthium* as restricting the leaching of inhibitors out of the seed and the entry of oxygen into the seed, where it could help destroy the inhibitor. An interesting variant of the inhibitor story was reported by Villiers and Wareing (1965); they found essentially no inhibitors in the dormant, dry seeds of *Fraxinus,* but when the seed imbibed water, large amounts of inhibitor were produced in the seed, thus apparently restricting germination.

The isolation of the growth inhibitor in dormant seeds and buds in Wareing's laboratory opened the way for its chemical identification, successfully achieved by Cornforth et al. (1965a), who established the structure of abscisic acid (see Fig. 8-6). The quantitative relationship between ABA levels and dormancy of *Fraxinus* seeds has been described by Sondheimer et al. (1968), whose data in Table 10-3 show that both the seed and the surrounding pericarp contain markedly larger amounts of ABA before chilling than after. Applications of ABA to various seeds have been shown to prevent seed germination (Eagles and Wareing, 1964) (See Fig. 8-1).

Table 10-2 *Xanthium* seeds develop thrust forces against the testa which are sufficient to rupture it in nondormant seeds but insufficient in dormant seeds. Passive forces were separated from active ones by anaerobic measurement; the force needed to rupture the testa was 67 g for large seeds and 56 g for small seeds (Esashi and Leopold, 1968a).

Germination phase	Force generated, g	
	Large seeds (nondormant)	Small seeds (dormant)
Passive imbibition:		
By axis	10	10
By cotyledons	20	10
Active enlargement:		
By axis	39	17
By cotyledons	15	4
Total force generated:	84	41

Table 10-3 The abscisic acid content of seeds and pericarps of dormant and nondormant *Fraxinus americana* (Sondheimer et al., 1968). Dormancy was broken by chilling the seed.

Plant material	ABA content, μM/kg
Seeds:	
Dormant	1.7
Nondormant	0.6
Pericarps:	
Dormant	2.8
Nondormant	1.8

Other growth inhibitors than ABA may be involved in the mechanism of dormancy; for example, Koller (1957) has found that the total osmotic value of the seeds of a desert *Atriplex* is sufficient to prevent its germination until enough rainfall has been experienced to elute these osmotica out. There are also several lines of evidence that phenolics, especially flavonoids, may be involved in the dormancy of some seeds (Phillips, 1961; Mikkelsen, 1966).

About 1960, some uneasiness arose about the idea of accounting for dormancy on the basis of growth inhibitors (see, for example, Buch and Smith, 1959). Frankland and Wareing (1962) were able to detect no change in inhibitor content of hazel or beech seeds with the cold treatment to break dormancy, and Kentzer (1966a) found that the inhibitors in dormant *Fraxinus* seeds were ineffective in inhibiting the growth

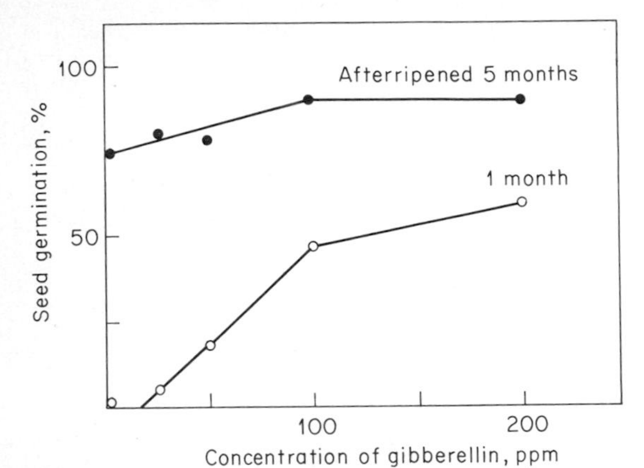

Fig. 10-12 The application of gibberellin to seeds of eggplant can remove dormancy, especially in seeds with limited amounts of afterripening (Suzuki and Takahashi, 1968).

of *Fraxinus* embryos. Meanwhile, Kahn et al. (1957) and Mayer and Poljakoff-Mayber (1957) reported that gibberellin applications can break the dormancy of lettuce seeds (see also Fig. 10-12). Gibberellins assayed in dormant and non-dormant seeds were found to rise with treatments which would break dormancy (Naylor and Simpson, 1961). Ross and Bradbeer (1968) found that during the cold treatment to break dormancy of hazel seeds there was a very large increase in GA content, the axis showing a hundredfold higher GA content than the cotyledons; this is consistent with the regulatory function of the axis on the germination processes of the storage tissues through the production of GA, a phenomenon discussed previously in this chapter. As with the growth inhibitors, some seeds may actually form the GA during germination, e.g., seeds of larch and pine (Fig. 10-13).

That cytokinins can stimulate seed germination has been known since Miller's report in 1956. The relief of dormancy by applied cytokinins is restricted to relatively few species of plants, and then only under some conditions (see, for example, Khan, 1971); e.g., lettuce seeds are relatively insensitive to cytokinin alone, but when the seed has been inhibited by ABA, cytokinin applications result in marked enhancement of germination (Khan, 1968). The ability of applied cytokinins to break dormancy is probably a consequence of their impressive ability to stimulate the enlargement of the cotyledons (Ikuma and Thimann, 1963; Esashi and Leopold, 1969a) and the presumed increase in the generation of force against the testa. While physiologists frequently infer that cytokinins may be involved in the natural control of seed dormancy, there are as yet no data indicating that they change in seeds in a manner related to the emergence from dormancy; in buds, however, there is some evidence that cytokinins increase as dormancy is lost (Domanski and Kozlowski, 1968). It is curious that the molecular requirements for cytokinin breaking of seed dormancy are quite different from the requirements for stimulating cell division (Kefford et al., 1965).

Auxins are ordinarily not present in dry seeds but are formed in early stages of the germination process (Fig. 10-14). The first auxin formed is probably released from a bound-auxin pool. While auxins perform many roles in the seedling, they probably are not involved in dormancy-control mechanisms (Poljakoff-Mayber et al., 1957).

Although ethylene was long ago shown to be effective in breaking seed dormancy (Vacha and

Fig. 10-13 During the germination of larch seeds, inhibitors are lost from the seeds and gibberellins accumulate. Inhibitors and giberellins assayed by oat-leaf bioassay (Michniewicz and Kopcewicz, 1966).

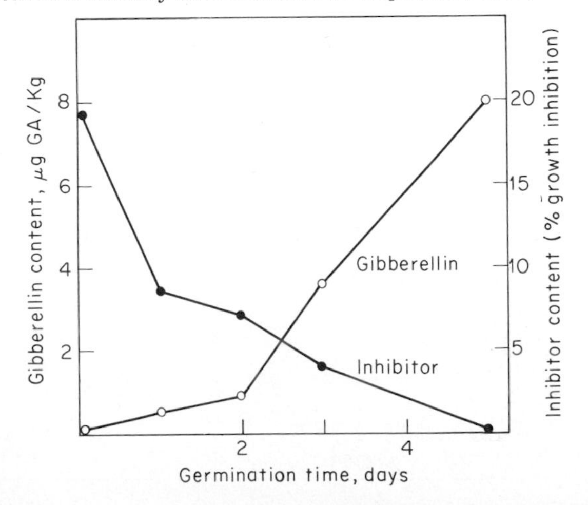

Harvey, 1927), it was ignored as a possible hormonal regulator of germination until 1964, when Meheriuk and Spencer reported that ethylene is formed by germinating oat seeds and Toole et al. (1964) found that it is formed by the peanut seed during germination. When the latter workers applied ethylene to their seeds, germination was enhanced. The formation of ethylene during germination time is illustrated in Fig. 7-1. In subterranean clover, the amounts of ethylene formed by the moistened seed are sufficient to stimulate the germination process (Esashi and Leopold, 1969b). Burdett (1972) has calculated that lettuce seed must experience an internal ethylene level of 0.74 nl/g of seed to emerge from dormancy. The principal locus of ethylene formation in germinating peanut seeds is the axis (Ketring and Morgan, 1969), again consistent with the idea that the axis provides the major hormonal signals for seed germination. Although this ethylene function is not reversed by CO_2 (Abeles and Lonski, 1969), it is known that CO_2 has a stimulatory effect on seed germination (Mayer and Poljakoff-Mayber, 1963); hence its suppression of the ethylene effect would be difficult to observe. Stewart and Freebairn (1969) have suggested that the inhibitory effects of high temperature on lettuce-seed germination may be due to the temperature interference with ethylene biosynthesis (see Fig. 7-3), as they could overcome the high-temperature dormancy by supplying ethylene.

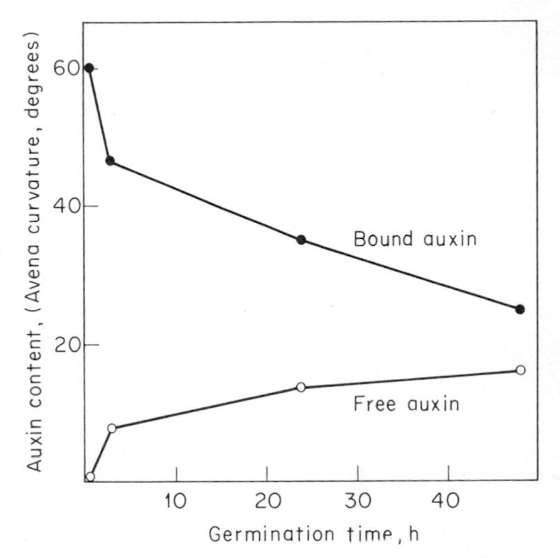

Fig. 10-14 During the germination of maize seeds there is an accumulation of free auxin and a decrease in bound auxins. Free auxin was taken to be the amount extractable with ether in 3 h, bound auxin from 3 to 45 h; assayed by oat-coleoptile curvature test (Hemberg, 1955).

Environmental Controls of Seed Dormancy

Temperature. Since seed dormancy serves to improve the passage of many plants through seasonal extremes of temperature, it is not surprising that temperature is a common environmental cue for emergence from dormancy. The requirement of some seeds for a low-temperature experience to break dormancy has been known for centuries. In some cases only a brief exposure to temperatures near freezing is needed to break dormancy; in others an extended period is needed; and in others dormancy is not actually broken until two winters have been passed in the ground (cf. Crocker and Barton, 1957). Temperatures near freezing are usually the most effective, though 10°C is often low enough for breaking dormancy. The agricultural practice of chilling seeds has been given the inappropriate and unphysiological name seed *stratification.*

Low temperature stimulates peach-seed germination, the effect increasing with the duration of the cold treatment from 4 to about 10 weeks. During a cold treatment there may be extensive changes in the distribution of food materials in the seed; e.g., *Heracleum* seeds translocate the bulk of their dry weight from the endosperm to the embryo during stratification (Stokes, 1952). The embryos of peony seeds have very low amino acid levels when dormant, but during the cold stratification they accumulate large amounts from the endosperm (Fine and Barton, 1958). In this case, gibberellin treatment, which breaks dormancy, also brings about the rise in amino acids in the embryo.

The low-temperature treatment can depress the growth-inhibitor content of the dormant seeds. Following the evidence of Hemberg (1949a) that inhibitors are involved in bud dormancy, Luckwill (1952) found that apple seeds are

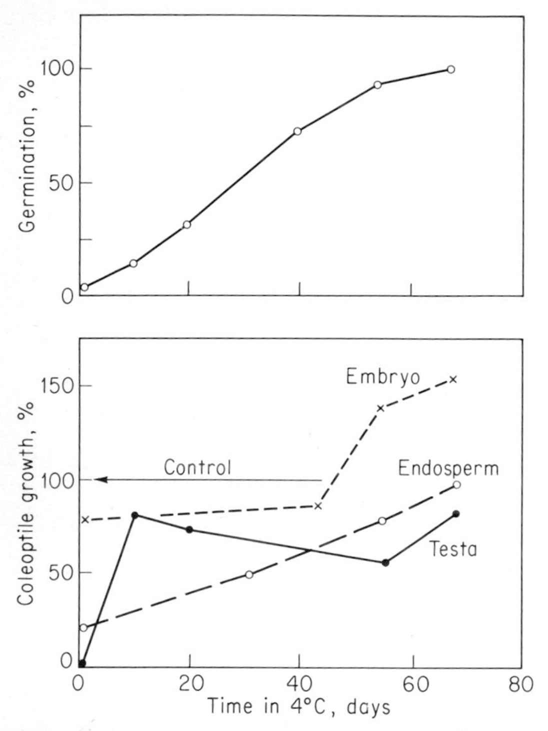

Fig. 10-15 During cold treatment to break dormancy of apple seeds, there is a decline in content of growth inhibitors, especially in the endosperm and testa. Ether extracts were tested for growth inhibition with oat-coleoptile sections, and germination percentages were determined using excised embryos (Luckwill, 1952).

greatly lowered in inhibitor content during cold stratification (Fig. 10-15). In this figure, the loss of dormancy during cold is shown by the increasing germination percentage, and the bioassay of the various seed parts indicates that a large amount of growth inhibitor initially present in the testa and endosperm disappears during the cold treatment. The data suggest that there may be an accumulation of growth-*promoting* substances in the embryo after about 50 days of chilling.

A comment about bioassays is in order. Most early experiments on growth inhibitors used the *Avena* coleoptile as the assay material, as in Fig. 10-15; the use of an assay material which is foreign to the biological event being studied leaves room for substantial errors in making deductions from the data. More modern experiments on seed dormancy would utilize embryos of the seed itself as the bioassay material (see Villiers and Wareing, 1965).

A remarkable variation of the low-temperature effect is the requirement of some seeds for an alternating-temperature experience. Morinaga (1926) first observed this and suggested that two temperatures are required for the mechanical modification of some limiting feature of the seed or seed coat. Increased percentage of germination or more uniform germination can often be obtained with alternating temperatures better than with any single-temperature treatment (Crocker and Barton, 1957). *Lycopus europaeus* shows a complete dependence upon a diurnal flux in temperature for germination, and the illustrative data in Fig. 10-16 show that full germination can occur in any daily flux greater than about 15°C. In this species, the daily flux is required for several hours (Thompson, 1969), but in some species it is satisfied in about 15 min (Taylorson and Hendricks, 1972). The effects of the daily changes can be summated by the seeds (Thompson, 1969).

Fig. 10-16 The seed of *Lycopus europaeus* requires a diurnal fluctuation in temperature for germination. The temperatures indicated were given for 16-h day and 8-h night for 7 days followed by a constant 25°C; any constant temperature from 6 to 34°C resulted in zero germination (Thompson, 1969).

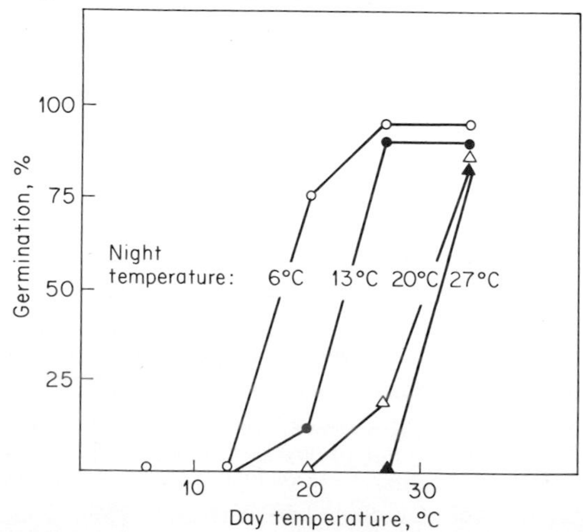

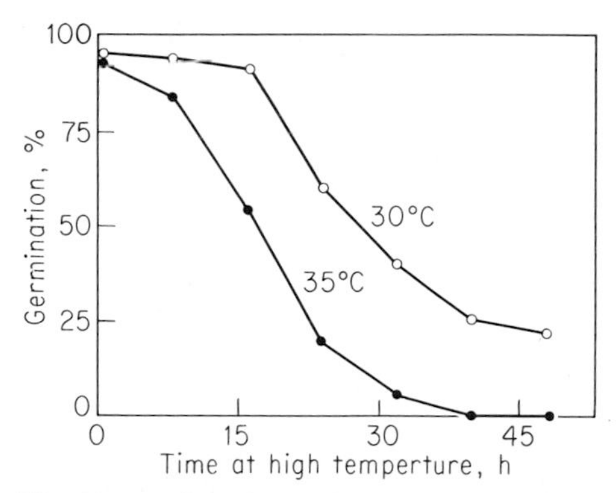

Fig. 10-17 Grand Rapids lettuce seed can be made dormant by exposure to 30 or 35°C. Tested for germination at 20°C (Toole, 1959).

Tager and Clark (1961) have reported that gibberellin applications can replace the temperature-flux requirement in *Asclepias fruticosa,* but after some exacting experimentation, Cohen (1958) concluded that the alternating temperatures more probably trigger some physicochemical or structural change than a change in any growth-regulating substance. Recall that Brown (1940) showed that elevations of temperature greatly increase the permeability of some seed membranes to gases. Toole et al. (1955) reported that KNO_3 can relieve the requirement for alternating temperatures in *Lepidium* seed, but the nature of this salt action is not known. An entirely different hypothesis has been suggested by Taylorson and Hendricks (1972); Isikawa and Fujii (1961) reported that red light can relieve the alternating-temperature requirement in *Rumex crispus,* and Taylorson and Hendricks suspected that the temperature flux could cause a cycling between the red-absorbing and far-red-absorbing forms of phytochrome. They found that far-red light can erase the promotive effects of alternating temperatures and suggested that either red light or a period of elevated temperature is needed to start a cycling effect between the two phytochromes.

High temperatures ordinarily increase the dormancy of seeds rather than improve the germination. Lettuce seed, for example, which is promoted in germination by moderately low temperatures, is made dormant by temperatures of 30 or 35°C (Fig. 10-17). This again is a quantitative effect, longer periods of heating resulting in deeper dormancy. The dormancy of lettuce seed imposed by high temperatures is relieved by light.

Light. The light requirement of many seeds may serve to bring about germination when the seed reaches the surface of the soil or when the soil or surrounding vegetation has been disturbed and the subsequent competition consequently lessened. It is a common requirement for the germination of many weed species.

The stimulation of germination by light is ordinarily quantitative, and Fig. 10-18 illustrates a light-dosage-response curve. In this illustration, there is also a temperature effect on dormancy, and if high temperatures were given before the light, the dormancy-induction effect of the temperature experience was erased by the light treatment. The quantitative effects of light on *Rumex* seeds have been reported by Isikawa and Fujii (1961), who found reciprocity between intensity and duration of light exposure.

The stimulation of germination of lettuce seed by light was studied in considerable detail by Flint and McAlister in 1937; they found that

Fig. 10-18 Light can overcome the dormancy of *Lepidum virginianum* seeds imposed by high temperature (35°C for 2 h) or can defer the inhibition by the same heat treatment given subsequently (Toole et al., 1955).

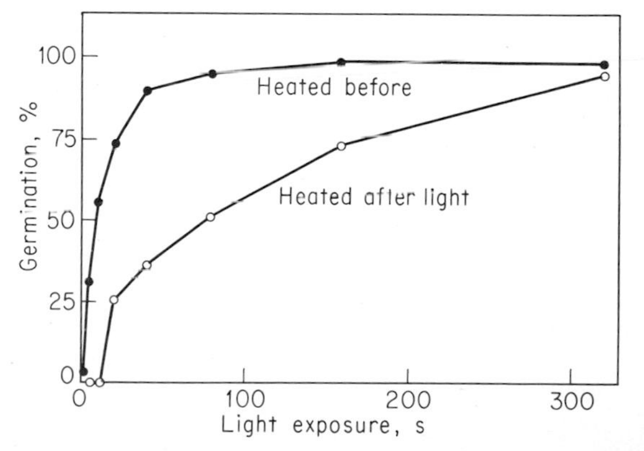

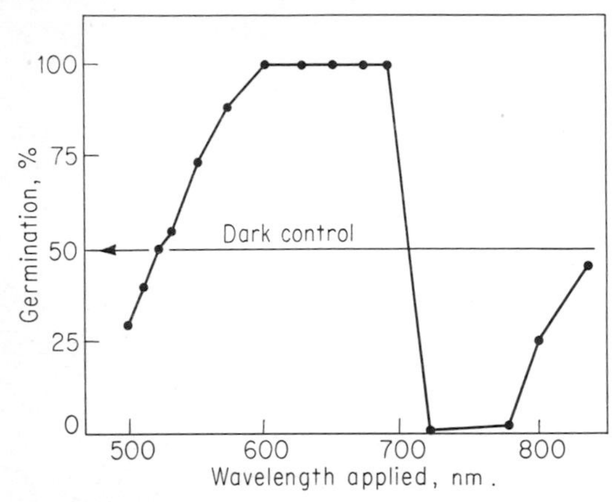

Fig. 10-19 Seeds of Arlington Fancy lettuce which are partly dormant can be stimulated to germinate by red light (600 to 690 nm) or inhibited by far-red light (720 to 780 nm) (Flint and McAlister, 1937).

red light is the most effective in breaking dormancy. Some of their data are shown in Fig. 10-19. In their wavelength studies they found that blue light and especially far-red light are very inhibitory of germination. The effects of red and far-red light on seed dormancy have since been found to be reversible, as shown in Table 10-4; this type of action implies that germination is controlled by a pigment which can exist in two interchangeable forms, a red and a far-red-absorbing form. That the responsible pigment is phytochrome is discussed in Chap. 15.

In contrast to the stimulatory effects of light on germination, numerous species are only inhibited by light (Mayer and Poljakoff-Mayber, 1963). *Citrullus* is curious in that the green and blue wavelengths of light are most inhibitory of germination, which does not indicate a phytochrome involvement.

The location of the light-sensitive system in the seed is not well known; in lettuce, the removal of the seed coat can relieve the light requirement (Evenari, 1957), which seems to imply that the pigment system resides in the seed coat. However, Ikuma and Thimann (1959) found light applied to the radicle end of the seed ineffective and light applied to the cotyledon end of the seed fully effective; they deduced that the cotyledon itself is the site of the light action. In *Citrullus*, the light-inhibiting action is localized in the radicle and not in the cotyledon (Koller et al., 1963).

A most surprising development in the light studies of dormancy has been that some seeds require not merely light but actually have a photoperiod requirement for germination. This was discovered simultaneously by Black and Wareing (1954) for birch seeds and by Isikawa (1954) for a wide selection of seeds, including some which have long-day requirements, e.g., *Eragrostis ferruginea*, and some with short-day requirements, e.g., *Veronica persica*. Some data illustrating the photoperiodic responses of seeds are given in Fig. 10-20 for *Begonia* seeds, which require a photoperiod of about 12 h or longer for germination and three or more repeated cycles. That there is a dark-period sensitivity for this photoperiod action is evident from the ability of a night interruption to depress the dark-period effect and permit germination as if in a long day (Fig. 10-21).

A clue to the possible action of light in regulating seed germination is the evidence of a strong effect of lactones on the light requirement. Nutile (1945), the first to observe this, found that coumarin applications to nondormant lettuce seeds make them strictly light-requiring. Similar effects have been found with other

Table 10-4 The stimulatory effects of red light and the inhibitory effects of far-red light on the germination of lettuce seed are entirely reversible. Light exposures of 1 min (red) and 4 min (far red) successively at 26°C (Borthwick et al., 1954).

Light treatment	Germination, %
R	70
R-FR	6
R-FR-R	74
R-FR-R-FR	6
R-FR-R-FR-R	76
R-FR-R-FR-R-FR	7

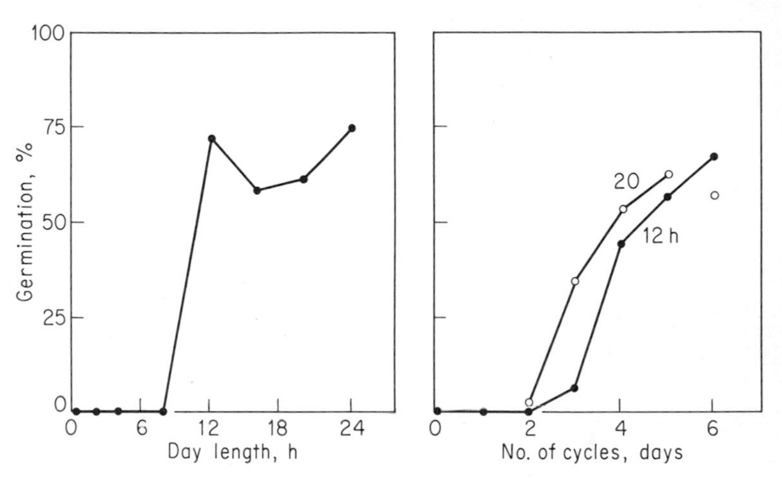

Fig. 10-20 The germination of seeds of *Begonia evansiana* is stimulated by a photoperiodic phenomenon; it requires photoperiods of 12 h or more, and the response is proportional to the numbers of long days given (Nagao et al., 1959).

lactones, e.g., rhamulosin (Mayer, 1964) and patulin (Berrie et al., 1967). One might infer that light regulates the availability of some substance which is sensitive to lactones or reacts with them.

This brief look at the light responses of seeds illustrates the diversity of light effects on dormancy. Some types of seeds are insensitive to light; others may be promoted or inhibited by it, depending upon the wavelength of light given; others have an absolute requirement for light; others for dark; and still others require a photoperiod. These effects of light are closely interwoven in some manner with temperature effects on seeds and with the growth substances involved in imposing or relieving dormancy.

Water. Dormancy of seeds can require the physical exclusion of water from the embryo; the effectiveness of hard seed coats in accomplishing this type of action has been illustrated in Figs. 10-10 and 10-11. In some desert plants, Koller (1957) found that the seed coats contain large amounts of some osmotic materials which can restrict the amount of water entering the embryo through osmotic means; exposure of the seeds to ample amounts of rain elutes the osmotic materials, and in this way germination can be synchronized with the onset of the rainy season.

Water can itself relieve dormancy in many types of seeds; the water elution of osmotic materials is paralleled by the effects of rain in relieving the dormancy of several types of desert plants. Went (1949) has shown that the relief of dormancy in many instances is due to the elution of growth inhibitors from the seed, especially from the seed coat.

Drying can relieve dormancy in some cases; the need for drying tomato seeds has already been mentioned. Lima beans are another familiar garden seed which requires drying. Klein and Pollock (1968) report that lima beans require drying down to less than 60 percent water content before they can germinate, and electron micrographs show that the drying period is associated with the dispersal of ribosomes away from the endoplasmic reticulum.

In some species, drying the seed actually imposes dormancy (Villiers and Wareing, 1965). In some cases it may bring about a coincidence between seed dormancy and the onset of a dry season (Longman, 1969).

A plethora of water can prevent the germination of many grains (Pollock et al., 1955), though this may not be a real dormancy but an exclusion of sufficient oxygen.

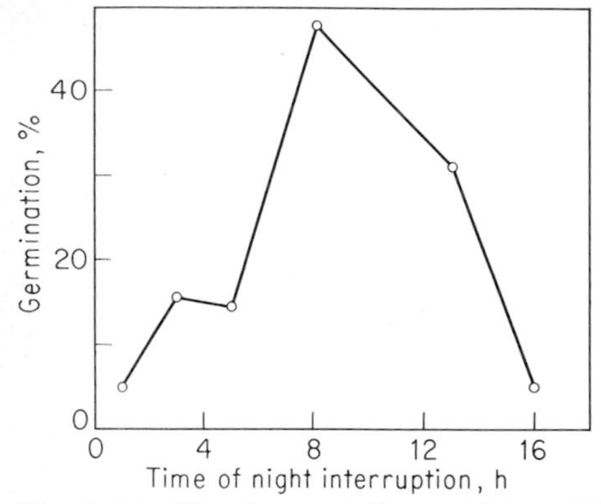

Fig. 10-21 The photoperiodic sensitivity of *Begonia* seeds is shown by the ability of a night interruption to cause germination even under a short-photoperiod regime; the light treatment (30 min) is most effective when applied in the middle of the 17-h night (Nagao et al., 1959).

Another effect of excessive water availability during germination is the development of injurious effects and reduced vigor of the seedling due to too rapid imbibition of water. These injurious effects are particularly noticeable in large-seeded plants such as the beans (Pollock and Manalo, 1970; Roos and Pollock, 1971).

Breaking Dormancy

The treatments utilized in breaking seed dormancy can be divided into four classes: mechanical, light, temperature, and chemical treatments.

Mechanical treatments. It is widely known that the seed coat can impose dormancy in many types of seed, due to impermeability to water or gases (especially oxygen), mechanical restriction of the embryo, or the presence of inhibiting substances in the seed coat. Mechanical removal of the seed coat can relieve all these limitations on germination. In horticultural practice, it is more common to weaken or rupture or puncture the seed coat, a treatment effective in most cases of seed-coat limitations of germination. Seed-coat *scarification* is accomplished by cutting a notch through the seed coat with a file or knife or shaking the seeds with an abrasive (Barton, 1947). Disruption of the coat can be achieved in some species by freezing the seeds or plunging them into liquid nitrogen, boiling them in water or alcohol, or, most commonly, soaking them in concentrated sulfuric acid (Isely, 1965).

Light treatments. Ordinarily there are simple illumination or photoperiodic exposures (Isely, 1965); the principal effects of light have already been discussed.

Temperature treatments. These are also standardized for various species and include low-temperature *stratification* or high-temperature treatments or diurnal cycles of high and low temperatures.

Chemical treatments. Treatments of this type are more diverse. We have already discussed the effects of the various growth regulators and oxygen on seed dormancy, and so further mention will be restricted to two compounds. Potassium nitrate is very effective in increasing the germination of seeds and is commonly included in the medium for tests of seed germinability (Isely, 1965). Hashimoto (1958) has suggested that this salt may serve to enhance the effectiveness of gibberellin in the germinating seed. Another chemical of importance is thiourea, first described as a germination stimulator by Thompson and Kosar (1939). This is probably the most widely effective chemical for breaking seed dormancy known, outside of the plant-growth regulators. Thiourea can relieve the light requirement of lettuce seeds and may act on the same site as the light regulation (Evenari et al., 1954; Evenari, 1957). In bringing about germination thiourea may cause an increase in gibberellinlike substances in the seed. It apparently does not cause a decrease in the inhibitory substances present (Wareing and Villiers, 1961).

BUD DORMANCY

The tendency of perennial plants to grow in cycles has been expanded in annual dormancy so that the seasons of environmental stress coincide with periods of inactive growth. The role of light in controlling the dormancy of trees was indicated long ago by Jost (1894) and more especially by Klebs (1914); that temperature experiences can break the dormancy of potato buds was reported by Müller-Thurgau (1880), and parallel work for the temperature regulation of tree dormancy was reported by Pfeffer (1904). The development of physiological information on dormancy since these early reports has been principally a quantification of the environmental cues perceived by the plant and the correlation of the environmental signals with regulatory controls at the level of plant hormones and nucleic acids, leading to the subsequent control of growth.

Entrance into dormancy involves not only the termination of active growth but also a morphologically distinctive state. In the formation of buds the elongation of the axis has been suspended, leaf enlargement has been suspended, and the shortened meristematic assemblage has been enclosed in distinctive scales. The dormant condition is associated with an inability to grow, a very low respiration rate (Pollock, 1960), and a low level of nucleic acids (see review by Vegis, 1964). The bud scales contribute to the inefficiency of the respiratory functioning in the bud axis, as evidenced by the fact that removal of the bud scales markedly increases the respiration (Pollock, 1953). The bud may require as much as 14 months for the completion of its development into a full component of leaf primordia enclosed by scales (Garrison, 1949), and during its dormant period the bud may continue to enlarge at a very slow rate and to develop new enclosing bud scales over a period of many years (Kozlowski, 1971).

It has already been mentioned that the dormant state may be one in which growth is blocked through a repression of the nucleic acid system (see Table 10-1). The onset of dormancy may result from functions which repress the nucleic acid system and the emergence from dormancy may result from functions which derepress this system. The information on dormancy presently available tells us a good deal about the quantitative and qualitative nature of environmental cues which can regulate dormancy and about the growth-regulator systems in the plant which may be involved in the dormancy system, but there is a gap in our understanding of how environmental cues bring about the changes in the repression state of nucleic acid systems.

There is some evidence that the onset of bud dormancy involves protein synthesis. Tubers of *Begonia evansiana* are induced to become dormant by the combination of red light and cool temperatures; if inhibitors of nucleic acid and protein synthesis are supplied during induction of dormancy, the tubers fail to reach the dormant state, as shown in Table 10-5. This evidence suggests that the environmental cue for onset of dormancy may be translated into the nucleic acid–directed synthesis of certain proteins which serve to carry out the program imposing dormancy.

Many species of trees can be brought into a dormant condition by short photoperiods of the sort that occur in late summer (Wareing, 1954), and many can be brought out of dormancy by long photoperiods. Woody plants which retain their foliage might be expected to respond to these photoperiods through some reaction in their leaves, but many species lose their foliage as they enter dormancy and lack the usual organs which we associate with photoperiodic perception. Wareing (1953) became interested in this phenomenon and found that the buds of several species of trees are capable of directly perceiving photoperiodic stimuli. For example,

Table 10-5 Inhibitors of nucleic acid and of protein synthesis can prevent the induction of dormancy in *Begonia* tubers. Dormancy was induced by 15°C exposure for 20 days; 0.3 m*M* solutions of inhibitors supplied during the 20-day period; percent germination measured after 70 days (Esashi and Leopold, 1969c).

Treatment	Tuber germination, %
Controls, 25°C	70
Induced at 15°C	4
Induced + 5-fluorodeoxyuridine	87
Induced + cycloheximide	87

buds of *Fagus sylvatica* are brought out of dormancy by long photoperiods, and removal of the outer bud scales considerably enhances the sensitivity of the buds to photoperiods.

Most photoperiodic phenomena are limited more by the dark periods than by the daily light periods in natural diurnal cycles, and a reliable test for this dark sensitivity can be made by applying a small amount of light in the middle of the night. Wareing (1953) found that the buds of *Fagus* are brought out of dormancy even on short photoperiods if such a night-interruption treatment was given. It is evident that all the means for photoperiodic controls are present in the bud scales of some trees and that both the daily light and dark periods are involved. A similar situation exists in the buds of tubers of *Begonia,* which are brought out of dormancy by long photoperiods; there too the buds are responsive both to long days and to short days with a night-interruption treatment (Esashi, 1961).

It is common for buds to be brought out of dormancy by low-temperature treatments. The requirement for low temperature is ordinarily satisfied during the winter or during seasonal cool spells; it has been known since the work by Müller-Thurgau (1880) with dormant potatoes and by Pfeffer (1904) with woody plants. The promotive effects of low temperature are ordinarily accumulated by the buds, and a quantitative increase in the growth of buds is obtained with increases in duration of the chilling treatment. An example with blueberries is given in Fig. 10-22. It is not uncommon for different types of buds to show different temperature responses, and in this case the flower buds require slightly less cold than the vegetative buds. The ability of apple trees to summate their cold experience was established by Chandler et al. (1937), who also showed that temperatures just above freezing are more effective in breaking dormancy than temperatures at or below freezing. This seems to imply that the cold requirement involves some metabolic action as well as a physical one.

Fig. 10-22 The buds of blueberry emerge from dormancy in response to increasing durations of low temperatures. Plants were left outdoors during winter for varying periods and were brought indoors to test for the ability of buds to begin growth (Darrow, 1942).

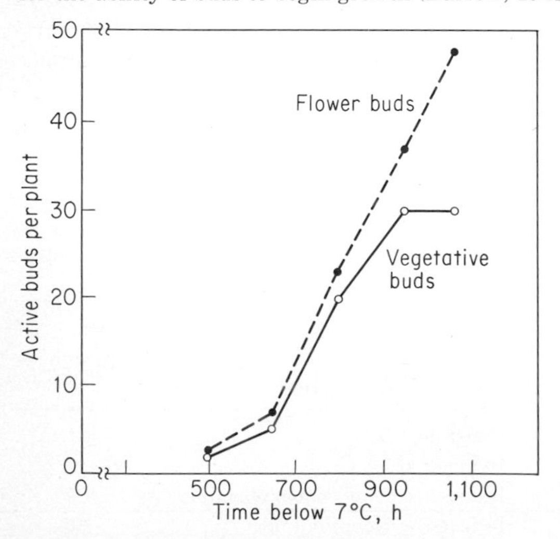

In some instances dormancy can be broken by heat treatment instead of cold, and some species can be brought out of dormancy either by heat or chilling. Molisch (1909) developed a warm-water treatment for breaking dormancy of leafless branches, using water between 30 and 40 °C. An example is the heat-treatment breaking dormancy of gladiolus corms (Loomis and Evans, 1928). Ecologically a high-temperature experience may break dormancy in situations where fire or exposure to intense sunlight has elevated the temperature.

Many different chemicals have been used successfully in breaking bud dormancy. The early work has been reviewed by Doorenbos (1953) and more recent work by Weaver (1972). Among the most generally effective chemicals are ethylene chlorohydrin and its more complex formulation as rindite.

Hormonal Control of Bud Dormancy

The concepts of hormonal regulation of bud dormancy have followed a sequence similar to that of seed dormancy. The first suggestion of a hormonal control was that auxin provides the hormonal inhibitory function (Michener, 1942);

however, the amounts of auxin found in dormant buds have never been high enough to support this suggestion (Samish, 1954).

An entirely different explanation arose from the finding (Hemberg, 1949*b*) that dormant buds of potato contain appreciable amounts of growth inhibitor and that the content of this inhibitor drops with treatments which remove the dormancy, e.g., rindite or ethylene chlorohydrin. Hemberg found the same situation in dormant buds of *Fraxinus,* and his sample data are shown in Fig. 10-23. He tested for the abundance of inhibitor in his bud extracts by adding increasing amounts of the extract to agar containing auxin, which he then analyzed by the *Avena* curvature test. The added extract was powerfully inhibiting when taken from dormant buds in October and was not appreciably inhibiting when taken from nondormant buds in February, even though no growth had yet begun.

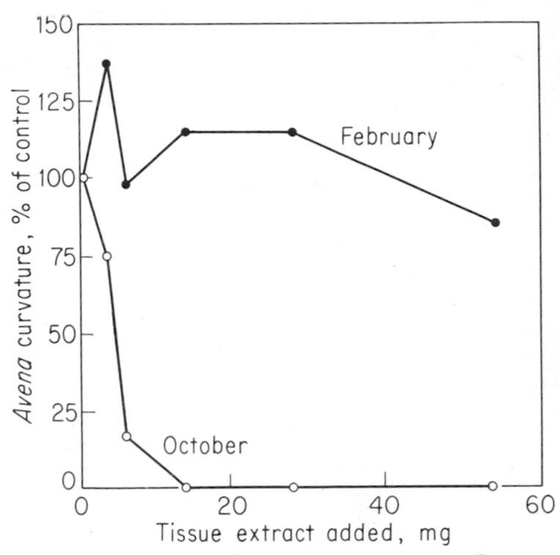

Fig. 10-23 Extracts of *Fraxinus* bud scales made in October contain large amounts of growth inhibitors, as indicated by the inhibition of the *Avena* response to auxin. In February the inhibitor content had diminished, and cuttings taken indoors then would grow (Hemberg, 1949*a*).

The concept of inhibitors serving as regulators of dormancy grew stronger with the accumulation of evidence that natural periods of dormancy are associated with high endogenous inhibitor contents. Illustrating this type of information are some data for *Acer pseudoplatanus* in Fig. 10-24, in which inhibitors are seen to appear in the leaves and the stem apices in late summer, followed by very substantial levels of the inhibitors in buds during the winter dormant season; the emergence from dormancy is associated with a decline in inhibitor levels in the buds in April. After several reports in which inhibitor levels were found to be high during the dormant periods, studies were carried out to determine whether the short photoperiod which induces dormancy would also induce the accumulation of inhibitors; several such studies in the later 1950s generally agreed that the environmental stimuli which bring on dormancy also bring on the elevated levels of growth inhibitors (Phillips and Wareing, 1959; Nitsch, 1957). In the buds of *Betula,* increases in inhibitor content are detectable after as few as two short photoperiods and decreases after two long photoperiods (Kawase, 1961).

Before the identity of the inhibitor in dormant buds had been established, it was found in numerous cases (as mentioned in Chap. 8) to have the characteristics of inhibitor β in paper-chromatographic separations (Nitsch, 1957). Varga (1957) suggested that phenolic compounds could account for the inhibitory action, but later work by Robinson and Wareing (1964) indicated that the principal component lacks an aromatic ring. Identification of abscisic acid as the principal inhibitor in dormant buds was made by Cornforth et al. (1965b) (see Fig. 8-6).

Essential to the inhibitor theory of bud dormancy is the demonstration that application of the inhibitor actually induces dormancy. The evidence with abscisic acid falls short of being unequivocal. Eagles and Wareing (1964) applied partially purified preparations of inhibitor β from dormant buds to seedlings of *Betula* and obtained the formation of dormant buds; after synthetic abscisic acid was available, similar experiments were carried out by El-Antably et al. (1967). Continuous application of ABA to the vascular system of leaves results in the termination of growth and the formation of terminal

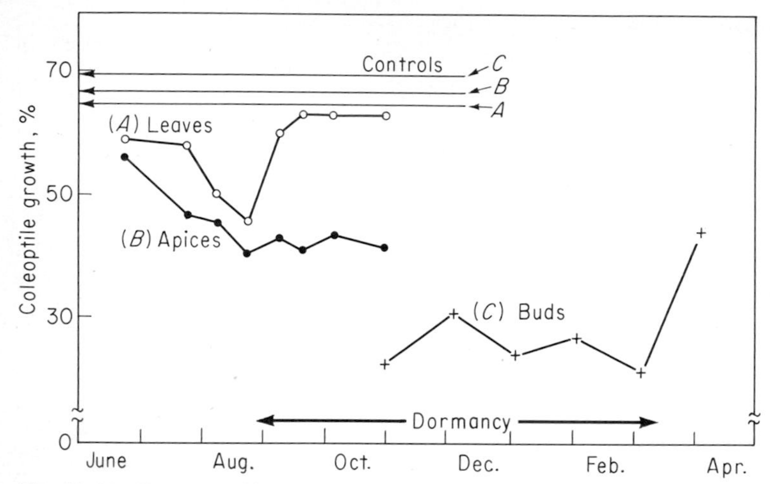

Fig. 10-24 Extracts of leaves, stem apices, and buds of *Acer pseudoplatanus* inhibit the growth of coleoptile sections. The content of growth-inhibiting substances increases as dormancy is entered and then declines again as dormancy is passed. Methanol extracts chromatographed in isopropanol-ammonia-water, and the inhibitor at R_f 0.7 followed by oat-coleoptile-section assay (Phillips and Wareing, 1958).

buds with the morphological features of dormancy in several woody species. The amounts of ABA required for this effect are very large, and when the ABA supply is withdrawn, growth is sometimes resumed. In species where there is no renewal of growth after ABA supply is stopped, chilling or kinetin treatment can restore growth.

In *Betula lutea,* which enters dormancy under short photoperiods with an associated large rise in inhibitor β content in its shoots (as in Fig. 8-10), specific examination of the ABA levels showed no increase with induction of dormancy (Loveys et al., 1974). Perhaps other types of inhibitors may be involved; Hashimoto et al. (1972) found a phenanthrene type of inhibitor involved in the dormancy of yam bulbils.

The photoperiodic induction of dormancy might be the result of the formation under short photoperiodic conditions of some inhibitor which is then translocated to the growing point and bud; the inference of a translocation of a hormonal regulator was markedly strengthened when Hoad (1967) found that samples of authentic phloem sap obtained from the honeydew of aphids on willow contain significantly more ABA under short photoperiods than under long ones.

Several lines of evidence suggest that bud dormancy is more complex than a simple regulation by inhibitors. There are instances in which the inhibitor level is high before the onset of dormancy (Blommaert, 1955) and others in which the inhibitor level does not decline when dormancy is broken (Blumenthal-Goldschmidt and Rappaport, 1965) although ABA may inhibit sprouting somewhat.

An interaction with gibberellins is suggested by several lines of evidence; applications of GA can break dormancy of potatoes (Brian et al., 1955) or woody plants (Vegis, 1964), and the levels of endogenous GA have been found to increase with chilling treatments (Smith and Rappaport, 1961). There are exceptions to the effectiveness of applications of GA in breaking dormancy and to the increases in endogenous GA with loss of dormancy (Rappaport and Wolf, 1969; Wareing, 1969). The stimulatory effects of GA on potato-bud sprouting are illustrated by the experimental results in Fig. 10-25; and in similar experiments GA is found to relieve the inhibitory effects of ABA.

While the early work on gibberellin effects on dormancy mainly involved breaking dormancy, in increasing numbers of cases gibberellin applications have been observed actually to induce dormancy, both in woody plants (Brian et al., 1959; Weaver, 1959) and in tubers (Nagao and Mitsui, 1959).

Although bud dormancy may be considered as a concerted interaction of inhibitory influences by ABA and promotive influences by GA (Michniewicz, 1967), scattered lines of evidence indicate that other plant hormones may also help regulate dormancy. Cytokinins are fairly effective in breaking dormancy in some woody plants (Benes et al., 1965), and they increase in buds of *Betula* and *Populus* during emergence from dormancy (Domanski and Kozlowski, 1968). There is a possibility that ethylene plays a role in dormancy regulation; there are many early reports of ethylene treatments breaking the dormancy of buds, corms, tubers, and woody plants (Pratt and Goeschl, 1969), and it has been established that ethylene is produced by tubers and that its production is enhanced by dormancy-breaking treatments with GA (Poapst et al., 1968).

Morphological Aspects of Dormancy

Bud scales are distinctive as morphological components of the bud, and, like the seed coat, they can play a major role in the dormancy mechanism.

A first regulatory function of bud scales involves respiratory activity and gas exchange. Pollock (1953) studied the respiration of maple bud scales and found that the scales respire markedly more actively than the enclosed meristematic tissue. When he removed the scales, respiratory activity of the meristem was markedly increased. He inferred that the scales impose respiratory limits on the meristematic tissue. His suggestion is supported by the reports that removal of bud scales can in some instances favor the breaking of bud dormancy (Abbott, 1970). Their function in photoperiodic regulation of dormancy has been mentioned (Wareing, 1954).

Scales may also contain rather large amounts of inhibitors which may participate in the inhibition of growth of the bud (Guttenberg and Leike, 1958). The flower buds of coffee tend to break dormancy in bursts after each rainstorm; Mes (1957) found that providing water by root irrigation does not break flower-bud dormancy but washing the plants with a garden hose does. He deduced that the rain serves to elute inhibitors from the buds, helping them emerge from dormancy.

Abbott (1970) studied the bud scales of apple as they relate to flower initiation and points out that buds arise from leaf primordia in which the leaf laminal part is aborted. Since it is the young leaf laminal meristem that provides a major source of gibberellin to the stem, the developing scale may be considered an inhibitory device. Abbott found that severing the blade laminal part from leaf meristems abruptly inhibited stem elongation. He suggests that bud scales act as mobilizing centers, depriving the enclosed bud meristem of substrates, and that the emergence from dormancy may involve the senescence of the bud scales and hence the relief of their mobilization action.

Fig. 10-25 Applications of gibberellin can break the dormancy of isolated buds from potato tubers (Rappaport et al., 1965).

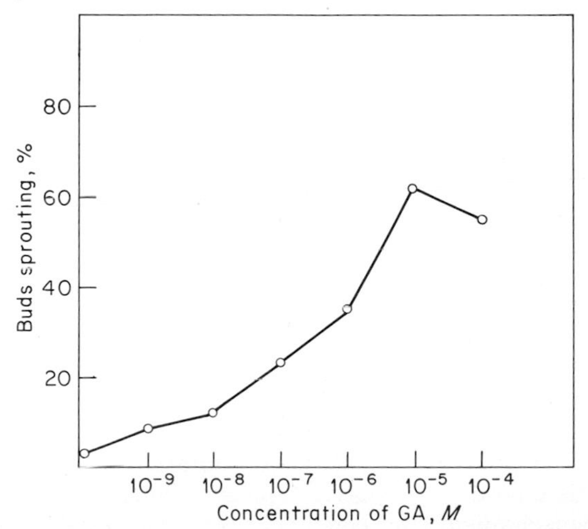

Fig. 10-26 Normal branch of *Potamageton obtusifolius* (*a*) and a turion, or winter bud (*b*). (*Redrawn from Fassett, 1960.*)

An interesting variant from bud dormancy is the *turion,* a dormant section of stem with appended leaves and terminal meristem which, in some aquatic species, breaks off and lies dormant in the water, ordinarily over the winter season. The formation of dormant turions is stimulated by short photoperiods (Czopek, 1962), and its dormancy can be relieved by applications of gibberellin or cytokinin (Czopek, 1964). This type of dormant shoot is shown in Fig. 10-26.

SUMMARY

The plant's ability to utilize the state of dormancy in meeting seasonal limitations of growth is a powerful adaptive capacity. There are many mechanisms in common with the regulatory systems in the dormancy of buds and seeds, especially the systems which seem to involve balances between inhibitory and promotive plant hormones, particularly ABA and GA.

While experimental information on nucleic acid activities in bud dormancy is rather limited, it seems reasonable to extend the concept of dormancy as a state of suppressed nucleic acid competence across both seed and bud types of dormancy. While the connections between the environmental cues and the hormonal and nucleic acid components of the plants are obscure, it still seems reasonable to presume that the environmental cues for dormancy lead to changes in the plant hormone components which in turn serve to switch the nucleic acid system on or off (Amen, 1968; Wareing and Saunders, 1971).

Physical limitations of growth, particularly by seed coats, may represent a dormancy mechanism of a different type entirely; restrictions of water or gases or physical restraint of the embryo may hold growth in check quite independently

Fig. 10-27 When the seeds of *Amaranthus retroflexus* are kept in moist medium at 20°C in the dark, they show a gradual increase in germination which becomes nearly complete in 60 months. Both seed lots shown here were completely dormant at the beginning of the test period (Crocker and Barton, 1957).

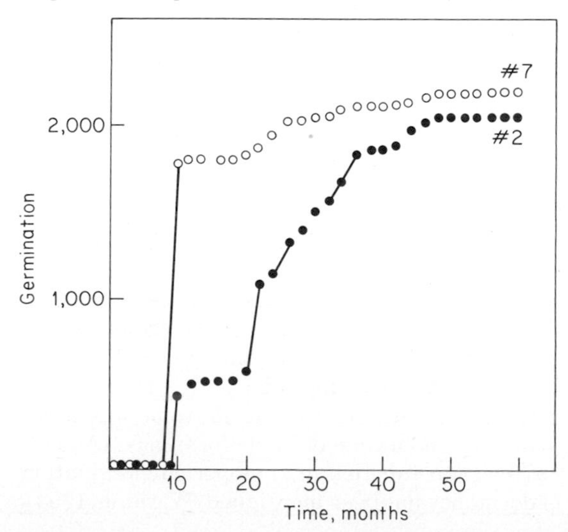

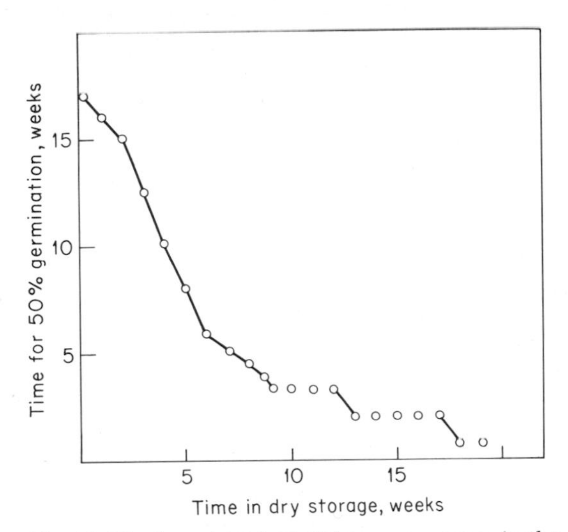

Fig. 10-28 A gradual loss of dormancy occurs in the seeds of *Impatiens balsamina* even during dry storage. Weekly tests of germination time show increasingly rapid germination after storage periods up to 19 weeks (Kroeger, 1941).

of the nucleic acid and hormone systems. Removal of the physical limitations may occur by the degradation of the seed coat by microorganisms; conversely, the embryo's ability to overcome the seed-coat limitation may be regulated by activities within the embryo itself; e.g., the embryo may secrete enzymes which can attack the coat, or it may develop sufficient thrust to burst through the seed coat by its own activities.

Dormancy serves the plant as an advantageous means of deferring growth; if the prerequisites for breaking dormancy are not met, the advantage could be turned into a loss. In the case of seed dormancy, at least, we have evidence that when the requirements for breaking dormancy are not met, many types of seeds gradually lose their dormant condition. This slow passage out of dormancy is very common, e.g., in seeds of the small grains and potato tubers. The gradual loss of dormancy occurs in seeds which remain in moist conditions (Fig. 10-27), and in many such instances there is a suggestion of an annual periodicity in the relief of dormancy. In Fig. 10-27, one can see a 12-month periodicity in germination. Loss of dormancy can also occur in many types of seed kept in dry storage (Fig. 10-28). *Afterripening* is the term applied to this deterioration of the dormancy mechanism in dry storage. The gradual emergence from dormancy can be interpreted as a type of fail-safe mechanism by which the plant will ultimately emerge even if the dormancy-breaking environmental cue is not experienced.

GENERAL REFERENCES

Briggs, D. E. 1963. Hormones and carbohydrate metabolism in germinating cereal grains. In B. V. Milborrow (*ed.*), "Biosynthesis and its Control in Plants." Academic Press, London, pp. 219–278.

Koller, D. 1972. Environmental control of seed germination. In T. T. Kozlowski (*ed.*), "Seed Biology." Academic Press, New York. Vol. 2, pp. 1–101.

Kozlowski, T. T. (*ed.*), 1972. "Seed Biology." Vol. 2, Academic Press, New York.

Nitsch, J. P. 1971. Perennation through seeds and other structures: Fruit development. In F. C. Steward (*ed.*), "Plant Physiology—a Treatise." Academic Press, New York, pp. 413–501.

Wareing, P. R. and P. R. Saunders. 1971. Hormones and dormancy. *Annu. Rev. Plant Physiol.* 22:261–288.

Woolhouse, H. W. (*ed*). 1969. "Dormancy and Survival." *Soc. Exp. Biol. Symp.* 23. 598 pp.

11 JUVENILITY, MATURITY, AND SENESCENCE

From observations on the developmental changes of plants as they pass through juvenility, maturity, and senescence, Goebel (1889) concluded that some plants have a type of heteromorphism between the juvenile and the mature states, which, unlike sexual dimorphism, exists as a sequence in the same organism, analogous to the juvenile-mature heteromorphism of many animals. The juvenile state has a distinctive morphology of leaves, stems, and other structures and is displaced by a morphologically distinctive mature state. After a plant has entered the mature state, it can be brought into flower by appropriate external cues.

Whereas the change from juvenile to mature conditions involves some alteration in morphology as well as a depression of growth rate, the change from mature to senescent conditions often involves a profound deterioration of growth rate associated with deterioration of various synthetic activities. In most perennials, the senescent condition of the plant is evident principally as a low growth rate, but in most annuals and biennials it is seen as a dramatic and rapid death, sometimes associated with brilliant colors.

JUVENILITY

With germination of the seed, most plants enter a state of vigorous vegetative growth during which they cannot be readily induced to a reproductive type of growth. One can visualize ecological advantages of a period devoted exclusively to rapid vegetative growth, in which the plant achieves a size which will be competitively stronger in the plant community. Physiologically, the juvenile state can be described as a period when the plant is capable of exponential increases in size, when flowering processes cannot be readily induced, and when the plant develops characteristic morphological forms (of leaves, stems, thorns, etc.).

It should be emphasized that juvenility is not necessarily devoid of flowering ability. Passecker (1949) noted that the morphological gradation from juvenile to mature forms of fruit trees is associated with a gradual increase in the ability

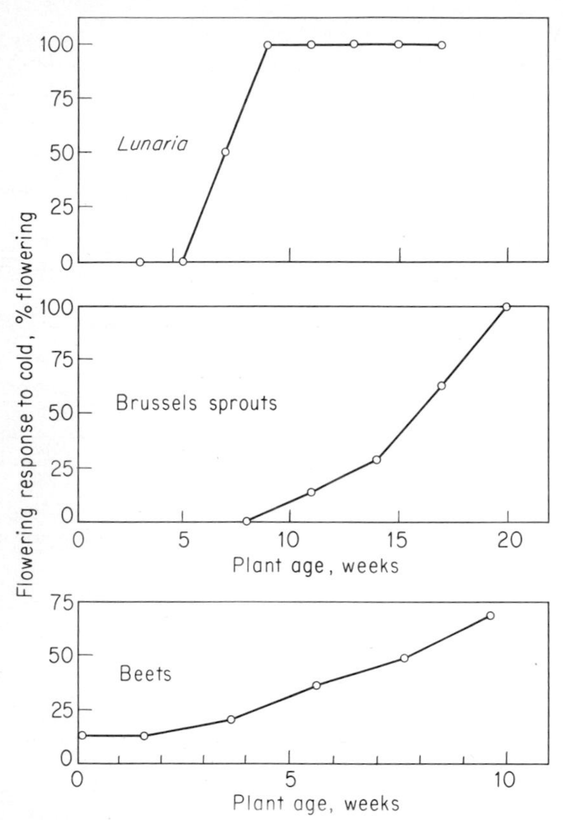

Fig. 11-1 Juvenility may be a period of complete inability to flower or of quantitatively poor response to flower stimuli. *Lunaria* cannot be induced by cold before 7 weeks of age (Wellensiek, 1958); brussels sprouts are not induced by cold before 11 weeks of age, after which there is a quantitatively increasing response (Stokes and Verkerk, 1951); and beets are simply increasingly responsive to cold as they pass out of juvenility (Wellensiek and Hakkaart, 1955).

of the tree to form flowers. De Muckadell (1954) emphasized that the morphological forms of trees are much better indexes of juvenility than the ability or lack of ability of the tree to flower.

Some examples of the relative facility with which plants may flower as they grow older are presented in Fig. 11-1. Each of these three species is induced to flower by low temperatures. *Lunaria* will not flower in response to cold treatment until it is about 7 weeks old, after which all plants will flower. Brussels sprouts likewise have a minimum age for flowering, but after that they show a gradually increasing responsiveness in numbers of plants responding to the cold. Beets, on the other hand, flower in response to cold at any age, but there is a gradually increasing responsiveness as the plants grow older, a quantitative rather than a qualitative juvenility with respect to flowering. These cases illustrate that a plant species may have a juvenile period in which the induction of flowering is completely obstructed or one in which flowering is only quantitatively impeded.

The vegetative condition is not exclusively a property of juvenility: a brussels sprout plant can pass far beyond its 11 weeks of juvenility and still not flower if it has not experienced an inductive cold condition.

Since flowering itself may not be a valid index of juvenility, the morphological expressions of juvenility may provide more meaningful information concerning the relative duration of juvenile periods in various plants. Juvenile leaf forms may occur only on the first node of the germinating seedling (as in bean) or on the first several nodes (pea, *Xanthium*). The narrow juvenile stems of brussels sprouts are formed for the first 11 weeks. Apple seedlings may continue the whiplike growth of the juvenile nonflowering state for 4 to 10 years. Juvenile types of stem growth, leaf form, and thorniness may persist for 10 to 15 years in *Citrus* and pecan trees, and flowering is absent or restricted during the time. A classic case of juvenility is the common ivy, *Hedera helix,* which retains the juvenile form of a creeping vine and palmate leaf for an indefinite time and only occasionally switches over to the mature form of an upright bush with entire leaves, after which it proceeds to flower and fruit. Some beech and oak trees may retain morphological features of juvenility for 60 years, and in fact in numerous horticultural plant materials, juvenility has become apparently fixed and permanent, as in the entire genus of *Retinospora,* which is a series of fixed juvenile forms of *Thuja* and *Chamaecyparis* (cf. Büsgen et al., 1929).

As the juvenile plant grows up, structural differences along its stem axis may reflect the gradual change from juvenile to mature types of growth. The presence of juvenile types of morphology at the base of an apple seedling and graded levels of increasing maturity up the tree were pointed out by Passecker (1949); dramatic evidence of the presence of transitional stages between juvenility and maturity in beech trees has been developed by de Muckadell (1959). Reproduced in Fig. 11-2 is Passecker's diagram of such a gradation up the tree, evident not only in the ability to flower but also in the leaf and stem forms and in the ability to transmit the physiological expression of juvenility to cuttings or graft sections taken from the various parts of the seedling. Cuttings from the base of a seedling may develop into new plants with juvenile leaf form and stem form, juvenile thorns (where present in juvenile wood), juvenile ease of rooting, and juvenile tendencies to retain leaves during the winter. In fact, then, the ontogenetic sequence of development from extreme juvenility to full maturity is preserved in a gradation from the base to the top of a mature tree. De Muckadell attributes this maturity gradient to the progressive aging of a meristem as it grows. Thus meristems left at the base of the tree retain the stage of juvenility which existed in the apical meristem at the time the laterals were established; the meristem which has grown entirely through to the top of the tree has aged in proportion to the extent of its growth. The most mature meristems are therefore at the apex of the tree and at the ends of its longest branches.

The gradation of maturity from the base to the top of a tree is illustrated in Fig. 11-3. Juvenile oak and beech trees tend to have a delayed abscission of leaves until spring. In older trees, the juvenile characteristic of leaf retention may be very evident in lower branches, even though the upper branches are mature (de Muckadell, 1954).

Juvenile Morphology

One of the most readily observed morphological expressions of juvenility is *leaf form* (Fig. 11-4). The unique leaf forms of new seedlings of conifers were pointed out by Beissner (1888) and detailed by Büsgen et al. (1929). For example, pine seedlings bear short juvenile needles in spirals instead of in fascicles of two to five elongate mature needles, and this leaf feature may persist for as long as 3 years of growth. Seedlings of *Thuja* have awl-shaped leaves for several years before changing over to the mature leaf form of appressed scales. These juvenile forms of *Thuja* and similar ones of the genus *Chamaecyparis* have become fixed in some plants by repeated propagation during the juvenile stage, as mentioned above. Among the angiosperms, juvenile leaves may be simple and mature leaves compound (bean, *Citrus*); rarely the reverse may be true (the phyllodium of koa). In the garden pea, the first juvenile leaves are reduced to scales. Increasing amounts of dissection or lobing of the leaves is often associated with increasing maturity. This has been described for cotton as a changing extent of leaf lobing during the development of the plant. The cotton seedling produces a simple entire leaf, and as the plant becomes mature, the leaf form gradually becomes palmate; after the plant has started to fruit, the leaves tend to return to an entire shape. Ideally one might find an entire series from juvenile to

Fig. 11-2 Diagrammatic representation of the gradient from juvenile tissues at the base of a fruit tree to mature tissues at the tip. (*From Passecker, 1949.*)

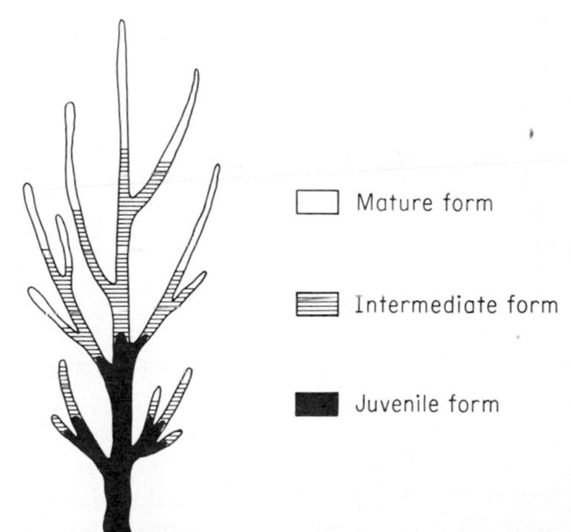

Fig. 11-3 The juvenile characteristic of delayed leaf abscission is evident in beech trees in the winter, the lower branches expressing the juvenile characteristics of abscission and the branch tips the mature characteristics.

mature and senescent leaf forms arrayed from the base to the tip of the cotton plant.

The usually simpler leaf shape and the faster growth rate of juvenile forms have led to the suggestion that the simple forms may result from the more rapid growth of juvenile leaves (Sussex and Clutter, 1960). This suggestion would be

Fig. 11-4 Contrasting leaf forms between juvenility and maturity: (*a*) the transition of leaves in *Juniperus virginianum* from the needlelike juvenile leaves on the left to mature, scalelike leaves of mature wood on the right; (*b*) three variations of leaves from juvenile wood of *Sassafras albidum* on the left compared with a leaf from mature wood on the right; (*c*) a leaf from the juvenile vine of *Hedera helix* on the left compared with a leaf from the mature shrub on the right.

(a)
(b)
(c)

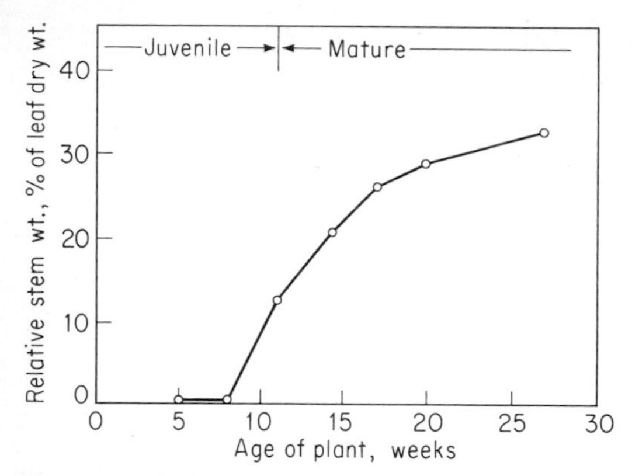

Fig. 11-5 As brussels sprouts pass from the juvenile to the mature form at about 11 weeks of age, the proportion of stem to leaves is markedly increased. These data are plotted as the ratio of stems to leaves in percent (Stokes and Verkerk, 1951).

conveniently compatible with the gradual transition from the formation of juvenile to more mature forms; experimental support for the idea remains to be developed.

Juvenile beech trees produce only shade leaves (Schramm, 1912), suggesting an adaptation of the seedling for growth in the shade, where beech seedlings generally become established.

In many species, the *stem* and its growing point exhibit morphological features of juvenility. Among woody plants, there is often a typical juvenile branching pattern, with long whiplike branches and a narrow branching angle (see, for example, Blair et al., 1956). The presence of thorns on the stem is characteristic of juvenile seedlings of *Citrus* and juvenile parts of many locusts. As the tree matures, the new growth at the apex and the longest branches ceases to develop thorns. Again, the degree of thorniness up the tree recapitulates the ontogenetic sequence of its changes from extreme juvenility to maturity.

The stem of ivy changes from a creeping vine habit to an erect shrub when it passes out of juvenility, as already mentioned.

A noteworthy change of stem morphology has been described for brussels sprouts (Stokes and Verkerk, 1951). Juvenile seedlings develop a narrow pointed stem with a thin apex; upon reaching maturity (11 weeks), the stem becomes wide and the apex broad and blunt. The rise in relative proportion of stem to leaf as brussels sprouts reach maturity is illustrated in Fig. 11-5. Enlargement of the apical meristem with maturity may be common to many herbaceous species (cf. Millington and Fisk, 1956).

Juvenile Physiology

Juvenile wood often roots much more readily than mature wood. This was noted by horticulturists before the concept of juvenility had become generally accepted (Gardner, 1929). Illustrative data are given in Table 11-1. Auxins can generally stimulate the rooting of cuttings, and Pilet (1958) observed more ready enzymatic oxidation of auxin in older pea tissues, suggesting that the superior rooting of juvenile cuttings may be due to a higher endogenous auxin content. Hess (1964), however, finding that auxin does not restore the readiness to root in mature ivy, proposed an explanation based on the decline of content of four rooting cofactors which he could extract and separate by paper chromatography.

The propagation of apples by rooting of cuttings is difficult, and English horticulturists sometimes maintain *stool beds* for propagation, where the apple plants are repeatedly cut back to force out juvenile basal shoots. The juvenile wood obtained in this way is used for making cuttings. Wellensiek (1952) has achieved a similar effect by continued debudding of small trees until juvenile shoots are sent out.

Propagation of woody plants by budding or

Table 11-1 The rooting of cuttings of *Hedera helix* shows a sharp decrease with maturity. This is expressed in cuttings from both dormant and growing plants (Hess, 1962).

	Rooting of cuttings, %	
	Dormant	Growing
Juvenile wood	83	100
Mature wood	0	17

cuttings provides some of the most significant evidence on the fixity of the juvenile morphology. Noting the gradient in degree of maturity from the base to the top of trees, de Muckadell (1954) made cuttings from graded locations up the tree and found that propagation from the juvenile base gives new plants which retain the juvenile features of the base. Those from the mature apex give new plants with the mature characteristics. This is true, for example, of the leaf retention in beech; propagules from the base retain their leaves through the winter. It is also true for the formation of thorns in locust; the propagules from the base develop thorns, and those from the apex are thornless.

De Muckadell (1959) has described other physiological features of juvenility of some woody plants, e.g., a greater tendency to develop autumnal colors and a greater shade tolerance.

The central theme of juvenile physiology is the relatively strong growth rate. In woody plants, increases in age are associated with decreases in stem-height increments, as illustrated for Ponderosa pine in Fig. 11-6. There is usually no sharp transition in growth rates between juvenile and mature trees but a gradual tapering of the growth rate. The morphological and other physiological characteristics of juvenility similarly taper off in a gradual transition. The most reasonable explanation of the declining growth rate between juvenile and mature stages is an increasing competition between growing points as the size of the tree increases. Wareing (1970) has suggested that besides the growth decline there are two other general changes with age, namely, the gradual loss of apical dominance and the gradual loss of geotropic orientation of the stems; he suggests that the increasing competition for nutrients within the plant may account for both these changes.

Modifying Juvenility

Efforts to modify the juvenile period can be grouped in two categories, efforts to intensify growth and pass through the juvenile phase more rapidly and efforts to retard growth so that flowering can be achieved earlier. The first experimental approach was facilitated by the discovery of photoperiodic controls of growth, and Wareing (1961) undertook to force the growth of birch trees under continuous long photoperiods. By this treatment, he was able to shorten the nonflowering period from the usual 5 or 10 years to less than 1 year; the deduction was that the size of the tree is an important factor in the passage out of juvenility. Pomologists, too, have noted that the passage out of juvenility may require the achievement of a certain size. For example, when various juvenile scions were grafted onto a common rootstock, the ability to flower was associated with the plant combinations which achieved the greatest size as stem diameter (Visser and De Vries, 1970).

A more common approach to the modification of juvenility is the application of treatments which slow down the growth rate. Plant breeders have dominated this type of effort, and graftage of buds from juvenile trees onto mature stocks has frequently been used in an effort to obtain early flowering. In some species, juvenile buds have proved obdurate to change by graftage,

Fig. 11-6 With increasing age, the growth in height of Ponderosa pines shows a decline from the rapid growth rates of juvenile trees to gradually lesser growth rates of mature and finally senescent trees (Meyer, 1938, for trees in site index 100).

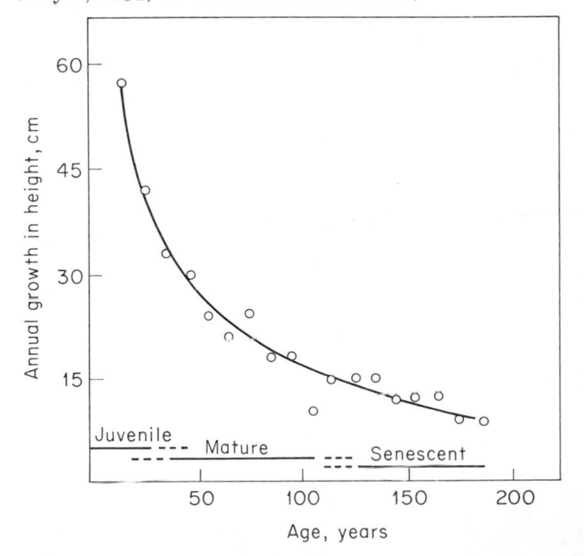

notably in the beech (de Muckadell, 1956) and brussels sprouts (Stokes and Verkerk, 1951). In other species some modification of juvenility has been achieved through grafting, as shown in Fig. 11-7. Slight increases of flowering of juvenile buds grafted onto mature wood have been reported for pecan (Romberg, 1944) and *Citrus* (Furr et al., 1947). Graftage has enhanced some juvenile features in mature buds grafted onto juvenile stocks, most notably in ivy (Doorenbos, 1954) and rubber (Muzik and Cruzado, 1958). In ivy, reversion to juvenility is most evident when only juvenile leaves are present on the graft combination and when temperatures are relatively high (Stoutmeyer and Britt, 1961).

In numerous species, gibberellin applications extend the juvenile period. This GA effect was first reported for ivy, which shows a temporary reversion from mature to juvenile morphology after GA application (Robbins, 1957). Similar enhancement of juvenility has been reported for *Citrus* (Cooper and Peynado, 1958), and the widespread effects of GA applications in reducing flowering of deciduous fruit trees have been interpreted as a similar effect (Luckwill, 1970). GA applications to *Ipomoea* likewise extend the juvenile period, as evidenced by the extension of the time over which juvenile leaf form is retained (Fig. 11-8). Luckwill (1970) has suggested that the rapid growth characteristics of juvenile trees bring about a high level of endogenous GA, and this condition may contribute to the nonflowering character of juvenility. It should be noted, however, that some fruit trees and members of some families of conifers are induced to flower precociously by applications of GA (Pharis and Morf, 1968). GA applications to *Kalanchoe* can induce flowering in otherwise juvenile plants (Wadhi and Ram, 1967). In brussels sprouts, the application of auxin can hasten the passage of juvenility (de Zeeuw and Leopold, 1955). Muzik and Cruzado (1958) have reported that extracts of juvenile rubber plants can enhance the juvenile feature of rooting, but it is not clear whether the effect is specifically on rooting or a modification of juvenility (cf. Hess, 1964).

Fig. 11-7 When buds from juvenile apple seedlings are grafted onto a dwarfing root stock (Malling IX), there is a consistent increase in the percentage flowering over those grafted onto a self stock (Visser, 1964).

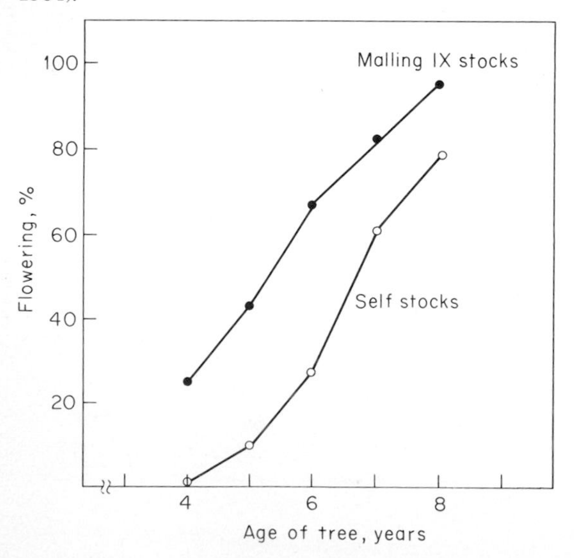

Pruning can enhance juvenility in many plant materials. Not only can pruning force out lower and hence more juvenile buds into growth, but it also appears to enhance the degree of juvenility expressed in growth. Many types of plants respond to pruning by at least temporary production of more juvenile leaf and stem forms.

Nature of Juvenility

What may constitute the endogenous regulation of the juvenile condition? The first thing we can say is that the duration of juvenility appears to be heritable. Visser (1965) has established this for fruit trees, and some of his evidence is reproduced in Fig. 11-9.

The description of the passage from juvenile to mature condition as a function of the overall growth of the apical meristem (de Muckadell, 1959) is interesting and helpful, but it does not make it easier to understand how pruning a plant can enhance juvenility or how the graftage of mature buds onto juvenile stocks can cause a regression to a more juvenile type of growth. Experiments on chemical modification of juvenility open up the interesting possibility that growth substances are involved in the control of these developmental changes and that the gradual shift from juvenile to mature growth may be due to an alteration in the proportions of endogenous substances.

In this connection some interesting and relevant experiments on nucellar buds of *Citrus* are highly suggestive. Swingle (1928) observed that nucellar buds germinate into juvenile seedlings, even though they arose from strictly maternal tissue which had budded itself into the embryo sac. In a later report (Swingle, 1932), he developed the concept that the passage of an embryolike experience in the embryo sac produced the juvenility feature, whether the embryonic cells arose from a true embryo or from a maternal nucellar bud. Detailed experiments by Frost (1952) and Cameron and Soost (1952) showed that the juvenile features of the nucellar seedlings are stable to graftage exactly like the juvenility of true seedlings. The early suggestion of Swingle (1932) that the embryo-sac experience stimulates a rejuvenescence in the de-

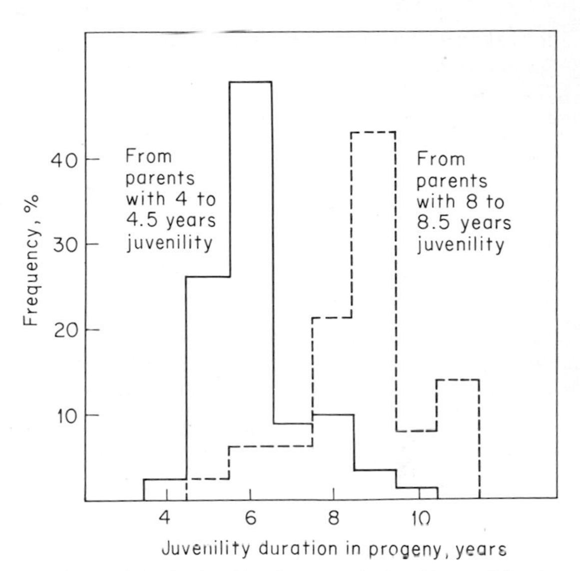

Fig. 11-9 The heritable characteristic of juvenility is seen in the extent of juvenile periods in apple seedlings from parents with relatively long juvenile periods and seedlings from parents with relatively short juvenile periods (Visser, 1965).

Fig. 11-8 Application of gibberellin to *Ipomoea caerulea* causes the retention of the juvenile narrow leaf form for an additional five nodes. Leaf shape is plotted as the ratio of width to length (Njoku, 1958).

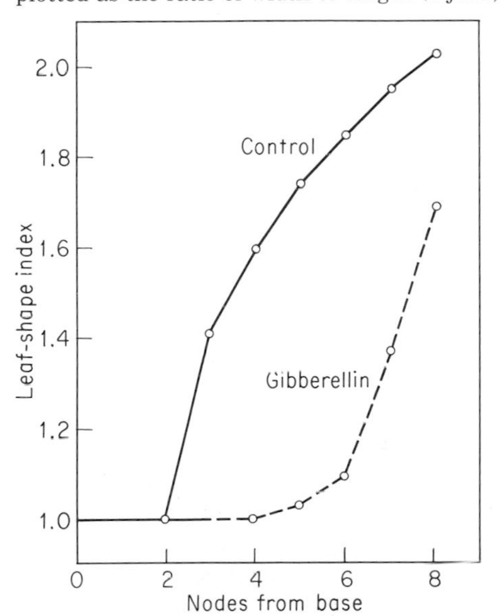

veloping ovary is a very attractive concept, with perhaps the further possibility of a chemical stimulation to the juvenile condition. Thus the meristem-aging concept of de Muckadell (1959) may represent a gradual dilution of a juvenility stimulus introduced at the embryo stage of development.

Wareing (1961) has suggested a distinction between *maturation,* which refers to fixed characters which are stable to grafting, and *aging* which refers to characters which are expressions of the functioning of the whole plant and hence modifiable with graftage. The concept of maturation by the gradual alteration of the genetic program in the meristem is supported by the work of Stoutemyer and Britt (1965), who grew juvenile and mature ivy tissues in cultures and observed that there was a markedly greater growth rate for the juvenile than the mature tissues. Many lines of evidence imply that the juvenile state is one in which there is a greater capacity for growth. The extent to which the greater growth is genetically programmed or represents interactions of nutrients and hormones from various parts of the plant remains to be assessed.

MATURATION

In some species with distinctive juvenile morphology, the transition from the juvenile to the mature condition is as gradual a change as the transition in growth rate illustrated in Fig. 11-6. The word mature is ordinarily used to refer to a state of development which is capable of flowering; in some species this capability is achieved only gradually, e.g., the increasing flowering of *Citrus* with age. Maturation, then, is achieved through a gradual transition of morphology, growth rate, and flowering capacity.

Some characteristics that change between the juvenile, mature, and senescent stages are fairly stable and only slightly altered by graftage of buds onto trees of another developmental stage; e.g., the juvenile leaf form and the capacity for flowering are relatively stable to graftage. Other characteristics that change with aging are more amenable to alteration, e.g., the reduction of growth and the weakening of apical dominance and geotropism. Surely a part of the change in growth characteristics is a consequence of the increasing competition between growing points as the tree becomes larger (Maggs, 1965).

The competition for nutrients between parts of the plant can be markedly intensified by the presence of fruits. In apple trees, for example, the diversion of increasing percentages of the total dry-weight increment into fruits is reflected in severe restrictions on the amounts of growth increment in the roots; this apparent competition effect is illustrated in Fig. 11-10. Since the roots are important as a source of growth hormones to the tops, competitive inhibition of the root system may have profound effects on the growth and vigor of the whole plant.

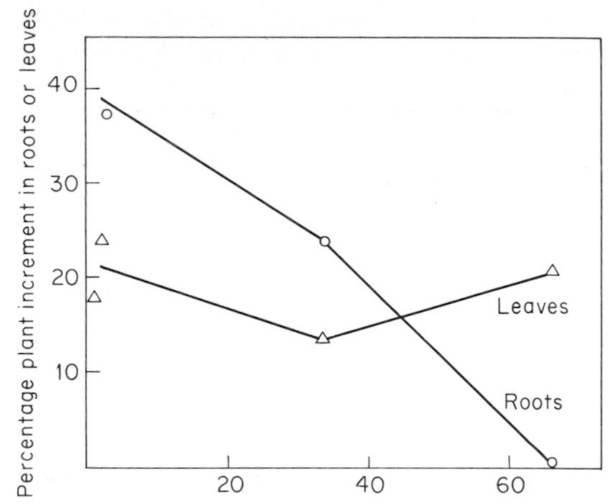

Fig. 11-10 When larger amounts of the total plant increment are diverted into fruits, there is a marked lessening of the plant increment invested in the roots; leaves do not show this lessening effect. Data for Worcester Pearmain apple grafted onto four different apple rootstocks (Barlow, 1970; data from Avery, 1969, 1970, and unpublished).

Quite aside from the changes in growth characteristics resulting from competition between branches or other parts of the plant, individual branches of fruit trees may pass through cycles of juvenile, mature, and then senescent characteristics. For example, a new and vigorous shoot on an apple tree may exercise a rapid, whiplike growth, with strong geotropic orientation and strong apical dominance for 1 or 2 years. As fruiting occurs on this branch, its growth rate is depressed, geotropism and apical dominance are depressed, fruiting increases further, and the entire branch becomes senescent. Forcing out a new shoot by removal of the branch will cause the entire cycle to be repeated.

Another cycle of change from juvenile to mature characteristics can be seen in the apical meristem. Heslop-Harrison and Heslop-Harrison (1970) have described the maturation of hemp as involving several changes in the apical meristem, including (1) a gradual increase in height of the meristematic dome (Fig. 12-1), (2) a gradual increase in density of RNA staining, and (3) a gradual increase in the activity of lateral meristems close to the apical dome. They find that each lateral passes through a transition from a more juvenile to increasingly mature anatomy, the cycle becoming shorter and shorter as the plant matures.

Still another cycle of aging occurs in the individual leaf; as each leaf unfolds, it enters a period of maximal synthetic activity, which then declines as successive leaves unfold. Illustrating the successive rise and fall of synthetic activity are data in Fig. 11-11 showing the abilities of pea leaves to synthesize nitrate reductase upon exposure to inorganic nitrate. Numerous other expressions of physiological activity follow similar time curves with the aging of leaves, including photosynthesis (Fig. 1-20), respiration, and photoperiodic responsiveness (Borthwick and Parker, 1940). The extent to which the apex of the stem serves to regulate these aging features in the leaves is worthy of note: removal of the terminal growing point above the leaf can restore many of the physiological functions mentioned, including photosynthesis (Das, 1968), activity of some enzymes (Varner et al., 1963), and photoperiodic responsiveness (Lam and Leopold, 1961).

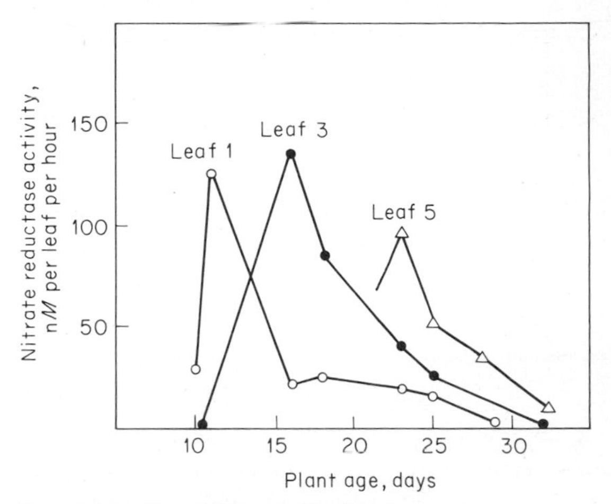

Fig. 11-11 The ability of a leaf to form nitrate reductase reaches a peak early in the life of the leaf and then becomes less as the leaf increases in age. These data are for three different leaves (numbered from the base) of pea plants, each given an exposure to nitrate to induce the formation of the enzyme 6 h before the enzyme levels were determined (Carr and Pate, 1967).

SENESCENCE

The deteriorative processes which naturally terminate the functional life of an organ, organism, or other life unit are collectively called *senescence.* In some plants this occurs as the gradual encroachment of deteriorative processes, as may be inferred from Fig. 11-6, and in other plants there may be a fairly abrupt deterioration leading to death.

Of course it is very difficult to state explicitly what processes naturally terminate the life of an organ or plant and what processes are simply reflections of the changes associated with the passage of time. Medawar (1957) has suggested that the term senescence be used to refer to natural changes toward moribundity and the term aging to refer to changes in time without reference to the natural development of death. Thus the loss of auxin responsiveness by pieces of tissue soaked in water is properly an aging phenomenon, whereas photosynthetic deterioration of leaves in the autumn is a senescence phenomenon. Senescence processes, then, may be considered as those aspects of aging which are involved in the natural expiration of life.

The central function of senescence is the facilitation of turnover. Evolution and natural selection are based on the turnover of organisms, and while most animals experience turnover as a consequence of predation or parasitism, plants are only rarely removed by a physical removal such as predation and in many instances achieve turnover through a programmed senescence.

But neither turnover nor senescence is limited to the organismic level. Modern experiments with radioactive isotopes show that there is a turnover in the substrates in an organism, a turnover of enzymes and nucleic acids, a turnover of organelles such as mitochondria and chloroplasts, in cells, e.g., phloem and xylem, and organs, e.g., leaves, flowers, and fruits. The existence of turnover at the species level was appreciated by Darwin, who remarked extensively upon the formation of new species and the disappearance of old ones in geologic time. There is extensive evidence for the concept of aging in plant species, with new, aggressive species showing strong adaptability to new environments and old species often showing more conservative, static distributions. To these latter the plant geographer applies the term *relic species* (Fernald, 1926).

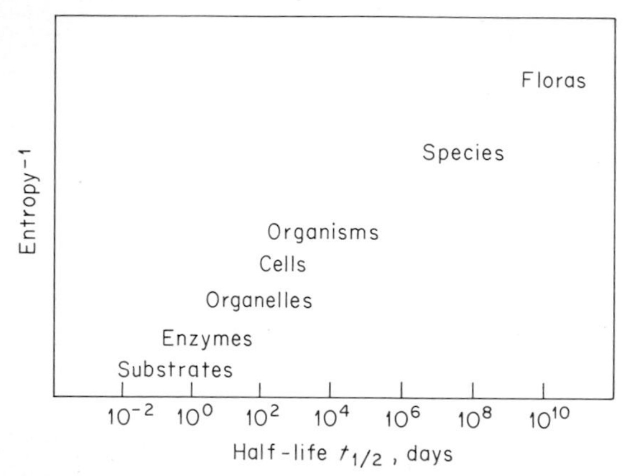

Fig. 11-12 A diagrammatic representation of the ranges of half-lives for various biological entities and components.

The concept of turnover can be applied to the collections of species that constitute floras: the floras of the Carboniferous period have been replaced by the more modern floras; the Alleghenian forest of magnolia, beech, and rhododendron which aggressively occupied most of North America and Greenland during preglacial times is now restricted mostly to the Alleghenian region of the United States, where it shows many conservative or relic characteristics.

The expected life duration of a biological entity can be expressed in a rough manner as the half-life, $t_{1/2}$, or the time over which half the members of a population can be expected to be removed. If one makes a rough approximation of the $t_{1/2}$ of common metabolic substrates in plants, a range of 10 to 100 min might be used, as illustrated in Fig. 11-12. Enzymes have a longer $t_{1/2}$, perhaps 5 to 100 h. Organelles such as mitochondria or chloroplasts vary in $t_{1/2}$, but a range of 10 to 100 days might be a reasonable approximation. Cells of the bark or of rapidly replaced parts might have a $t_{1/2}$ of 10 to 100 days and more persistent cells of 1,000 days. Organisms range commonly from 1- to 100-year lifetimes; species might be roughly approximated at 1,000 to 10,000 centuries; and finally, floras might be expected to last for 10 to 100 million years. As one plots out these comparative estimates of $t_{1/2}$ values, it seems that (at least for higher plants) the greater the degree of complexity (the lower the entropy), the greater its expected $t_{1/2}$.

If we accept turnover as a trancendent feature of biology, it follows that senescence might be observable in any organized biological unit.

Evolution has occurred as a product of natural selection operating upon a system of inherent variability between individual organisms. For evolutionary change to occur, it seems elementary to assume that the time scale for such changes must be an expression of the turnover of organisms. Thus it seems obvious that organisms with very long life spans can be expected to be relatively less adaptive in an evolutionary sense that those with short lifetimes. Among angiospermous plants life spans are widely diverse, ranging from the annual habit of many weeds to the biennial habit of forbs and the perennial habit of trees and shrubs, some persisting for 2,000 or even 5,000 years. Clonal species may persist over centuries; e.g., it is estimated that some clones of *Vaccinium* must have persisted for over 1,000 years (Darrow and Camp, 1945), and some of the clones of prairie grasses in the western United States may be the same as those which colonized the plains at the end of the glacial period 50,000 years ago (Hitchcock and Chase, 1931). In some species, sexual reproduc-

Fig. 11-13 Senescence may develop in several patterns in plants, such as overall death of the plant (*left*), the senescence of only the aboveground parts, the deciduous habit of leaf senescence, or the progressive senescence of leaves up the stem (*right*) (Leopold, 1961).

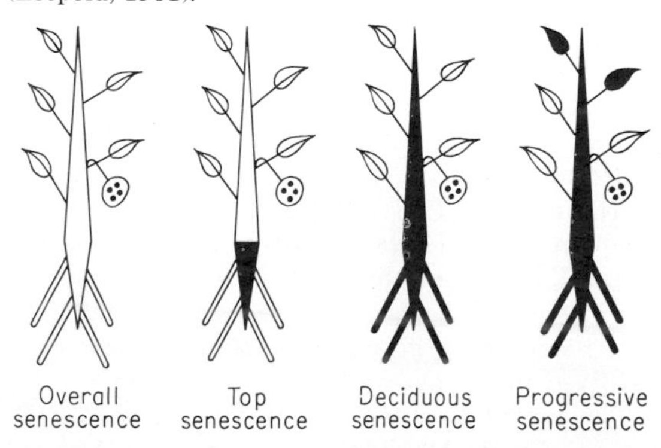

tion is displaced by such devices as parthenogenesis, and so the same genotype may persist over long periods of time, e.g., bluegrass.

Knowing that evolution advances on the basis of turnover of generations, one can expect with some certainty that in general there will be more adaptive characteristics in species with short $t_{1/2}$ values. At the species level, again there may be singular levels of persistence, as evidenced by the species of *Lycopodium* and *Equisetum,* which have persisted apparently unchanged since the Carboniferous period. Senescence, then, is an optional piece of physiological equipment, cutting off the life of some units very early and apparently omitted from the regulatory processes of others. When we realize that evolution advances on the basis of turnover of generations, it is not surprising that plants should have developed internally programmed senescence systems.

It may be helpful to describe the general types of senescence. In annual grains, the entire plant dies by some systemic function. In perennial herbs, the aboveground portions may die, but the root system and underground system remain viable, as indicated in Fig. 11-13. A less drastic development of senescence can be seen in the annual change in deciduous woody plants, in which all the leaves die but the bulk of the stem and root system remains viable. Still less drastic is the plant in which there is only a gradual progression of death of the leaves from the base upward as the plant grows. These types of senescence patterns (Fig. 11-13) illustrate the quantitative nature of senescence phenomena in plants and point up the close relationship between the senescence of leaves and the more elaborate processes of overall plant senescence. These various types of senescence may be quantitative expressions of similar physiological signals inside the plant.

Plant Senescence

The death of the whole plant commonly occurs in annual and biennial species upon completion of fruiting. Molisch (1938) suspected the reproductive activity itself of being lethal to the plant and found that removal of the flowers and fruits of several species of plants defers or prevents senescence (Fig. 11-14). Decline in growth rate is a common feature of the onset of plant senescence, and the removal of flowers and fruits can also restore the growth rate of some species, as shown in Fig. 11-15. Molisch suggested that the mobilization of nutrients out of the vegetative parts of the plant into the fruits causes senescence of the plant through a starvation process; while this is in part correct, there are some complications in this interpretation. For example, removal of flowers is noticeably more effective in deferring senescence than removal of fruits; furthermore, in male plants of spinach, the anthesis of male flowers causes plant senescence in much the same way, and senescence is deferred by removal of male flowers in a manner similar to that following removal of the female flowers and fruits (Leopold et al., 1959). The senescence of the whole plant seems to involve more complex interactions than the simple mobilization of nutrients from a plant with full synthetic competence.

Contributing to the senescence sequence is the deterioration of the photosynthetic effectiveness of the plant. Not only does each leaf lose photosynthetic abilities with age (Fig. 1-20; see also Fig. 11-20) but the whole plant shows a similar decline beginning at the time of flowering. This decline is illustrated for wheat plants in Fig. 11-16.

Another type of synthesis which may contribute in a major way to the regulation of senescence is the production of cytokinins by the roots. Mention has already been made of the suppression of root development by the fruiting activity of the plant (Fig. 11-10); and since the roots are a major source of cytokinins and other hormones for the whole plant, one might expect flowering to be associated with a sharp depression of cytokinins. Sitton et al. (1967) have analyzed the cytokinin content of root exudate of sunflower during the vegetative and flowering periods and find that the content falls dramatically as flowers begin to develop (Fig. 11-17). Since plant senes-

Fig. 11-14 The daily removal of flower buds as they appear on soybean plants results in a marked delay of senescence compared with untreated controls (Leopold et al., 1959).

cence ordinarily occurs after the completion of fruiting, the timing of cytokinin deficiency from the roots may seem at first too early to contribute to plant senescence; however, in numerous instances it has been observed that flowers themselves contribute to the development of senescence. It is an attractive possibility that flowering may lead to a restriction of the cytokinin supply from the roots, after which the cytokinins synthesized in the fruits exert an increasingly commanding effect on the translocation patterns in the plant. By this suggested scheme, completion of flowering and fruiting will leave the plant with no cytokinin supply from either the roots or the fruits, and mobilization of nutrients into the fruits will have depleted the vegetative parts.

The systemic development of plant senescence may have another contributing vector. According to Lockhart and Gottschall (1961), pea apices terminate growth even if all the flowers and fruits are removed from the plant; even if the apex of an older plant is grafted onto a younger plant, it terminates growth. These authors con-

Fig. 11-15 The growth rates of tomato plants decline after flowering and fruiting start and can be markedly restored by the removal of the flowers or fruits (Murneek, 1926).

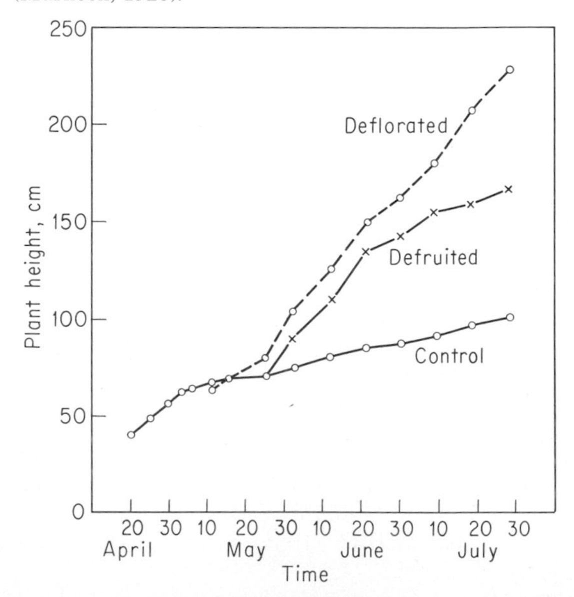

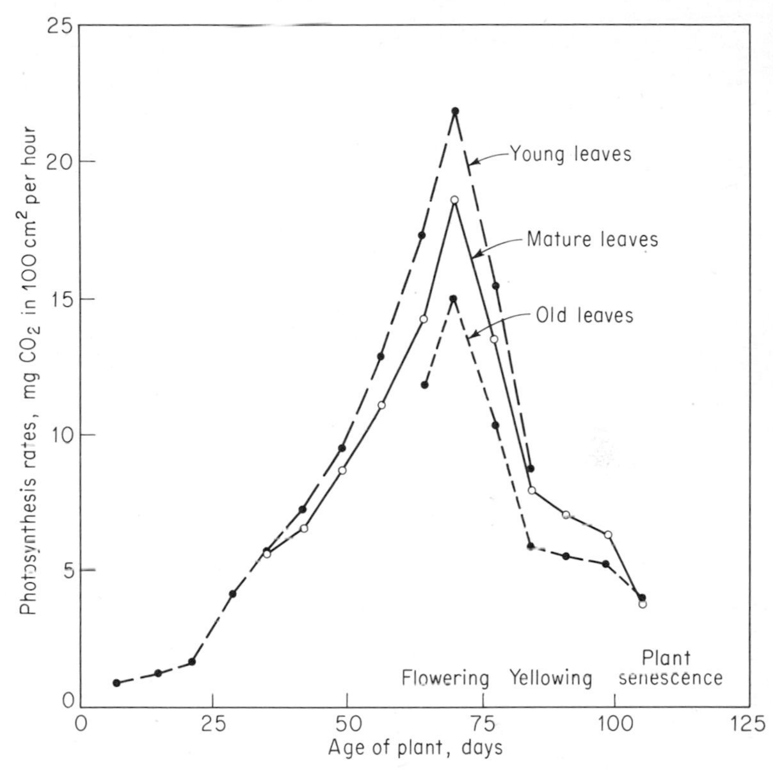

Fig. 11-16 With increasing age of the wheat plant, individual leaves show increasing photosynthetic rates until the time of flowering, after which the rates fall off rapidly, regardless of the age of the individual leaf. Photosynthesis measured under standard conditions of light and temperature (Singh and Lal, 1935).

cluded that there is a meristem senescence contributing to plant senescence, representing the consequence of some degenerative processes in the meristem itself. Marx (1968), also working with peas, showed that plant senescence is not closely correlated with flowering and fruiting and that certain cultivars continue to grow actively even after the completion of extensive amounts of fruiting.

The development of plant senescence can be modified by several factors. First, of course, there is a deferment of senescence obtained by removal of the flowers and fruits (Molisch, 1938). Several environmental factors which suppress normal plant growth enhance the tendencies to develop senescence, including the limitations of soil nutrients (Williams, 1936), a deficiency of water (Bakhuyzen, 1926), and excessive heat (Engelbrecht and Mothes, 1960). Ionizing radiations can promote aging, a response which has been shown very clearly for seeds (Sax and Sax, 1962).

There is evidence for several types of factors contributing to plant senescence: mobilization of nutrients by the reproductive parts, limitations upon the root system and the hormonal supply therefrom, deterioration of the anabolic effectiveness of the aerial parts, and a loss of growth capacity in the meristems themselves. The role of hormones in senescence will be discussed in connection with leaf senescence.

Leaf Senescence

The deteriorative processes in the leaf which lead to its senescence begin as soon as the leaf

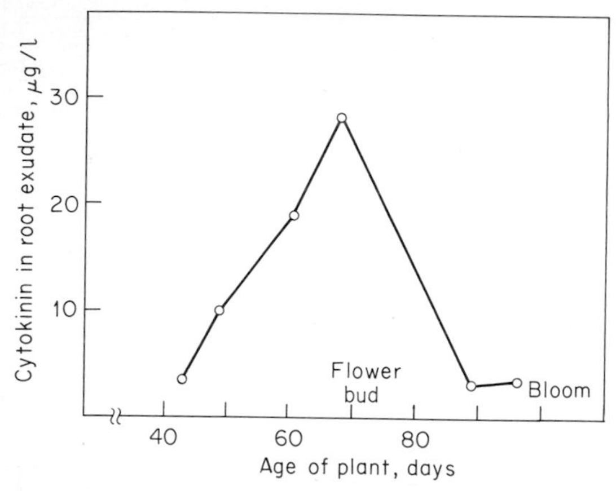

Fig. 11-17 An important supply of cytokinins for the plant comes from the roots, and at the time of flowering, the exudate from severed root systems shows a dramatic drop in cytokinin content. Each datum represents 100 plants, expressed as microgram equivalents of kinetin (Sitton et al., 1967).

has reached its full size. That there is a gradual decline in photosynthesis with leaf age has been known for many years (Rabinowitch, 1951), and the loss in synthetic capacity can proceed so far that there is an actual decline in dry weight of the leaf (Ballard and Petrie, 1936). It is also known that the stem apex may have deleterious effects on the integrity of the leaf (Petrie et al., 1939) and that the root may provide some substances which serve to extend its longevity (Chibnall, 1939).

The decline in photosynthetic rate starts soon after a leaf reaches full size, as can be seen in Fig. 1-20, along with apparent changes in respiratory activity. The pattern of declining photosynthesis is also seen in profile along the leaves of an individual plant, the youngest expanded leaves being the most effective and basal, or oldest, leaves the least effective (Sestak and Catsky, 1962). In synchronized *Chlorella* cultures a photosynthetic decline occurs with increasing age of the algal cells (Sorokin and Krauss, 1961). Respiratory activity during the life of the leaf commonly undergoes a gradual decline (Yemm, 1956); in some species an increase in respiratory rates at the time of leaf senescence resembles the respiratory climacteric shown by some fruits (Eberhardt, 1955; Yemm, 1956).

Chlorophyll and protein content fall rapidly as senescence develops, as illustrated in Fig. 11-18. It is possible that the decline of these two components may be structurally associated with the deterioration of the chloroplast, for Woolhouse (1967) has found that the most rapidly declining protein is the chloroplastic fraction I (RuDP carboxylase). In pea leaves the protein decline extends to nearly every component separable by gel electrophoresis (Carr and Pate, 1967).

The deterioration of synthetic activity during leaf aging leads one to expect a decline in RNA; in fact such a decline appears to be general (Bottger and Wollgiehn, 1958). In contrast with the simplification of protein profiles with senescence, there may be a marked increase in types of RNA species observable in the senescent leaf (Loening and Ingle, 1967). Von Abrams and Pratt (1968) have observed that in broccoli leaf disks the progress of senescence may be associated with a decline in RNA under some types of treatment but not under others; they suggest that RNA may not represent the basic control of the senescence processes. Wright et al. (1973) have found that some species of tRNA are lost during the senescence of pea leaves and cotyledons.

Fig. 11-18 The aging of bean leaves is associated with an extended decline in the protein and chlorophyll content. Data for primary leaves intact on the plant (Lewington et al., 1967).

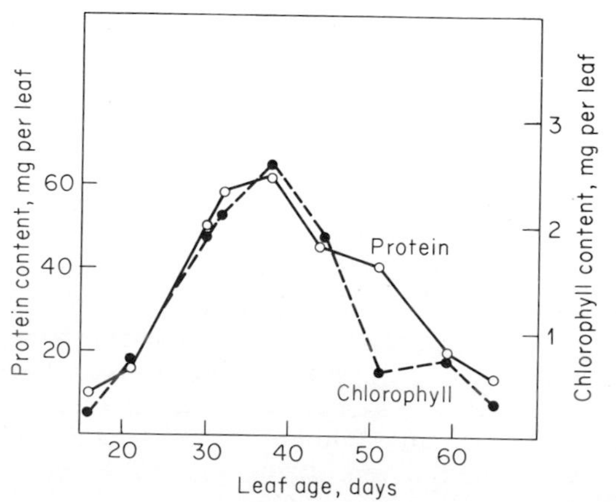

Several types of structural changes are associated with leaf senescence; at the ultrastructural level, as the leaf enters senescence, one of the first changes observed is a decline in the ribosomes and rough endoplasmic reticulum, after which the chloroplasts begin to show signs of deterioration. Markedly later are lesions in the nuclear membrane (Shaw and Manocha, 1965). Another type of structural deterioration is the increase in permeability of the leaf, first observed by Sacher (1957); later experiments by Eilam (1965) describe this type of change as an increase in the apparent free space of leaf tissues, as shown in Fig. 11-19.

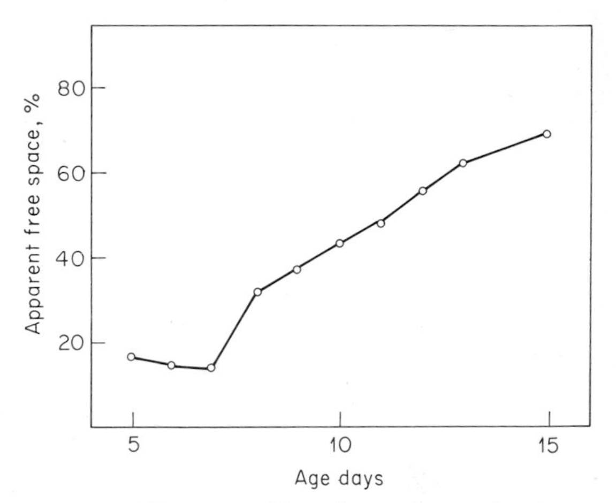

Fig. 11-19 The aging of bean leaves is associated with a rising apparent free space, which implies an increase in leakiness of membranes in the leaf. Apparent free space determined with radioactive sucrose; primary leaves of bean intact on the plant (Eilam, 1965).

Several environmental factors can alter the rate of leaf senescence, most notably, elevated temperatures, darkness, water deficit, and nutritive deficiencies. The action of heat in hastening senescence of tobacco leaves has been dramatically shown by Mothes and Baudisch (1958); the yellowing of the heat-treated leaves was associated with a lowered protein content. The hastening of leaf senescence by darkness was first described for tobacco leaves by Vickery et al. (1937), and most contemporary experiments on leaf senescence utilize darkness to hasten the development of senescent characteristics. The effects of light and darkness are probably complex; Goldthwaite and Laetsch (1967) found that poisoning photosynthesis can remove much of the senescence-retarding effect of light, but Haber et al. (1969) concluded from further experiments that an important part of the light effect is nonphotosynthetic. The experience of a rather brief water deficit can hasten leaf senescence (Gates, 1955), though the effect is less dramatic than darkness or heat treatments. Nutrient deficiencies, especially of nitrogen, have been known for many years to enhance leaf senescence (Williams, 1936).

A very useful device for deferring leaf senescence is the removal of mobilization centers in the plant. Mention has been made of the effectiveness of fruit removal in deferring plant senescence (Fig. 11-14), and this has been employed in some physiological experiments (Wareing and Seth, 1967); but a simple decapitation of vegetative plants can effectively suspend the senescence of leaves just below the point of cutting (Leopold, 1961). This provides a simple control for experimental measurements of senescence development (Woolhouse, 1967; Das, 1968). The extent to which leaf senescence can be prevented is evident from the data for photosynthesis of leaves from intact and decapitated bean plants in Fig. 11-20. Surely the growing apex exerts a strong influence on the progress of senescence of the leaves below it.

Regulatory Systems in Senescence

Each of the known plant hormones can have regulatory powers over the development of senescence. When cytokinins were found to defer senescence in *Xanthium* leaves (Richmond and Lang, 1957), for several years physiologists believed cytokinins to be distinctive in this ability; but in the middle 1960s it became clear that whereas some species are deferred in senescence by cytokinins, other species are deferred by gibberellins (Fletcher and Osborne, 1965) and still others by auxin (Osborne and Halloway, 1964). The other hormones, abscisic acid and ethylene, were found to accelerate the senescence of leaves of at least some species slightly (El-

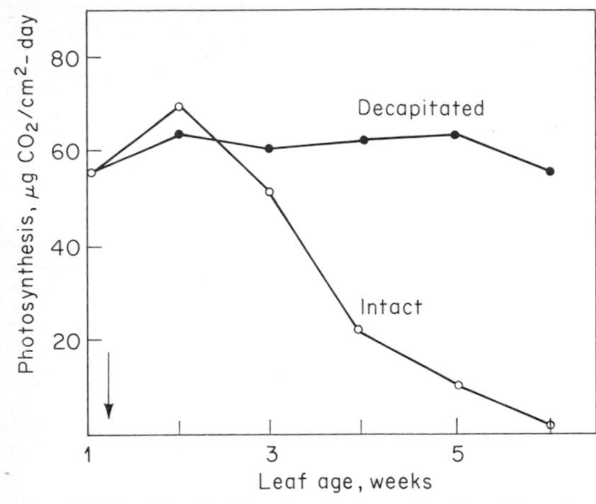

Fig. 11-20 The decline in photosynthesis with aging can be prevented by decapitation of the plant above the leaf in question. This implies a regulatory action by the growing point. Primary leaves of bean (Das, 1968).

Antably et al., 1967; Burg, 1968). There does not seem to be a real specificity for gibberellins or cytokinins for the deferral of leaf senescence; in *Rumex* leaves, for instance, deferral of senescence can be readily obtained through the application of gibberellin, but a similar effect can be obtained by adding cytokinin—though at 100 or 1000 fold higher concentration (Goldthwaite, 1972).

The deferral of senescence with hormone treatments results in an improvement of protein and RNA contents of the leaves, whether the deferral is done through cytokinin treatment (Osborne, 1962), gibberellin treatment (Fletcher and Osborne, 1965), decapitation of the plant (Mothes and Baudisch, 1958), or the rooting of leaves (Wollgiehn, 1961). The improvement following kinetin treatment is illustrated in Fig. 11-21. Experiments on the incorporation of radioactive amino acids into protein and of radioactive nucleotides into RNA established that increases are obtained when senescence is deferred, whether by cytokinin (Suguira et al., 1962), gibberellin (Fletcher and Osborne, 1965), or auxin (Osborne and Halloway, 1964).

Two interpretations are possible for the hormonal maintenance of protein and RNA levels in the leaf: the hormone may serve to stimulate the synthesis of these components, or, alternatively, it may serve to retard their decline. After the early experiments the first explanations were that the hormones did indeed act by stimulating RNA and protein synthesis; however, subsequent evidence has emphasized the possible role of retardation of catabolism. Kuraishi (1968) prelabeled the proteins of *Brassica*

Fig. 11-21 When excised *Xanthium* leaves are held in darkness, they become senescent in about 3 days, but the application of kinetin (40 mg/l) defers that senescence, as evidenced by the slower decline in chlorophyll, protein, and RNA content (Osborne, 1962).

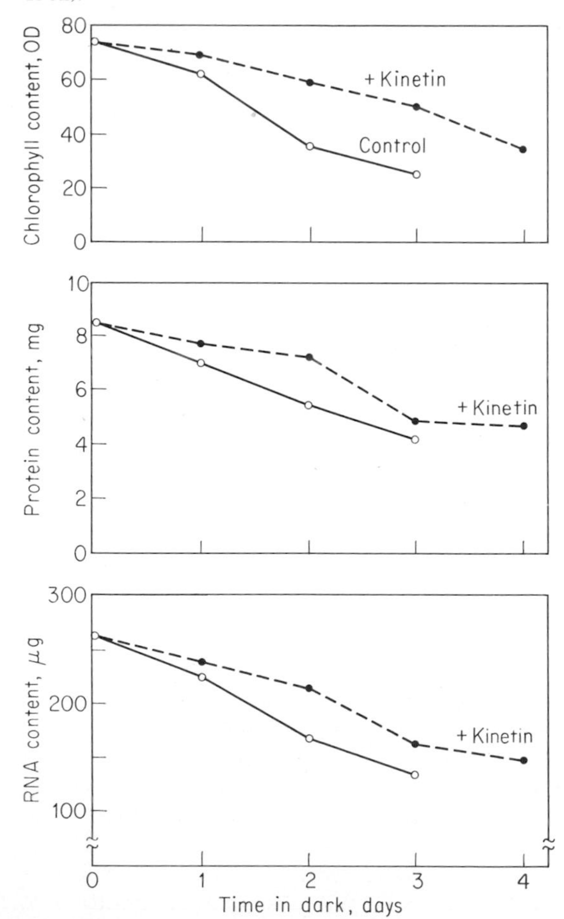

Table 11-2 Benzyladenine retards the breakdown of protein, as shown by the depression of the radioactivity lost from protein without changes in the specific activity. Corn-leaf disks pretreated with leucine*; benzyladenine (5 μM) added at time 0 (Tavares and Kende, 1970).

	Time after prelabeling, h		
	0	24	45
Loss of radioactivity from protein, %			
Control	0	28.4	49.6
BA	—	24.5	36.3
Specific activity of protein, cpm/μg:			
Control	53.3	57.2	51.0
BA	—	53.3	51.8

leaf disks with radioactive leucine and showed that the subsequent breakdown of protein is markedly deferred by cytokinin; he offered the interpretation of cytokinin regulation of senescence as a deferral of protein hydrolysis. In some similar experiments on corn-leaf disks, Tavares and Kende (1970) followed both the amount of label in the prelabeled protein and also the specific activity, and their data in Table 11-2 show that the cytokinin slows the loss of label from leaf protein without any associated increase in specific activity; this strongly supports Kuraishi's interpretation. Kende (1971) clearly pointed out the difficulties in interpreting experiments which show increased incorporation of label into protein and RNA as stimulations of synthesis. The cytokinin retardation of senescence has been shown to be associated with a lowering of both protease (Anderson and Rowan, 1965) and ribonuclease (Srivastava and Ware, 1965); an example of specific measurements of these two enzymes in tobacco leaves is given in Fig. 11-22. Martin and Thimann (1972) have added that the accrual of protease enzymes in the senescing leaf may involve the synthesis of the enzymes, since inhibitors of protein synthesis can inhibit the development of senescence in isolated oat leaves. In some species, however, inhibitors of RNA and protein synthesis accelerate leaf senescence (Mothes, 1964; Wollgiehn and Parthier, 1964).

Mobilization and Senescence

The effects of cytokinins, gibberellins, and auxin in deferring leaf senescence may be considered as simulating endogenous controls in the leaf; if they are applied to isolated leaf disks of appropriate species, they defer senescence. From experiments with whole plants, however, we can infer that there are systemic regulations involved in senescence as well. The removal of flowers or fruits or vegetative growing points can markedly defer the senescence of leaves, as can the development of roots on isolated leaves. One factor in the systemic or remote regulatory systems of senescence is directed translocation, or mobilization.

Precise experiments on mobilization were made possible by the pioneer experiments of Mothes (1959), who showed that localized applications of kinetin are not only effective in maintaining the green condition in the treated

Fig. 11-22 The deferral of leaf senescence by cytokinin may involve the lowering of activities of ribonuclease and protease. Tobacco leaves detached on day 6 and treated with kinetin as indicated (data of Balz, 1966, as cited by Kende, 1971).

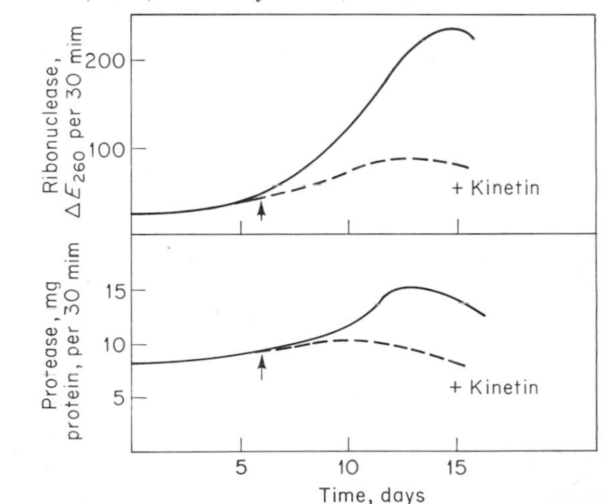

area but also mobilize tagged amino acids from remote parts of the leaf. Subsequent studies from his laboratory (Mothes et al., 1961) revealed that numerous types of solutes are subject to this mobilization action, including amino acids, sugars, inorganic ions, and some solutes not utilized in synthetic processes. Experiments by Gunning and Barkley (1963) established that in oat leaves there is a polarity of this mobilization effect, the base of the leaf being a much stronger mobilizing center than the apex. Müller and Leopold (1966) established that directed translocation of solutes by kinetin occurs as a movement through the phloem. Such an induced movement of nonutilizable solutes is consistent with the mass flow theory of phloem translocation. Leopold and Kawase (1964) induced senescence in leaves by the application of a cytokinin to remote leaves, presumably by drawing nutrients out of the untreated leaf.

The mobilization of solutes out of a leaf can markedly contribute to leaf senescence, just as the mobilization of nutrients into a leaf can defer senescence; this is a manner of correlative regulation which does not require the translocation of a specific chemical correlative carrier.

Theories of Senescence

There are no comprehensive theories of senescence in plants. The literature is filled with phenomena that relate to senescence, but a general theory has yet to be proposed.

The various types of evidence at hand show that most of the research reports fall into one or more of four categories:

1 Alterations in the nucleic acid and protein contents with senescence, including evidence that total nucleic acids and proteins decline (see, for example, Carr and Pate, 1967) and several reports of alterations in the species of tRNA (Bick et al., 1970; Venkataraman and de Leo, 1972). Even meristems of senescing plants can show these changes (Gifford and Tepper, 1962).

2 Changes in enzymic activities, e.g., increases in ribonucleases and proteases (see, for example, Srivastava, 1968; Martin and Thimann, 1972).

3 Alterations in hormone contents or the effects of applied hormones in altering senescence (see, for example, Sitton et al., 1967).

4 Structural alterations, particularly evidence of deteriorative changes in membranes and organelles (see, for example, Sacher, 1957; Shaw and Manocha, 1965; Poovaiah and Leopold, 1973).

Several underlying mechanisms can be visualized as triggers of the senescent decline. At the hormone level a trigger might include inadequate supplies of constructive hormones (cytokinins, gibberellins, auxins) or excessive amounts of deteriorative hormones with respect to senescence (ethylene, abscisic acid). At the enzymic level a trigger might include the deficiency of anabolic enzymes (respiratory enzymes) or the accumulation of hydrolytic enzymes (proteases, nucleases). At the substrate level a trigger might involve mobilization of nutrients and substrates. The two most likely means of changing these various mechanisms are through an alteration of the repression and derepression of the chromatin template, thus altering the types of nucleic acids and enzymes formed, or a structural alteration, e.g., breaking out of encapsulated enzymes (say, lysosome opening), altering homeostasis (increasing membrane permeabilities and hence loss of respiratory control), or forming new hormonal signals (the synthesis of ABA or ethylene following stress).

A working assumption might be as follows: the deterioration of cells will generally follow common symptomatic sequences, including deterioration of certain organelles and declines in proteins and nucleic acids. Whether the senescent deterioration is called forth by a programmed change in the genetic information on the chromatin, by the various types of stress, or by altered hormone profiles, the symptomology will be similar. From this point of view there need not be any specific hormonal signal provoking senescence; instead, any of several changes in nucleic acids, structure, substrates, or hormone profiles might trigger the complex set of deteriorative changes we call senescence.

Aging in Development

In the general view of plant development, there are numerous advantages inherent in the passage through successive stages of juvenility, with its aggressive vegetative growth, and maturity and senescence, with the mobilization of nutrients out of the vegetative parts into the fruits and seeds. The widespread occurrence of cellular and organ senescence shows plainly that the plant is equipped with devices through which senescence can be imposed in regular patterns. The extension of these same devices to the whole plant in overall senescence in some species seems natural. Stahl (1909) noted years ago that leaf senescence is somehow regulated by correlative events, for isolation of a part of a leaf from the rest of the plant defers its senescence. It is clear that in many ways senescence is a correlative event, regulated in part by the presence of flowers, fruits, apices, and roots.

GENERAL REFERENCES

Kozlowski, T. T. 1971. "Growth and Development of Trees." Vol. 1. Academic Press, New York. 443 pp.

Woolhouse, H. W. (*ed.*). 1967. "Aspects of the Biology of Ageing." *Soc. Exp. Biol. Symp.* 21. 634 pp.

12 FLOWERING

Flowering provides the plant with the mechanism for genetic outcrossing. The complexity of the developmental processes involved in the formation of flowers, fruits, and seeds is a tribute to the evolutionary values of sexual reproduction. How much simpler the plant organism would be if the vegetative components constituted the complete organism. And how diverse and complex are the regulatory systems and their environmental controls which direct the development of the elaborate and even rococo structures for reproduction.

In providing for sexual reproduction, flowering provides a means of securing a great variety of genetic recombinations by which new adaptations are made possible (Stebbins, 1957). These powerful advantages spring universally from the transfer of genetic materials between individuals, and it is reasonable to presume that plants have been under great and protracted evolutionary pressures toward improved synchrony of flowering between individuals, since only by the synchronous flowering of neighbor plants can outcrossing occur. The physiology of flowering is therefore mainly the study of how flowering is regulated and how synchrony is achieved through environmental cues and internal regulatory devices.

Beyond the central genetic advantage, the regulation of the timing of flowering offers several ecological advantages. In many species it permits the reproductive development to be completed before the onset of seasons of environmental stress. The existence of strictly vegetative periods in the growth of a young plant offers obvious advantages in the competitive interactions with other plants. Many perennial plants alternate vegetative with reproductive growth and are thus able to devote full energies to vegetative growth at some parts of the growing season. Some fruit trees have the ability to flower in alternate years, growing vegetatively in the intervening year.

The most conspicuous devices for achieving synchrony in flowering include the internal programming of earliness and the external cueing of flowering by such environmental features as light, photoperiods, and temperatures. A population of a given species in which the seeds germinate at a given time may achieve a synchrony in flowering time through their common earliness characteristics; in this respect, one could also include some germination controls as contributing to synchronous flowering. In some species a synchrony of anthesis is achieved by the environmental regulation of flower-bud dormancy; e.g., anthesis in coffee occurs in bursts which are triggered by rain.

Another widespread feature of the flowering process, in addition to synchronizing devices, is an increasing tendency toward flowering with plant age; while there are exceptions (Bhargava, 1964), plants generally advance toward the flowering condition with age. Not only is earliness of flowering a measure of such a tendency, but many species become increasingly responsive to environmental cues for flowering with age and may actually outgrow the need for photoperiod or vernalization stimulation. We have already suggested that ultimate independence from environmental cues is a type of fail-safe mechanism by which the plant ultimately succeeds in flowering even if the usual environmental cue fails to occur.

Fig. 12-1 The apical meristem of *Cannabis sativa* shows an increasing height of the meristematic dome during the aging of the plant and with the induction of flowering by short photoperiods. Apical meristems of male plants during the first plastochrons; profiles of the meristems are represented at the top of the figure (Heslop-Harrison and Heslop Harrison, 1970).

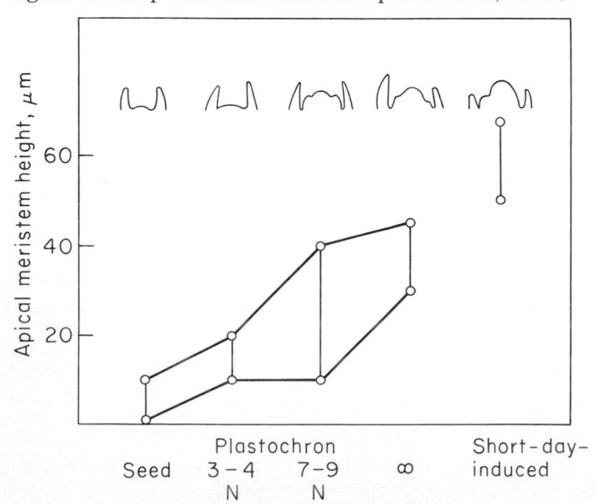

The gradual advance toward the flowering state is strikingly evident in some morphological changes in the apical meristem, as described for *Cannabis sativa* in Fig. 12-1. During the development from seed to mature plant the height of the meristematic dome increases, and at flower induction the dome becomes even more exaggerated. In fact, the suggestion has been made that the commencement of flower development may be triggered by the achievement of sufficient size in the meristematic dome (Evans, 1971). It is well documented that in some species the apical meristem must pass through a prefloral and then a floral stage in which there is a rise in synthesis of DNA and RNA and increased mitotic activity (Bronchard and Nougarède, 1970). The inability of juvenile plants to flower may well be a consequence of insufficient development of the terminal meristem (Heslop-Harrison and Heslop-Harrison, 1970).

The physiological controls of flowering may be exerted at any of several fairly definitive stages. Flowering may be regulated by the termination of the juvenile phase of development. Environmental cues may provoke the *induction* of the reproductive state in leaves, the *initiation* of floral meristems, the morphological *development* of flowers, or *anthesis* itself.

These reproductive developments are usually triggered by such environmental variables as temperature, light, and photoperiod. Since these variables change with the seasons, a programming of reproductive activities on a seasonal basis becomes possible.

EARLINESS

To the agronomist, the term *earliness* may refer to the calendar date of the first flowering or fruiting of a variety, but in the physiological sense it is usually applied to the morphological state. Thus greater earliness would refer to an earlier morphological differentiation of flowers and hence to the formation of the first flower primordia at a lower node.

In either context, earliness is an expression of the increasing tendency of plants to become reproductive with age. In some species, the arrival

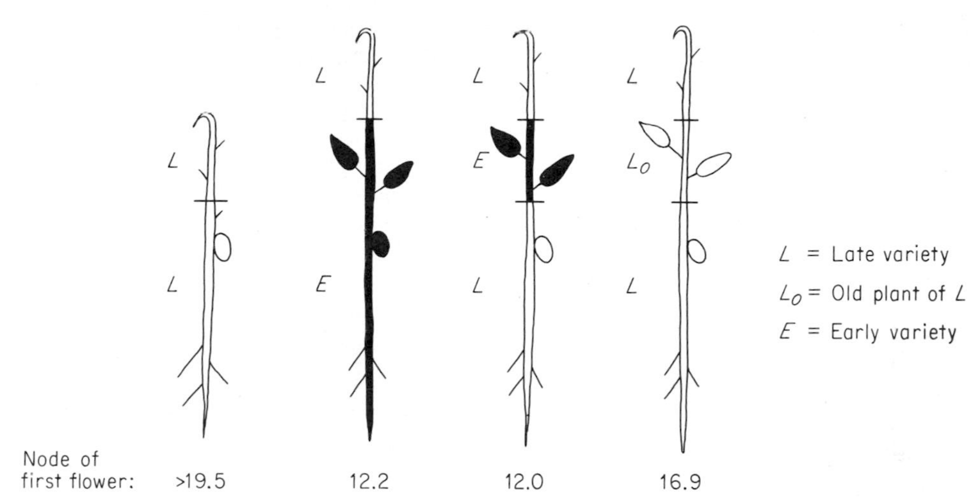

Fig. 12-2 The graftage of a late variety of peas (*L*) onto an early-variety stock (*E*) increases earliness, as indicated by a lower node of first flower. Increased earliness is also obtained if the early variety is used as an interstock or if an older piece of the late variety is used as an interstock. (*Adapted from Haupt, 1958.*)

at the reproductive state is apparently unalterably set, and treatments of the seed or the seedling have no effect on the earliness of flowering. The various species of bean are promiscuously flowering, and their earliness is apparently not alterable. The peanut (*Arachis*) is beyond any alteration of morphological earliness, for the first flower primordia are laid down at a standard node within the seed. Many species, however, are slower to develop their first flowers and are more amenable to environmental modification of earliness.

Developmental Factors

Working with peas, Barber and Paton (1952) attempted to analyze for earliness factors by graftage of early and late varieties. They found that an early-variety scion grafted onto a late-variety stock flowers at a higher node, i.e., is later. They suggested that some factor for lateness is supplied by the root system of the late varieties.

Early and late peas were grafted together to define the location of the inhibitory effect of the stock plants more precisely. Haupt (1958) found that a late scion grafted onto a leafy early stock flowers early (Fig. 12-2). When the early stock was reduced to an interstock with leaves, the increase in earliness still obtained. The stock and interstock effects require the presence of leaves, indicating that a promoter of flowering may come from the leaves of early varieties (Fig. 12-2).

A dramatic case of flower promoters apparently arising from leaves is *Ipomoea;* the Jersey type of sweet potato is stubbornly vegetative when grown in the United States, but Lam and Cordner (1955) were able to bring it into flower by graftage onto readily flowering stocks of other species of *Ipomoea.* Flower formation in the sweet potato scion continued only as long as leaves remained on the stock plant. There are species in which this leaf requirement does not hold, however (Wellensiek, 1966).

Promoters and inhibitors of flowering may be effective internal regulators of the earliness of some plants. Attempts to extract the promoters and inhibitors of earliness have ordinarily been unpromising (Moore and Bonde, 1962). With peas the inhibitor action can be simulated with GA (Moore and Bonde, 1958).

In a more general way, internal factors for earliness can be associated with increases in plant age and increases in plant size. Experimentally, evidence on these factors can best be col-

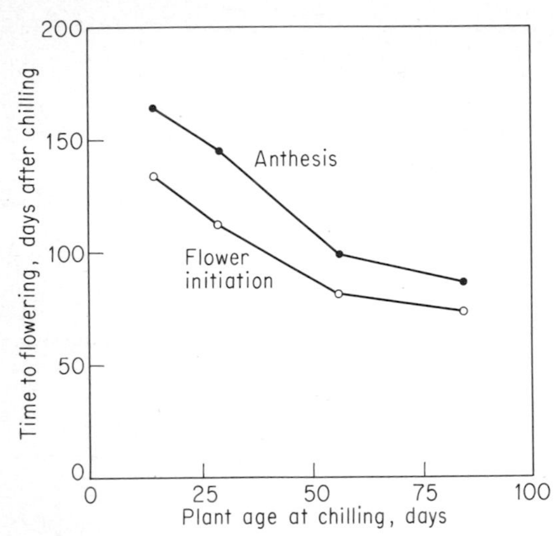

Fig. 12-3 The ability of celery plants to flower in response to a low-temperature treatment increases as the plants grow older, as evidenced by the shorter time to flower initiation and to anthesis after the completion of the cold treatment (4 weeks at 8°C) (Pawar and Thompson, 1950).

lected with plant species which respond to flower-stimulating treatments such as temperature and photoperiod; consequently the results obtained may be somewhat peripheral to the question of inherent earliness, but a few examples will show the age and size effects on flowering. In celery plants, the responsiveness to low temperature increases with the age of the plant, both flower initiation and ultimately anthesis being achieved sooner after low-temperature induction of flowering (Fig. 12-3). In experiments with cabbage seedlings of uniform age, Boswell (1929) was able to show that the proportion of plants flowering is directly related to the size of the plant; he asserted that a minimum stem diameter must be achieved before flowering can occur. This is very similar to the minimum size required for the passage out of juvenility (Fig. 11-5).

In experiments with pepper, Rylski and Halevy (1972) have observed that there is a gradient in the readiness with which lateral buds can flower, basal nodes being slowest to flower when forced out by pinching and more apical buds being fastest to flower under similar conditions. Such a gradient is reminiscent of the gradient of juvenility and maturity along the axis of some trees (Fig. 11-2).

Environmental Factors

The ability of temperature treatments to shift earliness was noted long ago by Sachs (1872), and since then many plants have proved to be so modifiable. Low temperatures commonly increase earliness, vernalization of winter grains being a well-known example. The flowering of some biennials in response to low-temperature experience is another example, though the term earliness applies more precisely to the spontaneous tendency of indeterminate plants to flower, and many biennials require a low temperature for flowering.

The tomato ordinarily sets its first flower buds at about the twelfth to the seventeenth node, depending on the earliness of the variety. Lewis (1953) found that by holding newly emerged seedlings at a low temperature, the first flower

Fig. 12-4 The earliness of Ailsa Craig tomato plants is increased by 2 to 5 weeks of low temperature (16 to 10°C) given at the cotyledon-expansion stage. The increased earliness is indicated by the lower node of first flower. If the plants are grown at 10 to 16°C, exposure to high temperature (27°C) causes a delay in flowering (Calvert, 1957).

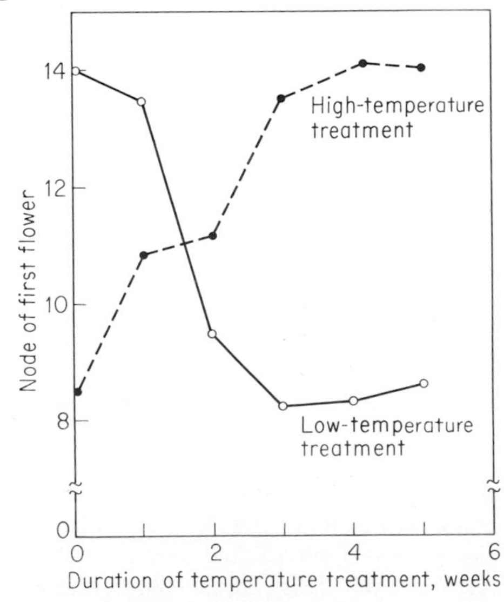

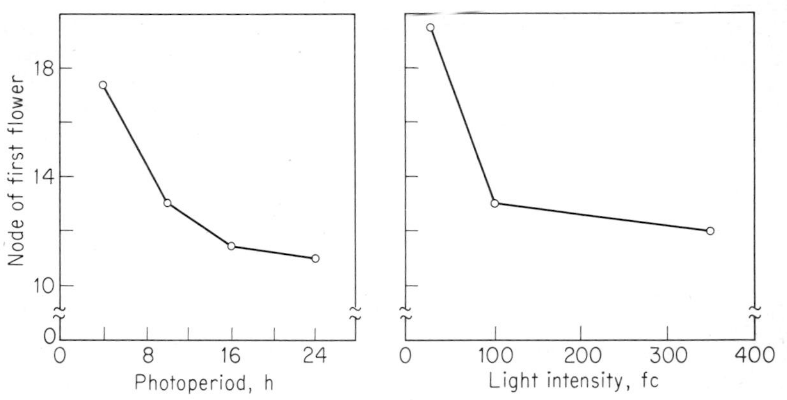

Fig. 12-5 The flowering of pea (Kleine Rheinlanderin) is similarly increased in earliness either by increases in photoperiod or by increases in light intensity under constant photoperiod. Cotyleodons were removed to increase the light sensitivity. Experiment at left was with 350 fc; at right, with continuous light (Haupt, 1958).

could be brought down to a lower node. This effect is illustrated by some data of Calvert (1957), shown in Fig. 12-4, concerning exposure of seedlings to various durations of low temperature (16 to 10°C). Three weeks of such treatment lowered the first flower almost six nodes. Seedling experiences at high temperatures (27°C) had the opposite effect, shifting the first flower to a higher node.

The ability of light to alter earliness was dramatically revealed by Garner and Allard (1920) with the discovery of photoperiodism. When plants flower in response to long or to short days, this does not necessarily mean that the influence is photoperiodic; in numerous cases flowering can be altered simply by the amount of light rather than by periodic effects of light and dark interactions. Effects of the quantity of light on earliness are revealed by light-intensity experiments, as illustrated by some data for peas (Fig. 12-5). Plants in which the cotyledons had been removed flowered at a lower node when longer day lengths were given, and the same effect was achieved if the light intensity was increased in a constant day length of 24 h.

Growth Rate

In many perennials, the differentiation of flower primordia in the meristems occurs when rapid stem growth has ended. A dramatic example of the occurrence of flower initiation at the time of growth termination can be seen in some data of Nasr and Wareing (1961) for black currant (Fig. 12-6). Such a correlation is probably common to many fruit trees.

Ringing the bark of fruit trees can often be effective in increasing earliness, usually bringing about flowering on trees of apple and *Citrus* 1 or 2 years earlier than the 4 to 15 years needed for flowering (see, for example, Harley et al., 1942). The resulting wound slows translocation of sap out of the branch so treated, retarding growth and damming nutrients and growth substances in the branch. The Peruvian cube tree grown in California never flowers naturally, but ringing can cause flowering at will (Cooper et al., 1945). The leaves produce a positive stimulus for flowering, since removal of leaves above the ring erases the flowering tendency (Magness et al., 1933). Bending stems out of vertical orientation is also effective in retarding growth and enhancing earliness in woody fruits. Such bending occurs naturally in older branches as a result of fruit loads, and pruning practices take advantage of the abundant flowering of pendant branches (Christopher, 1954). Artificially bending branches down can force many woody plants to flower earlier (Longman, 1961). In several species of trees it has been found that bending the branches leads to a marked increase in the endogenous ethylene levels (Leopold et al., 1972).

An ingenious device for experimentally retarding growth to assess its effect on flowering has been utilized by Kojima and Maeda (1958). Embedding radish seedlings in gypsum plaster to restrain growth, these workers found a pronounced increase in the flowering tendencies of the restrained plants. Other growth-retarding treatments enhancing flowering include the application of the growth inhibitor maleic hydrazide or sugar solutions of high osmotic values.

After an unexpectedly strong fumigation of a greenhouse, Fisher and Loomis (1954) noted that soybean plants were forced into flowering; similar effects could be obtained with pinching (cutting off the tip of the stem) or even partial leaf-removal treatments, which inhibited stem elongation. They proposed that there is an antagonism between vegetative and reproductive growth, a concept seriously considered also by Wellensiek's group in the Netherlands (Wellensiek et al., 1954; cf. also Carr and Ng, 1955).

Many chemical treatments of plants can induce flowering. Among the first of these was the auxin induction of flowering in pineapple (Clark and Kerns, 1942). This treatment is associated with a temporary suspension of leaf growth, perhaps related to the ethylene produced (Burg and Burg, 1966c). Auxins are also effective in encouraging the flowering of the litchi tree in Hawaii (Nakata, 1955), and this too is associated with an overall inhibition of vegetative growth. Many other growth-retarding treatments are similarly effective, including withholding irrigation water to suppress vegetative growth (Nakata, 1955). Naphthylphthalamic acids have been employed by Teubner and Wittwer (1955) to force early and abundant flowering of tomatoes; this treatment, which parallels the effects of cold treatment of seedlings shown in Fig. 12-4, appears to act through an inhibition of vegetative growth. De Zeeuw (1956) showed that the chemical treatment is ineffective if the young expanding leaves of the tomato are removed, and these pinching treatments cause increases in flowering similar to those obtained by the chemical application.

Fig. 12-6 In black currant, stem-length increases are completed in mid-August, and the appearance of flower primordia follows in the next 4 weeks (Nasr and Wareing, 1961).

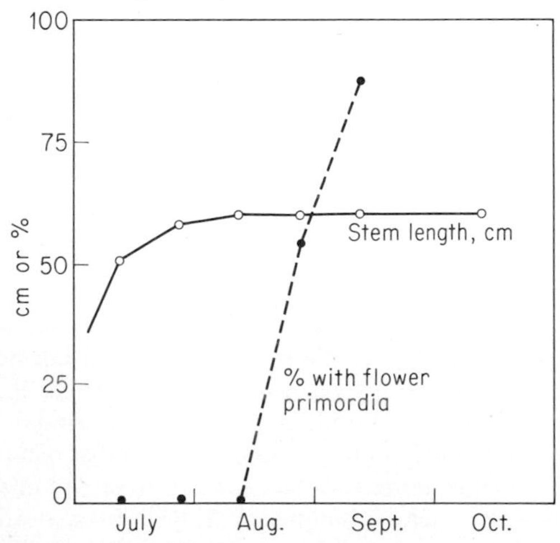

The increases in tomato earliness obtained by leaf removal were described by de Zeeuw (1954). Subsequent studies reveal that early varieties of tomato are not made earlier by such pinching treatments but all the late varieties tested are made earlier (Heinze, 1959). These observations are consistent with the possibility that lateness in tomatoes may be due to a flowering inhibition in the late varieties associated with the activities of vegetative growth.

The gradual amplification of the flowering tendencies of plants with age can result in varying degrees of earliness between species and varieties. The evidence concerning the endogenous factors influencing earliness suggests that both promotive and inhibitory influences are acting on the potential of the plant to flower. The ability of various environmental factors such as temperature, light, and photoperiod or of mechanical and chemical treatments to alter the earliness of a plant can be tentatively interpreted as relating to the relative expression of these endogenous factors. The similarities of the environmental factors which can alter earliness to those which can actually induce flowering (temperature, photoperiod) strongly suggest that the same mechanisms are involved in the earliness and the flower-induction processes.

PHOTOPERIODISM AND FLOWERING

The idea that flowering might be controlled by a hormonal stimulus, first proposed by Sachs (1882), appears to be borne out by the more recent finding that the stimulus is mobile in the plant (Cajlachjan, 1936). The mobility of the stimulus really became clear only after the discovery of the effects of day length on flowering by Garner and Allard (1920), day lengths being most effectively perceived by the leaves of plants. The flowering stimulus must move then from the leaves to the meristem, where the morphological change is initiated.

That day lengths can control reproductive activities was touched upon by several plant physiologists, especially Klebs (1918), who actually succeeded in causing the flowering of *Sempervivum* by illumination during the night; but it remained for Garner and Allard (1920) really to establish the concept of day length as a control of seasonal events—or photoperiodism.

The seasonal changes in photoperiods provide a precise and reliable cue of the seasons; for biological systems in the middle latitudes, day lengths oscillate between about 9 and 15 h of light per day, the oscillation increasing in amplitude with increasing latitude and decreasing toward the equator to an almost uniform day length at the equator itself (Fig. 12-7). It is evident that day length is a more effective signal in the middle latitudes than in the tropics, and in fact it is probably more common as a physiological cue to plants in the middle latitudes.

An interesting means of representing the photoperiod and temperature features of different climates devised by Ferguson (1957) is shown in Fig. 12-8, where the day lengths are plotted against mean monthly temperatures. The annual climatic cycle of seasons appears as an orbit through these two environmental coordinates. Associated with the small photoperiod changes in the tropics is a small temperature flux. The increasing seasonality of the climate with increasing latitude is evident as the increasing range of climatic orbit. Not only is the magnitude of the seasonal cue larger in higher latitudes, but the need for seasonal changes in plant development is often greater.

Photoperiodic Classes

Garner and Allard (1920) recognized three classes of photoperiod responses: short-day plants (SDP), long-day plants (LDP), and indeterminate plants. Among SDPs and LDPs, some species may be strictly controlled in flowering by photoperiods, and many others may be only quantitatively accelerated or delayed in flowering by the photoperiods. The difference is sometimes quite blurred: some plants are strict SDPs as seedlings, becoming only quantitatively accelerated by short days as they grow older and finally indeterminate with further aging.

Some common examples of strict SDPs are the fall- and winter-flowering species of tobacco, *Xanthium,* ragweed, soybeans, *Pharbitis,* and *Poinsettia.* Some common strict LDPs include *Hyoscyamus,* spinach, plantain, winter barley, and *Lolium temulentum.* The distinction between the classes lies in their responses to the length of the night period, not to any special range of day lengths. Thus, SDPs require a dark period of a greater than threshold length for flowering, whereas LDPs are prevented from flowering by dark periods of greater than threshold length. The short days of late summer provide the SDP with adequate night length for flowering, and the long days of early summer provide the LDP with a brief enough night to permit flowering (Fig. 12-9). The major photoperiodic influence of the night period, discovered by Hamner and Bonner (1938), will be discussed after mentioning two other photoperiodic classes.

Some species originally considered to be SDPs were later found to be unable to flower under continuous short-day conditions (Dostal, 1950). Sequential experience of long days followed by short days is necessary for *Bryophyllum* and some other crassulacean species, some lilies,

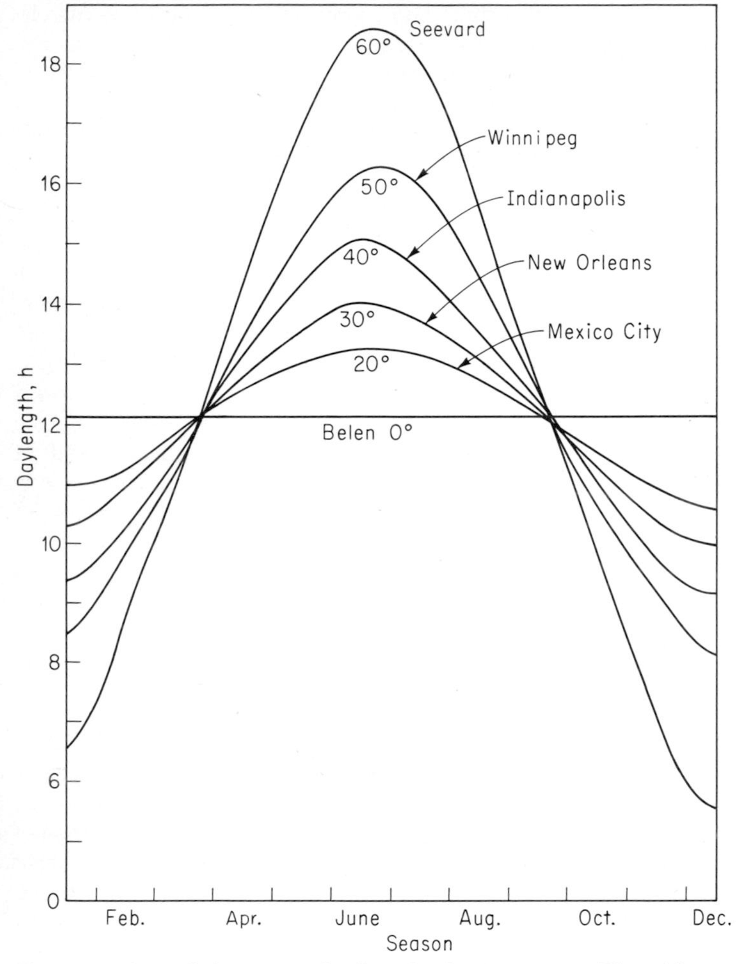

Fig. 12-7 Annual changes in day length, showing the increases in amplitude with increasing latitude. Plotted here are the times from sunrise to sunset for six different locations.

and *Cestrum.* Conversely, some species require short days followed by long days, e.g., *Campanula medium, Scabiosa succisa,* and the common *Trifolium repens.* The existence of plants requiring multiple photoperiod experiences suggests that the flowering stimulus may involve multiple regulators or hormones. We shall return to this issue later.

Physiologists have concentrated their experiments concerning photoperiodism mainly on the regulation of flowering; however, photoperiods control a whole spectrum of other developmental events (including flower development, sex expression, tuber and bulb formation, growth rate, dormancy, and germination). The great span of photoperiodic regulations may mean that photoperiodic cues lead to the formation of separate hormones which specifically regulate each of the developmental functions or, alternatively, that photoperiodic cues alter the interactions of a common set of hormones and thus regulate developmental functions. With respect to flowering, the problem of photoperiodic control might be restated to ask whether regulation is achieved through the formation of a specific flower-inducing hormone or through changes in proportions of the common plant hormones.

The genetics of photoperiodic classes has received little attention; in some cases the inheritance of the photoperiodic class appears to be controlled by a single recessive gene (Lang, 1948), but more often it appears to be the product of multiple interacting genes (Goodwin, 1944; Griesel, 1966).

Graftage experiments between species in the different photoperiodic classes show that similar stimuli are common to the various types; grafting a flowering LDP to a SDP, e.g., *Hyoscyamus*

Fig. 12-8 The annual climatic cycle can be expressed as the seasonal changes in day length and temperature. The climate for any one place is represented as an orbit of these two parameters. Sample climatic orbits are shown for five locations from 6 to 59°N latitude (*after Ferguson, 1957*).

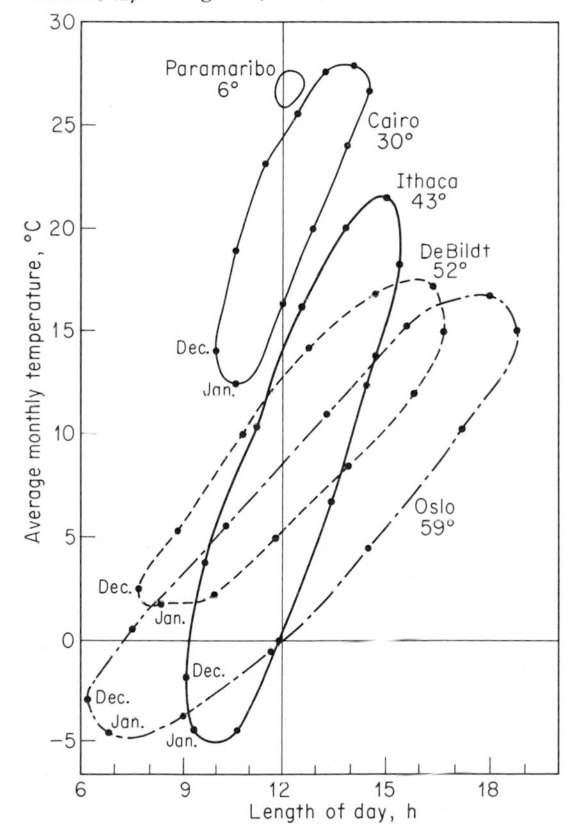

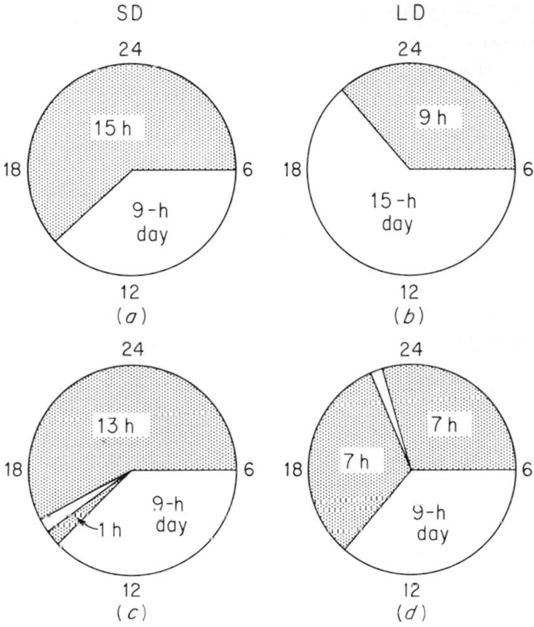

Fig. 12-9 Some patterns of day and night periods on a 24-h cycle that invoke short-day responses (*a*) and (*c*) or long-day responses (*b*) and (*d*) in photoperiodism. (*a*) and (*b*) A short day and a long day. (*c*) A short day with a light break near the start of the night evokes a short-day response and (*d*) with a light break near the middle of the night it evokes a long-day response.

to tobacco, can cause the SDP to flower (Melchers and Lang, 1941). The converse graft pairing a flowering SDP to a LDP has induced flowering in the LDP, but with more difficulty (Lang and Melchers, 1943; Zeevaart, 1957). Indeterminate species induce flowering in either LDP or SDP graft partners (Melchers, 1937). A surprising feature of the graftage results is that flowering may be induced in some graft pairings even when the *donor* plant is not flowering; for example, Melchers (1937) was able to induce flowering in *Hyoscyamus* by graftage to the SDP tobacco, whether the latter was induced or not; again, Cajlachjan (1965) has elaborated the same type of experiment, obtaining flowering stimulus from a nonflowering graft partner.

Numbers of features of the photoperiodic classes imply that the regulation of flowering may involve multiple interacting regulators:

the requirement of some plants for different sequences of photoperiods—long days followed by short days (LSDP) or short days followed by long days (SLDP)—the lack of simple hereditary characteristics of the photoperiodic classes, and the stimulation of flowering across grafts from nonflowering donor plants. On the other hand, several features of the photoperiodic classes suggest that there may in fact be a flowering hormone, e.g., the stimulations of flowering across grafts between the different classes, the flowering of parasites such as dodder when it is parasitizing flowering plants from any photoperiodic class (Frattianne, 1965), and occasionally successful extractions from flowering plants of materials which can induce flowering in noninduced plants (see, for example, Lincoln et al., 1961).

Fig. 12-10 Expressed as the percentage of plants flowering, the photoperiodic induction of the LDP or SDP acts nearly like an all-or-none response. One inductive cycle will usually induce all *Xanthium* or *Lolium* plants to flower. Species requiring several days for induction may show a limited quantitative response, but all plants become flowering with a few inductive cycles (Evans, 1960; Lang and Melchers, 1943; Schwabe, 1959).

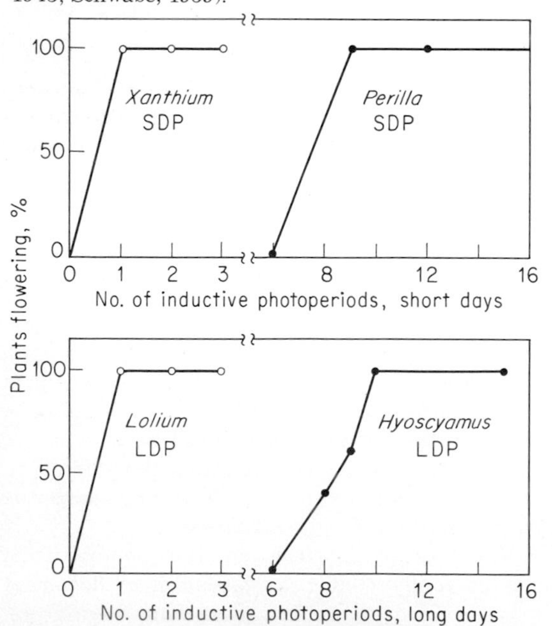

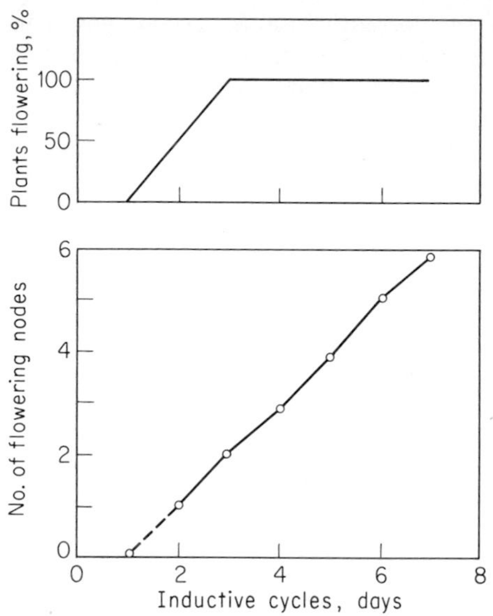

Fig. 12-11 Expressed as the numbers of flowering nodes per plant, the photoperiodic induction of soybean appears as a quantitative response, with increasing numbers of nodes with flowers. The percentage of plants flowering is still essentially an all-or-none response (Hamner, 1940).

Photoperiodic Processes

The two principal means of studying the characteristics of the photoperiodic flower stimulus are examination of the kinetics or quantitative characteristics of photoperiodic induction and separation of the floral induction into component or partial processes. The kinetics of the flowering processes can be measured using such criteria as the percentage of plants flowering after an inductive treatment, the rate of development of the flowers as evidenced by the progress of changes in the meristem, or the number of days required for flowers to become visible. In a few species, e.g., *Hyoscyamus,* the number of inductive photoperiods may be reflected in a quantitative increase in the number of plants which initiate flowers, as shown in Fig. 12-10. More commonly, when a sufficient number of inductive cycles has been given, all the individuals so treated become flowering, e.g., *Xanthium pensylvanicum, Lolium temulentum*). A quantitative increase in numbers of flowers can often be obtained with increases in number of photoperiods, even though the percentage of flowering is more nearly an all-or-none response (Fig. 12-11). Neither of these parameters has been widely accepted for quantitative or kinetic measurements (Lang, 1952).

If repeated inductive photoperiods are given, one would expect increasing amounts of the flowering stimulus to accrue in the plant and be reflected in increased rates of floral development. Using stepwise changes in the meristem to measure the rates of development, and presumably the amount of the flowering stimulus, Salisbury (1955) made quantitative comparisons like those shown in Fig. 12-12. This morphological measure of flowering rate has provided a basis for kinetic analysis and has also made possible much progress in understanding the sequence of partial processes constituting the photoperiodic phenomenon.

The light period. At least a part of the light-period influence on photoperiodism is photosynthetic. The minimal light intensity for an effective light period is usually about 100 fc (Hamner, 1940), a value lying in the range at which photosynthesis would pass the compensation point. In addition, it has been found that depriving the leaves of CO_2 during the light period erases its effectiveness (Parker and Borthwick, 1950) and supplying plants with sugar solutions in darkness has occasionally proved successful in bringing about flowering (Takimoto, 1960). Part of the sugar effect may be through enhancement of translocation out of the leaf (Carr, 1959).

If the light period were simply a photosynthetic one, the quality of the light applied during the light period should alter photoperiodism only as it alters the photosynthetic activity. In comparisons of light qualities during the day, Stolwijk and Zeevaart (1955) found some striking differences between photosynthetic effects and photoperiodic effects. For example, red light given for 16-h photoperiods did not induce flowering in *Hyoscyamus* but in combination with blue or far red it did. Oda (1962) found that blue light during the day made *Lemna* change from a SDP into an indeterminate plant. Hillman (1966) found that if red or blue was applied at the end of the photoperiodic day, he could erase or reinstate the photoreversibility of a night-interruption reaction in *Lemna*. Hillman (1969) has averred that the only generalization possible from the experiments on light quality during the day is that red light has markedly different actions than blue or far-red light. Whether phytochrome is responsible for these effects is not clear.

The dark period. One of the most productive findings by researchers in photoperiodism (Hamner and Bonner, 1938) is that the dark period is the critical part of the 24-h photoperiodic cycles, which they deduced from the simple experiment of applying a brief light period in the middle of the dark period and showing that the SDP *Xanthium* is prevented from flowering even under short-day conditions.

The effectiveness of light exposures in disrupting the effectiveness of the dark period has opposite actions on SDP and LDPs. Placed on short days, an SDP will flower, and a night interruption will prevent flowering. An LDP

Fig. 12-12 A quantitative expression of flowering intensity is the rate of flower-primordia development. The stage of development in *Xanthium* is seen to advance more rapidly under continuous inductive cycles than after only 2 short days. Floral stages range from a convex apical meristem (stage 1) to clusters of meristematic florets (stage 8) (Salisbury, 1955).

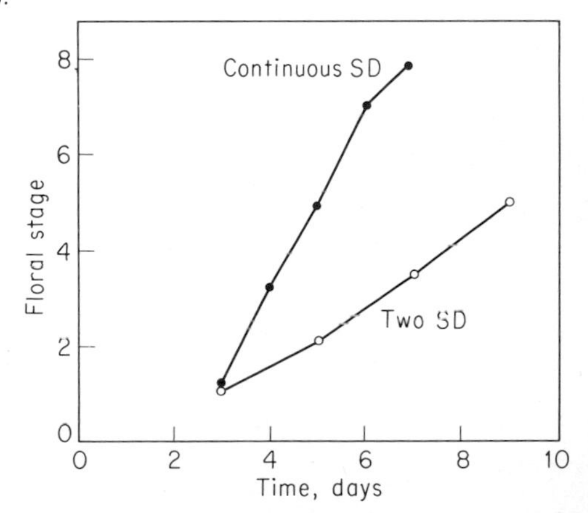

placed on a short day will not flower, and a night interruption will induce it to do so (Fig. 12-9).

Knowledge of the night-interruption effect led to one of the major breakthroughs in our understanding of photoperiodism: establishment of the action spectrum, and then identification of the pigment for the night-interruption effect. Applying light of various wavelengths to the leaves of SDPs in the middle of the long dark periods permitted a comparison of the relative effectiveness of the various parts of the spectrum. The results obtained are shown in Fig. 12-13 for *Xanthium*. Similar action spectra have been obtained for each of the photoperiodic plants tested, showing a maximal effect in disrupting the night period by red light in the region of 640 and 660 nm. Following this came the unexpected finding that far-red light is effective in erasing the red-light action, also represented in Fig. 12-13. The resolution of these two opposite light actions came with the demonstration by Hendricks et al. (1956) that light has reversible effects on a pigment which can assume two forms, a red-absorbing P_r and a far-red-absorbing P_{fr} form. This pigment was subsequently isolated by the group at Beltsville, Maryland, that had proposed its existence (Butler et al., 1959) and was

Fig. 12-13 Action spectra for the effects of light on the photoperiodic night. The light interruption of the night is most effective with red light (620 to 660 nm), and reversal of the red-light effect is most effective with far-red light (725 nm); data for *Xanthium* (Hendricks and Borthwick, 1954).

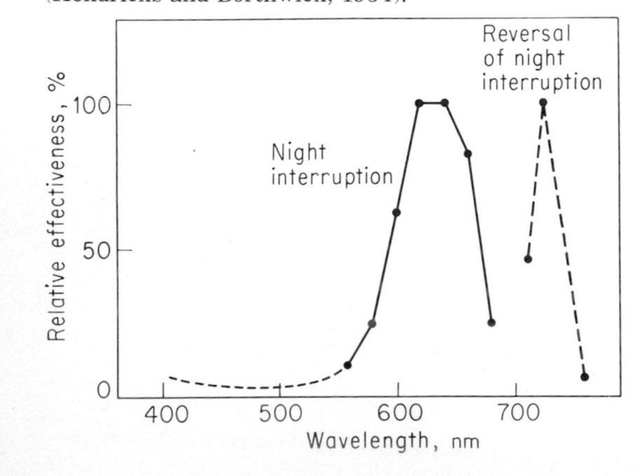

named *phytochrome*. The pigment and its physiological actions are discussed in Chap. 15. For the present discussion, we can presume that the effect of a night interruption is dominated by red light, converting P_r into P_{fr}; red light can enhance the levels of P_{fr} in the leaf, P_{fr} being detrimental to the progress of the night reactions (Hendricks and Borthwick, 1965).

Is it possible to distinguish between separate reactions or partial processes going on during the dark period? If the critical length of the dark period is due to the necessity for the completion of a sequence of two or more steps, it may be helpful to apply modifying treatments at various times during the night to look for selective changes in effectiveness of the treatment. The first factor so examined was the interruption of the night by light. When light is applied at various times to measure its comparative effectiveness, it reaches its maximal effectiveness approximately at, or shortly before, the critical number of hours of darkness needed for an effective night. For example, for *Kalanchoe* Harder and Bode (1943) found a night interruption to be most inhibitory of flowering when applied between the fifth and seventh hours of the night (Fig. 12-14), and Salisbury and Bonner (1956) found that for *Xanthium* it is most inhibitory at the eighth hour or shortly before (Fig. 12-14). Each of these species requires about 9 h of darkness for the induction of flowering.

Early interpretations of night-interruption curves like those in Fig. 12-14 suggested that a timing reaction going on in the dark period required the phytochrome to be in the P_r form before the timer could be started and that it was linear, like the linear escape of sand in an hourglass or egg timer. According to this explanation, effectiveness of night interruptions in the middle of the night was greatest because the linear timer mechanism in the leaf was complete by then and the threshold night duration had been accomplished. If this were the correct interpretation, applications of far-red light at the onset of the dark period should convert the phytochrome to the P_r form promptly, starting the linear timer in the leaf sooner and enhancing the flowering response of a SDP; surprisingly,

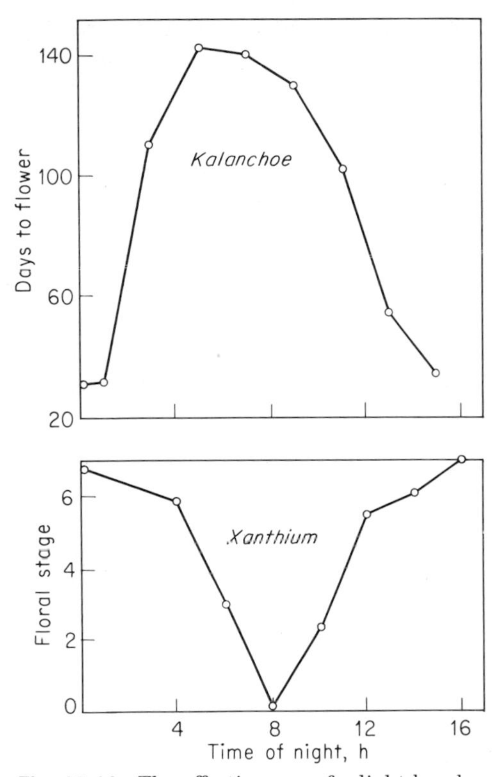

Fig. 12-14 The effectiveness of a light break applied during the photoperiodic night measured as the delay of the time to flower in *Kalanchoe* (*above*) or as the inhibition of flower stage of *Xanthium* (*below*). In each case a standard light dose was applied at various times during the 16-h night (Harder and Bode, 1943; Salisbury and Bonner, 1956).

though, a far-red treatment at the onset of the long-day period actually inhibits the flowering of *Pharbitis* (Nakayama, 1958).

An alternative explanation of the night-interruption curve arose from experiments of diurnal cycling. Bünning (1948) suggested that many plants experience diurnal cyclings between a photophile phase (one enhanced by light) and a skotophile phase (one that requires darkness and is suppressed by light). The applicability of his cycling concept to photoperiodic induction was greatly extended when Carr (1952) designed experiments to test the effects of night-interruption treatments over a period longer than 24 h; he reported the surprising result that if the dark period was very long and the night-interruption range could thus be extended, there would be periods of inhibition of flowering for interruptions in the skotophile, or "night," section of the diurnal cycle and promotions of flowering for interruptions during the photophile, or "day," section of the cycle. His approach has been utilized by several subsequent researchers, as illustrated in Fig. 12-15, which shows that the cycling between inhibitory and promotive effects is on an approximately 24-h cycle and that each inhibitory sector of the curve may be interpreted as a repetition of the night-interruption curve in Fig. 12-14. Plants grown in complete darkness can be entrained into this cyclic responsiveness by light of very low intensity (about 20 fc, according to Marushige and Marushige, 1966).

The 24-h cycling of photoperiodic responsiveness resembles the diurnal cycling of numerous other physiological activities, e.g., leaf movements, leaf acidity, stomatal opening and closing, and flower movements (Bünning, 1963). Several components of nucleic acid synthesis are likewise diurnal, as are mitotic frequencies in some instances (Evans, 1971).

A cardinal feature of the diurnal, or circadian

Fig. 12-15 The effects of a night-interruption treatment, when applied at different intervals during a long dark period, describe a circadian periodicity of effects on flowering, with periods of flower inhibition alternating with periods of slight flower promotion. Data for *Pharbitis* plants given one 72-h dark period (Takimoto and Hamner, 1965).

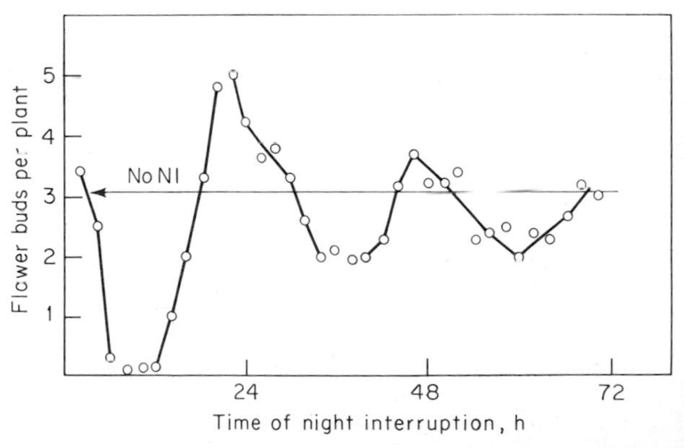

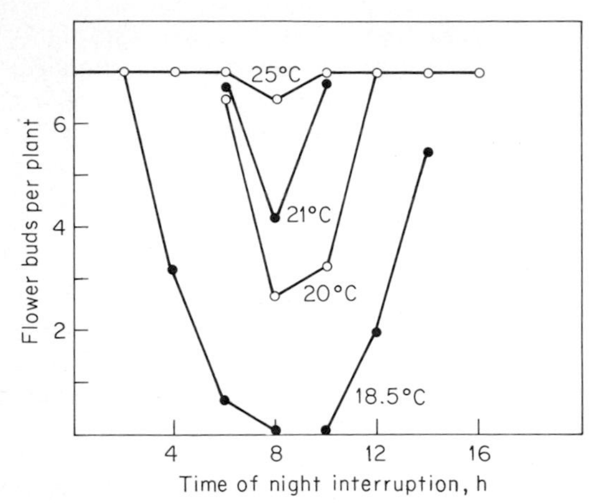

Fig. 12-16 Whereas temperatures can markedly alter the photoperiodic effectiveness of a long night, the applications of night interruptions (5 min red light at various times during the night) indicate that the timing reactions are relatively insensitive to temperature differences. Data for *Pharbitis* at four night temperatures (Takimoto and Hamer, 1964).

(approximating the day), rhythm is that it is temperature-compensated; i.e., the endogenous rhythm is not speeded or slowed by temperature differences to the extent that would be expected of an enzymic reaction (Bünning, 1963). The relatively small effect of temperature on the timing of the night-interruption effect is illustrated by the experiment in Fig. 12-16, where the maximal effectiveness of the night interruption remains at about the eighth hour of the dark period, over the temperature range from 25 to 18.5°C.

With the circadian rhythm as a basis, the night-interruption curves are explained as reflections of the plant's oscillation between a skotophile and a photophile phase rather than any involvement of a linear timer. Some species of photoperiodically sensitive plants show only a weak cycling of night-interruption responses (see, for example, Moore et al., 1967), but this may be a reflection of substrate deficiency accruing in the protracted darkness, for Cumming (1967) has shown that there is a marked improvement of the cycling in darkness when the plants are provided with various sugars.

The present explanation of the night-interruption curves follows the hypothesis of Könitz (1958), that phytochrome in the far-red absorbing form P_{fr} is inhibitory of the skotophile phase and essential for the photophile phase. Red light converts phytochrome into P_{fr}, which is inhibitory or promotive depending upon the oscillation phase.

In SDPs the regulating principle formed in the dark period has been referred to as the *flowering hormone,* but of course in LDPs the product of the dark period is an inhibitor of flowering; hence the term *regulating principle* is preferred.

Translocation. In addition to the photoperiodic processes leading to synthesis of a regulating principle, there is a requirement for its translocation out of the leaf. That the flower stimulus may move in the carbohydrate-translocation system was first suggested by Stout (1945), and subsequent studies have lent much support to the idea. The beneficial effects on flowering from sugars added to an induced leaf have been attributed to the provision of better translocation activity in the phloem (Carr, 1959).

The timing of translocation out of the leaf can be observed by the simple device of removing the induced leaves at various time intervals after completion of an inductive night (Imamura, 1953), but for actual calculations of velocities of translocation longer distances must be traversed. Some ingenious experiments on this question (Imamura and Takimoto, 1955) used two-branched *Pharbitis* plants with a single leaf on one branch and a single bud on another, timing the arrival of the flower stimulus in terms of the node in the remaining bud at which flowering first occurred. Using several variations of this experiment, they found velocities of translocation of about 2 to 4 mm/h. King et al. (1968) found much higher velocities for that species, and in their experiments the flower stimulus moved with very much the same velocity as the photosynthate from the same leaf. In the LDP *Lolium,* however, they found velocities of about 2 cm/h for the flower stimulus and 77 to 105 cm/h for photosynthate. Evans (1971) has suggested that in some cases the flower stimulus may move by a system separate from the phloem translocation of photosynthates.

Lam (1965) has found that the flower stimulus will move through the plant in different patterns according as mobilization centers are left on the plant or cut off. The magnitude of mobilization centers may account in part for the differences in translocation velocity found by different workers, but it seems unlikely to account for velocity differences between the flower stimulus and photosynthate.

The Photoinduced State

The alteration of the leaf by photoperiodic induction is an interesting subject of argument and a crux of the problem of photoperiodic regulation of flowering. Once a leaf has been photoperiodically induced, we must presume that it proceeds to export substances capable of altering the ontogenetic development of the apical meristem; the duration of this export might tell us something of the nature of the induction.

The ability of the leaf to export substances which evoke the change in the meristem to a floral one can be tested either by grafting the leaf to a noninduced stock or by forcing out buds at the axil of the leaf; in either case the leaf serves as a nearly exclusive source of substrates for a newly expanding meristem. Reversion to vegetative growth by plants which have been once photoperiodically induced does not necessarily imply that the leaves have lost the ability to export the flower-evoking stimulus, for the development of noninduced leaves below the developing apex can interfere with the stimulus reaching the apex (Moskov, 1941).

Graftage experiments have shown that the leaves of some species such as *Xanthium* remain induced for only a few days (Carr, 1959). In *Pharbitis,* a leaf which has had relatively few inductive photoperiods also produces flower stimulus for only a few days, but leaves that have had many inductive short days remain induced for many days (Imamura, 1953). *Perilla,* too, remains induced for a few days or for several months, depending upon the number of inductive short cycles it has been given (Zeevaart, 1957). Collectively, the experiments on grafting suggest that leaves may "remember" the photoperiodic induction treatment for a few days or for extended periods, perhaps for the life of the leaf with heavily inductive treatments.

The same question about the length of persistence of the induced state in leaves can be approached by measuring the ability of single induced leaves to provide flower stimulus to the bud at the opposite side of the same node (Cajlachjan and Butenko, 1957). With this method, it is found that *Perilla* leaves show a declining effectiveness in evoking flower development. As shown in Fig. 12-17, after inductive short days are terminated, the leaves evoke declining numbers of flowers; with leaves given 19 or 15 short photoperiods, there is no persistence of the induced state after the short days are stopped; with 22 or 25 inductive short days, the induced state is lost after 20 or 45 days, respectively. Zeevaart (1969) has stated that if the leaves had been given even more inductive cycles, the induced condition probably would have continued for the functional life of the leaf.

A reasonable conclusion from all these experiments seems to be that in some leaves the photoinduced state is very transient (as in *Xanthium*); in other species persistence is related to the extent of photoperiodic induction and ranges from essentially none after few photoinductive cycles to a prolonged persistence if enough photoinductive cycles are given (as in *Pharbitis* and *Perilla*).

For the concept of flowering induction as a change in the genetic program through the derepression of *flowering genes* (Zeevaart, 1964a), it would be appealing for the induced state to be irreversible; Zeevaart (1969) asserts without qualification that the induced state in leaves is irreversible. Nevertheless, both graftage experiments and bud-forcing experiments show that only under the conditions of very strong photoperiodic induction is the induced state truly persistent, and even then only in some species (see, for example, Zeevaart, 1958). The

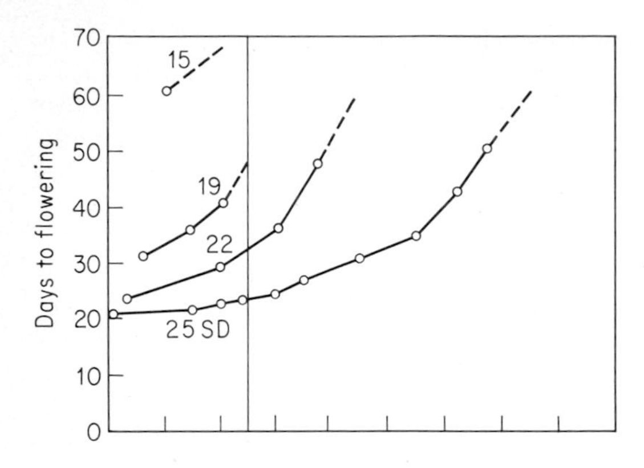

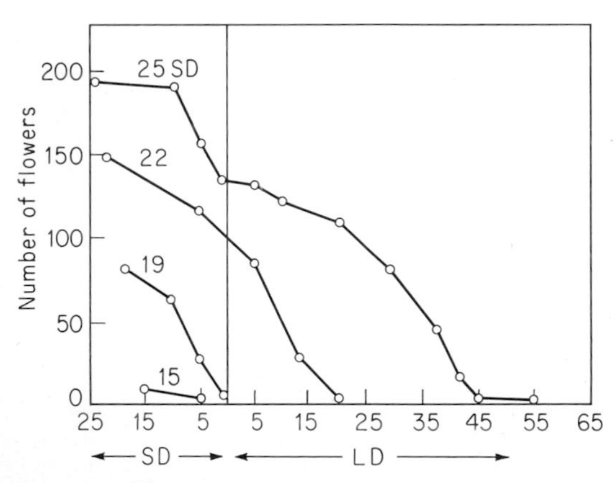

Fig. 12-17 After 15 to 25 short days, the leaves of *Perilla* may become decreasingly effective as suppliers of the flower stimulus. In paired-leaf experiments one leaf was induced; after various intervals of time the opposing leaf was removed, and the flowering of its axillary bud was used as an indication of the flower stimulus still coming from the induced leaf. As long as it remains on the plant, the opposite leaf prevents the short-day leaf from inducing its axillary bud. With time, the induced leaf becomes less effective as a flower stimulator, indicated by longer times to flowering (*above*) and by fewer flowers induced (*below*) (Lam and Leopold, 1961).

photoinduced state can be a transient condition in the leaf and in many instances is.

A remarkable example of the transient condition is found in the plants which can become sources of the flower stimulus indirectly. In experiments with *Xanthium,* Lona (1946) found that young leaves on plants induced not by photoperiodic treatment but by graftage to photoperiodically induced parts could act as donors of the flower stimulus when grafted onto yet another vegetative receptor plant. Zeevaart and Lang (1962) showed such indirect induction of *Bryophyllum* leaves, and Wellensiek (1966) found the same for *Silene armeria.* In these species, the flower stimulus may be propagated by leaves which were not photoperiodically induced but only indirectly brought into the induced state.

Another question is whether photoinduction involves the formation of stimulatory substances (as a flower hormone) or inhibitory substances. The evidence for the existence of stimulatory substances is clear in the quantitative experiments already described, but there is also evidence that inhibitors of flowering are involved in photoperiodism. Moskov (1936) noted that in chrysanthemum, a SDP, leaves which received only long days inhibited the flower initiation. While numerous photoperiodic species show such inhibitions by uninduced leaves, other species like *Xanthium* may not (Hamner and Bonner, 1938). Finding that removal of all leaves would permit flowering of the LDP *Hyoscyamus,* Lang and Melchers (1943) suggested that the photoperiodic induction specifically involves the removal of the inhibition by uninduced leaves. A similar situation has been reported for the SDP strawberry (Guttridge, 1959). It has even been possible to define the photoperiodic requirements for the formation of a flower inhibitor in *Xanthium* leaves (Lincoln et al., 1956) and *Salvia* leaves (Bhargava, 1963).

Wellensiek (1959) has identified the inhibitory state with the light period in SDPs and suggests that darkness nullifies the inhibitory state formed in light. In support of this idea is the fact that some SDPs flower more readily under weak light; and in fact at lower intensities the threshold photoperiod for flowering of SDPs may be considerably lengthened (Krumweide, 1960) or even removed (de Zeeuw, 1953). Furthermore, low temperatures applied during the day period can promote flowering of *Perilla* or *Pharbitis* (Wellensiek, 1959; Ogawa, 1960) or even induce flowering of *Xanthium* on long days (de Zeeuw,

1957). Perhaps low temperatures suppress formation of flowering inhibitors during the light period. How these features might be applied to LDPs is an interesting problem yet to be developed experimentally.

The photoperiodic control of flowering has provided the most important experimental tool we have to develop our understanding of the physiology of flowering. Cyclic experiences of light and darkness can bring about the reproductive state, and consecutive processes which require either light or darkness must proceed separately to accomplish this change.

Exploitation of photoperiodism in the analysis of flowering has developed principally along two lines: analysis of the light-sensitive changes which occur in the night period, especially with respect to the pigment involved, and analysis of partial processes through kinetic experiments with physical or chemical treatments selectively to alter some parts of the sequence of events in the photoperiodic cycle. While the pigment involved in the dark-period regulation has been extracted and the movement of the flowering stimulus can be followed, the nature of neither the intermediate steps nor the stimulus has been established.

VERNALIZATION AND FLOWERING

Temperatures provide another salient climatic cue for plants, a ready signal of seasonal changes. Many species are induced or promoted to flower by low temperatures, especially many biennials and perennials. Less common are species caused to flower by high temperatures, e.g., annuals such as spinach, rice, China aster, and *Rudbeckia*.

The promotive effects of low temperatures on flowering are termed *vernalization,* a word coined by Lysenko (1928). For this discussion the term will include the actual induction of flowering in species which require low temperatures, e.g., cabbage, celery, and beet, as well as the hastening of flowering in species which are only quantitatively promoted by low temperatures, e.g., the winter grains, lettuce, and radish. In each of these, the flowering responses to temperature can be considered as an ecological coordination of flowering times, providing the associated advantages discussed earlier.

To illustrate the quantitative and sometimes qualitative nature of the vernalization response, some data on vernalization of four species of *Lolium* are given in Fig. 12-18. The flat curve for *L. temulentum* shows a lack of any apparent vernalization response, and the curves for *L. multiflorum* and its hybrid with *L. perenne* show quantitative responses to vernalization, i.e., shortening of the time required for the appearance of flowers. The hyperbolic curve for *L. perenne* illustrates an actual requirement for low temperature and the quantitative nature of the effects of further extensions of low-temperature experience beyond the essential minimum for flowering.

The classical case of vernalization is the hastening of flowering in winter grains; the quantitative nature of this response is illustrated in Fig. 12-19. These curves indicate that the low-

Fig. 12-18 The relative responses of four genetic strains of *Lolium* to vernalization. *L. temulentum* shows no response; *L. multiflorum* and its hybrid with *L. perenne* show quantitative responses; and *L. perenne* shows a complete requirement for vernalization. Growth conditions: 23°C during the day; 17°C at night; long photoperiods (Evans, 1960).

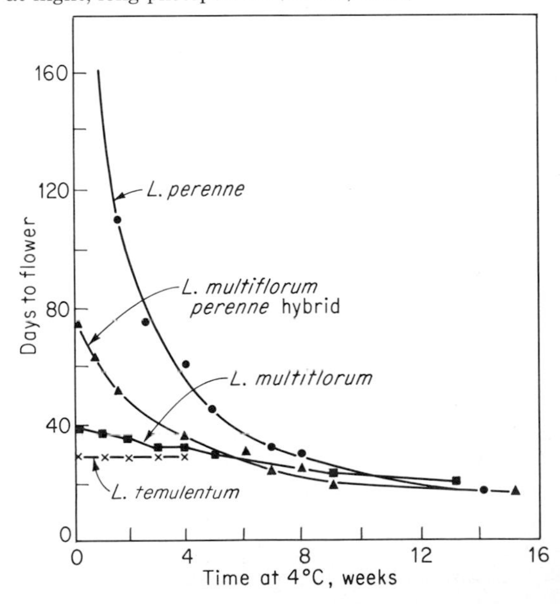

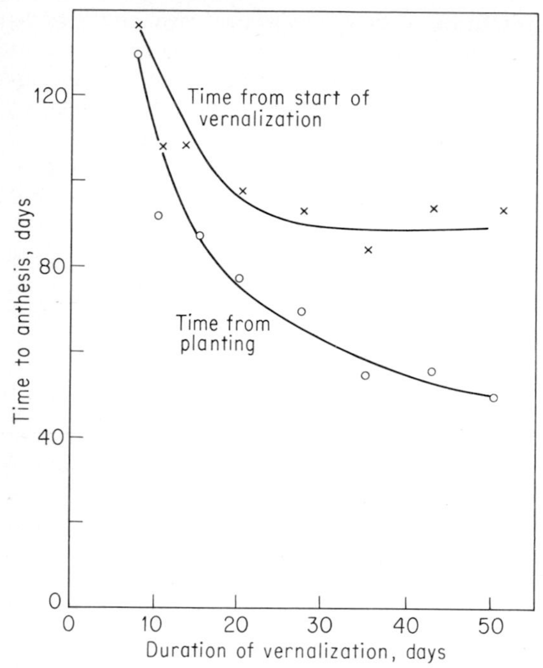

Fig. 12-19 The quantitative response of Petkus winter rye to vernalization. The time to anthesis is decreased by vernalization whether one measures from the start of the cold treatment (*above*) or from the time of planting out after the cold treatment (*below*) (Purvis and Gregory, 1937).

temperature experience really does advance the time of flowering, whether it is plotted as time from the start of vernalization or as time from planting out after completion of the vernalization treatment.

Kinetics of Vernalization

Low-temperature effects are obtained in some species when the moistened seed is chilled; in other species they are obtained only when growing plants are chilled. Some species respond readily to chilling at either stage.

Vernalization of whole plants was discovered simultaneously for two horticultural crops, cabbage and celery (Boswell, 1929; Miller, 1929; Thompson, 1929). All these reports noted that as plants become older they respond more readily to a low-temperature stimulation. This is frequently true for whole-plant vernalization; and some data shown in Fig. 11-1 are representative of such a changing sensitivity. *Lunaria* (Fig. 11-1) has a juvenile period during which low temperatures are without effect. Other species such as beet (Chroboczek, 1934) become increasingly sensitive with age but lack a completely insensitive juvenile period. Evans (1960) describes the temperature response as an exponential function, the slope of which shifts with increasing plant age.

Instead of increasing vernalization responsiveness with age, in some species the best response is obtained by chilling the seeds. Perhaps the vernalization of seed was first noted by Gassner (1918), who found that by moistening seeds of winter grains and keeping them in the cold before planting out he could make them flower as spring varieties. Lysenko (1928) described vernalization of seeds as a general phenomenon, and the entanglement of this seeming alteration of genetic characteristics with political ideology has had a sad and curious history (Huxley, 1949).

In species where both the seed and the whole plant are capable of responding to vernalization these two responsive periods may be separated by a stage of markedly less sensitivity. A striking example of these two stages in *Arabidopsis thaliana* has been described by Napp-Zinn (1957), and some of his data are plotted in Fig. 12-20. If vernalization is started within the first day or two after moistening the seed, flowering is induced very early. If the plant has several days in which to start growth before the vernalization is started, flowering is less readily induced. A gradually increasing responsiveness to vernalization then becomes evident in plants of increasing age.

The possibility that rapid growth may be antagonistic to the flowering response has occurred to many investigators, and even in the earliest reports on low-temperature induction of flowering, attempts were made to establish whether the low-temperature effect might be simply a suppression of growth (Thompson, 1929). Cholodny (1936) suggested a general explanation of vernalization based on the suppression of vegetative growth. Numerous types of evidence deny that the growth check is the essential part of the vernalization (Thompson, 1929; Gott et al., 1955), but in some special cases there is reason to believe that the growth check is specifically involved in the response; e.g., in tobacco Steinberg (1952) achieved the same increases in earliness with low-temperature treatments and with transplanting, pruning, or other treatments which would check the growth of the plant. Tomato responses to low temperature may be a similar case (Lewis, 1953; cf. Fig. 12-4), but most examples of vernalization cannot be explained in this way.

Many species which respond to vernalization are also sensitive to photoperiodic stimulations of flowering. The most common combination is that of low temperature and LDP, e.g., in many Cruciferae and other biennials (beet and the winter grains). But some species may be LDPs and induced by high temperature (spinach), SDPs and induced by low temperature (chrysanthemum), or SDPs and induced by high temperature (China aster). The interactions of these two environmental cues are either supplementary or complementary. Thus low temperatures may quantitatively displace the critical photoperiods without altering the critical day length needed (Cathey, 1957); or low temperatures may entirely replace a photoperiod requirement (Koller and Highkin, 1960).

The actual temperature experienced greatly alters the effectiveness of any vernalization treatment, and for most species temperatures in the range of 0 to 5°C are optimal for vernalization. The temperature range for Petkus winter rye is plotted in Fig. 12-21, illustrating that in this species not only is the temperature range of 0 to 5°C effective but that there is a real effectiveness of temperatures as low as −5°C. One is curious to know what reactions in the plant might be taking place better at subfreezing temperatures than at 15°C, for example. In other species the vernalization effectiveness is rather sharply limited at 0°C, e.g., the Japanese radish (Kimura, 1961).

After the completion of vernalization, warm temperatures can have a depressing effect on flowering earliness. Purvis (1948) noted that with later planting dates of Petkus rye there was a requirement for a somewhat greater vernalization time to achieve the same degree of earliness, which she attributed to the warmer temperatures of the soil at the time of planting.

Devernalization

The fact that high temperatures after vernalization can nullify the effect was perhaps first noted in celery by Thompson (1929) and by various horticulturists working with the low-temperature responses of biennials (Miller, 1929; Chroboczek, 1934). The recognition of devernalization as a physiologically significant event, however, was made by researchers at the Imperial College, London, working with the vernalization of winter grains (Purvis and Gregory, 1945). When vernalized grains were exposed to 35°C for even 1 day, the vernalization effect was erased. The plants were not damaged since devernalized plants could be revernalized effectively by repetition of the cold treatment. The same reversible devernalization could be obtained with anaerobic conditions (Gregory and Purvis, 1937).

Fig. 12-20 The effectiveness of a standard vernalization treatment of *Arabidopsis thaliana* changes with stage of growth. Vernalization applied as a seed treatment or at various ages during seedling growth reveals a period of marked responsiveness at the seed stage and then low sensitivity until the seedling has developed for a few weeks. The flowering response is plotted as the days to flower after a 38-day period at 2°C. Unvernalized plants did not flower (Napp-Zinn, 1960).

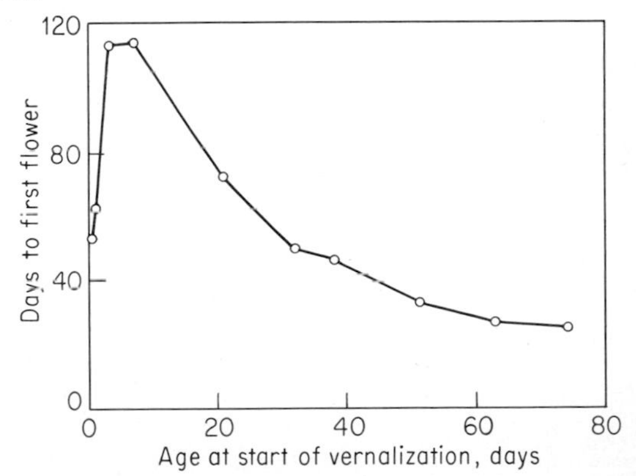

This finding implies that an unstable product formed during the cold experience is metabolized away if high temperature or anaerobiosis is experienced before the product establishes its effect. This hypothetical product of vernalization is not known.

A horticultural application of the devernalization effect is the control of flowering of onions. This crop is properly a biennial, and the young bulb which is formed the first year can be induced to flower by vernalization during the winter (or in cold storage). For the production of large bulbs in the second year, flowering is undesirable. After it was learned that high temperatures prevent flowering (Heath, 1943), this response was developed as a commercial devernalization treatment for onion sets after storage at low temperatures (Lachman and Upham, 1954).

Devernalization in chrysanthemums can be achieved with low light intensities or by decapitation and forcing out basal buds (Schwabe, 1954). High temperatures are ineffective in this species.

Fig. 12-21 A comparison of the relative effectiveness of various temperatures in vernalizing seed of Petkus winter rye. Curve at right is for 42 days vernalization, dissected 133 days later (Purvis, 1948); curve at left is for 45 days vernalization, dissected 49 days later (Hänsel, 1953). Optimum temperatures seem to be between +7 and −3°C. Flower "score": meristem as a single ridge = 10; meristem as a double ridge = 20; stamens differentiated = 30; spike emergence = 40; anthesis = 50. Subsequent days each = 1.

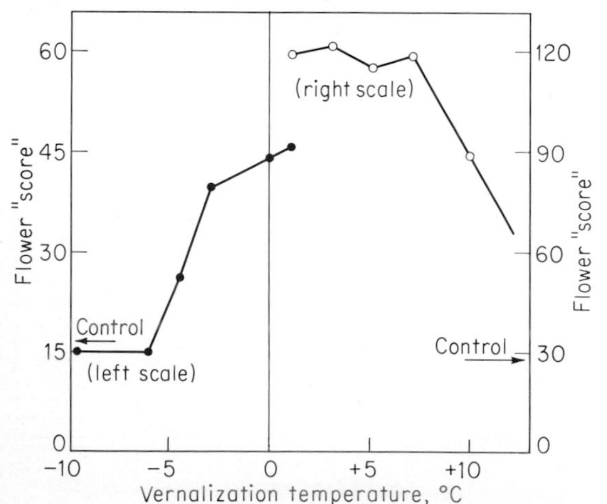

The Vernalization Stimulus

Unlike the photoperiodic stimulus to flowering, the vernalization stimulus is perceived ordinarily by the apical meristem (Chroboczek, 1934). In a series of clear-cut experiments, Gregory and de Ropp (1938) found that the excised embryo of winter rye can be vernalized and that even a fragment of the embryo growing point can still be effectively vernalized in tissue culture.

Physiologists were satisfied that the apical meristem is in fact the locus of vernalization until Wellensiek (1961) reported that the young expanding leaves of *Lunaria* can be vernalized and further established (1962) through some ingenious experiments that leaves can receive the vernalization stimulus provided they are experiencing active cell division. Thus the more specific generalization seems to be that dividing cells are the site of vernalization, a requirement the meristems are well qualified to meet.

Once a meristem has been vernalized, all growth that develops from that source is vernalized (Schwabe, 1954). This suggests a metastable modification of some self-replicating component in the meristematic cells. There is no experimental evidence in the literature of a declining effectiveness of vernalized meristems, though in vernalizable perennial species, where growth ceases each year following reproduction, reversion can occur through the regrowth of new buds which are nonvernalized (Schwabe, 1954).

Unlike the photoperiod stimulus, the product of vernalization seems to be quite immobile. When active meristems are vernalized, inactive meristems on the same plant may remain unvernalized. Graftage experiments have not shown a translocation of the product of vernalization.

Attempts to stimulate or modify vernalization with various growth substances and metabolic modifiers have not been very helpful, e.g., the apparent stimulations of vernalization with auxins by Cholodny (1936) and others. With the discovery of gibberellins, a new hope sprang up that a substance might be involved. Several laboratories reported that at least some of the vernalization effects could be obtained with gibberellin applications to seeds (Lona and Bocchi, 1956; Purvis, 1960), and in certain biennial species a complete replacement of the vernalization effect could be obtained by gibberellin application to the plants (Lang, 1957). This complexity of responses suggests two things: (1) at least some of the vernalization effects may be related to the natural gibberellin substances in the plants, and (2) not all the vernalization responses are a result of the same physiological mechanism, since a substance which could replace the cold for one species would ostensibly be expected to replace it for all species.

The salient features of the vernalization effects on flowering seem to be that dividing cells can perceive a low-temperature stimulus which alters the morphological expression of growth for a protracted period of time. The ability of tissues to "remember" a temperature experience and to remain altered in a metastable manner is a prime example of one of the most remarkable problems in developmental physiology. That the change in metastable state should be provoked by temperatures near or even below freezing does not seem to imply that the perception of the temperature experience involves a synthesis of a new substantive material but rather that some physical change occurring during the cold experience persists in subsequent cell lineages.

SUBSTANCES REGULATING FLOWERING

The study of developmental processes should ultimately provide an explanation of physiological events on a biochemical basis. The studies of earliness, photoperiodism, and vernalization all hold some promise of a more complete biochemical understanding of the flowering processes. The existence of natural growth substances which regulate the quantity and quality of growth offers interesting possibilities of explaining (at least partially) such developmental events as flowering.

Growth substances have already been found to play important roles in the endogenous control of flowering. Historically, each of the major explanations of flowering has been founded on the particular aspect of plant physiology under most rapid development at the time. Thus, the first attempt at a general explanation of flowering was on the basis of the carbohydrate-nitrogen ratio (Kraus and Kraybill, 1918), at a time when nutrition was the main concern of plant physiologists. Later as auxins came to the forefront, it was proposed that they were the controlling entities in flowering processes (Liverman, 1955). With the establishment of the gibberellins as natural growth regulators in plants, they too were invoked as endogenous controls of flowering —and with the most substantial experimental basis (Lang, 1957). As plant physiologists focused on nucleic acids, they in turn have been used to explain flowering (Zeevaart, 1962). This historical mirroring of the special interests of physiologists does not detract from the relevance of the resulting information to flowering, but it does account for the spurts of progress in localized areas of physiological thinking.

Physiological information on the effects of growth substances on flowering has passed through an empirical period which showed that in some conditions each of the known hormones can regulate flowering in some plant species. Efforts have been made to link the effects of auxin, gibberellin, ethylene, and abscisic acid on flowering with endogenous changes with photoperiod and vernalization; while there are some promising correlations of changes in these hormone levels with flower-initiating events, generally the results are not encouraging. For example, while gibberellins play a role in the flowering of many LDPs, it has not been established that the long-day effects can be generally accounted for on the basis of increased gibberellin levels in plants. A more realistic position might be that each of the known plant hormones can regulate flowering under some circumstances (and in some species); perhaps the control of flowering by photoperiods or vernalization may occur through multiple interactions between these and possibly other regulatory principles.

Gibberellin

Finding that gibberellin applications can cause rosette plants to elongate in a manner suggestive of bolting led quickly to investigations of the possibility that gibberellins could cause flowering in rosette plants. In fact, it is widely effective (Lang, 1957). The effects are dramatic (Fig. 12-22) and widespread among many species (Wittwer and Bukovac, 1957).

An extensive array of cold-requiring species is brought to flowering by gibberellin applications, including the biennial *Hyoscyamus,* the biennial cruciferous species (cabbage, turnip), and other cold-requiring species (beet, carrot, endive, parsley). Though not all cold-requiring plants can be induced to flower with gibberellin, the list is impressive.

The relation to photoperiod requirements is more complicated. Numerous LDPs have been induced to flower with gibberellins, though it may be more accurate to say that gibberellin may increase the tendency of LDPs to flower instead of entirely replacing the long-day requirement (Chouard, 1960). Among the LDPs gibberellin has been shown to cause or promote flowering in spinach, dill, lettuce, stock, radish, *Rudbeckia, Hyoscyamus,* and others. Many species of LDPs which can be stimulated to flower by GA have a rosette habit of vegetative growth. For a time it was thought that the stimulations of flowering were closely tied to stimulations of stem elongation (Burk and Tso, 1958), but some flower stimulation is quite separate from the growth stimulation (see, for example, Cleland and Briggs, 1969). Gibberellin applications will replace the long-day requirement of LSDPs such as *Bryophyllum* and *Coreopsis* (Harder and Bunsöw, 1956).

While most gibberellin stimulations of flowering have been reported for LDP and vernalizable species, some important stimulation effects have also been observed for SDPs. Application of gibberellin can induce flowering in the strict SDP *Pharbitis* at near-threshold photoperiods (Ogawa and Imamura, 1958) or in *Xanthium* plants treated also with an extract from flowering plants (Carr, 1967). But in some SDP species, applications of gibberellin can bring about flowering over a very wide range of photoperiods (Lona, 1962). In many SDPs gibberellin only inhibits flowering (see, for example, Nitsch, 1968).

If endogenous gibberellins do participate in the control of flowering, one would hope to find increases in the gibberellin content of plants upon induction with photoperiods or low temperatures. Efforts to find such endogenous changes have met with mixed results. Lang (1960) found very small increases in GA after low-temperature induction of biennial *Hyoscyamus,* and very small increases were found in winter wheat upon vernalization (Suge and Osada, 1966). Photoperiodic induction of LDPs has provided only slightly more convincing increases in gibberellin with induction of spinach (Radley, 1963), *Ribes nigrum,* and strawberry (Wareing and El-Antably, 1970) and *Nicotiana* (Cajlachjan and Lozhnikova, 1964). Stoddart (1966) has suggested that gibberellin is metabolized more rapidly and to different end products under short days than long; Wareing and El-Antably (1970) have suggested that the lowering of gibberellin content in several species under short photoperiods may regulate induction through a combined action with increasing abscisic acid contents.

Evans (1971) concludes that the presence of gibberellins is a probable prerequisite for the evocation of flowering at the meristem in all plants and that the gibberellin level is often limiting to flowering in LDPs. He suggests, too, that vernalization may induce a greater capacity for gibberellin synthesis but an actual increase may not be evident until later in the progress of the plant's growth. Gibberellin applications that do not stimulate flowering may be interpreted as cases where gibberellin content is not limiting to flower evocation.

Auxin

This growth substance was reported to induce flowering in the pineapple by Clark and Kerns (1942); while this response has proved very useful in the production of pineapple, it is probable

Fig. 12-22 *Silene armeria,* which ordinarily requires long days for flowering, is induced to flower in short-day conditions (9 h) with applications of gibberellin. From left to right, plants were given 0, 2, 5, 10, 20, or 50 μg per plant each day (Lang, 1957).

that the regulation of flowering is due not to auxin but to an effect of the ethylene produced for about a week following the auxin treatment (Burg and Burg, 1966c).

Auxin does have an effect on the photoperiodic induction processes in leaves. This was first described as an inhibition of flowering in SDPs when auxin was applied to leaves (Thurlow and Bonner, 1947). Salisbury (1955) suspected that the auxin was inhibiting some aspect of the induction in leaves, and he carried out some careful timing experiments in which auxin was applied to the leaf of *Xanthium* at various times during the inductive dark period and the subsequent light period. These experiments show that the inhibition is much more pronounced for application during the dark period (Fig. 12-23), and he deduced that auxin interferes with dark-period reactions (Salisbury and Bonner, 1956).

If auxins play an important role in the photoperiodic induction of leaves, one would expect marked changes in auxin content with photoperiodic treatments. Attempts to find such changes have given little or no support to the idea that auxins play such a role (Cooke, 1954).

Other Chemical Regulators

Each of the plant hormones can alter flowering in at least some situations. Besides the striking effects of gibberellin in photoperiod and vernalization processes and the less powerful effects of auxin, cytokinins can promote flowering of several SDPs (Ogawa, 1961; Nitsch, 1968) or actually induce flowering of some SDPs (Maheshwari and Venkataraman, 1966) or some LDPs (Michniewicz and Kamienska, 1965). Beever and Woolhouse (1973) report that photoperiodic induction of the SDP *Perilla* is associated with a striking increase in cytokinin levels in the xylem sap coming from the roots. Applications of cytokinin plus GA is especially effective in some SDPs (Pharis, 1972). Applications of abscisic acid, too, can promote or induce flowering in some SDPs (El-Antably and Wareing, 1966) and can inhibit flowering in at least one LDP (Evans, 1966) but this inhibitor does not show changes with photoperiod treatment other than diurnal rises each day and declines at night (Zeevaart, 1974). Ethylene has been known to be capable of stimulating flowering in pineapple for many years (Rodriguez, 1932) and has recently been found to promote flowering in a few other species (Nitsch, 1968).

Since graftage work indicated that there might be a universal flowering hormone among higher plants, many attempts have been made to extract such a material out of induced leaves and other plant parts. There were occasional early reports of successes, but only rarely have successful experiments been repeatable. The first convincing success with plant extracts was a report by Lang et al. (1957) that extracts of fruits of *Echinocystis* induced flowering in biennial *Hyoscyamus,* undoubtedly because of the gibberellin content of the extracts. Since that time, Harada and Nitsch (1961) have obtained extracts from cold-requiring plants and LDPs which induced flowering in a manner suggestive of gibberellin, though the material in their extract did not have all the properties of gibberellin itself. Tomita (1959) has also obtained an extract which hastens flowering of small grains in a manner suggestive of a gibberellin.

Fig. 12-23 The inhibitory effect of auxin on photoperiodic induction of flowering in an SDP is most pronounced when the auxin is applied to the leaf during the inductive night. *Xanthium* plants given one long night and mM NAA applied at intervals during the night or subsequent day; flower stage determined after 14 days (Salisbury, and Bonner, 1960).

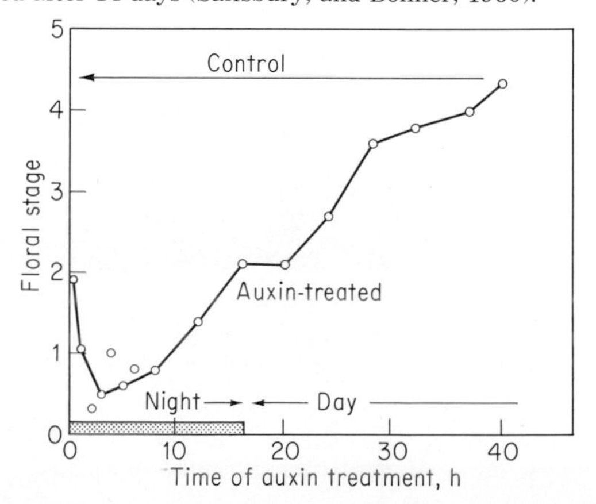

In several instances, extracts of gibberellin-like substances from SDPs are effective in inducing flowering in SDPs. Some remarkable results by Bünsow et al. (1958) and Ogawa (1962) have involved the induction of flowering in *Pharbitis* and *Bryophyllum,* and an extract of *Xanthium* (Lincoln et al., 1961) has shown some effectiveness in inducing flowering in that same species. The *Xanthium* extract is markedly more effective in stimulating flowering of some species when applied along with exogenous gibberellin (Carr, 1967; Hodson and Hamner, 1970).

The possibility that these several extracts may yield a widely effective flowering hormone is somewhat dimmed by the reports of effective stimulations of flowering by diverse chemicals; e.g., in tissue cultures, sucrose can substitute for the photoperiod requirement of the SDP *Pharbitis* (Takimoto, 1960) or of the LDP *Sinapis alba* (Deltour, 1970). Other compounds which might be considered as endogenous and which have been found to stimulate flowering include vitamin E (Michniewicz and Kamienska, 1965), and some mixtures of nucleic acids (Cajlachjan et al., 1961), and uridine (Tomita, 1964) or uracil (Cummins, 1969). One should deduce from this that the successful extraction of some material from plants which would stimulate flowering in others does not necessarily indicate that one has separated the flowering hormone or even a factor which regulates flowering in vivo.

Numerous synthetic compounds can inhibit flowering, including especially the growth retardants (Baldev and Lang, 1965). It seems likely that growth-retardant effects may be a consequence of the depression of gibberellin synthesis, since gibberellin applications can ordinarily restore flowering (Zeevaart, 1966), although there are some exceptions (Cleland and Briggs, 1969). The photoperiodic reactions in the leaf appear to be unaltered by the growth retardants (Evans, 1964b).

GENETIC MECHANISMS OF FLOWERING

The idea of the programming of flowering through repression and derepression of genetic information was approached experimentally by Salisbury and Bonner (1960) and Heslop-Harrison (1960). In each case, inhibitors of nucleic acid synthesis were utilized in an effort to alter flower induction. Salisbury and Bonner (1960) tested the relative effectiveness of 5-fluorouracil (5-FU) applications at various times during the photoperiodic induction and found that flowering is most inhibited by applications at the beginning of the inductive dark period, as shown Fig. 12-24. They suggest that nucleic acid synthesis is critical both to the induction processes in the leaves and to the conversion of the apical meristem to flowering. Heslop-Harrison (1960) applied 2-thiouracil, which did not cause the growth inhibitions characteristic of 5-FU, and yet effectively inhibited the flowering of hemp. From his experiments he suggested that nucleic acid synthesis is involved in the conversion of the apex and not in the induction of leaves.

The flowering processes can be interfered with by inhibitors of nucleic acid synthesis or inhibitors of protein synthesis (Hess, 1961; Collins et al., 1963), and these inhibitors can act against flowering in SDPs or LDPs (Zeevaart, 1964a; Evans, 1964a), against flowering in vernalizable species (Suge and Yamada, 1965), or in the autonomous flowering of peas (Mitra and Sen, 1966). Later work has been in general agreement that the locus of nucleic acid involvement in flowering is at the apical bud (Zeevaart, 1964). Histochemical examination of the terminal meristems during photoperiodic induction has revealed increases in the intensity of RNA stains and protein-staining activity (Gifford, 1964) and ultrastructural changes occurring as soon as 3 h after the end of an inductive night (Gifford and Stewart, 1965). Attempts to find altered nucleic acid profiles after induction have led to mixed results. Some small changes have been reported in the nucleic acid base ratios after induction (Hess, 1961; Yoshida et al., 1967), whereas in others only quantitative changes have been reported (Ross, 1962; Cherry and van Huystee, 1965). Evans (1971) notes that the experimental difficulties in detecting changes in the apical meristems are enormous, since the apex itself is probably less than 1 percent of the terminal bud usually employed for such experi-

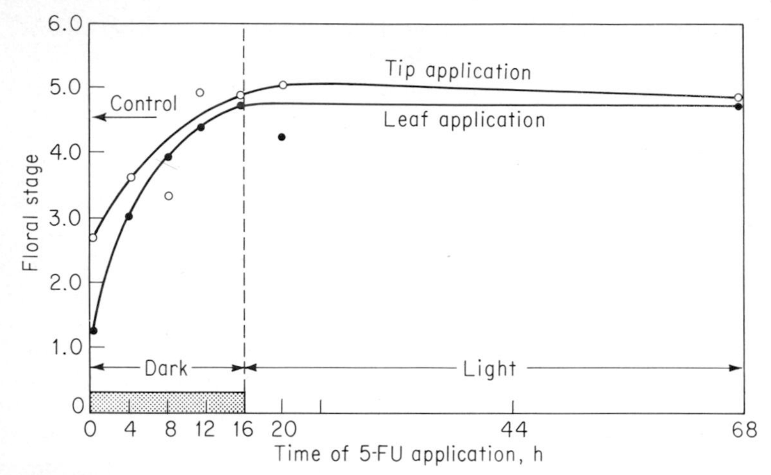

Fig. 12-24 5-Fluorouracil is most effective in inhibiting the flowering of *Xanthium* when applied at the beginning of the inductive dark period. Similar inhibitions are obtained whether the inhibitor is applied to the leaf or to the growing point at 2 mM (Salisbury and Bonner, 1960).

ments, and that the associated leaf primordia represent a very large background, making it very difficult to detect changes in extractable nucleic acids.

Some evidence has indicated that after vernalization new species of proteins may be formed in the embryos of small grains (Nitsan, 1962; Teraoka, 1967), and in a few instances some nucleotides enhance flowering when applied to plants or tissue cultures (Tomita, 1961; Suge and Yamada, 1963).

It seems altogether reasonable to presume that the onset of flowering involves an alteration in the genetic instructions being translated into RNA and then protein synthesis. It has been suggested that this involves the synthesis of specific mRNA (Galun et al, 1964), but regulation could be achieved through other RNA components, too (Cherry, 1970). Umemura and Oota (1965) have added an interesting further detail: the LDP *Lemna gibba* quickly reverts to vegetative growth after the long photoperiods are stopped, and the inhibition of flowering with nucleic acid inhibitors is no longer effective when they are added after the end of the inductive photoperiods. Umemura and Oota suggest that the nucleic acids which bring about flowering must be quite short-lived.

An interesting suggestion has been made by Bernier et al. (1967) that the conversion to the reproductive state may involve coordination of several rhythmic processes in the meristem. In *Sinapis alba* they find daily peaks in mitoses in the middle of the night. RNA and protein synthesis may show cyclic peaks in the morning, DNA synthesis at night (Evans, 1971); the diurnal changes in effectiveness of various inhibitors of flowering may well be related to these cyclic events in the apex. Jacqmard (1968) found that GA applications strikingly increase mitotic activity in the apical bud of *Rudbeckia.*

As hormone systems frequently involve steroids, the possibility that flower induction involves the formation of a steroid in the induced leaf has been tested by Bonner et al. (1963). A complex fat (SKF 7997) known to inhibit steroid synthesis in animal systems applied to leaves of *Xanthium* during short-day induction yielded marked inhibitions of flowering. This effect was specifically located in the leaf, applications to the stem tip being without apparent effect. The inhibitor was most effective when applied just before the inductive long night, and Bonner et al. interpreted this as indicating an inhibition of dark reactions. The inhibition was not reversed by the application of steroids, but the possibility of a steroid hormone is nevertheless attractive. This same inhibitor has been found to effectively inhibit the leaf-induction processes in the LDP *Lolium temulentum* (Evans, 1964a).

At present four major facts about the nature of the system regulating flowering are at hand:

1 Graftage experiments and translocation experiments indicate that a mobile stimulus is supplied by photoperiodically induced leaves.

2 Vernalization experiments indicate that a nonmobile stimulus occurs in the meristems.

3 The ability of each of the known plant hormones to stimulate or to inhibit flowering indicates the possibilities for control of flowering in at least some instances through the interactions of these natural regulators.

4 There is an apparent requirement for nucleic acid synthesis in the meristem for evocation of the flowering response.

Numerous lines of evidence suggest that multiple factors may be involved in the control of flowering in the plant: the existence of flower inhibitors as well as flower promoters, the requirements of some plants for multiple photoperiodic experiences and others for sequential photoperiod and temperature experiences, and the ability of several natural or synthetic growth regulators to stimulate flowering in certain instances. For such a central and important developmental function as flowering, higher plants may well have evolved multiple types of control systems through which flowering can be achieved.

DEVELOPMENT OF THE FLOWER

Sexuality

A physiological device which provides for outcrossing—in addition to regulations of flower timing such as earliness, photoperiodism, and vernalization—is the separation of female and male parts into different flowers on the same plant (the monoecious habit) or on separate plants (the dioecious habit). Genetic control of the dioecious habit is achieved by an X and a Y chromosome in a manner similar to that in man, XY combinations being male and XX combinations female (Westergaard, 1948). The monoecious habit, at least in some plants, is an expression of a modified gene on the X chromosome (Janick and Stevenson, 1955).

Maleness and femaleness can usually be altered by environmental variables such as temperature, photoperiod, and nutrition or by the application of growth substances. Heslop-Harrison (1963) points out that in many species with unisexual flowers, each flower has the rudiments of both sexes in the meristem, and unisexuality occurs as a result of the suppression of some of these parts; in some species, e.g., hemp, corn, the Cucurbitaceae, and the unisexual trees, the flower meristem does not contain both male and female rudiments, but the sexual parts are developed during the growth of the flower bud. Galun et al. (1962) have been able to steer the development of excised floral buds of cucumbers into male or female development through the application of gibberellin or auxin, respectively.

Among monoecious plants, there is usually a characteristic distribution of the male and female flowers over the plant. To cite two examples, in the squash plant there is a period in which the young plant produces only male flowers, followed by hermaphroditic flowers, female flowers, and finally parthenocarpic female flowers (Fig. 12-25). In the Scotch pine the sequence is almost reversed; after completion of the juvenile period of growth, the young tree first produces female flowers, and then as the tree ages, male cones are borne in the lower branches, especially in suppressed laterals (Wareing, 1958b).

Monoecious plants can respond to numerous environmental variables with altered sex expression; the nutrition, photoperiod, and temperature experiences are most common factors. In some species, high nitrogen levels encourage maleness (cucumber), in others femaleness (*Begonia*); likewise long photoperiods encourage maleness in some species (especially SDPs) and femaleness in others (especially LDPs) (Heslop-Harrison, 1957b). Low temperatures, especially low night temperatures, encourage femaleness

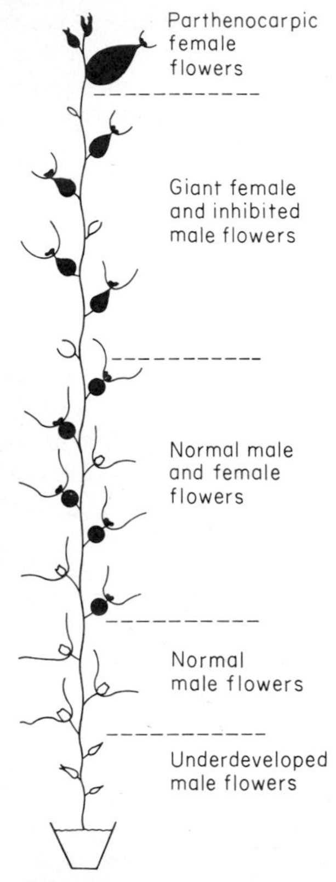

Fig. 12-25 In the acorn squash there is a sequence of development of flowers, with the first flowers being underdeveloped male flowers, followed by male flowers, and then increasingly female flowers, until parthenocarpic female flowers are produced (Nitsch et al., 1952).

in numerous species such as spinach, hemp, and cucurbits, as illustrated in Fig. 12-26.

There is a wide range of hormonal effects on sexuality. Auxin was the first hormone found to have regulatory effects on sex; Laibach and Kribben (1950) reported the stimulation of female flower production by auxin application to gherkin. Heslop-Harrison (1956) obtained female flowers on genetically male plants of hemp following auxin application. Gibberellin, first found to stimulate male flower production in cucumber (Galun, 1959), can induce male flower formation on genetically female plants (Mitchell and Wittwer, 1962), but in at least some species gibberellin stimulates femaleness (Bose and Nitsch, 1970). Cytokinins can induce the formation of female flower parts on genetically male grape (Negi and Olmo, 1966). Ethylene can stimulate femaleness in cucumber plants in a manner parallel to the effect of auxin (Robinson et al., 1969). Rudich et al. (1972) report that female flower buds of cucumber produce more ethylene than bisexual buds do.

Like the regulation of flowering itself, the control of sexuality may well be achieved in the plant in part by interactions of several endogenous growth hormones. Atsmon et al. (1968) measured the gibberellin content of strains of cucumbers which were genetically female and found these plants to have lower gibberellin contents than plants which can develop male flowers. Bilderbach (1972) has suggested that cytokinins may be essential for the development of carpels in *Aquilegia* based on flower tissue-culture work. The intensity of the photoperiodic induction of flowering can also serve to regulate the sexuality of the flowers developed; e.g., *Xanthium* plants given minimum short-day induction produce only male flowers, whereas with several

Fig. 12-26 The formation of female flowers in acorn squash plants is markedly enhanced by lower night temperatures. These data, for plants on an 8-h photoperiod and 26°C day temperature, show that lowering the night temperature to 20°C results in the formation of more female flowers at lower nodes than at 26°C (Nitsch et al., 1952).

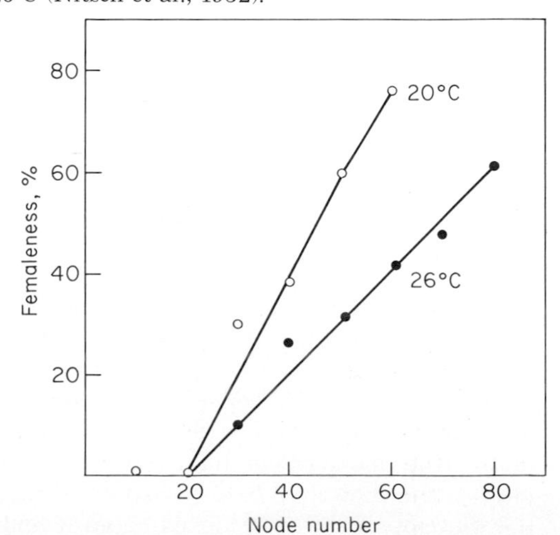

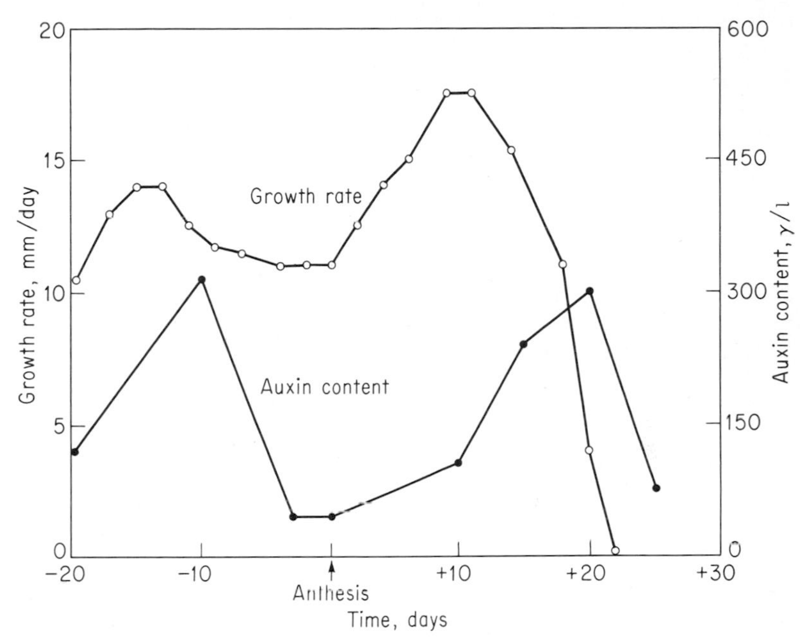

Fig. 12-27 The growth rate of the *Fritillaria* pedicel peaks before anthesis, declines as the flower opens, and rises again when fruit growth proceeds. The diffusible auxin content also shows a two-peaked curve, but it does not entirely coincide with the growth peaks (Kaldewey, 1957).

inductive cycles they produce hermaphroditic flowers. Increasing femaleness with increasing photoperiodic induction is frequently observed (Heslop-Harrison and Heslop-Harrison, 1958). Thus sexuality is often very closely linked to the intensity of the floral induction processes in the plant.

Separation of the male and female parts can be achieved by genetic separation of the sexes onto different individual plants or by separate differentiation associated with the autonomous flowering program in the plant or environmental cues. Separation of the sexual parts represents one more complication in the regulatory systems within the plant, and again the combined array of the common plant hormones appears to play an important part in this regulatory function.

Growth of the Flowers

An appurtenance to flower growth is the development of the pedicel. It is apparently dominated by stimuli produced in the developing flower bud, for the pedicel growth rate appears to reflect the production of growth substances within the bud (Katunsky, 1936). Kaldewey (1957) and Zinsmeister (1960) have plotted the growth rates of the pedicels of *Fritillaria* and *Cyclamen,* and sample data in Fig. 12-27 show a double growth peak. Most rapid growth occurs just before flower opening; there is a drop in growth during bloom and another peak just after fruit-set. Each of the two growth peaks was associated with peaks of auxin production by the flower, and the depressed growth period in *Fritillaria* was associated with the production of an inhibitor as well. The stamens appear to be the source of the inhibitor, which has a pronounced effect on the development of a hook in the pedicel during the period of bloom.

Within the flower, several floral parts have distinctive effects on the growth of the whole. Marrè (1946) observed that the removal of stamens from young, developing flowers of several genera results in a striking decline in the mobilization of starches into the developing flower

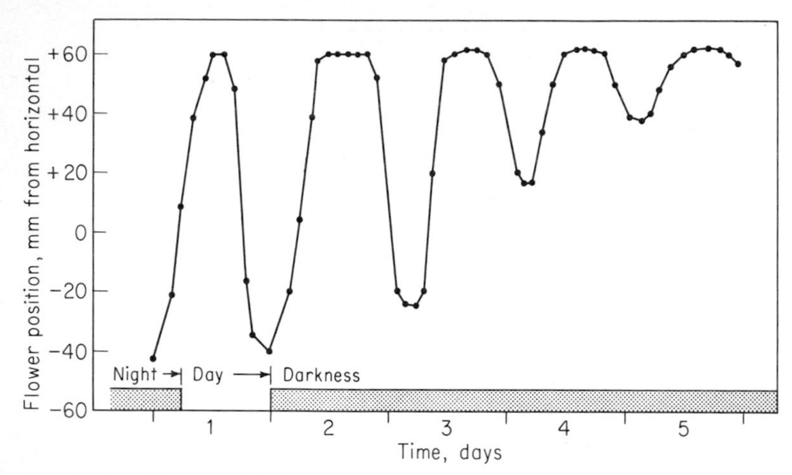

Fig. 12-28 Flowers of *Kalanchoe* have a diurnal pattern of movement, being elevated in daytime and drooping at night. Endogenous rhythms are evidenced by the continuing diurnal movements even after the plant has been placed in continuous darkness (Bünsow, 1953).

and a cessation of mitotic activity in the developing ovary. Kaldewey's work (1957) implicated the stamens in the production of the inhibitor involved in the bending of the pedicel during bloom. The ovary, too, plays a critical role in flower development and is a rich source of auxin (Katunsky, 1936). Ovary removal during development ordinarily leads to floral abscission. Laibach and Troll (1955) find that in *Coleus* this abscission-preventing action is limited to the stigma of the ovary. In their experiments removal of the stigma led to prompt flower abscission, and neither auxin nor pollen applications were effective in replacing the stigma in preventing abscission. More often, however, it has been found that auxin applications to flowers retard their abscission (Roberts and Struckmeyer, 1944).

Flower Movements

Flowers are capable of several types of movements, some turning toward the sun, e.g., *Helianthus,* and some being elevated during the day and drooping at night or opening and closing in a diurnal fashion.

To illustrate the sun movement, Shibaoka and Yamaki (1959) have performed experiments on the east-west movements of *Helianthus* and describe them as a growth phenomenon of the stem. It appears from their experiments that the growth differential is regulated by the auxin content of the sides of the stem toward or away from the sun. There is a diurnal periodicity involved, however, since removal of the plants to continuous darkness does not terminate the east-west movements for several days.

Up-and-down movement is common to many flowers and is a function of the growth of the pedicel. In *Kalanchoe* (Bünsow, 1953) the movement is repeated daily during the period of anthesis. Like the east-west movement, the elevation of the flowers shows a diurnal periodicity, since placing the plant in continuous darkness does not terminate the movement for several days (Fig. 12-28). The experiments of Kaldewey (1957) with *Fritillaria* indicate that the raising of the flower is regulated by auxin and inhibitor content of the pedicel.

Another movement of flowers is the opening and closing of the corolla. In some species this movement is repeated daily, usually with the flower opening in the day and closing at night (Bünsow, 1953). The night jasmine (*Cestrum nocturnum*) shows the reverse cycle of opening at night and closing during the day (Fig. 12-29)

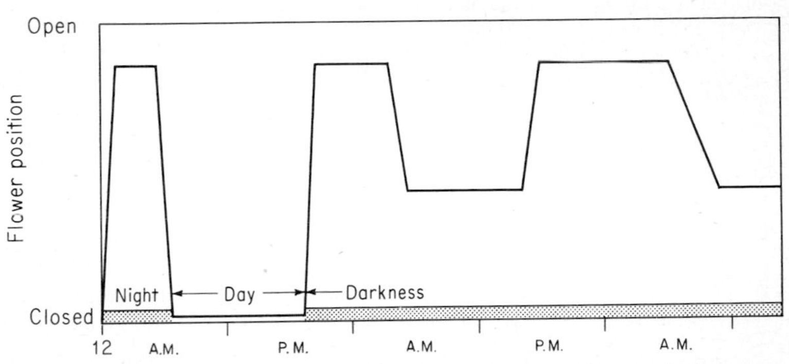

Fig. 12-29 The diurnal cycle of flower opening and closing of *Cestrum nocturnum* may continue even when the flowers are placed in continuous darkness (Overland, 1960).

and also has a parallel diurnal cycle of fragrance emission and nectar secretion. Goldschmidt (1968) has suggested that auxin regulates the opening movements of *Citrus* flowers.

Flower closure commonly takes place after pollination. In *Portulaca* the closing movement occurs 4 h after pollination, as illustrated by the pollination-timing experiments in Fig. 12-30. Here the closure appears to be controlled by the stigma, for its removal causes closure at any time. Many species respond to pollination by corolla wilting rather than closure (see, for example, Hsiang, 1951).

In contrast to the relatively slow movements

Fig. 12-30 Flowers of *Portulaca grandiflora* close about 4 h after pollination. For example, pollination at 7 A.M. (P_7), just after the flower has opened, results in closure between 11 and 12 A.M. Pollination at 1 P.M. (P_1) results in closure between 4 and 5 P.M. (Iwanami and Hoshino, 1963).

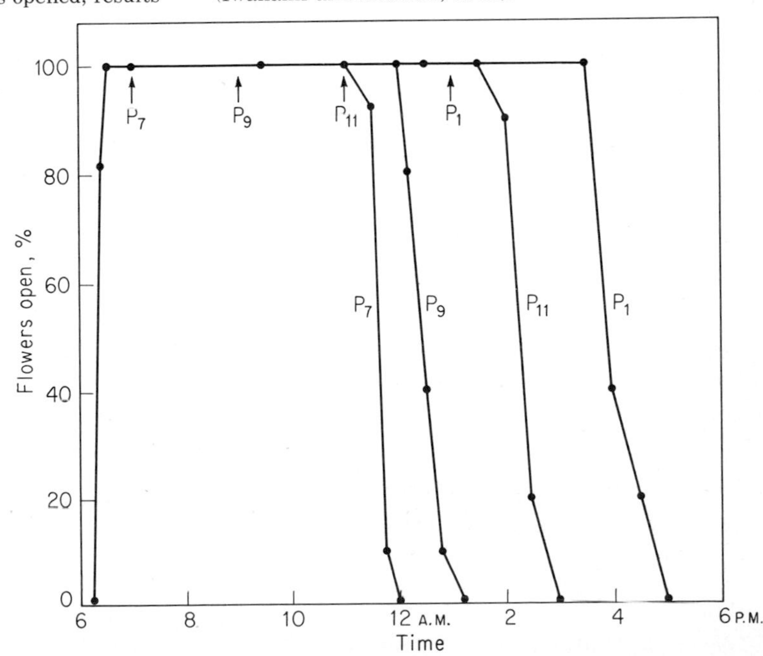

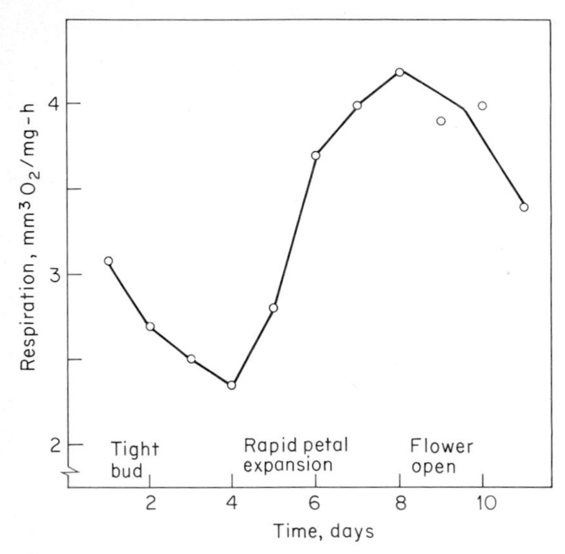

Fig. 12-31 The respiratory rate of the rose flower declines up to the time of rapid petal expansion, after which there is a dramatic rise lasting until the flower is entirely open (Siegelman et al., 1958).

of flowers and flower opening and closure, some species show rapid thigmotropic movement of stamens and styles. When visiting insects touch the stamens of some flowers, they touch off a rapid bending response which may move the anther sacs toward the body of the insect. In some species, e.g., *Sparmania* or *Berberis,* the bending is in the direction of the center of the flower; in *Portulaca,* however, the bending can occur in any direction according to the side stimulated by the insect touch (Sibaoka, 1969). The stamen movement is completed in only a few seconds. The stigmatal lobes of some species likewise show a rapid movement when stimulated by insect touch; e.g., *Mimulus* and *Incarvillia* (Sibaoka, 1968).

Growth Substances in Flowers

The production of relatively large amounts of auxin by the flower bud has been mentioned. In addition to large amounts of auxin produced in the ovary and in the pollen, there is evidence that a substantial amount of auxin is produced in the petals of some flowers during the limited period of flower opening (Takeyosi and Fujii, 1961). It appears that auxin produced in flower parts may prevent abscission of the flower.

Auxin can also have a strong influence on developmental activities taking place in the flower. A dramatic stimulating effect of auxin on the development of the orchid embryo sac has been described by Heslop-Harrison (1957). He found that auxin introduced to the orchid ovary by pollination triggers the entire development of the embryo sac. Until the flower is pollinated or supplied with an external source of auxin, the orchid embryo sac does not develop beyond the single-cell stage.

The stimulating effects of gibberellins on flower development have been described by Wittwer and Bukovac (1957). Vasil (1957) has demonstrated that auxins, gibberellins, and kinins are all needed for the maturation of the anther sacs of onion flowers. In the *Bougainvillea,* applications of gibberellin actually retard the development of the inflorescence, and the growth retardant CCC promotes it (Hackett and Sachs, 1968).

The respiratory behavior of the flower during its period of growth and senescence is sometimes dramatic. Siegelman et al. (1958) followed the respiratory rates of the petals of developing rose flowers and found a declining respiration during the early stages of flower-bud growth, followed by an impressive rise during the period of rapid petal expansion, as shown in Fig. 12-31. From the time of full bloom, the respiration rate tends to fall, and the longevity of the flower after this time is apparently closely associated with the decline in respiratory ability (Coorts et al., 1965). The sharp respiratory peak just before full bloom appears to be a common feature of flowers (Ballantine, 1966; Carfantan, 1970).

As a product of what must be a very complicated set of developmental controls, the flower is a fairly static organ as far as growth is concerned, but it has a potential for enormous growth in its ovary once the stimulus for fruit-set has been received.

GENERAL REFERENCES

Evans, L. T. 1971. Flower induction and the florigen concept. *Annu. Rev. Plant Physiol.* 22:365–394.

Evans, L. T. 1969. "The Induction of Flowering." Cornell University Press, Ithaca, N.Y. 488 pp.

Schwabe, W. W. 1971. Physiology of vegetative reproduction and flowering. In F. C. Steward (*ed.*), "Plant Physiology, a Treatise." Academic Press, New York. 6A:233–412.

13 FRUITING

By definition, an angiosperm is a plant which bears its female parts (and its subsequent seeds) within a pericarp, or enclosing tissue. This enclosing tissue is the site of most controls over the processes of fruiting; it is the tissue on which pollen must germinate, the tissue through which the pollen tube must grow to reach the ovules, and the tissue which ordinarily forms the fruit. The story of fruiting is dominated by the physiological events in the pericarp and accessory tissues surrounding the ovules and later the seeds.

POLLINATION

Pollen Physiology

Two morphological types can be recognized among angiosperm pollens: a binucleate type, with relatively simple requirements for germination and growth, and a trinucleate type, with apparently much more complicated requirements (Brewbaker, 1959), represented diagrammatically in Fig. 13-1.

Binucleate pollen is characteristic of many relatively primitive orders, e.g., the Magnoliales, Ranales, and Liliales, and of the common tree families including the Juglandaceae, Betulaceae, Fagaceae, and Urticaceae. Some relatively advanced taxa have this pollen type, including the Leguminosae. These pollens experience one mitosis after the microspore stage, having then a loose vegetative nucleus, which gradually degenerates, and a generative nucleus surrounded by a distinct unit of cytoplasm. This generative cell advances behind the tip of the pollen tube as the germination and growth proceed, and several hours after germination a second division takes place, at which time the nuclei are ready to fertilize the embryo sac.

Trinucleate pollen is characteristic of many of the more advanced orders and families, including the Gramineae and Juncaceae, the Compositae, and the Rubiales, Caryophyllales, and Polygonales. This pollen experiences two mitoses after the microspore stage; when the pollen is shed, it has one vegatative nucleus and two generative nuclei. Since trinucleate pollens are rarely capable of germination on artificial media,

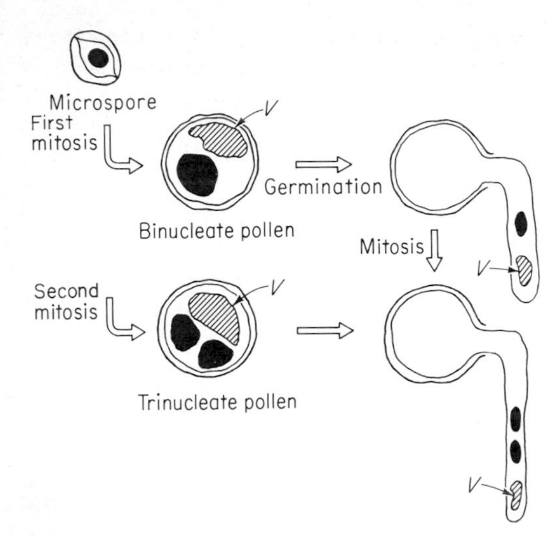

Fig. 13-1 Mitosis in the microspore can produce binucleate pollen (*above*), which completes its last mitosis after germination, or trinucleate pollen (*below*), in which the last mitosis has been completed before pollen germination. Vegetative nuclei are labeled *V*; generative nuclei unlabeled. (*Adapted from Brewbaker, 1959.*)

they are difficult to study. Apparently they have some very specialized requirements for germination and growth, and little is known of their physiology. Brewbaker (1959) cites the trinucleate pollens as the ultimate in gametophytic reduction, completing the last gametic mitosis even before the pollen is shed from the parent plant.

The germination of pollen is usually immediate upon placement upon a receptive stigma or agar medium. In some species respiratory activity begins with a few minutes and rises over a period of 2 h (Tupy, 1962); in others there is a marked depression of respiratory activity after about 30 min during the emergence of the pollen tube (Dickinson, 1966). Studies with radioactive nucleotides show that RNA synthesis begins within 15 min; while actinomycin D blocks this incorporation, it does not prevent the initial stages of pollen germination. Therefore the essential nucleic acids for germination are presumably present and ready to go (Mascarenhas, 1966). Both the generative and the vegetative nuclei are active in RNA synthesis (Young and Stanley, 1963). Protein synthesis begins even more rapidly, being detectable after only 2 min of germination (Mascarenhas and Bell, 1969).

Ultrastructural studies of pollen germination and growth reveal that in the few minutes before emergence of the pollen tube, there is a great activation of Golgi bodies, which then release enlarged vesicles. The growth of the pollen tube involves bursting these vesicles at the surface of the pollen-tube tip (Fig. 13-2) and releasing the pectinaceous contents into the pollen-tube wall (Rosen et al., 1964). That the vesicles blebbing off the Golgi bodies do contain pectins has been demonstrated by the use of pectinase on the electron microscope sections (Dashek and Rosen, 1966). The vesicles apparently break in the zone of tube growth (the first 5 μM), and their contents are contributed to the pollen-tube wall.

It should be mentioned that the germination of the pollen is associated with a rapid release of proteins into the germinating medium (or the stigma); numerous enzymes have been indicated as being present in the secreted material (Stanley and Linskens, 1964). Such proteins have obvious possible roles in the self-incompatibility reaction, not to mention hay fever and other pollen allergies.

The nutritive requirements for binucleate types of pollen are simple (Brink, 1924). This heterotrophic organism can be readily germinated and grown on agar or in water, with surprisingly high concentrations of sucrose (5 to 30 percent). The supplied sugar maintains an osmotic equilibrium in the pollen tube, without which the cells become inflated and burst (Fig. 13-3); the sugar in many instances also serves as a nutritive substrate for the metabolism of the cell (O'Kelly, 1955; but cf. Visser, 1955). Profound benefits of added boron were first reported by Schmucker (1933). Some representative data for the growth responses of pollen tubes to boron are shown in Fig. 13-3.

Like root growth, the growth of pollen tubes is markedly stimulated by calcium (Mascarenhas and Machlis, 1962). A concentration-response curve is shown in Fig. 13-4. It is striking that the optimum concentration is very high indeed,

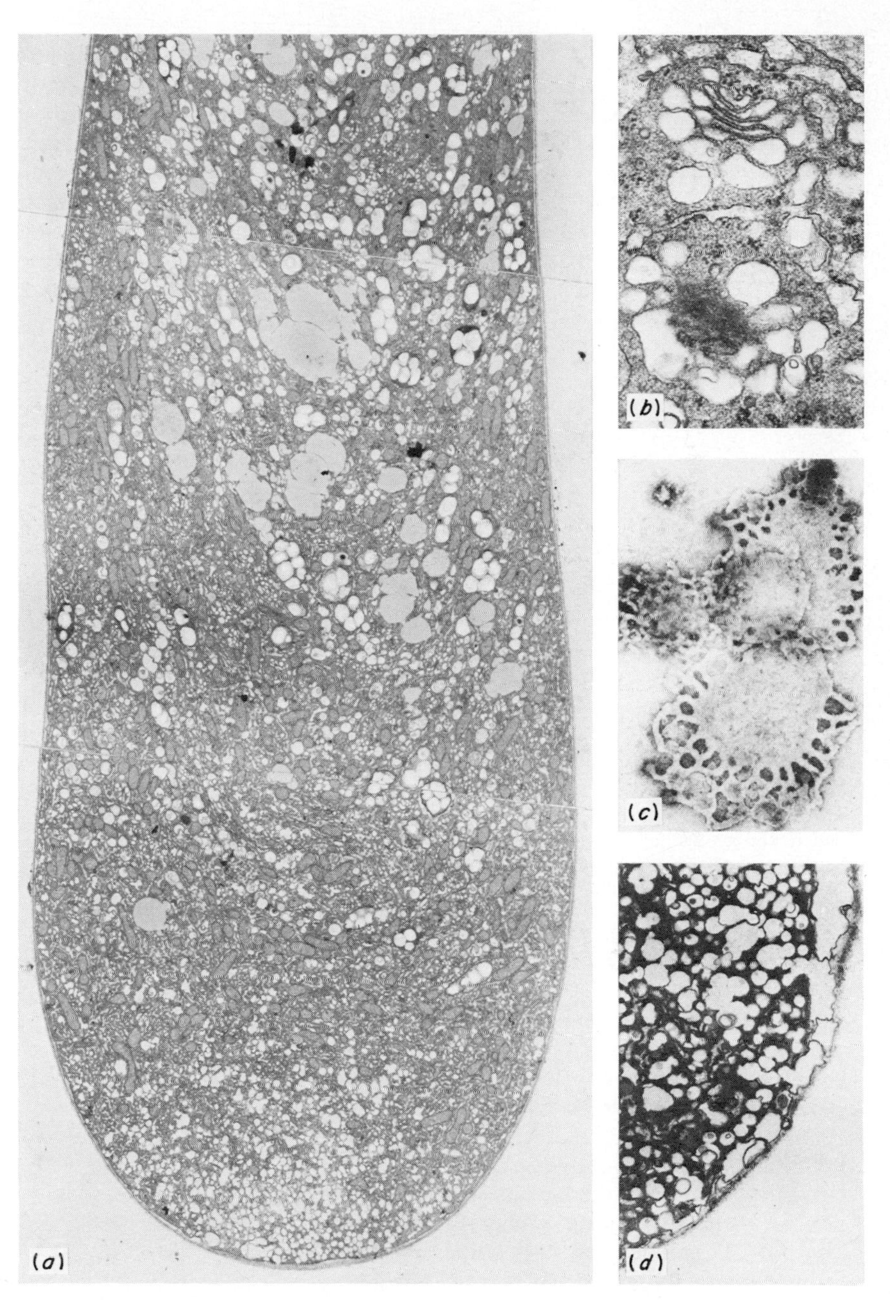

Fig. 13-2 Electron micrograph of a pollen tube (*A*), showing the abundant Golgi bodies and the numerous vesicles which apparently release their contents into the wall at the tip region of the pollen tube. Details of Golgi bodies in cross section (*B*) and longitudinal section (*C*) and a detail of the tip, showing vesicles opening into the tube wall (*D*). (*Courtesy of D. J. Morré.*)

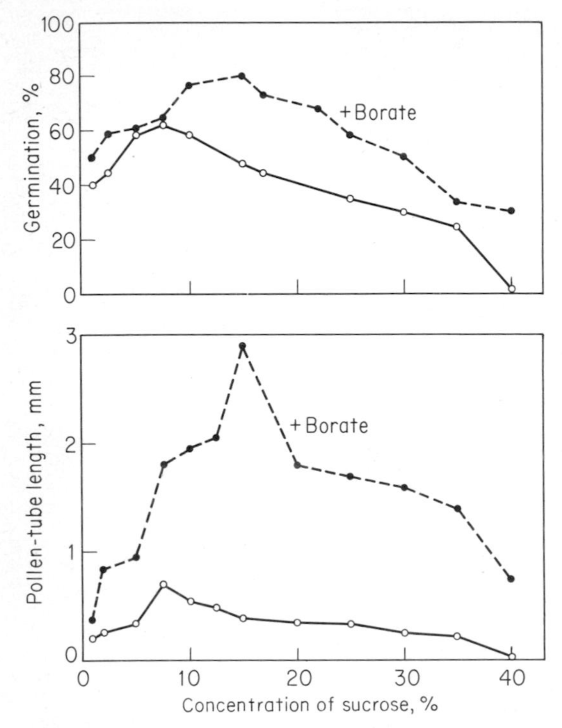

Fig. 13-3 The beneficial effects of sucrose in the germination medium for *Cucumis melo* pollen are evident both in the improved percentage germination and in the growth of the pollen tube. Addition of borate makes the sucrose effect more extensive. Germination for 4 h at 36°C (Vasil, 1960).

being 1m*M* or more. The suggestion has been made that the tropistic growth of pollen tubes to the ovule may be in response to a gradient of calcium content from the stigma to the ovule.

Stimulatory effects of CO_2 have been reported both for pollen germination (Sfakiotakis et al., 1972) and pollen tube growth (Nakanishi et al., 1969). These effects are not attributable to relief of any ethylene inhibitions.

While pollen contains substantial amounts of auxin (Muir, 1947) and gibberellin (Barendse et al., 1970), pollen germination and tube growth are not appreciably affected by the known plant hormones. This property, too, they share with roots. However, for many years there has been experimental evidence for naturally occurring promoters and inhibitors of pollen-tube germination and growth (see, for example, Brink, 1924; Iwanami, 1957). Experimentally, the presence of such growth substances is shown by the tendency of pollen tubes to grow faster or slower on nutrient agar in the presence of style tissue or style extract. Logically one would expect pollen-tube growth promoters to be involved in the chemotropic direction of the tube into the ovule and growth inhibitors to be involved in incompatibility reactions between pollen and styles or stigmas.

Growth-promoting substances may be present in the pollen itself and may help to account for the common beneficial effects of abundant pollination as contrasted with pollination of a style by only a few pollen grains. This population effect, first recognized by Brink (1924), has been quantitatively described by Brewbaker and Majumder (1961). Placing pollen at different densities on a nutritive agar, they observed increasing percentages of pollen germination with increasing densities of pollen. That this effect is due to the abundance of a germination promoter in the pollen is evident from the fact that the addition of a water extract of pollen results in maximal germination at every pollen density (Fig. 13-5).

The stigmas may provide substances of a bene-

Fig. 13-4 Calcium ions have a very beneficial effect on the growth of pollen tubes, which continues up to concentrations above 1 m*M*. *Antirrhinum* pollen grown on 20% sucrose, 0.1% yeast extract, 0.01% boric acid, and 1% agar for 8 h; calcium added as $CaCl_2$ (Mascarenhas and Machlis, 1964).

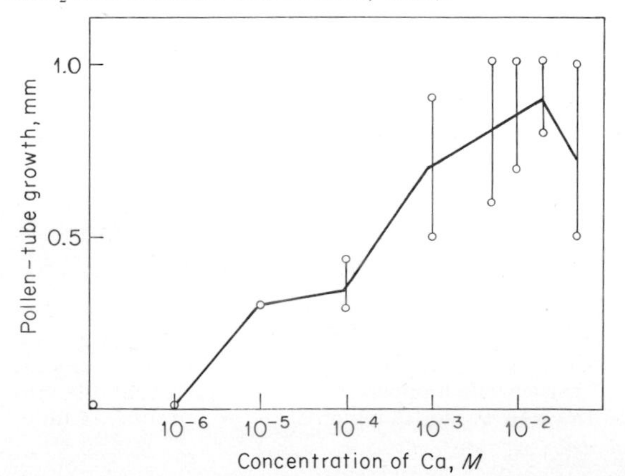

ficial nature, including secretions which may promote germination of the pollen or limit the tendency of pollen to become inflated and burst. These may be actual droplets of liquid which are secreted by the stigma upon maturation, as in the Liliaceae and Solanaceae, or when the pollen lands, as with grasses (Wantanabe, 1961), or they may be released by rupture of a covering stigmatic film when disturbed by a visiting insect, as in many legumes (Elliott, 1958).

Outcrossing Mechanisms

The principal benefits provided by sexual reproduction are the crossing of separate genomes and the consequent effects on vigor and genetic adaptability. For these benefits to be realized in plants, the ovary must be fertilized with pollen from other plants. The near proximity of pollen from the same flower gives self-pollination much the higher probability unless there are special mechanisms to ensure outcrossing. The mechanisms may be positional, e.g., dioecious plants which have male and female floral parts on separate plants or monoecious plants which have the sexes separated into different flowers or separated by the more subtle devices of heteromorphic variabilities in relative lengths of styles and anther filaments. Another common mechanism involves a differential ripening time for pollen and style. A dramatic illustration of differential timing is the avocado flower, which opens on one day when its female parts are receptive; the flower closes and then reopens the next morning, when only its male parts are ripe (Peterson, 1955). Any of these positional or timing modifications can encourage outcrossing. Another set of outcrossing mechanisms involves incompatibility reactions between the pollen or pollen tube and one of the ovary parts: stigma, style, or ovule. These incompatibility mechanisms can suppress or prevent the effectiveness of pollen in the ovaries of the same plant.

Experiments on the nature of self-incompatibility indicate that in each case the ovary imposes inhibitory actions upon the effectiveness of the pollen through suppression of pollen germination, inhibition of pollen-tube growth,

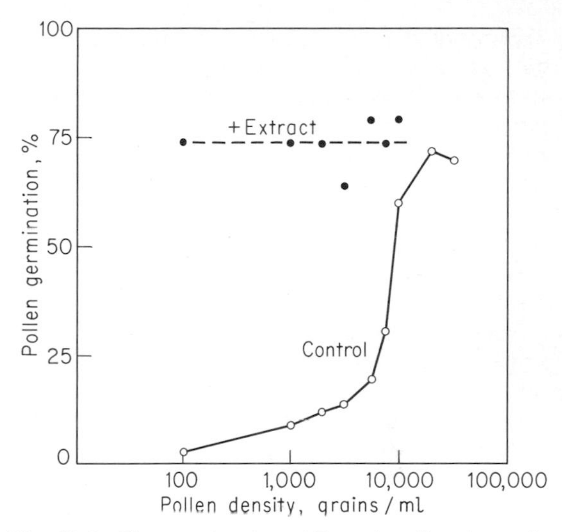

Fig. 13-5 The germination of *Petunia* pollen is much improved by higher densities of pollen. A water extract of pollen can provide maximal pollen germination at any of the tested pollen densities. On 10% sucrose agar with 100 ppm borate (Brewbaker and Majumder, 1961).

or limitation of the fertilization of the ovule itself. Lewis (1949) estimates that over 40 percent of angiosperm species have self-incompatibility mechanisms.

Self-incompatibility by prevention of pollen germination is most characteristic of the species with trinucleate pollen. Some scattered instances of incompatibility due to failure of the pollen tube to make entry into the stigma are known (Smith and Clarkson, 1956); this failure may be enzymatic, since the proteins released by the germinating pollen contain enzymes which serve to dissolve the pectinaceous film over the stigma in the Cruciferae (Kroh, 1964). Through the clever device of grafting excised style tips together, Hecht (1964) has shown that *Oenothera* self-incompatibility involves the failure of pollen tubes to pass through the stigma even when the pollen tubes approach the stigma from the stylar side. Most attention has been given to cases of incompatibility by inhibition of pollen-tube growth in the style. This inhibition is commonly a relative slowing of the selfing pollen tubes in comparison with the crossing tubes. Emerson (1940) first described this differential slowing of the selfing tubes, and representative data are given in Fig. 13-6. A clever extension of this type of experiment was made by Straub (1946), who pollinated excised stylar tips of *Petunia* and allowed the pollen tubes to grow through the sections and out the severed end into a nutrient agar, where they could be observed. Selfing pollen showed a markedly slower growth rate after passing through the styles, in contrast to crossing pollen.

Even moderate differences in growth rate of selfing and crossing pollen increase the probability of outcrossing enormously. Bateman (1956) showed that a slightly slower tube-growth rate by self pollen can result in over 90 percent cross-fertilizations of the ovules when mixtures of self and crossing pollen are applied to *Cheiranthus* stigmas.

The mechanism of the retardation of pollen-tube growth may be through the formation of inhibitors of the tube growth by the ovary (Straub, 1947) or through an inhibition by immunological reactions of the tube to the stylar tissue (Lewis, 1952). Examination of the antigenic reactions in rabbit serum has shown that incompatible crosses of *Oenothera* do in fact give antigenic responses, whereas compatible crosses do not (Makinen and Lewis, 1962). A dramatic means of bypassing the incompatibility mechanisms in the stigma and style has been used for members of the Papaveraceae: this involves the injection of a slurry of pollen in a weak borate solution directly into the ovary, where the pollen germinate and set seed (Kanta and Maheshwari, 1964).

Fig. 13-6 When *Oenothera organensis* pollinations are made in incompatible crosses, the pollen-tube growth rate is relatively small, as indicated by the high frequency of short pollen tubes in the styles compared with pollen tubes in compatible styles. A summary of 2,500 measurements, 3 to 10 h after pollination (Emerson, 1940).

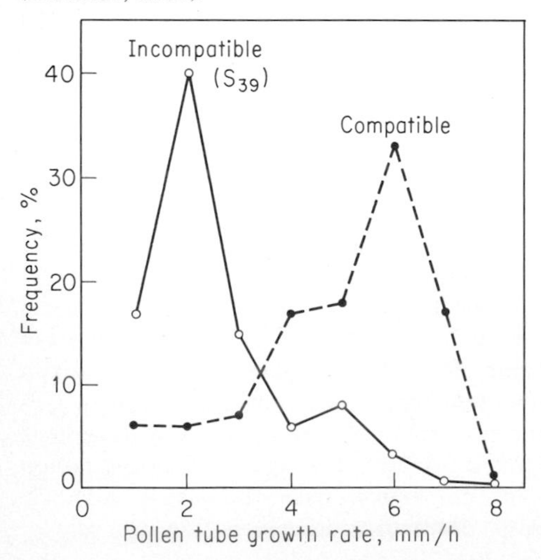

Another site of incompatibility control is at the ovule itself; for example, in *Citrus* flowers, incompatible pollen germinates normally on the stigma and grows through the style normally but fails to penetrate the ovule (Ton and Krezdorn, 1966).

Genetic incompatibility in some species such as *Oenothera* and *Petunia* is under the control of multiple alternative alleles, of which two are present in the diploid stylar tissue and one in the haploid pollen tube. If the latter is identical with either of the stylar alleles, incompatibility will result (Emerson, 1940). Within the species of *Oenothera organensis,* 45 compatibility alleles have been identified. In that species, then, any given individual will be incompatible to itself and to only 1 of 45 other possible genetic combinations within the species (Emerson, 1940). In red clover there are up to 88 compatibility alleles (Pandy, 1967).

A genetic model has been suggested to account for the incompatibility mechanism. If one assumes that each allele for incompatibility directs the formation of a protein which can form dimers, and if the monomer from the style reaches a similar monomer in the pollen tube, a dimer may be formed which is itself inhibitory of pollen-tube growth (Lewis, 1964; Ascher, 1969). A highly useful finding is that the incompatibility mechanism is sometimes removed by low- or high-temperature experiences (Kwack, 1965; Ascher and Peloquin, 1966); this fact together with the ineffectiveness of incompatibility mechanisms in some cases before bud opening and after aging of the flower (el Murabaa, 1957) suggests that the formation of inhibitors may be involved.

In the event that crossing fails, there may be real benefits to the plant in settling for self-pollination. Kerner (1895) has described numerous mechanisms by which selfing can be achieved by flowers in which crossing has failed. Among them are devices which bring the stigma to the anther of the same flower or, conversely, the anther to the stigma. Species of *Epilobium* show continued growth of the style as the flower ages, and if crossing has not been achieved, the curving growth of the style ultimately brings it in contact with the neighboring anther and the flower is selfed. In *Digitalis* the old corolla is abscised with its anthers attached, and as it falls from the flower, the anthers sweep over the stigma and selfing results (Meeuse, 1961). In *Portulaca* the anthers grow gradually closer to the stigma as the flower ages, and if insect pollination has not been achieved by the end of the flower's life, the anthers actually touch the stigma and selfing results (Iwanami and Hoshino, 1963). The slower growth of selfing pollen through the style (Bateman, 1956) and the increasing compatibility with selfing pollen as a flower ages (el Murabaa, 1957) offer other fail-safe devices for achieving self-fertilization if crossing pollen does not reach the ovule.

In some species selfing is entirely the rule; the best-known example of this *cleistogamous* type of flower is the violet, in which the pollen germinates while still in the anthers, growing through the anther wall to reach the style even at an early bud stage. The cleistogamous flower ordinarily does not expand its corolla (Uphof, 1938).

The physical mechanisms by which pollination is accomplished make a fascinating and often amazing compilation (Faegri and van der Pijl, 1966). Insect pollination is nearly restricted to the angiosperms and ranges from relatively crude "mess and spoil" activities by insects, to highly elegant devices to place the pollen on the visiting insect, e.g., the formation of pollen baskets over the entryway to the flower which drop pollen on the entering insect, pollen guns which shoot balls of pollen at him, or dehiscence mechanisms which are set off by the vibrations of the insect's wings. An astonishing device is that of the *Ophrys* orchid, in which the flower is shaped like the female of the hymenopteran wasp that visits the flower and on which the wasp performs the act of copulation with the flower, pollinating the female floral part as he does so (Baker, 1963). Wind pollination is common in most gymnosperms and in many angiosperms, of course. Wind-pollinated species produce enormous quantities of pollen per ovule (Faegri and van der Pijl, 1966). Their special adaptations are less elaborate, e.g., the production of markedly larger stigmas or stigmas which are hairlike and serve to screen pollen out of the passing air, explosive pollen-release mechanisms which serve to increase the pollen range, and winglike projections on the pollen grains which may serve to increase their air buoyancy. Water pollination occurs in some aquatics and may involve the release of anthers onto the surface of the water, where the female flowers may mechanically catch them through a funnel of the water meniscus, or floating pollen which is similarly trapped or may sink into funnel-shaped ovaries underwater.

The success of the angiosperms in achieving outcrossing through the various mechanisms of timing of flowering and through the self-incompatibility mechanisms and the employment of zoological agents in transferring pollen may well account for their very rapid evolutionary development in geological time.

FRUIT-SET

With the successful pollination of the flower, a burst of growth of the erstwhile ovary occurs, and development of a fruit begins, usually with a simultaneous wilting and abscission of the petals and sometimes of the stamens. These changes, which mark the transition of the flower into a young fruit, are called fruit-set.

Flower Fading

Upon pollination, most flowers either close their corollas (see Fig. 12-30) or more generally experience a rapid fading and collapse of the petals and sometimes other floral parts. Floral collapse upon pollination is sometimes associated with a marked burst of respiratory activity (Hsiang, 1951) and of ethylene production (Akamine, 1963). In *Vanda* orchids, a striking ethylene biosynthesis results from pollination. It is known that ethylene has an autocatalytic effect on its own biosynthesis in flowers (Burg and Dijkman, 1967), which seems to account for the effects of ethylene in shortening the storage life of cut flowers (Smith and Parker, 1966).

Receptivity

The capacity of the flower to set fruit often depends on the receptivity of the female parts to the pollen. The receptive condition may last for only a few hours, as in the mango, or for over a week, as in the tomato. In some species the receptive condition of the ovary is indicated by the exudation of a viscous material on the stigma which may hold the pollen and perhaps nourish it as well.

Receptivity of the ovary frequently begins before the flower bud is opened, and in many garden crops the incompatibility reactions are undeveloped for the few days just before flower-bud opening; this feature is of considerable use to the plant breeder, who can effect inbreeding by dissecting flower buds and applying the pollen inside. Changing receptivity with time is illustrated by the experiments with brussels sprouts in Fig. 13-7. This crop is somewhat self-incompatible, and from these data it can be seen that although self-pollination can be effected from 2 to 5 days before or after bud opening, it is quite unsuccessful on the day the flower opens. Cross-pollination, on the other hand, can be effected over the entire 5 days before bud opening to 5 days after bloom.

Abscission of the flower frequently terminates the period of female receptivity. In the tomato flower the ovule remains viable even after abscission, for abscised flowers can be cultured on agar and successfully induced to set fruit (Leopold and Scott, 1952).

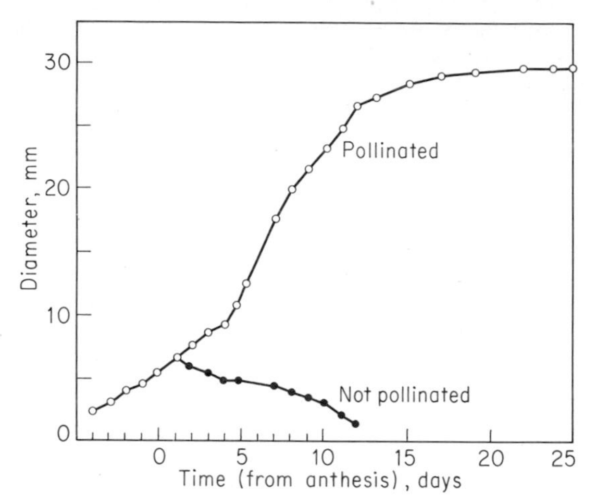

Fig. 13-8 Growth of the ovary or fruit of *Cucumis anguria* shows a rapid rise after pollination or an actual decline if pollination is not achieved (Nitsch, 1952).

Fig. 13-7 The successful setting of seed in brussels sprouts is greatest for cross-pollinations at the day of flower opening; selfing, however, shows an incompatible reaction at the time of flower opening, and better set can be achieved if pollen is inserted into the bud 2 or 3 days before opening or if pollination is done 2 or 3 days after anthesis (el Murabaa, 1957).

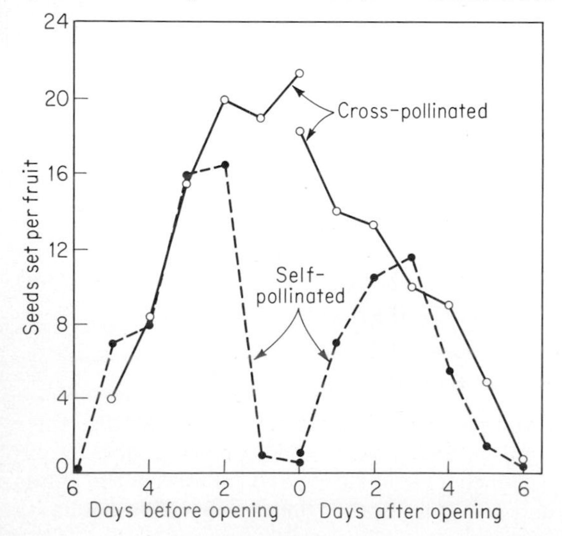

Growth Relations

With successful pollination, a burst of ovary growth follows. Germination of the pollen is often the catalyst of ovary growth; in many fleshy fruits the increase in ovary growth occurs before there has been sufficient time for the growth of pollen tubes into the ovules (Yasuda, 1934). The stimulus of ovary growth is illustrated by some data of Nitsch (1952) for cucumber, which show (Fig. 13-8) a small ovary growth before pollination and then a marked increase in growth after pollination, whereas unpollinated flowers actually decrease in ovary size. In other species the ovary may be static in growth until pollinated.

The extent of the growth stimulus may be correlated with the pollen-population phenomenon; i.e., with increasing density of pollination there may ensue a greater growth response of the ovary. In tomatoes, for example, Verkerk (1957) has described markedly faster fruit-growth rates for more heavily pollinated ovaries. In the passion flower, Akamine and Girolami (1959) observed that heavier pollinations result in increases of fruit-set, an associated markedly greater seed number, and an ultimately greater fruit size. In many plant-breeding programs where small amounts of pollen must be used, a carrier pollen of another species or spores of *Lycopodium* are often utilized to help create the pollen-population effect.

Pollen is a rich source of auxin, and in fact many of the earliest studies used water extracts of pollen as an auxin supply. Fitting (1909) observed that pollen extracts applied to orchid flowers can cause wilting and abscission of the petals and swelling of the ovary in a manner suggestive of fruit-set. Laibach (1932) assayed the auxin content of many species of pollen and reported quite large amounts in some species. Suspecting that the growth stimulation following fruit-set could be related to auxin, Gustafson (1936) discovered that fruit-set of many species can actually be induced with auxin treatments, substituting completely for the pollination, particularly in the fleshy fruits of many solanaceous and cucurbitaceous species.

The auxin stimulus involved in fruit-set comes not only from pollen but also from the ovary. Pollination results in a stimulation of auxin formation in the ovary; Muir (1942) found a sudden formation of auxin in the ovary of tobacco flowers within 2 days of pollination (Fig. 13-9). In contrast, unpollinated ovaries persisted in having only small auxin contents. In later work (1951) Muir established that this ovary auxin represents a stimulated synthesis of auxin by some factor from the pollen. The synthesis of auxin does not engulf the whole ovary at once but follows a progressive pattern strikingly similar to the progress of the pollen tubes down the style. In some perceptive experiments, Lund (1956) described this correlation in tobacco. His data show that 20 h after pollination, auxin synthesis occurs principally at the stylar tip; at 50 h the synthesis has moved to the base of the style, and at 90 h, when the pollen reaches the ovule, the ovary base is the site of most auxin synthesis. Lund chromatographed his auxin extracts and identified indoleacetic acid and tryptophan as the principal forms showing the increase, though later Paleg and Muir (1959) also found an unidentified neutral auxin which increased initially after pollination.

After Gustafson (1936) discovered that some species of fruits can be set with auxins, it became evident that the property is not shared by most fruits. While auxins are effective in causing set of tomato, pepper, eggplant, tobacco, holly, okra, figs, and numerous relatives of the

Fig. 13-9 Diffusible auxin obtained from the ovaries of tobacco flowers shows a striking increase after pollination. Diffusates for 3 h assayed by *Avena* curvature test (Muir, 1942).

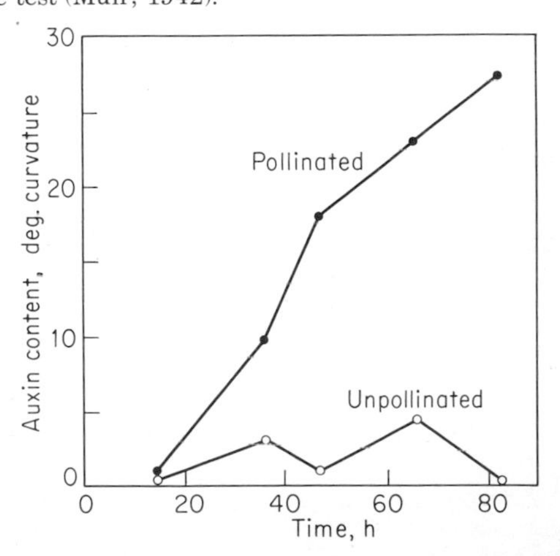

cucumbers and melons, an estimated 80 percent of horticultural species cannot be set with auxins.

Soon after gibberellins assumed a position of popularity among physiologists, it was discovered that they too can induce fruit-set in tomatoes (Wittwer et al., 1957). But in addition, applied gibberellins can cause fruit-set in many species not set by auxin, especially grapes (Weaver and McCune, 1958), many stone fruits (Crane et al., 1960), and apples and pears (Luckwill, 1959). Dennis and Nitsch (1966) found that gibberellins are present in enlarging apple fruits, and gibberellins extracted from young fruits can induce parthenocarpic fruit-set in other apple flowers (Dennis, 1967). There are numerous other reports of the occurrence of gibberellins in newly set fruits of various kinds.

Cytokinins bring about fruit-set when applied to figs (Crane and van Overbeek, 1965) and improve the setting of pollinated cucurbits (Jones, 1965). Dramatic increases in fruit-set and fruit growth of grapes have been obtained by dipping flower clusters into solutions of cytokinins (Weaver et al., 1966), which are present in very young fruits of several kinds (Steward and Simmonds, 1954; Goldacre and Bottomley, 1959), but published evidence that they act as a natural regulator of fruit-set is lacking.

It is noteworthy that in numerous species, outcrossing provides markedly better stimulations of fruit-set than selfing (see, for example, Akamine and Girolami, 1959; Bergh et al., 1966); in avocado orchards, trees of slightly different genetic background can markedly increase fruit-setting on adjoining trees.

Parthenocarpy

From a general biological viewpoint, it seems surprising that some plants can experience fruit-set and develop normal (though seedless) fruits without any fertilization of the ovule. As students of natural selection we expect major developmental processes to be associated with at least some beneficial outcome for the species, but parthenocarpic fruit does not seem to offer the plant any benefit whatsoever, appearing more like a biological accident.

Parthenocarpy is of widespread occurrence, especially among species which have large numbers of ovules per fruit, e.g., bananas, pineapples, tomatos, melons, and figs. Many seedless horticultural varieties of these fruits are exclusively parthenocarpic. One can not equate parthenocarpy with the presence of numerous ovules, for many-seeded fruits are not necessarily capable of parthenocarpy and even some single-seeded fruits can be parthenocarpically set (see, for example, Hartmann, 1950).

Several types of parthenocarpy occur, some requiring pollination and others not. Parthenocarpy may occur as (1) fruit development without any pollination, (2) fruit development stimulated by pollination but proceeding to full development even without the pollen tube's ever reaching the ovule and effecting syngamy, or (3) seedlessness as a result of the abortion of the embryo before the fruit reaches maturity. Illustrative of the first case is the occasional complete parthenocarpy of tomatoes, peppers, pumpkins, and cucumbers and the consistently seedless *Citrus*, banana, and pineapple. Parthenocarpy stimulated by pollination without syngamy is illustrated by the parthenogenetic species of *Poa;* fruits of triploid plants which are therefore genetically sterile also fit into this second group of parthenocarpy. The type which follows from embryo abortion is common in some cherries, peaches, and grapes. A curious type of intermediate parthenocarpy is that of the *Gingko*, in which fruit development is stimulated by pollination but the pollen tube experiences a suspended growth until the fruit is ripened and has fallen to the ground; the ripe fruit has developed without syngamy, but pollen-tube growth begins again, and syngamy does occur some months later.

Special environmental conditions are frequently associated with parthenocarpic fruit development. It often occurs following the extensive periods when normal fruit development has been prevented, as in plants which have failed to set fruit over an extended time. Osborne and

Went (1953) were able to induce parthenocarpy in tomatoes with low temperatures and high light intensities, conditions under which pollination is poor. Nitsch et al. (1952) found similar induction of parthenocarpy in cucumbers at short photoperiods and with low night temperatures. In some species which are reluctant to exercise parthenocarpy, seed abortion can be induced by frost or low temperature (apples and pears) or even by fog (olives), and parthenocarpic fruits sometimes result.

After his discovery that auxins can induce parthenocarpy, Gustafson (1939) examined the ovaries of seeded and seedless species, seeking a relationship between the auxin content and the natural inclination toward seedlessness. Some of his data for the auxin contents are given in Table 13-1 from which it appears that among the citrus fruits and grapes the seedless varieties have appreciably higher auxin contents than the seeded varieties. Luckwill (1957) has suggested that parthenocarpy may represent a state of auxin autotrophy in the ovaries.

The physiological basis for parthenocarpy remains obscure. It occurs most commonly among fruits with large numbers of ovules, suggesting that the ovule may provide some of the chemical constituents which stimulate fruit-set and fruit growth. That the ovules are sites of auxin synthesis in some fruits seems reasonable (see, for example, Nitsch, 1950), and the fact that auxin applications can bring about fruit-set enhances the possibility that auxins are critical in fruit-set. However, many fruits do not so respond to auxins, and so there must be a further component of the fruit-set stimulus in the plant. Stephen's (1958) suggestion that factors favoring fruit-set come from the roots of fruit trees may be interpreted as implying a role of gibberellins and cytokinins.

Nitsch (1970) has taken the position that the fruit is an aggregation of numerous types of tissues, which may be expected to require separate regulatory signals. Crane (1965) was able to induce parthenocarpy in figs with auxins, gibberellins, or cytokinins, and he suggested that the central requirement for set might be the triggering of mobilization activity in the fruit; thus any of the mobilizing regulators would suffice. In some species, auxin applications can result in

Table 13-1 Auxin content of the ovaries of unopened flowers, comparing seeded with seedless varieties (Gustafson, 1939).

	Extractable auxin, μg/kg fresh weight, *Avena* curvature assay	
	Seeded varieties	Seedless varieties
Orange:		
Satsuma	—	1.01
Washington	—	0.73
Robertson	—	1.16
Paper rind	0.58	
Valencia	0.58	2.39
Lemon:		
Eureka	0.43	0.78
Grape:		
Thompson	—	2.74
Black manulka	—	1.30
Muskat	0.34	

initial fruit-set, but the subsequent development of the fruits fails without proper pollination, as in olive, hops, avocado, mango, and corn. In special cases the application of auxins improves fruit-set when natural set has been poor, e.g., frosted flowers of apple or pear.

The auxin forcing of fruit-set has achieved transient commercial use in the tomato industry at times when natural fruit-set is poor. There has been a more extensive commercial application of auxins for fruit-set in grapes.

Setting fruits with auxin commonly brings about an abortion of the developing embryos which may have already been fertilized. When auxins have been applied to tomato flowers in the field, seedless fruits commonly result even though there may have been some fertilization. The effect of auxin in causing embryo abortion has been nicely defined by some data of Luckwill (1953) in connection with studies of how auxin applications to apple fruits can cause fruit thinning. In the data of Fig. 13-10, auxin applied at the time of petal fall resulted in a striking increase in the abortion of the embryos.

On the other hand, applications of growth regulators have sometimes been effective in actually increasing the seed yield after pollination of difficult crosses. For example, Zafar (1955) applied naphthaleneacetic acid to potato seedlings, and the flowers that were subsequently formed were markedly slower in abscission, permitting much better seed set with pollination. Enhancement of fruit-set by the applications of cytokinins along with pollination of emasculated flowers (Jones, 1965) may be in part a direct effect of the growth regulator on fruit-set, but it may also be in part a suppression of ethylene effects on the flower; the injury associated with emasculation of the flower may be expected to produce ethylene, and cytokinins are known to have a suppressive action on ethylene (Burg and Burg, 1968a).

Fig. 13-10 When apple flowers are sprayed with auxin (40 mg/l naphthaleneacetic acid) at petal fall, a dramatic increase in the frequency of seed abortion in the fruits follows, resulting in a thinning action on the fruits; Crawley Beauty apple (Luckwill, 1953).

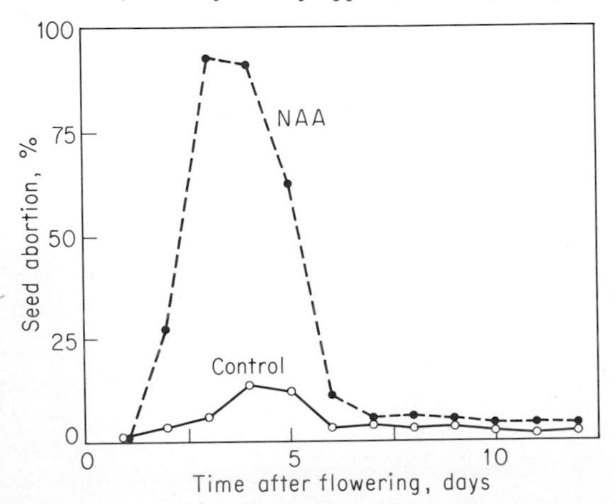

Limitations of Fruit-Set

Fruit-set may often limit the productivity of fruit crops. Three general categories of limited fruit-set can be recognized, those due to limited pollination, limited nutrients, and the precocious abscission of flowers and newly set fruits. Some plants set fruits from only a very small proportion of flowers, perhaps only 1 flower in 100 or more, e.g., such many-flowered racemes as mango, macadamia nut, and litchi. The mango may have as many as 6,000 flowers on a single panicle, of which only perhaps 2 to 4 will produce fruit (Singh, 1960).

Lack of pollination can arise from several causes. In tomato, which is usually a self-pollinating plant, weak light intensities or low temperatures can alter the flower structure so as to prevent the dehiscence of the anthers when the stigma is growing through the anther ring (Howlett, 1936). In other crops, pollination may be limited because of very short periods of female receptivity, as in the mango (Spencer and Kennard, 1955) and many legumes. Again, the pollen which does reach the stigma may be ineffective. Pollen sterility develops readily in the tomato with excessively low or high temperatures (Smith, 1932). Rain on the flowers can cause pollen frustration, illustrated by the experiments in Fig. 13-11. When a simulated rain was applied at various intervals after pollination of passion flowers, within the first hour the rain destroyed

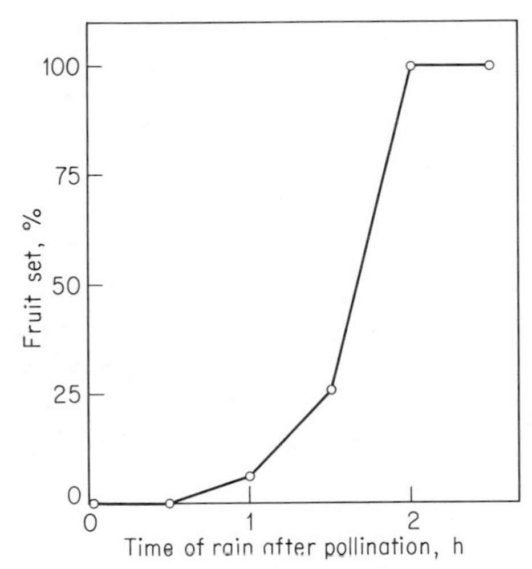

Fig. 13-11 The set of passion flowers is prevented if rain falls on the style in the first hour following pollination. Data for cross-pollinations of 30 to 50 flowers per time (Akamine and Girolami, 1959).

the effectiveness of the pollen by causing it to burst, just as bursting occurs in media of too low osmotic concentration. Chemical treatment to cause pollen sterility shows possibilities for preventing unwanted fruit-set (Eaton, 1957; Moore, 1959).

Early observations of fruit-set in cotton led Mason and Maskell (1928b) to suggest that there might be nutritive limitations of fruit-set. Eaton and Ergle (1953) later denied the idea since the foliar applications of organic nutrients did not improve cotton fruit-set. While the techniques of foliar applications of nutrients generally do not improve fruit-set, several lines of evidence clearly support the concept of nutritive limitations. In tomato, Leopold and Scott (1952) found that mature leaves are essential for the setting of flowers on the plant; they could substitute for the leaves by supplying various organic nutrients to excised flowers and obtain good fruit-set. Murneek (1927) established that the presence of fruits on the spider plant (*Cleome*) effectively limits further fruit-set. Van Steveninck (1957) found that the deterioration of fruit-set with time in yellow lupine can be averted by removal of fruits already set (Fig. 13-12). Perhaps a similar competition for nutrients is responsible for the abortion of the ovules in the stylar end of bean fruits (Gabelman and Williams, 1962), a competition effect which becomes more severe under conditions of water stress. In some species such as grape and cotton, girdling a branch can markedly improve fruit-set, presumably through interference with the translocation of nutrients out of the branch.

Flower abscission frequently limits fruit-set in tomato. Because the flowers are abscised in a viable condition (Leopold and Scott, 1952), any treatment which defers abscission may be effective in increasing fruit-set. In cyclamen, abscission follows promptly if the ovary ceases growth, and there the application of the auxin naphthaleneacetamide defers flower abscission and improves fruit-set probably by temporarily sustaining ovary growth (el Murabaa, 1957).

The moment of fruit-set is a time of dramatic change; the collapse of the flower and the sudden onset of growth of the ovary constitute the changeover from a flower to a growing fruit. The sudden increase in growth should be expected to involve sudden increases in synthesis of protein and RNA. Fuller and Leopold (1972) have examined these changes with pollination of

Fig. 13-12 The presence of set fruits on yellow lupine inhibits the fruit-set of subsequently opening flower whorls. If the whorls on which set has been obtained are cut off, full fruit-set is restored (van Steveninck, 1957).

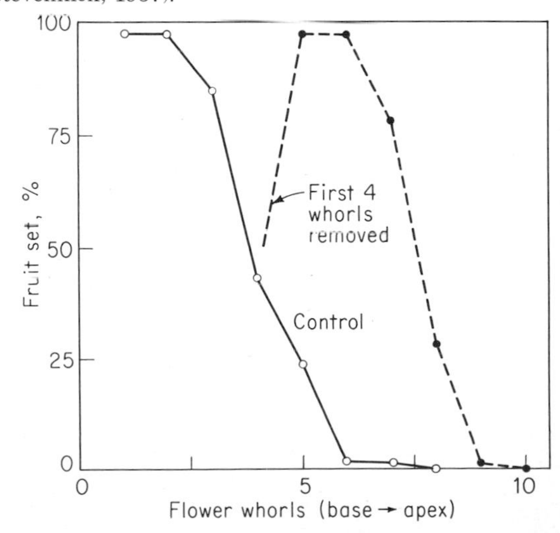

cucumber and find a large increase in RNA synthesis in the ovary within 12 h after pollination. Pollen germination occurs within minutes after placement on the stigma, and since about 16 h is required for pollen-tube growth to the ovules, the signal for RNA synthesis apparently arises from the passage of the pollen tubes through the style. Tupy and Rangaswamy (1973) have found marked increases in rRNA after pollination of Nicotiana ovaries.

In conclusion, although fruit-set is a basic step in the life cycle of higher plants, at present very little is known about its physiological control mechanisms.

FRUIT GROWTH

The tissues which enclose the ovules of angiosperms include, of course, the pericarp, a thin membranous envelope around the seeds in some fruits, a fleshy envelope in others, and a stony shell around others, plus accessory tissues. These latter vary, but in some species the enclosing tissues may be (1) only the ovary tissue, or gynaecium (as in the tomato and stone fruits), (2) the gynaecium plus receptacle tissue (as in the apple), (3) the combination of numerous gynaecia into a single fruit (strawberry, raspberry), or (4) the combination of numerous flowers into a common fruit (pineapple). Bollard (1970) stated that the central theme of fruit growth seems to be the mobilization of substrates into the various tissues associated with the ovules.

The tissues enclosing the ovule, for which the angiosperms have been named, have been elaborated into a wide diversification in various fruits. Even in the gymnosperms the naked megasporangium has been slightly elaborated in the *Ginkgo* into a simple fleshy enclosure; in *Ephedra* the enclosing tissue is slightly more elaborate, including perianth tissue. In the angiosperms, the female structure is enclosed in more elaborate tissues.

Besides the morphological structures for the development of fruits, the angiosperms have two other physiological capabilities which may have been crucial in the evolution of elaborate fruits: they have evolved the capability for very strong directed translocation or mobilization of foodstuffs, and for a type of organ senescence which involves softening and sweetening changes. We see parallel mobilization into tubers, bulbs, and corms as well as into angiosperm fruits; and we see also parallel organ senescence which involves respiratory and mellowing changes in angiosperm leaves (autumn colored leaves, the "ripening" of tobacco) with some of the respiratory and hydrolytic changes we normally associate with fleshy fruits. Thus the angiosperms are uniquely capable of mobilizing nutrients into large and sometimes elaborate fruits and of changes in quality which make them effective in the dispersal of seeds through their appeal to various animals.

In the physiological literature, almost exclusive attention has been paid to the growth and development of the fleshy fruits since these are of special horticultural interest; but there is scarcely any information on the development of the majority of fruits, which are not fleshy.

Fruit-Growth Rates

The growth of the fruit has two main components: the growth of the pericarp (or enclosing tissues) and the growth of the tissues derived from the fusion of pollen nuclei with the embryo-sac nuclei. Each of these categories may have, in turn, separate components, including the soft mesocarp and the stony endocarp, the embryo and the endosperm. Each of the component tissues may have distinctive periods of rapid and slow growth.

The fertilization process in angiosperm species involves the union of one of the germinal nuclei from the pollen tube with two, three, or four of the polar nuclei of the embryo sac to form the endosperm and the union of the other pollen nucleus with the egg nucleus to form the zygote, from which the embryo will develop. The fruit may therefore be composed of diploid ($2N$) tissues from the maternal parent, 3 to $5N$ tissue from the endosperm, and diploid tissue of the new embryo. Structurally, the fruit may be mostly parental tissue (strawberry), mostly endosperm tissue (the small grains), or mostly embryo tissue (Compositae).

Many fruits have growth patterns of the simple sigmoid type common to most cells, tissues, and organisms, starting with an exponential increase in size and then slowing in a sigmoid fashion (Fig. 13-13). This type of growth curve is common to the apple, pineapple, strawberry, pea, tomato, and many other fruits. A second group of fruits has a more complicated growth curve, involving two periods of growth increases with a period of slow or suspended growth in between, also represented in Fig. 13-13. The double growth curve is common to probably all the stone fruits, e.g., peach, apricot, plum, and cherry, and to some nonstony fruits, e.g., fig, grape, and currant. The growth curves do not seem to be distinctive for the different morphological types of fruits, for berries, pomes, and simple and accessory fruits occur in each type.

The double growth curve may be expressed in overall fruit growth and in separate growth curves for such internal parts of the fruits as the embryo, endosperm, and fleshy or stony parts of the fruits. Carr and Skene (1961) have found that in bean the entire fruit apparently shows a simple sigmoid growth curve, whereas the seeds develop with a double growth curve. A classic illustration of the separate periods of growth of the various parts of the fruit is the study by

Fig. 13-13 The growth curves for fruits are commonly of two types: sigmoid curves, as for apple, or double sigmoid curves, as for the cherry.

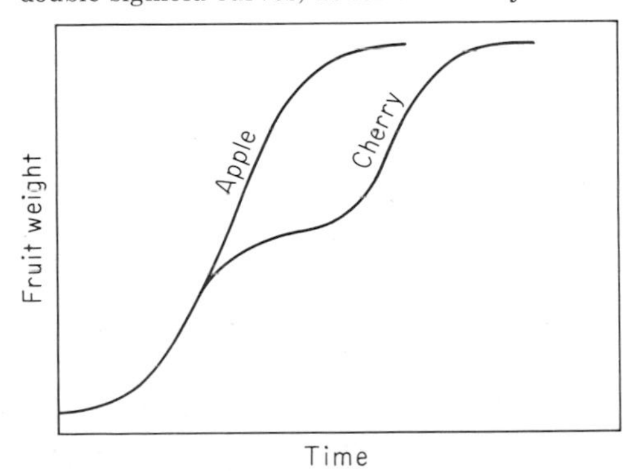

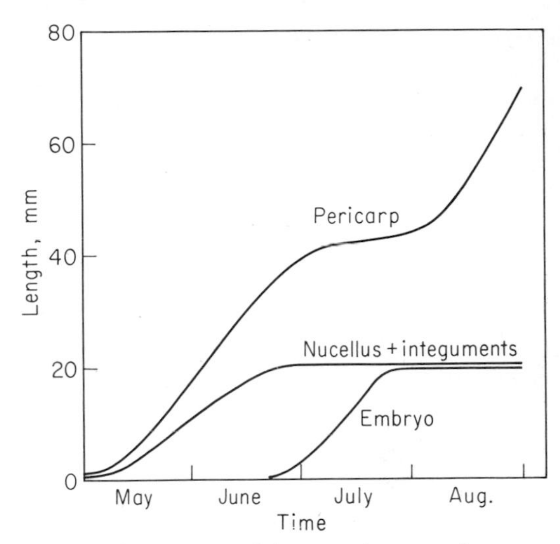

Fig. 13-14 Comparison of the growth curves for various parts of the fruit of Elberta peach shows that the double growth curve is expressed by the pericarp tissues and that the time when the first growth period ends is approximately the time of completion of nucellar growth and commencement of embryo growth (Tukey, 1933).

Tukey (1933) of the selective growth of peach-fruit parts (Fig. 13-14).

The enlargement patterns of fruit growth reflect both cell-enlargement and cell-division activity. There is wide variation in the extent to which cell division participates in fruit growth, ranging from cases in which cell division (except in the embryo and endosperm tissues) has been completed at the time of pollination (*Ribes*, *Rubus*) to cases in which there is a brief period of cell division just following pollination (tomato, *Citrus*, cucurbits, apple, *Prunus*) or extended periods of cell division after pollination (strawberry).

Mobilization

It is unnecessary to point out that the fruits which contain large amounts of food materials must import them from other parts of the plant. The means by which such a mobilization of substances occurs is poorly understood. Mobilization can permit fruit growth at the expense of materials in the leaves, as the data of Fig. 13-15

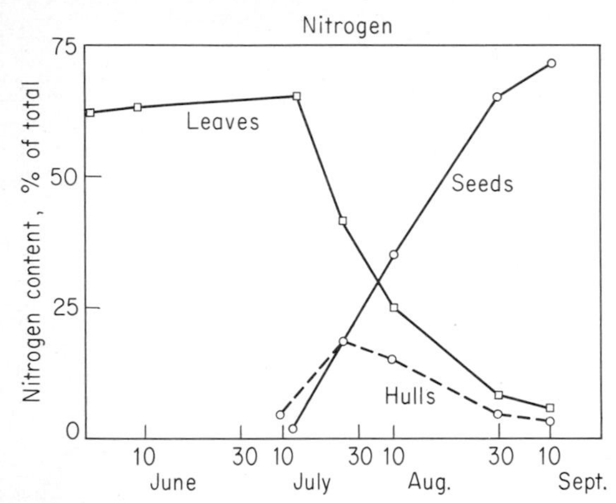

Fig. 13-15 Mobilization of nitrogenous materials from the leaves into the hulls and then into the seeds during fruiting of *Vicia faba* (data of Emmerling, 1880; cited by McKee et al., 1955).

illustrate. One can see changing centers of mobilization during the development of the pea fruit, the hulls enlarging in the early stages and shrinking later. The growth of separate parts of a fruit can involve shifting mobilization patterns between the parts, and nutrients may be moved consecutively from part to part, following the mobilization shifts.

The question should be asked: Which organs supply the nutrients for fruit growth? Often the nearest leaves or other photosynthetic tissues are principal sources from which the mobilization draws; e.g., in small grains even the awns of the flower head provide over 10 percent of the fruit dry weight, and the rest of the flower spike provides an additional 30 percent (McDonough and Gauch, 1959). Such determinations can be nicely made by radioactive labeling of the separate contributing plant parts.

As increasing numbers of fruits draw upon the nutrient supplies of the plant, competitive limitations on the growth rates of the fruits begin to set in. In tomato, Verkerk (1957) has shown that the fruit size achieved declines with increasing numbers of seeds developing on the plant. More simply, this can be expressed as a decrease in fruit size with increasing number of fruits developing on the plant, as shown in Fig. 13-16. The mobilizing forces existing between fruits have been mentioned in Chap. 3.

Fruit Size

The ultimate fruit size may be correlated with cell size in some species. For example, Tukey and Young (1939) found that in cherries the fruit size seems to be determined by the cell sizes achieved in the fleshy parts of the fruit. In other fruits the number of cells is related to ultimate size, as Bain and Robertson (1951) have shown for the apple (Fig. 13-17). The apple also achieves a considerable amount of size through the development of intercellular spaces during the second half of fruit enlargement. This growth phenomenon is evident from the data in Fig. 13-18, which show that during the last half of growth, fruit volume increases more rapidly than fruit weight. In the finished fruit, 25 percent of the apple is air space. The white refractive quality of apple flesh is related to the resulting cell-air interfaces. In contrast to apple, the grape shows a greater increase in weight than in volume during the later periods of fruit growth (Coombe, 1960), indicating an enhanced deposition of solids instead of the opening of air spaces.

From these brief comments it can be seen that

Fig. 13-16 Competitive actions between fruits are indicated by the decreases in fruit size with increased numbers of fruits present on tomato plants (unpublished data of Hafen and Leopold).

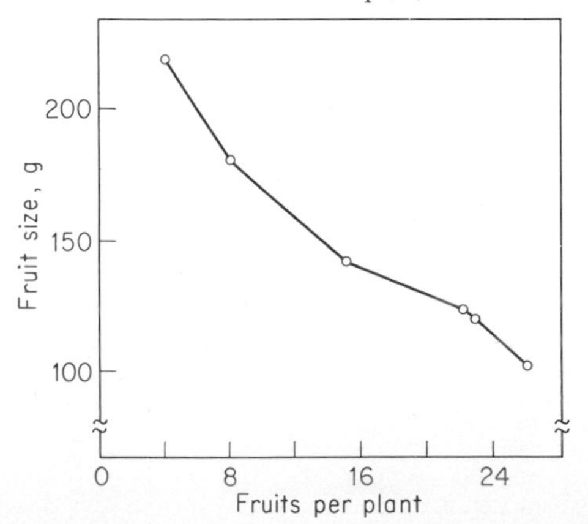

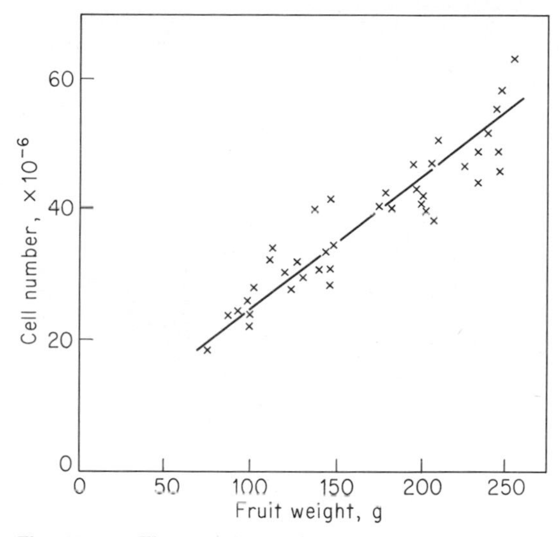

Fig. 13-17 The correlation between fruit weight and cell number in Granny Smith apples from a single tree (Bain and Robertson, 1951).

growth may be an expression of a wide variety of events, from the development of air spaces to the loading into the fruit of sugars without corresponding volume increases and from cell division to cell enlargement (and of course associated various types of tissue differentiation); finally, there may be preferential growth of any of several or successive morphological parts of the fruit.

Role of Seeds

There is a puzzling inconsistency about the evidence for the role of seeds in fruit growth. Abundant evidence indicates that seeds regulate many aspects of fruit growth, and yet in numerous species seedless fruits can grow in a manner very similar to seeded fruits. The concept of seeds as sources of regulatory substances for fruit growth is based on three types of evidence (Nitsch, 1963):

1 The removal of the fertilized ovules frequently terminates the growth of the fruit.

2 The geometry of the fruit frequently reflects the distribution of seeds inside, regions containing seeds outgrowing those parts without seeds.

3 In many fruits there is a strong correlation between the numbers of seeds in the fruit and the overall fruit size achieved.

The role of seeds in fruit development was described by Nitsch (1950) for the strawberry, a convenient fruit for such a study since the seeds are all on the surface. Removal of the seeds prevents the growth of the fleshy fruit, and in fact trimming all the seeds from two sides prevents growth there, resulting in disk-shaped fruits, as shown in Fig. 13-19*b*. Nitsch provided some evidence that the seeds stimulate fruit growth by providing auxins, for an auxin paste could adequately substitute for the seeds on the fruit (Fig. 13-19*d*). Nitsch also noted that in strawberry the fruit size becomes proportional to the number of seeds developing on it.

Fig. 13-18 During the growth of the apple, there is a greater increase in fruit volume than in fruit weight, indicating the development of air spaces as a part of fruit growth (Bain and Robertson, 1951).

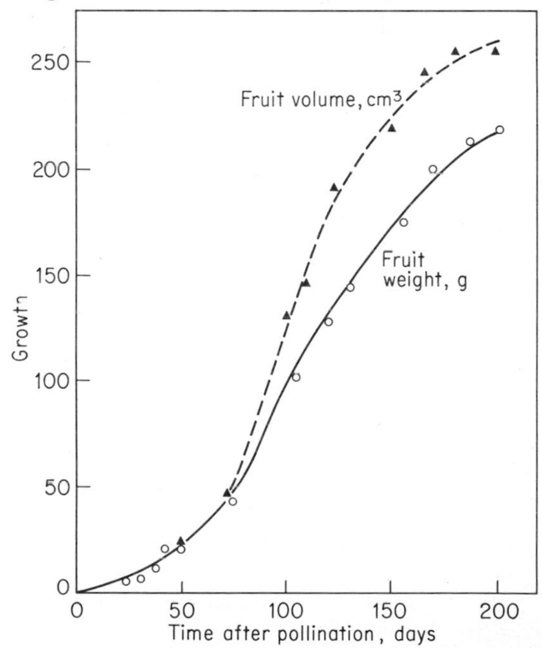

Fig. 13-19 Removing the developing achenes on two sides of the strawberry fruit (*b*) results in a disk-shaped fruit; removing all achenes (*c*) suppresses fruit growth; and removing all achenes but supplying the fruit with an auxin paste (*d*) can restore normal growth (Nitsch, 1950).

The distribution of seeds often markedly alters the shape of the fruit. For example, if a pear has seeds only on one side, that side of the fruit develops preferentially, resulting in a distorted asymmetrical fruit (Luckwill, 1959). Many parthenocarpic fruits are shaped differently than seeded fruits of the same variety; e.g., fruits of some squash varieties are oblong when seedless and pear-shaped with seeds. Pears, too, are oval when parthenocarpic and develop the typical pear shape only when seeds are present.

There are many correlations between the numbers of seeds and fruit size. Some representative data for the apple are presented in Fig. 13-20; the correlation is consistent with the concept of seeds as a source of growth-stimulating signals in the fruit (see also Fig. 13-19).

In fruits which show the double growth curve, the period of suspended growth occurs at the time of maximal growth of the seeds in the fruits (see Fig. 13-14). This situation suggests a possible competition between the growth of the fruit and the growth of the seeds, and in fact in the Early Richmond cherry the feature of earliness is correlated with the abortion of the embryo at the beginning of the period of suspended growth; the cherry then shows almost no period of suspended growth and finishes off early, as shown in Fig. 13-21.

Seeds, then, appear to provide strong influences on the growth of fleshy fruits, ranging from strongly promotive to inhibitory effects, depending upon the stage of fruit growth measured. How seedless fruits compensate for the missing seeds is not clear, but there is some evidence that other parts of the fruit take over the synthesis of growth substances.

Growth Substances

When Nitsch (1950) found that auxin paste would act in place of missing seeds in the growth of strawberry fruits, he naturally raised the question whether endogenous auxin is responsible for fruit growth. Analysis of the auxin content of strawberry fruits at various stages of development encouraged him to believe it is, for he found that a large production of auxin occurs in the first 2 weeks of fruit growth, when rapid growth is getting under way (Fig. 13-22). He found large increases of tryptophan in the fleshy parts of the fruit and obtained evidence that the seeds provide factors necessary for the conversion of the tryptophan into auxin, presumably thus stimulating the fruit growth.

The strawberry growth rate is sigmoid; what is the situation in fruits with a double sigmoid growth curve? From studies of the black currant, Wright (1956) found that by paper chromatography he could separate three auxins from the developing fruit: a neutral one which peaks with the first sigmoid growth phase, an acidic one which peaks more during the second growth phase, and a third which is apparently more related to fruit abscission than to growth. In this case, the double growth curve correlates with a double peak of auxin production.

The correlation of fruit growth with endogenous auxin became clouded, however, when both grape (Nitsch et al., 1960) and fig (Crane et al., 1959) yielded evidences of an auxin increase only for the first growth period; the second period of fruit growth was not associated with a rise in auxin content. Coombe (1960) interpreted the situation in grape as evidence that the first growth period is controlled by auxin whereas the second is a consequence of an osmotic accumulation of carbohydrates. But in peaches even the first phase of growth lacks an associated rise in auxin (Stahly and Thompson, 1959); instead, the auxin content rises perceptibly only during the period of suspended growth. Three auxins were detected, and each showed the same characteristic peak at the period of least growth.

While the possibility of explaining fruit growth as an auxin effect grew dim, another substance entered as a possible factor in fruit growth, gibberellin. Besides the observations that gibberellin treatments can stimulate fruit growth, already mentioned, Coombe (1960) detected measurable amounts of endogenous gibberellins in the first growth stages of Seedless Emperor grape, and Jackson and Coombe (1966) found gibberellin content to rise to a peak in apricot fruits during the first period of growth. In bean fruits Skene and Carr (1961) have found peaks of gibberellin content both early and late in the fruit-growth period (see Fig. 13-23). To complicate the picture, Raju and Das (1968) have found a two-peaked curve for gibberellin content in the developing pepper fruit even though this fruit has only a single period of growth (sigmoid pattern). Before we conclude with Crane (1969) that gibberellin levels in fruits just do not correlate well with growth rates, it should be pointed out that developing seeds are notable sources of gibberellins. Extraction of developing seeds often yields rich supplies of gibberellin (Skene and Carr, 1961); nevertheless, in seedless fruits, gibberellin is formed in the developing fruit (Weaver and Pool, 1965, for grape; Hayashi et al., 1968, for apple).

Turning from the unconvincing evidence of gibberellin control, we should ask whether cy-

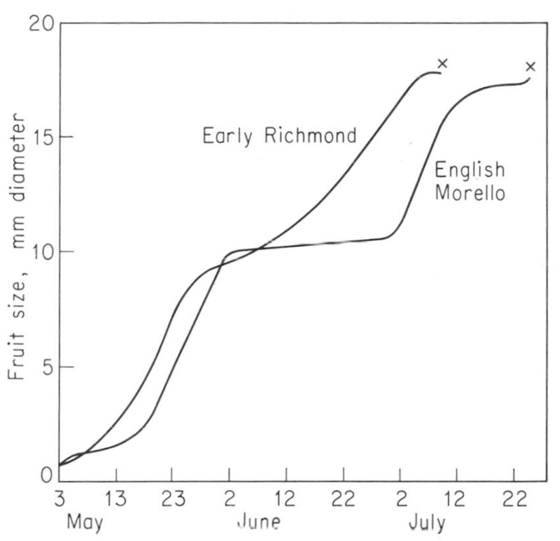

Fig. 13-21 Growth curves of two varieties of cherry, showing the rather extended period of suspended growth in the late variety English Morello and the almost absent period of suspended growth in Early Richmond (Tukey, 1934).

Fig. 13-20 The correlation between the size of apple fruits and the numbers of seeds per fruit (Visser, 1955).

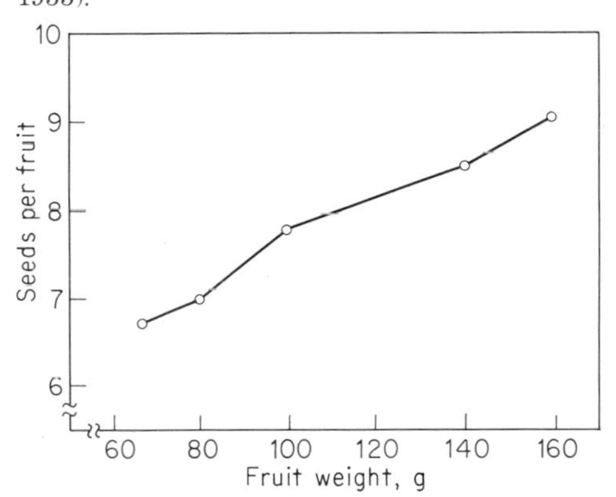

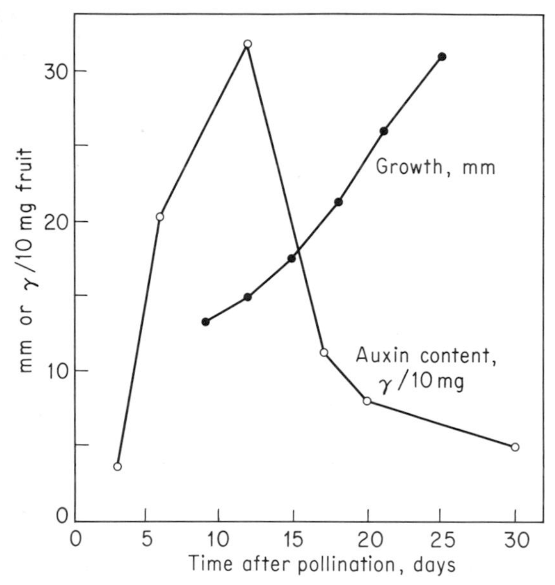

Fig. 13-22 The growth of the strawberry fruit is associated with a burst of extractable auxin in the fruit during the first 2 weeks after pollination (Nitsch, 1950).

tokinins may serve as fruit-growth regulators. The occurrence of cytokinins in developing fruits, especially during the early periods of fruit development, is fairly well established in the literature (Goldacre and Bottomly, 1959, for apple; Nitsch and Nitsch, 1961, for tomato). In apple fruits, the seeds appear to be central sources of cytokinin (Letham and Williams, 1969). Applications of cytokinins can bring about large increases in fruit growth in some varieties of grape (Weaver and van Overbeek, 1963). The possible correlation of cytokinin levels with fruit growth has not been well documented.

The possibility that growth inhibitors may play a regulatory role in fruit growth was first brought up by Thompson (1961), who found an inhibitor in the receptacle of unpollinated strawberries but not after pollination when fruit growth had started. Gabr and Guttridge (1968) have shown ABA to be present in such material. Kentzer (1966b) has observed an inhibitor to accumulate at the time of completion of fruit growth in *Fraxinus*, implying that an inhibitor could be involved in the termination of fruit growth; but the role of inhibitors in fruit growth remains speculative.

Ethylene may be a normal component of enlarging fruits even before the start of the climacteric respiratory rise (Burg and Burg, 1965a), but the possibility that ethylene plays a role in the growth of some fruits was not seriously investigated until Maxie and Crane (1968) found that the application of ethylene to figs during the suspended period of fruit growth leads to dramatic increases in fruit growth. They showed that the reports of stimulated fruit growth by application of auxins to figs could probably be accounted for as a consequence of the production of ethylene following the auxin treatment.

Looking back over the various lines of evidence that each of the known plant hormones can

Fig. 13-23 The growth of seeds in bean fruits compared with the gibberelin content. The two peaks of growth roughly correspond in time to two peaks in gibberellin level, as determined by wheat-leaf bioassay (Skene and Carr, 1961).

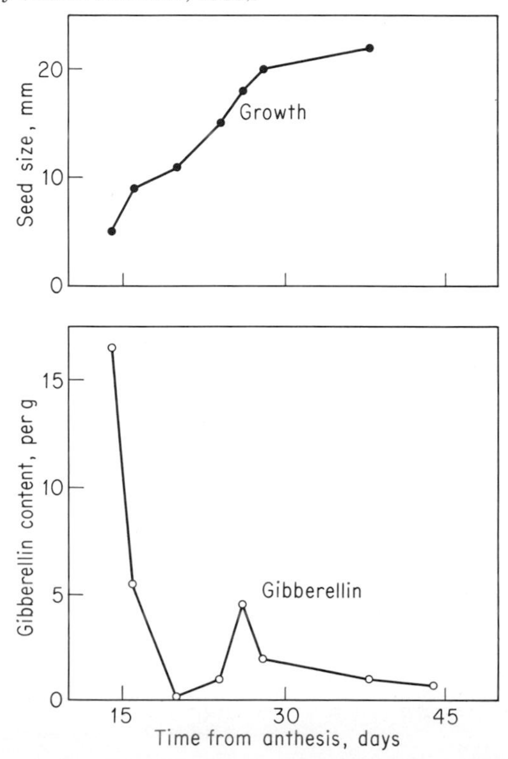

participate in the regulation of fruit growth, particularly the evidence that certain hormones may be correlated with growth in certain fruits at some times but not at others, one must set aside any tendency to think of a specific hormone as the principal regulator of fruit growth. If hormones do regulate fruit growth, one hormone may regulate growth at one time, and another at another time, or in another species; or alternatively regulation might be a result of the interactions of multiple hormones in any given fruit. This possibility of interactions has become important since Bradley and Dahmen (1972) found that peach mesocarp tissue in culture responds poorly to auxin, cytokinin, or ethylene but splendid growth is obtained when all three are supplied. One may recall, too, that blurring interactions seem to apply to the hormonal regulation of stem growth of other organs. Just as the evolution of fruits has been a complex of participating tissues from different morphologic origins, so also the evolution of regulatory systems in fruit growth are complex and diverse, utilizing any of the known hormone systems or perhaps combinations of several.

Little has been done on the physiological functions underlying the growth processes in fruits; Antoszewsky (1970) was unable to find evidence of changes in osmotic pressure or water potential during the growth of strawberry fruits, although there was a continuing mobilization of phosphorus and sugars by the growing cells. The extent to which wall softening may participate in fruit growth has not been investigated; the formation of intercellular spaces during apple enlargement (Fig. 13-18) indicates that losses of cell-cementing properties may be involved in the growth of some fruits.

FRUIT DROP

Fruit drop is a widespread horticultural problem. Not only does abscission occur at the time of normal fruit ripening, but in many fruits it occurs shortly after pollination and fruit-set or during the period of preferential growth of the young embryos in the seeds. A dramatic case of fruit drop is the *Macadamia* nut (Fig. 13-24),

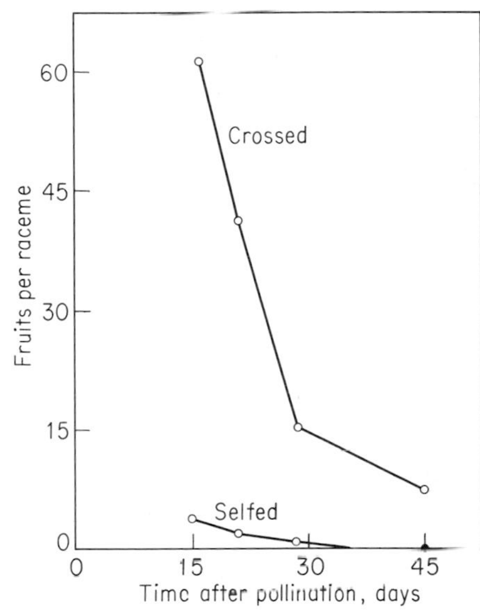

Fig. 13-24 Fruit drop in the *Macadamia* nut during the period of fruit growth results in the loss of a large percentage of the fruits (Urata, 1954).

where ordinarily more than 90 percent of the young fruits may be shed during fruit growth; fruits resulting from self-pollinations are in this case even more susceptible to drop, and ordinarily all the selfed fruits may be lost. In apple production, two periods of precocious fruit drop are recognized: *early drop,* which occurs between the period of initial swelling of the ovary and the start of endosperm development, and *June drop,* which occurs later during the period of rapid embryo development (Luckwill, 1953).

The presence of auxins may be the major deterrent to the abscission of fruits. The simplest assumptions suggest that when the auxin content of the fruits becomes low, fruit drop follows. Luckwill's (1953) data with apple support this scheme, both early drop and June drop being associated with periods of lesser auxin content in the seeds. A few years later, Wright (1956) assayed for the auxins in currant fruits, using both the conventional straight-growth test and an abscission-inhibition test employing *Coleus* petioles. On his chromatograms of fruit extracts, he found one spot which was quite effective in inhibiting abscission but which did not show growth-stimulating activity common to the usual auxins. He correlated the quantities of this auxin in currants with the periods of suppressed fruit drop and suggested that there might be some auxins in control of fruit growth and quite different ones limiting fruit drop. This situation remains unique to currant fruits up to the present and whether the retention of fruits is commonly controlled by auxin in the usual sense or by such abscission inhibitors as Wright describes remains to be ascertained.

Precocious fruit abscission can be relieved in many species by the application of auxin sprays, and it is likely that the natural drop relates at least in part to a low endogenous auxin content in the fruits.

Two other regulators that might be expected to be involved in fruit drop are ethylene and abscisic acid. Lipe and Morgan (1972) have detected substantial increases in ethylene production by cotton fruits during the first few days after fruit-set, and again a large production just before fruit dehiscence. Using another variety of cotton, Davis and Addicott (1972) have found large increases in ABA in the fruits occuring during the period of fruit dehiscence (Fig. 13-25). Each of these abscission-stimulating hormones may be participating in the fruit drop and dehiscence processes. The use of spray chemicals to stimulate fruit abscission is growing for many horticultural crops with the increasing need for mechanical harvesting; compounds which increase the ethylene levels are the most effective (Bukovac et al., 1969).

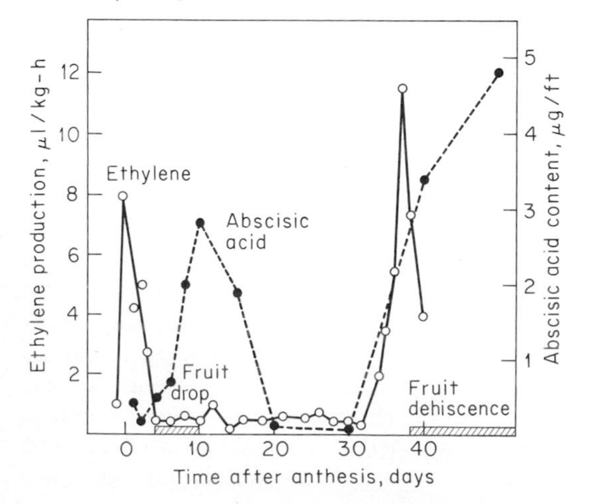

Fig. 13-25 A comparison of the changes in ethylene production of an Acala cotton fruit and the content of abscisic acid in Stoneville cotton fruits with fruit drop and dehiscence. Ethylene production increases before the drop of young fruits, and ABA content rises shortly afterward. Ethylene production precedes the time of fruit dehiscence, and ABA content rises during dehiscence (Lipe and Morgan, 1972; Davis and Addicott, 1972).

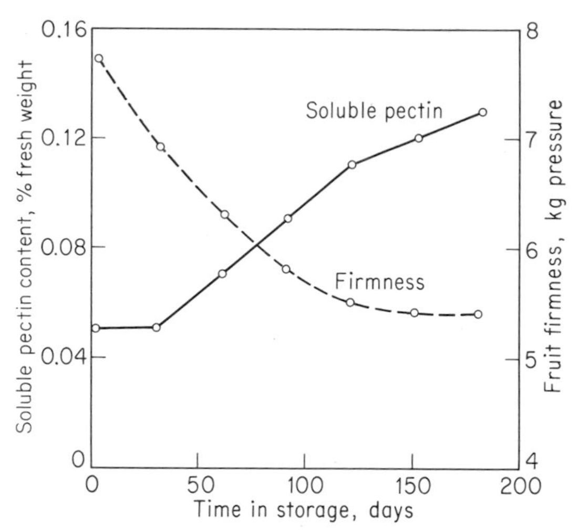

Fig. 13-26 During the ripening of apple fruits, the declining firmness is associated with a rise in the content of soluble pectins. Delicious apples in 2°C storage (Gerhardt and Smith, 1946).

FRUIT RIPENING

As the fruit reaches the end of its growth period, it may undergo some characteristic qualitative changes, which are collectively referred to as ripening. Often this includes a softening of the fleshy fruits. In some cases ripening is stimulated by picking the fruit; for example, in California the ripening of avocados characteristically does not occur until after the fruit is picked; in many other fruits ripening is hastened by picking, e.g., banana, apple, papaya.

To clarify terms, the word *maturation* refers to the processes associated with a fruit reaching full size, and *ripening* refers to the processes which qualitatively transform the mature fruit.

Changes with Ripening

The general changes associated with ripening, including softening of the fruit flesh, hydrolytic conversions of storage materials in the fruit, and changes in the pigments and flavors, can be attributed to the energy provided by respiratory activities. This concept, first specifically proposed by Biale (1950), is now well documented. Especially relevant is evidence that respiratory inhibitors can effectively prevent ripening (see, for example, Marks et al., 1957). Channeling the respiratory energies into the ripening processes involves marked changes in the enzymic components of fruits during ripening; a central theme in this subject is the formation of new enzymes and the marked alteration of the quantities of other enzymes.

Softening is one of the most dramatic changes with the ripening of fleshy fruits. From the widespread reports of changes in cell-wall-degrading enzymes, it appears that wall degradation is a major part of fruit softening. In some fruits, considerable contributions to softening occur through the hydrolysis of cell contents, e.g., the hydrolysis of starches in squash and fats in avocado.

Softening is ordinarily measured by the force needed to press a plunger a given distance into the fruit; changes during the ripening of apples are illustrated in Fig. 13-26. The softening is interpreted as the solubilization of pectic substances from the middle lamellae and hence an associated rise in soluble pectins. The solubilizing of the pectic substances may occur through an increase in the methylations of the galacturonic acids or through a shortening of the polygalacturonide acid chains or both.

Illustrating the formation of enzymes which may degrade or solubilize the pectins or cellulose components of the cell walls during ripening are data in Fig. 13-27; in dates the early stages of ripening involve the appearance of polygalacturonase and cellulase. In the mature green fruit these enzymes cannot be detected, from which we can infer that the ripening process may involve the new formation of these enzymes. Polygalacturonase appears to play a role in the softening of several types of fruits (Hobson, 1964; Zauberman and Schiffman-Nadal, 1972).

Hydrolytic changes during ripening usually lead to the formation of sugars. Various fruits show widely different rates and extents of such hydrolytic activities; e.g., the banana ripens extremely fast, and the hydrolysis of starch is precipitous (Fig. 13-28). Apple fruits are more gradual in hydrolytic as well as in other ripening actions, and citrus fruits such as orange and lemon are ponderously slow, sometimes taking months. Hydrolytic activities can give rise to increases in sugars, not only from starches but also from fats (Beevers, 1961).

In contrast to other food reserves, proteins are actually synthesized at increasing rates during the ripening process in apples, as shown by Hulme (1939, 1954). This important exception to the hydrolytic tendencies will be discussed below in connection with the mechanism of ripening.

Quality changes in ripening fruits include

Fig. 13-27 With the onset of ripening of date fruits, there is a marked appearance of cell-wall-degrading enzymes, including polygalacturonase and cellulase; these may be inferred to play a role in the softening of the fruit during the time of sweetening, as indicated by the rise of invertase (Hasegawa et al., 1969; Hasegawa and Smolensky, 1970, 1971).

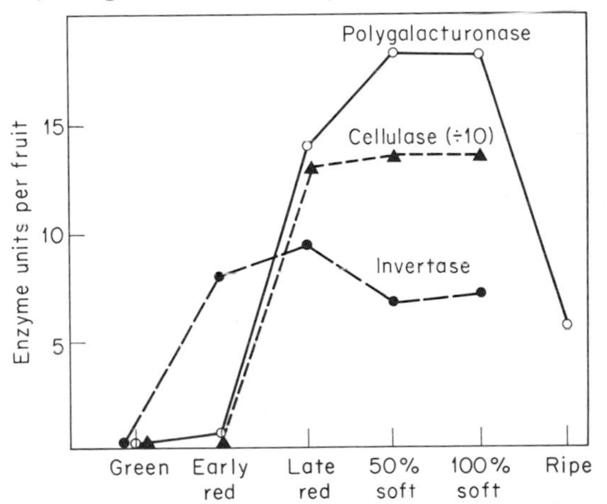

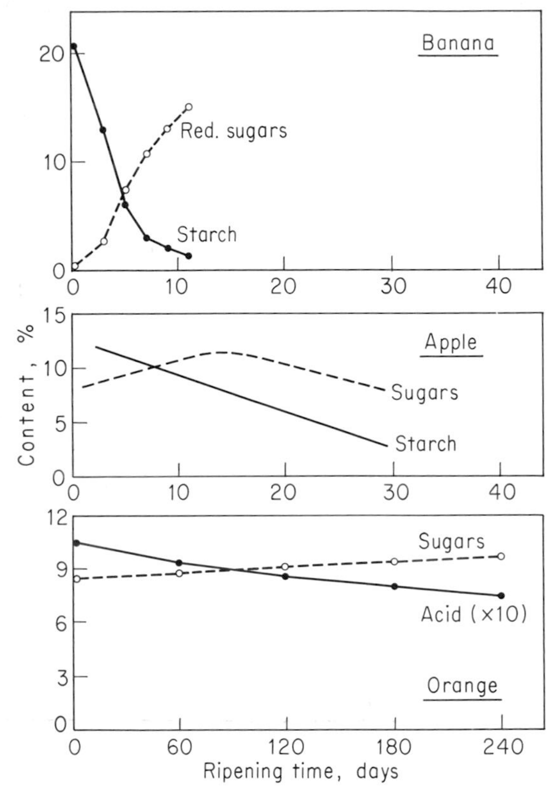

Fig. 13-28 Fruit-ripening processes may be very rapid, as in banana; moderate, as in apple; or slow, as in orange. The most rapid ripeners are usually associated with a hydrolysis of starch or other reserve material (von Loesecke, 1950; Smock and Neubert, 1950; and Stahl and Camp, 1936).

marked changes in pigmentation, production of flavor materials, and usually depletion of astringent substances. The pigment changes during ripening of three fruits are shown in Fig. 13-29 to illustrate the usual drop in chlorophyll content and the formation of carotenoids. The loss of chlorophyll from fruits may be synchronous with ripening (as in banana), or it may occur only in the earliest stages of ripening (in the orange) or, rarely, after other indexes of ripening have been passed (in some pears). The coloration of the ripening fruits may be a consequence of the formation of carotenoid pigments (as in the orange) or a consequence of the disappearance of chlorophyll with little or no net formation of carotenoids (as in the banana).

The pigment changes occur mainly in the chloroplasts, of course, converting them from green chloroplasts with grana into chromoplasts with more disperse thylakoid membranes. Examination of these changes at the ultrastructural level reveals that there is an extensive synthesis of new thylakoids, and Spurr and Harris (1968) have concluded that a dynamic synthesis of membranes occurs during the conversion in pepper fruits.

Changes in enzyme components of the fruit are, of course, involved in the altering pigmentation during ripening; one of the most general changes is an increase in chlorophyllase (Looney and Patterson, 1967).

The pigments of some fruits are limited to carotenes (papaya) and of others to anthocyanins (strawberry); in some the pigments are restricted to the shell (apple), and in others they permeate the entire fleshy part (peach). The anthocyanin or flavone pigments may be formed in response to sunlight (Dugger, 1913), and the action spectrum for the pigmentation of apples indicates the involvement of phytochrome (cf. Siegelman and Hendricks, 1958). Sugars may accentuate the tendencies to form pigments (Thimann and Edmondson, 1949). These considerations, plus the fact that pigmentation may be strongly altered by temperature (Denisen, 1951) in a manner distinctly different from ripening changes, indicate the color changes may not be proportional to fruit ripening.

The development of flavor substances in fruits remains for future research, and surely the advent of gas chromatography will allow rapid advances in this area. In apples numerous esters, aldehydes, and ketones have been identified (cf. Ulrich, 1958), plus a series of saturated and unsaturated hydrocarbons (Meigh, 1959). Siegelman and Hendricks (1958) have made the surprising observation that light inhibits the formation of some flavor substances in apples, apparently diverting the necessary substrates over to the formation of pigments instead.

The decline of astringent materials such as

phenolics (Reeve, 1959) is common to pomological fruits, but little is known of the significance of the changes.

Lest the description of fruit ripening take on the aspect of a deterioration, it should be stressed that the ripening fruit is ordinarily an effective synthetic organ. The many enzymes involved in the ripening changes must be synthesized in part by the ripening fruit; the strong effectiveness of protein synthesis in apples during ripening has been well documented by Hulme (1954). Likewise, RNA synthesis is very active in the early stages of ripening (Hulme et al., 1971).

Fig. 13-29 Pigment changes during fruit ripening ordinarily involve a decline in chlorophyll and transformations of carotenoids. The banana does not change in total carotenoids; the apple carotenoids rise as the chlorophyll declines; and the orange carote noids; the apple carotenoids rise as the chlorophyll declines; and the orange carotenoids increase considerably later (von Loesche, 1950; Workman, 1963; Miller et al., 1940).

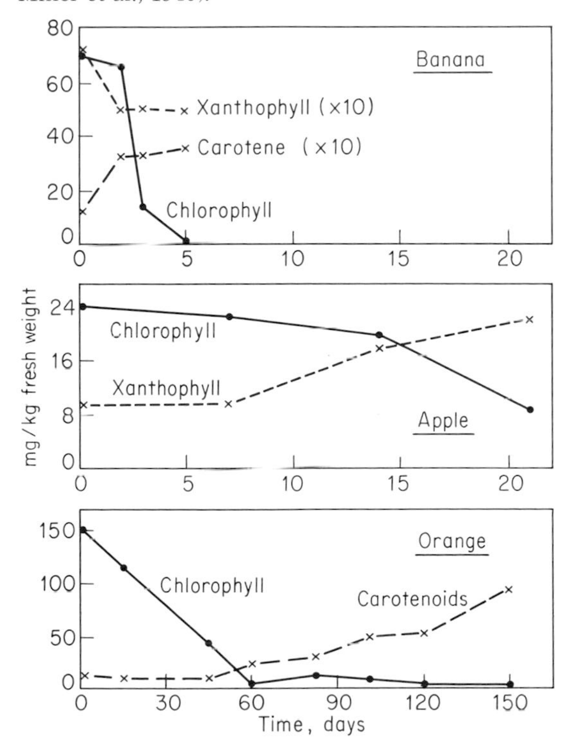

The mitochrondria of the ripening fruit have very strong respiratory control (Ku et al., 1968), and their good metabolic activity persists until the later stages in fruit senescence—after the passage of the respiratory climacteric (Bain and Mercer, 1964).

The Respiratory Climacteric

From studies of the ripening of apple fruits, Kidd and West (1930) discovered that ripening is associated with spectacular changes in respiratory rates, including a lowering of respiration in the mature fruit, followed by a large increase in respiration during the time of ripening. After reaching a climacteric peak, respiration falls off again as the fruit begins to enter a senescent decline. These workers also established that the effectiveness of refrigeration in prolonging the storage life is associated with a suppression of the intensity of this respiratory climacteric and prolongation of its duration.

In many fleshy fruits the climacteric peak occurs at the time of optimal eating quality, as in pears; in others it slightly precedes this optimum (apple, banana). In tomatoes the climacteric is reached well before the fruit is fully ripe. Biale (1950) has suggested that the climacteric is associated with the hydrolysis of food reserves in the fruit, pointing out that fruits which do not ordinarily experience a climacteric after harvest, e.g., orange, lemon, and fig, do not experience extensive hydrolytic activities (see Fig. 13-28).

The climacteric may occur rapidly, as in the avocado and banana; at intermediate rates, as in pear, mango, and apple; or not at all, as in orange and lemon. Some comparative climacteric curves at room temperatures are illustrated in Fig. 13-30; in each case there is a distinct respiratory decrease in the mature fruit, followed by the rise to the climacteric peak and then a drop again.

In contrast to the climacteric fruits and the fruits like *Citrus* that maintain a steady respiration during ripening, some fruits show only a decline in respiration during the ripening period. Examples include the pepper (Howard and

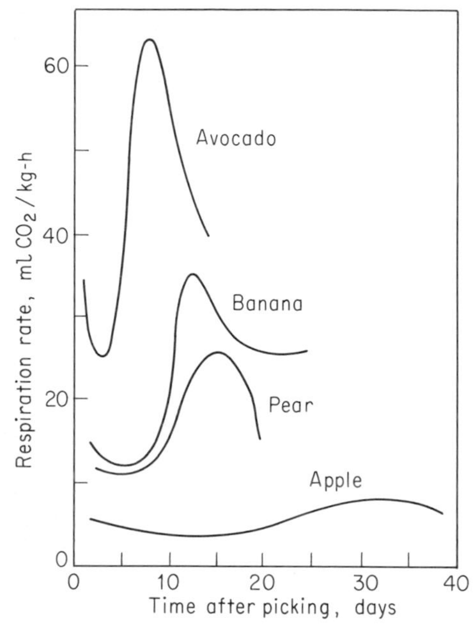

Fig. 13-30 The rate of ripening processes is reflected in the intensity of the respiratory climacteric, the fastest-ripening avocado showing the greatest climacteric peak, and banana, pear, and apple showing lesser peaks, in that order (Biale, 1950).

Yamaguchi, 1957) and the peanut (Schenk, 1961).

Most fruits which experience the climacteric do so whether they are picked or ripen on the tree; however, some fruits, especially the avocado in California, will start the climacteric only after being picked. Like ripening processes in general, picking hastens the climacteric rise of many fruits (Smock, 1972). If a fruit has passed the climacteric peak before being picked, its respiratory pattern will be only a steady decline, a feature which is useful for determining the state of ripening of picked fruits (Kidd and West, 1937).

The original observations of Kidd and West (1930) that low temperatures suppress and defer the progress of the climacteric appear to apply to fruits generally. In some fruits too low a temperature will erase the ability to proceed with the climacteric (e.g., pear; Hansen and Hartmann, 1937).

Knowing of the climacteric, Kidd and West (1933) were able to develop storage techniques which utilize controlled atmospheres—low oxygen and elevated carbon dioxide levels. Such treatments can effectively prevent the development of the climacteric rise and result in much improved storage qualities. Controlled atmosphere storage has come into widespread use in recent years, especially in apple storage (Smock, 1958). Storage of fruits in polyethylene bags can roughly accomplish some of the same effect since the plastic can lower the oxygen and elevate the carbon dioxide levels around the fruit (Smock and Blanpied, 1958).

The abilities of some gases to stimulate the climacteric form the basis of an ancient Chinese custom of ripening fruits in rooms where incense was being burned. Kerosene stoves were used to improve the coloring of lemons in California in the 1920s, and Denny (1924) found that unsaturated hydrocarbons in the fumes are responsible. Trying ethylene as a simple model, he found that this gas can strongly promote ripening activities. Gane (1937) specifically established that ethylene stimulates the climacteric rise; fruits which had reached the climacteric peak were insensitive to the ethylene treatment. He further established that ripe bananas stimulate the climacteric in the same manner as ethylene and that ethylene can be identified in the emanations from the ripe fruit (Gane, 1935).

After the discovery that ethylene can trigger the climacteric rise, 30 years passed before precise evidence was brought to bear on the question of whether ethylene serves as a ripening hormone. With the development of gas-chromatographic measurements of ethylene, Burg and Burg (1962) were able to detect a marked rise in ethylene formation in fruits either just at or just before the onset of the climacteric rise (Fig. 13-31); they made quantitative experiments on the effectiveness of ethylene in inducing the climacteric and found that fruits exposed to higher concentrations of ethylene experience greater respiratory climacteric peaks and the climacteric develops in a shorter time, as illustrated in Fig. 13-32. Depriving banana fruits of ethylene by placing them under reduced pressure results in a remarkable blockage of ripening (Burg and Burg, 1965a). Supporting the interpretation that the partial vacuum prevents ripening by depriving the fruit of ethylene is the evidence that introducing pure oxygen into the partially evacuated atmosphere does not restore the ripening activities, but introduction of some ethylene with the oxygen does (Burg and Burg, 1966a). On the basis of these experiments Burg and Burg assert that ethylene is in fact a ripening hormone.

The ability of ethylene to induce a respiratory climacteric is not exclusive to climacteric fruits. Nonclimacteric fruits such as oranges and lemons can be forced into a climacteric respiratory peak by exposure to ethylene during the ripening period (Denny, 1924). Some fruits such as strawberry are singularly insensitive to ethylene with respect to ripening (Matson and Jarvis, 1970).

Fig. 13-31 When gas chromatography was used, the rise in respiration during ripening was found to be associated with a concomitant rise in ethylene production in mango and to be preceded by a burst of ethylene production in the banana. Fruits of Kent mango, 24°C; Gros Michel banana, 16°C (Burg and Burg, 1962).

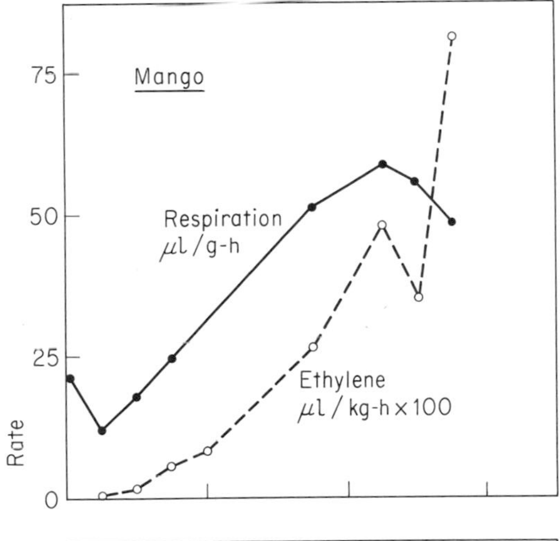

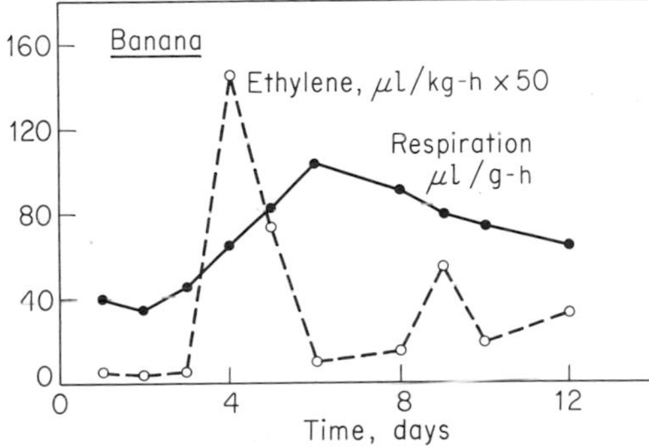

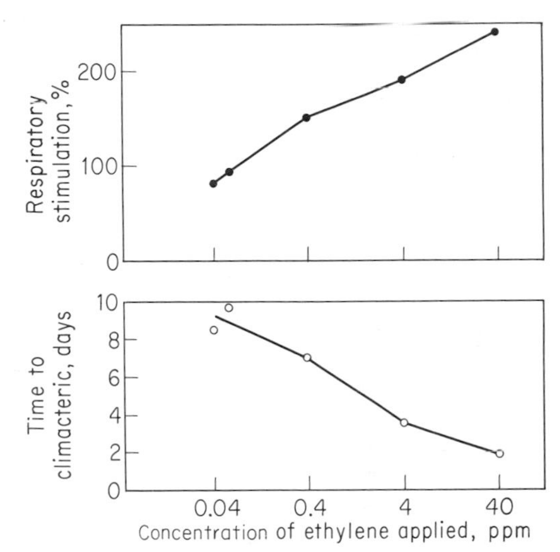

Fig. 13-32 Ethylene treatments of Kent mangoes can quantitatively regulate the climacteric achieved during ripening, as shown by the shortening of the time to climacteric peak (*below*) and by the greater magnitude of the respiratory peak achieved (*above*) with increasing concentrations of ethylene applied for 2 days at 24°C (Burg and Burg, 1962).

Ripening Mechanisms

The essential elements in the fruit-ripening complex might be considered to be (1) a respiratory source of energy for ripening, providing the fuel, so to speak, for (2) the synthesis or formation of new sets of enzymes, and (3) the actions of these enzymes in bringing about the changes characteristic of ripening such as softening, pigmentation, and quality changes in the fruits. Early theories which attempted to explain ripening were all concerned with the respiratory aspects; only fairly recently has the concept of ripening become concerned with the unleashing or derepression of a distinctive genetic program for ripening.

The first theory to account for ripening was advanced by Kidd (1934). Following in detail the changes in constituents of apples during ripening, he observed that when the climacteric rise begins, fructose starts to disappear from the cytoplasm of the cells and the respiratory rise continues as long as fructose is being supplied from the vacuoles to the cytoplasm. He proposed, therefore, that the climacteric is a consequence of the metabolism of a special substrate, an *activated fructose.* Biale (1950) has noted the possibility that such a role could be taken by a phosphorylated fructose. At present there is only scattered evidence that sugars or any other substrates serve as critical substrates for ripening; Hulme and Neal (1957) reported that malate can hasten ripening of apples, Rakitin et al. (1957) found a similar effect in persimmon with methanol, and Hobson (1965) has hastened the ripening of tomatoes with methionine. The methionine effect might be due to an enhanced biosynthesis of ethylene rather than a metabolic utilization of the methionine.

A second theory was proposed by Wardlaw and Leonard (1936) from their studies of gas-exchange experiments with several tropical fruits. They noted that with increases in size, fruits develop physical limitations to the free exchange of carbon dioxide and oxygen; as these limitations become acute, CO_2 accumulates inside the fruit and oxygen levels become very low. They proposed that the respiratory climacteric is an anaerobic type of respiratory shift. Careful measurements of the internal gases revealed that such accumulations of CO_2 and depletions of O_2 do occur, and, as illustrated in Fig. 13-33, these changes coincide with the development of the climacteric. In further support of their idea that the climacteric rise is a type of anaerobiosis, they noted that some fruits actually ripen from the center outward (Leonard and Wardlaw, 1941) and that elevations of temperature which magnify the climacteric also enhance the state of anaerobiosis inside the fruit. While their measurements were fine for the techniques available at the time, their theory seems to be deficient in two respects: (1) exposing climacteric fruits such as apple or avocado or banana to elevated carbon dioxide depresses the climacteric and retards ripening (Blackman and Parija, 1928); (2) peeling or slicing avocado or banana fruits and thus bringing about a lowering of internal carbon dioxide and elevation of oxygen levels in the tissues enhance respiration and hasten ripening (Ben-Yehoshua et al., 1963).

Fig. 13-33 The respiratory climacteric in banana fruits is associated with large changes in the internal atmosphere in the fruit, including a marked depression of O_2 content and a rise in CO_2 content (Leonard and Wardlaw, 1941).

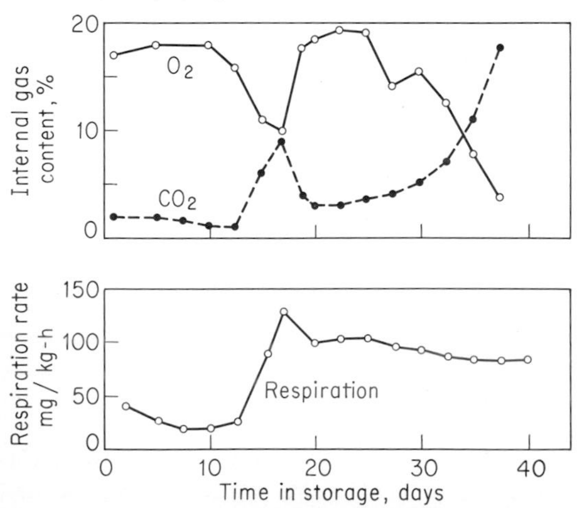

Twenty years passed before another theory of ripening was proposed. As our knowledge of phosphorylations as the principal means of minting the currency of metabolic energy grew, the possibility arose that the availability of phosphate acceptors might naturally limit metabolic rates. Pearson and Robertson (1952) first tested the possibility in ripening apples by measuring the response of fruits at different stages of ripening to additions of the uncoupling agent dinitrophenol. This poison increases respiration by increasing the availability of phosphate acceptors at the expense of phosphate esters. They found that it evokes large respiratory increases in preclimacteric but not climacteric or postclimacteric fruits. Millerd et al. (1953) further developed the idea in studies of ripening in avocado; they confirmed the dinitrophenol response in that fruit and found that extracts of climacteric fruits can cause respiratory uncoupling in mung bean mitochondrial preparations.

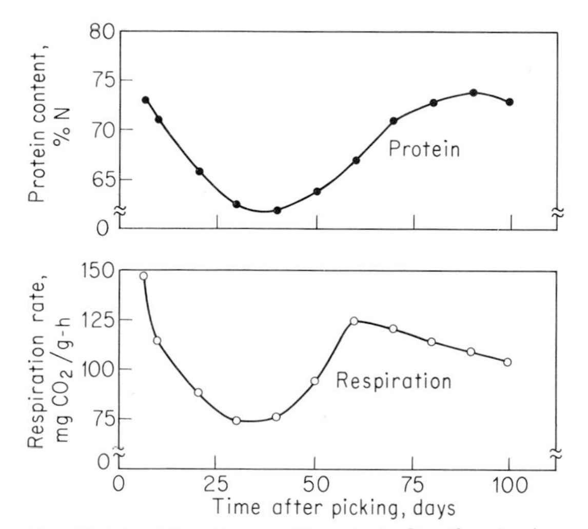

Fig. 13-34 After the preclimacteric dip, the rise in respiration is also associated with a rise in protein content in Bramley's Seedling apple, indicating an enhanced synthesis with the respiratory burst (Hulme, 1954).

They suggested specifically that the respiratory rise is a consequence of the accumulation of naturally occurring uncoupling poisons which obstruct the effectiveness of oxidative phosphorylation. The respiratory rise would then be interpreted as a racing of the respiratory motor because of the uncoupled phosphorylation.

The uncoupling theory of ripening implies that with the climacteric rise there develops a depreciation of available energies for biosynthesis, and even before the theory was proposed, Hulme (1939, 1954) had noted that ripening in apples is associated with considerable increases in protein synthesis. The rise in protein content during ripening is illustrated in Fig. 13-34. The uncoupling theory is more directly denied, however, by experiments which show that uncoupling agents such as dinitrophenol actually block the ripening of tomato fruits (Marks et al., 1957). Furthermore, studies of the respiratory control of mitochondrial preparations from ripening fruits show that oxidative phosphorylation actually increases during ripening, as shown for avocado fruits in Fig. 13-35. If uncoupling were occurring, one would expect the levels of ATP in the fruit to fall, and they actually increase during the ripening of melons (Rowan et al., 1969). In some experiments using the oxygen electrode as a measure of respiratory control, Wiskich et al. (1964) showed that avocado mitochondria maintain a strong respiratory control up to the time of the climacteric peak. It appears, then, that the respiratory climacteric, instead of being due to an uncoupling of oxidative phosphorylation, is a period when there is a close coupling of phosphorylation and oxidation.

Another theory of ripening has been suggested by Sacher (1959, 1966). He found that solutes leak with increasing ease from banana fruit slices as ripening proceeds and proposed that ripening and the associated respiratory rise are related to membrane deterioration. His experiments were criticized by Burg et al. (1964) on the basis that the amounts of solutes in the fruit tissue were also increasing during ripening and could account for the leakage of solutes. Sacher, however, showed that the apparent free space increased, using radioactive solutes

Fig. 13-35 The respiratory climacteric in avocado fruits is reflected in an increase in oxidative phosphorylation, as indicated by the rise in the P/O ratio; pyruvate as substrate (Romani and Biale, 1957).

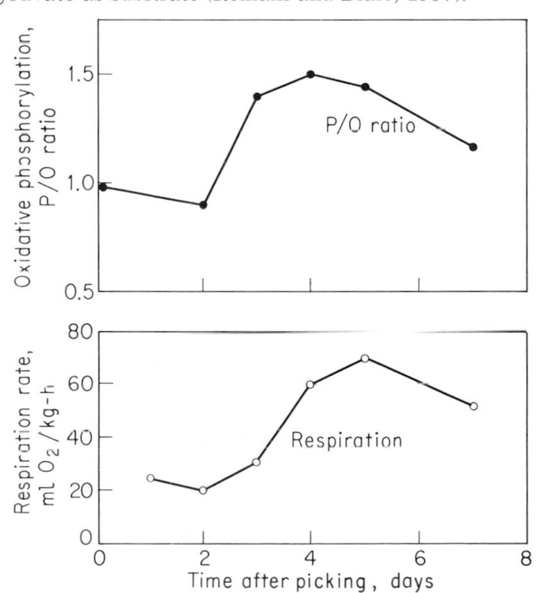

rather than endogenous solutes; his experiments showed a rise in apparent free space beginning even before the start of the respiratory climacteric, thus offering new promise as a possible cause of the climacteric rise. The question has been reexamined by Brady et al. (1970), who reported that the apparent free space does not increase until after the respiratory climacteric rise has started, and they, too, take the position that the respiratory rise cannot be a consequence of the increase in leakiness of the fruit cells.

A modern concept of ripening on the basis of the synthesis of new enzymes involved in the ripening process began to emerge with the work of Richmond and Biale (1966), who obtained increases in the incorporation of radioactive amino acids into the protein of avocado fruits during the first half of the climacteric rise; they suggested that the many enzymes known to arise during ripening would logically spring from this activity of protein synthesis. Frenkel et al. (1968) expanded this point of view into an overall concept of ripening, showing that in pear fruits, the blockage of protein synthesis by cycloheximide effectively prevented ripening. They separated protein components by gel electrophoresis and found that radioactive amino acids are incorporated into many more protein components during the early climacteric rise than before the rise begins; again, the biosynthesis of new proteins for ripening processes is implied. Hulme et al. (1971) have extended the incorporation experiments to show that ripening apples also show an enhanced RNA synthesis, and Marei and Romani (1971) have reported marked increases in the incorporation of radioactive uridine into each of the RNA fractions of fig fruits following the stimulation of ripening with ethylene.

Considering the picture of ripening as a respiratory source of energy which can be applied to the synthesis of new proteins (or enzymes) or to the changes in the fruit which may be directed by these new enzymes, one might ask how closely linked the respiratory activities are with the actual changes in terms of softening, pigmentation, and quality changes in the fruit. Until the late 1960s, every evidence showed a close linkage between the respiratory climacteric and overt changes involved in ripening; if one slowed down ripening by low temperature or by controlled gas atmospheres, one observed similar slowing of the respiratory climacteric. If one shuts down the metabolic production of energy, e.g., with dinitrophenol, arsenite, or fluoroacetate, ripening is ordinarily terminated (Marks et al., 1957; Ben-Yehoshua, 1964). Inhibiting protein synthesis with cycloheximide, Frenkel et al. (1968) showed that the softening and pigmentation changes of ripening can be terminated without any depression of the respiratory climacteric, as shown in Fig. 13-36. A loose connection between respiration and ripening has also been shown by Dostal and Leopold (1967), who reported that whereas the ethylene stimulation of tomato ripening is associated with a hastening of the respiratory climacteric, the application of gibberellin can defer ripening, e.g., the development of red color, without any apparent alteration of the climacteric. In two types of experiments, it has been found that ripening can proceed even though the climacteric has been markedly suppressed; in pears, Hansen and Blanpied (1968) were able to induce ripening in relatively immature fruits by brief exposure to ethylene without the development of the usual respiratory climacteric; in tomatoes, Pharr and Kattan (1971) could almost suppress the respiratory climacteric by placing the fruits in a slow airflow system without apparently interfering with the ripening processes of softening or color development. It seems reasonable to believe, then, that respiration provides the energy for the synthesis of new enzymes and for the action of these enzymes in bringing about the ripening changes but that the linkage of respiration to these activities is fairly weak; treatments which markedly depress some of these activities may not necessarily have comparable effects on others. It is well established, also, that the ripening process is not energized by a catabolic decline in the respiratory system but at least the early part of the ripening process requires an orderly and effectively controlled respiratory system.

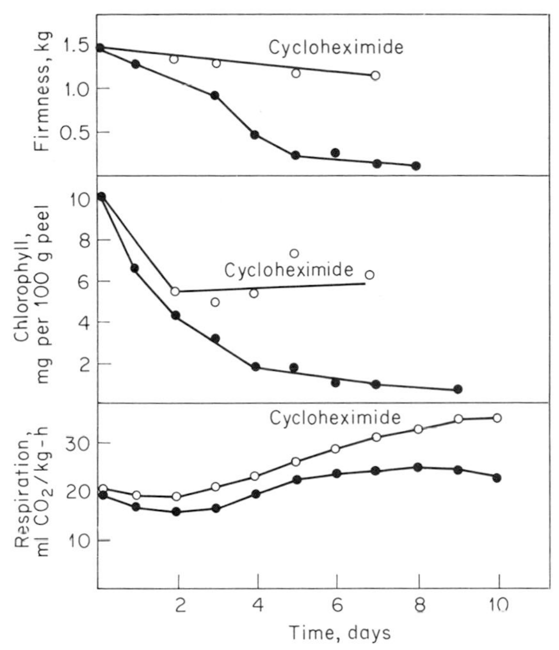

Fig. 13-36 The application of the poison cycloheximide to Bartlett pears can block the ripening processes, as measured by loss of firmness and of chlorophyll, without blocking the respiratory climacteric (Frenkel et al., 1968).

Hormonal Controls of Ripening

The ability of a fruit to become ripe is closely related to some changes experienced at maturity. Massive doses of ethylene can bring about ripening changes in immature fruits (see, for example, Lyons and Pratt, 1964b), but as a fruit reaches maturity, or full size, its responsiveness to the ethylene trigger is enormously increased. In many species, picked fruits ripen much more readily than fruits on the tree, and this has been the basis for the suggestion that some (hormonal?) factor supplied by the tree defers ripening (Biale, 1960).

Turning to the effects of the known hormones on fruit ripening, we see the same theme repeated as for many other phases of plant development: various plant hormones can serve to regulate fruit ripening in at least some instances. Early experiments with auxins indicated that they stimulate ripening (Clark and Kerns, 1943), but these observations are no doubt reflections of the auxin-stimulated ethylene formation (Maxie and Crane, 1968). In later experiments (Vendrell, 1969) auxin applications were quite effective in deferring the ripening of bananas and pears, and Frenkel (1972) has suggested that the onset of ripening in several types of fruits may be a loss of hormonal control by auxin. The concept of an auxin deferral and ethylene promotion of ripening finds a parallel in the interactions of these hormones in abscission. Gibberellin was first reported to retard fruit ripening by Coggins and Lewis (1962), and the effect of this hormone is widespread, having been shown now for many species (Dilley, 1969). Cytokinin effects on ripening are not well known, and only preliminary reports of inhibitions of ripening have been made (Abdel-Gawad and Romani, 1967). Abscisic acid has been reported to increase with maturity and during ripening of strawberries (Rudnicki et al., 1968). Ethylene of course is known to play a critical and triggering role in the onset of ripening of climacteric fruits (Lyons et al., 1962). While the effects of some of the hormones are known in only a sketchy and preliminary way, it appears that in at least some kinds of fruits each of the hormones may have some regulatory potential.

Ethylene is the hormone with the most powerful regulatory role in ripening. Whereas earlier experiments suggested that ethylene might best be described as a product of the ripening changes (Hansen, 1942), more recent experiments have established that a rise of ethylene occurs at the onset of the climacteric rise and can be assigned the role of the trigger of ripening (Burg and Burg, 1962). In some fruits the production of ethylene is predominantly during the early part of the climacteric rise, and in others it continues to rise all during the climacteric, as shown in Fig. 13-31. Burg and Burg (1966a) have pointed out that the onset of ripening is associated not only with a rise in the ability to biosynthesize ethylene but also a marked increase in ethylene responsiveness. One might reasonably expect the ethylene stimulation of ripening to be engendered by two types of effects: (1) ethylene may be bringing about the formation of new types of enzymes in the fruit (as shown, for example, by Riov et al., 1969), and (2) ethylene may be stimulating its own biosynthesis, as described in Chap. 7.

Ethylene treatments can induce some leaves to experience a respiratory climacteric (Herrero and Hall, 1960) and to develop many of the pigment changes commonly associated with fruit ripening.

It is well known that unsaturated gases other than ethylene are produced by ripening fruits (Meigh, 1959), and such gases apparently may have stimulatory properties not unlike those of ethylene (Maxie and Baker, 1954). Ethylene production in apples, however, is a thousandfold more intense than the production of other volatile unsaturated hydrocarbons, and so ethylene appears to be the dominant stimulant in this group.

In an interesting experiment carried out by McMurchie et al. (1972) propylene was utilized to stimulate the ripening of various fruits; thus they could separately measure the gas utilized to induce ripening and the ethylene biosynthesized by the fruit. The impressive result was that climacteric fruits such as banana respond to the ripening stimulus by biosynthesizing large amounts of ethylene, whereas nonclimacteric fruits such as the citrus fruits respond to the ripening stimulus without any ethylene formation. This may indicate that climacteric fruits are those which show the autocatalytic effects of ethylene on ethylene biosynthesis.

Ethylene is not a universal trigger of fruit ripening. Some nonclimacteric fruits are quite impervious to ethylene regulation, e.g., the strawberry (Matson and Jarvis, 1970).

The effects of gibberellin in the regulation of ripening deserve special attention. Its effectiveness in blocking the pigment changes with ripening (Coggins and Lewis, 1962) has led to commercial use of gibberellin in citrus culture. Its effectiveness in the deferral of ripening of other fruits (tomato, banana, mango, pear, apricot) suggests the possible interpretation that gibberellin acts in fruits in indirect opposition to ethylene, as shown in several growth and enzyme systems (Scott and Leopold, 1967). It has already been mentioned that gibberellin can alter pigmentation changes without altering the respiratory climacteric. The gibberellin deferral of ripening can be overcome by applications of ethylene (Russo et al., 1968). Wade and Brady (1971) assert that deferral of the pigment changes in ripening bananas can be achieved with applications of gibberellin or auxin or cytokinin.

In conclusion, ripening appears as an unveiling of enzyme systems which bring about the alterations of the fruit, and respiratory energy must be provided both for the synthesis of the new enzyme systems and for their actions in ripening. Finally, hormonal regulation may be involved in the change from a mature fruit resistant to ripening to one which is responsive to ripening signals and then finally, of course, in the ripening signal itself.

GENERAL REFERENCES

Hulme, A. C. (*ed.*). 1970. "The Biochemistry of Fruits and Their Products." Academic Press, London. 620 pp.

Sacher, J. A. 1973. Senescence and postharvest physiology. *Annu. Rev. Plant Physiol.* 24:197–224.

Spurr, A. R. 1970. Morphological changes in ripening fruit. *HortScience* 5:33–35.

14 TUBER AND BULB FORMATION

RELATION TO GROWTH

ENVIRONMENTAL EFFECTS

NATURE OF THE STIMULUS

TUBER AND BULB GROWTH

TUBER RIPENING

TUBER REUTILIZATION

SUMMARY

The fruit is not the only organ with mobilization properties so strong that it swells into an inflated storage structure. The swellings of stems into tubers or of leaf bases into bulbs or of roots appear to be different morphological manifestations of the same ontogenetic end. In each case a powerful set of mobilization activities builds up a large storage of carbohydrates or fats, and the location of this mobilized depot is an organ in which the polarity has been degraded.

The formation of tubers and bulbs has several interesting analogies to the flower and fruiting activities of plants. In addition to the powerful mobilization actions which are shared with numerous fruits, tubers and bulbs may be formed in response to an induction phenomenon which occurs in the leaves and is transmitted to the part which will do the swelling. There follows a morphological differentiation and then growth of the storage organ; to continue the analogy, the tubers and bulbs, like some fruits, then experience a ripening phenomenon. The storage organs other than fruits occur as *tubers* in potatoes and jerusalem artichokes, *bulbs* in onions and certain grasses and sedges, or simply as enlarged basal regions in the *stem*, e.g., the corms of gladiolus and the marvelously enlarged stem base of kohlrabi, or finally, as enlargements of the *root* itself, e.g., radish, carrot, and beet.

How the swelling organ develops is not well understood. In the potato tuber, the first morphological beginning is the lateral enlargement of cortical cells immediately behind the tip of a stolon; lateral enlargement is then extended through the mitotic activity of cortical cells, and the mitotic activity continues throughout the progress of tuber growth. The internodes and nodes of the stolon are retained during the tuber development, as illustrated in Fig. 14-1.

The formation of tubers normally occurs at the tips of ageotropic stems or stolons, the lack of polar elongation being there associated with recumbent stem growth. The association with stolons, however, is not consistent, for in *Begonia evansiana* tubers are formed on short lateral branches along the upright stem. The tuber retains the nodes as "eyes," with a leaf bract or scar and subtended bud; the expression

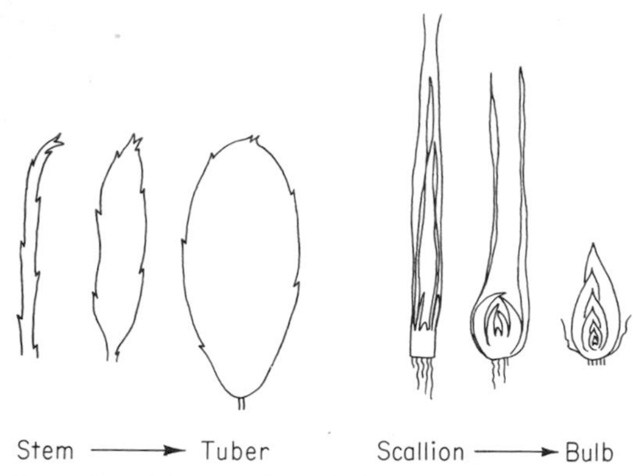

Fig. 14-1 The morphological changes involved in tuber formation (*left*) include an inflation of the stem through a swelling type of growth. In the formation of a bulb (*right*), the inflation is not of the stem but of the bases of the young leaves.

of apical dominance by the terminal eye over the other eyes of the tuber is a natural carryover from the stem situation.

The formation of an onion bulb is a consequence of the mobilization of carbohydrates into the bases of very young leaves. As described by Heath and Holdsworth (1948), there is a cessation of growth of the apical meristem and roots, a cessation of cell division generally, and an investment in a lateral swelling type of growth by the young leaves (Fig. 14-1). When the bulb starts to sprout, elongation and mitotic activities are restored in the newest leaf primordia, which were enclosed inside the bulb scales, and the enlarged scales deteriorate as they yield their stored carbohydrates.

Morphologically, the formation of storage organs is associated with a strong localized lateral enlargement of cells and the deposition in these cells of storage polysaccharides, usually starch and sometimes inulin. So the deposition of starch becomes a central part of storage-organ formation.

Assuming that sucrose is the principal carbohydrate being translocated to the tuber, the synthesis of starch from sucrose will be the biosynthetic system involved. While the biosynthesis of starch is not well understood, it apparently proceeds through the formation of glucose nucleotides (Leloir et al., 1960, 1961). The arriving sucrose is presumed to be acted upon by sucrose synthetase:

$$\text{Sucrose} + \text{uridine diphosphate} + \text{ATP} \leftrightarrows \text{uridine diphosphate glucose} + \text{fructose} + \text{ADP}$$

The uridine diphosphate (UDP) glucose is the presumed immediate precursor of starch, being incorporated onto the end of a starch polysaccharide chain by amylose synthetase:

$$\text{UDP glucose} + (\text{glucose})_n \rightarrow \text{UDP} + (\text{glucose})_{n+1}$$

The fructose formed by the sucrose synthetase step can be phosphorylated by ATP, converted to glucose 1-phosphate by hexokinase, and then made into UDP glucose through the utilization of one more ATP, thus providing another glucose unit for starch synthesis. The conversion of one sucrose molecule into two amylose links in a starch grain uses up two high-energy phosphate bonds:

$$\text{Sucrose} + \text{UDP} + 2\text{ATP} \rightarrow 2 \text{ amylose links} + \text{UDP} + 2\text{ADP} + 2\text{P}_i$$

Edelman (1963) has pointed out that this is a very expensive synthesis with regard to the energy consumed; for the synthesis of two starch linkages of 4300 cal each, the plant expends the high-energy sucrose linkage (6,600 cal) plus two ATP linkages ($2 \times 7{,}700$ cal), which gives an efficiency of only about 40 percent. He points out, however, that it is unlikely that this low efficiency is significant in the economy of the plant, since one respired hexose unit can generate between 30 and 40 high-energy phosphate linkages, and in the tuber itself there is hardly a shortage of hexose units.

Besides amylose synthetase, two other enzymes are important to starch synthesis. Some glucose polymer must be present before amylose synthetase can build further glucose blocks; in potato there is a starter enzyme, called D enzyme, which can make some primer out of simple saccharides such as maltotriose (Edelman, 1963). Branching of the amylose chain occurs through the action of Q enzyme, which in potato moves blocks of about 20 glucose units off the end of the amylose chain and inserts them on the side, thus initiating a branch. Branching is of great importance: it increases the loci on which starch synthesis can occur and is the major factor determining the physical characteristics of the starch polymer and the starch grain.

The physical growth of the starch grain has been studied with radioactive tracers by Badenhuizen and Dutton (1956), who have dramatically shown that the insertion of hexose into the starch grain occurs predominantly at the periphery of the grain; thus there is a layering of the starch, the outermost layer being the most recent deposition and the most readily hydrolyzed off. Brief exposure of the leaves of potato to $^{14}CO_2$ permits the observation that the starch deposition occurs most actively in the periphery of the tuber and most particularly at the distal end, where cell division is most active. Similar experiments on inulin deposition in the tubers of jerusalem artichoke reveal the same pattern (Edelman and Popov, 1962).

When tuber formation begins, Werner (1935) has observed that the potato plant shows marked accumulations of starch systemically, in leaves, stems, and stolons; Wellensiek (1929) noted that pinching off the growing points of the stems can enhance tuber formation and interpreted this as a response to the increased carbohydrate level in the plants when the growing points are removed. Mes and Menge (1954) have shown that potato stems in tissue culture form tubers in response to high levels of sucrose in the culture medium. Edelman (1959) has offered the interpretation that the formation of the tuber is a consequence of the accumulation of sugars in the plant under the pressure of CO_2 and light from the outside; he describes the accumulation of storage carbohydrate as an expression of excretion rather than thrift.

RELATION TO GROWTH

It seems general that the formation of storage organs, including tubers, bulbs, corms, and enlarged stems or roots, is associated with the suppression of elongation growth. This can be illustrated by the data in Fig. 14-2, where the period of potato-tuber growth is seen to coincide with a period of no increase in leaf area and even a decrease during the later weeks. With bulb formation in onions, there is a systemic suppression of the above-ground parts (Heath and Holdsworth, 1948) and especially of root growth (Sideris, 1925). Suppressed growth activities are associated with the formation of corms (Mansour, 1968), tuberous roots (Moser and Hess, 1968), and enlarged stems (Selman and Kulasegaram, 1967). That pinching off the growing points can stimulate tuber formation has already been mentioned; tuberization can also be enhanced by the application of growth regulators which serve to inhibit growth, including cytokinin (Asahira and Nitsch, 1968), ethylene (Catchpole and Hillman, 1969), abscisic acid (El-Antably et al., 1967), and growth retardants (Nagao and Okagami, 1966; Moser and Hess, 1968).

ENVIRONMENTAL EFFECTS

Like most developmental processes, the formation of tubers or other storage organs can be variously regulated by photoperiods, light, and temperature. That photoperiods can influence

Fig. 14-2 The start of tuber development in potato plants is associated with the cessation of the growth of new leaves (unpublished data of Borah; cited by Ivins and Milthorpe, 1963).

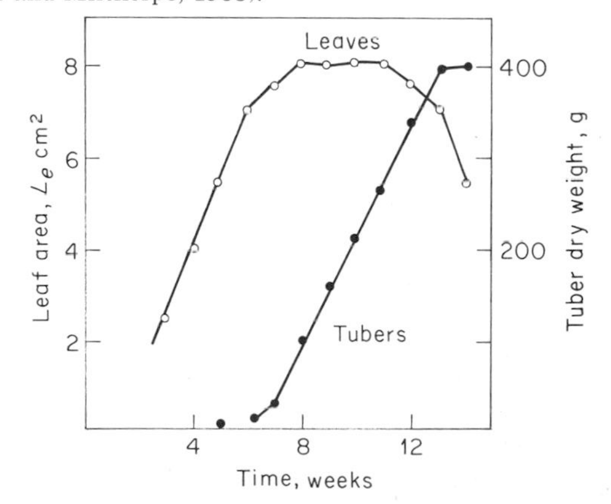

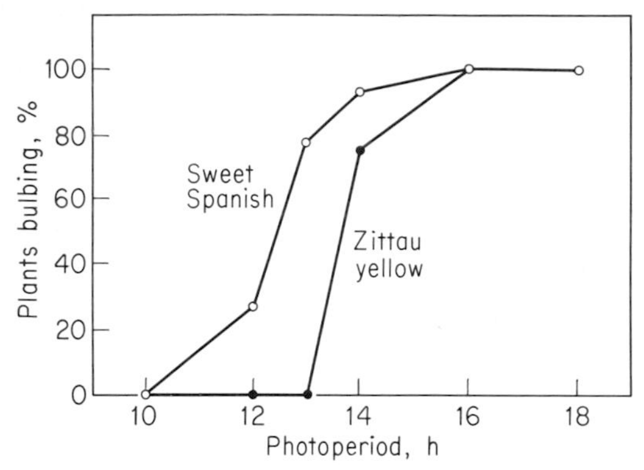

Fig. 14-3 Various onion varieties show differences in the long-day requirement for bulb formation, Sweet Spanish onion forming bulbs under somewhat shorter photoperiods than Zittau Yellow (Magruder and Allard, 1937).

the formation of potato tubers was first noted by Werner (1935), who observed that potato plants form tubers more readily under short photoperiods than long. In some potato varieties this photoperiodic effect may be a strict control (Madec and Perennec, 1962). In *Begonia,* aerial tuber formation may likewise be strictly controlled by short photoperiods (Esashi, 1960).

The research workers at Beltsville, Maryland, who discovered photoperiodism found that onion-bulb formation is controlled by photoperiods in a wide selection of cultivated varieties (Magruder and Allard, 1937). Bulb formation is accelerated by long days in the varieties tested, some varieties forming bulbs under day lengths as short as 12 h and others only at longer minimum day lengths (Fig. 14-3).

The short-day stimulation of tuber formation is shared by potato, *Begonia,* some *Dahlia, Helianthus tuberosus, and Phaseolus coccinius.* The long-day stimulation of bulb formation is shared by the onions, early garlic, the stem storage organs of Chinese radish, and some iris. Many of these photoperiod responses were noted by Allard and Garner (1940).

The photoperiod control (or influence) on tuber and bulb formation embraces all the classical features of photoperiodism as worked out for the control of flowering. The leaf is the photoperiod-perception organ; a stimulus is translocated from the leaf to the responding organ; quantitative responses are obtained in terms of the new organ formation and development; and night interruption may be effective in converting a short-day into a long-day type of stimulus. Some of these features are illustrated in Figs. 14-4 and 14-5.

The role of leaves in the perception of day lengths for tuber formation in *Helianthus tuberosus* was noted by Hamner and Long (1939). More precise experiments on the leaf as the perceiving organ have been published for *Begonia* by Esashi (1960). He compared the relative effectiveness of leaves of different ages and areas, finding that the youngest expanded leaf is the most photoperiod-sensitive, just as it is the most sensitive to other photoperiod effects. That the photoperiodic perception for bulbing is also by leaves has been adequately demonstrated by Heath and Holdsworth (1948), who removed the leaf blades of onions and showed that the photoperiod sensitivity is lost until new blades have developed.

In addition to the strict photoperiod requirement for storage-organ formation in some species (*Begonia*), there are quantitative photoperiod effects in others. A clear-cut example is the *Dahlia,* which forms tubers at any day length but with much greater tuber development under short photoperiods. Again the conditions which gave the greatest tuber development were as-

Fig. 14-4 Tuber formation on aerial nodes of *Begonia evansiana* requires photoperiods with a night length of 11 h or longer. The stimulus is quantitatively increased with increasing numbers of short days. The stage of tuber formation is taken to be the product of the apex height and width upon dissection (Esashi, 1961).

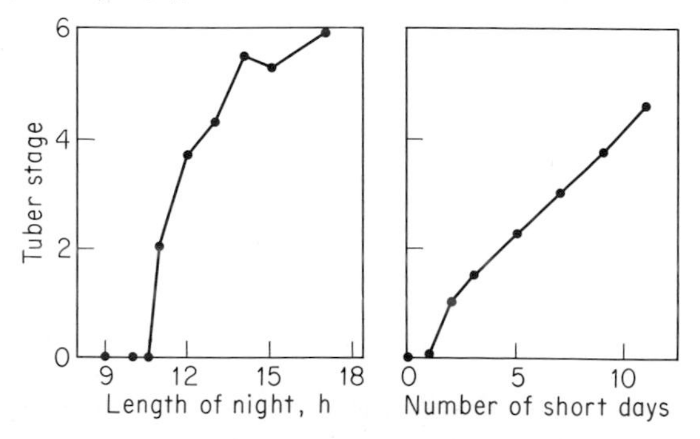

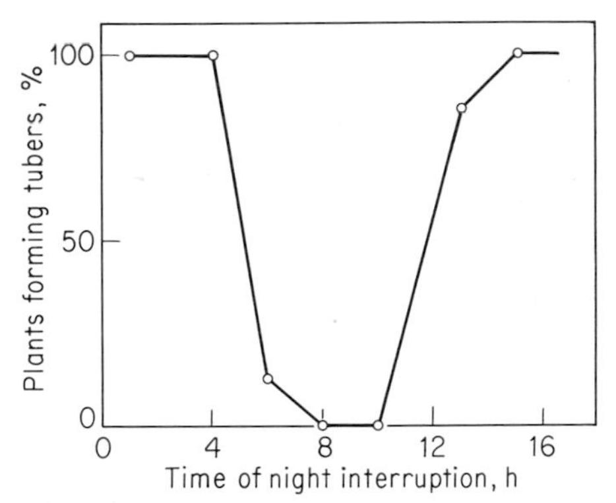

Fig. 14-5 Night-interruption treatments can prevent tuber formation in *Begonia* on short photoperiods (8 h), and the light is most effective in the middle of the long dark period. Light of 45 fc was applied 1 h in each of 20 nights (Esashi and Nagao, 1958).

sociated with the least growth of shoots. A light period which interrupts the night on a short photoperiod depresses tuber development and stimulates the growth of the shoots.

Quantitative effects of light on tuber development can be seen in some varieties of potato. For example, Fig. 14-6 shows that the extent of tuber development is depressed or eliminated at low light intensities.

Temperature regulation of storage-organ formation is known in a few cases. For example, in Fig. 14-7 it is seen that *Freesia* forms corms preferentially at low temperatures, conditions under which the growth of stems is suppressed. In *Begonia evansiana* tubers will form even without short photoperiods if low night temperatures are given (Esashi, 1964; Esashi et al. 1964). In the Japanese artichoke higher temperatures are needed for the formation of stolons, after which lower temperatures will bring about tuber formation (Lagarde, 1971).

NATURE OF THE STIMULUS

Studies of the tuberizing stimulus were greatly facilitated by Barker's (1953) finding that tubers can be formed from potato meristems in tissue culture. This technique has been developed in numerous laboratories (e.g., Mes and Menge, 1954; Gregory, 1956; Chapman, 1958) with considerable benefit. Even from the initial report, it was clear that any node or meristem of potato can form tubers; the property is not distinctive to the underground stolons. But such tuber formation is characteristic of tissues taken from short-day-induced stems; pieces taken from noninduced potato plants are very slow to produce tubers in culture. Chapman (1958) noted that if a two-branched potato is given inductive short days on one branch only, cuttings from that branch readily form tubers and cuttings from the uninduced branch do not.

The stimulus to produce tubers moves characteristically down the plant, as shown not only by Chapman's (1958) experiment but also by the observation that a girdle or other interference with downward transport in the stem leads to tuber formation just above such an obstruction (van Schreven, 1949). The tuber stimulus moves readily across a graft (Gregory, 1956) parallel to the flowering stimulus. After photoperiodic induction, the tuber stimulus will persist in cuttings for a limited time (Gregory, 1956), but reversion to the noninduced state occurs eventually (Chapman, 1958; Esashi, 1960).

The various types of evidence for the exis-

Fig. 14-6 Effects of light intensity on tuber development are shown for the potato grown under three different light intensities (Bodlaender, 1963).

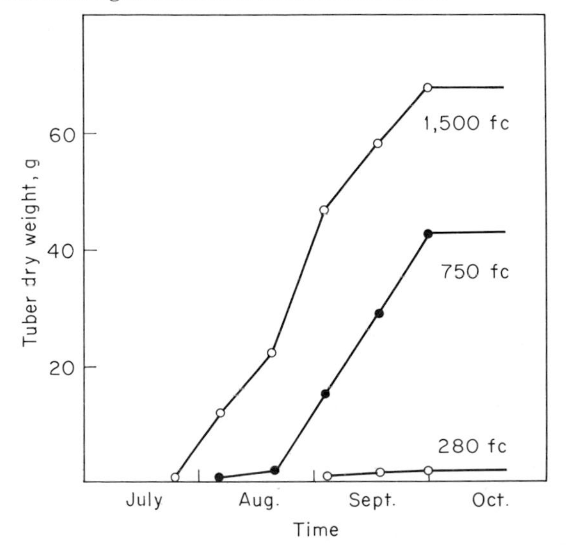

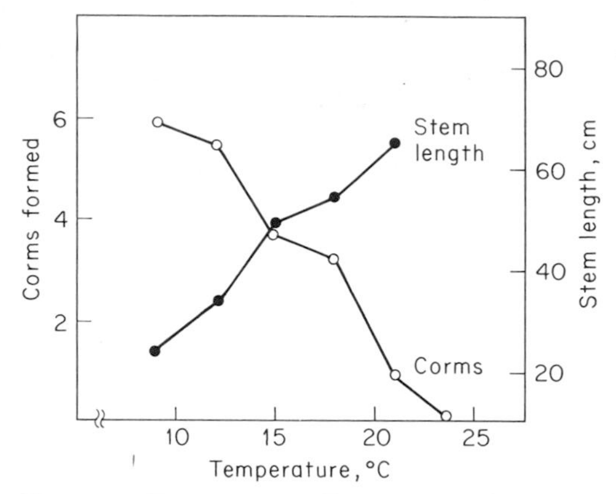

Fig. 14-7 Temperature effects on corm formation for *Freesia,* showing greater development of corms at lower temperatures, and associated depression of the development of stems (16-h photoperiods) (Mansour, 1968).

tence of a tuber-forming hormone are strikingly parallel to those for a flowering stimulus: a photoperiodic event in the leaves forms the stimulus; it shows the features of a threshold day length, night-interruption sensitivity, reversion of the leaf to the noninduced state, translocation of the stimulus across a graft, and quantitative response to numbers of photoperiods at a remote meristem.

Wellensiek (1929) made the suggestion that the stimulus for tuberization might be sugar itself, and Mes and Menges (1954) provided some impressive support for this idea when they found that sucrose can stimulate tuber formation by potato-stem pieces in tissue culture. Gregory (1956) repeated the tissue-culture experiments and found that if no photoperiodic induction for tuberization has been given, high sucrose levels in the culture medium stimulated only stem growth; he asserted, therefore, that sugar is not itself the tuberization stimulus.

Graftage experiments provide the strongest evidence for the existence of a tuberizing hormone, as they do for the existence of a flowering hormone. Gregory (1956) grafted induced potato scions onto noninduced stocks and showed decisively that the tuberizing stimulus can carry the signal from the induced to noninduced parts; furthermore, the stimulus moves in a polar manner toward the base of the plant. Madec (1963) provided further support for the concept of a specific tuberizing stimulus in that tomato scions grafted onto potato rootstocks do not lead to tuber formation but potato scions do. Nitsch (1966) found that in jerusalem artichoke, not only can the tuberizing stimulus cross a graft from scion to stock, but an ordinary sunflower scion, with no photoperiodic sensitivity as far as we know, could under short photoperiods provide the tuberizing stimulus to an artichoke stock. Madec (1963) has made some promising progress in extracting a tuberizing stimulus from induced shoots of potato. Sap pressed from induced plants was placed on uninduced plants and produced a tuberizing effect (along with an inhibition of growth). Further characterization of this extractable stimulus should be promising.

Each of the plant hormones has been claimed to be involved in the formation of storage organs in some instance. Auxins were first implicated in the tuberization of potato (Craniades, 1954; van Schreven, 1956) and soon thereafter in the bulb formation in onions (Clark and Heath, 1959). Esashi et al. (1964a) noted that the induction of tuberization in *Begonia* is inhibited by applications of auxins and that the auxin content of leaves declines during short-day induction treatments. Subsequent experiments (Sano and Nagao, 1970) revealed a marked increase in auxin oxidase with the inductive short-day treatment. In all reports on gibberellin, this hormone appears to inhibit storage-organ formation, including the potato (Okazawa, 1960), kohlrabi (Selman and Kulasegaram, 1967), and *Begonia,* (Cho, 1970), and tuber-inducing short photoperiods can lead to rapid decline in gibberellin content in potato plants (Railton and Wareing, 1973). Cytokinins give marked promotions of tuberization in *Begonia* (Esashi and Leopold, 1968b) and potato (Palmer and Smith, 1969) and of stem swelling in kohlrabi (Selman and Kulasegaram, 1967). While cytokinins are readily obtained by extraction from tubers of potato and *Begonia* (Tizio, 1966; Esashi and Leopold, 1968b), the stimulation of tuber growth obtained by applied cytokinins is principally a

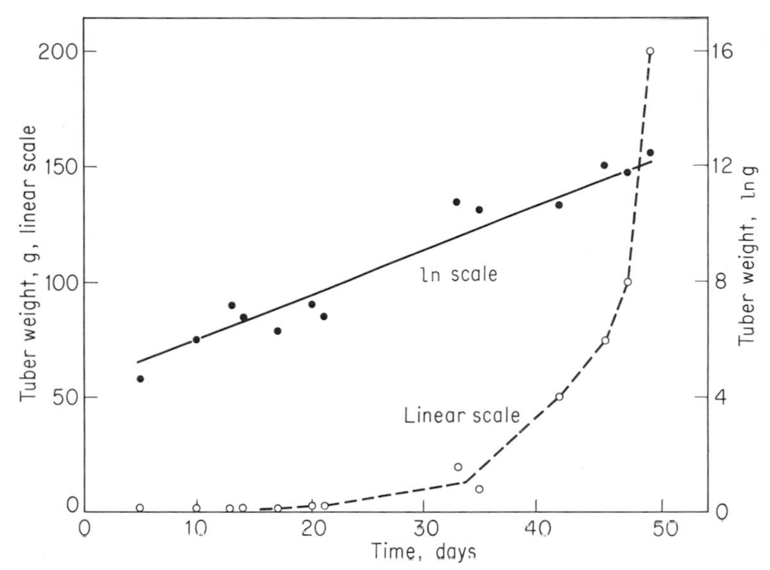

Fig. 14-8 The tuber weights of Cobbler potatoes increase in approximately an exponential manner, as shown by the plot on a logarithmic scale (Plaisted, 1957).

stimulation of cell enlargement in the tuber, whereas normal tuber growth involves cell division. Stimulations of tuberization have also been reported with ABA (El-Antably et al., 1967) and ethylene (Catchpole and Hillman, 1969), and inhibitor β (ABA?), which appears to be present in large amounts in the leaves of potato during the period of tuberization (Ivins and Milthorpe, 1963). Biran et al. (1972) have found that during the short-day induction of tuberization in *Dahlia* plants, there is a threefold increase in ethylene production by the plants before the onset of tuber development.

TUBER AND BULB GROWTH

The progress curves for tuber growth show an exponential character (Fig. 14-8). Thus, the great bulk of the tuber-filling activity occurs very near the end of the growth season. In potato, the tuber grows by roughly equal activities of cell division and enlargement. In the development of the bulb of the onion, however, there is apparently no cell division (Heath and Holdsworth, 1943), and growth is simply by swelling of the leaf cells already there.

An interesting feature of tuber growth is the facility with which materials can be moved from one developing tuber to another. The number of developing tubers can actually decrease during the growing season (cf. Hardenburg, 1949), and the entire contents of some tubers can be transferred into others. The translocation from a parent tuber into many peripheral tubers is illustrated by the work of van Schreven (1956). Also, it is not uncommon for potato vines to start growing again late in the season after a killing frost, and this growth is at the expense of the tuber crop.

TUBER RIPENING

Unlike the dramatic ripening changes common to the fleshy fruits, ripening transformations in potato tubers are relatively subtle. In terms of tuber constituents, ripening is associated with a marked *drop* in sugar content, as shown in Fig. 14-9. Associated with the sugar decline is an increase in the starch content, and, reminiscent of the ripening of apples and some other fruits, there is also an increase in protein content. Respiration is apparently steady, and no evi-

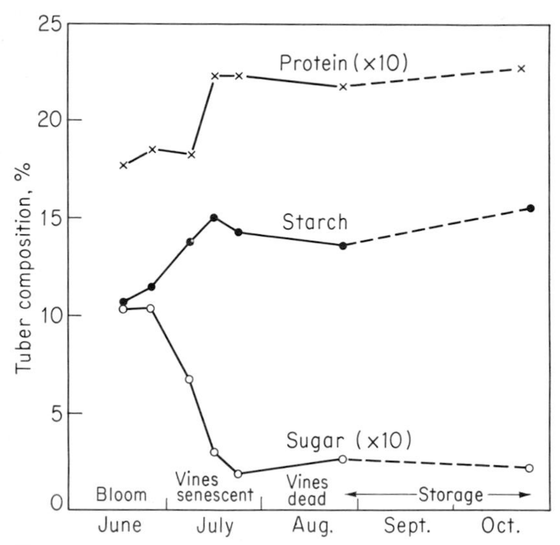

Fig. 14-9 Tuber ripening takes place at vine senescence, at which time there is a marked increase in starch and protein content and a drop in sugar content of the tubers; Irish Cobbler potatoes (Appleman and Miller, 1926).

dences of a climacteric have been reported, although at least in iris bulbs ethylene can markedly stimulate the respiratory rate (Kamerbeek and Verlind, 1972).

As the tubers ripen, the vines frequently undergo an active senescence, dying back to the ground. The ripe tuber possesses superior properties for storage, having a markedly thicker skin and forming suberin layers over bruises more readily than an unripe tuber. Ripened tubers are also of superior quality for the production of potato chips.

TUBER REUTILIZATION

Obviously the tubers and bulbs make an excellent propagation material, having generous reserves of carbohydrate and meristems. Even before the tuber or bulb is removed from the parent plant, the carbohydrate reserves may be mobilized out of the storage organ. The resorption of potato tubers has been mentioned. In some dramatic experiments of van Schreven (1956), potato tubers sent out shoots in the dark which formed myriads of new tubers, mobilizing the carbohydrates to fill them from the parental tubers. Like the mobilization of carbohydrates into and out of some fruits or parts of fruits (Fig. 13-15), the assimilation and dissimilation processes are not well understood.

When tubers are utilized as planting material, the depletion of carbohydrate reserves progresses steadily in time and, as shown in Fig. 14-10, removal of the growing points from the shoots generated by the tuber terminates depletion of the tuber reserves.

The utilization of the tubers involves, of course, the hydrolysis of the carbohydrate polysaccharides in the tuber; Jefford and Edelman (1960) have followed the hydrolysis of the inulin in jerusalem artichoke tubers during the period of growth after planting, and their data show the impressive accumulation of fructose while the larger sugar polymers decline.

The hydrolysis of starch and inulin occurs through enzymatic pathways different from their deposition. Edelman (1963) points out that the utilization of different enzymic systems in the

Fig. 14-10 After planting, jerusalem artichoke tubers experience a general hydrolysis of carbohydrate reserves, as illustrated by the decline in dry weight (as percent of initial weight); that this hydrolysis is directed by the growing points is indicated by the cessation of weight loss when the shoots are removed from the new plant (Edelman, 1963).

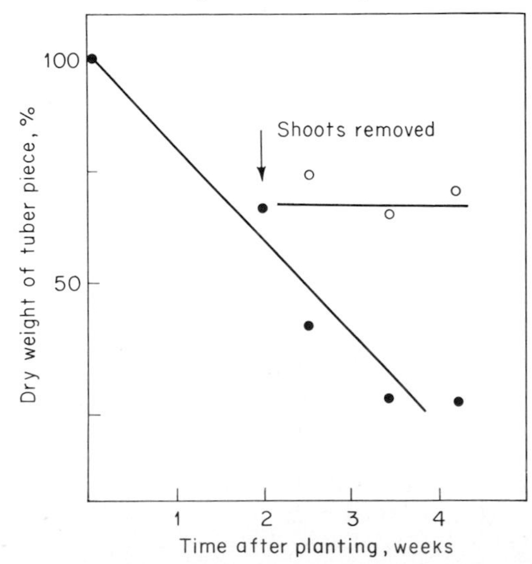

breakdown of storage polysaccharides offers some real advantages in terms of the control of the two types of process. The hydrolysis of starches involves several types of amylase. In potato tubers, a well-known sweetening reaction occurs when the tubers are stored at low temperatures. This phenomenon of sweetening has been attributed to the activity of phosphorylases (Arreguin-Lozano and Bonner, 1949) and of invertase (Pressey and Shaw, 1966). The work on phosphorylase indicates that inhibitors of phosphorylase formed in the tubers at high storage temperatures suppress the progress of sweetening; however, Porter and Rees (1954) were unable to show such an inhibitor related to temperature experience of the tubers.

The dormancy and sprouting of tubers and bulbs and corms is dealt with in Chap. 10.

SUMMARY

The formation of tubers, bulbs, corms, and other storage organs appears to be further dramatic expression of the capability of higher plants for mobilization phenomena. While the mechanism of mobilization activities is not well understood, the involvement of several of the growth hormones can be inferred; there are some bits of evidence that each of the known plant hormones may be involved in the regulation of tuber and bulb formation, some hormones having promotive effects and some having inhibitory effects on the storage-organ formation. Evidence for the possible role of a tuberizing hormone comes from the photoperiodic induction of tuberization and from the demonstration that a hormonal stimulus moves across grafts. The tuberizing stimulus is another regulatory principle which moves in a polar, basipetal direction. The possibility that there is a specific tuberizing hormone or a bulb-forming hormone again raises the question whether regulatory controls in the higher plant are achieved by separate hormonal regulators for each of the developmental steps or by interactions of the common plant-growth substances.

GENERAL REFERENCES

Burton, W. G. 1972. The response of the potato plant and tuber to temperature. In A. R. Rees et al. (*eds.*), "Crop Processes in Controlled Environments." Academic Press, London, pp. 217–233.

Heath, O. V. S. and M. Holdsworth. 1948. Morphogenic factors as exemplified by the onion plant. *Symp. Soc. Exp. Biol.* 2:326–350.

Ivins, J. D. and F. L. Milthorpe. 1963. "The Growth of the Potato." Butterworths, London.

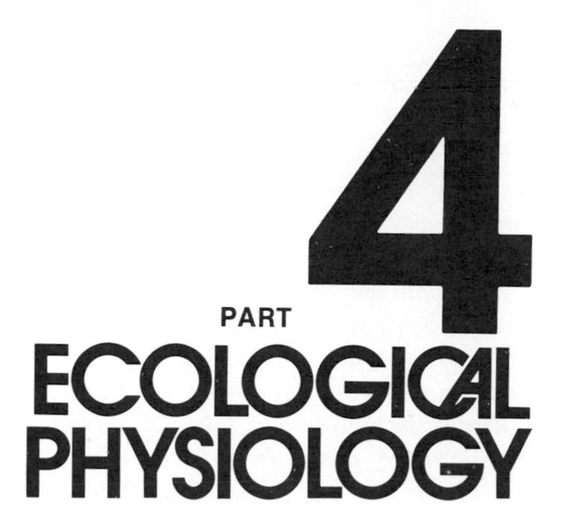

PART 4 ECOLOGICAL PHYSIOLOGY

LIGHT

TEMPERATURE

WATER

All living things must adapt to their surroundings in order to survive and reproduce. Natural selection, the inexorable force behind this adaptation, provides organisms with an enormous variety of stratagems for utilizing local resources. Every organism can therefore be viewed as a microcosm within an enormously greater ecosystem, and to gain some appreciation of the whole, we must first understand its parts, i.e., individual plants. Adaptive mechanisms with ecological significance become, then, our primary concern.

In the final analysis, everything is energy, and all biological events involve energy change. Plant communities are coupled to their environment via energy flow, *a concept that provides a unifying thread for us to grasp as we explore dynamic interdependence within communities of living things.*

The major environmental inputs that drive terrestrial systems include solar radiation, temperature, water, and nutrients. Coupled with these environmental constraints are the plant's own internal regulators, whose control is modified according to environmental conditions. These controls are analogous to feedback mechanisms in model systems which portray the plant taking stock of itself. One concern of ecological physiology therefore is to understand these control signals: What does a plant perceive within itself or its environment that leads to feedback and hence to a modification of growth?

Competition in pure stands such as agricultural communities can be especially intense because all plants express identical demands. If supplies of water, nutrient, or light are insufficient, or if temperatures are unfavorable for growth, all members will survive but they will not thrive. If environmental stresses intensify, all individuals will eventually die because their uniform genetic composition demands the same environmental fulfillment at the same time. Pure stands can also suffer such a fate under natural conditions. Replacement will then occur by new dominants which are more attuned to local fluctuations in environmental conditions.

Natural plant communities, however, are more commonly characterized by genetic diversity *than* uniformity. *Even within a stand of the same species some individuals will be physiologically more drought- or heat-tolerant, have a better ca-*

pacity to acquire nutrients from impoverished soils, or be capable of compensating for low light intensity. Such individuals are more likely to endure environmental limitations, will come to occupy that particular niche, and will eventually constitute an ecological race.

Driven by economic expedience, agriculturalists modify the environment to achieve optimum crop performance. They relieve limitations on plant growth and so maximize yields (at least within the genetic limitations of their crops). In nature, however, species do not necessarily occur in those habitats which promote optimal growth because competition in favorable locations is often too intense. Ecological races emerge whose physiological adaptations enable them to grow, or at least survive, under less conducive conditions, where they are able to compete successfully against plants which are less endowed with these adaptations.

Inadequate illumination, temperature extremes, moisture stress, or nutritional deficiency—all constitute major limitations to plant growth and development. In response to such environmental stresses higher plants have evolved with a bewildering range of morphological and physiological systems which enable them to avoid or endure rugged conditions. Related systems which promote maximum utilization of environmental resources confer strong competitive abilities in favorable locations and deserve equal emphasis. These mechanisms which enable plants to thrive or at least survive have profound ecological consequences and constitute the subject of ecological physiology.

If we attempt to collect the various effects of environmental factors on the plant into general categories, we can recognize four types of physiological alterations:

1 *The environmental action may bring about alterations of components or structures in the plant, e.g., the greening and expansion of leaves by light.*

2 *The environmental variable may alter the rates of synthetic processes in the plant, e.g., the effects of light and carbon dioxide on photosynthesis.*

3 *The environmental factor may bring about the onset of new processes, serving as a cue for some aspect of regulation of growth or development.*

4 *The environmental variable may impose limits on the success or distribution of the plant.*

The concept of environmental variables serving as cues for plant regulation implies that plants can perceive and then in a sense remember an environmental event for some time. The persistence of the plant's reaction may be measured for comparative purposes in terms of its half-life, $t_{1/2}$. *Thus the* $t_{1/2}$ *for an alteration of photosynthesis by a change in light intensity may be in the order of seconds. The* $t_{1/2}$ *of other responses to light may be extended to a matter of minutes, e.g., the alteration of stomatal opening or the alteration of chloroplast position by light. A* $t_{1/2}$ *of several hours may be observed for plant responses to tropistic stimuli. The alteration of flower induction by photoperiodism may have a* $t_{1/2}$ *of several days or weeks. And finally, environmental experiences may bring about metastable changes in the plant, changes which persist until another environmental cue is experienced. Examples of such on-off environmental switches are the regulation of dormancy, dwarfism, and vernalization. The ability of the plant to respond with a range of persisting aftereffects gives a whole new dimension to environmental regulation of the plant: regulation in time.*

The close match between sensing mechanisms within a plant and the characteristics of environmental variables deserves mention. The major light responses of plants have action spectra which peak within a range of wavelengths corresponding to maximum incident sunlight. Photoperiodic reactions are instigated by light receptors sensitive enough to recognize sunset and sunrise but not sensitive enough to be influenced by moonlight; the extent of low-temperature experience needed for the relief of dormancy in apples is close to the accumulated cool weather of temperate climates during winter. This coincidence of sensing systems and environmental conditions provides further evidence of how evolutionary forces can shape biological machinery and will be a recurring theme in our discussions of ecological physiology.

15 LIGHT: AN ENVIRONMENTAL FACTOR IN ECOLOGICAL PHYSIOLOGY

Ecosystems represent basic self-sufficient communities within the biosphere which are wholly sustained by solar radiation. While these integrated complexes operate within environmental constraints set by temperature, water, and nutrition, the tempo of their existence ultimately depends upon how effectively they utilize sunlight.

ENERGY SUPPLY

Solar energy spans a wide spectrum of electromagnetic radiation from the hertzian band (which includes radio waves down to 150 kH_z and corresponding to a wavelength of several hundred meters) through visible and ultraviolet up to cosmic radiation of unimaginable energy (10^{22} H_z) capable of destroying any form of life.

Generation of this solar energy has naturally aroused intense interest, and physicists have indulged in heated conjecture about the underlying reactions. Earlier proposals, e.g., stellar contraction and radioactive decay, are clearly inconsistent with present knowledge. Solar energy arises from thermonuclear reactions similar to the fusion process utilized so spectacularly in the hydrogen bomb. The solar reaction similarly involves conversion of hydrogen into helium and is probably catalyzed by carbon and nitrogen. Energy is produced in accordance with Einstein's famous equation $E = mc^2$ (Robinson, 1966). This thermonuclear reaction then produces energy at the expense of matter, but our sun is not about to be extinguished for want of fusible material; the sun should continue its emission for at least another 10^{11} years (Robinson, 1966).

Solar radiation entering our earth's atmosphere undergoes major modification. Atmospheric constituents absorb radiant energy at specific wavelengths in a selective way according to energy requirements for electronic transition, vibration, and rotation. Absorption of ultraviolet energy results from electron transitions in the atomic and molecular forms of oxygen, nitrogen, and ozone. Absorption coefficients are large, and in some ultraviolet bands virtually no radiation reaches the earth's surface. By contrast, infrared radiation is absorbed in our atmosphere

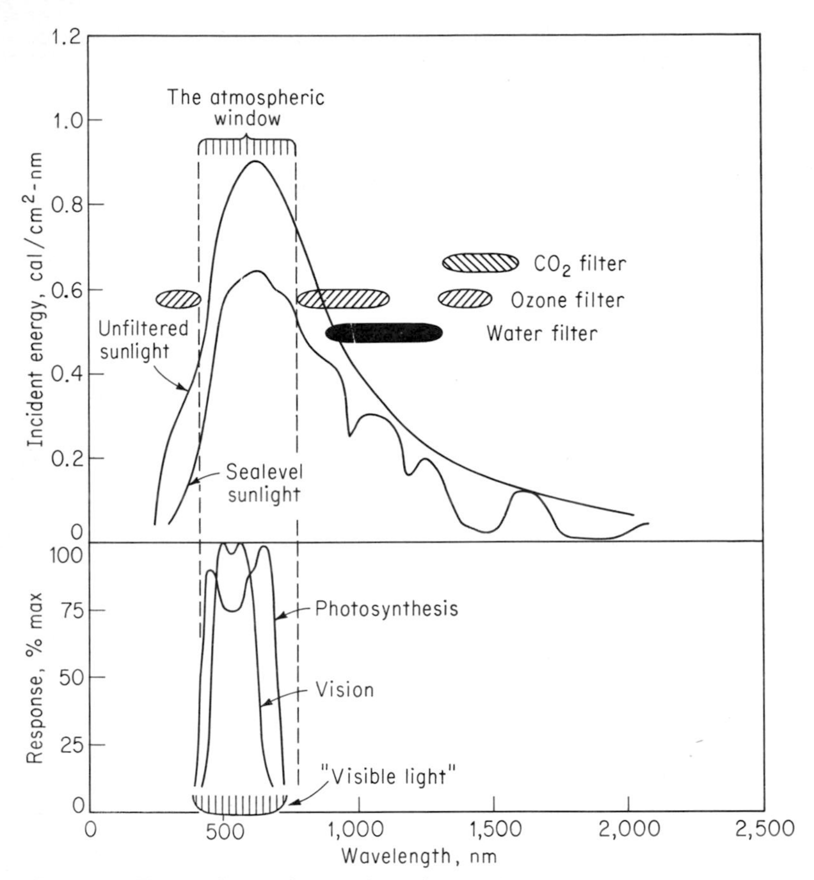

Fig. 15-1 The incident radiation from the sun is most abundant in the wavelengths used for such biological reactions as photosynthesis and vision.

by polyatomic molecules such as H_2O, CO_2, and O_3 due to alterations in their vibrational and rotational states. Significantly, the biosphere which has developed under this selective umbrella is acutely sensitive to ultraviolet and infrared radiation. Gates (1963), for example, stresses that ozone absorption in our planet's stratosphere shields organic complexes at the earth's surface from dissociation by ultraviolet radiation which would otherwise rupture the C—O, C—H, and O—H bonds essential for plant and animal life.

Solar energy entering our atmosphere is scattered by molecules and dust particles to form diffused skylight (as opposed to direct solar radiation). When these particles responsible for scattering are microscopically small (radius less than one-tenth of radiant-energy wavelength), the scattering coefficient is inversely related to wavelength (*Rayleigh's formula*). Consequently the scattering coefficient for blue light is greater than that for red light. Accordingly, the skylight produced by air-molecule scattering has a blue color. On the other hand, scattering due to larger particles (*Mie scattering*) is independent of wavelength, so that transmitted and reflected sunlight still looks white to our eyes.

Solar radiation is attenuated by the earth's atmosphere in a highly selective way. Carbon dioxide, ozone, and water in the atmosphere remove substantial amounts of energy in infrared wavelengths (between 850 and 1,300 nm) while absorption of ultraviolet, due specifically to ozone, produces an abrupt termination of energy reaching the earth's surface at around 290 nm. The relatively simple gaseous composition of the earth's atmosphere, however, does give rise to an extremely narrow slot in the entire electromagnetic spectrum where very little absorption occurs. Sunlight can therefore stream toward the earth's surface through the convenient frequency gap in the visible region (Fig. 15-1). Similar gaps in the absorption band associated with vibrational and rotational transitions in atmospheric gases at infrared wavelengths allow reradiation of surface energy back to space.

Our atmosphere thus provides a window through which shine visible wavelengths, and life forms in the biosphere are a direct consequence of this selective absorption. Ultraviolet and infrared wavelengths can be highly damaging to biological materials. Wavelengths below 400 nm are absorbed in a nonspecific fashion by common biological constituents, and since the energy level per quantum increases with decreasing wavelength, these short wavelengths dump rather heavy packages of energy onto cellular components. Molecular disorganization with attendant biological injury can result. With wavelengths beyond the visible region energy absorption again becomes damaging because cellular water absorbs infrared radiation (Figs. 15-1 and 15-2). These water molecules amount to more than mere cogs in the cell's metabolic machinery; water represents a universal cradle for biological macromolecules, and its absorptive properties predispose living cells to damage by infrared irradiation.

Between these two potentially harmful categories of radiation stands the visible spectrum. Its wavelengths are selectively absorbed by numerous types of carbonaceous molecules, especially by some with nitrogen and metallic components through which electron resonance is enhanced. Strict selectivity in light absorption then promotes selective translation of its energy into organized biochemical systems, a basic essential for light utilization.

At the earth's surface, visible wavelengths

Fig. 15-2 Sunlight is altered in quantity and spectral composition during its passage through the earth's atmosphere and vegetative cover. (*Modified from Gates, 1965.*)

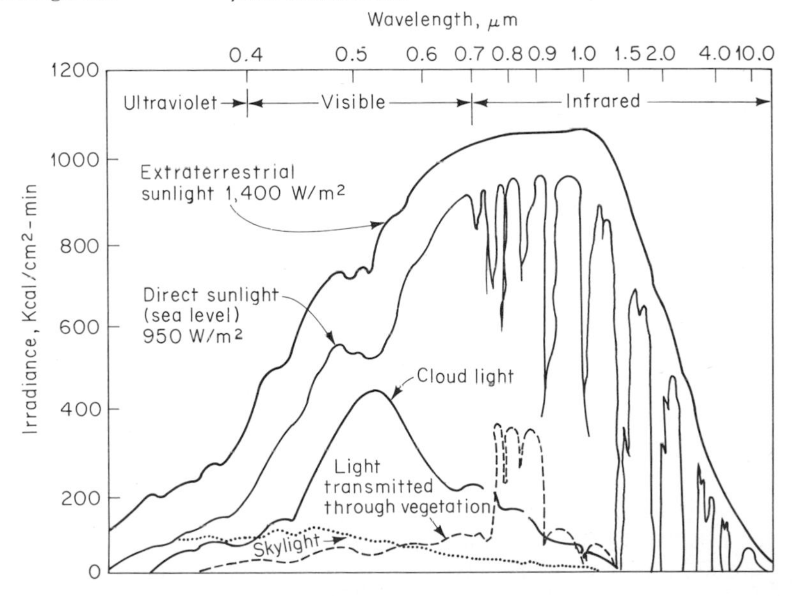

represent about 50 percent of total solar radiation (Fig. 15-2). It is probably no coincidence that biological systems make most effective use of this particular band of radiation. Selective absorption by vegetation is clearly evident in Fig. 15-2. Even cloud light, which is of course more diffuse than direct solar radiation, offers almost exclusively visible radiation. The wavelength band most effective for photosynthesis straddles the same band as that eliciting the greatest response in animal vision (Fig. 15-1). Within this span of wavelengths are the action spectra for the principal light-regulated systems of biology, including plant tropisms, seed germination, photoperiodism, and pigmentation responses in plants and animals. Each of these responses is mediated by pigment systems which effect selective guidance of solar energy into organized chemical systems: synthetic systems in photosynthesis, biochemical triggering systems in vision, tropisms, photoperiodism, and pigmentation responses to light. The pigments may be unsaturated carbon skeletons, like the carotenoids; carbon associated with nitrogen, as in the flavins; or carbon associated with nitrogen and metals, as in chlorophyll. Through selective absorption by biological pigments, sunlight streaming through the atmospheric window both sustains and regulates the biosphere below. Biochemical evolution has undoubtedly favored this synchrony of available and utilizable solar energy.

PLANT REACTIONS

The amount of energy available for photosynthesis or photoregulation depends upon interception by the leaf and attenuation within it. Higher plants are reasonably similar with respect to their energy absorption despite variation in chlorophyll concentration above a certain minimum (2 to 3 mg/dm^2; Shul'gin and Kleshnin, 1959). A simple model showing the fate of unidirectional incident energy falling on a leaf is given in Fig. 15-3. E_i, the incident energy, makes an angle of incidence θ with the leaf surface. E_r is reflected radiation, and E_a and E_t represent the radiant energy absorbed by and transmitted through the leaf, respectively. Light intensity at the leaf surface equals $E_i \cos \theta$. Radiant energy absorbed by the leaf can be further apportioned as

$$E_a = E_{\text{rad}} + E_{\text{H}_2\text{O}} + E_p$$

where E_{rad} represents energy which heats the leaf and becomes reradiated, $E_{\text{H}_2\text{O}}$ signifies energy used to vaporize water, and E_p is energy used in photosynthesis. This energy budget within a leaf is discussed in Chap. 16; our present concern is reflection and transmission. The reflection coefficient changes greatly with alteration in θ, and this departure is further modified according to spectral composition (Howard, 1966; Woolley, 1971). Since a plant leaf is optically heterogeneous, rays of light encountering such a medium undergo multiple refraction, reflecting at different angles, and so contributing to scattering and secondary absorption within the leaf.

Cellular structures in a leaf are large compared to wavelengths of visible radiation. Typical cell dimensions of say 15 by 20 by 60 μm for palisade cells are orders of magnitude greater than wavelengths in the visible spectrum, but organelles in these cells present a different picture. Each photosynthetic cell may contain between 50 and 150 chloroplasts. These organelles are about 5 to 8 μm in diameter and roughly 1 μm thick. Granal stacks in these plastids are still smaller (Fig. 1-1), amounting to a length of 500 nm and thickness around 50 nm. These dimensions approach the wavelength range of visible radiation and may produce a lively optical interaction in illuminated leaves because scattering or diffraction takes place when light encounters structures whose dimensions are comparable to its wavelength. Cellular dimensions are too large, but chloroplast and grana dimensions are conductive to scattering.

Operative pigments in chloroplasts include chlorophyll (65 percent), carotenes (6 percent), and xanthophylls (29 percent); percentage distribution is of course highly variable (Gates et al., 1965). Chlorophyll *a* and chlorophyll *b* are most frequent in higher plants, but altogether about 10 forms have been isolated, each with a unique absorption spectrum. When such a pigment absorbs radiant energy, there is a displacement of pi electrons in a resonance system through the pigment molecule. This activated state may last for only a fraction of a second, and then the energy corresponding to its displacement is given up. Such activation energy can be yielded in several ways: by reemission of light (fluorescence) or of heat energy, by transmission of the activated state to another molecule, or by consumption of this energy in some biochemical event. Light-activated plant pigments employ all these methods in dissimilating activation energy.

Absorption spectra and pigment behavior depend upon the pigment's surrounding medium (Loomis, 1965). In organic solvents, chlorophyll and its derivatives display a brilliant red fluorescence, which accounts for 25 to 30 percent of absorbed quanta, but in the cell this value drops to 2 to 3 percent because the activation energy becomes usefully coupled to other pigment systems and biochemical processes.

In photosynthesis, as in other light-driven reactions, absorption and action spectra do not show precise coincidence. The discrepancy between chlorophyll absorption and photosynthesis action spectra is largely due to contributing actions of accessory pigments such as carotenoids. This general broadening in an action spectrum bears some resemblance to light absorption by the intact leaf (Fig. 1-9).

The sharp absorption peaks which characterize organic solutions of pigments are not evident in intact leaves for a variety of reasons. In particular, attenuation of light within a leaf occurs over a much broader spectral region due to scattering by organelles in the leaf, e.g., mitochondria, ribosomes, nuclei, starch grains, and other plastids.

The spectral composition of incident and transmitted solar radiation (Fig. 15-2) demonstrates how plants have become magnificently adapted to their radiation environment. They absorb efficiently in those regions of the spectrum where energy is readily usable and poorly in the near infrared, which serves to minimize heat load, but do absorb far infrared to become efficient

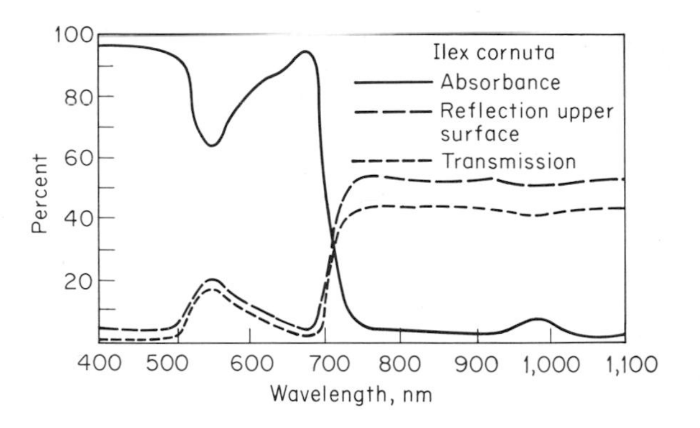

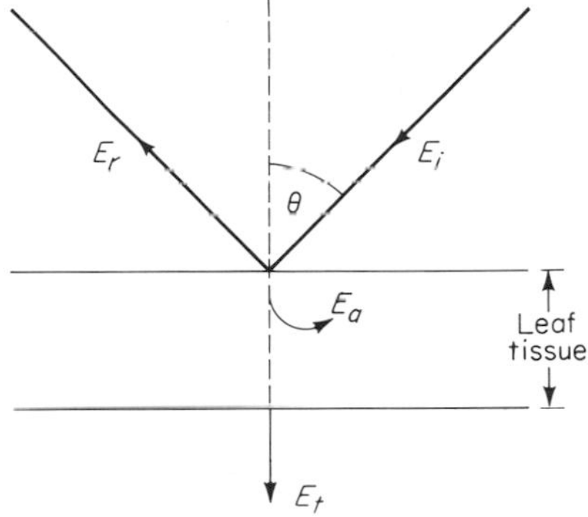

Fig. 15-3 Solar radiation E_i striking a leaf surface is reflected E_r, absorbed E_a, or transmitted E_t. The relative magnitude of these quantities depends upon leaf optics and spectral composition of incident light, but absorbance is generally greatest in red and blue regions, while green light is preferentially reflected or transmitted (Gates et al., 1965; Kriedemann et al. 1964).

radiators. Further anatomical subtleties of leaves enhance their utilization of absorbed sunlight. Leaves are optically heterogeneous with interfaces between cell walls and air spaces offering ideal sites for internal reflection. At such interfaces there is a change in the index of refraction from 1.33 for liquid water to 1.00 for air contained by intercellular spaces (Willstätter and Stoll, 1913). This remarkable property of leaves provides an efficient internal reflection at each interface. A ray of light is therefore likely to encounter several internal reflections before transmission or reflection back outside. Photosynthetic prospects for the single leaf and for the community at large are thereby improved, especially at high light intensity, where reflectivity assumes greater prominence (Kharin, 1965; Tageeva and Brandt, 1961).

PHOTOSYNTHETICALLY ACTIVE RADIATION

Photosynthesis is undoubtedly the giant among all biological light reactions, but the efficiency of this process on an ecological scale is pitifully low. The supply of solar energy is out of all proportion to its photosynthetic utilization. The full extent of solar energy pouring onto our planet staggers the imagination. Can one conceive of 10^{21} kcal of energy being invested in every acre of the earth's surface over every year of its existence? In electrical terms, that amounts annually to 3.5 million kilowatt-hours per acre. Top-quality agriculture might fix 5 percent of this energy resource; natural communities are even lower.

Radiation effective in photosynthesis accounts for about half the solar energy reaching the earth's surface, so that photosynthetic efficiency is doubled if we limit considerations to visible light. Loomis and Williams (1963) prepared such a balance sheet in their estimate of maximum productivity by a corn crop (Table 15-1).

In budgetary terms, only 12 percent of the energy received by the community has been retained in the community, and even this potential value is well in excess of values encountered under field conditions (Wassink, 1959). Low conversion is primarily attributable to the low efficiency with which the plant's photosynthetic apparatus converts light energy into chemical energy. The high light-quantum requirement for photosynthesis is the largest single cause of this inefficiency, but the problem at least is common to all plants, so that ecological advantages will be gained from other adaptive features of the plant's photosynthetic system.

Radiation climate within a plant community is characterized by chaotic short-term variation (including sun flecks) superimposed on a diurnal cycle whose duration and amplitude (in intensity) vary with season. Added to these variations in intensity are quality differences in incident sunlight according to weather conditions and time of day (direct sunlight, diffuse skylight, and cloud light are compared in Fig. 15-2). Condit and Grum (1964) also describe spectral changes according to solar altitude and atmospheric conditions which lead to small-scale variations in light quality.

Radiation within the community undergoes further changes in both quality and intensity. Figure 3-17 illustrates the exponential decline in intensity with depth, but spectral composition also changes. Leaves transmit principally green and far-red wavelengths, whose photosynthetic effectiveness is less than other components of the visible spectrum (Fig. 1-9). Resultant alterations in relative levels of red and far-red light within a community are therefore worth considering in connection with the regulation of growth processes. Our immediate concern, however, is photosynthetically active radiation

Table 15-1

Total solar radiation per day, cal/cm^2	500
Total quanta (400–700 nm), microeinsteins/cm^2	4,320
Albedo loss (shortwave reflection 8.3%), microeinsteins/cm^2	360
Inactive absorption losses (10%), microeinsteins/cm^2	432
Total quanta (400–700 nm) available for photosynthesis, microeinsteins/cm^2	3,528
Carbohydrate produced (assuming quantum requirement = 10 einsteins/mol), μmol/cm^2	353
Total respiratory losses (assuming 33% of fixation), μmol/cm^2	116
Net carbohydrate production, μmol/cm^2	237

(PhAR) and plant responses to changes in light climate.

Low Light Intensity

When intensity limits full expression of a plant's photosynthetic potential, morphological adaptation can maximize interception of incident radiation. Leaves and leaflets on Leguminosae, for example, show coordinated movements which place leaflets parallel to the sun's rays under high insolation but perpendicular to incident light at low insolation. Such a *day-sleep* phenomenon improves light utilization within a legume sward by modifying community structure according to available light. Under high insolation, upper leaf layers remain light-saturated despite their vertical orientation, while lower layers of horizontal leaves enjoy better exposure at the same time.

Leaf movements can also occur in response to the sun's daily passage across the sky; e.g., at any time of day, the leaves of *Malva neglecta* are always perpendicular to the solar beam (Yin, 1938). The great majority of plants show some leaf movement in response to solar direction and intensity, and even those without joints adjust their laminae as long as their petioles are still elongating. Final position then depends upon prevailing light and plant response. In shade plants, leaves are generally in one layer and at right angles to incident light. In deep shade leaf mosaics develop which maximize light interception but minimize mutual shading. Young *Acer* seedlings on the forest floor show this highly effective and aesthetically pleasing leaf arrangement. Formation of larger thinner leaves confers additional ecological advantage in deep shade, but only certain categories of plants are capable of benefiting from this particular adaptation (Jackson, 1967; Hughes, 1965) (see Chap. 3).

Leaf mosaics also form in aquatic communities, but the control system depends upon differential extension of petioles. Water lily petioles continue to grow as long as any part of the leaf blade is submerged. If a young leaf comes to rest on top of an older one, its petiole starts to elongate and continues to do so until the older leaf is exposed again (Funke, 1938). Available space obviously limits this form of expansion.

While plant morphology in particular, and community architecture in general, govern the *interception* of solar radiation, physiological processes in the leaf govern its *utilization.* Leaves on shade plants, for example, have at their disposal an impressive array of adaptive responses to low light intensity. Mesophyll cell size is reduced and laminar surface enlarges. Chloroplasts become rearranged in photosynthetic tissues, and in the moss *Funaria,* they line upper surfaces of cells under low light but align along vertical walls when sunlight is strong (Zurzycki, 1953). Even the chloroplasts themselves modify their light-harvesting equipment. Ballantine and Forde (1970) demonstrated a change from the rudimentary grana of plastids in soybeans grown under high light intensity (220 W/m^2) to well-formed grana in chloroplasts of mesophyll cells when plants grew under low light (90 W/m^2)

Photochemical adaptation in shade-tolerant ecotypes described earlier might well find some correlation with granal development, as shown in Fig. 15-4. In shade plants of Australian rain forests this adaptive feature in the form of granal development is greatly accentuated and leads to high photochemical efficiency (Goodchild et al., 1972). Such handsome grana might well equip shade leaves with higher photochemical efficiency, but these components are themselves prone to damage by high light intensity. When shade ecotypes of *Solidago virgaurea* are grown in stronger irradiance, their photosynthetic apparatus deteriorates in association with a loss of photosystem II activity (Björkman, 1968). Significantly, photosystem II activity has shown some measure of correlation with granal development in other land plants and algae (Andersen et al., 1972, and literature cited). Corn leaves, with their obligate need for high light intensity in order to saturate their photosynthetic apparatus and with little capacity for shade adaptation, do show some reaction to reduced light intensity. In full sunlight the chloroplasts of mesophyll cells are filled with a dense granular structure, but in

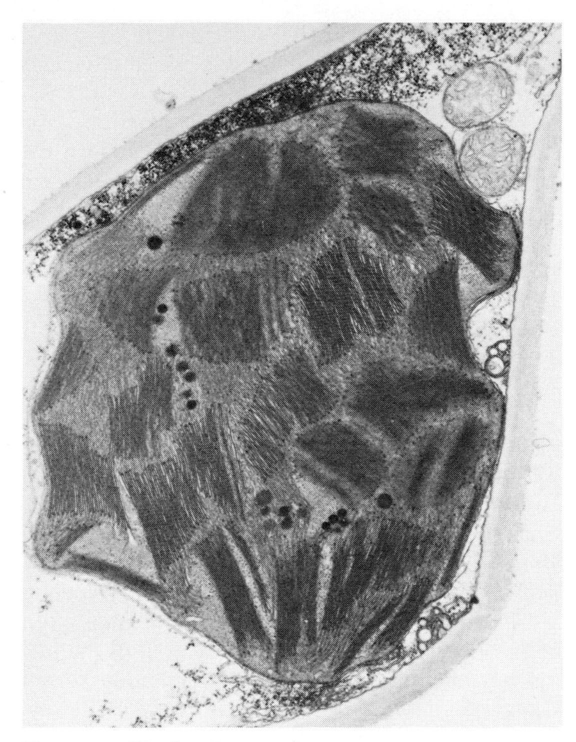

Fig. 15-4 Shade-tolerant plants can accentuate granal development in their chloroplasts when exposed to low light intensity. Increased photochemical activity (especially in photosystem II) is associated with this structural alteration. Chloroplast shown here is from a shade leaf of the rainforest plant *Alocasia macrorrhiza*. 17,480 (*Courtesy of D. Goodchild.*)

plants grown under 33 percent full sun this material adopts a lamellar structure (Osipova and Ashur, 1965).

Comparative photosynthesis in sun leaves and shade leaves is summarized in Fig. 15-5. These data resemble those of Wassink et al. (1956) for photosynthetic adaptation in *Acer pseudoplatanus* but cover a wider species range. In Fig. 15-5, sun leaves were taken from *Phaseolus vulgaris, Lycopersicon esculentum, Helianthus annuus, Nicotiana tabacum, Glycine max, Ricinus communis,* and *Gossypium hisutum;* shade leaves were taken from *Oxalis rubra, Saintpaulia ionantha, Philodendron cordatum,* and the ferns *Dryopteris* sp. and *Nephrolepis exaltata.*

Most cultivated annuals and biennials bear typical sun leaves (upper curve in Fig. 15-5). Their photosynthetic apparatus is light-saturated at one-fourth to one-fifth full sunlight, with an absolute capacity well in excess of that for shade leaves (C_4 plants are of course even more impressive; see Fig. 1-11). Shade leaves, on the other hand, are much thinner and have lower rates of dark respiration and a lower value for light compensation. Although light-saturated rates of photosynthesis are lower than values for sun leaves, their survival in deep shade is certainly enhanced by these other components. Respiratory losses in particular can be a primary factor in shade adaptation, since both herbs and tree seedlings which are recognized as shade-tolerant also embody this feature of respiratory readjustment to light intensity. Shade-intolerant species lack this capacity (Grime, 1965; Loach, 1967; Larcher, 1969). This distinction becomes especially apparent at elevated temperatures, where shade-intolerant species are "pressured" into high rates of respiration at the expense of survival but shade-tolerant species show only moderate response even at 35°C (Fig. 15-6). Grime (1965) suggested that his shade plants

Fig. 15-5 Shade leaves generally show lower values for dark respiration, light-compensation point, and photosynthetic capacity than sun leaves. Average data from seven species of sun plants and five species of shade plants. (*Adapted from Böhning and Burnside, 1956.*)

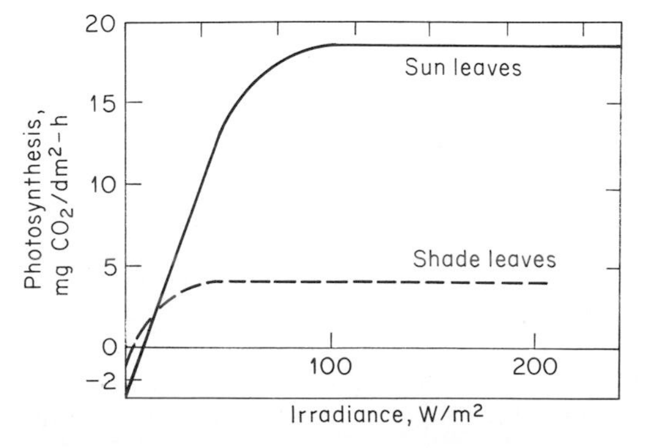

were metabolically slow, and the low Q_{10} in Fig. 15-6 certainly attests to lethargy. Ecological races of *Lolium perenne* which have successfully adapted to the hot climate of Algeria are similarly less sensitive to temperature than their genetic counterpart from Denmark, where respiratory economy offers no selective advantage (Eagles, 1967).

Shade tolerance represents more than simply detuning. Shade plants are undoubtedly conservative, and substrate losses are minimized, but there is also a positive side to the ledger. Photochemical efficiency contributes to this balance. Gross differences between sun and shade leaves studied by Böhning and Burnside (1956) masked a subtle difference in the early part of the light-response curves which is a key to success in deep shade. Wassink's data on shade adaptation in *Acer pseudoplatanus* (Wassink et al., 1956) show the distinction, as do Sparling's (1967) values for woodland herbs in Ontario and Björkman's measurements on *Solidago virgaurea*. In every case, plants capable of shade adaptation develop a higher photochemical efficiency, which is expressed by a steeper slope in the early phases of their light-response curves. Björkman and his colleagues analyzed this adaptation in great detail; some data are given in Fig. 15-7. Subsequent work has been devoted to this fragile photochemical distinction that confers an advantage in deep shade (compare *C* and *D* in Fig. 15-7) but suffers in strong light (compare *A* and *B* in Fig. 15-7).

Björkman (1966) concludes from his work on sun and shade races of *Solidago virgaurea* that the light-absorbing system in its photosynthetic apparatus is more efficient in shade races than in sun races when both are grown in *weak* light. When a shade race is grown in *strong* light, photochemical function is severely impaired, probably because photosystem II is inactivated (granal disorganization may be implicated; cf. Fig. 15-4). Sun races of *Solidago virgaurea* show no detrimental effect of strong light during growth; moreover, their light-saturated rates of photosynthesis are substantially greater if plants are grown in strong light because of their inherent capacity for RuDP carboxylase formation. Shade races cannot match this capacity and therefore operate at a distinct disadvantage in exposed situations.

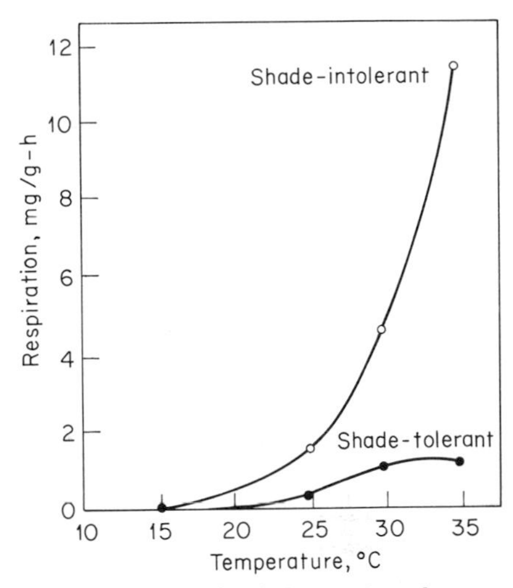

Fig. 15-6 Respiratory losses by disks cut from leaves of a shade-tolerant tree (*Pachysandra*, spp.) are minimal despite high temperature. Corresponding tissue from a shade-intolerant tree (*Paulownia tomentosa*) shows an exponential rise in respiration with increased temperature (Grime, 1965).

Species which become established in a shaded niche either *endure* low light due to a combination of morphological and physiological adaptation (described above) or *avoid* deep shade by escaping toward sunlit regions at an early stage of their vegetative existence. Anatomical specialization in the form of tendrils, coiled stems, or adhesive pads aids this escape. Tendrils are superbly adapted to this role, even to the point of disregarding unsuitable anchorages and then coiling faster in light than in darkness so that the parent vine gets better illumination (Jaffe and Galston, 1968).

High Light Intensity

Intense solar radiation can be an asset or a hazard depending on species and location. Under desert conditions, for example, there is no selective advantage in achieving total interruption

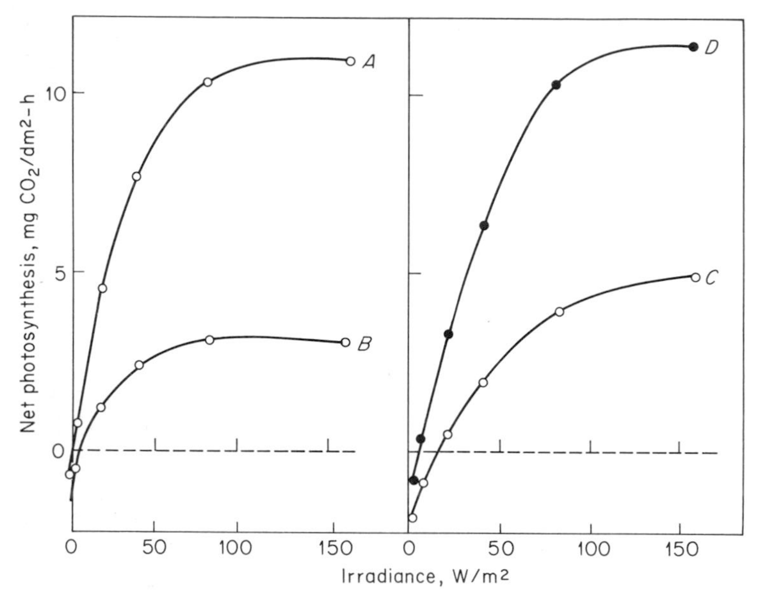

Fig. 15-7 Leaves of *Solidago virgaurea* from shaded habitats are prone to damage by high light intensity. A = shade-adapted; B = after transfer to strong light; C = grown in strong light; D = after 7 days in weak light. $A \rightarrow B$ represents photosynthetic deterioration in shade-adapted plants under the influence of strong light. $C \rightarrow D$ shows photosynthetic restoration of a shade-tolerant plant after transfer from strong back to weaker light. (*Adapted from Björkman and Holmgren. 1963.*)

of incident radiation because water supply frequently limits growth, strong irradiance is of little use photosynthetically and simply contributes to the plant's heat load. In this situation leaves do well to avoid direct exposure. Eucalyptuslike leaves, which hang vertically and characterize Australia's arid regions, or pine needles, which facilitate heat dissipation, both achieve this end. Other special cases include the so-called compass plants *Sylphium laciniatum* and *Lactuca scariola.* Leaves on these herbs adopt a vertical orientation in a north-south plane so that noontime exposure is at an absolute minimum, maximum exposure occurring in early morning and late afternoon. The compass plants commonly grow on steppe or prairie, where they extend well above other species without shading them unduly. When grown near solid structures that cast deep shadows, compass plants orientate their leaves along a plane of average highest light intensity.

Vertical orientation of foliage is thus a means of minimizing heat load under strong irradiance but at the same time providing sufficient illumination to saturate the leaf's photosynthetic apparatus for much of the day. Flat leaves, on the other hand, can reduce their energy absorption with epidermal hairs or other surface coverings (waxes), which increase reflectivity especially at high angles of incidence (discussed earlier in this chapter). A corrugated epidermal surface amounts to a further anatomical refinement which accentuates diffused reflectance and so benefits adjacent foliage as well as the exposed leaf.

Strong sunlight which gets absorbed despite this armory of protective stratagems is mainly wasted. Bonner (1961) suggests that chloroplasts are saturated at intensities around one-twentieth of full sun because their light-harvesting capacity is well in excess of their ability to convert this absorbed energy into a chemically usable form. This limited ability stems from their biophysical makeup: Emerson and Arnold (1932) were first to show that a photosynthetic unit contains about 2,500 chlorophyll molecules for every center within the plastid capable of reducing CO_2. The chloroplast with large photosynthetic units is therefore highly efficient at low light intensity, as discussed previously in connection with shade plants, but under strong illumination most energy goes to waste because the photosynthetic unit is too large; i.e., the light-harvesting system captures energy much faster than its utilization at energy-transforming centers. We might predict that a smaller photosynthetic unit should offer greater efficiency in strong sunlight; virescent mutants of cotton, tobacco, and pea plants show this characteristic. Their chloroplasts have few grana (in direct contrast to chloroplasts from shade-adapted leaves, Fig. 15-4), but instead they contain a high proportion of stroma lamellae (Benedict et al., 1972, and literature cited). Virescent mutants are most readily distinguished as seedlings, because foliage on mature plants reverts to normal in terms of chlorophyll content and grana formation. As immature seedlings, these mutants still achieve normal rates of photosynthesis under full sunlight, but their pale leaves transmit more energy than wild types, so that their heat load is reduced. Ecological consequences of light absorption without commensurate heat accumulation are self-evident. If shade-adapted plants are deliberately exposed to strong light, their photosynthetic apparatus is damaged (Fig. 15-7), whereas ecotypes that emerge at lower latitudes or in exposed situations respond to strong light and eventually achieve photosynthetic rates higher than shade-adapted races. The behavior of *Solidago virgaurea* has already been considered in detail (Björkman, 1966), while grasses such as *Lolium perenne* and *Dactylis glomerata* and the perennial needles of spruce are known to behave in a similar fashion (Wilson and Cooper, 1969; Eagles and Treharne, 1969; Katrushenko, 1965).

Despite their adaptation to exposed habitats, these species still fail to make maximum use of incident radiation; their leaves are light-saturated at intensities well below full sunlight. C_4 plants, on the other hand, *do* make effective use of strong light (see Figs. 1-11 and 1-25) and become fiercely competitive in favorable locations where dominance is a direct function of the plant's capacity to assimilate CO_2 and utilize its photosynthate for leaf expansion and vertical extension. If a C_4 plant's photosynthetic characteristics are coupled with stoloniferous or rhizomatous growth plus abundant and easily disseminated seed, this species becomes highly competitive, especially under intense solar radiation. Its evolutionary emergence and continued dominance at low latitudes follow almost automatically.

While C_4 plants are taxonomically diverse, species within the genus *Sorghum* show further differentiation in photosynthetic capabilities according to origin and so bear some analogy to physiological races within C_3 plants discussed earlier and to the upward readjustment in photosynthetic activity of leaves on well-spaced soybean plants under field conditions (Beuerlein and Pendleton, 1971). For example, Downes (1971) has shown that *Sorghum arundinaceum,* which originated in the low-light environment of equatorial forests, is capable of faster photosynthesis and wider stomatal aperture if conditioned to half sunlight rather than full sunlight. By contrast, species from high-radiation environments (*Sorghum bicolor*) display greater photosynthetic capacity when maintained in full sun.

Rapid assimilation confers decided ecological advantage on C_4 plants, and in mixed stands of C_4 grasses and C_3 legumes, grasses inevitably come to dominate. Comparative data on their light response and temperature sensitivity given in Fig. 15-8 elucidate this dominance. Grass leaves are amenable to illumination from either side, and with equal effectiveness, while their photosynthetic system is still not light-saturated at intensities approaching full sunlight. Quantum efficiency in grass leaves at this strong irradiance is almost twice that recorded for legumes under similar conditions (average $\phi = 0.10$ and 0.06 mol CO_2 per einstein, respectively). C_4 grasses therefore possess a combination of leaf anatomy, photochemical equipment, and biochemical machinery which fits them superbly to tropical regions with a hot and humid atmosphere where intense solar radiation is superimposed on an abbreviated growing season.

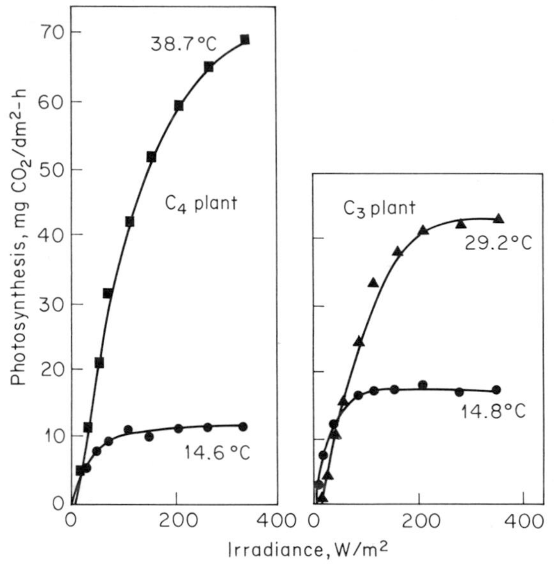

Fig. 15-8 C_4 plants generally show greater photosynthetic capacity and higher temperature optima than C_3 plants. These light-response curves compare *Pennisetum purpurem* (elephant grass, a C_4 plant) with *Vigna luteola* (a tropical legume, a C_3 plant). This advantage is lost at low temperature (Ludlow and Wilson, 1971).

Light Duration

Sustained bright light inhibits photosynthesis in shade-adapted plants (Björkman, 1968; Holmgren, 1968), whereas fluctuations in light intensity provide some opportunity for repair (Kok, 1956*a,b*) as well as favoring higher photosynthetic yield (CO_2 fixed per quantum of absorbed energy). Short intervals of light and dark, of 1 s or less, can enhance energy conversions because dark reactions concerned with CO_2 assimilation can run to completion and thereby restore the photosynthetic apparatus to its full efficiency at the start of each new light phase. This response is a feature of both algae and higher plants (Steeman-Nielsen, 1957). Enhanced yield is evident in Fig. 15-9 where vine-leaf photosynthesis is expressed as a function of light intensity for both continuous and intermittent illumination (0.05 s light, 0.20 s dark). Predictably, intermittent light has no effect on photosynthetic yield in weak light, corresponding to the linear portion of the continuous-light-response curve (Fig. 15-9), because photosynthesis is limited by absorbed light and not by CO_2-fixing capacity, i.e., dark reactions. In this instance, irradiance had to exceed about 50 W/m² (one-tenth full sun) before dark reactions imposed any limitations. Quantum efficiency under intermittent light was 0.04 mol CO_2 per einstein, compared with 0.02 under continuous illumination. The intermittency factor (Rabinowitch, 1956) was therefore 2.0. This difference emphasizes a leaf's potential ability to make more effective use of brief flecks of direct sunlight which fall on leaves within a canopy, i.e., sun flecks, than an equivalent energy supply from continuous sunlight, and variations of this periodicity do occur in nature. Leaves in natural communities experience a constantly changing light climate where diffused light is supplemented by occasional sun flecks. In erect crops such as corn, sun-fleck contribu-

Fig. 15-9 Photosynthetic yield in *Vitis vinifera* (CO_2 fixed per unit of absorbed light) is higher under intermittent light than continuous illumination. Duration of light and dark cycles of 0.05 and 0.20, respectively, so that completion of dark reactions allowed more efficient utilization of bright flashes (Kriedemann et al., 1972).

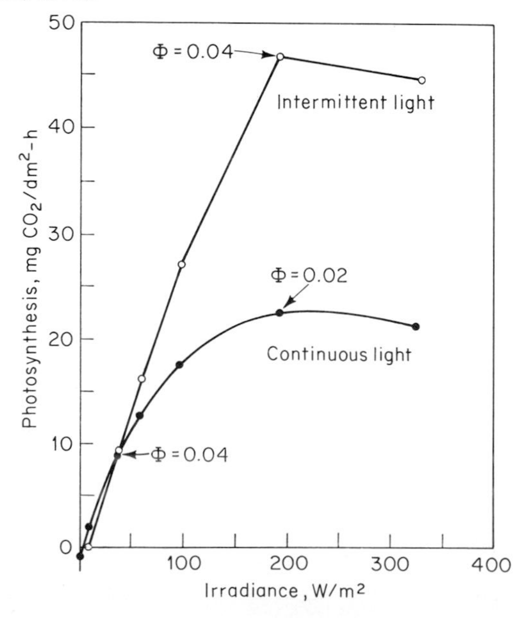

tion to canopy photosynthesis is substantial although its significance in bromegrass-alfalfa canopies is negligible (Norman and Tanner, 1969). Well-ventilated tree canopies do admit sun flecks with a high frequency, and Evans (1956) estimates that sun flecks occupy 20 to 25 percent of forest floors in Nigeria, where this form of illumination accounts for 80 percent of incident radiation at noon.

The quantitative significance of sun flecks for photosynthesis is difficult to assess, and estimates differ sharply (Anderson, 1964), but in well-ventilated canopies of *Betula, Eucalyptus, Acer, Populus,* and some *Vitis* species sun flecks occur with a frequency and duration that matches the capacity of leaves to use them with maximum effectiveness. Brief periods of illumination on the floor of dense forests should help offset respiratory losses and so contribute to long-term maintenance of shade-adapted tree seedlings that would otherwise perish (Ino, 1970; Kriedemann et al., 1972).

Short-term fluctuations undoubtedly enhance photosynthetic yield, but total plant growth is of greater ecological concern. Variable shading can be highly advantageous to some species, especially shade-tolerant ones, and in a substory crop such as *Coffea arabica* growth is undiminished by variable shade which allows rapid fluctuations in light intensity across the entire population as well as over the surface of individual leaves (Huxley, 1969). Beneficial effects of short-term variation in light intensity have also been demonstrated under controlled conditions by Lemon and Bogacheva (1957), who grew cucumber, tomato, and radish under a movable bank of fluorescent tubes (0.8 m^2) which swung back and forward over the crop. Incident light varied between 6 and 600 fc and optimum growth occurred under a 6- to 7-s periodicity. Cucumber productivity was increased by 50 percent, while radish growth almost doubled. Tomatoes were less responsive, although variable illumination did elicit some increase (10 percent) over steady conditions.

Fluctuations in lighting of longer periodicity (cycles lasting from 1 min to about 1 h) lead to *reduction* in photosynthetic yield (Garner and Allard, 1931; Rabinowitch, 1956). Whereas short cycles allow dark reactions of photosynthesis to run to completion, longer cycles up to 1 h lead to induction phenomena which then occupy most of the subsequent light period. This initial cranking up of the photosynthetic machinery was discussed in Chap 1. Both biochemical and stomatal components are involved. When fluctuations of this order begin to vary in both amplitude and frequency, the plant experiences climatic "noise," and its photosynthetic performance drops even further. Impaired stomatal function appears to be a contributing factor (Aufdemgarten, 1939; Evans, 1963). Photosynthetic readjustments to changed conditions are therefore not immediate; the plant operates with some measure of inertia and tends to take up environmental fluctuation. One outcome for overall growth is that plants perform as near-perfect integrators of photosynthesis rather than light. McCree and Loomis (1969) expounded this principle in computing rates of photosynthesis in fluctuating light which simulated natural conditions.

Diurnal or seasonal changes in light climate also influence plant growth, but here the outcome is generally more favorable. Uninterrupted strong light which supports rapid photosynthesis seems to exhaust the plant, whereas some diurnal change in temperature or diurnal reduction in light intensity offers relief (Hillman 1956). Went (1958) attributed this steady rundown over the course of a sunny day to an accumulation of photosynthetic products, a hypothesis already discussed in detail by Neales and Incoll (1968) and referred to in Chap. 1. Bleaching of leaf pigments due to strong light (full sunlight) has also been suggested by Nutman (1937) and Böhning (1949) as a probable cause, while a steadily mounting reaction to unfavorable leaf-water potential as the day advances could be a contributing factor in both herbaceous plants and woody perennials (Bormann, 1953; Kozlowski, 1957; Kriedemann and Canterford, 1971).

Sustained photosynthesis under continuous illumination will obviously favor both productivity and competitive ability, especially within the Arctic Circle, where plants can experience 24 h of illumination at certain times of year. In wild birch, for example, photosynthesis in regions around 67.5°N lat. continues around the clock while the sun does not set. Under this moderate irradiance, both cultivated plants (potato, cereals, beans) and wild species, regardless of photoperiod grouping, photosynthesize 24 h/day in the arctic environment (Kislyakova, 1960). Young aspen trees (*Populus tremuloides*) behave similarly under controlled conditions, showing greatest carbon gain under continuous illumination of moderate intensity (Bate and Canvin, 1971).

Growing seasons are obviously abbreviated at high latitudes, and plants clearly benefit from their capacity to utilize continuous light, but the duration of conditions favoring growth is also affected by altitude. Plants show clear adaptation to the greater insolation of high altitudes in the form of epidermal coatings which enhance reflectance, but properties of their photosynthetic apparatus are also altered. Altitudinal races of *Typha latifolia* provide an illustration. McNaughton (1967) isolated chloroplasts from two ecological races of *Typha,* a high-altitude population from Wyoming and a maritime population from California. The ability of these extracts to carry out the Hill reaction showed negative correlation with growing-season duration. The high-altitude race had twice the photochemical activity of the population from California. Admittedly, Hill activity does not necessarily control productivity or competitive ability, but higher photochemical activity was certainly compensating in some way for an abbreviated growing season.

Light Quality

The photosynthetic effectiveness of visible radiation varies according to wavelength (action spectrum Fig. 1-9b), but the chemical identity of fixation products also depends upon the spectral composition of incident light. Exposed foliage therefore receives a more effective balance of solar radiation and should also generate a different range of photosynthetic products compared with shaded leaves, which experience a greater proportion of transmitted radiation enriched in wavelengths corresponding to green (and sometimes blue) light (Anderson 1964). Accordingly, we would expect the biochemical nature of a leaf's fixation products to vary according to time of day and position within the community.

Work in the Soviet Union (Voskresenskaya, 1950) gave an early indication of light-quality effects on photosynthetic products. Leaves were first infiltrated with KNO_3 in the hope of relieving any possible limitation nitrogen supply might have imposed on protein synthesis and then illuminated by red or blue light of similar quantum flux density. Leaves added more dry weight in red light than in blue light, as would be expected from the relative quantum effectiveness of these two wavelength categories (McCree, 1971, and Fig. 1-9b), but in red light, 68 percent of the added dry matter was due to carbohydrates compared with 42 percent in blue light. Subsequent experiments by the same investigator confirmed that red light favors carbohydrate accumulation, and related work on rice leaves in India (Das and Raju, 1965) has revealed that blue light stimulates accumulation of protein and other noncarbohydrate substances.

Diurnal effects on fixation products (Table 15-2) could also depend in part on light quality because Horvath and Feher (1965) showed an enhanced carbohydrate buildup in stems of *Phaseolus vulgaris* exposed to green light. Under natural conditions, the chemical distribution of ^{14}C-fixation products similarly depends upon environmental factors. In leaves of *Zygophyllum rosovii* (a steppe plant) and *Lidelofia stylosa,* ^{14}C products were displaced in favor of sugar formation as the day progressed (Table 15-2).

PHOTOMORPHOLOGICALLY ACTIVE RADIATION

Regulatory effects of light on plant form and development (photomorphogenesis) are of pervading importance both for individual organisms and for their relationships with other members of their community. Ecosystems are both sustained and regulated by light-driven biochemical reactions. Just as a plant's photosynthetic activity equips it with the capacity for storing the fuel that runs the plant machinery, the

Table 15-2 $^{14}CO_2$ fixation products vary according to time of day. ^{14}C sugar formation is enhanced as the day advances (Filippova, 1957).

	Time of experiment	^{14}C absorbed by leaf, cpm dried-leaf preparation	Soluble sugars, % of $^{14}CO_2$ absorbed
Zygophyllum rosovii	6 A.M.	170	13.5
	9 A.M.	2,130	16.5
	7 P.M.	340	30.3
Lidelofia stylosa	6 A.M.	320	30.6
	9 A.M.	2,840	83.5
	7 P.M.	690	94.0

photomorphogenic systems provide means through which the plant can adapt to environmental phenomena.

Pigment Systems

Light can force numerous functional changes in the plant through the direct activation of plant components by photons; examples are the damage to plants due to direct photodestruction of nucleic acids, damage to metabolic systems through absorption of ultraviolet light, or destruction of auxin by the direct absorption of radiation by the auxin molecule. In most ecological situations, such direct photoalterations of nonpigment plant components are rare. By far the most widespread regulatory actions by light involve the absorption of light by pigment systems.

There are three major pigment systems which function in response to PmAR in higher plants, namely phytochrome, the carotenoids, and the flavins. The absorption spectra of the two forms of phytochrome, of β-carotene and riboflavin, are illustrated in Fig. 15-10. The spectra illustrate again the clustering of biologically important light phenomena in the region of the spectral window of the atmosphere (Fig. 15-1). The regulatory functions of phytochrome include the regulation of photoperiodism, dormancy, and numerous morphological features related to etiolation. The functions of carotenoids include most probably and most importantly the regulation of phototropism of etiolated seedlings. Flavins may contribute to the third positive phototropic curvature of seedlings (Zenk, 1964), and they may also participate in various phytochrome functions under higher light intensities through the transfer of light energy from the blue to the red wavelengths (Shropshire et al., 1964).

The mechanism through which light may introduce alterations in regulatory systems is most commonly an alteration of *cis* to *trans* carbon-carbon double bonds. Just as the perception of light in animal vision involves photoalteration of a *cis* bond in retinal to a *trans*, so also it is probable that the light regulation of phototropism may occur by a *cis* to *trans* alteration of a bond in carotene. A similar possibility for the phototransformation of phytochrome from the P_r to the P_{fr} form would be a *cis* to *trans* shift of a double bond between the third and fourth

Fig. 15-10 A comparison of the absorption spectra for three types of pigments known to participate in photomorphogenic reactions: Riboflavin (Rbf), β-carotene (Car), two forms of phytochrome (P_r and P_{fr}).

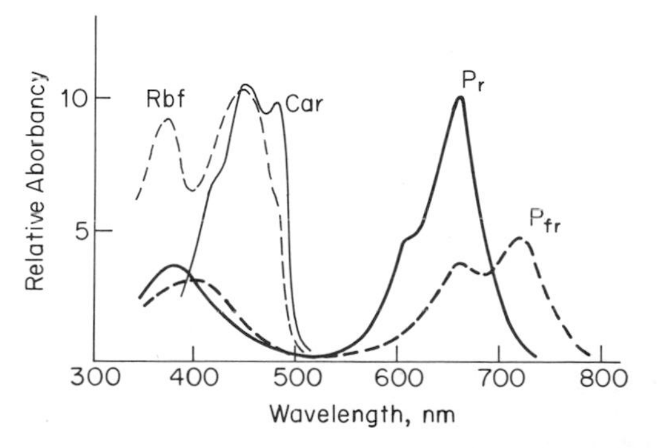

pyridine rings of the tetrapyrole structure in the chromophore. Another possibility is a shift of a proton between the two ends of the tetrapyrole structure of phytochrome (Fig. 15-11). A third possibility for light alteration is an oxidation-reduction reaction within the pigment; this remains a possibility for phytochrome, and may be the nature of riboflavin responses to light.

The kinetics of the pigment change must be presumed to be instantaneous; likewise, in at least some instances the photomorphogenic change may begin at once when light is applied, or in other instances after 10 or 20 min (Meijer, 1968).

In a simple photosystem, the number of photons required to make each molecule of pigment change should be the same; in this case, then, one would expect the extent of the biological response to be constant if the same number of photons were applied, even over a wide range of radiation densities. This gives us the reciprocity test of any photoresponse: if the same total amount of radiation is applied over a range of intensity time variations, then the biological response should be uniform, if there is a single controlling photoreaction. As an example of this reciprocity test, data of Briggs (1960) are shown in Fig. 15-12 for the phototropic curvature of *Avena* coleoptiles over a range of light intensities. In the upper graph, it can be seen that as long as the total light dose was 1,000 mcs, the curvature response remained at about 30°, reciprocity holds, and it is probable that a single pigment system is driving the reaction. In the lower graph, under a total light dose of 10,000 mcs the curvature is not constant, and reciprocity does not hold.

Phytochrome. The late date for the discovery of phytochrome was due to the fact that the pigment was not visible in plants. But the finding by Hendricks and coworkers at Beltsville, Maryland, that several photomorphogenic responses of plants had action spectra which peaked in the red region (Parker et al., 1946), and then that many of these actions of red light were reversible with far-red light (Borthwick et al., 1952),

Fig. 15-11 The photoinduced change of P_r into the biologically active form P_{fr} is probably associated with a structural alteration in the chromophore of phytochrome involving a shift in two hydrogen atoms (Hendricks, 1968).

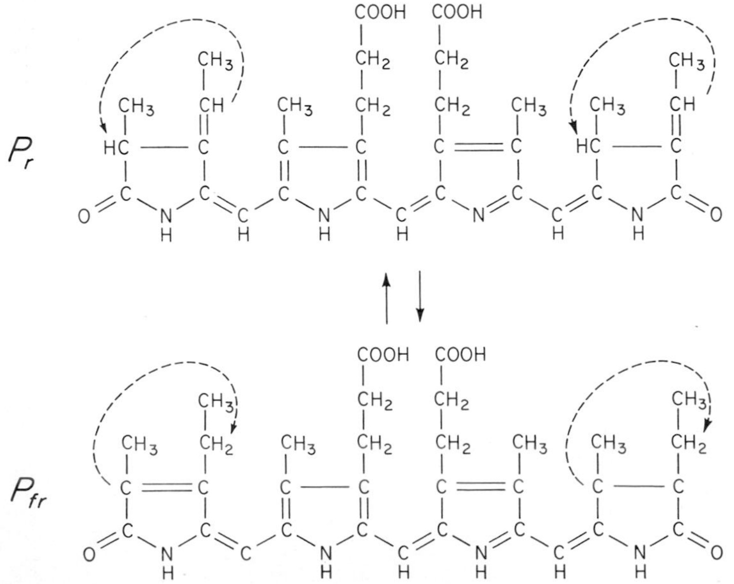

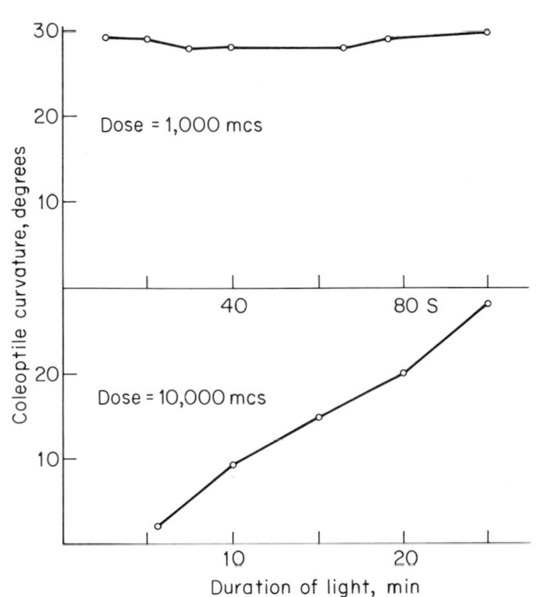

Fig. 15-12 The phototropic response of oat coletiles shows reciprocity when intensity × duration equals 1,000 mcs (above), but does not show reciprocity when the light dose equals 10,000 mcs (below) (Briggs, 1960).

opened the way for a systematic search for some pigment which would show interconvertible red and far-red absorbing forms. Here was a case of inductive reasoning which predicted a plant pigment which nobody could see. Final resolution of the search came with the isolation of such an interconvertible pigment, which was then given the name of phytochrome (Butler et al., 1959).

Measurement of phytochrome in vivo was made possible by designing a spectrophotometer which can separately apply monochromatic light to plant tissue and then rapidly scan the spectrum for changes in absorptive properties (Birth, 1960). Using this spectrophotometer, Butler et al., (1959) found a spectral shift (Fig. 15-13) in absorption by etiolated plant parts following red or far-red illumination. Alkaline extracts of light-sensitive plants (commonly made at pH 7.8; cf. Briggs and Rice, 1972) yield a solution which has similar spectral shifts. Siegelman and others subsequently extracted phytochrome from seedlings, achieved extensive purification, revealed its identity as a chromoprotein, and showed that the shift in chromophores results in blue or green color depending on molecular forms (cf. Fig. 15-14). Exposure to red light (approximately 660 nm) preferentially activates the red absorbing form P_r and converts it to the far-red absorbing form (P_{fr}). Exposure to far-red light (approximately 730 nm) has the reverse effect (see Figs. 15-11 and 15-13). Etiolated cereal seedlings are favored for investigations of this type because they contain far more phytochrome than light-grown seedlings and because the absence of chlorophyll simplifies spectral analysis.

This photoreversibility is seen in the characteristic action spectra for numerous responses of plants to weak light, an example being the regulation of leaf enlargement shown in Fig. 15-15.

Fig. 15-13 Responses to red light which are reversed by far-red light may be related to a difference spectrum. Absorbance of a phytochrome extract from corn shoots is shown after red and after far-red irradiation, and the difference spectrum is represented below (Butler et al., 1959).

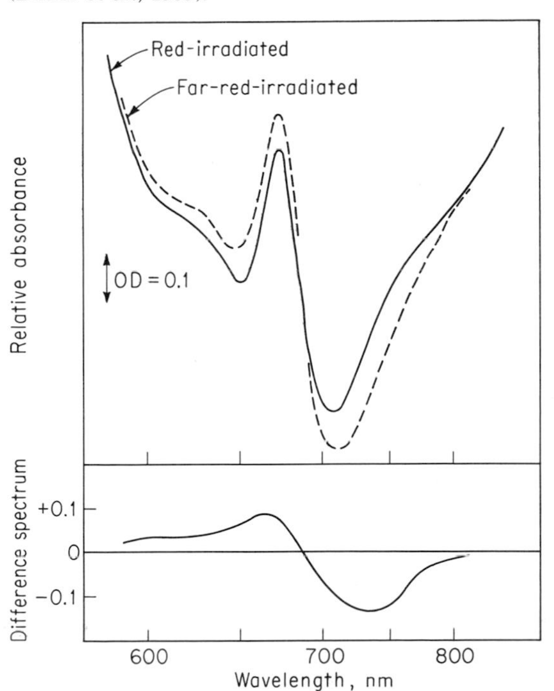

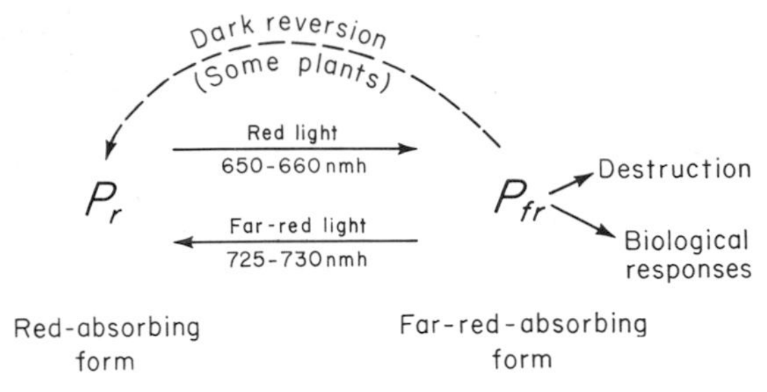

Fig. 15-14 Exposure of phytochrome to red light (650 to 660 nm) preferentially activates the red-absorbing form P_r and converts it to the far-red-absorbing form P_{fr}, which may then be utilized in biological responses or destroyed; in some instances it may be converted back to P_r in a dark reversion. Far-red light (725 to 730 nm) reverses the red-light effect.

Endogenous levels of phytochrome are extremely low, estimated at about 10^{-7} M in maize (Siegelman and Butler, 1965). The highest amounts have been found in meristematic tissues (Hillman, 1967), and especially in parsnips, artichokes, cauliflower, and brussels sprouts (Hillman, 1964).

There have been wide discrepancies in the molecular weights reported for phytochrome, and this is probably due to the ease with which it is attacked by proteases. An estimate of 120,000 has been made by Gardner et al. (1971), based on isolations which obviated the earlier problems.

The structure of the chromophere of phytochrome has been established in general terms, although the exact nature of structural differences between P_r and P_{fr} is still the subject of some speculation. One proposal is shown in Fig. 15-11. Despite this uncertainty, everyone at least agrees that phytochrome is a tetrapyrrole and that intermediate forms between P_r and P_{fr} do exist (Lipschitz et al., 1966). These intermediates may represent different forms of the protein component of the phytochrome system, although P_r and P_{fr} almost certainly represent two distinctive conformational states of the protein. This protein has in fact been credited with enzyme activity (Tezuka and Yamamoto, 1969), which raises other intriguing prospects for light regulation of metabolic processes.

The conversion of phytochrome between the P_r and P_{fr} forms can be driven by light, of course. It was soon discovered that the conversion between P_{fr} and P_r could occur in darkness in some plants, as is indicated in Fig. 15-14. This dark reversion may be substantial in some green plants (Butler et al., 1963), but probably not in most grass seedlings (Hillman, 1967). There are temperature sensitivities here, too; low temperatures may hold the P_{fr} from reverting to P_r in the dark (Klein et al., 1967).

An important aspect of the phytochrome dynamics in plants is the well-established fact that the P_{fr} form is ordinarily subject to destructive forces in the plants. This destruction, especially evident in grass seedlings, has a Q_{10} of about 3 (Pratt and Briggs, 1966), requires oxygen (Butler and Lane, 1964), and shows zero order kinetics, so it is probably enzymatic (Pratt and Briggs, 1966). P_{fr} is very easily destroyed by sulfhydryl poisons, and is readily denatured by urea (Butler et al., 1964). Examination of the kinetics of the red and far-red responses has shown that the P_{fr} is the biologically active form of phytochrome, and the P_r form apparently is not involved in most light reactions (Hendricks and Borthwick, 1965).

Fig. 15-15 The stimulation of leaf enlargement is most effective for red light, and the red-light effect is reversed by far-red light; etiolated bean seedlings (Downs, 1955).

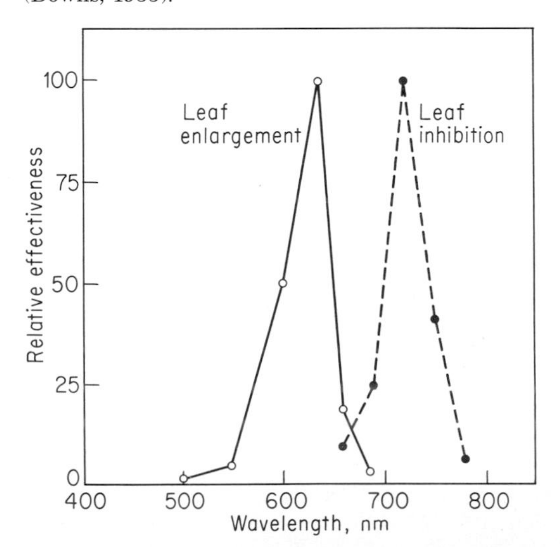

Therefore, the regulatory functions of phytochrome can be passed over either by the passage of time in darkness (in cases where the dark reversion to P_r occurs), or by the passage of time in darkness where enzymatic destruction of P_{fr} occurs, or by exposure to light which would convert the P_{fr} into P_r.

Attempts to correlate the amount of phytochrome which was in the P_{fr} form with the photomorphogenic effects of light ran into many complications, and it soon became evident that there were few good correlations between the measured proportion of phytochrome conversion and the biological response obtained; the deduction was that there must be pools of phytochrome in the plant, only some of which may be serving in regulatory roles whereas others are not (Hillman, 1967).

Phytochrome is an exceptionally sensitive photoreceptor and can trigger growth responses following exposure to momentary flashes of weak red light despite its low concentration in plant tissues (about 0.1 percent of total protein). Dark-grown *Avena* seedlings, which are particularly sensitive to light, can be used to illustrate effects from low irradiance. As little as 9.1 μcal/cm^2 of red light (623 nm) is enough to retard growth of the first internode by about 10 percent due to effects on meristematic activity (Goodwin and Owens, 1951). By comparison, clear moonlight represents 4 μcal/cm^2-min, of which 5 percent would occur as red light in the region of 623 nm. Less than 3 h of exposure to clear moonlight should therefore produce a measurable reduction in internode growth on etiolated *Avena* seedlings.

High-Energy Reactions

While the early studies of phytochrome reactions had repeatedly indicated a photoreversibility between red and far-red treatments, there were also reports of light-regulated plant functions which had somewhat different characteristics: no reversibility between the red and far-red, peaks of activity in the red and the blue with peak effectiveness at about 710 nm (instead of the phytochrome peaks at 660 and 725), and requiring higher light levels than the common phytochrome systems. These light systems have been given the name high-energy responses, or HER. A sample action spectrum for a HER response is given in Fig. 15-16, in this case for the inhibition of the elongation of the hypocotyl of *Sinapis* seedlings. Many HER effects are related to photoregulation of seedling responses; thus there is a HER control of hypocotyl elongation in *Sinapis* (Mohr, 1959a), of hair formation in the epidermis of the hypocotyl (Mohr, 1959b), a stimulation of opening of the hypocotyl hook in lettuce seedlings (Mohr and Noble, 1960), and a stimulation of anthocyanin development in the hypocotyl (Vince and Grill, 1966). In each case, these controls combine a peak effectiveness of red light (710 nm) with an effectiveness of blue light—especially in the region of 430 nm.

There seem to be two options by which to account for the HER phenomena: they may be due either to a single pigment with sensitivity to light in both the blue and the red regions, or to the interactions of two pigments, one absorbing the blue and another absorbing the red wavelengths. Examination of Fig. 15-10 suggests that either of these explanations might be invoked using pigments we already know of in the plant. A combination of either a flavin or a carotenoid

Fig. 15-16 The action spectrum for the inhibition of hypocotyl elongation in lettuce seedlings, showing maxima in the blue region, suggestive of the action spectrum in Fig. 15-18 but with another peak of effectiveness in the red region. Light effectiveness is expressed on a comparative basis with 447 nm (Hartmann, 1967).

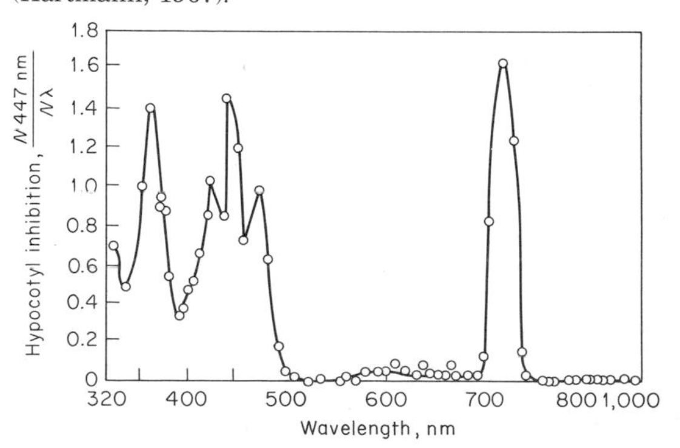

with phytochrome could provide such an action spectrum. Alternatively, phytochrome itself has absorbance characteristics in both wavelengths. One will note at once that neither form of phytochrome has an absorption peak at 710 nm where the HER systems characteristically peak; but if the characteristic high light dose needed for HER is due to an activation of the phytochrome pigment instead of an interconversion effect, then it might be expected that the red region of the action spectrum might peak at the isobastic point—where the two absorption curves (P_r and P_{fr}) cross—and this is approximately 710 nm (Siegelman and Hendricks, 1958).

An interesting alternative suggestion has been made by Hartmann (1966); he proposes that the HER action spectrum may be entirely due to phytochrome, and that the peak of activity at 710 nm may be due to the maintainance of ratios between P_r and P_{fr}, with 710 nm being able to drive the phytochrome conversions in either direction, thus maintaining a given ratio between the two forms. He has done experiments on a HER response, the inhibition of hypocotyl growth in *Sinapis* seedlings, and applied light of two different wavelengths, 658 to drive P_r to P_{fr} and 768 nm to drive P_{fr} to P_r. As illustrated in Fig. 15-17, the plants showed no responsiveness to either alone, but, with increasing amounts of illumination at 768 nm (increasing the proportion of phytochrome in the P_r form), then growth was inhibitable with 658 nm light (returning the P_r in part to the P_{fr}, bringing the ratio into the regulating state).

Fig. 15-17 Hypocotyl lengthening in lettuce seedlings can also depend upon the combined effects of light at two different wavelengths. Irradiation at 658 nm becomes effective in eliciting a quantitative response only when superimposed over light at 768 nm. (Hartmann, 1967).

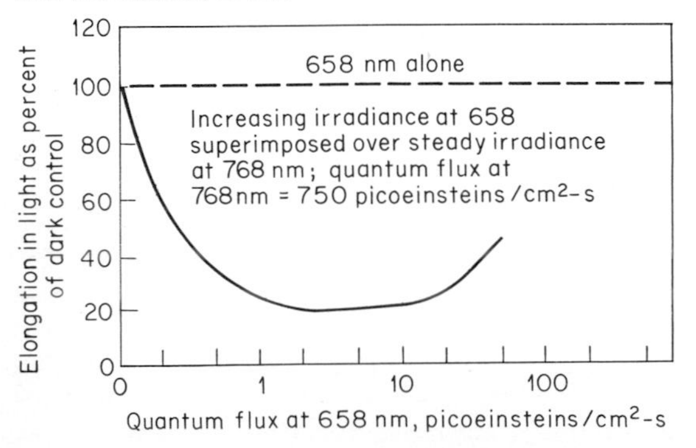

Mohr has found that reciprocity does not hold for the HER response of anthocyanin formation in the far-red light; this could indicate that the photocontrol does not involve a single pigment, or alternatively, that a single pigment has more than one effect, such as pigment conversion and pigment activation.

Photomorphogenesis

One of the changes in plant form given earliest attention by plant physiologists was that of phototropism. As quantitative experiments have been developed with increasing precision (Figs. 9-15 and 9-16), a precise action spectrum has been worked out as in Fig. 15-18. One can see a maximal effectiveness of blue light (440 nm), with shoulders of activity at wavelengths somewhat blending the characteristics of shoulders of β-carotene and riboflavin (cf. Fig. 15-10). In most solvents, β-carotene does not have a shoulder of absorbance in the region of 370 nm, but the absorbance shoulders in the region between 400 and 500 nm strongly suggest β-carotene. Hager (1970) has provided the ingenious suggestion that the peak at 370 may be a consequence of the combined action of polar and nonpolar solvents on the carotenoid at its site in the cell.

The list of photomorphogenic influences on the etiolated seedling is enormous. The elongation of the seedling stem is inhibited by light (Blaauw, 1915); some species are most inhibited by the phytochrome system, others by the HER system (as in Fig. 15-16). In grass seedlings, such an inhibition is most manifest in the mesocotyl, which is inhibited most effectively by red light at about 650 nm (Weintraub and McAlister, 1942). Associated with the inhibition of stem elongation, red light can also stimulate the commencement of leaf enlargement; this is another red-light effect reversible with far-red light (Parker et al., 1949), and is illustrated in Fig.

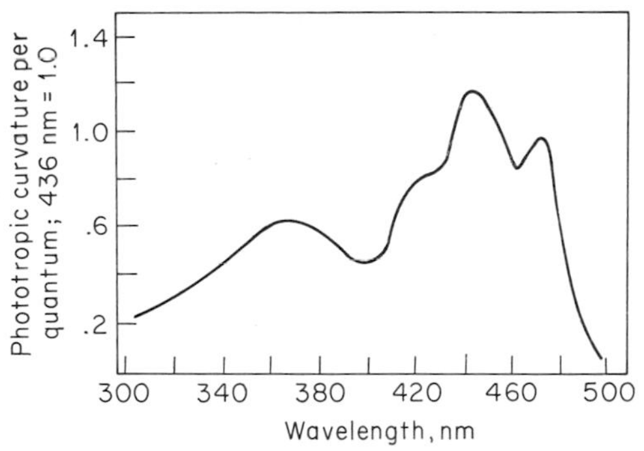

Fig. 15-18 Action spectrum for the first phototropic curvature of *Avena* coleoptiles, showing the combination of a peak at about 370 nm like riboflavin, and the peaks at about 420, 450, and 470 like β-carotene (Curry, 1969).

15-15. The complementary action of light in depressing stem and enhancing leaf growth is shown in Fig. 15-19. In the case of grass seedlings, a central component of the stimulatory effect of light on leaf expansion is the stimulation of leaf unrolling; this event involves a light stimulation of gibberellin release from some bound form (Loveys and Wareing, 1971). In many dicot seedlings, the etiolated seedling carries its terminal growing point through the soil in the form of a hook, either a hypocotyl hook or an epicotyl hook (see Fig. 10-6); another photomorphogenic effect is the removal of this hooking condition as illustrated in Fig. 15-20. This event involves a light suppression of ethylene biosynthesis (Goeschl et al., 1968). Numerous other morphological changes can be induced in seedlings by light, including stimulation of formation of epidermal hairs (Mohr, 1959a), the differentiation of the epidermis itself (Borgstrom, 1939), the stimulation of the biosynthesis of anthocyanins (Withrow et al., 1953) or of betalains (Wohlpart and Mabry, 1968), the induction of polarity in germinating spores of ***Equisetum*** (cf. Fig. 9-1) (Haupt, 1957), or the regulation of the conversion of fern filaments into prothalli (Mohr and Ohlenroth, 1962). Perhaps effects of light in potentiating other morphogenic responses should be mentioned as well; small amounts of light can enhance the geotropic responses of the *Avena* coleoptile (Blaauw, 1961), or of the roots of maize (McNitt et al., 1974); red light can likewise enhance the subsequent phototropic responsiveness of maize coleoptiles to subsequent blue light (Briggs, 1963). And finally, red light can enhance the subsequent photosynthetic efficiency of leaves (Zucker, 1972).

In dense communities, growth regulation due to subtle differences in PmAR with depth through the canopy is discernible, despite generally high levels of radiation. Leaves transmit much far-red light compared with red and blue light, so that shaded organs experience much more far red compared to the red and blue of exposed foliage. Consequences for plant morphology are seen in Kasperbauer's (1971) work on dense plantings of tobacco. Partly shaded plants in a row grow much taller than plants on the outside of the community, and these inside plants have longer internodes with no axillary branches and develop light green thin leaves. This growth reaction can be simulated under controlled conditions if individual plants are exposed to far-red light at the close of each photoperiod. There are dramatic cases of photomorphogenic regulation of development, including most especially the light involvements in dormancy (Chap. 10) and photoperiodism (Chap. 12).

Fig. 15-19 Light applied to etiolated pea seedlings causes an inhibition of internode growth and a stimulation of leaf enlargement (Parker et al., 1949).

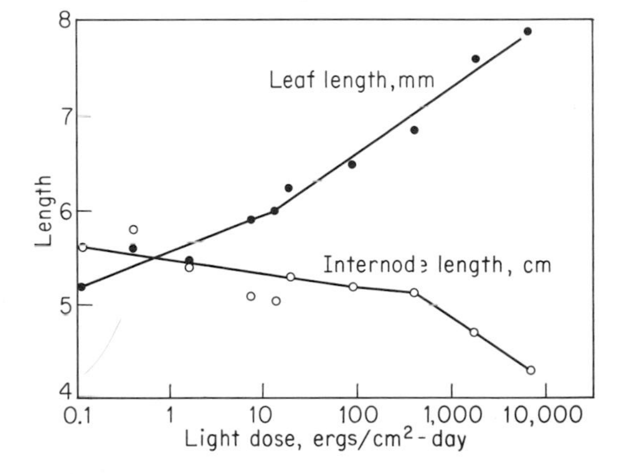

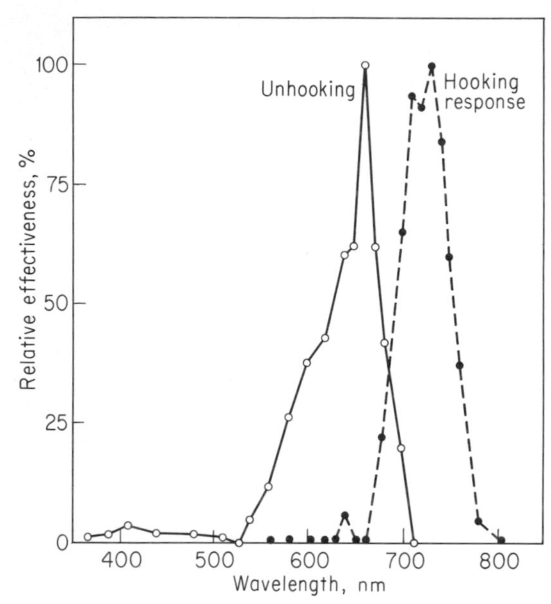

Fig. 15-20 The action spectrum for removing the hypocotyl hook from etiolated bean seedlings shows the peak effectiveness of red light; the red-light effect is removed by subsequent far-red light (Withrow et al., 1957).

Mode of Action

The mechanism of photomorphogenic controls in plants may be through light alterations of hormonal components, alterations of enzymic components (this could be through an alteration in the repression or derepression of genetic information), or through alterations of membrane characteristics (particularly of membrane permeability or bioelectric charge).

Hormone Regulation

Some hormones can substitute for phytochrome-mediated reactions, even in darkness, which lends force to the idea that hormones are directly involved in relaying photomorphogenic phenomena. A relationship between red light and gibberellin biosynthesis is already well established. Reid et al. (1968), for example, demonstrated how exposure to red light leads to a rapid increase in endogenous gibberellins in etiolated leaves of barley (*Hordeum vulgare L.*). Further experimentation with metabolic inhibitors then revealed how both RNA and protein synthesis are prerequisites for the red-light induction of gibberellin synthesis (Reid and Clements, 1968). The association between leaf unrolling and a rapid increase in endogenous gibberellins following exposure to red light was investigated further with etiolated wheat leaves by Loveys and Wareing (1971*a,b*). These workers encountered a faster increase in gibberellin-like substances following exposure to red light and with a concomitant fall in the level of bound gibberellins during early phases of the reaction. Since neither AMO-1618 nor actinomycin D had any effect on this initial production (following a 5-min exposure to red light), the initial rapid increase was attributed to release of gibberellin from a bound form. Subsequent production of gibberellin in these wheat leaves (30 min exposure to red light) did involve synthesis, rather than simply solubilization, and coincides with previous observations by Reid et al. (1968) on barley leaves.

Cytokinins are also subject to rapid fluctuation following exposure to red light. Van Staden and Wareing (1972), working with seeds of *Rumex obtusifolius,* reported an increase in endogenous cytokinins within only 10 min of irradiation.

Yet another hormone, but one with a totally different character—ethylene—appears to be involved in phytochrome-mediated reactions. As discussed in Chap. 7, ethylene participates in a host of physiological reactions, but the actual production of this hormone can be under phytochrome control. Kang and Burg (1973) have described such a situation in *Amaranthus,* where a system involving ethylene in anthocyanin production is regulated by phytochrome.

Light can cause the formation of growth inhibitors in some cases; Blaauw-Jensen (1954) has reported the photoconversion of chlorophyllide into a growth inhibitor. Abscisic acid has been found to be synthesized more rapidly during the daylight hours than in darkness in at least some plants (Zeevaart, 1971).

Enzymic regulation. The two most definite cases of light regulation of enzymes are the

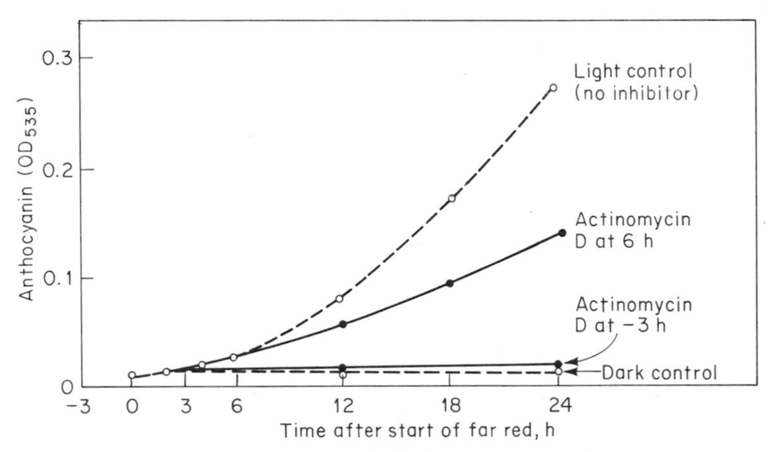

Fig. 15-21 Synthesis of anthocyanin in mustard seedlings (*Sinapis alba*) is influenced by far-red light. When actinomycin D is added before the onset of far-red light (−3 h), no anthocyanin synthesis occurs in response to the light. If antibiotic is added during or after the lag-phase (6 h following the start of illumination), anthocyanin synthesis still proceeds. (*Adapted from Lange and Mohr, 1965.*)

stimulation of phenylalanine ammonia lyase and nitrate reductase (Zucker, 1972). Each of these enzymes shows diurnal cycling with higher levels of activity during the daylight hours. Further, in each case the rise of enzymic activity during exposure of plants to light is prevented by inhibitors of protein synthesis such as cycloheximide. The diurnal cycling effects are complex, however, for in both of these cases there is some involvement of inhibitors or inactivation of the enzyme in darkness, and light may be serving in part to alter these processes.

An inhibitor of chymotrypsin accumulates in the dark and is dispelled by light (Ryan and Housman, 1967). Red light can enhance the formation of several enzymes involved in photosynthesis (Zucker, 1972), and in the case of etiolated grasses, red light leads to a marked increase in the starch hydrolytic activity in the unrolling grass leaves (Klein et al., 1963).

Mohr (1966a,b) has proposed that in the photomorphogenic responses, light may serve to bring about the derepression of some parts of the genetic information, thus leading to the new synthesis of enzymes involved in anthocyanin synthesis, or growth etc.

One piece of interesting evidence which relates to Mohr's view comes from work on anthocyanin synthesis (Lange et al., 1967). Etiolated seedlings of *Sinapis alba* will not maintain a steady rate of anthocyanin synthesis unless P_{fr} is constantly produced. Once the light source is removed, the rate of synthesis declines, but it is immediately restored under illumination following 12 h of darkness. No secondary lag occurs, but cycloheximide or puromycin (inhibitors of protein synthesis) inhibit this secondary rise in the rate of anthocyanin synthesis (Fig. 15-21). Actinomycin D proved less effective as an inhibitor of this second phase than in inhibiting the initial rise.

A difficulty with the theory that photomorphogenic controls are achieved through a derepression of certain genetic information lies in the timing of the responses to light. As mentioned previously, light can bring about an inhibition of growth within 1 minute, or in the cases of other wavelengths, after 10 or 20 min (Meijer, 1968). In higher plants, this is a very short time for the completion of the formation of a new *m*RNA and the synthesis of new enzymes (cf. Evans and Ray, 1969). In some cases, the light effect is lost as soon as a supply of P_{fr} is not at hand, and considering the relatively slow rates of turnover of enzymes in plants, this rapid loss of light influences would not be expected if genetic controls were the basis of the light effects.

Membrane Permeability

Hendricks and Borthwick (1967) proposed that many short-term responses mediated by phytochrome are highly suggestive of changes in the biophysical properties of membranes under the influence of P_{fr}. Flowering in *Pharbitis nil,* for example, is suppressed following irradiation with red light for only 3 min in the middle of an inductive dark period (Fredericq, 1964), due presumably to the formation of P_{fr} over this short time. Leaf unrolling in etiolated cereal seedlings provides a further illustration of rapid effects: the stimulus for unrolling is under phytochrome control (Virgin, 1962) and is transmitted from an irradiated to a nonirradiated part of a leaf in only 20 s or even less (Wagné, 1964).

Direct evidence of permeability changes mediated by phytochrome comes from the research of Jaffe and Galston (1967). Leaflets of *Albizzia* and *Mimosa* normally remain open during the day but close in darkness. However, if phytochrome is converted to P_r by far-red irradiation at the end of a photoperiod, the leaflets remain *open* in *darkness*. Subsequent irradiation with red light promotes formation of P_{fr}, and leaflets then begin to close within a few minutes. Closure is accompanied by the release of solutes from the cut ends of detached, but functional, pinnae, Jaffe and Galston (1967) interpreted these events in terms of a rapid permeability change due to production of P_{fr} under red light.

Tanada (1968) has also reported fast responses to irradiation under the influence of phytochrome. Barley root tips will adhere to negatively charged wet glass if phytochrome is in its P_{fr} form but are released in far-red light. This photoreversible adhesiveness changes within 30 s of illumination and requires the presence of IAA, ATP, ascorbic acid, and Mn^{2+}, K^+, and Cl^- ions. Omission of any ingredient erases the effect.

Root tips of mung bean also attach themselves to a glass surface (previously charged with phosphate ions) under the influence of red light, although this system differs from barley in some respect. While red light has a direct effect on attachment in both systems, *prior* irradiation of seeds or seedlings with red light promotes attachment in mung bean but diminishes attachment in barley (Tanada, 1972). Subsequent work (Tanada, 1973) revealed a fascinating hormonal involvement in this general response which again differs between the two systems. In mung bean, IAA causes detachment, while ABA leads to attachment; barley root tips were the exact opposite in their response to both IAA and ABA. Gibberellic acid, ethylene, and cytokinins had no effect in either system.

Adhesion is thought to depend on charge relationships between root tips and the underlying glass. Tanada (1968) has proposed that root tips possess a net positive charge over their surface when phytochrome is in its P_{fr} form.

Phenomena like the control of leaflet movement in *Albizzia* or *Mimosa,* root-tip adhesion, and rapid-growth inhibition (Meijer, 1968) are highly suggestive of rapid changes in biophysical properties that would hinge on membrane characteristics and lend support to the Hendricks and Borthwick (1967) view of phytochrome action. Other categories of plant responses of a more substantive nature, which result in the expression of new physiological capacities within the plant relating to growth and development, necessitate an additional theory for phytochrome action.

RHYTHMS

Many photoresponses in plants become so enmeshed with the daily march of environmental factors that they become self-sustaining. Even under constant illumination, plants that have been tuned to natural conditions will show persistent rhythms such as opening and closing of leaves and flowers. Rhythms of this sort have been recognized for at least two centuries, and probably the earliest report comes from Androsthenes, who noted daily movements of plant leaves during his travels with Alexander the Great. The ability of light to regulate such a rhythm was described by Candolle as early as 1832, while Sachs (1872) provided information

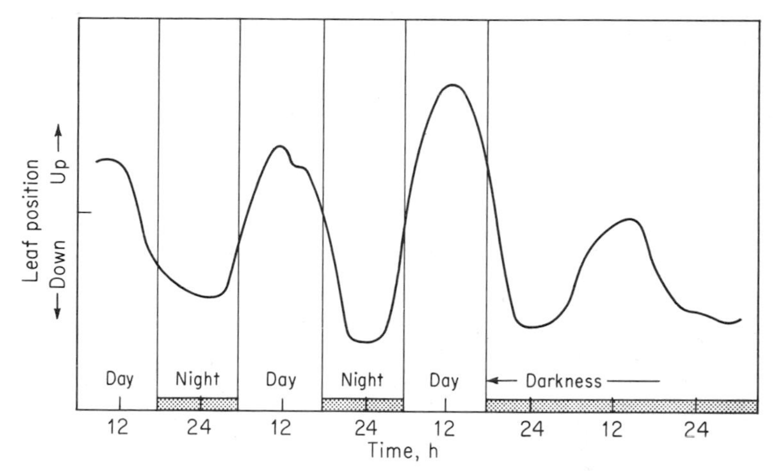

Fig. 15-22 The diurnal movements of *Albizzia* leaves elevate the leaves in the morning and lower them again in the evening. If they are deprived of light, the movement is repeated for a limited time and on a limited scale (Pfeffer, 1915; from Bünning, 1959).

on growth responses. He noted how some plants grow more rapidly at night and deduced that light could have specific inhibitory effects. Nevertheless, daily growth does show a component of diurnal periodicity, as illustrated by data from Ball and Dyke (1954), who observed a diurnal oscillation of growth rate in oat coleoptiles even under uniform conditions in darkness. Oscillation was initiated at the time of removal from light to darkness.

Diurnal periodicity in leaf movements was also noted by Pfeffer (1873). Leaves became elevated to a horizontal position in the morning, but were lowered to a "sleep" position in the evening. Such movements are synchronized by light; illumination during the night causes leaves to resume their day position. In constant darkness leaf movements persist for a few days as though daily light-dark cycles were continuing. *Albizzia* leaves show this same phenomenon (Fig. 15-22), Bünning and Lörcher (1957) examined this light action in qualitative terms and concluded that red light encourages the rhythm whereas far-red light erases the effect. This situation does vary with species and conditions, because Holdsworth (1959) and Fortanier (1954) have both found that in some species either the red effect is not reversible with far-red light or the response is driven better by blue than by red light.

Light-regulated rhythms with a periodicity of about 1 day, i.e., circadian rhythms, are now recognized in plants, algae, and protozoa. They include leaf and flower opening, stomatal movement, photosynthetic capacity, meristematic activity, root growth, and ion uptake; even the plant's capacity to respond to photoperiodic stimuli is mediated by phytochrome. Each function is triggered by light and is regulated in its periodicity by the duration of illumination. These circadian rhythms become so entrained in the day-night cycle that they persist for several days despite removal to steady conditions.

SUMMARY

The enormous scope of light influences on plants, from substrate synthesis to the regulation of growth and developmental sequences, establishes light as the most powerful single environmental force. The quality of sunlight has become precisely mirrored in the spectral sensitivity of many plant processes, while the diurnal rhythm of night and day is reflected in periodicities of plant responses, e.g., timing mechanisms for photoperiodic reactions and the sustained circadian rhythms of growth and metabolism. Major regulating actions of light fall within the spectral areas of photosynthesis (PhAR), red–far-red phytochrome effects, or the blue-light effects thought to be mediated by carotenoid-flavin types of photoreceptors (PmAR). These key light-response systems have developed over the course of evolutionary history to regulate an enormous array of growth and developmental functions ranging from rhythmic movements of stems, leaves, and flowers to the synchronization of complex steps which coordinate growth and reproduction.

GENERAL REFERENCES

Briggs, W. R., and H. V. Rice. 1972. Phytochrome: chemical and physical properties and mechanism of action. *Annu. Rev. Plant Physiol.*, 23:293–334.

Clayton, R. K. 1971. "Light and Living Matter." Vol. 2: The Biological Part. McGraw-Hill Book Co., New York., 243 pp.

Cumming, B. G., and E. Wagner. 1968. Rhythmic processes in plants. *Annu. Rev. Plant Physiol.*, 19:381–416.

Hillman, W. S. 1969. The physiology of phytochrome. *Annu. Rev. Plant Physiol.*, 18:301–324.

Mitrakos, K., and W. Shropshire (*eds.*). 1972. "Phytochrome." Academic Press, London. 631 pp.

Mohr, H. 1972. "Lectures on Photomorphogenesis." Springer-Verlag, Berlin. 237 pp.

Zucker, M. 1972. Light and enzymes. *Annu. Rev. Plant Physiol.*, 23:133–156.

16

TEMPERATURE: AN ENVIRONMENTAL FACTOR IN ECOLOGICAL PHYSIOLOGY

Although biologically important reactions are limited to a small temperature range, plants display remarkable thermal resilience by occupying sites as diverse as Antarctica at one extreme and Death Valley, California, at the other, they can operate only within close limits of heat content. If temperatures are too low, biological reactions are stifled by inadequate energy, while the complex structures of proteins become disrupted by temperature extremes in either direction.

Within the overall span of temperatures from absolute zero, where molecular motion ceases, to the highest temperatures where atoms exist intact (> 10,000 K), biological activities are limited to an exceedingly narrow range of 50°C from 273 to 323 K (0 to 50°C). Limited on the lower side by the freezing point of water and on the upper side by protein denaturation, life is squeezed into a tiny slot, about 0.5 percent of the range over which atoms can exist (Fig. 16-1).

Temperature extremes within the biological range exert an unremitting selective pressure, and individuals succeed or perish on the basis of their capacity to either avoid or endure temperature extremes thanks to a combination of their morphology and physiology. Small leaves and photosynthetic stems help the desert plant avoid lethal heat load and yet achieve some photosynthetic gain with an absolute minimum of water loss, while bud scales and autumn leaf fall encourage low-temperature survival. Special mechanisms of frost and heat hardiness (there is some analogy in their physiology) may then be coupled with morphological adaptation to produce highly resilient organisms of great diversity; e.g., plants can avoid temperature extremes through dormancy.

Under a temperature regime that favors plant growth and development, other environmental factors such as light, water, and nutrients will shape the community and determine its species composition, but temperature can have regulatory influences which become translated into selective pressures. Plant responses to temperature and mechanisms which serve to protect plants against unfavorable temperatures can be viewed against this background.

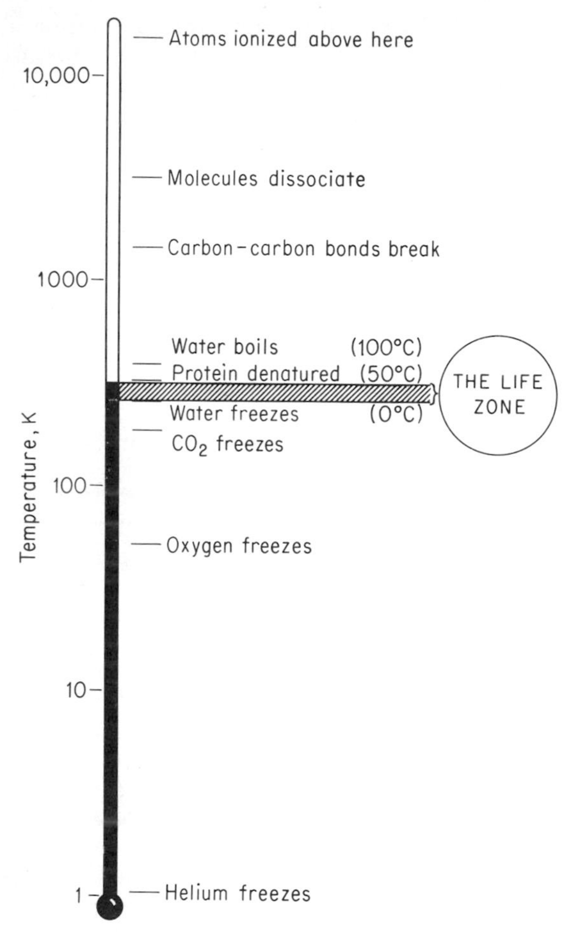

Fig. 16-1 Biological processes are restricted to a very narrow sector of the overall temperature range known to exist.

THE PLANT'S HEAT BUDGET

At any temperature above absolute zero, molecules are in continuous motion. This motion is accentuated in proportion to the amount of heat supplied, and the energetics of these molecular or atomic units is manifested in plant temperature.

Following this idea of motion, we can imagine heat in terms of energy associated with disordered motion. It can be generated by the physical impact of one mass against another, as when a hammer strikes an anvil and results in more disordered energy. Electromagnetic waves also project energy, so that solar radiation pouring onto an object generates more disordered energy in the absorbing system and increases its heat load. Since the energy associated with each photon is directly related to the frequency of its own radiation (hence inversely proportional to its wavelength), a given photon flux of light will dump more energy onto a plant than the same flux of infrared radiation. As shown in Fig. 16-2, most solar energy beyond the visible spectrum is either reflected or transmitted through the leaf, so that heating results principally from the absorption of photosynthetically active radiation. A leaf may absorb anything between 75 and 90 percent of incident energy in this wavelength band (400 to 700 nm), of which one-fifth becomes effective for photosynthesis; the other 80 percent must be dissipated. A leaf's heat budget provides an indication of how successfully this organ can intercept solar radiation to satisfy its photosynthetic requirements while minimizing heat load.

Single Leaves

Plants are coupled to their environment via energy flow, and leaf surfaces represent transducers in this circuit. A plant's heat budget therefore depends on events within the leaf-air interface. This site is now considered in detail.

The energy budget of an individual leaf takes the form

$$Q_{abs} = Q_{rad} + Q_{cond} + Q_{conv} + Q_{evap} + Q_{stor} + Q_{met}$$

The Q's simply represent energy in one form or another translated to the common denominator of calories per square centimeter per minute. Individual terms are as follows:

Q_{abs} = total incident radiation absorbed by leaf's surface

Q_{rad} = amount of energy reradiated as longwave radiation from the leaf's surface

$Q_{cond} + Q_{conv}$ = energy exchanged between leaf and ambient air by combined conduction and convection

Q_{evap} = energy consumed in transpiration, i.e., latent-heat exchange

Q_{stor} = net gain of energy during transient periods when more energy enters the leaf than is leaving; temperature would then rise; during steady-state conditions this term would be zero

Q_{met} = amount of energy comsumed by photosynthesis or generated by respiration (i.e. metabolic components)

While Q_{met} is of immense significance physiologically, this term is quantitatively small and of little consequence for the leaf's overall equation for heat exchange. Under steady-state conditions Q_{stor} is zero, so that both terms can be safely ignored in our analysis of the leaf's heat-budget equation.

Irrespective of location, over an extended period any organism will lose as much energy as it gains; otherwise it would get hotter and hotter or cooler and cooler. In either event the organism would die. Given the hostile environments where plants live (from hot springs to the Antarctic landscape), resident organisms must still achieve a suitable energy balance, otherwise they simply could not exist. Regardless of energy-flux distribution within this equation, the budget must balance, so that in the final analysis all components are countable. This principle of energy conservation provides a sound basis for our analysis of how plants interact with their energy environment in a variety of situations.

Plant leaves obtain energy from the surrounding atmosphere and from solar radiation. Their spectral characteristics (Fig. 15-3) enable them to absorb photosynthetically active radiation efficiently while showing weak absorption of near-infrared wavelengths, which account for a substantial quantity of energy within the solar spectrum. By discarding this radiation (Figs. 15-3 and 16-2) many plants avoid lethal temperatures. Leaves, however, are efficient absorbers of far-infrared energy, but since solar radiation is low in these wavelengths after its passage through the atmosphere, heat gain from this form of energy is of little consequence. Ironically, the reradiation of far-infrared energy actually cools plants, because their leaves both absorb and radiate these wavelengths with a high degree of efficiency. Considerable energy derived from absorption of visible radiation is then reradiated as far infrared.

Absorption depends on leaf orientation, surface characteristics, and geometry. The temperature of a large, flat, dark object held at right angles to incident light is closely coupled to the solar intensity. On the other hand, cylindrical objects such as tree trunks, twigs, or pine needles show far less dependence. Vertically oriented leaves therefore offer an ecological advantage under high insolation because their heat load is minimized (Fig. 15-3). Tender foliage on tea

Fig. 16-2 Leaves preferentially absorb radiant energy at wavelengths corresponding to maximum insolation. The relative levels of absorption reflection and transmission are all wavelength-dependent. Infrared (800 nm) is poorly absorbed due to the high reflection and transmission. (*Adapted from Rabideau et al. 1946; Curtis and Clark, 1950; and Wolpert, 1962.*)

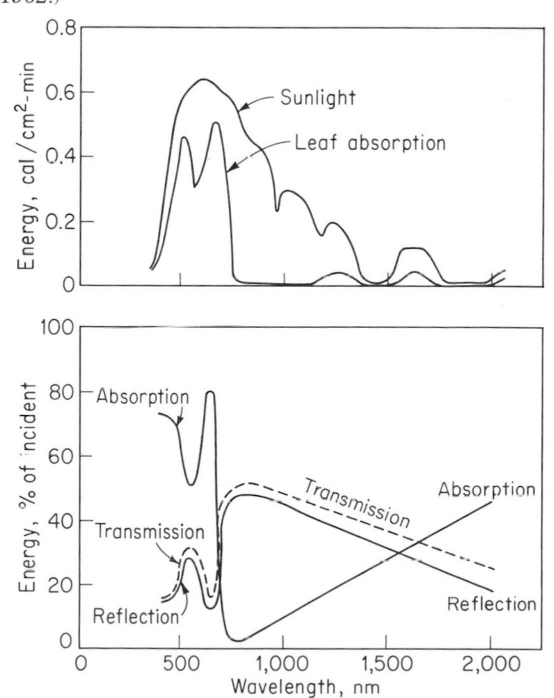

plants illustrates such an effect: despite their interception of strong sunlight, exposed tea bushes yield less because their foliage gets too hot. Shade can reduce leaf temperature from 40 or 45°C in horizontally exposed foliage to about 30 to 32°C, that is, close to ambient-air temperature. Tea bushes with vertically oriented foliage perform much better under Indian conditions because their leaves are cooler although still adequately illuminated (Hadfield, 1968).

Reflectance characteristics of leaf surfaces obviously influence absorption. While all leaves tend to reflect or transmit energy between 750 and 1,000 nm (near infrared), differences emerge with respect to reflectivity of visible radiation. In a comparison of mesophytic plants with plants adapted to desert or subalpine conditions, Billings and Morris (1951) found that adapted species reflect more light in the visible region (400 to 700 nm). Stanhill et al. (1966) subsequently measured the reflection coefficient of visible radiation, i.e., albedo, throughout the day for a number of vegetative associations. Their data are shown in Table 16-1. Desert vegetation was predictably the most reflective. Surface coatings or anatomical features which lend a glassy or silvery appearance to many desert plants minimize temperature increase by favoring reflection over absorption.

Leaf temperature is also coupled in a positive sense with air temperature. This interaction depends upon free or forced convection, so that leaf shape and size influence the outcome. Large convection coefficients, i.e., tight coupling, are characteristic of small objects; their surface temperatures depend closely on air temperature. Tiny dissected leaves on arid-zone plants, pine needles, or small insects are in this category. Solar radiation is of relatively less significance in determining their heat budget than it is for large organisms, e.g., banana leaves or elephants!

Table 16-1 The reflection coefficient of visible radiation (albedo) for different plant communities varies according to habitat. Desert vegetation typically reflects a greater proportion of incident sunlight than mesophytic communities (Stanhill et al., 1966).

Surface	Albedo (seasonal mean)
Pine forest	0.123
Open oak forest	0.176
Coastal sand-dune shrubs	0.195
Desert and wadi vegetation	0.373
Fishpond	0.115

Energy gain depends upon interception and absorption of solar energy, plus some adjustments to ambient temperature, while energy loss depends upon a combination of conduction, convection, reradiation, and transpirational cooling, so that leaf temperature becomes an expression of energy distribution as described by the leaf's heat-budget equation.

Leaf temperature is the whole crux of the leaf's energy balance because the direction and magnitude of heat exchange which comes from convection reradiation and transpiration are all dictated by a combination of temperature and atmospheric conditions. Leaves are in effect forever striving to balance their heat-budget equations by balancing transpiration against leaf temperature. These parameters are known to be functions of air temperature, irradiance (especially energy absorbed), wind speed, relative humidity, leaf dimensions, and diffusive resistance. Hence leaf temperature and transpiration are dependent variables which relate in turn to at least seven other independent variables—all operating *simultaneously*. Understandably, plants do not maintain a thermodynamic equilibrium with their environment. Even under reasonably steady conditions, energy intercepted by the plant becomes sensible heat and sets up new gradients. At night, atmospheric and leaf temperatures still differ because longwave radiation is being exchanged. The leaf's heat budget therefore remains dynamic and seldom becomes static.

Given such a complex situation, it is little wonder the scientific literature on environment-plant relationships which purports to analyze leaf temperatures is often highly confused. Gates and coworkers at the Missouri Botanical Garden have made a detailed analysis of heat exchange by plant and animal systems, and their exhaustive studies, published in a series of papers since

1962, have substantially clarified the situation (Gates, 1968, and literature cited). Leaves or leaf replicas were held in wind tunnels and other controlled situations. Basic data on environment-leaf interactions were obtained and subjected to mathematical analysis. Models so constructed were verified by field observation.

Gates (1969) summarizes the energy budget of a leaf in precise physical terms as follows:

$$Q_{abs} = \epsilon\beta\, T_l^4 + K\left(\frac{V}{D}\right)^{1/2} (T_l - T_a) + L\frac{sd_l\,(T_l) - \text{r.h.}\,sd_a(T_a)}{r_l + r_a}$$

where ϵ represents emissivity of the leaf surface, β is a radiation constant, K is another physical constant derived from wind tunnel experiments, V stands for wind velocity in cm/sec, D is leaf width in cm, T_l is leaf temperature and T_a air temperature, L represents the latent heat of vaporization of water (in this case 580 cal/gm at 30°C); $sd_l(T_l)$ and $sd_a(T_a)$ stand for saturation vapor densities at the temperature of leaf and air respectively; r.h. indicates atmospheric relative humidity while r_l and r_a represent leaf (principally stomatal) and boundary layer resistance respectively.

Gates and coworkers (Gates, 1969) then developed a computer simulation of the leaf's activities and solved the heat-budget equation for temperature and transpiration over a wide range of environmental conditions and leaf parameters. Figure 16-3 provides one such illustration, where transpiration has been computed in terms of leaf temperature and diffusive resistance for a range of air temperatures. Other environmental factors such as solar radiation, wind, or relative humidity can be examined the same way. The particular leaf referred to in Fig. 16-3 measures 5 by 5 cm, and its diffusive resistance is variable. Energy is being absorbed at the rate of 1.0 cal/cm²-min, and atmospheric humidity is 50 percent (relative humidity). Wind is blowing across the leaf at 200 cm/s, and air temperature is varying. In other words,

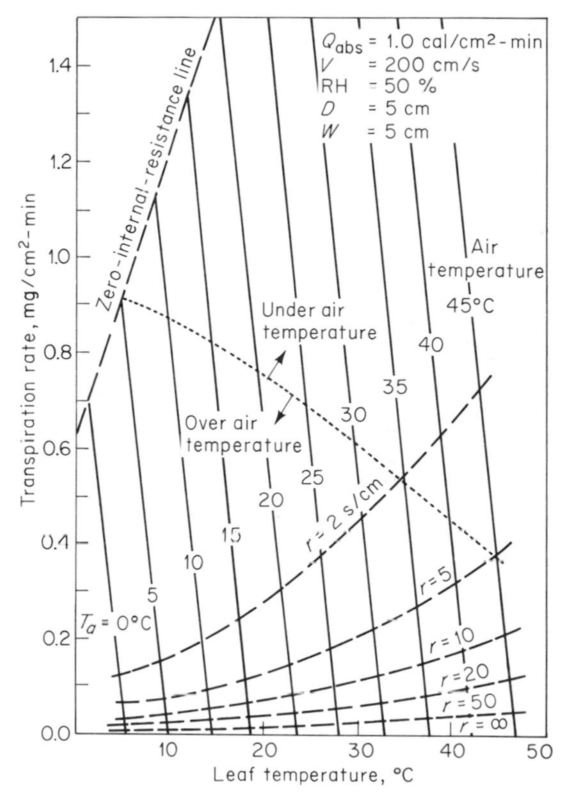

Fig. 16-3 Transpiration is a function of leaf temperature but varies according to leaf size, stomatal resistance, and atmospheric conditions. For this hypothetical leaf (5 by 5 cm), surface temperature is always greater than air temperature under cool conditions, but at high air temperature the leaf can be cooler if transpiration is sufficiently high (Gates, 1968).

we are considering an average leaf on a clear sunny day.

The effect of leaf temperature on transpiration is particularly evident at low leaf resistance (stomata wide open), and under the conditions in this figure, the leaf will be cooler than ambient air only when low leaf resistance favors rapid transpiration. Such a leaf (say r = 2 s/cm) will show a 50 percent increase in transpiration from about 300 to 450 μg/cm²-min when leaf temperature rises from 20 to 30°C. If air temperature rises from 20 to 30°C, this same leaf will change from 22 to 31°C. By contrast a poorly transpiring leaf (r = 20 s/cm) will rise from 23 to 33°C.

Table 16-2 Temperatures of small leaves on many desert plants in Utah are within 3°C of air temperature despite intense solar radiation and close proximity to an exceedingly hot soil surface. *Opuntia,* with its large blades and virtually zero transpiration, becomes substantially hotter (Gates et al., 1968).

		Temperature, °C	
Surface measured	Time	Air	Leaf
Artemesia tridentata	0830	24	24
	1000	30	32
	1500	35	37
Pinus edulis	0830	24	23
	1100	30	33
	1300	32	34
Opuntia sp.	0830	24	34
	1000	30	45
	1030	30	46
Soil	0830	24	36
	1000	30	48
	1100	30	50
	1400	32	60

Gates and coworkers applied their detailed mathematical models to other aspects of energy relationships between leaves and their environment. Wind speed and leaf size are of ecological interest. A large leaf, say 20 by 20 cm, really suffers under high insolation on hot humid days at low wind speed because it relies heavily on transpirational cooling. If stomata close due to temporary moisture stress and leaf resistance increases to about 20 or 50 s/cm, leaf temperature can easily climb to 50°C. Under desert conditions, where air temperature might readily exceed 45°C, broad-leafed mesophytes would undoubtedly perish. Small leaves should offer a sharp contrast according to Gates' mathematical model, and field observations confirm predictions that their temperatures will be very close to ambient for just about all manner of environmental conditions (Table 16-2). Small leaves would therefore confer a decided ecological advantage under hot, arid conditions by avoiding lethal temperatures due to excessive heat load. The *Opuntia* (Table 16-2) offers an exception, in that its leaves endure, rather than avoid, high temperatures. The performance of the small-leafed desert plants is all the more remarkable in view of their proximity to the hot ground. Soil surfaces a few centimeters away reach 60°C!

In general, large and fully exposed leaves are from a few degrees up to 20°C higher than air temperature, whereas shaded leaves would be close to air temperature or slightly below it. Figure 16-4 illustrates this principle. In the horizontal and exposed leaf, interception of solar radiation dominates the leaf's heat budget; hence the sudden temperature reduction in shade (Fig. 16-4). Transpirational cooling has an added effect under these conditions. *Mimulus cardinalis* provides an illustration of leaf-temperature reduction due to transpiration (Gates et al., 1964). This plant grows in moist situations and has particularly low leaf resistance, which favors rapid transpiration. As a consequence, even fully exposed leaves maintain their temperature below ambient.

The significance of transpirational cooling as opposed to other forms of heat dissipation (conduction, convection, and reradiation) must be put into perspective. It is a common misconception, held by physiologists, ecologists, and agriculturalists alike, that leaves transpire in order to stay cool. Granted, the ability to transpire will make a difference in leaf temperature, but this principle applies to environmental situations that encourage rapid transpiration from broad leaves with low stomatal resistance. Sunflower (*Helianthus annuus*) leaves are a case in point. Martin (1943) has demonstrated a leaf-temperature depression of 8°C due to transpiration at a rate of 996 mg/dm^2-h. Conversely, stomatal closure should lead to temperature increase under equivalent conditions, a prediction borne out by Slatyer and Bierhuizen (1964b) when transpiration suppressants increased cotton-leaf temperature by as much as 9°C. Cook et al. (1964) provided an illustration of this same effect on tomatoes, where leaf temperature was measured under illumination following treatment with sodium azide. This inhibitor blocked stomatal opening in the light, and leaves were 4°C hotter than control foliage.

If water supply limits plant performance, as it usually does in arid regions, leaf resistance approaches high values (100 to 200 s/cm), and other forms of heat dissipation must predominate. The shape and orientation of thin organs, e.g., pine needles, phyllodes, cladodes, or highly dissected small leaves, in combination with surface coatings help minimize light interception and serve to maintain leaf temperature close to ambient-air temperature. Heat dissipation then depends primarily on conduction, convection, and reradiation.

Operationally, small leaves favor close correlation with air temperature. The importance of *conductive heat loss* was first pointed out by Brown and Wilson in 1905. The flat, small leaves used in their experiments were ideally suited to this form of heat loss. The leaf loses heat to air molecules in a boundary layer immediately adjacent to its surface; convection then aids the withdrawal of this warmed air. Wind amplifies the whole effect by lowering boundary-layer resistance (Fig. 1-19). Parlange and Waggoner (1972) have provided a visual demonstration of this effect on reed leaves. In a highly original piece of work, forced convection of heat was observed under natural conditions by painting the leaves with liquid crystals that provide a measure of temperature without any disturbance to airflow. Temperature differences up to 15°C were observed between leading and trailing edges of nontranspiring painted leaves. By contrast to the elongate reed leaf, a broad circular leaf is distinctly unsuited to convective heat loss. Vogel (1970) demonstrated this effect with leaf replicas (copper plates) in a wind tunnel. Regardless of wind speed or orientation with respect to airflow, a circular plate offers the worst possible shape for heat dissipation. A lobed structure, on the other hand, showed substantial convective cooling irrespective of wind speed or orientation. The natural occurrence of deeply lobed leaves illustrates the ecological significance of these physical principles.

If transpiration is impeded by high stomatal resistance and conductive loss is minimized due to large surface area and meteorological conditions which lead to a thick boundary layer, the leaf can still lose heat by *radiant exchange.*

All surfaces lose heat by radiation, depending upon their emissivity (dull vs. polished) and according to the fourth power of their absolute temperature (referred to in the heat-budget equation). Curtis (1936) provided an early illustration of radiant exchange when he positioned a cool blackbody in the vicinity of illuminated leaves; their temperature immediately fell by 2 to 3°C. In absolute magnitude, this effect is small, but it does illustrate a principle which is going to assume much greater significance under desert or alpine conditions where leaf temperature can be high and the clear sky serves as a blackbody which absorbs radiant energy (Fig. 16-5). Leaves exposed to a cold night sky easily fall to 5 or 7°C below air temperature, whereas cloud cover reflects this energy back and radiant cooling is suppressed. Consequently, plants are less prone to frost injury on cloudy nights than on crisp clear nights. Cold soil, on the other hand,

Table 16-3 A warm root system enhances every facet of vine growth between bud burst and fruit-set on rooted cuttings of grape (Woodham and Alexander, 1966).

Root temp., °C	Flowers per inflorescence	Fruit-set, %	Shoot fresh wt, g	Root dry wt, g	Shoot-root ratio, g dry wt
11	1,299	2	1.8	3.0	0.96
20	1,539	10	19.8	6.2	1.21
30	1,768	23	74.2	8.5	4.13

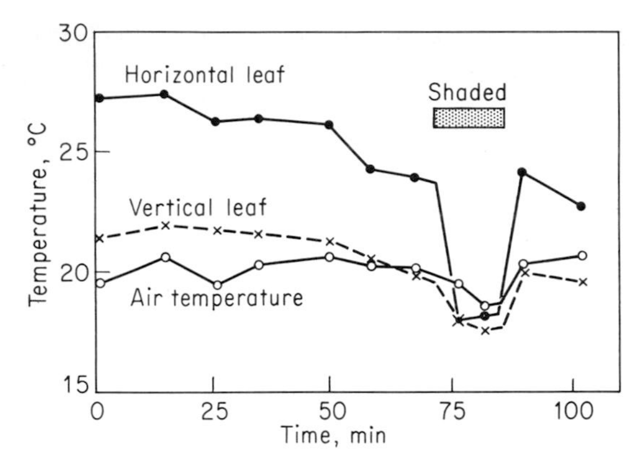

Fig. 16-4 In sunlight, a horizontal apple leaf is more than 7°C warmer than ambient air. Shading or vertical orientation reduces the temperature difference between leaf and air (Curtis and Clark, 1950).

serves as an acceptor for radiant heat so that leaf temperature is reduced even further (Geiger, 1957).

Greenhouse plants are similarly deprived of radiant cooling. Since glass reflects the infrared energy radiated from leaves and other objects inside the enclosure, incoming energy is trapped as sensible heat and produces the *greenhouse effect.* Heat budgets of plants in growth chambers are similarly displaced because a hot bank of electric lights is substituted for the cool sky. Physiological responses to temperature in growth chambers can therefore be expected to differ from field conditions because of this difference in radiant cooling.

Plant Communities

Taking the simplest situation of short, dense vegetation which is growing actively and is well supplied with water and nutrients, Lemon (1963) estimates that 75 to 85 percent of incoming radiation absorbed during the day is used to evaporate water (either from leaf or soil surfaces), 5 to 10 percent is absorbed by soil and becomes sensible heat (i.e., the ground gets warmer), another 5 to 10 percent is transferred as sensible heat to the atmosphere by convective processes (i.e., the air warms up), while no more than about 5 percent goes into photosynthesis.

Evaporative heat flux accounts for most energy exchange in a well-watered situation (Fig. 17-2), but other terms in the community energy balance assume dominance under less mesic conditions, especially the exchange of sensible heat between the community surface and surrounding air. During daylight hours, bare soils become markedly heated, reaching temperatures much higher than the surrounding atmosphere (Table 16-2). In sharp contrast to foliage (Fig. 16-2) dark soils reflect only 7 to 10 percent of incident radiation; the remainder is all absorbed, and the soil surface sometimes becomes hot enough to result in plant damage. "Burning" can occur on black muck soils, where reflection is low and absorption high; onion seedlings commonly suffer in this respect in spring, before the seedlings are large enough to shade the soil and produce a more favorable microclimate.

Greatest diurnal variation therefore occurs at the surface of dark soils in fallow areas (light dry soils are more reflective); some data in Fig. 16-6 illustrate this principle. Air temperatures above a fallow soil are compared with air temperatures above ground in a crop of sugar cane. Conduction and convection increased air temperature by 7°C above the bare ground. By contrast, the presence of vegetation displaced the

Fig. 16-5 On a clear night, tomato-leaf temperatures drop below air temperature because the night sky acts as a black body and absorbs radiant heat from the leaf. Note the temperature rise when cloud cover suppresses this radiation or when ice forms within the leaf tissue. (Shaw, 1954).

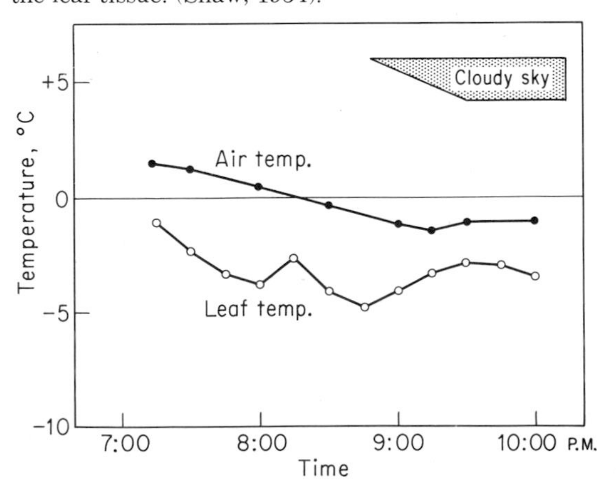

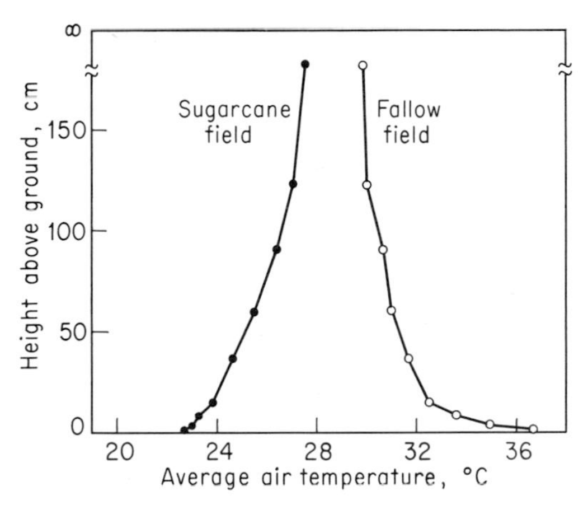

Fig. 16-6 Air temperatures are higher above fallow ground than within a standing crop; foliage intercepts radiant energy and minimizes the soil-air temperature differential. Air temperature in the weather box was 29°C (Geiger, 1957).

system's heat balance to such an extent that air within the crop was 7°C lower than ambient atmosphere. Not only does vegetation have a moderating influence on heating due to insolation, but heat losses from the soil at night are also suppressed.

Leaf tips are especially effective in reradiating infrared radiation; they might heat during the day, but then they become cooling surfaces at night (Fig. 16-5). As leaves cool, dew formation is encouraged, which then contributes to the plant's water economy—a significant effect in semiarid regions. Clumps of foliage are especially effective as centers for dew formation, and the compact foliage in tree crowns or trellised grape vines offers this specific advantage.

By virtue of their efficient nighttime cooling, foliage tips are also susceptible to frost damage. Radiant cooling can force an exposed leaf several degrees below air temperature, so that exposed foliage at the top of a plant may actually freeze, even though air temperature is still above 0°C.

PHYSIOLOGICAL RESPONSES

Plant Growth

Since growth is an expression of enzyme-catalyzed reactions, we would expect a plant's temperature-response curve for growth to bear a close analogy to the shape of an enzyme-response curve. There should be a rapid rise in the lower temperature ranges (0 to 15°C) followed by a steady increase above 15°C to an optimum value somewhere between 20 and 30°C. Supraoptimal values should lead to a decline which gains speed as the thermal death point approaches (about 50°C). Pea roots and corn coleoptiles (Fig. 16-7) behave in this fashion (at least between 12 and 35°C), while leaf expansion and relative growth rate in whole plants also coincide with prediction (Figs. 3-3 and 3-4). As with enzyme activity, plant growth is more than doubled by a 10°C rise in temperature, although this response is diminished at higher temperature ranges.

Physiological concepts of plant response to temperature were greatly amplified by F. Went and coworkers when the Earhart Laboratory was operative in California. His first major contribution related to night temperatures and their importance for whole-plant growth. The response of tomato plants to diurnal fluctuation com-

Fig. 16-7 Pea roots and corn coleoptiles show a steady response in their growth to temperature increase between 12 and 30°C. Note the large Q_{10} between 12 and 22, and the smaller Q_{10} between 20 and 30°C (Leitsch, 1916; Lehenbauer, 1914).

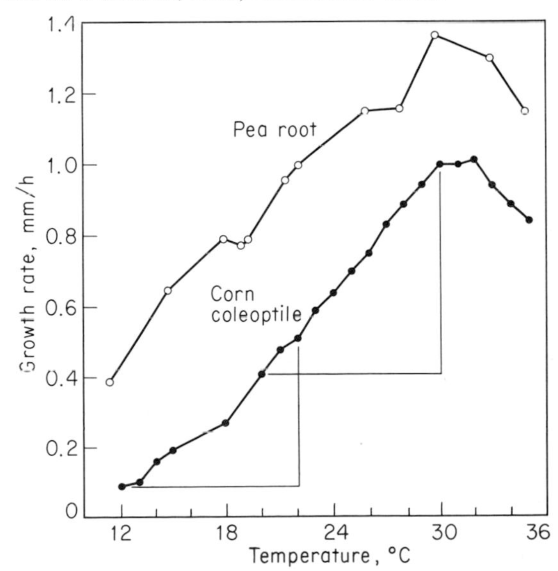

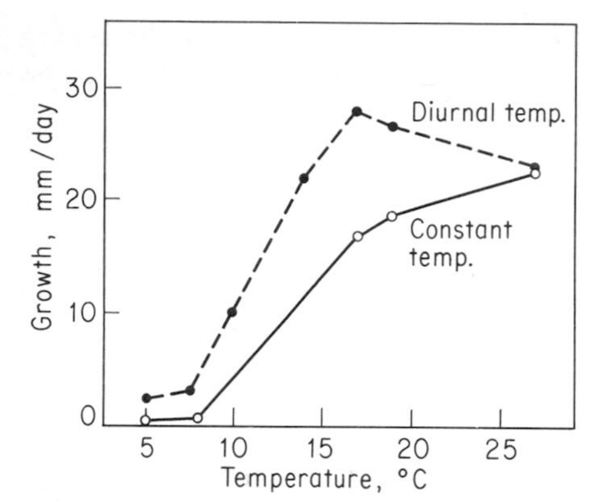

Fig. 16-8 The beneficial effects of lowered night temperature. Growth rates of tomato plants were superior when a day temperature of 26°C was alternated with lower night temperatures, for example, 17°C, than in the absence of a day-night differential (Went, 1948).

pared to steady day and night temperatures is shown in Fig. 16-8. Plants held in a constant environment at 17°C grew less satisfactorily than plants which experienced a day-night differential. Went applied the term *thermoperiodism* to this type of response, which seems to characterize whole-plant growth. In fact, a plant's requirement for day-night differential becomes accentuated as the organism enlarges. Dorland and Went (1947) showed, for example, that pepper plants 24 days old and held in a 26°C day temperature, showed optimum growth around 25°C night temperature, but by 96 days their preferred level was 12°C.

Horticulturalists have long realized the advantages of lowering greenhouse temperature at night; depressed nyctotemperature is especially beneficial when light intensities have been low during the day. Went (1945) described this interaction of daytime light intensity and optimum night temperature for the tomato. At high light intensity (full sun) the optimum night temperature for stem elongation is at least 25°C compared with about 8°C for the lowest intensity (8 percent full sun).

Carbon dioxide enrichment places a totally different complexion on temperature-growth relationships. Calvert (1972) discusses such effects on the basis of his work at the Glasshouse Crops Research Institute in England, where the benefits of CO_2 enrichment were enhanced by elevated night temperature.

Calvert obtained a 90 percent increase in the productivity of early tomatoes, due principally to an increase both in number of fruit originally set and their eventual size. Krizek et al. (1972) provide further information on the response of some F_1 hybrid annuals to day-night temperature differentials and conclude that higher temperature regimes than are traditionally used in controlled environments may be employed provided other environmental factors such as light, CO_2, relative humidity, and nutrient supply are not imposing any limitation on plant growth.

Stem growth is only one of many plant reactions to lowered night temperature; the quality of growth, earliness or intensity of flowering, and fruit development can all benefit (Went, 1957). In guayule, for example, night temperature has a profound effect on rubber production; synthesis of this material is optimized between 4 and 7°C but virtually ceases above 15°C (Went, 1953).

Temperature effects on plant growth were analyzed in Chap. 3 in terms of leaf expansion and net assimilation rate, but metabolic responses to day-night differentials were not considered. Went (1948) noted that lower night temperatures result in higher sugar content in many plants and suggested that sugar translo-

Fig. 16-9 Low night temperature leads to preferential root growth in potato and tobacco with the result that the top-to-root ratio declines. Pea plants were unaffected (Went, 1957).

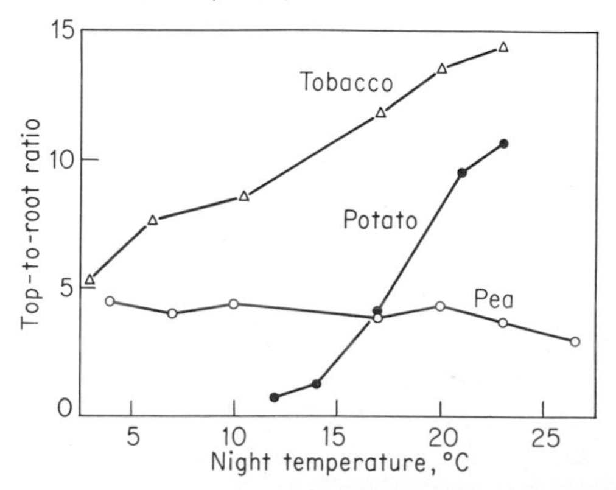

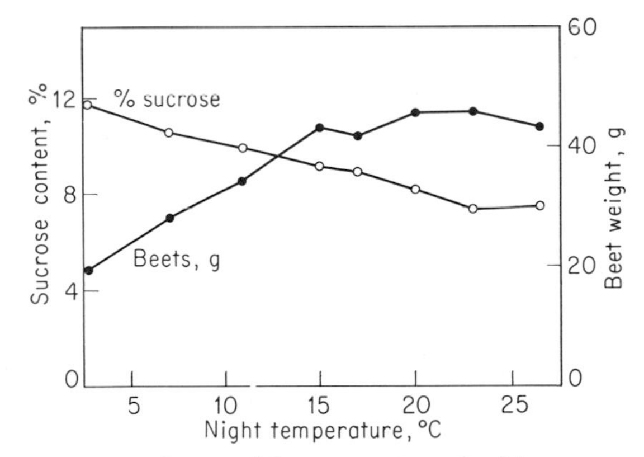

Fig. 16-10 Lower night temperatures lead to enhanced sugar content in sugar beet roots, although the consequences of this effect for total dry matter productivity tend to be offset below 15°C by a decline in root growth (Ulrich, 1952).

cation must be favored under these conditions. However, experimental evidence which has accumulated on this point indicates that translocation is most favored by temperatures around 25°C (Whittle, 1964; also Fig. 2-21), so that some other explanation is needed to account for growth responses to temperature differentials. Went was originally encouraged to make this suggestion by the enhancement in root growth that occurred when night temperature was lowered. This response is illustrated in Fig. 16-9 for three species. As root growth is favored over stem growth, the top-to-root ratio declines. Tobacco and potato (both members of the Solanaceae) are particularly sensitive in this respect. Pea is included as an illustration of one plant which shows no alteration in growth distribution according to night temperature.

In root crops such as potato, night temperatures also influence yield (Went, 1959). Sugar beet provides a further illustration, lower night temperatures increasing sugar content (Fig. 16-10). Nevertheless, when night temperatures were depressed below about 15°C in those particular experiments, there was a reduction in root growth which offset any yield response due to higher sugar concentration.

Stimulatory effects of low temperature on growth can be attributed to a systemic plant response rather than to a direct stimulation of a given plant organ (reviewed by Nielsen and Humphries, 1966). The systemic nature is readily evident in Fig. 16-11, which shows some of Richardson's (1956) data for maple roots. By applying controlled temperatures either to the whole plant or to aerial organs he observed the expected improvement in root growth at lower temperature. When temperature treatments were applied specifically to the roots, low temperature had the converse effect: growth was impaired. Nelson (1967) has also demonstrated such an effect on cotton plants whose roots were held at 12, 18, or 24°C. The diminution he observed in water uptake and leaf expansion on plants with cold roots (12°C) was attributed to a reduction in their permeability. Another woody perennial (*Vitis vinifera* L.) behaves similarly (Table 16-3). These plants were grown in nutrient culture under greenhouse conditions, so that aerial organs experienced favorable conditions for growth while their root systems were held at 11, 20, or 30°C. The benefit these vines derived from a warm root system is self-evident. Every component of vegetative and reproductive growth was favorably affected. Such an amplification in

Fig. 16-11 Beneficial effects of low night temperature on growth of maple roots is not a localized effect on the roots themselves because this enhancement in growth occurs only when low night temperatures are applied to the aerial organs (Richardson, 1956).

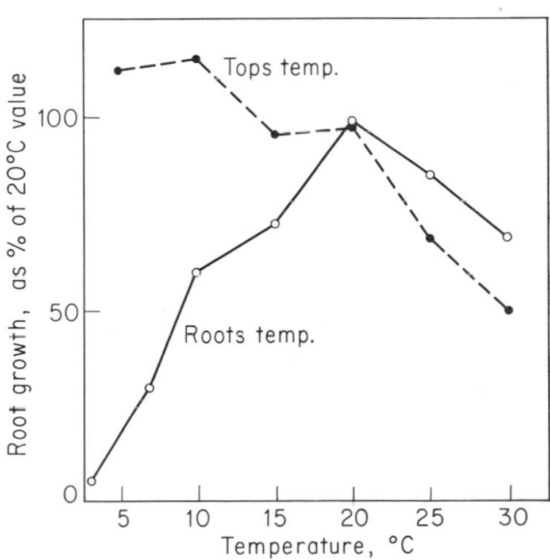

plant growth is highly suggestive of hormonal control with some regulatory compound being formed in the roots and eliciting responses throughout the plant. The results shown in Table 16-3 prompted Skene and Kerridge (1967) to make a detailed analysis of root-derived cytokinins under top-temperature–root-temperature regimes that offered sharp contrasts in vine performance. Root exudates showed quantitative and qualitative differences in cytokinin composition as detected with soybean-callus assay. For both root temperatures, major activity on paper chromatographs appeared between R_f0.6 and R_f0.8; the major disparity occurred at R_f0.1 to 0.2; roots held at 30°C showed no cytokinin activity in this region.

Temperature environments which do not fluctuate can restrain growth. This effect was illustrated by Highkin (1958) with peas, which showed a 20 percent reduction in growth at 10°C compared with values recorded under a diurnal variation in temperature. Even more intriguing was the transmission of this inhibition to the next generation of peas, and the effect became accentuated as subsequent generations of peas were grown in this static environment. Three generations under alternating temperatures were required before full potential was restored. Highkin and Lang (1966) subsequently demonstrated residual inhibition of a similar character in pea plants maintained at 23°C. Their growth rate and seed production, together with other parameters of their performance, differed significantly according to the temperature they experienced during germination (ranging from 3 to 31°C). Some herbicides and growth retardants show a similar persistence in their effects, which can extend from one to as many as three generations (Chap. 18).

Fig. 16-12 A comparison of three methods for predicting harvest time for Marquis wheat at one location (Moro, Oregon). *A*: number of days; *B*: day-degrees above 0°C; *C*: day-length-degrees times photoperiod (Nuttonson, 1948).

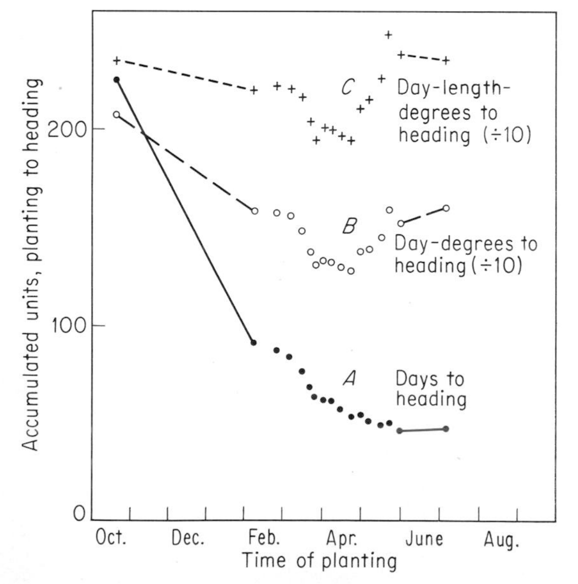

Heat Units

Growth of many field crops is roughly proportional to ambient temperature, and this correlation has been used to predict a probable harvest time for several crops. Such predictions would greatly facilitate many agricultural and horticultural operations. Calculation of daily accruals of heat units was first attempted by Candolle (1855) and has been modified extensively since that time. The number of days to crop maturity varies enormously, as indicated in Fig. 16-12 for the time to heading in Marquis wheat planted on various dates within a single region. Using Candolle's method of summing the mean temperature for each day, one can estimate the crop's progress with more precision. This type of heat unit, *day-degrees,* lowers variability to some extent, but Nuttonson (1948) made an additional refinement where day-degrees are multiplied by the number of daylight hours. Cumulative effects of temperature and light duration are therefore taken into account. This type of calculation is shown in Fig. 16-12 as day-length-degrees. The relative precision of these three methods is indicated by the variability around the median in each case, there being a ±66 percent variability in days, a ±29 percent variability in day-degrees and a ±12 percent variability in day-length-degrees. Went (1957) nevertheless offers some cautionary advice: heat-unit methods provide for greater precision at one location than in diverse regions because variability due to edaphic factors or other climatic components is at a minimum.

If growth rates are integrated over the course of a growing season (Chap. 3), maturity can be predicted with even greater certainty. Medcalf (1949) adopted this approach for pineapples growing in Hawaii. He measured the rate of growth of pineapple plants at each temperature in the growing range and multiplied the number of hours of each day at each temperature by the relative growth rate of the plant at that temperature. Crop maturity was predicted in this way with a high degree of precision.

While environmental factors that have a direct influence on plant growth vary in the same direction as temperature, this system of heat units will work. Its temperature range is confined, and predictions are most realistic for species whose development is not closely regulated by photoperiod. Heat units take no account of sequential temperatures, which many species require for successive steps in their development.

Plant Development

The ordered sequence of events generally referred to as development is not simply the result of autonomous processes governed solely by genetic constitution. Environmental factors, especially light and temperature, interact with the plant's genetic controls to elicit developmental changes. Some Dutch research was among the first to describe such a temperature pattern. Blaauw and coworkers in Holland have published work of fundamental importance over nearly 40 years on the temperature requirements of 23 different bulbous species which vary according to developmental stage (Hartsema, 1961). They combined the use of temperature-controlled rooms with morphological criteria on rates of development. Data given in Fig. 16-13 illustrate the sequence of temperature requirements during bulbing and subsequent development in tulips. When these bulbs are lifted from the ground in summer, their initial requirement is a rather high temperature, which favors

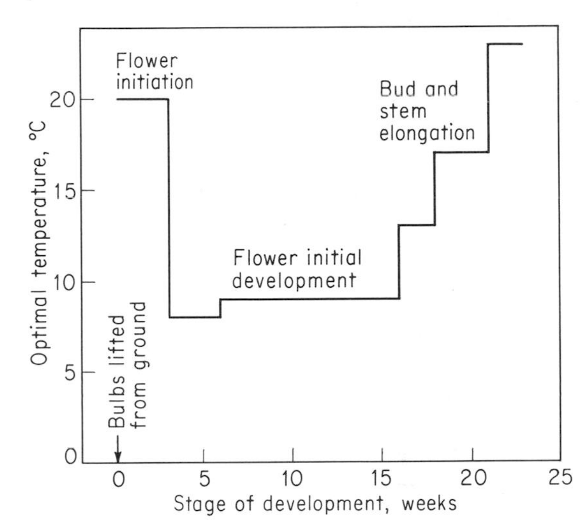

Fig. 16-13 In Copland tulip, the optimal temperature for development shifts with consecutive stages from a conspicuously low temperature for the development of flower initials to a much higher temperature for successive stages of elongation (Hartsema et al., 1930).

formation of new flower initials, but 3 weeks later, as the flower initials develop, a substantially lower temperature is required. Eventually, after 16 weeks, temperature optimum rises gradually as buds and stems elongate to form new aerial organs.

A plant often does not proceed to the next stage of development until a particular temperature requirement has been satisfied. Biennial plants, for example, require cold for flower induction, and the flowering stock (*Mathiola*) usually begins to flower after experiencing winter temperatures. Post (1935) established this response as a causative relationship by showing that stocks are *unable* to flower until they have experienced temperatures below 15°C. His data represent one of the first demonstrations of a specific cold requirement for flowering. The plants show a complete shift to the flowering condition when transferred to a warm greenhouse once they have experienced 3 or more weeks of cool temperatures.

While low temperatures can favor the onset of flowering, high temperatures may prevent flowering. Devernalization (Chap. 12) illustrates this principle. In other instances, the experience of high temperature can erase flowering tendencies and maintain the plant's vegetative condition. Sugar beets, for example, are induced to flower by low temperatures but remain vegetative if they are stored at 23°C for 3 weeks subsequent to their cool treatment (Stout, 1946). Similarly, onion bulbs are induced to flower by low temperature, and the subsequent removal to high-temperature storage will restore all plants to a completely vegetative condition (Lachman and Upham, 1954).

Temperature regulation of plant development is also expressed in dormancy and dwarfism. High temperatures (50 to 100°C) induce dormancy in the seeds of *Digitalis* and *Sporobolus* (Siegel, 1950) (see also Fig. 10-16). In lettuce, dormancy induced by high temperature was completely erased by brief exposure to red light, thereby implicating phytochrome as the regulatory system. Dwarfism represents a totally different measure of response to temperature; some novel effects were first observed by Tukey and Carlson (1945b) and Flemion and Waterbury (1945). Peach seeds held in warm temperatures during the first weeks of germination produce dwarfed plants which may retain this condition for over 10 years (Pollock, 1962).

Temperature effects on plant metabolism are virtually instantaneous, or else the biochemical shift will hold for as long as the temperature is maintained. But how can temperature influences on such plant functions as flowering, vernalization, and dwarfism persist long after the treatment is withdrawn? There appears to be no basis for assuming that differences in availability of substrates resulting from temperature experience can account for the protracted effect on plant development. What alternatives exist?

The altered developmental state of vernalization persists in all meristems derived from a vernalized meristem, and because dwarfism persists in all meristems on a dwarf shoot, the most reasonable assumption is that the temperature experience causes a change in the metastable state of some regulatory system within cells which is then self-propagating. Persistence in time suggests a specific involvement by nucleic acid metabolism.

These control systems almost certainly involve growth substances. Many cold-requiring steps, for example, can be satisfied by a gibberellin treatment. Lang (1956) discovered that gibberellin applications will substitute for cold induction of flowering, while Flemion (1959) has demonstrated that gibberellin also replaces the cold requirement during peach-seed germination, which results in normal rather than dwarf plants. Gibberellin can also overcome seed dormancy (Kahn et al., 1957), as illustrated in Fig. 10-12. Seeds rendered germinable by light are normally made dormant again by a 35°C treatment. Seeds treated with gibberellin will continue their germination despite high temperature (Fig. 16-14).

If growth substances induce a plant response despite countervailing temperatures, some change in the endogenous hormones would be expected to occur during the more normal sequence of events. Lang and Reinhard (1961) obtained tentative evidence that gibberellin content of *Hyoscyamus* increases with the low-temperature induction of flowering, but the precise role of these substances in temperature responses is still being established.

Metabolic Adjustments

Since temperature is an expression of disordered energy, lowered plant temperatures will be associated with lowered molecular reactivity. Within limits set by the cell's thermal resilience, the converse will be equally applicable. A convenient term for describing how reaction rate is influenced by temperature is the *temperature coefficient*, Q_{10}. According to the van't Hoff rule, $Q_{10} = (k_{t+10})/k_t$, that is, the ratio of a reaction rate at one temperature divided into the rate at 10°C higher. It is relatively low (1.2 to 1.4) for physical reactions that depend upon molecular diffusion or photochemical processes. Enzyme-catalyzed reactions typically show a higher Q_{10}, values ranging from 1.3 to 5 but commonly close to 2. As heat is applied to substances involved in a physical process, the total increase in disordered energy pushes up reaction rates in relation to the change in absolute temperature. By contrast, as heat is applied to reactants in an enzyme-catalyzed process, an increase in disordered energy leads to much larger increases in the frequency with which reactants achieve the necessary activation energy for enzymatic catalysis, so that reaction rate rises more nearly exponentially.

If Q_{10} was constant over any range considered, then reaction rate is exponential and for any temperature interval $t_2 - t_1$:

$$\ln Q_{10} = \frac{10}{t_2 - t_1} \ln \frac{k_2}{k_1}$$

Arrhenius then developed a more precise formulation of temperature relations where

$$\ln k = \frac{-E}{R}\left(\frac{1}{T} + C\right)$$

in which k is the velocity constant, R the gas constant, T the absolute temperature (K), C the constant of integration, and E the energy of activation. The extra disordered energy a system needs to acquire before a reaction can proceed is called its *activation energy;* it can be derived from experimental data by plotting ln k against $1/T$,K. If the system's reaction rate conforms with the Arrhenius analysis, a straight line is obtained—for example, see Fig. 16-18, lower plot. This analysis seems more appropriate to cell-free systems intended for characterizing enzymes over the biokinetic range because the integrated functions of higher plants, e.g., respiration and photosynthesis, tend to yield a smooth curve rather than a straight line (Went, 1961). This departure stems from the nature of metabolic events which embody both enzymatic and physical components. As the temperature of plant materials becomes elevated, enzymatic activities are preferentially increased, and physical steps such as gaseous diffusion often become biologically limiting. As temperatures are lowered, physical processes are very slightly restricted, whereas enzymatic steps become increasingly limited by biological processes.

Temperature changes not only influence reaction rate but also lead to physical alterations in tissues (especially membrane function), which can in turn alter metabolic directions. The solubility of two basic ingredients for biochemical activities, O_2 and CO_2, is greatly altered by temperature changes. As temperature is lowered, substantially more CO_2 and O_2 can be held in solution in plant cells. Since CO_2 solubility increases more sharply with temperature decline, there may be a preferential depression of those sectors of metabolism which are involved in the production of CO_2. The surge in respiratory CO_2 evolution that occurs as plant temperature is adjusted upward seems to be an expression of this solubility effect, although Forward (1960) is inclined toward a more physiological explanation.

While low temperatures depress these terminal aspects of metabolism, at the same time they enhance the availability of oxygen, provided of course that gaseous diffusion into the sap does

Fig. 16-14 Seeds of *Lepidium virginianum* require light for their germination, but after illumination they can be made dormant again by high temperature. Gibberellin treatment 2 mM breaks dormancy without being temperature-reversible. These seeds were maintained at 20°C except for the period at 35°C (Toole and Cathey, 1961).

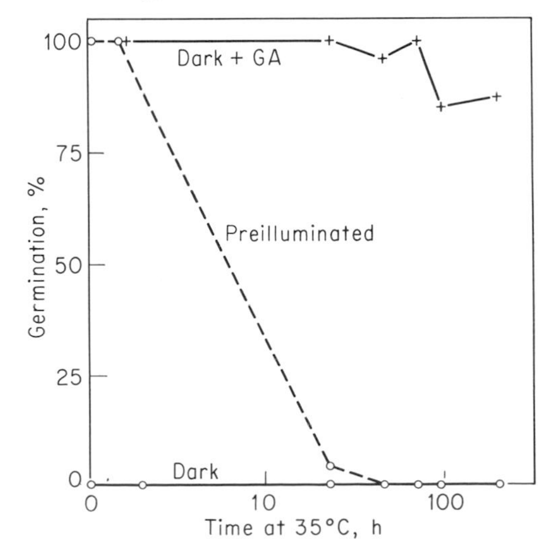

not become limiting. Low temperature therefore offers a beneficial situation for energy-producing aspects of metabolism but a restraint on CO_2 generation. The promotion of organic acid metabolism in the dark (when CO_2 is not being diverted into photosynthesis) by low temperatures illustrates this improvement in metabolic efficiency under cool conditions. Crassulacean plants embody this characteristic; malic acid tends to accumulate under mild conditions but is consumed at high temperatures (Brandon, 1967). Decarboxylation of malate in other acid-containing tissues such as grape berries is also favored by elevated temperatures (Kriedemann, 1968b; Kliewer, 1964).

In general, we can visualize effects of low temperature in terms of alterations in metabolic rate of equilibrium position, and starch metabolism provides an example. Temperature effects on starch hydrolysis in artichoke leaves (Fig. 16-15) have been measured by Wassink (1953), who found the usual rise in hydrolytic activity with increasing temperature to be modified by a secondary peak of activity below 13°C. Effects of temperature on the equilibrium position of starch synthesis vs. hydrolysis may be related to changes in leaf acidity according to temperature.

Fig. 16-15 Starch hydrolysis in *Helianthus tuberosus* leaves is greatly accelerated by temperatures above 25°C but shows a secondary peak just above 0°C. Starch hydrolysis is represented by relative values for iodine before and after $6\frac{1}{2}$ h at temperatures indicated; data are plotted as differences in values before and after treatment (Wassink, 1953).

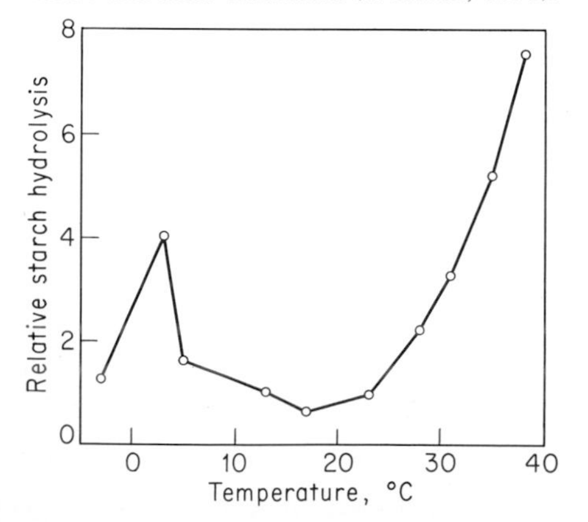

Plant tissues commonly show increases in sugar concentration at temperatures below about 10 or 15°C. A well-known case is that of potato tuber, which becomes undesirably sugary when stored between 0 and 5°C (Müller-Thurgau, 1882).

Another category of energy-rich storage compounds, the lipids, is also sensitive to temperature. Buttrose and Hale (1971) have described a situation where surplus photosynthate in grapevine leaves is stored either as starch or lipid, the balance depending upon temperature. Their data, summarized in Table 16-4, emphasize the preponderance of lipid at 35°C, which seems to have occurred at the expense of starch accumulation. These changes are not expressed in surface waxes but are associated with structural alteration in chloroplasts and an increase in leaf chlorophyll (Table 16-4). Effects of temperature on chloroplast structure similar to those described by Buttrose and Hale (1971) have been reported by other workers (Klein, 1960; McWilliam and Naylor, 1967; Ballantine and Forde, 1970), so that temperature effects on chloroplast structure and function occur in widely divergent species. C_4 plants with their highly specialized leaf anatomy and metabolic compartmentation have an obligate need for high temperature if their full potential for growth is to be expressed. Their chloroplasts show structural disruption under cool conditions (10°C), and their competitive advantage over C_3 plants would be lost in such an environment (Taylor and Craig, 1971; Taylor and Rowley, 1971).

PHYSIOLOGICAL ADAPTATION

A plant's metabolic responses to temperature embody instantaneous effects, e.g., enzyme activity, membrane permeability, and substrate concentration, and cumulative reactions, whereby the plant "remembers" a temperature experience and adapts its physiology accordingly. Such adjustments may lead to *hardening* and enable the plants to endure more intense conditions of heat or cold. Not all categories of higher plants are capable of such *hardening off* and are therefore limited, geographically, in their

Table 16-4 Storage products in grapevine leaves vary according to temperature. Vines grown at 18°C formed starch, whereas lipid predominated at 35°C (Buttrose and Hale, 1971).

Day temperature, °C	Starch, % dry wt	Total lipid, % dry wt	Total chlorophyll, mg/dm^2
18	23.3	5.9	2.8
25	10.9	9.8	4.7
35	1.3	16.1	5.6

natural occurrence. This hardening process, which enables plants to avoid or endure unfavorable temperatures, is clearly of great ecological significance and depends upon subtle adjustments in their physiology—our next concern.

Thermal Injury and Hardening

It is common knowledge that ice melts at 0°C and that water freezes to form ice, but this does not usually occur at 0°C; some degree of supercooling before freezing is the norm. Freezing is possible only when conditions favor the birth of a microscopic ice crystal known as a *nucleus.* This process of nucleation is immediately followed by crystal growth and constitutes freezing. Absolutely pure water can form crystals spontaneously, but this does not happen until about −40°C. Such conditions simply do not occur in nature because water is invariably in contact with cell surfaces and other matter as well as holding in suspension material which aids nucleation, but some degree of supercooling before freezing always occurs due to biological materials in solution.

When fully turgid plant tissue is exposed to freezing temperatures under laboratory conditions (Fig. 16-16), the material itself shows a steady drop in temperature to a value well below 0°C, generally referred to as the *undercooling point.* Once ice-crystal formation is initiated, tissue temperature jumps to a higher value as heat is liberated (latent heat of freezing) and may remain fairly steady before resuming its steady decline with time. The extent to which this initial freezing point is depressed below 0°C depends on dissolved solutes, colloids, and some forms of binding between H_2O molecules and cellular constituents. In curve I of Fig. 16-16, the first freezing point is barely detectable because the normal fresh leaf has a minimum of free water in its intercellular spaces. Following one cycle of freezing and thawing, the undercooling point is lowered, and freezing shows up as an abrupt jump in the curve. Presumably, the previous cycle of freezing and thawing resulted in water redistribution which led to increased solute concentration in cells and more free water in intercellular spaces.

Once ice forms in spaces between cells, as generally happens in nature, water potential in that region is lowered and H_2O molecules show a net migration from the protoplast and through the semipermeable plasmalemma toward regions of crystallization. Solute concentration in the cells is thereby increased, so that their prospect of suffering from ice formation is automatically diminished. In addition, the plasmalemma discourages entry of ice crystals from outside the cell to seed the aqueous phase in the protoplast. Leaves of many temperate plants such as *Rhododendron* offer further compartmentation with respect to ice formation. Frozen leaves *look* wilted but resume normal life after thawing out again. Similarly, *Camellia* leaves become semitransparent but recover without damage after thawing. This type of alteration in physical appearance is due to formation of *frostblaze,* whereby ice-crystal formation is localized between readily cleaved layers of tissue and thereby rendered harmless. Montemartini described the phenomenon in frozen *Buxus* leaves as early as 1905 (see Steiner, 1933).

When rapid cooling occurs under laboratory conditions, intracellular ice can be formed, as

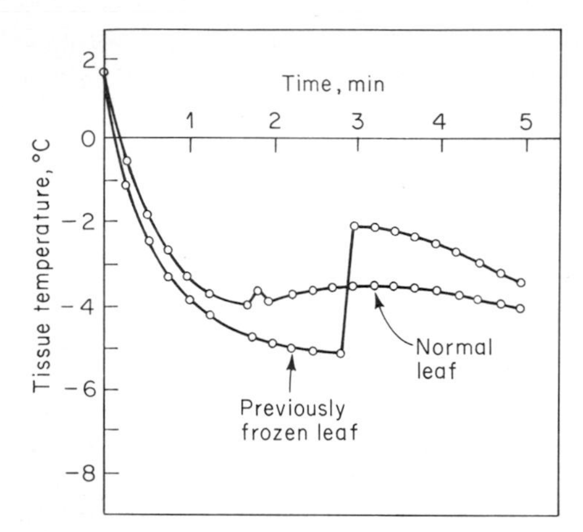

Fig. 16-16 Plant material held in a freezing mixture shows a steady decline in temperature until the moment of ice formation within the tissue. Release of heat causes a momentary rise in temperature. If the tissue is thawed and refrozen, the freezing curve shows a more accentuated rise; *Buxus microphylla* (Hatakeyama and Kato, 1965).

opposed to the extracellular formation of ice in nature. Freezing then occurs in a sudden wave, cell by cell. Tissue is inevitably killed; injury results from ice-crystal formation in the protoplast. Cytoplasm becomes structurally disrupted, and protein denaturation results. Levitt (1956) also suggests that membrane laceration would occur [cellular inclusions such as starch grains may play some role in this damage (Siminovitch et al., 1953)]. A rapid freeze of unhardened tissue would therefore offer the most extreme situation (Fig. 16-17).

Tissue damage is revealed by a loss of differential permeability to vital stains, inability to show plasmolysis and deplasmolysis, or a sudden change in electrical conductivity (Aronsson and Eliasson, 1970). Using these indexes, it is possible to demonstrate that injury may also occur during thawing (Fig. 16-17). Rapid thawing is especially deleterious; frozen tissue brought up to 2°C in 30 to 60 min showed a completely different degree of damage from tissue warmed up within a few seconds (Levitt, 1957*a*). Immediately after thawing, water becomes redistributed within the tissue, and sudden turgor changes have deleterious effects on membrane integrity. This disruption is offset by external osmotica. Cabbage leaves frozen to −7°C were almost uninjured if they were slowly thawed and then exposed to sugar solutions be-

Fig. 16-17 Low-temperature damage depends upon both the rate of temperature change and the tissue condition. *(a)* Lethal effects of low temperatures on cabbage leaves were accentuated by rapid *freezing* although hardening conferred resistance during either rapid or slow freezing. *(b)* By contrast, rapid *thawing* caused extensive damage, even to hardened tissue, whereas slow thawing resulted in minimal injury. Hardening was induced by exposure to low temperature (2 to 5°C) (Levitt, 1957).

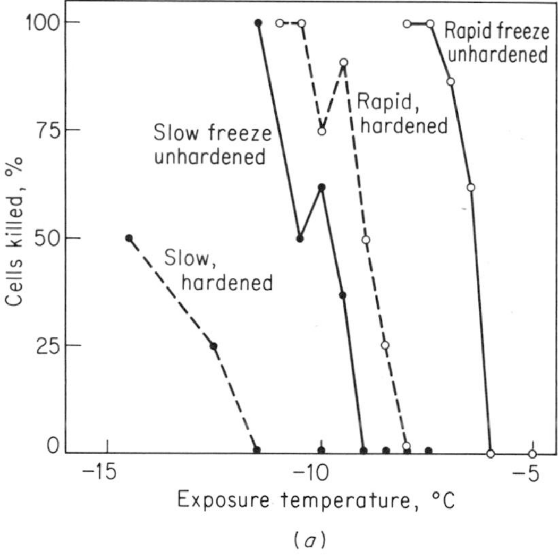

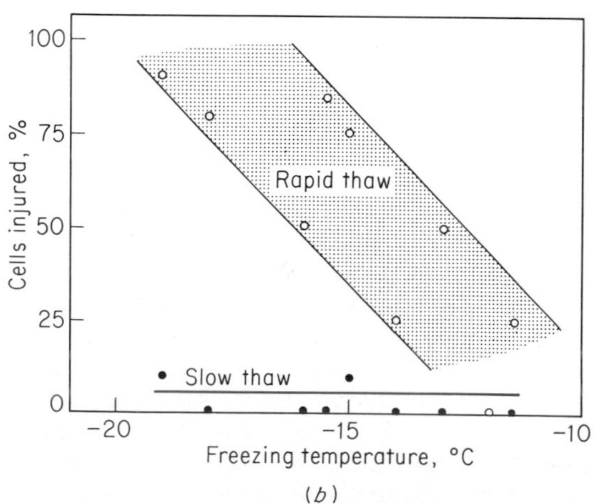

tween 0.5 and 1.25 *M*. Even with rapid thawing, damage was substantially lower if tissue was immersed in glucose solution.

Any of these situations where the protoplasm is subject to physical stress will be conducive to injury (Levitt, 1957*a*). Rapid freezing or thawing could mechanically abrade the protoplasm and disrupt physical integrity. Abrasion seems to involve interposition of ice crystals and simultaneous mechanical deformation of the cytoplasm as its aqueous matrix is moved into and out of such crystals. Placing cells in an osmotic medium after thawing slows down the rate of reentry or at least limits final turgor within cells, which in turn cushions stresses on the protoplast.

Formation of ice crystals at subfreezing temperatures undoubtedly disrupts plant tissue, but many tropical and subtropical individuals are injured by a day or so of chilling temperatures (0 to 10°C). Mature green avocados, bananas, papayas, and pineapples, for example, cannot be stored under such conditions without tissue deterioration. Chilling injury has long been associated with a decrease in protein level and a matching increase of amino acids; Levitt (1969) provides evidence that protein hydrolysis associated with chilling injury has been increased by a factor of 4 to 9 times the normal rate. Since a plant's metabolism depends upon enzyme activity, protein hydrolysis will disturb metabolic activity. Membrane integrity is also lost at higher temperatures in these chilling-sensitive tropical plants than in temperate and hardened species. Lyons et al. (1964) have associated this aspect of chilling sensitivity with inflexible mitochondrial membranes. In their experience, cold-tolerant species had more flexible membranes (composed of a higher proportion of unsaturated fatty acids), which enabled organelles to shrink or swell according to changes in temperature and osmotic conditions. Subsequent work of Lyons and Raison (1970) showed a change in metabolism with lowered temperatures in *unhardened* plants (Fig. 16-18).

While chilling injury is generally characterized by a time-temperature relationship, even brief frosts can prove lethal to tropical plants.

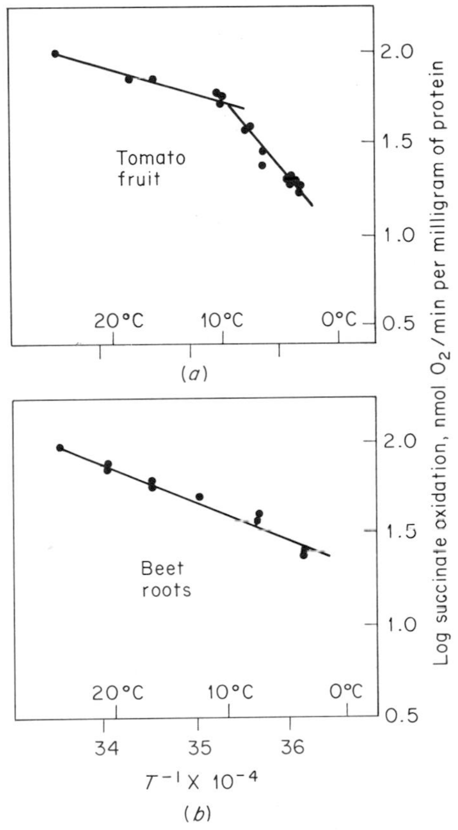

Fig. 16-18 Respiratory data, presented here as Arrhenius plots, demonstrate a differential response to temperature in mitochondria isolated from chilling sensitive tissues *(a)* compared with chilling-resistant tissues *(b)*. Discontinuity in the plots of chilling-sensitive tissues suggests that reduced temperature has affected some property of the mitochondrial membrane; lipoprotein complexes have probably undergone a phase transition (Lyons and Raison, 1970).

This factor imposes strict limitations on the geographic distribution of such plants or necessitates heavy expenditure on frost-protection systems when commercial expedience puts these sensitive plants beyond natural environmental boundaries. Cool storage of tropical produce involves similar problems.

In sharp contrast to tropical and semitropical plants such as sunflower, soybean, and *Begonia*, temperate crops, including cabbage and wheat, and a wide range of deciduous and evergreen trees can become hardened off and are then more resistant to frost. The extent to which hardening can protect tissues is truly dramatic. In Fig. 16-17, hardened cabbage tissue withstood temperatures of nearly −15°C. In an extreme case, Sakai (1956) was able to lower hardened mulberry leaves to the temperature of liquid nitrogen (−210°C) without killing them, provided that the temperature was lowered slowly for the first 30°C below freezing.

Frost-hardy tissues maintain two lines of defense. Their primary line stems from a protoplast which resists mechanical stresses. If proteins avoid intermolecular S—S bonds (Levitt, 1969; Andrews and Levitt, 1967) and increase their measure of bound H_2O molecules (Levitt, 1959*b*), the protoplasm remains more ductile when subject to frost (or high-temperature dehydration). The secondary line of defense adopted by frost-hardy tissue complements the first line by helping to reduce stresses actually imposed upon protoplasmic material. This reduction is

Fig. 16-19 During the frost season in Minnesota, black locust bark accumulates a relatively high sugar content at the expense of starch reserves. High solute concentration favors frost resistance (Siminovitch et al., 1953).

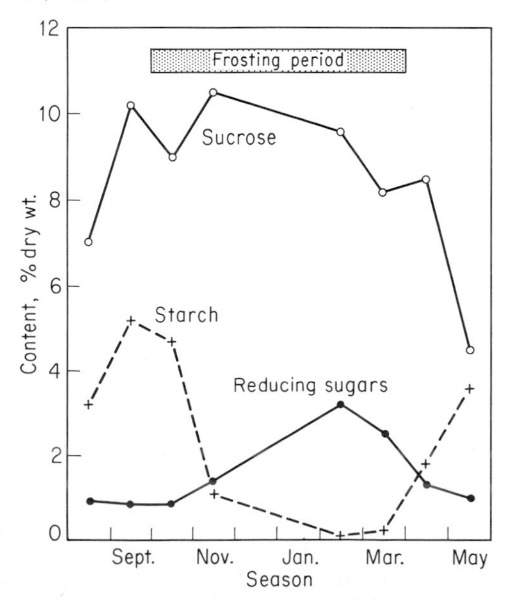

achieved by a combination of small cells, flexible cell walls, permeable membranes, and a high solute concentration in small vacuoles. Membrane permeability, in particular, contributes to frost resistance by allowing a reversible influx of solutes which forestalls excessive dehydration and shrinkage when ice forms in spaces between cells. The freezing resistance of cell-free systems can also benefit from a similar mechanism (Williams and Meryman, 1970).

In actively growing tissue, the protoplasm of individual cells becomes highly viscous and inelastic on dehydration, whereas in dormant cells it remains more ductile. Accordingly, the first stage of hardening coincides with a cessation of growth. If CO_2 assimilation is sustained during this time, sugars will accumulate or insoluble carbohydrate reserves become converted into solutes that accumulate within the vacuole leading to an increased osmotic potential.

During these preliminary stages of hardening there are generally some characteristic alterations in carbohydrate components. Changes observed by Siminovitch et al. (1953) on locust bark are representative of changes associated with hardening. In Fig. 16-19 there is a pronounced drop in starch content but an increase in simpler sugars with the onset of winter (reminiscent of changes in *Helianthus* tuber, Fig. 16-15). Such adjustments in starch or sugar levels contribute toward increases in cell-sap osmotic pressure during hardening. In addition, Siminovitch et al. (1953) suggest that starch disappearance is a critical factor in hardening and support their case with some striking photomicrographs which show how starch grains contribute to frost injury through their abrasive action in stressed cytoplasm (Siminovitch and Briggs, 1954).

A plant's capacity for cold acclimation by means of these biophysical mechanisms is clearly under genetic control (cf. tropical and temperate species), but this capacity is fully expressed only after specific interactions with photoperiod and temperature. The seasonal pattern of cold resistance in the living bark of *Cornus stolonifera* (Fig. 16-20) illustrates the sequence of events which lead to cold acclimation in this woody shrub, which grows widely in

North America. The first stage of acclimation is associated with growth cessation and resting bud formation; autumn coloration would be another indication in deciduous trees and shrubs. During this first phase, shortening photoperiod leads to formation of a *hardiness-promoting factor,* which moves from leaves to overwintering stems through the bark. Weiser (1970) summarizes the evidence for this view, which is based on experiments involving defoliation, girdling, grafting, and split-plant studies. Especially pertinent to this concept is the observation that leaves of a hardy genotype which have been photoinduced to produce the factor can enhance the cold acclimation of a branch on a less hardy genotype when the two are grafted together. The chemical identity of this signal becomes a fascinating question. Irving (1969) has detected increased levels of abscisic acid and decreases in gibberellin in *Acer negundo* which seem closely related to hardening. Levels of abscisic acidlike inhibitors bore closer connection to the onset of hardening and were highest under short days or long days with cold (5°C) nights.

The second stage of acclimation (Fig. 16-20) is triggered by frost and does not involve photoperiod or translocatable factors. The evidence is as follows. A plant exposed to frost and long days during this phase will still become hardy; by contrast, one experiencing only short days but no frost will not proceed beyond the first stage. Temperature experience rather than photoperiod is therefore the crucial factor for completion of stage 2. A wealth of physiological changes occur at this time which equip the cell's protoplasm with the resilience referred to previously, including increased membrane permeability, reduced hydration, and structural alterations to proteins.

The third stage of acclimation is induced by subzero temperatures (−30 to −50°C) and equips hardened twigs with an ability to endure temperatures down to −196°C. This last phase is generally regarded as physical, rather than physiological, and Weiser (1970) favors the notion that cellular water becomes bound with such tenacity that it resists dehydration and reduces the amount of water available for destructive crystallization. This ultimate hardiness is easily lost, and the rapid fluctuations normally encountered in the midwinter hardiness of stems and buds probably arise out of changes in this third phase of resistance.

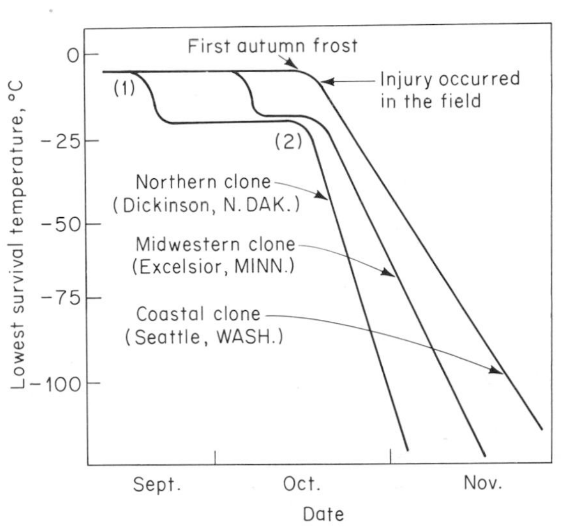

Fig. 16-20 Seasonal patterns of cold resistance in the living bark of *Cornus stolonifera* vary according to climate race and local conditions. The first stage of acclimation (1) is associated with growth cessation and the formation of a *hardening factor* under shortening days. The second stage (2) may then proceed, whereby resistance is intensified by exposure to low temperature. Northern and Midwestern clones were more responsive to changing seasonal conditions than the coastal race (Weiser, 1970).

Hardiness to frost, heat, and drought are closely correlated. When plants become frost-hardy due to combined effects of growth cessation and physiological responses at a cellular level, they are automatically heat- and drought-hardy as well. The plant's ability to endure tissue desiccation forms a common denominator for these three conditions. As in low-temperature hardiness, some species acquire a hardiness to high temperatures by experiencing them. For example, avocado tissues which had experienced a brief heat treatment at 50°C (2 to 3 min) become decidedly hardened against damage from heating up to 55°C (Schroeder, 1963). This form of hardiness may result from an increase in cy-

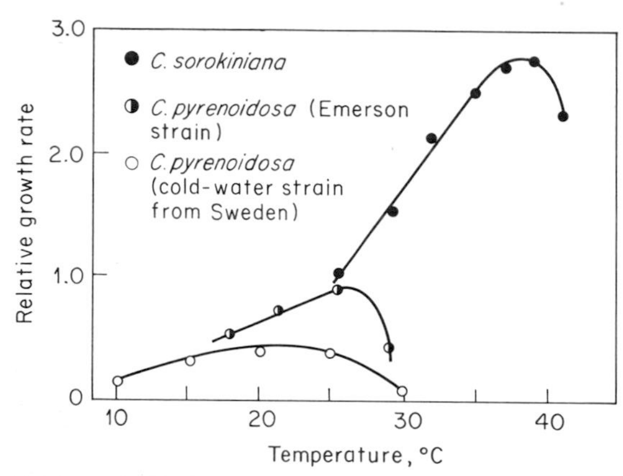

Fig. 16-21 High-temperature limitation to growth varies widely between species. Growth rates of three types of *Chlorella* demonstrate such disparity. The Emerson strain showed growth inhibition above 25°C, whereas a high-temperature strain (*C. sorokiniana*) achieved optimum growth near 40°C (light intensity was 1,600 fc) (Fogg, 1969).

toplasm viscosity and associated changes in bound H_2O within the cells, as mentioned in connection with low-temperature hardening (Levitt, 1956).

Variations in resistance to hot conditions seem related to morphology. Small dissected organs minimize heat load (Gates, 1968), while the succulent habit of other desert plants, e.g., the Crassulaceae, enable them to endure high temperature (Sachs, 1864). Lange and Schwemmle (1960) tested this proposition experimentally. They grew *Kalanchoe* under different photoperiodic conditions, which altered its degree of succulence, and subsequently found a close correlation between heat resistance and succulence under various day-length conditions and different leaf ages.

Despite a variety of adaptive morphology and physiological mechanisms which can minimize heat load and water loss at high temperature, thermal injury does occur. Protein denaturation is involved (Sachs, 1864), while toxic principles are also generated and translocated from local lesions (Yarwood, 1961). This observation on the migration of chemicals from zones of injury to healthy tissue suggests that damage results from the deteriorative effects of protein denaturation and toxic products, which then contribute to further damage.

To counteract this prospect of high-temperature lesions due to inactivation of key metabolic processes or end-product accumulation, Bonner (1957) proposed a "chemical cure of climatic lesions." His idea was to introduce essential metabolites which become deficient at unfavorable temperature. Bonner (1943) had reported that *Cosmos* plants showed low-temperature lesions that were chemically reparable by thiamine, while Galston and Hand (1949a) were able to associate high-temperature damage in peas with impaired synthesis of adenine. Langridge and Griffing (1959) then extended this hypothesis by growing 43 different ecological races of *Arabidopsis thaliana,* taken from around the globe, at 25, 30, and 31.5°C. Eight races were found to show a disproportionate decline in growth at 31.5°C while five races displayed morphological symptoms of high-temperature damage. Three of these five races gave significantly increased growth at 31.5°C if supplemented with vitamins, yeast extract, and nucleic acids, respectively. Biotin was revealed as a specific stimulant of growth by two races at 31.5°C and also prevented formation of temperature lesions. Cytidine fulfilled this same function in a third race.

In this context of tissue deterioration at high temperature, Englebrecht and Mothes (1960) found that leaf yellowing due to heat treatment can be prevented by kinetin applications. They attributed this chemically imposed heat resistance to the kinetin stimulation of protein synthesis, an effect which they had reported previously (Mothes et al., 1959). Kuraishi et al. (1966) subsequently demonstrated a related effect in pea plants, where benzyladenine treatment enhanced frost hardiness within a few days of four successive foliar applications.

To recapitulate, damage to plants by either high or low temperatures involves the imposition of stresses on the cytoplasm. Development of hardiness appears as a modification of the cytoplasm and encompassing membranes which offsets rapid fluxes of water molecules (a principal source of cytoplasmic stress) due to temperature extremes. An additional effect of high temperature is protein denaturation, and the formation of toxic substances may be an associated event. Under extremes of temperature, structural damage to the cytoplasm is a major component of the forces which impose temperature limits on plant growth and survival.

Ecospecies Physiology

Species vary in their sensitivity to temperature, and even within a single taxonomic group races emerge which differ enormously in both their endurance and developmental responses to seasonal patterns and extremes of temperature. Widely diverse phyla exhibit this ability. In the green alga *Chlorella,* for example, the growth of ordinary strains is inhibited above 25°C compared with a high-temperature strain, found by Sorokin (1960), whose growth is optimal at 40°C (Fig. 16-21) (Sorokin and Krauss, 1962). Certain blue-green algae endure even higher temperatures and obviously thrive in hot springs (50 to 75°C), e.g., at Yellowstone National Park (Brock, 1967).

Higher plants also reveal distinctions in this respect, though of a more subtle character, which influence their geographic distribution. Went (1957), for example, represented seasonal changes in day and night temperatures as shown in Fig. 16-22. The annual march of diurnal temperatures occurring naturally at Pasadena, California, is shown as a narrow ellipse in the figure. As an illustration, night temperature in January averages about 7°C, and day temperatures are around 14°C; during the year these parameters rise to a peak in August but come down again by the end of the year. A plant like *Zinnia* has a range of temperature requirements indicated by the dotted circle *Z,* and since seasonal changes meet these requirements between June and September, *Zinnia* grows well in Pasadena. The China aster (*C*) requires lower day and night temperatures, but these requirements are still met in spring and fall. *Saintpaulia (SP)*, on the other hand, requires a high night temperature which is *not* met in Pasadena, and this species does not thrive there.

The ecological consequences of temperature also becomes expressed along climatic gradients of latitude and altitude. Alpine timberlines provide a clear illustration of a vertical gradient which imposes sharp discrimination against growth and reproduction or establishment of trees as opposed to shrubs and herbs. Daubenmire (1954) offers his *heat-deficiency hypothesis* to account for such transition zones. Curiously enough, timberlines tend to coincide with the 10°C isotherm, but air temperature per se is unlikely to limit tree growth because pine trees and other arboreal species in the association can photosynthesize at reasonable rates despite prolonged periods of low temperatures (Pharis et al., 1967; Scott et al., 1970). Soil temperature, on the other hand, is a more likely determinant. The growth of tree seedlings would be restricted by a cold root zone (discussed previously); consequently, these young trees become more susceptible to other environmental pressures than dwarf perennials of the alpine tundra (Spomer and Salisbury, 1968).

Latitudinal gradients in seasonal temperatures also dictate plant growth and community structure. Such controls are clearly evident in the life cycles of forage species in western Europe, including *Lolium perenne, Dactylis glomerata, Phalaris tuberosa,* and *Trifolium subterraneum.*

Fig. 16-22 The seasonal change of diurnal temperatures for Pasadena, California, intercepts the optimal temperature ranges for *Zinnia* (*Z*) and China aster (*C*) but not for *Saintpaulia* (*SP*). The former species thrive at that location but not the latter (Went, 1957).

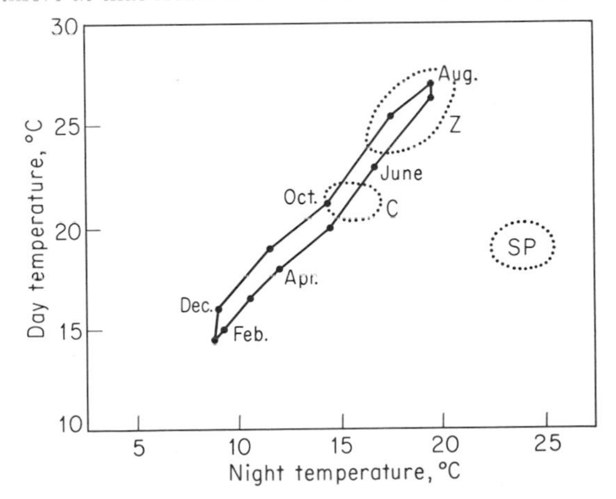

These plants have become commercially significant not only in Europe but also in North America and Australasia. One particular climatic transect, from Mediterranean regions to northern Europe, shows a pronounced gradient in the seasonal distribution of temperature, rainfall, and day length. *Lolium* species have developed ecological races to suit these conditions. Their temperature optima for photosynthesis show a clear distinction on the basis of geographic origin (Fig. 16-23). The Mediterranean population of *Lolium perenne* combines a higher optimum temperature for photosynthesis with a lower Q_{10} for respiration. The additive effect of these two responses in the Mediterranean race would favor its growth over that expected from the Danish population at higher temperature (Eagles, 1967a).

Temperate species, particularly those from regions with a pronounced Mediterranean climate, have become adapted to summer drought and winter growing season; this involves active growth at moderately low temperature. As indicated in Fig. 16-23, 10 to 15°C provides optimum conditions for photosynthesis. In sharp contrast to this situation, tropical grasses and related C_4 genera that have become adapted to an abbreviated growing season under the higher temperatures and solar radiation of lower latitudes are decidedly sensitive to cool conditions.

Fig. 16-23 Temperature has somewhat different effects on races of rye grass of Danish and Algerian orgin (*solid and broken lines,* respectively). The more northern race has a lower temperature optimum for photosynthesis and a lower respiratory rate at lower temperatures (Eagles, 1967*a*).

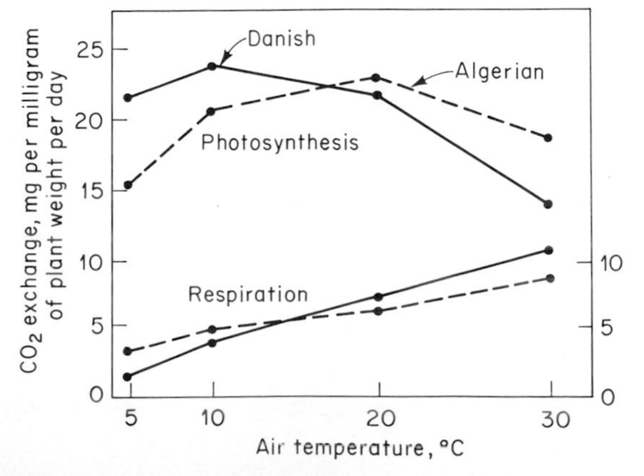

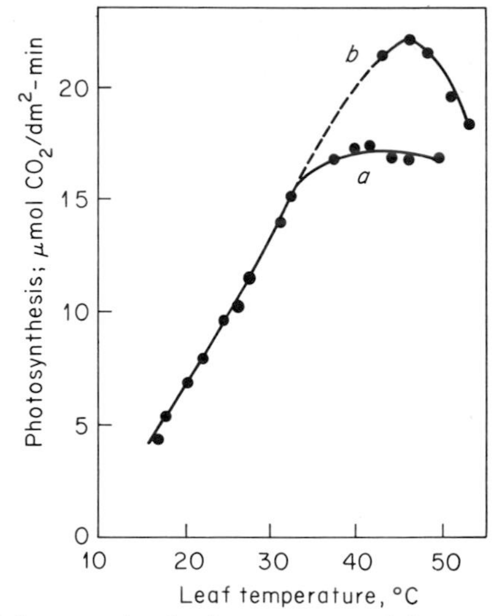

Fig. 16-24 A low-growing herbaceous perennial from Death Valley, California, shows a remarkable adaptation to its habitat. Optimum temperature for photosynthesis in this C_4 plant is around 47°C. Full sunlight is required for maximum performance. (*a*) 1.07 cal/cm^{2}min; (*b*) 1.33 cal/cm^{2}min (Björkman et al., 1972).

A combination of low temperature (10°C) but high light intensity (160 W/m^2) imposes climatic *stress* on these individuals. Taylor and coworkers in New Zealand have documented the responses of some C_4 plants including *Zea mays, Sorghum bicolor, Paspalum dilatatum,* and *Amaranthus lividus,* to such conditions (Taylor and Craig, 1971; Taylor and Rowley, 1971; Taylor et al., 1972). The photosynthetic responses of these plants are reminiscent of chilling injury described previously in connection with tropical and semitropical plants. The chloroplasts become disrupted, and chlorophyll synthesis is arrested. Yellowing tips on *Paspalum* leaves are in fact an early visual indication of their unfavorable reaction to winter conditions under a Mediterranean type of climate.

C_4 plants respond favorably to warm conditions, and a virtual absence of photorespiration, in concert with their other biochemical specialization, gives them a higher optimum temperature for photosynthesis (Fig. 15-8). These characteristics promote ecological dominance in the tropics, where environmental conditions are highly conducive to rapid growth of C_4 plants (Chap. 15), but the hot, arid environment imposes different demands. It favors crassulacean plants with some measure of succulence which helps them endure high temperature (discussed earlier) and is usually combined with an inverted stomatal rhythm which contributes toward a greater economy of water use (Chap. 17). Other categories of desert plants rely on photosynthetic specialization within small and dissected organs or stem tissues which allow CO_2 assimilation but encourage heat dissipation. Chlorophyllous stem tissue and leaves of the small leguminous tree *Cercidium floridum,* for example, make it ideally suited to the washes and sandy areas of deserts in California and Arizona, where it abounds (Adams et al., 1967; Adams and Strain, 1968).

Compared to this situation, where arid-zone species are characterized by their endurance of unfavorable conditions rather than their capacity for heavy production, the summer performance of a small herbaceous perennial on the floor of Death Valley, California, seems almost incredible. This plant's photosynthetic activity reaches a peak at temperatures that would prove lethal for most mesophytes; predictably, it is a C_4 plant. The remarkable performance of this perennial herb *Tidestromia oblongifolia* is illustrated in Fig. 16-24. Björkman and his colleagues from Stanford conducted their measurements in the field by enclosing aerial organs of single plants in a small, air-conditioned assimilation chamber. Photosynthesis showed peak rates around 47°C, and under these conditions assimilation increased linearly with solar radiation up to *full* sunlight. The water economy of this plant was equally impressive. Its root system explored the soil to a depth of 40 cm and provided sufficient moisture to sustain high rates of transpiration, 12.6 g/dm^2-h, that is, an order of magnitude higher than most mesophytes. Gas exchange could not be maintained at this rate for extended periods because the plant would ultimately lose turgor. Nevertheless, by mid-afternoon at the experimental site, leaf water potential commonly fell to around −25 bars and yet leaf resistance was only 5 s/cm. Most plants would show complete stomatal closure and a virtual cessation of photosynthetic activity at such low potential (Chap. 17). Björkman et al. (1972) conclude that despite searing heat on the floor of Death Valley, photosynthesis in *Tidestromia oblongifolia* will often be limited by light intensity.

SUMMARY

Over the course of their evolutionary history, plants have come to occupy every reasonably stable niche on the earth's surface, regardless of its temperature regime. Species may not necessarily thrive under conditions of extreme heat or cold, but at least they exist, and this is due to combinations of morphological features and physiological mechanisms which equip them with a remarkable degree of thermal resilience.

Extreme temperatures impose a heavy selective pressure, and successful individuals embody intricate systems which enable them to avoid or endure unfavorable conditions. In more moderate situations, temperature assumes a different role; it governs biological tempo. Ecosystems are driven by solar radiation, but temperature becomes the regulator of energy flow within the biosphere by dictating the rate and often the direction of biological reactions, due to its influence on developmental sequences.

GENERAL REFERENCES

Evans, L. T. (*ed.*). 1963. "Environmental Control of Plant Growth." Academic Press, New York. 449 pp.

Gates, D. M., and L. E. Papain, 1971. "Atlas of Energy Budgets of Plant Leaves." Academic Press, London. 277 pp.

Geiger, R. 1957. "The Climate Near the Ground" (trans. M. N. Stewart). Harvard University Press, Cambridge, MA. 482 pp.

Lyons, J. M. 1973. Chilling injury in plants. *Annu. Rev. Plant Physiol.,* 24:445–466.

Rees, A. R., K. E. Cockshull, D. W. Hand and R. G. Hurd (*eds.*). 1972. "Crop Processes in Controlled Environments." Academic Press, London, 391 pp.

17 WATER: AN ENVIRONMENTAL FACTOR IN ECOLOGICAL PHYSIOLOGY

Water is not only the most abundant single substance in the biosphere but probably is the most remarkable as well. Extraordinary physical properties characterize this liquid: it has the greatest surface tension, highest specific heat, biggest latent heat of vaporization, and, except for mercury, the strongest thermal conductivity of any known liquid. Water molecules are more than simply cogs in the plant's metabolic machinery; they are an integral part of living systems, and at the ecological level they represent a major force in shaping climatic patterns. Water is essential for life in both the biochemical and biophysical senses, and its influences are both internal and environmental. Since more than 95 percent of the earth's surface experiences some limitation on agricultural production because it is either too hot or too dry, or too wet, water availability has ecological consequences on a global scale.

BIOPHYSICAL ASPECTS

Physicochemical Properties

Hydrogen bonds, reinforced by dipole forces, endow water with physical properties that would not be expected on the basis of the atomic components of H_2O. Among these are a high specific gravity and a very high specific heat. Not many natural substances require 1 cal to increase the temperature of 1 g of material by 1°C. Another peculiar feature of water is its high heat of vaporization; over 500 cal is required to convert 1 g of liquid water into its gaseous phase at 100°C. Such an enormous requirement ties up large amounts of heat, contributing both to climatic stability and to heat budgets of individual organisms.

A fourth property of water with biological consequences is its high heat of fusion: 80 cal must be removed from 1 g of water for it to change from liquid to solid at 0°C. This property enhances the thermostatic effects of water by releasing large amounts of heat as ice is formed. Frost protection in horticultural situations which is afforded by overhead sprinklers relies on this feature of heat liberation. Heat released during this phase change is also evident in freezing curves for plant tissues (Fig. 16-16).

Water in Plants

Higher plants, like other living creatures, have evolved with biochemical systems that operate in an aqueous medium. Cytoplasmic materials are cradled in a matrix of water molecules which amount to 80 to 90 percent of fresh weight for fully hydrated tissue. This water is much more than simply a solvent for metabolites; it is a structural component of proteins and nucleic acids. Within this aqueous medium, hydrogen bondings between water and the electronegative sites on protein molecules are of paramount importance. The coupling is weak and transient, but water molecules become structurally integrated with proteins, by hydrogen bondings, especially with nitrogen and oxygen positions. If these links are restricted, protein structure is destroyed and the system is no longer viable. Temperature limits in the biosphere are due principally to this factor. The force of hydrogen bonding decreases with increasing temperature, and some proteins begin to denature above 30°C as the tenacity of hydrogen bonding diminishes. Conversely, as temperatures fall below 0°C, water molecules become rearranged into a relatively static crystalline lattice with drastic consequences for protein structure. Even though large energies for hydrogen bonding still exist, the static bondings between water molecules in the crystalline lattice limit the availability for hydrogen bondings with proteins, and protein structures face denaturation.

Biological macromolecules cannot exist in their complicated secondary foldings or spiraled helixes without the support of hydrogen bonding within their water matrix. If this bonding is restricted (as in partly desiccated tissue), the macromolecules collapse in a state of denaturation. The water matrix is also critical for metabolic activity because its properties of hydration permit ready reactivity between molecules in solution as well as between enzymes and their substrates. The state of hydration of the photosynthetic enzyme RuDP carboxylase, for example, is a prime determinant of this enzyme's reactivity (Fig. 17-1). IAA oxidase in plant tissues also demonstrates a clearly defined pattern of metabolic effectiveness and water potential in situ (Darbyshire, 1971*a,b*).

Water is also a reagent in biological reactions; it is a substrate for photosynthesis but a product of respiration. The surge in respiratory activity commonly associated with plant-moisture stress (Kaul, 1966) has been suggested to entail beneficial effects on cellular moisture status as water molecules are generated metabolically (Genkel et al., 1967).

In biophysical terms, water molecules have additional properties which hold significance for the plant's aqueous phase: they lack a ready means of attraction or movement. The molecule is small, and its electric charges are not large enough to permit easy movement by gradients in electric potential, while its ready resonance characteristics accentuate its slipperiness when water molecules have to be moved. In sharp contrast to the anions and cations taken up by plants, water moves readily only along gradients in free energy as manifest in the diffusive activities of its own molecules. Osmotic influences or tensions can generate such a gradient. Suggestions of metabolic water uptake have been made at various times (Spanner, 1952), but overwhelming evidence favors gradients in free energy as the primary motive force behind water movement (see, for example, Levitt, 1954).

The semipermeable nature of plant-cell membranes leads to compartmentation of nutrient elements or other osmotically active material. If the plant's moisture status is favorable (demand due to transpiration is not exceeding water supply from the soil), turgor develops. Resultant pressures have desirable consequences, as in growth (Burström et al., 1967; Ray et al., 1972), or highly undesirable consequences, as in fruit splitting (Frazier and Bowers, 1947; Considine and Kriedemann, 1972). Cell division and subsequent vacuolation depend upon freely available water to act as a hydraulic force so that a plant's water status is immediately reflected in processes depending upon positive turgor. Extension growth in shoots and roots, leaf expansion, organ enlargement, and stomatal function are all clearly coupled to moisture status. (Kaufmann (1969) has in fact utilized this feature of response in fruit volume to follow changes in moisture status of orange trees.)

Biochemical processes are equally sensitive to tissue hydration, and except for specialized organs such as seeds, a reduction of only 20 to 25 percent below maximum hydration is sufficient to cause metabolic disaster for many enzymes.

Water vapor and ice are chemically pure, but liquid water has an extreme dielectric constant and consequently is never pure under natural conditions. Water does in fact dissolve just about anything to some degree, but fortunately organic polymers show particularly low solubilities. Dissolved solutes and the adsorption of water molecules onto cell surfaces and colloids disturb the water's liquid structure. In thermodynamic terms, free energy is lowered, and if pure free water is taken as the reference state, where water potential is zero ($\psi = 0$), plant water will have a potential less than zero. Plant ψ is probably positive only during guttation or related phenomena. ψ has dimensions of energy per unit volume, e.g., ergs per cubic centimeter, but is usually expressed in bars or atmospheres.

The water potential of a solution is quantitatively equivalent to its osmotic potential, but in plant tissues, the differential permeability of limiting membranes, combined with resilient cell walls, leads to some measure of turgor pressure which makes a positive contribution to water potential. In the terminology of Meyer and Anderson (1952), this situation is summarized as DPD = OP − TP, that is, diffusion-pressure deficit is the result of osmotic influences (which lower diffusive energy) and turgor pressure (which raises diffusive energy). The term DPD has been abandoned for a number of reasons, particularly since diffusion pressure is no longer used in physical chemistry or thermodynamics. Furthermore, diffusion is no longer considered the sole mechanism by which water moves along gradients in free energy or chemical potential. Component potentials in plant-water relations can now be written as $\psi = \psi_p + \psi_s$, where ψ_p stands for a pressure potential arising from turgor (TP in the previous equation) and ψ_s represents solute or osmotic potential (less than free water and therefore negative). A matrix potential ψ_m and gravitational potential ψ_z (both negative) can be included in this expression of plant water potential if desired. Since gravitational influences ψ_z which lower the capacity of water to do work are insignificant compared to other factors, ψ_z is usually disregarded. However, matrix potential ψ_m must be considered; this term accounts for water molecules that are retained by surface forces associated with the cytoplasmic matrix and structural members. In meristematic tissue, which is typically nonvacuolated, much of the water will be retained by matrix forces, and comparatively little free solution will exist. As tissues lose water, both matrix and osmotic forces

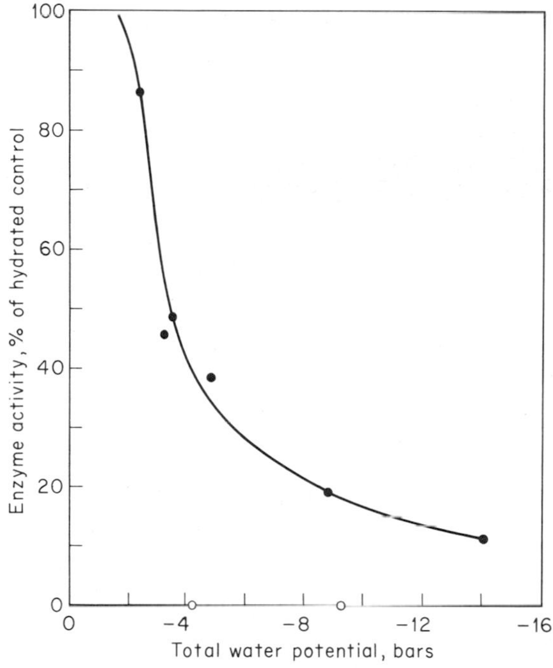

Fig. 17-1 The photosynthetic enzyme ribulose diphosphate carboxylase from commercial sources shows a marked decline in activity as hydration is lowered. Water potential of enzyme preparations was manipulated physically on a pressure membrane. (*Adapted from Darbyshire and Steer, 1973*).

will increase, and in structural components such as cell walls, the matrix forces increase rapidly as solid surfaces are drawn together. Water loss from the system is progressively reduced as ψ diminishes and constitutes an important mechanism in offsetting dehydration under dry conditions. In woody plants, leaf ψ may fall to -100 bars in desiccated material, of which ψ_m would be a significant component, whereas crop plants more commonly endure values as low as -10 to -20 bars (Slatyer, 1967).

The aqueous phase of plant tissues clearly amounts to a dynamic rather than static entity, and it fulfils both biophysical and biochemical roles. Water molecules act as structural components of macromolecules and as metabolites in biochemical reactions. At a cellular level, water provides a suitable medium, and not simply a solution, for physiological events. By virtue of its unique combination of physical properties, water helps to buffer plant tissue against fluctuations in the external environment. Both physical integrity and metabolic reactivity therefore depend upon the energy status of these molecules, which in turn finds expression in a collective term, the plant's *water potential.*

Fig. 17-2 A diagrammatic representation of the energy budget at the earth's surface. During the day (*a*) evaporation accounts for most solar energy absorbed; (*b*) it also makes a substantial contribution to heat dissipation at night (Tanner and Lemon, 1962).

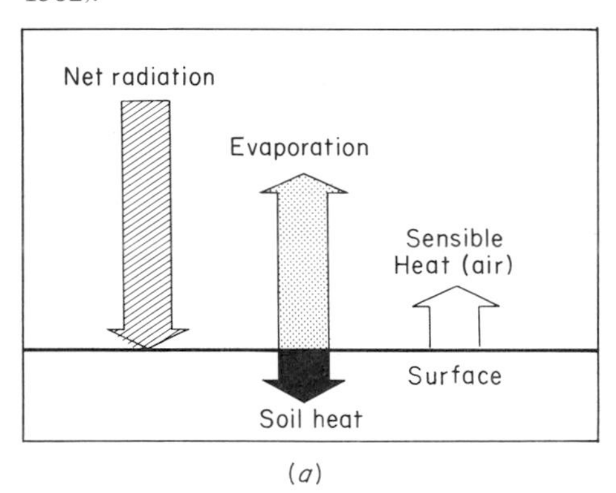

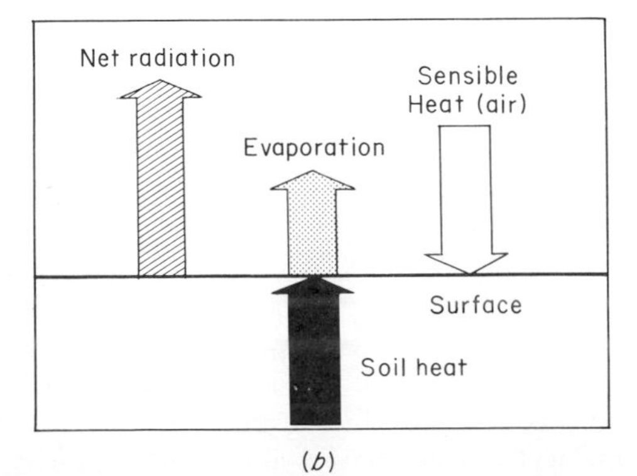

Water in the Biosphere

Less than 10 percent of the earth's surface is readily available for cultivation by man; the remainder is too steep, too rocky, too salty, too hot, too cold, or too dry. Within these categories, water is the outstanding environmental deficiency which limits plant growth. We are confronted by a paradox: water is by far the most abundant single substance in the biosphere, and yet plant growth in terrestrial habitats is more likely to be limited by water than by any other factor. This disparity between supply and demand hinges on distribution and condition of available water. The earth's oceans and terrestrial reservoirs contain 1.5 billion cubic kilometers, but most of this water (97 percent) is held in oceans whose solute concentration makes them "too dry" to supply the needs of higher plants (Penman, 1970). Since most mesophytes (wilting point around $\psi = -16$ bars) have a higher water potential than seawater ($\psi \approx -20$ bars), they are unable to extract moisture from that source. The presence of other dissolved salts, especially chlorides, heightens the unsuitability of seawater or saline groundwater for most plants, although halophytes have become specifically adapted to this situation. Specialized organs, e.g., bladders and salt glands, on leaves perform the dual function of excreting undesirable solutes from the plant body (Thomson et al., 1969; Ziegler and Lüttge, 1966, 1967) and generating a water-potential gradient between root zone and transpiring surfaces which favors moisture absorption (Scholander et al., 1962).

In climatic terms, "water is the driver of

nature" (Leonardo da Vinci); water has a moderating influence on temperature fluctuations because of its unique physical properties, which endow a large body of water with tremendous thermal inertia. Climatic differences between maritime and continental regions are due principally to this fact. Its high heat of vaporization makes a further contribution to temperature moderation at the earth-air interface by virtue of the energy consumed in evaporation. This phase change accounts for a substantial fraction of energy exchanged at the earth's surface during both day and night (Fig. 17-2).

As a climatic force, the influence of water extends well beyond the earth's surface. Atmospheric water filters solar radiation (Figs. 15-1 and 15-2), while surface moisture buffers temperature changes due to diurnal and seasonal fluctuations in transmitted energy. The huge volume of water on our planet (equivalent to a layer 1.5 mi deep over its entire surface) can store prodigious amounts of heat. These same characteristics of high specific heat and high heat of vaporization which make water a good climatic thermostat also apply to temperature regulation in plants and animals, both of which utilize evaporative cooling and derive substantial benefit from a high moisture content, which helps them buffer their own temperature against fluctuations in environmental conditions.

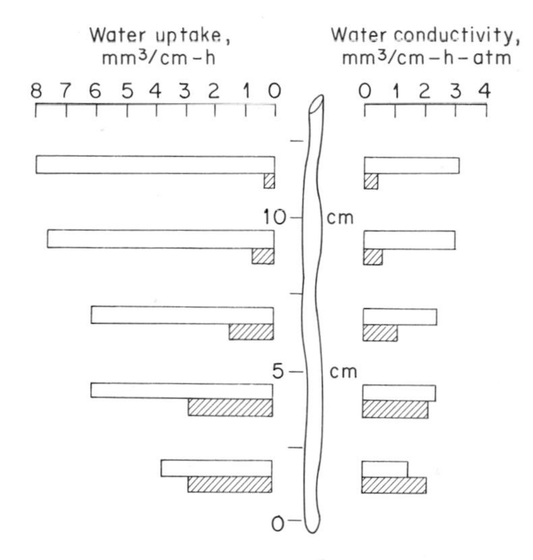

Fig. 17-3 Water uptake *(left)* and water conductivity *(right)* are shown for different zones along a single root from a broad-bean plant. At low transpiration *(solid bars)* tension within the root zone is moderate, and most uptake occurs just behind the root apex. With increased transpiration *(open bars)* tension within the root alters the absorption pattern (Brouwer, 1965).

PLANT-WATER RELATIONS

Transpiration lowers water potential ψ at evaporative sites within leaves, and this effect is immediately translated to the root system via tensions in the plant's vascular system. If leaf ψ is to be maintained, this demand for water must be satisfied continuously; otherwise moisture stress develops in foliage, stomata close, and assimilation is adversely affected. An actively growing plant therefore needs to maintain liquid-phase continuity from soil water, through its vascular system, and all the way to evaporative sites in leaves. Such a system will be immediately responsive to plant needs (Spomer, 1968), and the question of water supply to meet this demand can be viewed in terms of a water-potential gradient favoring moisture flow which has to negotiate a series of resistances within the soil-plant-atmosphere continuum.

Water Uptake

Aerial organs are capable of absorbing water from a humid atmosphere or liquid film, but for most practical purposes the root system accounts for virtually all the water and nutrients which enter a plant. A host of anatomical and physiological studies (Kramer, 1956; Esau, 1960; Brouwer, 1965) have shown that the zone of most active water absorption lies just behind the root tip (Fig. 17-3). This zone coincides with the normal region of root-hair development. Such epidermal outgrowths obviously enable the root system to explore a greater volume of soil; mycorrhizal associations may fulfil a similar function, but the absolute magnitude of their contributions is difficult to establish under natural conditions (Bowen and Rovira, 1969).

Water uptake by roots is greatly affected by temperature, and this has been taken to imply some metabolic involvement although the real explanation probably lies in membrane permeability. Discontinuities in relationships between temperature and permeability probably arise from phase shifts in membrane lipids (Ginsburg and Ginzburg, 1971), and hence the passage of water molecules.

Water-potential gradients arising from transpiration are generally assumed to be the driving force behind water absorption by the root system. However, we must acknowledge the rapid movement of ions into roots, which would cause some adjustment of osmotic potential with attendant effects on water movement. The effect tends to be significant for cell enlargement, but the transpiration stream is dominated by gradients in ψ (Slatyer, 1967). Figure 17-4 illustrates how transpiration can generate a gradient in water potential that is followed by water uptake within the root system.

In dry soil a vapor gap may form between root surface and soil particles so that water uptake would have to occur across a vapor phase. Soil water potential drops exponentially with decrease in soil-water content (Fig. 17-5), so that transpirational pull becomes decreasingly effective in satisfying the plant's water require-

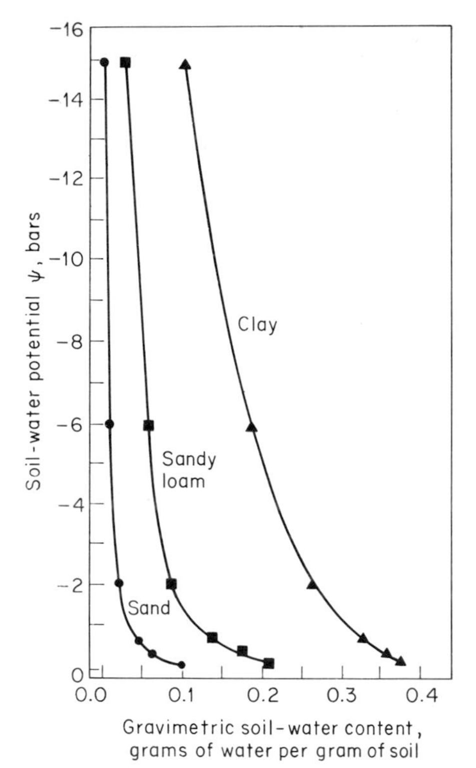

Fig. 17-5 Clay soils have a greater water-holding capacity than sandy soils comprising larger particles, but in both situations water is held with progressively stronger forces as the soil mass dries out (Slatyer and McIlroy, 1961).

Fig. 17-4 Diurnal rise in transpiration precedes the rise in water uptake, illustrating how the reduction in water potential due to evaporative losses from the leaves is subsequently transmitted to the root zone; ash trees (Kramer, 1937).

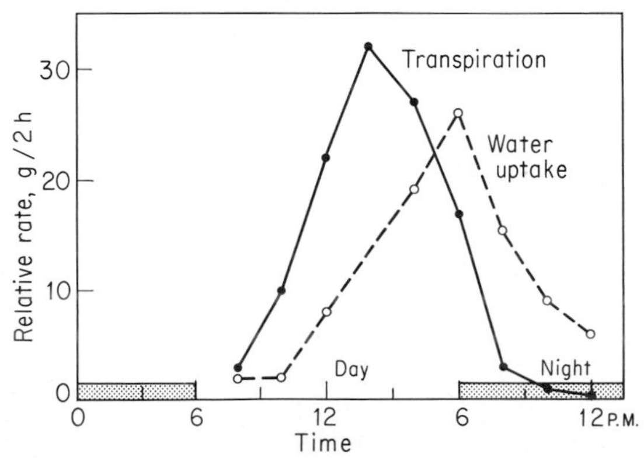

ments under drought conditions. As water potential drops, root tips ultimately lose contact with soil particles. Philip (1958) made such a prediction after examining the kinetic properties of water and solute uptake by roots.

Studies at the Auburn Rhizotron (root-observation chamber at Auburn University, Alabama; Taylor, 1969) have provided a vivid demonstration of diurnal changes in root extension and contraction in a fascinating time-lapse movie (Huck et al., 1970). Root extension ceased during the day, and subsequent contraction broke contact with soil particles, but after sunset, turgor was restored and extension growth started again. The exact sequence of events varies with species, soil conditions, and evaporative demand above ground.

Water Movement

Plant stems fulfil both structural and hydraulic functions. Their vascular tissues must cope with prodigious longitudinal flow of xylem sap over great distances (100 m in some species). Such structures must possess high conductivity, as shown in Table 17-1. Lianas, with their long open vessels, are noteworthy for low resistance to flow, whereas conifers, whose vascular system comprises short thin tracheids, offer much greater resistance (see Heine, 1971, for a detailed discussion). Given these requirements, a simple network of open pipes is still not good enough, and plants have evolved with a vascular system which can have a high conductivity (Table 17-1) plus an ability to accommodate interruptions in hydraulic continuity (Fig. 17-6). Scholander et al. (1957) demonstrated this ability in *Tetracera*. Stems of this liana were examined for tension on the xylem-sap column using a simple lateral punch manometer. As the stem was severed and its cut end placed in water, sap tension above the cut fell almost to zero (Fig. 17-6). If the dish of water was withdrawn and air was pulled into the xylem by transpiration, tension soon rose steeply; but when water was resupplied, tension again diminished. Surprisingly, water flow proceeded as it had before these operations despite the introduction of air bubbles. Scholander et al. (1955) performed similar experiments on grape-

Table 17-1 The flow of xylem sap is facilitated by high conductivity—a consequence of large-diameter conduits with few transverse septa. Long open vessels of a tropical liana offer a sharp contrast to the trachieds of conifers in this respect, and sustain vastly different rates of sap flow (derived from Huber 1956).

Plant material	Specific conductivity, cm^2/s-atm
Roots;	
deciduous trees	10–160
Stems:	
Deciduous trees	1.6–3
Conifers	0.6
Lianas	7–36

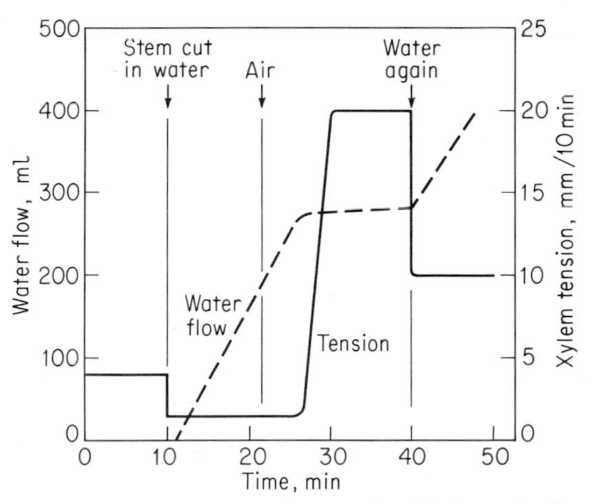

Fig. 17-6 Water flow and the hydraulic forces in the xylem of *Tetracera* stems vary with the water supply. When a stem is cut under water, tension is low and water flow progresses nicely; when the stem is taken out of water, tension rises as air enters the xylem elements. When the stem is placed in water again, flow is restored and tension in the xylem abates, despite the discontinuity (Scholander et al., 1957).

vine stems. With short lengths, water movement regained its original rate, although recovery was incomplete in longer stems (> 10 m). These experiments imply that water movement up a plant stem is not a simple flow through open tubes because the introduction of air bubbles does not break the flow but merely sets a requirement for higher tensions before flow can proceed again at full speed. Overlapping cuts in a trunk of *Pinus resinosa* made by Postlethwait and Rogers (1958), plus those of Mackay and Weatherley (1973) on *Acer* and *Gossypium,* confirm this ability to conduct water despite apparent gaps in liquid-phase continuity. However, crosswalls, pits, tyloses, and other anatomical components of xylem tissue have very low values for air entry, so that bubbles introduced by cutting do not spread extensively and water can move around overlapping cuts or from moistened zones at the base of Scholander's liana stem and so maintain liquid-phase continuity. These observations meet Preston's (1958) criticism against Dixon's (1914) cohesion theory that flow continues even though a high proportion of larger vessels may contain air at any one time: first, not all vessels are needed, and second, liquid-phase continuity is met by lateral migration of the water column.

Dixon's cohesion theory for water flow depicts a direct transmission of moisture tension from sites of transpiration (leaves) to zones of absorption (roots) via xylem conduits inside the plant's vascular network. Xylem sap continuity within this tortuous pathway, plus a high tensile strength within individual columns, are prerequisites. This theory has become a historic part of plant-water relations but does not adequately explain how substantial tensions can be developed and sustained within the vascular cylinder (Scholander, 1972) which enables water to rise above a height equivalent to zero atmospheric pressure (approximately 10m). Plumb and Bridgman (1972) have further propositions on the role of microfibrils on xylem cell walls in aiding this ascent. The microfibrils are depicted as supporting columns of water inside the xylem via the formation of a gel-like structure within individual conduits. Despite such sophisticated concepts, one problem remains: How do vessels in tall trees refill once they have been ruptured? In herbaceous plants or shrubs nocturnal improvement in moisture status could lead to refilling, with root pressure as the driving force. But these pressures develop only 1 to 2 bars, and then only with wet soil ($\psi_{soil} \approx 0$); root pressure could not possibly refill vessels above 10 to 20 m on tall trees in dry ground. Gibbs (1958) has provided some intriguing data which relate to this question. He showed that water content in trunks of tall trees decreases during summer to a minimum in autumn but increases by the following spring. It seems that vessels which become drained do not refill that same season even though soil moisture is adequate. Although new water-filled vessels are laid down during the growing season and contribute to the conducting system, once empty, they do not refill until the next spring.

In contrast to plant roots, stem permeability is not sensitive to metabolic inhibitors or to temperatures above freezing (Zimmermann, 1964). Since water ascends the xylem via the lumen of its individual elements (Fig. 2-6) and need not traverse cellular membranes, this independence is to be expected. Scholander et al. (1961) froze a section of liana stem in dry ice, and water movement ceased, due presumably to cavitation in all vessels. Upon thawing, normal absorption and transfer were restored, confirming the absence of even indirect metabolic involvement in maintaining the transpiration stream.

The actual flow of water movement through a stem will obviously vary according to plant demand and soil supply, i.e., the ψ gradient, and resistance to flow. In the equation

$$f = \frac{\psi_{\text{roots}} - \psi_{\text{leaf}}}{R_{\text{root}} + R_{\text{stem}} + R_{\text{leaf}}}$$

the resistance of each component depends upon overall dimensions of the organ in question, as well as specific conductivity (Table 17-1), and in this type of assessment stem resistance R_{stem} is often the smallest single component in the plant's overall resistance to liquid flow (Table 17-2). Speed of movement, which is substantial, was measured on a spruce tree as early as 1937 by Hüber and Schmidt. They warmed sap in a localized zone and followed the movement of this heat pulse with a series of thermocouples above and below the warmed zone.

The water moved at 120 cm/h at midday, when evaporation demand was greatest, dropping to less than 10 cm/h at night. Sap ascent at night

Table 17-2 The total resistance to water flow in the plant is relatively small in the leaves and stem compared with the roots (calculated from Jensen et al., 1961).

	Resistance to water flow, 10^2 atm-sec/cm^3	
Plant component	Sunflower	Tomato
Leaves	2.8	1.2
Stem	5.5	1.5
Roots	6.5–9.2	3.6–6.2
Whole plant	14.7	8.9

refills some of the elements drained during the day and also contributes to restoration of leaf water potential ψ_{leaf}.

Water gains entry into leaves via petioles or leaf sheaths and moves in the liquid phase from vein endings to evaporative sites, principally the substomatal cavity. Crowdy and Tanton (1970) have tracked this movement with Pb-EDTA chelate and confirm that cell walls are an important pathway (see Figs. 2-4, 2-6). Many thousands of individual xylem elements terminate in each square centimeter of leaf surface, and the leaf's vascular system is characterized by the close spatial relationship of mesophyll cells and vascular strands. This distribution system is so effective that very few leaf cells are separated from a vascular element by more than two to three other cells. Vascular bundles are enclosed in bundle-sheath cells, so that no part of the vascular system is directly exposed to intercellular air; water supply to mesophyll cells benefits accordingly. This fine network of vascular elements in a leaf presents a relatively higher resistance to the flow of liquid water. Cox (1966) estimated that leaf resistance R_{leaf} is about twice that of roots plus stem, $R_{root} + R_{stem}$, although Jensen et al. (1961) obtain a lower value (Table 17-2). In any event, the pathway of movement in leaves is substantially different from that in other organs, because cell walls figure so prominently (Strugger, 1949; Gaff et al., 1964). Russell and Woolley (1961) calculate that the ratio of water flow in walls compared with a vacuole-to-vacuole pathway is about 50:1. This pathway offers little hindrance to water flow, and leaf transpiration is primarily regulated by resistances in the vapor phase (Slatyer, 1967). It is important to note this present distinction between resistance to gaseous diffusion and liquid flow; while stomata exert a primary control over transpiration, they do not represent the major resistance to water movement in a plant, which resides in the liquid and not in the gaseous phase (Levitt, 1966).

Xylem sap ascends the stem in a passive response to evaporation demand. This movement of water, in itself, is of no direct advantage, but plants benefit in terms of uptake and translocation of nutrients. As water enters the root system, it sweeps in large amounts of dissolved solutes. Hylmo (1953) has demonstrated (Fig. 17-7) that calcium uptake is linearly related to transpiration over a wide range in external concentration. Such movement of water in the liquid phase can also favor redistribution of solutes in the plant. Martin et al. (1967) report instances in apple trees under moisture stress where fruits act as reservoirs for transpiring leaves and the withdrawal of water from fruits is accompanied by a migration of calcium from fruit to leaves.

Apart from its contribution to a leaf's heat budget under certain conditions (Chap. 16), transpiration affords no additional benefits to higher plants and can be viewed as an inevitable consequence [or prerequisite? (Walter and Stadelmann, 1968)] of their adaptation to terrestrial habitats over the course of evolutionary history.

Transpiration

Water evaporates from leaves (transpires) for the simple reason that relative humidity inside a leaf is higher than atmospheric moisture levels outside. Brightly illuminated leaves are general-

Fig. 17-7 The uptake of water can be associated with the entry of some solutes. In pea plants, for example, increasing transpiration resulted in greater uptake of calcium ions. A linear relationship was found for each of three concentrations of $CaCl_2$ (Hylmo, 1953).

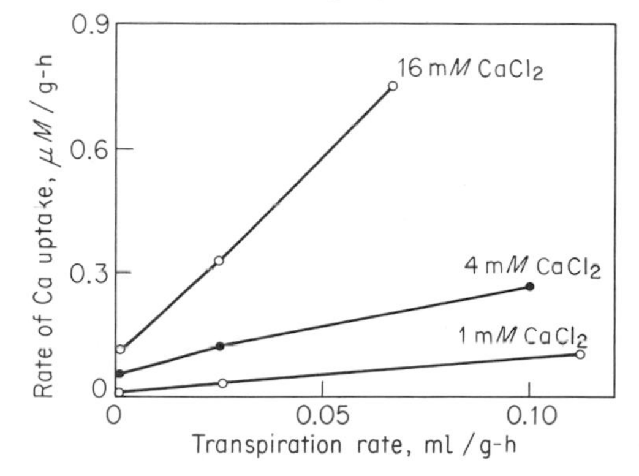

ly warmer than their ambient atmosphere (Chap. 16), so that a gradient in water-vapor pressure from leaf to air will exist even during humid weather. Under moderate environmental conditions, say 35 percent relative humidity and air temperature 25°C, water-vapor content would be 8 mg/l. For a leaf at this same temperature, moisture level within its air spaces would approach 30 mg/l (allowing for some leaf-air temperature differential). This sharp difference in water-vapor concentration, 30/8 mg/l, occurs across a few millimeters immediately adjacent to the surface of a transpiring leaf. Such a steep gradient obviously favors rapid transpiration, but a leaf does not behave like a free-water surface because freely transpiring leaves lose water principally via stomata and the combined effects of their density and aperture constitute a diffusive resistance (Fig. 1-19) which can in large measure regulate transpiration.† Despite stomatal resistance, the steep gradient in vapor pressure from leaf to air, which is well in excess of the concentration gradient for CO_2 uptake, favors a rate of transpiration well in excess of CO_2 fixation. Transpiration commonly ranges from 500 to 2,500 mg H_2O/dm^2-h (although higher values can occur; Chap. 16) whereas CO_2-fixation rates are essentially two orders of magnitude lower at 5 to 25 mg CO_2/dm^2-h (Table 1-1). Meyer and Anderson (1952) state that many herbaceous plants transpire an amount of water equal to several times their own volume in a single day; somewhere in excess of 99 percent of water absorbed by the plant would be lost in transpiration while less than 1 percent would be retained in the plant.

This large disparity between water vapor lost and CO_2 fixed leads to a heavy water requirement if dry matter productivity is to be maintained.

†The nature and derivation of stomatal diffusive resistance have been discussed in Chap. 1; the combination of boundary-layer and stomatal resistances exercises close control over transpiration. While water-vapor loss from leaves resembles diffusion through multiperforate septa in some respects, the so-called *perimeter law* appears to be a totally inadequate description founded on misconceptions (Cowan and Milthorpe, 1968).

When a plant community is adequately supplied with water and nutrients, both water use and dry-matter production depend upon receipt and utilization of solar radiation. These two plant parameters will then be coupled to some degree; their relationship has been termed *transpiration ratio,* i.e., grams of water lost per grams of dry matter produced. This concept was introduced at Rothamsted (England) in 1850 by Sir John Lawes as an index of water-use efficiency (Monteith, 1966). It has since been modified to the form of a *production ratio* (reciprocal of transpiration ratio), which is derived meteorologically from potential photosynthesis and potential transpiration. Growth and water consumption are usually proportional in a temperate climate, but the relationship breaks down under hot, arid conditions or strong solar radiation because the mechanism for photosynthesis becomes light-saturated whereas the forces driving transpiration do not. Understandably, production ratios will then vary with climate, and values for a grass sward drop from $\frac{1}{500}$ under relatively overcast conditions in Denmark to $\frac{1}{2500}$ in the West Indies (Stanhill, 1960). The value is likely to decline even further under more arid conditions.

Since evaporation accounts for most of the solar radiation absorbed at the earth's surface (Fig. 17-2), plant transpiration is closely coupled to diurnal variation in solar radiation (Fig. 17-8). In this situation (Tempe, Arizona) a sward of Sudan grass (irrigated) is experiencing strong wind under arid conditions, and heat dissipation due to transpiration is in excess of incoming solar radiation. This discrepancy, which stems from the combined effects of strong wind, high temperature, and low relative humidity, is accentuated by mowing the stand adjacent to the site of measurement. These well-watered plants are in fact behaving rather like wicks. Their internal evaporative surfaces within leaves are a clear order of magnitude greater than leaf area (Esau, 1965; Turrell, 1965), which is in turn greater than land area (leaf-area index, LAI, expresses this ratio of leaf area to land surface, Chap. 3). The relative levels of plant transpiration and soil evaporation therefore vary with crop density; Brun et al. (1972) provide quantitative information on this point by separating out the plant and soil components of evapotranspiration from soybean and sorghum fields. The proportion of transpirational water was clearly correlated with plant density; transpiration accounted for 50 percent of total evapotranspiration at LAI = 2 but as much as 95 percent at higher plant density (LAI = 4).

In summary, evapotranspiration from a plant community expressed in terms of land area can easily exceed the value from bare wet soil or free water. The greater evaporative surface (wick effect) more than compensates for the additional resistance in the water-vapor pathway (Slatyer, 1967).

If soil moisture is insufficient or the root system inadequate to meet the continuous daily demands of evapotranspiration, stomatal controls come into operation and leaf characteristics rather than weather conditions dominate water-vapor flow. The effect of leaf resistance in changing relationships between transpiration and leaf temperature has already been summarized (Fig. 16-3); this resistance is a direct consequence of stomatal function.

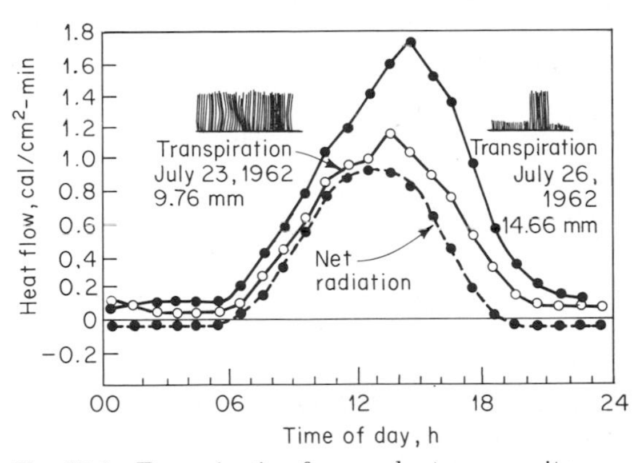

Fig. 17-8 Transpiration from a plant community accounts for the bulk of absorbed radiation but also depends on other meteorological conditions. This sward of Sudan grass growing in Arizona showed strong evaporative demand which resulted in more heat being dissipated than was obtained from net radiation. Transpiration (hence heat flow) at the test site was further accentuated by mowing surrounding regions (van Bavel et al., 1963).

STOMATAL PHYSIOLOGY

In furnishing the needs of plant metabolism (light, water, CO_2, nutrients) a leaf is faced with a dilemma: it must assimilate atmospheric CO_2, but at the same time it must minimize evaporative losses. Some regulatory system is clearly needed, and specialized epidermal organs (stomata) fill this role (Fig. 17-9).

Although botanists have observed stomata for the past 300 years, the complete mechanism of stomatal movement remains unexplained. Malpighi (1674) must be credited with probably the earliest description of stomatal function and its relation to the ascent of sap and subsequent transpiration. His charming explanation (Thomas, 1965) was

that part of the nutrient sap which has entered the roots . . . at length slowly reaches the leaves by way of the woody vines. This is necessary so that the sap may linger in the adjacent vesicles . . . and be fermented. In this the warmth of the surrounding atmosphere is of no little assistance for it helps . . . to evaporate that which is of no service. For this purpose nature has provided the leaf with numerous special glands or bellows for the sweating forth and gradual elimination of moisture so that the sap being thereby condensed may be more readily digested in the leaves. . . .

Structure and Function

We can summarize a multitude of characteristics all emphasizing that guard cells are not ordinary epidermal cells, despite their location (Fig. 17-9). They have distinctive morphology and anatomy, cytoplasmic inclusions, and vacuolar properties plus a capacity to alter their volume in response to light and CO_2. Their metabolic complement is analogous to that of a mesophyll cell, and they contain chloroplasts, although these are smaller and fewer than those in a mesophyll cell. These smaller chloroplasts, however, are exceptionally rich in starch, as is readily demonstrated with the standard iodine

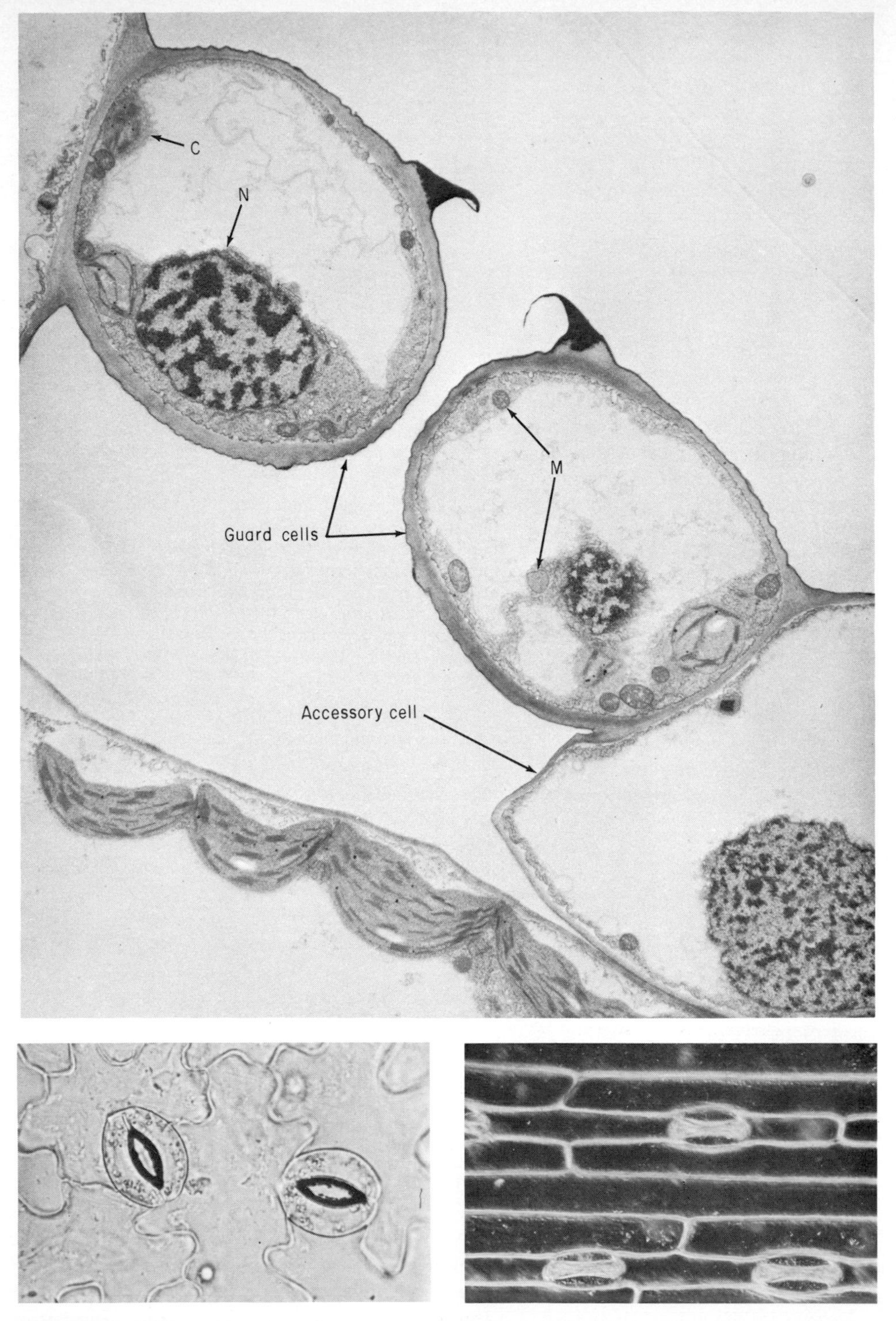

Fig. 17-9 The stoma of tobacco, showing guard cells with prominent nucleus (*N*), chloroplasts (*C*), and microbodies (*M*). The thinner walls of accessory cells, which can be seen here, contribute to shorter accessory-cell reaction time to changes in leaf potential (Fig. 17-11) (Pallas and Mollenhauer, 1972). Inserts: stoma of *Vicia faba* (left) and *Bromus uniloides* (right), x400 (*photos by B.R. Loveys*).

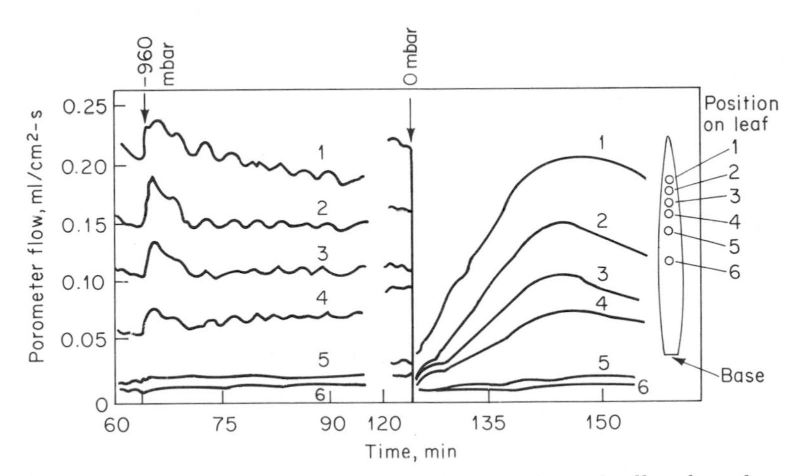

Fig. 17-10 The stomatal resistance of a detached maize leaf shows an almost instantaneous response to changes in leaf water potential ψ_{leaf}. The leaf base was supported under deionized water, ψ_{leaf} was initially pulled down (65 min) by applying vacuum to the leaf's water supply. Stomata showed a momentary opening as the turgid guard cells enlarged in response to a reduction in the lateral constraints imposed by the slower-reacting epidermis. Conversely, the release of vacuum (125 min) elicited a temporary reduction in stomatal aperture. (*Adapted from Raschke, 1970.*)

reaction on an epidermal strip. Unlike other epidermal cells, the guard cells contain a wealth of enzymes, including phosphorylase, ATPase, and peroxidase, and they are able to accumulate ions such as Ca^{2+} or K^+ against a concentration gradient when sufficient ATP is available. The nucleus of the guard cell is also more prominent and the osmotic pressure of its vacuolar contents is higher than in adjacent epidermal cells (as determined by the plasmolytic method). With the stomatal pore open, guard-cell osmotic pressure is commonly between 13 and 18 bars, with most measurements around 15 bars (Meidner and Mansfield, 1968). It is perhaps significant that the total soil-moisture level commonly regarded as wilting point is also 15 bars (Fig. 17-16).

Stomata come in a variety of shapes, sizes, and location, but in each case, increased turgor in the guard cells opens the stomatal pore. In the elliptically shaped guard cells of most dicotyledons, the uneven thickening of the inner and outer walls causes an increasingly curved shape as the guard cell yields to the increased turgor. This principle can be readily demonstrated by attaching a piece of adhesive tape to a sausage-shaped balloon: when the balloon is inflated, it becomes curved.

The stomata of graminaceous plants are dumbbell-shaped; here an increase in the volume of the guard cells at each end of the pore causes mutual repulsion, and the opposing sides of the stomatal pore move apart as the guard cells enlarge. Electron micrographs of *Avena* stomata by Kaufman et al. (1970) help illustrate this principle. An increase in turgor pressure results in even greater enlargement by the adjacent subsidiary cells and the rest of the epidermis (cf. Fig. 17-9). Elastic subsidiary cells also offer less impedance to guard-cell expansion, which favors sensitivity in stomatal movements.

Guard cells plus subsidiary cells, which we can regard collectively as the stomatal apparatus, constitute part of the hydraulic system of the leaf and show an immediate response to changes in its hydrostatic pressure. In the experiment in Fig. 17-10, vacuum was applied to the water supply of a detached leaf exposed to the light. The stomata, which were already open, immediately opened further in typical hydropassive movement, cf. Fig. 17-11, and in distal portions of the leaf stomatal aperture showed cyclic oscillations, which seemed related to an overcompensation for the water lost following the initial opening reaction. Once vacuum was released (see Fig. 17-10), the stomata closed abruptly as lateral pressure from epidermal cells was reimposed, before returning to a steady-state aperture comparable to that at the outset of the experiment. Raschke established that the stomata showed a lag time of only 0.1 s before responding to either a positive or a negative change in pressure; this emphasizes the close coupling of the guard cells to changes in hydraulic pressure of the leaf's vascular system.

Opening and closing reactions of the stomata, particularly the hydropassive opening and subsequent closure that follows excision, can be viewed in these same terms. If the xylem is under tension when a leaf is severed from the plant, there will be a sudden release of this tension at the moment of excision, making water freely available to the leaf. For a brief period the accessory cells and adjacent epidermal cells actually gain turgor. This should produce a slight initial closing reaction (Fig. 17-11) before the hydropassive opening that occurs when accessory cells and remaining epidermal cells become flaccid as they lose water. Since their loss in turgor occurs more quickly than that of the guard cells, they offer less constriction and the stomatal pore actually enlarges (hydropassive opening, Fig. 17-11). As the guard cells and the leaf generally lose turgor, the stomatal pore shows a final and decisive closure.

Fig. 17-11 Changes in stomatal aperture following excision of a bean leaf reveal an initial closing reaction followed by a more definite hydropassive opening as accessory cells lose turgor faster than guard cells. The second closing reaction after 3 min. coincides with a general decline in ψ_{leaf} and in guard-cell turgor. (*Adapted from Meidner, 1965.*)

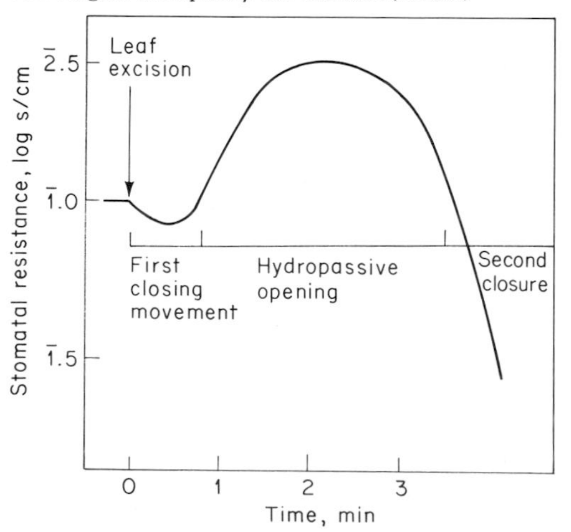

Environmental Controls

If we view stomata as regulating the passage of CO_2 into the leaf and set aside transpiration as an unavoidable and often embarrassing consequence of opening, stomatal responses to light and CO_2 fit a logical pattern. Under illumination, photosynthesis calls for CO_2; and as CO_2 in the substomatal cavity is depleted, the stomata open. Conversely, when external CO_2 supply exceeds consumption by the leaf, this leads to increased levels in the substomatal cavity and the pore closes but will reopen as CO_2 is depleted. Regardless of whether the CO_2 is consumed by photosynthesis or by the crassulacean acid metabolism of succulents (Neales, 1970), lowered CO_2 tension will elicit the same opening response.

The time taken for stomata to react to light can vary from many hours (especially in older leaves) to a few minutes, as in *Xanthium pensylvanicum* (Fig. 17-12). While a threshold intensity of light must be exceeded to induce opening, stomatal aperture can then show a progressive increase with light intensity up to the levels required for photosynthetic saturation. Figure 17-13 illustrates this type of response for a mature citrus leaf, but it does not necessarily typify the stomatal behavior of leaves on rapidly growing herbaceous plants or of younger citrus leaves, where stomatal aperture can reach a maximum at light intensities far below the level required to saturate photosynthesis.

The role of CO_2 in regulating stomatal movement was anticipated in something like an inspired guess by Scarth in 1932, and subsequent work by Heath and colleagues at Reading (England) established what must now be regarded as a tenet of stomatal physiology: in either illuminated or darkened leaves, and in-

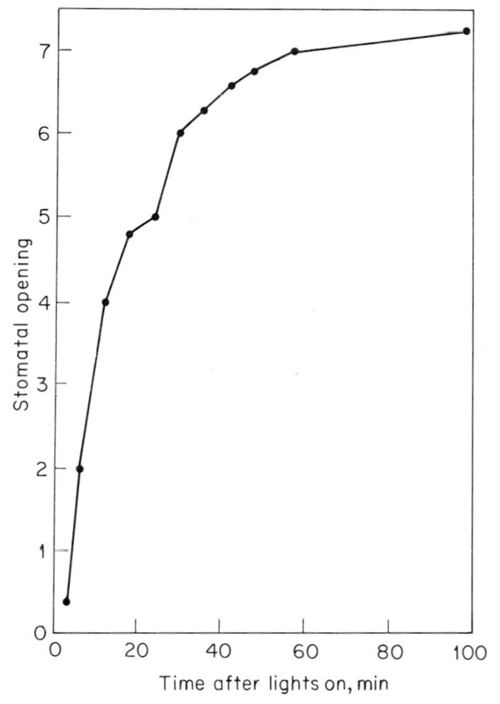

Fig. 17-12 Stomatal aperture in leaves of *Xanthium pensylvanicum* approaches its maximum within a few minutes after illumination. Irradiance was 80 W/m²; the preceding dark period was 8 h (Mansfield and Heath, 1963).

dependent of the presence of chlorophyll, increased CO_2 levels induce closure whereas depleted CO_2 causes opening.

This closure at elevated CO_2 has unfortunate consequences from the horticultural point of view because it limits the prospects for increased productivity from CO_2 enrichment. Provided the improved diffusion gradient more than offsets the reduction in stomatal aperture at elevated CO_2, some improvement in photosynthesis and more especially in water-use efficiency should occur with CO_2 enrichment. This was the effect observed by Moss et al. (1961) on corn plants when the CO_2 content of the air was increased from 310 to 575 ppm. Despite a 23 percent reduction in transpiration, photosynthesis was increased by 30 percent.

Stomatal opening is affected by light quality as well as intensity, although the action spectrum for opening departs from that for photosynthesis. In particular, blue light is far more effective than red in promoting stomatal opening (Mansfield and Meidner, 1966). This discrepancy between blue and red light is accentuated in corn, where Raschke (1967) found that leaves exposed to a CO_2-free atmosphere require 7 times the amount of energy at 769 nm as at 439 nm. When one is confronted with such a CO_2-independent blue-light effect, the conclusion is inescapable that the mechanism is not simply photosynthetic and that carotenoids are involved in the initial absorption.

One clearly recognizable internal control that seems to operate on the stomatal mechanism is an opening-closing stimulus. The apparent transmission of an opening stimulus is seen in variegated leaves, where the stomata in green areas open normally and those on nearby unpigmented regions also tend to open as if they "felt" an opening signal from the neighboring green parts. This could reflect a gradient in CO_2 concentration within the variegated leaf, but such a pos-

Fig. 17-13 Stomatal resistance and photosynthetic rate in orange leaves show related changes according to light intensity (Kriedemann, 1971.) Leaf resistance declines as stomata open.

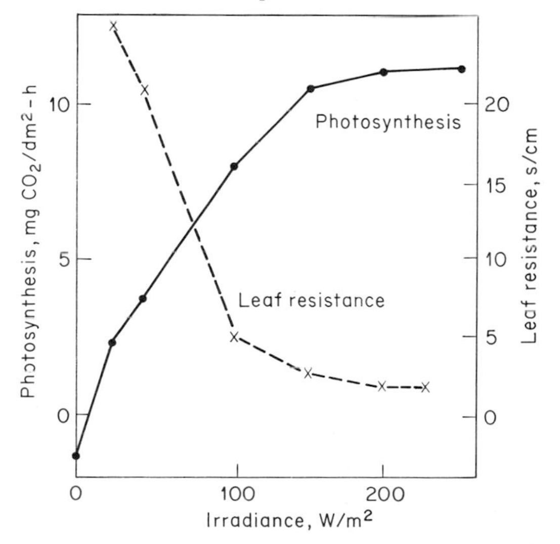

sibility was rejected by Heath and Russell (1954), who observed stomatal opening as much as 1.7 cm from an illuminated portion of a wheat leaf held in a CO_2-free airstream. The case for a translocatable opening-closing stimulus is thereby strengthened.

Stomatal Mechanisms

Irrespective of their size, shape, or location in the leaf, stomata open when their guard cells achieve a turgor pressure which is sufficiently greater than the accessory cell's turgidity. The central question in stomatal action is how this turgor is generated in the guard cells. Four systems may be envisioned: (1) active absorption of water, (2) passive absorption of solutes followed by passive absorption of water, (3) active absorption of solutes followed by passive absorption of water, and (4) formation of solutes within the guard cells followed by passive absorption of water [this corresponds to the classical Scarth (1932) theory]. The primary contentions of this earlier theory were that osmotically active materials were derived in situ, and that environmental factors such as light and CO_2 concentration influenced stomatal aperture via their effects on the levels of these solutes. Provided leaf moisture status was adequate, a higher concentration of osmotically active material would foster an increase in guard-cell turgor, and hence stomatal opening. Guard-cell photosynthesis was reguarded as a potential source of soluble material, while the attendant reduction in CO_2 concentration within the substomatal cavity was thought to favor a conversion of starch into sugar due to local changes in guard cell pH. The direct effects of atmospheric CO_2 concentration on stomatal aperture, independent of light (and therefore photosynthesis) were explained in terms of this pH effect on starch/sugar interconversion.

Levitt (1967) has made a penetrating analysis of the pros and cons for all these possibilities, concluding that "of the four conceivable methods of achieving an increase in turgor pressure leading to stomatal opening, the only one that appears possible is the formation of solutes in the guard cells followed by the passive absorption of water." This is essentially the old Scarth theory, i.e., starch in the guard cell is converted into sugar, which leads to passive absorption of water, increased turgor, and stomatal opening. Supposedly, the phosphorylase that catalyzes this conversion is sensitive to CO_2-concentration changes in the substomatal cavity via changes in the pH of the guard-cell sap; e.g., higher CO_2 would induce lower pH, which favors starch formation and therefore stomatal closure. Conversely, low CO_2 concentration leads to higher pH, enhanced conversion of starch into sugar, and hence stomatal opening. Light would therefore cause opening due to photosynthetic reduction in CO_2 concentration within the leaf. CO_2 also alters tissue permeability (Glinka and Reinhold, 1972), but the significance of these effects has yet to be considered for stomatal physiology.

Although attractively simple in concept, this theory based on starch-sugar interconversion is nonetheless inadequate on the following counts:

1 The leaves of some plants (onion) have guard cells devoid of starch, and yet they work just as well as those of other plants.

2 The CO_2-concentration change to which stomata are sensitive, for example, 0.03 to 0.01 percent, is insufficient to induce a major alteration in guard-cell pH and certainly far less than the change from pH 5.0 to 7.0 observed during stomatal opening by Scarth and Shaw (1951).

3 The starch-sugar interconversion is too slow to account for observed responses (a vexing question and perhaps still not resolved).

4 Why should blue light be so much more effective than red light in inducing stomatal opening if simple photosynthetic sugar formation (starch-sugar interconversion) is involved?

In a closely reasoned paper, Levitt (1967) dealt with these and other objections, concluding that the case against the classical theory is by no means proved and proposing a new scheme analogous to the Scarth hypothesis except that carboxylic acid level rather than CO_2 concentration is regarded as the decisive factor. This modified theory accounts for virtually all the known stomatal responses but is deficient in that the biochemical nature of the starch-sugar conversion is not specified. The classical theory implies that phosphorylase catalyzes both synthesis and breakdown of starch. However, the celebrated work of Leloir's group in Argentina emphasizes that any scheme of reversible starch formation is inconsistent with the general situation for starch synthesis, which goes via sugar nucleotides (Chap. 14) (de Fekete and Cardini, 1964). Phosphorylase may of course be involved in starch breakdown, but this leaves us with only one-way stomatal control, obviously an unsatisfactory situation.

Instead of prolonging this debate over the classical hypothesis, some additional data on potassium-ion accumulation renew our interest in the third possibility for a stomatal mechanism, namely, active absorption of solutes followed by passive absorption of water. Potassium ions, plus a counterion, could be the solute in question, because we now have convincing evidence that potassium ions enter the guard cells in association with the light-stimulated opening of the stomatal pore. Part of the traditional wisdom of stomatal physiology was that guard cells respond selectively to different ion species and to concentration (Iljin, 1957), but there is now evidence for a key role for potassium ions. Fujino (1967) summarized his experiments (published only in Japanese since 1959), which subscribed to this view, but he held reservations whether enough potassium ion is absorbed during illumination to account quantitatively for the necessary turgor change. On the other side of the Pacific, Fischer and Hsiao (1968) estimated, from some completely independent research, that light-induced potassium-ion uptake is in fact sufficient to account for stomatal opening. They suggested that "the absorption of extracellular solutes such as K^+ may be the primary mechanism for stomatal opening. . . ." Subsequent work from Hsiao's laboratory at Davis, California, localized the guard cell as the specific site of potassium-ion accumulation. Humble and Hsiao (1970) treated epidermal strips of *Vicia faba* with sodium cobaltinitrite stain and neatly demonstrated a rise in potassium content in the guard cells of illuminated tissue. This localization of potassium ions has been confirmed with the electron-probe technique by Sawhney and Zelitch (1969) and by Humble and Raschke (1971) (Fig. 17-14).

Since selective absorption of potassium ions by the guard cells occurs only under illumination, the driving force is likely to come from light-generated energy and presumably ATP produced inside the guard cell. The studies of Humble and Hsiao (1970) suggest that photosystem I and cyclic electron flow could supply the necessary energy because diuron (DCMU) had no appreciable effect on opening in their experiments, whereas DCMU in combination with carbonylcyanide-*m*-chlorophenyl hydrazine (Cl-CCP), commonly regarded as an uncoupler of photophosphorylation, completely inhibited opening. ATP generated this way has been implicated in ion uptake by other leaf tissues under illumination (see Nobel, 1970), so that we can justifiably speculate that the epidermal strip from *Vicia faba* also works this way.

Given such a mechanism for K^+ uptake by illuminated guard cells, and assuming that it applies equally well to intact plants and to epidermal strips, we need to explain additional facets of the guard cell's physiology. Why should CO_2-free air lead to the same K^+ influx and opening reaction as illumination? What anion accompanies the K^+ absorbed into the guard cell, or at least how is the K^+ influx balanced electrochemically? There is no succinct answer, but some data from Nobel (1970) hold promise for an eventual explanation. Using pea leaves, he established that the light-dependent uptake of K^+ ions depends upon the pK of the accompanying organic acid; the lower the pK, the lesser the uptake. In his experiments (pH 6.3) $K^+HCO_3^-$ favored K^+ uptake over that from any other K^+ salt, such as acetate, formate, glycolate, or pyruvate. The contrast between HCO_3^- and salts of inorganic acids with low pK was especially striking; K^+ uptake from K^+Cl^- was only 7.5 μmol per gram of fresh weight per hour, whereas

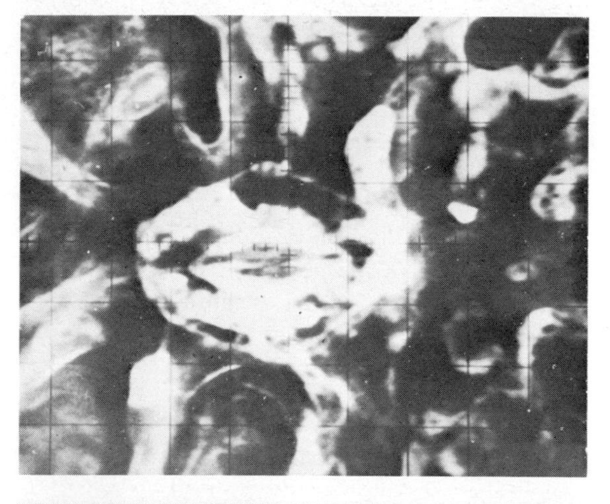

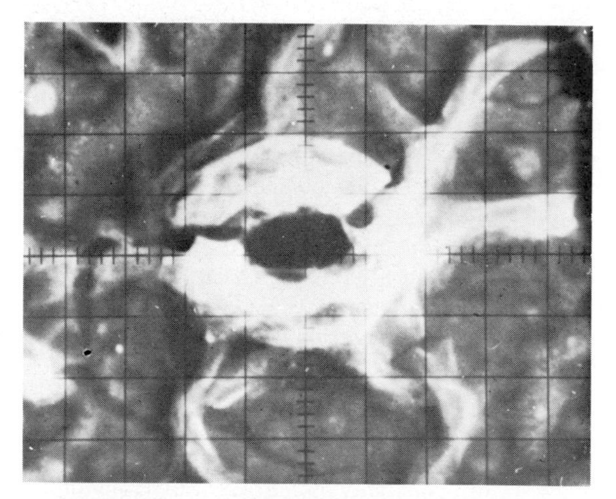

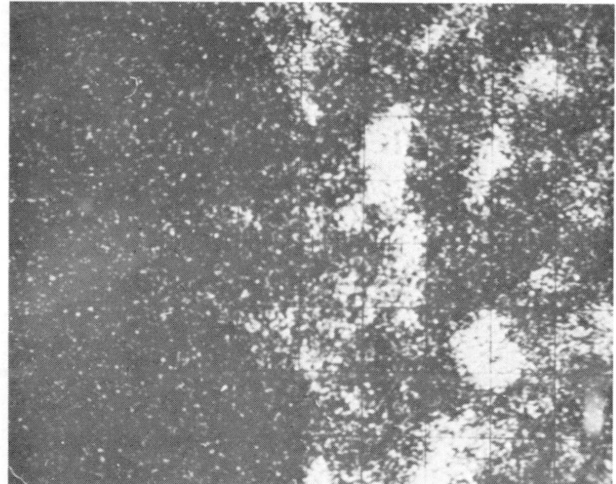

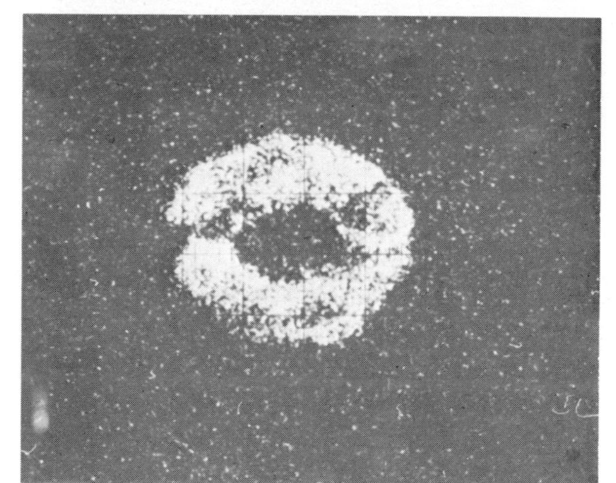

Fig. 17-14 The concentration of potassium into guard cells as stomata open is illustrated with the electron-probe microanalyzer. When a stoma of *Vicia faba* is closed (*upper left*), there is no apparent concentration of potassium (*lower left*), whereas the opened stoma (*upper right*) shows an impressive concentration of potassium in the guard cells (*lower right*) (Humble and Raschke, 1971).

K^+ uptake from $K^+HCO_3^-$ solution was increased to 56 μmol/g-h (fresh weight).

The presence of HCO_3^- in the guard cell or extracellular solution would presumably stimulate the uptake of K^+, compared with the effect of other anions, and this HCO_3^- might even be the hitherto unknown counterion for K^+ absorption. Moreover, different aspects of stomatal function call for different sources of osmotically active material. Rapid stomatal responses would require sudden changes in osmotic potential of guard cells as provided by potassium-ion flux, whereas sustained opening would then depend on a shift in starch-sugar interconversion initiated by the alteration in potassium-ion concentration.

Hormonal Implications

Stomatal behavior is closely coupled to leaf water potential (Fig. 17-10), but guard-cell operation is not purely hydraulic. The stomatal apparatus shows differential permeability and is under hormonal control, and evidence presented below implicates at least two broad categories of endogenous regulators: (1) abscisic acid and related inhibitors, which are associated with stomatal closure (Loveys and Kriedemann, 1974); and (2) cytokinins, which tend to encourage opening (reviewed by Kuiper, 1972). Stomata appear subject to control by a feedback mechanism involving the opposing action of these two hormones. Stomatal opening is encouraged by cytokinin in the range 10 to 0.05 μM (Kuiper, 1972), while abscisic acid causes closure within this same concentration range (Tucker and Mansfield, 1971, Kriedmann et al., 1972a) and offsets the cytokinin effect on opening (Cooper et al., 1972) Water balance in animals can be regulated by a hormone (vasopressin) which controls membrane permeability, so that stomatal regulation may bear some analogy to the animal system, ABA and cytokinins taking the place of vasopressin.

If ABA and cytokinins were instrumental in the regulation of stomatal function via changes in ionic flux, a tight coupling between permeability and hormonal status would be needed. This suggestion gains support from Glinka and Reinhold (1971, 1972), who demonstrated that ABA raises osmotic permeability (solutes) and diffusional permeability (H_2O) of carrot root cells, and from Mansfield and Jones (1971), who found a reduction in potassium-ion concentration guard cells of *Commelina communis* following ABA treatment.

Since moisture stress has such profound effects on endogenous levels of ABA and cytokinins, stomatal responses will depend upon changes in leaf water potential per se, plus attendant changes in hormone balance.

ECOPHYSIOLOGY

Water resources impose inescapable constraints within the biosphere, and in most places where plants manage to grow, their performance hinges on water supply. Ecological success generally depends on how efficiently plants can acquire and utilize their local supplies. Acquisition of water by the plant thus necessitates absorptive organs with maximum surface area and a water potential lower than surrounding media (soil or atmosphere). Under arid or saline conditions, this requirement imposes an unremitting selective pressure, and ecological races have evolved which differ widely in their capacity to generate or endure low water potential. By contrast, a rain forest or an aquatic community presents an entirely different picture. Nevertheless, terrestrial communities encounter water deficiency more commonly than sufficiency, and their existence is inevitably characterized by sustained, or at least transient, moisture tension.

Water-use efficiency in higher plants is inherently low (although C_4 plants are superior to C_3 in this respect), and since transpirational losses are an unavoidable corollary of photosynthesis, the leaves of most plants sometimes encounter a hostile environment where evaporative demand greatly exceeds supply. Crassulacean plants minimize this problem with their inverted stomatal rhythm, in which stomata are closed during the day and opened at night (Fig. 17-15). In other higher plants, their first line of defense for maintaining water status under drought conditions is stomatal closure. Even with stomata firmly closed, transpiration does not cease completely because epidermal surfaces are not absolutely impervious to water despite cuticular layers, wax platelets, and a variety of cellular outgrowths which all contribute a higher diffusive resistance.

Unless plant-moisture status is restored, the stress in a droughted plant will intensify and the organism must be equipped with some mechanism for preservation as a functional unit capable of regeneration (as in perennials) or else escape via immediate reproduction. Seeds, spores, and other propagules develop as a form of latent life which endures inordinately greater levels of desiccation than the parent organism. Plants achieve drought resistance by either system, and their ecological significance can be seen both in short-term reactions to moisture stress and in the evolution of races with adaptive mecha-

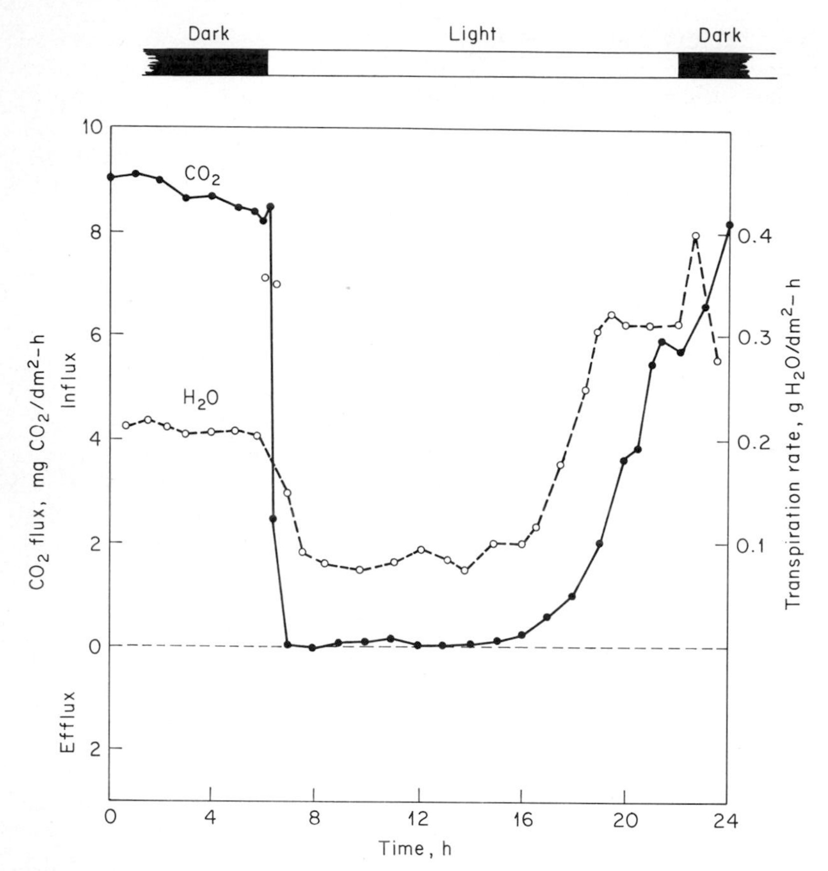

Fig. 17-15 Fluxes of water vapor (◦) and carbon (•) over a 24-h cycle of light and dark reveal an inverted stomatal rhythm in *Agave americana*. Nocturnal stomatal opening minimizes evaporative losses and confers an ecological advantage under dry-land conditions (Neales et al., 1968).

nisms which enable them to avoid or endure low water potential.

Moisture Stress

A plant's reaction to moisture stress embodies both physical and physiological components. The physical side is illustrated schematically in Fig. 17-16. [Actual data of Gardner and Nieman (1964) on pepper plants show close agreement.] Soil is initially moist ($\psi_{soil} = 0$) and as the soil mass dries out, there is a progressive decline in its water potential. Associated curves show diurnal changes in water potential at root surfaces within the soil ψ_{root} and the moisture status of foliage ψ_{leaf}.

By day 4, ψ_{leaf} drops below the wilting line for a substantial period and simulates the temporary wilt that is commonly observed on broad-leafed plants around midday (pumpkins and some other cucurbits growing outdoors wilt dramatically). By day 5 (Fig. 17-16) ψ_{soil} has also fallen below the line taken as permanent wilting point, so that plant turgor cannot possibly be restored that evening, as occurred previously, and the plant will now remain wilted until soil moisture is recharged.

A droughted plant's capacity to absorb soil

moisture is naturally dictated by the tensions it can endure within its own tissues, and these may be well in excess of 15 bars (Fig. 17-17), but water absorption is further impaired by retarded root growth (Fig. 17-18). This reaction sets in at surprisingly high potential ($\psi_{soil} = -1$ to -2 bars) and means that the plant's root system no longer explores a given volume of soil with the same effectiveness. Tissue shrinkage, known to accompany the cessation in cotton-root extension (Huck et al., 1970) accentuates the plant's difficulty in meeting transpirational demands because vapor gaps develop between root surfaces and soil particles.

If the plant's demands for water are not met, turgor is lost and leaves (especially old ones) adopt a wilted appearance. ψ_{leaf} at this stage is commonly about -15 bars in herbaceous crop plants and compares with ψ_{soil} at permanent wilting point in Fig. 17-16, but species differ enormously in their physical expression of moisture stress. Tomato plants, for example, first show permanent wilting at -19 bars, whereas values for cotton and privet are -38 and -48 bars, respectively (Slatyer, 1957). Other woody

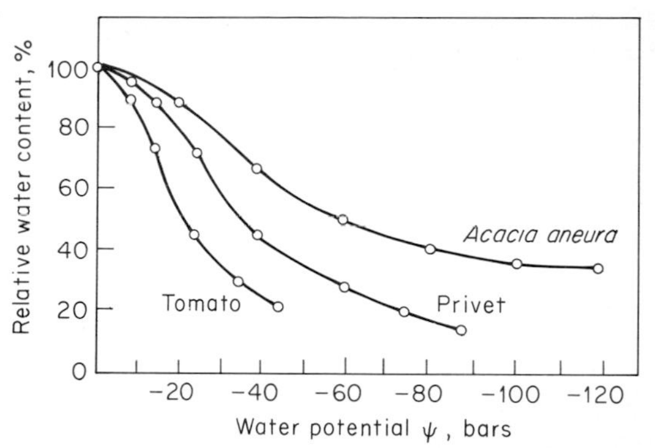

Fig. 17-17 A mesophytic species such as tomato shows a relatively small change in leaf water potential for a given loss in water content, compared with the xerophytic acacia. The capacity to generate low potential represents a distinct advantage in the arid zone (Slatyer, 1960*b*).

Fig. 17-16 A schematic representation of changes in leaf water potential ψ_{leaf} root-surface water potential ψ_{root}, and soil-mass water potential ψ_{soil}. A hypothetical plant starts off in wet soil ($\psi_{root} \approx 0$), and the moisture status of soil and plant declines as transpiration proceeds. The horizontal dashed line indicates the value for ψ_{leaf} where wilting occurs (Slatyer, 1967).

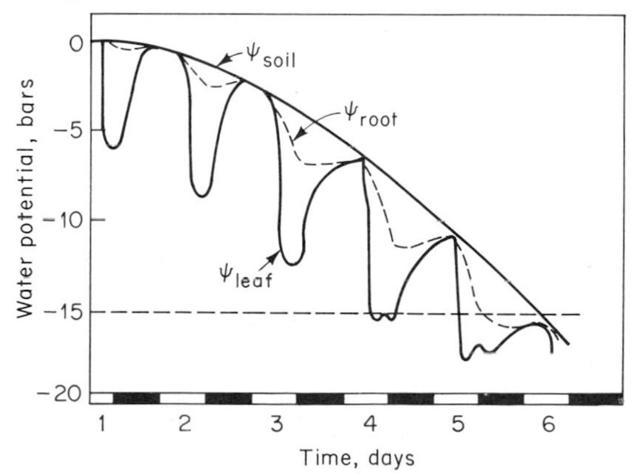

perennials are even more impressive with a desert shrub (*Acacia aneura*) developing a stress of 120 bars ($\psi_{leaf} = -120$ bars). The ability of these phyllodes to endure such a low water potential (Fig. 17-17) has ecological benefits and relates to a structural resilience which offsets their physical collapse even at exceedingly low water potential.

Before any visual expression of moisture stress (wilting), physiological reactions to moisture stress have already occurred. Stomatal aperture, which is so closely coupled to ψ_{leaf} via guard-cell turgor (Fig. 17-10), will have already diminished (minimum reached between -10 and -15 bars), so that transpiration will fall accordingly. Other turgor-dependent processes, e.g., leaf enlargement, are even more sensitive (Boyer, 1970; Hsiao et al., 1970) and come to a complete halt at potentials around -4 bars (see also Figs. 3-5 and 3-6. This hydraulic coupling between guard-cell turgor, transpiration, and ψ_{leaf} is so tight that concurrent cyclic oscillations in all three parameters (Lang et al., 1969) can be sustained for several hours under constant environmental conditions (Barrs, 1971). These oscillations are initiated when water loss at peak transpiration is just in excess of the leaf's capacity to meet evaporative demand; stomata

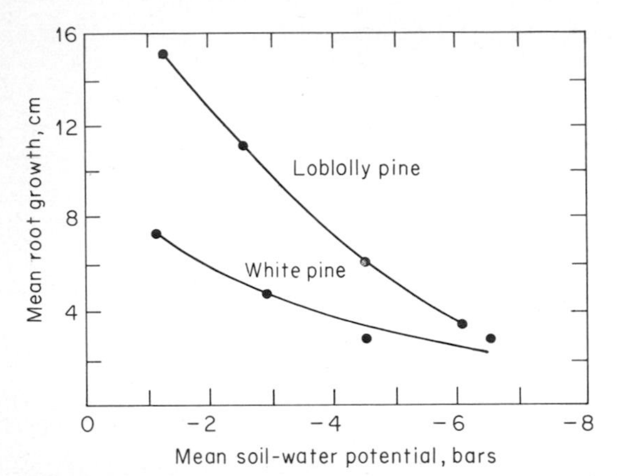

Fig. 17-18 Root growth of pine seedlings is adversely affected by low water potential in the rooting medium. Roots were measured as they grew out of a soil mass into humid air; treatments lasted 20 days (Kaufmann, 1968).

close in response to this mild stress, ψ_{leaf} is then restored, stomata reopen, and as transpiration picks up, the closing cycle is instigated once more (Kriedemann, 1968; 1971). Changes in ψ_{leaf} during the course of these cyclic oscillations are probably sufficient to displace the relative levels of abscisic acid and cytokinins within foliage, and in view of their regulating influence on stomatal behavior, they may also contribute to stomatal oscillations.

How far ψ_{leaf} must fall before stomatal closure is triggered varies with species (Kriedemann and Canterford, 1971) but is also time-dependent, as shown in Fig. 17-19. The upper curve represents behavior of leaves on long shoots excised in the vineyard. Changes in ψ_{leaf} following excision were followed on successive leaves along the shoot with a pressure chamber (Scholander et al., 1965) while photosynthesis was monitored continuously on adjacent foliage. At the first suggestion of photosynthetic decline, this same leaf was immediately removed for ψ_{leaf} determination. Pooled data (upper curve in Fig. 17-19) reveal a sharply defined cutoff. Photosynthesis was virtually unaffected until ψ_{leaf} fell to between -13 and -15 bars. When similar vines were grown in containers and allowed to drought gradually, a totally different relationship was obtained (lower plot, Fig. 17-19). Photosynthesis declined linearly with diminishing ψ_{leaf} and reached a minimum at wilting point. This time factor in the equilibration between ψ_{leaf} and photosynthesis implies the existence of physiological components which override purely physical factors in a leaf's photosynthetic reaction to moisture stress.

A further suggestion of metabolic involvement in plant-water relations is shown during recovery from moisture stress. When vines referred to in Fig. 17-19 (lower curve) were rewatered, photosynthetic activity did *not* recover despite restoration of favorable water status ($\psi_{\text{leaf}} \approx -2$ bars) that same day. Such *aftereffects* of moisture stress are widely recognized, and the related stomatal behavior of tobacco and broad bean has been considered in detail by Fischer et al. (1970). They reported a depression in stomatal response to light following moisture stress; the aftereffect was proportional to leaf water deficit immediately before rewatering. The effect is also shown in Ashton's (1956) data for sugarcane, where several days elapsed before photosynthesis showed full recovery from a brief wilting experience. Schneider and Childers (1941) had previously observed that photosynthetic recovery sometimes requires a week. The inhibition of leaf expansion during moisture stress (referred to above) also persists well after ψ_{leaf} is restored (Gates, 1955a,b) and demonstrates how the degree of inhibition, which might persist for several days, also has implications for ultimate leaf size and hence plant growth (see also Chap. 3).

Virtually every facet of the plant's physiology is dislocated to some degree during moisture stress and subsequent recovery. On the metabolic side, Magness et al. (1933b) observed that wilting in mature leaves is associated with carbohydrate depletion due to mobilization and export, while leaf senescence follows even though water supply is later restored. According to Gates and Bonner (1959), growth inhibition after wilting is attributable to enhanced RNA destruction; the protein decline observed by Mothes (1928) on older leaves senescing because of stress might well have been related to this disruption of RNA metabolism. (Tissue hardening in response to moisture stress also involves nucleic acid metabolism, discussed later.)

Biophysical responses to moisture stress stem from changes in root permeability, which then affect restitution of ψ_{leaf} after rewatering (Boyer, 1971*b*), and from a loss in hydraulic continuity between soil and leaves. Stomatal function becomes impaired, and photosynthesis declines due to changes both in gaseous diffusive resistances (Troughton and Slatyer, 1969; Redshaw and Meidner, 1972) and in metabolic systems concerned with CO_2 fixation (Boyer, 1971*a*; Stewart and Lee, 1972). Photochemical efficiency is also affected (Nir and Poljakoff-Mayber, 1967), while increases in respiration and photorespiration contribute to the diminution in net photosynthesis (Brix, 1962; Heath and Meidner, 1961).

Water status and hormone physiology are inextricably related, and alterations in ψ_{leaf} evoke such rapid changes in endogenous regulators that it is surprisingly difficult to distinguish between direct physical effects of changes in ψ_{leaf} per se and those mediated via a shift in hormone levels. Stomatal reactions are a case in point. Excised leaves typically show brief hydropassive opening followed by complete stomatal closure within a few minutes (Fig. 17-11). ψ_{leaf} commonly falls to between −10 and −15 bars and agrees well with field observations, where stressed plants cease photosynthesis at about this same value for ψ_{leaf} (Fig. 17-19). It is not unreasonable to conclude that stomatal closure is a direct consequence of reduced ψ_{leaf} and to exclude metabolic involvement. However, endogenous levels of ABA can double within a few minutes of leaf excision (Loveys and Kriedemann, 1973), and even turgid leaves undergo stomatal closure following changes in ABA concentration of this order (Kriedemann et al., 1972a). Hence the question: To what extent are stomatal reactions to ψ_{leaf} purely physical, and to what extent physiological?

Hormonal disturbances clearly make contributions to the aftereffect of moisture stress whereby stomatal function and other turgor-dependent processes such as leaf expansion are not restored for several days after rewatering. This pattern of decline and subsequent recovery of gas exchange, plus associated changes in ψ_{leaf} and ABA-like inhibitors, is shown in Fig. 17-20. Irrigation was withheld from this group of potted grapevines for 6 days before rewatering.

By day 6 (Fig. 17-20) stomata were firmly closed, and leaf resistance (r_l, Chap. 1) varied from 50 to 100 s/cm (cf. Fig. 1-23). Leaf water potential had fallen to −13 bars by day 6 (laminae were flaccid), and the concentration of endogenous ABA-like inhibitors had increased by a factor of 44. Since endogenous levels of ABA need only double to initiate stomatal closure, the metabolic effects of a fourty-four-fold increase on stomatal physiology must have been profound. Equally remarkable is the precipitous drop in endogenous inhibitors when vines were rewatered, and although ψ_{leaf} was restored that same day, photosynthetic recovery took much longer. The aftereffect of moisture stress is therefore not directly attributable to current levels of inhibitors during recovery, but the massive accumulation by day 6 is likely to have impeded the synthesis of photosynthetic enzymes in common with other proteins formed *de novo,* and the slow turnover of RuDP carboxylase (Peterson et al., 1973) would then protract photosynthetic recovery. ABA inhibition of total RNA synthesis and of RNA polymerase activity in corn coleop-

Fig. 17-19 Vine-leaf photosynthesis shows a rapid decline once ψ_{leaf} falls to between −13 and −15 bars on excised shoots (*upper curve*). If moisture stress is imposed more gradually (over a period of days rather than minutes), photosynthetic activity becomes adjusted to ψ_{leaf} and a linear relationship is developed (*lower curve*) (Kriedemann and Smart, 1971).

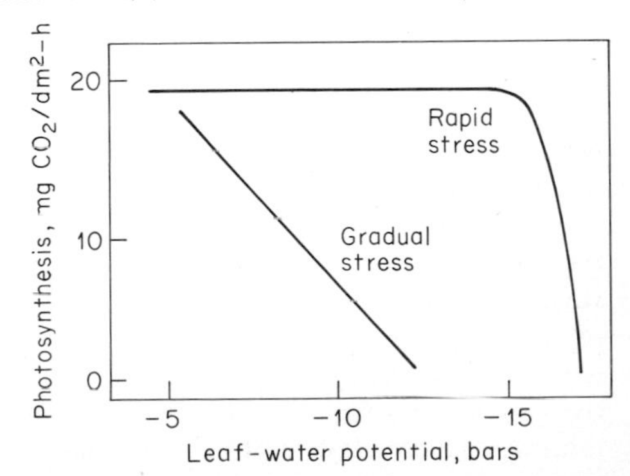

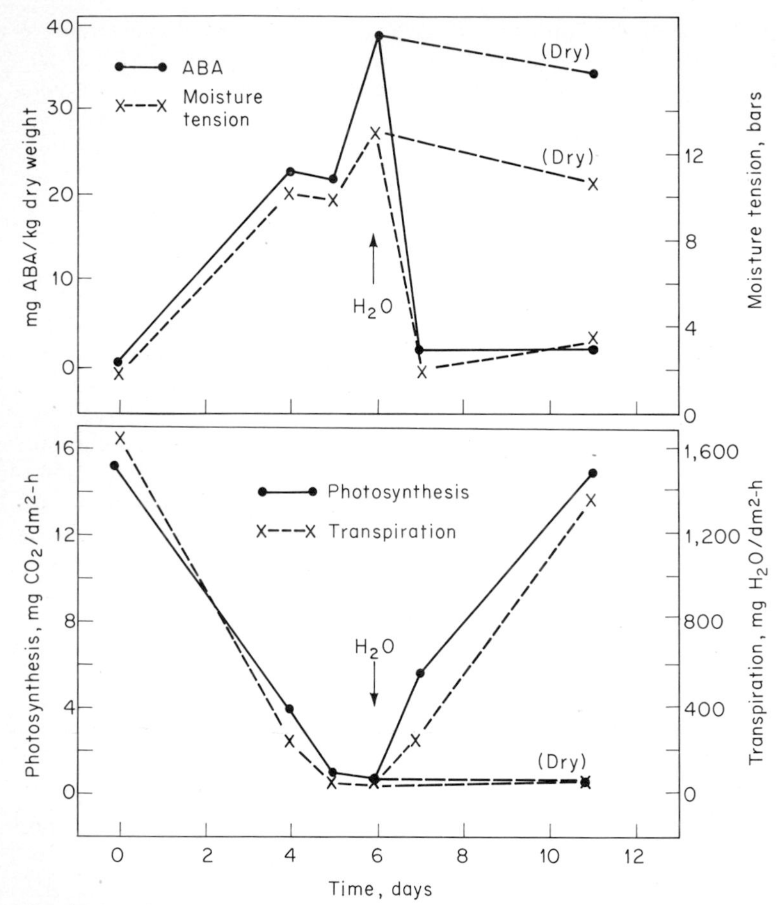

Fig. 17-20 Grapevines deprived of irrigation show a progressive increase in moisture tension with an associated increase in ABA-like inhibitions. Peak values ($\psi_{leaf} \approx -13$ bars; ABA increased 44-fold) coincide with stomatal closure (*lower curves*). Once vines are irrigated, leaf-water potential is restored almost immediately and ABA levels fall precipitously, but stomatal function remains impaired for some days (Loveys and Kriedemann, 1973).

tiles reported by Bex (1972*a,b*) confirms this view on the mechanism of aftereffects and helps explain previous observations (Gates and Bonner, 1959) on moisture stress interfering with nucleic acid metabolism.

Physiological consequences of moisture stress frequently relate to abscisic acid (Mizrahi and Richmond, 1972) but are not solely attributable to this source. Moreover, hormonal participation in stomatal physiology is not necessarily limited to environmental stress. A variety of manipulative treatments, including shoot-tip removal, stem cincturing, fruit excision, and even changes in day length, can evoke alterations in stomatal behaviour that show close correlation with changes in endogenous levels of ABA and related inhibitors (Loveys and Kriedemann, 1974; Loveys et al., 1974).

Cytokinins were shown earlier to influence stomatal behavior, and because of their close involvement in protein synthesis and other aspects of metabolic regulation (Chap. 6), cytokinins should be considered within the present context of plant-water relations. Stomata on tobacco plants treated with kinetin appear less responsive to moisture stress, and this pretreatment accentuates plant-water deficits in saline media (Mizrahi et al., 1971). ABA pretreatment has the opposite effect on stomatal sensitivity to ψ_{leaf} and alleviates moisture stress under saline conditions; i.e., stomata close more readily, and the tobacco plant's moisture status is preserved. Other work at the Negev Institute for Arid Zone Research in Israel has also established that roots of droughted plants supply less cytokinin to aerial organs and that leaves on such plants also generate more ABA in situ. Both factors would accentuate stomatal sensitivity (Ben-Zioni et al., 1967; Itai and Vaadia, 1971) and would contribute to the maintenance of leaf turgor.

These two categories of growth regulators clearly interact in their influence on plant-water relations. The stomatal-hormonal interrelationships in a wilty mutant of tomato (*flacca*) provide a better idea of how control systems can operate. Tal et al. (1970) have a wilty mutant of the normal tomato variety Rheinlands Ruhm which suffers from sluggish stomatal closure under moisture stress and consequently wilts in situations where any normal tomato plant would be able to shut its stomata and remain turgid. If the mutant is sprayed with ABA, it reverts to normal behavior, showing less tendency to wilt (Imber and Tal, 1970; Tal and Imber, 1972). In the untreated condition, *flacca* has a lower concentration of endogenous ABA-like inhibitors —hence its insensitivity to moisture stress (Tal and Imber, 1970). But an additional control enters the picture because reciprocal grafts between normal and mutant tomatoes revealed a root factor from *flacca* that *promoted* stomatal opening in normal plants. This information, combined with bioassays (soybean callus) of leaf extracts from *flacca* showing higher cytokinin activity, emphasizes the dual nature of hormonal controls over this stomatal response (Tal et al., 1970). Hormonal interaction was confirmed in experiments where continuous ABA application to *flacca* produced normal stomatal behavior in association with lowered levels of endogenous cytokinins. Tal and Imber (1971) then went on to examine hormonal effects on root resistance because plant-moisture status obviously depends upon the combined effects of root permeability and stomatal diffusive resistance. Their attention had been directed to root permeability by observations that decapitated *flacca* plants produce a lower volume but higher concentration of root exudate than normal tomatoes. Data are summarized in Table 17-3. Kinetin and ABA have opposing effects on volume and concentration of exudate. The low volume of exudate characteristic of *flacca* was even lower during kinetin treatment; i.e., root resistance was increased in both normal tomatoes and the wilty mutant.

The overall consequences of these hormonal interactions can be summarized as follows. Cytokinins decrease stomatal resistance but increase root resistance to water absorption so that a plant becomes prone to moisture stress. (The normal decline in cytokinins in droughted plants might even represent a form of self-protection.) Abscisic acid counteracts this cytokinin effect by increasing both root permeability and stomatal resistance; hence plant turgor is favored. (The massive rise in endogenous ABA during moisture stress forms a natural corollary to the cytokinin response mentioned above.)

These two forms of hormonal regulation (root permeability and stomatal physiology) therefore equip the plant with additional means of buffering its own system against fluctuations in evaporative demand and of adaptation to moisture stress.

Adaptation to Stress

Plants which successfully occupy arid regions are termed *drought-resistant,* a form of adaptation that can stem either from a capacity to maximize water uptake but minimize loss so that tissue hydration is maintained, or from an ability to survive tissue desiccation. The ability of a plant to withstand low water potential within its tissues is termed *hardiness* and can be regarded as one specific part of the more general question of drought resistance. Thus, drought resistance (or *drought tolerance*) indicates some ability to

Table 17-3 Root resistance dictates the volume and concentration of root exudate. A wilty mutant (*flacca*) of the normal tomato Rheinlands Ruhm has high root resistance and is characterized by lower endogenous ABA. Exogenous ABA offsets this effect, whereas kinetin accentuates it in both varieties (Tal and Imber, 1971).

	flacca		Rheinlands Ruhm	
Hormone treatment	Exudate volume, ml per 100 mg root dry wt per day	Exudate osmotic pressure, atm	Exudate volume, ml per 100 mg root dry wt per day	Exudate osmotic pressure, atm
0	2.6	0.89	5.0	0.73
Kinetin	1.6	0.96	1.8	1.14
ABA	6.5	0.62	10.2	0.74

grow in a dry environment, whereas drought hardiness is the ability to endure tissue desiccation.

Structural changes during tissue dehydration in naturally adapted organs are most conveniently observed in seeds. Even in moist fleshy fruits, e.g., tomato or watermelon, mature seeds have a remarkably low water content, despite their high osmotic potential. This automatically implies some physiological mechanism which actively excludes moisture from seeds during their development but allows reimbibition after replanting. McIlrath et al. (1963) followed this loss of water from tomato seeds during fruit ripening and found that the seeds actively lose about 50 percent of their free water during this period. Even seeds placed in distilled water became dehydrated to a similar extent (Abrol and McIlrath, 1966).

Water exclusion during seed development is again an apparent gradient in water potential within the ripening fruit and must rely on metabolic processes, whereas water loss from the droughted plant is entirely physical, and tissue desiccation occurs when systems regulating the plant's water economy are inadequate to cope with the demand-supply situation. As water content declines, water potential naturally falls, but the organism stands a much better chance of retrieving a favorable water balance by generating a disproportionately lower potential for a given drop in moisture content. Desiccation-resistant plants show this characteristic, and data in Fig. 17-17 illustrate the contrast with tender herbaceous species. *Acacia aneura,* which can be regarded as a xerophyte (Slatyer, 1960*b*), not only withstands extraordinarily low ψ_{leaf}, but its own moisture loss is also more effective in generating a reduction in ψ_{leaf} and so obtaining moisture from its environment. This combination endows the acacia with a strong competitive advantage over other shrubs, e.g., privet, and especially over herbaceous plants, e.g., tomato, in a dry situation.

Some South African plants adjacent to the Namib Desert (precipitation nil) are even more remarkable with respect to desiccation resistance. Gaff (1971) describes a number of angiosperms whose leaves survived equilibration over concentrated sulfuric acid. Tissue equilibration was extremely slow, and minimum moisture content was not achieved until 4 to 8 weeks had elapsed (water must have been tightly bound, and diffusive resistance would have been high). Rehydration was much faster (completed after 24 h immersion under water), and physiological activity was restored within a single day.

Whether seeds or photosynthetic organs possess this hardening mechanism, some underlying principles are common to both. Several studies reviewed by Kuiper (1972) show interrelationships between environmental adaptation and membrane characteristics. In wheat plants, for example, adaptation to drought leads to greater membrane flexibility and is associated with a decreased mobility of cellular water because water molecules become more firmly bound within the cytoplasm. Changes in protein hydration as tissue water potential falls (Table 17-4) emphasize the tenacity with which cytoplasmic water is retained. These changes in tissue hydration lead to conformational alterations in membrane components and hence permeability. One additional metabolic shift with consequences at a cellular level is in protein synthesis. Henckel (1970) has shown that ^{15}N becomes incorporated into protein much faster in drought-resistant than susceptible wheat plants and probably helps offset the disabling effects of moisture stress on protein synthesis in drought-susceptible varieties of the same species.

A plant's potential for developing drought hardiness also finds expression in its capacity to modify nucleic acid composition (Kessler, 1961), in particular the RNA base ratio; protein synthesis would be modified accordingly. Kessler and Frank-Tishel (1962) demonstrated such an effect in a hardy species of the Oleaceae, which was interpreted as a protective reaction in drought resistance. In their hardy olive, GC/AU ratios of RNA formed during hardening increased from 1.07 to 1.38 in leaves suffering a 40 percent water deficit. In a nonhardy species (*Ligustrum*) the base ratio remained unchanged during a gradual onset of moisture stress, and these plants showed no physiological adjustment indicative of hardening.

Regardless of these metabolic changes, the cytoplasm of hardened plants must also undergo concurrent changes in physical properties in order to cope with mechanical stresses that accompany desiccation and remoistening (Iljin, 1957). As plant tissues dry out, cells collapse and the protoplasm is pulled inward by a shrinking vacuole and outward by the resisting cell wall to which it adheres. If walls are rigid, strong tensions develop, which lead to destruction. Cells that do survive are subject to new mechanical stresses upon remoistening because the vacuole swells more quickly than the cytoplasm. Iljin's (1957) view is supported by observations that cell diameter may be reduced to 10 or 20 percent of normal values during desiccation, that torn protoplasm can be seen adhering to cell walls in stressed tissue, and that pretreatment with plasmolyzing solutions releases contact between cell wall and protoplasm and improves drought hardiness of cells. Similarly, small cell size, small vacuoles, and a high concentration of cell sap all increase hardiness.

Within the cytoplasm itself Levitt (1962) proposes that "injury is due to an unfolding and therefore a denaturation of the protoplasmic proteins. This results from the formation of intermolecular —SS— bonds induced by the close approach of the protein molecules due to . . . dehydration. . . ." Levitt (1962) originally offered this explanation in connection with frost resistance, but it is considered equally applicable as a mechanism behind drought resistance (Gaff, 1966).

Plants can also achieve drought resistance without becoming hardened by simply avoiding desiccation. They need to combine mechanisms for water conservation with systems for water acquisition. Where water supply is limited by solute concentration in the root medium, halophytes have emerged with the capacity to draw water from a root zone that proves too "dry" for most mesophytes. Seawater (potential = −20 bars) is in this category, but some mangroves,

Table 17-4 Desiccation lowers the hydration of structural protein. The mean death point of about −90 bars corresponds to approximately 2.3 water molecules per amino acid residue, although minimal hydration was reached at higher potential (compare Fig. 17-1). Present data were obtained by equilibrating acetone-precipitated structural protein to various water potentials by vapor exchange for 1 week (Gaff, 1966).

Water potential, bars	Number of H_2O molecules per nitrogen atom in protein
0	6.3
− 9	5.5
−18	2.7
−37	1.9
−46	1.9
−65	1.7
−86	1.7†

†Equivalent to approximately 2.3 per amino acid residue.

which inhabit estuaries, have developed salt glands on their leaves to establish a ψ gradient which favors uptake. This process relies primarily on the active secretion of salts (Scholander et al., 1962) whose composition varies according to solutes in the root zone (Thomson et al., 1969). The physiology of salt glands is especially intriguing because they endow halophytes with their capacity to develop exceedingly low ψ_{leaf} in localized regions which is then transmitted to their roots without desiccating adjacent tissue. Salt glands on mangrove leaves become covered by sodium chloride crystals on sunny days, but their excretion continues despite direct contact with this concentrated brine. These glands are all the more remarkable because their secretion is not driven by evaporation. Solutes continue to exude, even under a film of oil, provided leaf hydration is maintained. Their salt-exuding mechanism has become the subject of some detailed electron microscopy (Levering and Thomson, 1971; Ziegler and Lüttge, 1966) and microautoradiography (Ziegler and Lüttge, 1967).

While halophytes cope with saline media by active excretion of undesirable solutes, small amounts of sodium chloride actually encourage the growth of an arid-zone shrub (*Atriplex halimus*) under low humidity (Gale et al., 1970). Solute accumulation in small leaf bladders offsets moisture stress when evaporative demand becomes excessive (low humidity) but confers no benefit at high humidity; in fact growth is inhibited.

Salts secreted by leaf glands or accumulated in bladders contribute toward ψ_{leaf} by favoring water uptake from the root zone, but localized reduction in ψ_{leaf} due to solute accumulation at evaporative sites in leaves also lowers the vapor-pressure gradient from leaf to air and so reduces transpiration. Whiteman and Koller (1964) refer to this beneficial effect in the halophyte *Reaumuria hirtella*, which grows in the Negev Desert, Israel. In common with other halophytes, this plant tolerates ψ_{leaf} down to -300 bars!

Few mesophytes could withstand such low potential, and even if they could, any benefit would be limited to transpiration control. There would be little prospect of their recharging leaf moisture at night from atmospheric sources because stomata would be closed. Crassulacean plants, on the other hand, possess an inverted stomatal rhythm plus dark CO_2 fixation (Fig. 17-15), so that they enjoy the related benefits of minimal transpiration and the possibility of rehydration at night. Plants with crassulacean acid metabolism are superbly adjusted to desert conditions; this form of adaptation largely explains their success in the arid zone (Loomis et al., 1971). During the course of their evolution, these plants developed the capacity for nocturnal CO_2 assimilation (Fig. 17-15), instead of synchronizing their assimilation with light-driven photosynthetic reactions which would necessitate opening during the day (Neales et al., 1968). Despite their inverted rhythm, the leaves of *Agave americana* referred to in Fig. 17-15 show normal stomatal behavior with respect to CO_2 concentration (i.e., a drop in substomatal CO_2 concentration triggers opening; Neales, 1970). In ecological terms, their nocturnal opening coincides with minimum leaf-air vapor-pressure gradients so that transpiration is reduced and water-use efficiency is at a maximum.

Stomatal resistance complements lowered vapor-pressure gradients in controlling transpiration, and xerophytes also utilize this principle in minimizing evaporative losses. Whereas mesophytes have stomata level with their epidermal surface, species adapted to arid situations have sunken stomata, which, combined with waxy coatings and abundant epidermal hairs, contribute to high leaf resistance (see Fig. 1-19). Rapid stomatal closure at the first sign of impending moisture stress would offer additional economy, and sorghum, widely regarded as more drought-resistant than corn, incorporates this effect (Sanchez-Diaz and Kramer, 1971). Consequently, corn loses more water before wilting than sorghum and its recovery is slower; sorghum plants dampen fluctuations in internal moisture status, so that growth and productivity benefit accordingly.

Water economy in perennial shrubs or succulents of the arid zone is also favored by their severe reduction or virtual elimination of leaf surfaces (Grieve and Hellmuth, 1968). Spherical or cylindrical morphology reinforce this economy by minimizing surface-volume ratio; the reduction in size of assimilatory organs makes a greater measure of convective cooling possible, and their heat budget no longer relies on transpirational cooling to the degree required by broad horizontal leaves (Chap. 16). Succulents make a further contribution to their own moisture status by developing water-storage tissues within their own "leaves" or stems.

Desert ephemerals represent one category of arid-zone plant which relies on an abbreviated life cycle for continued existence within the ecosystem. In contrast to perennial plants of the arid zone, these ephemeral herbs show neither drought resistance nor hardening mechanisms during their vegetative phase; they escape moisture stress by compressing their developmental cycle. Ecological success depends on synchronizing germination with conditions that offer the best prospect for completion of the life cycle, and soil water is the operative factor. Germination is delayed by chemical inhibitors in their seed coats (Chap. 10) which require substantial rainfall for their elution. Brief showers or intermittent rain which is insufficient to recharge soil water and provide adequate reserves for plant growth is also insufficient to elute the inhibitor.

Both regenerative and reproductive phases of plant development are influenced by water supply. Salter and Goode (1967) cite numerous instances where reproductive development in both herbaceous and perennial plants is triggered by moisture stress. Hormonal changes (ABA and cytokinins) known to accompany even brief reductions in water potential are possibly involved, but this mechanism, which is of significance for both ephemerals and perennials, remains obscure.

SUMMARY

Higher plants have emerged on the evolutionary time scale with a remarkable variety of mechanisms for adapting growth and development to their resources of water. In aquatic situations, special mechanisms for root aeration are combined with leaf characteristics (no cuticle) that favour substrate exchange with their aqueous environment; in salt-marsh plants (e.g., mangroves), the aeration problem becomes compounded by a need to eliminate solutes that accompany the transpiration stream, and aerial organs are suitably equipped (salt glands). Mesophytes, on the other hand, usually enjoy a greater abundance of available moisture in their root zone, but water-use efficiency can then be a major factor determining ecological success. Here we see the refinement of stomatal mechanisms to embody both hydraulic and hormonal control systems; this emphasizes the degree to which such higher plants are attuned to their aqueous resources.

While stomatal control over transpiration represents a dynamic form of adaptation in mesophytes, more static forms exist in xerophytes whose gross morphology and fine structural details promote resilience. Despite an aerial environment which can be acutely unfavorable, arid-zone plants have evolved that are particularly successful at reducing their evaporative losses by minimizing heat load but maximizing diffusive resistance. Coupled with these morphological adaptations is the emergence of biochemical patterns in crassulacean plants, where nocturnal assimilation confers a decided ecological advantage in the arid zone. If these adaptive features fail to meet environmental demands and vegetative organs cannot harden enough to endure desiccation, reproduction is instigated and provides the species with yet another line of defense.

GENERAL REFERENCES

Dainty, J. 1969. The water relations of plants. In M. B. Wilkins (*ed.*). "The Physiology of Plant Growth and Development." McGraw-Hill Book Co., Maidenhead, England. Pp.421–454.

Hsiao, T. C. 1973. Plant responses to water stress. *Annu. Rev. Plant Physiol.*, 24:519-570

Kozlowski, T. T. (*ed.*). 1968-1972. "Water Deficits and Plant Growth." Vols. 1, 2, and 3. Academic Press, New York.

Kramer, P. J. 1969. "Plant and Soil Water Relationships: A Modern Synthesis," McGraw-Hill, New York. 482 pp.

Meidner, H. and T. A. Mansfield, 1968. "Physiology of Stomata." McGraw-Hill Book Co., Maidenhead, England. 179 pp.

Slatyer, R. O. 1967. "Plant-Water Relationships." Academic Press, London. 366 pp.

PART

5 CHEMICAL MODIFICATION OF PLANTS

CHEMICAL REGULATION OF GROWTH AND DEVELOPMENT

Since we know that the natural control of growth and development of plants is achieved principally through the endogenous chemical messengers, or hormones, it would be reasonable to expect application of the same chemical species to permit an orderly modification of the plant. While it is often possible to alter the plant with such applications, they are surprisingly ineffective in several instances. When auxin was the only known plant hormone, many botanists expressed discontent or uneasiness with the hormone concept since applications of indoleacetic acid had little apparent effect on plants beyond causing abnormal growth at very high concentrations. It was only after synthetic analogs of auxin were discovered that chemical modification of plants began to emerge as a technology. The chemical structures of synthetics which show activity in altering the hormone systems are sometimes astonishingly different from the natural hormone, reminding one of the striking contrast between the synthetic diethylstilbesterol and the natural animal hormone estrone.

Armed with the knowledge that there are at least five hormone systems in plants, that each has fairly definable regulatory effects, and that synthetic analogs of each of them are available, one might hope that intelligent selection of the synthetics would permit fairly precise chemical regulation of growth and development; but even with these prerequisites it has often proved difficult or impossible to achieve chemical regulation by the logical selection of synthetic chemicals. Presumably this is due to the complex interactions that occur between plant-hormone systems (as roughly illustrated in Fig. 9-30); elevated concentrations of one may bring about sharp adjustments in the synthesis or destruction of other hormones. But also, the difficulty of achieving regulation may be related to the extent to which multiple hormones are involved in endogenous controls over growth and development processes (cf. Table 18-1).

While progress in the technology of chemical regulation of plant performance has been slow, hindered by the delicacy of hormone interactions, the buildup of pressure from the agricultural industry for development of chemical modification has been enormous, due to the greater need for agricultural products and the growing necessity of producing them with less and less manpower. Since about 1965, there has been an enormous expansion of the chemical devices available for agricultural production, and with the intensifying pressures for increased production and avoidance of hand labor, this expansion will surely continue.

18 CHEMICAL REGULATION OF GROWTH AND DEVELOPMENT

While knowledge of endogenous regulatory systems in plants was rapidly expanding in the 1940s and 1950s, the applications of chemicals to plants were predominantly as selective herbicides. Beginning in the 1960s, however, an increasingly effective technology of chemical regulation of plants evolved. The rapid development of this technology has been made possible partly because of the greater understanding of endogenous regulatory systems, partly because of the growing pressures on agriculture to provide alternatives to the use of increasingly expensive labor, and partly because of the existence of extensive chemical screening systems developed by agricultural chemical firms as an outgrowth of the herbicide industry.

A surprising aspect of the chemical regulation of plants has been the emerging realization that instead of separate hormonal systems regulating separate developmental processes, each of the known types of hormones can participate in the regulation of almost any phase of plant growth or development. If one makes a checklist, like Table 18-1, listing the various phases of plant development and marking off each of the hormones known to alter that developmental process in at least some plant species, one finds that just about any developmental process can be altered by any of the hormones under some circumstances. It appears that developmental processes may be under simultaneous control by numerous regulatory factors in the plant.

Important for chemical regulation is the readiness with which exogenous chemicals can be introduced into most plants—through the leaves, the stems, roots, flowers, and fruits. The ready entry of chemicals into plants may be a carryover from the primitive condition; plants may never have entirely lost their abilities for heterotrophic existence and the capacity for utilizing exogenous materials.

ENTRY OF CHEMICALS

The intact plant can be thought of as a mosaic of cells in a matrix of mainly aqueous wall and intercellular spaces. The watery matrix provides channels through which solutes can generally

Table 18-1 The diversity of effects of the five known types of plant hormones is illustrated for 12 different growth and developmental processes. The existence of a known effect of each hormone on a developmental process is indicated by an x; the absence of such a mark does not imply no hormonal effect on that developmental process but the absence of such a report in the literature (modified from Leopold, 1972).

	Known regulatory activities for:				
Developmental process	Auxin	Gibberellin	Cytokinin	Abscisic acid	Ethylene
Dormancy		x	x	x	x
Juvenility	x	x			
Growth rate	x	x	x	x	x
Flower initiation	x	x	x	x	x
Sex determination	x	x	x		x
Fruit-set	x	x	x		x
Fruit growth	x	x	x	x	x
Fruit ripening	x	x	x		x
Tuberization	x	x	x	x	x
Abscission	x	x	x	x	x
Rooting	x	x	x		x
Senescence	x	x	x	x	x

pass. Uptake of solutes is possible through the root, the foliage, the stem, and other surfaces of the plant; the various sites of entry of chemicals do have differences, however, in their ability for screening solutes. Starting with the root, entry through the mucilaginous sheath surrounding the root tip is very easy for apparently any chemical which can exist in the aqueous soil solution, and there is ordinarily no screening at this uptake site. Entry into leaves, on the other hand, occurs through various layers of wax, cuticle, and cell wall, and through stomata, and at this site there can be some remarkable screening of solutes. Some of the most impressive examples of selective entry are among the herbicides; e.g., among the triazine herbicides, simazine is unable to enter the foliage of most plants, whereas closely allied atrazine penetrates leaves easily. One might presume the readier entry of atrazine to be related to its greater water solubility, but for dactal, a substituted urea with solubility comparable to simazine, entry into leaves is very easy. Herbicide research has established that seedling grasses may take up important amounts of herbicides through the mesocotyl and the first node; in the soil these stem structures are more readily penetrated by herbicides than adjacent underground leaf structures are (Prendeville, 1968).

Leaf Absorption

Pioneer work on the entry of chemicals into leaves was done by Aslander (1927) in connection with the development of sulfuric acid as an herbicide. Using surprisingly modern types of analysis, he found that the herbicide enters chiefly through the stomata, since the first cellular damage is localized in the mesophyll cells adjoining the stomata. He noted that entry is rapid and not averted by an artificial rain even 1 h after application. Elevated temperatures increase the herbicide effectiveness, while dry conditions lower the effectiveness. The species selectivity of the herbicide can be accounted for on the basis of selective entry.

The time patterns of entry of chemicals into leaves are diverse (Fig. 18-1), ranging from rapid but brief uptake for 2,4-D at 32°C to prolonged but slow uptake at lower temperatures. A carrier or wetting agent can enormously extend the uptake period. Actually, the diversity of uptake-time-course curves like those in Fig. 18-1 are probably due to physical variations at the leaf surface. When time-course studies are made under conditions where there is no drying or other shift in physical characteristics, the time course is linear (Jyung and Wittwer, 1964; Yamada et al., 1965).

It was once asserted that uptake of chemicals through leaves is an active metabolic process (Jyung and Wittwer, 1964), but it is probable that the involvement of metabolism is secondary and related to the fate of the solute after uptake has been completed. The concentration curves for uptake show a linear increase with increasing concentration (Jyung and Wittwer, 1964) or a steeply rising curve on a logarithmic scale (Fig. 18-2), as would be expected for diffusion.

Fig. 18-1 The time course of absorption of 2,4-D into bean leaves. Elution of unabsorbed 2,4-D was measured after applications of 67 μg to a primary leaf. Carbowax (0.5 percent) extended the absorption time from below 10 to over 72 h, and elevated temperatures increased the absorption rates (Rice, 1948).

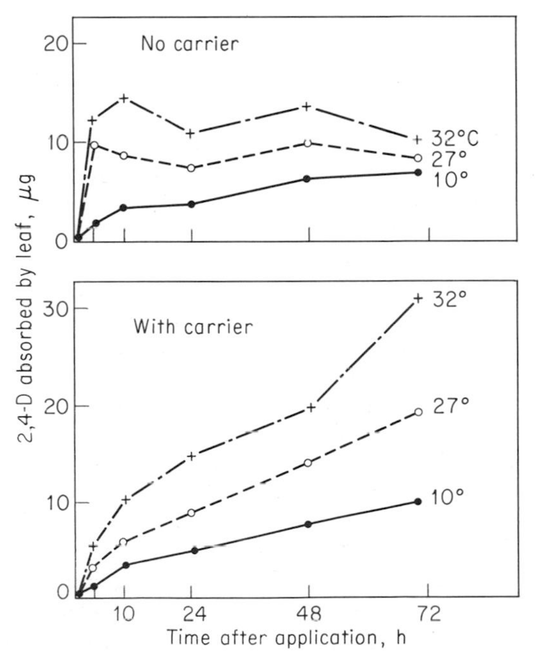

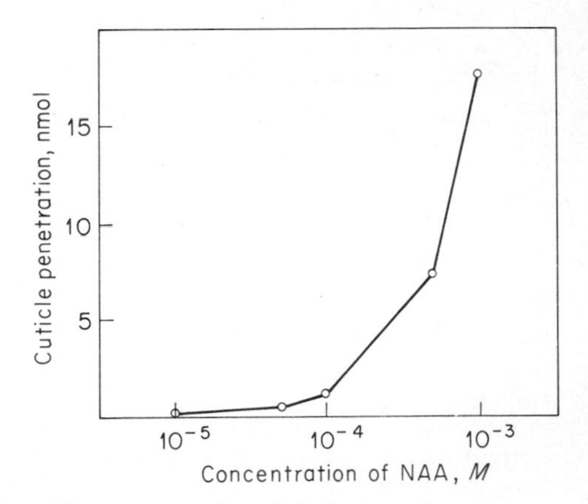

Fig. 18-2 The passage of naphthaleneacetic acid through the isolated cuticle of pear leaves shows a marked dependence upon the concentration of solute applied; penetration in 24 h (Norris and Bukovac, 1969).

The first barrier to entry of chemicals through leaves (Fig. 18-3) is the cuticular wax, which is composed of fatty alcohols and esters. Waxes are solids at ordinary temperatures and show decreasing solubilities with increasing chain lengths in the range of 24 to 36 carbons. They are synthesized in the cytoplasm (Chibnall and Piper, 1934), and in some cases they are exuded at the leaf surface through minute pores in the cuticle; in other cases they move in a more diffuse pattern to the surface, where they are deposited (Hull, 1970). The *blooms* on cabbage leaves, grapes, and apple fruits are common examples of extruded waxes, and in some leaves they are prolific enough to yield commercial amounts of waxes, e.g., carauba and candelilla wax from the wax palm and a xerophytic euphorb shrub, respectively. In some plants, the wax layer may make up to 15 percent of the leaf dry weight (Eglinton and Hamilton, 1967). Using a clever technique for isolating cuticles, Skoss (1955) was able to show that the permeation of the cuticles by aqueous materials can be almost

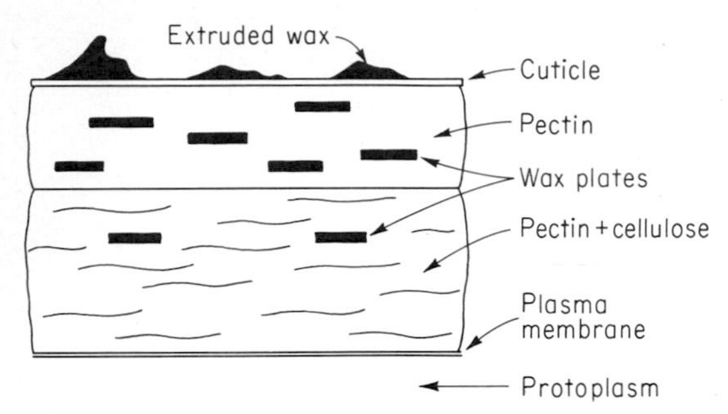

Fig. 18-3 Components of an epidermal cell wall, showing extruded wax overlying a cuticle, then cell-wall layers of pectin and of pectin plus cellulose with some wax plates included. Inside the cell wall is the protoplasmic membrane enclosing the protoplasm. (*Modified from Schieferstein and Loomis, 1959.*)

entirely prevented by the wax components. The density of the wax layer may increase markedly with age of the leaf (Skoss, 1955), and in some species the composition of the wax layer may change without any change in the total amount per unit leaf area (Albrigo, 1972). Waxes are deposited in a variety of patterns (Fig. 18-4), including plates, granular, rodlike, and membranous layers, and it has been suggested that the actual form of the wax deposits may be a function of the patterns of solvent evaporation during their deposition (Juniper, 1959).

The second layer of the leaf to be penetrated by applied chemicals is the cuticle. This varnish-like skin of polymerized fatty acids, esters, and soaps is apparently formed by an oxidative drying process on the exposed cell-wall surfaces. The fatty components are synthesized in the cytoplasm, of course, and apparently move through the cell-wall region as minute oily droplets lacking any membranous envelope (Hull, 1970). In very young leaves, the cuticle may form uniformly over stomata as well as epidermal cells and then rupture at the stomatal sites during leaf enlargement (Albrigo, 1972). The cuticle retains enough acid end groups to show a fairly strong ionic binding for cations (Yamada et al., 1966); anions show less ion exchange with the cuticle, and nonionic solutes such as urea are markedly more permeable through the cuticle, showing no apparent binding. Cuticles tend to increase in thickness with leaf age (Fig. 18-5) and with higher light intensities, and they form more extensively on the upper epidermis than the lower. The cuticle has a rather spongelike feature, which makes it contract under dry conditions and expand into a looser structure under moist conditions, thus facilitating uptake of solutes. Leaf hairs are likely to have minimal cuticles over their basal cells when the hairs are still young, and these hair bases form a ready site of chemical entry into leaves (Haberlandt, 1914; Dolzmann, 1964). The same is true of the cells lining the substomatal pores (Scott, 1950).

Inside the cuticle is the cell wall, usually composed of an outer pectin layer and an inner secondary wall layer of cellulose fibers embedded in pectin and other noncellulosic polysaccharides. The wall is quite permeable to water unless the frequency of wax plates in it is high. The greater frequency of waxes in the outer layer may result in a gradient of increasing polarity from the cuticle to the inside of the cell wall (van Overbeek, 1956).

The last barrier which an applied chemical must pass is the protoplasmic membrane, or plasmalemma. While there is considerable argument about the basic architecture of the membrane, it is principally composed of a double layer of fatty acids, or phospholipids, with the hydrophilic acid or phosphate groups oriented toward the two surfaces of the membrane and protein components either sandwiching over the fatty layer or variously embedded in it or both (Korn, 1966; Vanderkooi and Green, 1971). The fatty components of the membrane exert the greatest limitation on the passage of solvent and solutes, and Horowitz and Fenichel (1965) have calculated that the diffusion coefficient for a membrane made up of just the fatty layers without protein units embedded in it would be about 1,000 times less than for a layer of vulcanized rubber of the same thickness (100 Å).

Passage through the membrane can occur by diffusion through pores in the fatty layer, by

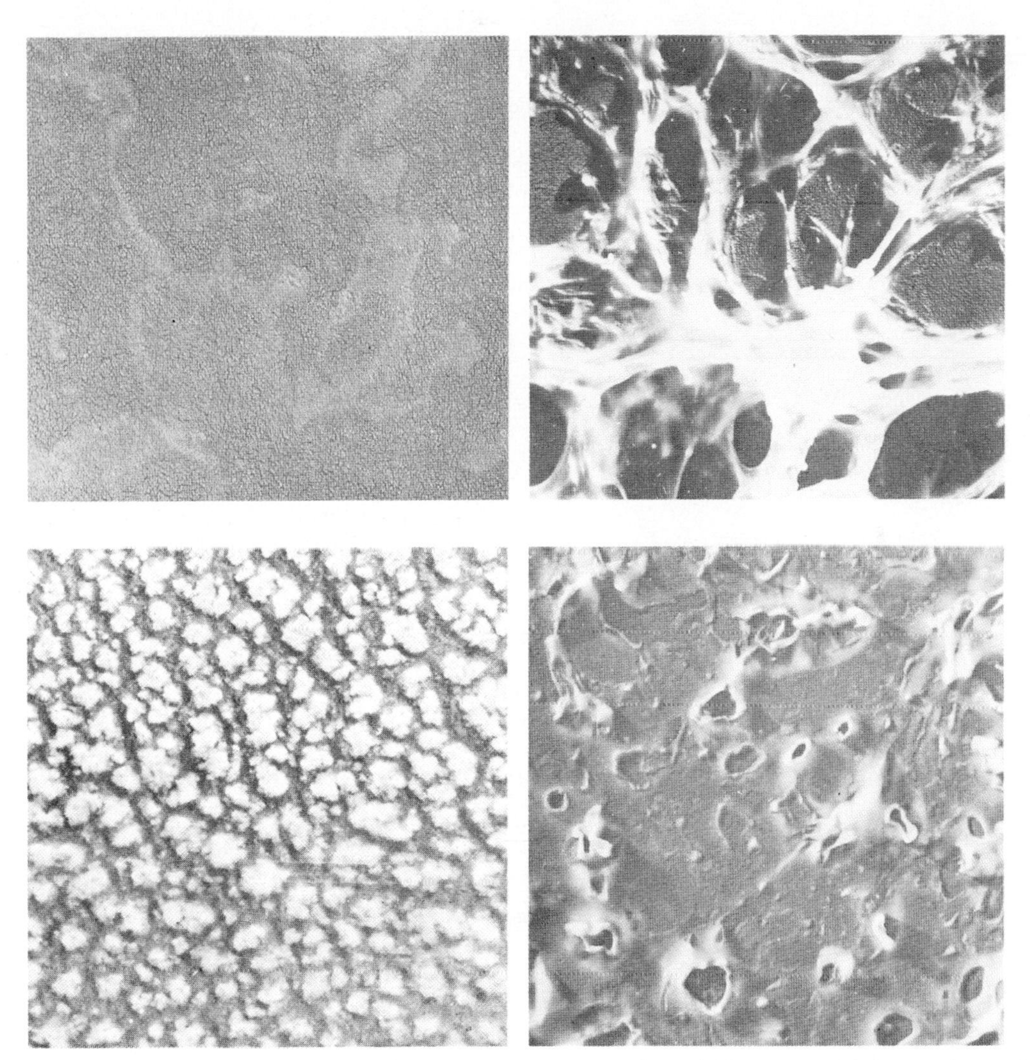

Fig. 18-4 Electron micrographs of the wax extrusions of leaves (X12,000): *Asplenium,* almost lacking wax; *Picea,* showing nets of wax; *Peperomia,* showing rods; and *Nicotiana,* with heavy plates of wax (Müller et al., 1954).

solubilization into the fatty layer, or by attachment to proteinaceous carrier sites within the membrane. Therefore entry through the membrane can be influenced by the size, shape, or charge of a molecule, its solubility in the fatty membrane layer, and any molecular features which would alter its ability to become attached to carrier sites in the membrane.

Major progress in understanding the proteins involved in the permeation of membranes by solutes has been made possible by the isolation of the actual transport proteins from some cell

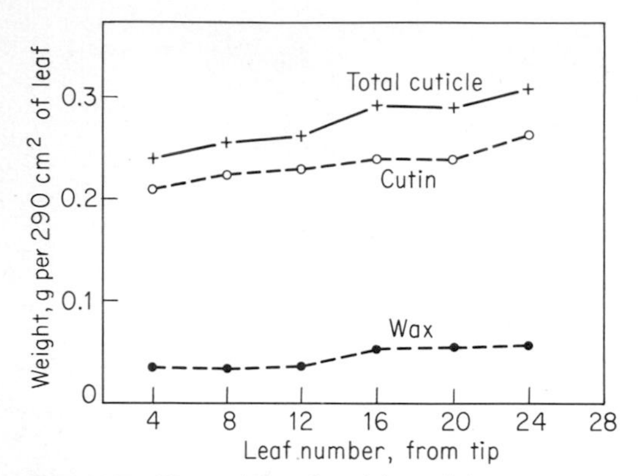

Fig. 18-5 The weight of cuticle and its two components, cutin and waxes, increases with leaf age. Weights of samples retted from leaves of *Hedera helix* at various nodes from the tip down (Skoss, 1955).

membranes (Pardee, 1968) and by the incorporation of model carriers into synthetic membranes which impart carrier properties to the membranes (Läuger, 1972).

Electron micrographs of cell walls have indicated that in addition to plasmodesmata, which provide connecting channels between adjoining cells, tiny pores, or ectodesmata, may permeate the walls of epidermal cells (Lambertz, 1954; Franke, 1961). These pores may serve as convenient channels for permeation of epidermal cell walls by aqueous materials, though the possibility remains that they are artifacts in the photographic preparations.

In view of the structure of the cell wall, penetration through the wall by chemicals and solvents can apparently be via three pathways:

1 Hydrophilic materials can enter through the aqueous phases from the acidic components of the cutin to the pectins and the water-permeable protoplasmic membrane.

2 Lipophilic substances can enter through the fatty components (waxes, cutins, and the lipoidal layer of the membrane).

3 Chemicals with both hydrophilic and lipophilic end groups can enter along the interphases between the two types of wall component.

This last feature can be illustrated by the auxins, with their lipophilic ring and hydrophilic acid side chain. Drying the epidermis can greatly retard entry of the materials with hydrophilic tendencies because of the shrinking of the aqueous phases of the wall and the decreasing wettability of the cuticle surface. There is actually a diurnal cycle of wettability; it is greater at night and less with the drying actions of the day (Fogg, 1947).

Stomata can provide a means of entry for some applied materials and not others. Calculations of the relations between stomatal aperture sizes and movement of volatile sprays into leaves led Turrell (1947) to the deduction that volatiles should be able to move into the leaf through stomatal openings. The accuracy of these deductions was slow to be grasped, but it is now clear that the stomata are major pathways of entry of such materials as lipophilic dyes, auxin esters, and volatile oil herbicides (Minshall and Helson, 1949; Currier and Dybing, 1959; van Overbeek and Blondeau, 1954). There are apparently two vectors of solute entry relating

Fig. 18-6 The entry of the dye fluorochrome into leaves of *Pyrus communis* is related to the concentration of a surfactant (Vatsol OT) and the state of opening of the stomata. Entry of 0.17 percent fluorochrome for 6 min before washing and measuring entry by fluorescence in ultraviolet light (Dybing and Currier, 1961).

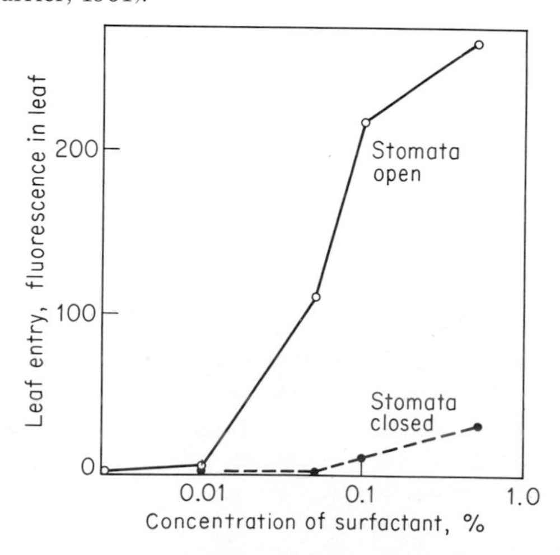

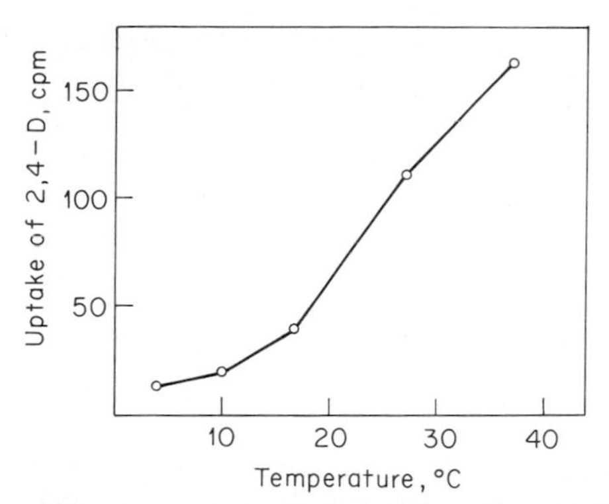

Fig. 18-7 The uptake of 2,4-D by bean leaf disks shows a fairly steep temperature dependence, with a Q_{10} of about 2 in the lower ranges; radioactive 2,4-D at 100 ppm for 8 h (Sargent and Blackman, 1962).

to the stomata. (1) Entry may occur through the stomatal opening, as is apparently the case in the case of fluorochrome. Entry through the stomata is enhanced by the presence of a surfactant carrier, as illustrated in Fig. 18-6. (2) There may be readier entry of solutes through the cuticle in the region immediately surrounding the stomata without specific involvement of the stomatal pore (Holly, 1964). These two vectors are rather difficult to separate, since the more volatile materials, e.g., volatile esters of herbicides, not only enter the stomatal openings but also penetrate the cuticle around them. Among the materials that readily enter leaves through the stomata are volatile oils. At least some of the air-pollution components which damage plants are presumed to enter the leaves through open stomata since closure can greatly lessen the damage (Ledbetter et al., 1960; Eagle and Gabelman, 1966); however, experiments with ozone and peroxyacetylnitrate as smog components indicate that entry of these pollutants is independent of the stomatal opening or closure (Dugger et al., 1962).

Many environmental factors can modify the entry of chemicals, of which the most powerful are probably the extent to which the leaf surface remains wet after the spray application and the temperature and light conditions after application. It is reasonable to presume that the most rapid entry of chemicals as in Fig. 18-1 occurs over the period of time when the applied drops of water remain on the leaf surface. Factors like the presence of wetting agents (Figs. 18-1 and 18-6), high relative humidity, and large droplet size in the applied spray increase entry principally through improvement of the persistence of the water solvent on the leaf surface. The effects of temperature on entry are illustrated in Fig. 18-7. The steepness of the temperature response suggested to Sargent and Blackman (1962) that entry involves some active process or processes; since the uptake measurements were made after 8 h time, it is possible that metabolic alterations of the solute after entry may have been contributing, rather than simple entry of the solute. Light sometimes has a stimulatory effect on entry, as illustrated in the data for 2,4-D entry into *Coleus* leaves shown in Table 18-2. The pH of the spray solution also has a marked effect on entry, most particularly for dissociable compounds such as acids. Illustrating the pH effect on uptake are data for uptake of two acidic substances in Fig. 18-8. The poor entry at high pH conditions may be a consequence of the ion's being caught up by the anion-exchange properties of the cell walls and cuticle. Nonionizable solutes show little alteration of entry by the pH of the solution (e.g., maleic hydrazide; Crafts, 1961). The entry into older

Table 18-2 The uptake of 2,4-D by *Coleus* leaf disks is increased by light, especially when uptake is through the abaxial surface of the leaf (Sargent and Blackman, 1962).

	Uptake of 2,4-D, cpm	
	Abaxial surface	Adaxial surface
Dark	15	16
Light	381	34

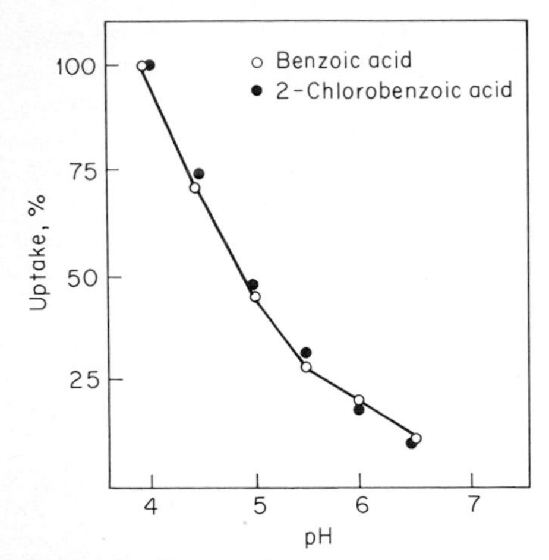

Fig. 18-8 The pH dependence of uptake by pea-stem sections is strikingly similar for two different benzoic acids. The pK_a of benzoic acid is 4.21, and of 2-chlorobenzoic acid is 3.98 (Venis and Blackman, 1966). Uptake at pH 4 is taken as 100.

leaves can be markedly less easy than into young leaves. With increasing age, leaves often become more resistant to uptake of applied chemicals (Ahlgren and Sudia, 1964), and this phenomenon, as well as the readier entry through the abaxial or lower leaf surface, may reflect in part the relative amounts of cuticle.

Root Absorption

There are less effective barriers against the entry of water-soluble chemicals into roots than into leaves. Roots sometimes show some selectivity, but for the application of organic chemicals through the soil solution the selectivity is usually sufficiently weak to be of no concern to people involved in soil applications of herbicides or growth regulators (cf. Woodford, 1958).

The root tip, where most entry of water and solutes occurs, is enclosed in a mucilaginous sheath which provides an aqueous matrix continuous with the apoplast or the aqueous phases of the cortex, including especially the cell walls and intercellular spaces. These spaces are dominated by pectins, hemicelluloses, and celluloses, all of which provide fixed carboxyl groups and consequently make rather good cation exchangers. The entry of solutes into these free spaces is of course passive and shows marked differences between cations and solutes with negative or no residual charges. The exchange capacity for cations can be seen in experiments like that shown in Fig. 18-9, where radioactive-strontium uptake is followed in time; when the roots are placed in an unlabeled-strontium solution, a large proportion of the label initially taken up is lost. There is very little anion-exchange capacity in the root free space (Elzam and Epstein, 1965). Nonionized solutes show a similar free entry into the free space of the root (Barber and Shone, 1966).

Through the exchange capacity of the mucilaginous sheath at the surface, roots can pick up insoluble cations, e.g., insoluble forms of iron, by contact exchange with the soil particles (Jenny, 1961).

The fixed negative charges in the root free space were once considered to be a part of the mechanisms by which cations enter the root (Briggs et al., 1957), but Epstein et al. (1962) have used divalent cations such as calcium to satisfy the fixed negative charges of barley roots and have shown that the entry of monovalent

Fig. 18-9 The uptake of strontium by barley roots shows a rapid initial phase for about 20 min, followed by a slower, steady uptake rate. When the roots were transferred from the 1 mg/1$^{89}SrCl_2$ to unlabeled $SrCl_2$, considerable elution of the label out of the roots resulted, indicating a freely exchangeable form (Epstein and Leggett, 1954).

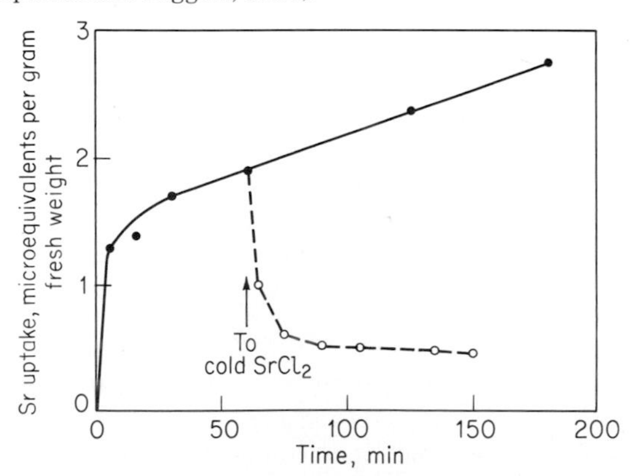

cations can then occur without any apparent binding to the negative charges in the root free space.

Of course the pH of the ambient solution will markedly alter the cation-exchange properties of the root free space, and at low pH there will be markedly less cation binding (Shone and Barber, 1966). The similar pH curves for uptake of acids with different pK_a values suggest a major effect of pH on the root exchange capacity.

While any solute can enter the free space of the cortex, entry into the more intimate parts of the root, the stele, and the vascular system leading up into the rest of the plant is restricted by the endodermis (van Fleet, 1961). Two types of evidence, however, suggest that the passage of chemicals from the soil solution into the transpiration stream may not be necessarily restricted by the endodermis: (1) there is a movement of solutes into the transpiration stream which does not require metabolic involvement but is pulled along by transpiration, and (2) there is a movement of solutes into the transpirational stream when the solute is supplied in relatively high concentrations. Both these phenomena are evident in the experiments of Kylin and Hylmo (1957) on sulfate uptake by wheat plants. Dumbroff and Pierson (1971) have noted that there are relatively poorly developed regions of endodermis at the point of each xylem strand and that passive movement of solutes might occur there; and one might suggest in addition that a strong transpirational pull might magnify the movement of water through these passage cells. From an operational point of view, we must try to integrate the fact of facile entry of chemicals from the soil solution into the root without any apparent screening with the fact of restricted entry past the endodermis and the strong selectivity of the plasmalemma. Kylin and Hylmo (1957) offered the picture of a passive uptake through the root, which is driven by transpiration, combined with an active uptake through the cells enclosing the stele, which is driven by metabolism.

Various lines of evidence indicate that metabolism is critical to some aspect of the uptake

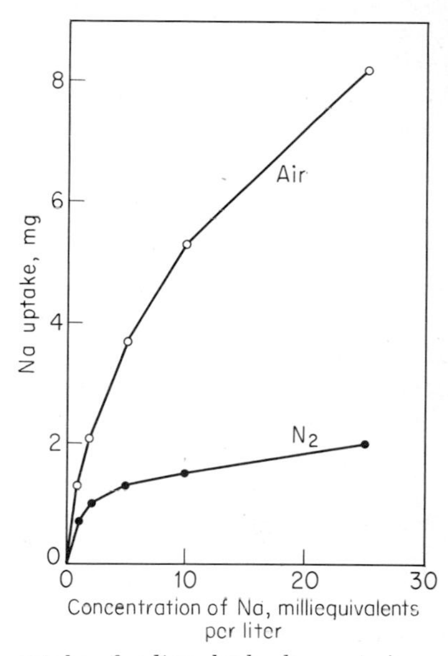

Fig. 18-10 The uptake of sodium by barley roots is strongly suppressed by a nitrogen atmosphere, indicating a respiratory involvement in uptake; after 3 h in $Na^{24}Cl$, roots were rinsed three times before counting (Epstein and Hagen, 1952).

of solutes. The uptake of ions by excised roots is reduced under anaerobic conditions (Fig. 18-10). The uptake of many solutes is inhibited by metabolic poisons, and the temperature effects on uptake show Q_{10} values up to 2.8, indicative of enzyme-catalyzed processes. Energy production for the movement of ions across the plasmalemma is presumed to spring from ATP, and Hodges et al. (1972) have provided evidence that ATPases related to ion transport are located on the plasmalemma.

While roots may not screen the entry of solutes, cases are known in which one compound can be held back in the roots whereas a closely related analog can move readily along the transpiration stream into the tops. For example, Crowdy and Rudd-Jones (1956) noted some striking differences in the movement of various sulfa drugs through the root system, even though they all entered the root easily. Many herbicides are translocated poorly after uptake from the soil solution because of their retention in the roots, e.g., 2,4-D in grass roots (Crafts, 1959*a*) and the triazine herbicide simazine in corn (Harris et al., 1968). The same explanation applies to the entry of some systemic insecticides through the roots (Mitchell et al., 1960).

Entry Rates

When plant cells are immersed in solutions containing inorganic or organic materials, one can perceive two stages in the kinetics of entry: an initial rapid entry for about 20 min, followed by a steady though slower uptake rate. This was described for the entry of inorganic ions into roots by Epstein and Leggett (1954), as illustrated in Fig. 18-9. They interpreted the first rapid uptake as a free exchange, since they found that removing the roots from solutions of a radioactive ion to similar molarities of a nonradioactive ion results in the elution of most of the label that had entered in the initial phase; they estimated the amounts of ion that had been physiologically fixed as the amount not eluted in this manner. The two phases of uptake are also found with the entry of materials into leaves. For example, Johnson and Bonner (1956) followed the kinetics of uptake of 2,4-D by oat-coleoptile sections and found precisely the same sequence.

While the rapid, reversible uptake of solutes appears to involve the movement of solutes into free space, ion-exchange phenomena, or perhaps adsorption phenomena, the steady-state uptake occurring after 30 min in Fig. 18-9 involves metabolic events, since they can be selectively inhibited by metabolic poisons (see, for example, Johnson and Bonner, 1956).

In anion uptake, there is often little difference in the amount of anion eluted out of the root with water and with equimolar solutions of unlabeled anion (see, for example, Elzam and Epstein, 1965). Knowing that the free space has relatively strong cation-exchange properties and weak anion-exchange properties, one might equate the amounts of ions exchangeable with unlabeled solutions with the amounts held in the free space by the ion-exchange capacities.

The entry of nonionizable compounds is least affected by the pH of the solution in which it is applied (e.g., indoleacetonitrile, Anker, 1958; coumarin, Audus, 1949). However, some nonionizable solutes have an improved entry at pHs near neutrality (e.g., maleic hydrazide, Crafts, 1959*b*), which would logically be attributable to the greater permeability of the cuticle with dissociation.

There can be a foreshortening of the entry from a sprayed solution due to drying of the spray droplets. Thus, instead of entry continuing over a period of days, it may be terminated in a few hours (Fig. 18-1). For this reason, the addition of solubilizing or carrier materials to the solution can be especially important.

Carriers in aqueous spray solutions, first utilized by Mitchell and Hamner (1944), may be of three general types: surfactants, humectants, and oil emulsions. All three may function as wetting agents to increase the solvent continuum between spray droplet and leaf surface. Surfactants include the propylene glycols, detergents, and a variety of other carriers which may increase the wettability of the leaf surface. Humectants include glycerin, calcium salts, and other additives which attract water to the drying spray droplet. Oil emulsions solubilize the more lipophilic materials, which then are suspended in water by the formation of emulsions. All these carriers increase the spreading properties of the droplet and the ability of the droplet to wet the leaf surface, and they can improve entry through spreading action and/or prolonging the drying period for the droplet. Effects of carriers in prolonging the period of chemical entry into leaves are well illustrated in the data of Fig. 18-1; effects in facilitating the spreading action through the stomata are illustrated in Fig. 18-6.

Environmental factors, especially temperature and relative humidity, can have strong influences on the entry of applied sprays. The increases in uptake with increases in temperature have been mentioned. This appears to hold for both organic and inorganic materials (Sargent and Blackman, 1962; Barrier and Loomis, 1957). A dramatic case of humidity affecting entry is that of maleic hydrazide. As illustrated in Fig. 18-11, entry of the potassium salt is almost obliterated at humidities below 75 percent. The more humectant diethanolamine salt enters more readily at each of the humidities.

Apparent Free Space

As chemicals enter the root from the soil solution, a considerable proportion may be eluted out again by washing (Fig. 18-9). Quantitative measurements of the amounts of ions that could be eluted from roots led Hope and Stevens (1952) to the concept of an apparent free space as a region of the root which was freely accessible to soil solution and solutes and from which an active accumulation of nutrients could proceed. Calculations of the extent of this accessible space on the basis of the amounts of ions which can be eluted after exposure of roots to known molarities of the ion indicate that apparent free space may normally constitute from 13 to 23 percent of the root volume (Butler, 1953). Subsequent investigations indicated similar extents of apparent free space for numerous plant and animal cells (Kramer, 1957). To accommodate such an extensive space, Hope and Robertson (1956) suggested that the cytoplasm is all freely accessible to the soil solution and that the vacuole is the site of final accumulation of ions into a state from which they are not readily eluted. Kramer (1957) suggested the organelles of the cytoplasm, e.g., the mitochondria, as additional accumulation sites.

The morphological extent of free space is not experimentally established, however. Calculations of the root volume freely accessible to chemicals may well be deceptively high, for adsorptive phenomena can be expected to concentrate the chemicals, e.g., the exchangeable materials adsorbed onto cell walls, and so the calculations of the volume of apparent free space on the basis of the solute concentration may be too large. Scott and Priestley (1928) proposed that the soil solution can freely move in the cell walls, which would provide the equivalent of a free space from the epidermis to the endodermis of the root. Strugger (1949) assumed that the cell walls account for much of the water and solute movement in roots.

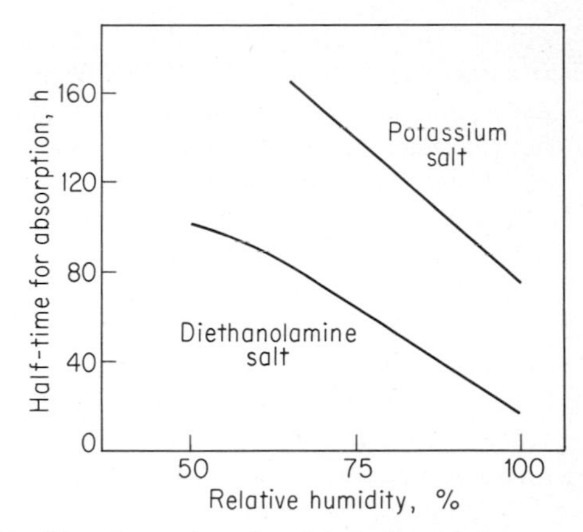

Fig. 18-11 The absorption of maleic hydrazide by tomato leaves is strongly influenced by the relative humidity. After application of 1 ml of 1,600 mg/1 solution of the potassium or diethanolamine salt, the time required for absorption of half the chemical was found to be markedly shorter at higher humidities; absorption measured as the amount not eluted with a 200-ml wash of detergent solution (Smith et al., 1959).

In an effort to pin down the location of the apparent free space more precisely, Crowdy and Tanton (1970) allowed leaf disks to take up lead salts into the free space and then precipitated the lead with H_2S; the precipitate was located specifically in the cell-wall regions (Fig. 2-6). It is widely believed now that free space is ordinarily located in the wall region of healthy, young tissues (Butler, 1953; Epstein, 1972).

MOVEMENT OUT OF PLANTS

That chemical materials can move out of plants has been known for many years; de Saussure (1804) was probably the first to note this type of event, and more recently an apparent excretion has been reported for nitrogenous compounds (Virtanen et al., 1936), carbohydrates and organic acids (Knudson, 1920), and exogenous chemicals (Linder et al., 1957). Studies of herbicide uptake led to the finding that 2,4-D can move out of cotton roots into the ambient solution (Crafts and Yamaguchi, 1958). More dramatic excretion was reported for methoxyphenylacetic acid (Linder et al., 1957); bean roots excreted as much as 20 percent of the amount taken up.

The movement of materials out of plants can occur in two ways: a simple diffusion, e.g., leaching of some materials out of leaves or roots (Tukey, 1970), or an active excretion. There are several lines of evidence for the active excretion of exogenous chemicals, including a high temperature coefficient (Saunders et al., 1966) and a requirement for oxygen (Foy et al., 1971).

Among the herbicides reported to be excreted from roots, there is a striking chemical specificity; e.g., among the benzoic acids, Foy et al. (1971) have shown that dicamba (a dichlorinated methoxybenzoic acid herbicide) applied to the leaves of bean plants is excreted from the roots, whereas tricamba (a trichlorinated methoxybenzoic acid) is not excreted. A series of analogs of the benzoic acid series was tested, and the property of excretion was not clearly related to precise chlorine-substitution positions.

All the excreted herbicides were found to be unaltered in structure. The ability of the excreted chemical to affect the growth of neighboring plants has been demonstrated convincingly in several instances (Foy et al., 1971).

Evidence for excretion of exogenous materials out of stems (Venis and Blackman, 1966) and leaves has been reported (Saunders et al., 1966), and this activity, too, shows remarkable chemical specificity. As illustrated in Fig. 18-12, 2,4,5-trichlorophenoxyacetic acid is excreted from cotton leaf disks, whereas the unsubstituted phenoxyacetic acid is not. This exit of the herbicide was also apparently an active process, judging from the high-temperature coefficient and the sensitivity to metabolic inhibitors.

Fig. 18-12 The time curve for uptake of 2,4,5-T by cotton leaf disks shows an evident excretion of the auxin after about 2 h; the unsubstituted phenoxyacetic acid is not excreted; each acid at 15 μM (Saunders et al., 1966).

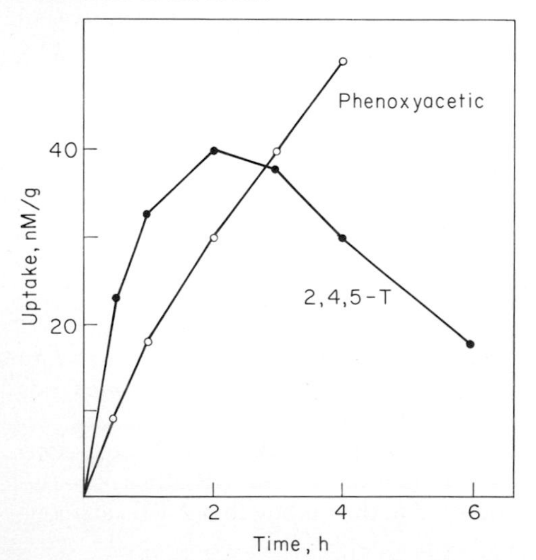

TRANSLOCATION

There appear to be four principal pathways for the translocation of materials after uptake by the roots or leaves. Movement may occur (1) in the xylem along with the transpirational stream, (2) through the phloem or other cells such as ray parenchyma, (3) through the cell walls, or (4) through the intercellular spaces.

The upward movement through the xylem was described by Hitchcock and Zimmerman (1935) for some organic compounds which were applied to the soil solution; the movement of inorganic nutrients through such a pathway was already well known. Wiebe and Kramer (1954) established that a wide variety of radioisotopes is translocated most effectively from the subterminal region of the root, where the xylem begins, and Weaver and DeRose (1946) found that steam killing a sector of stem still permits the upward movement of organic materials in the xylem from the soil solution into the foliage.

The downward movement in the phloem was clearly established for applied dyes by Schumacher (1933). This is the principal pathway of movement of materials applied to the leaves (Hay, 1956). For herbicidal substances, the translocation may be of relatively short duration (Hay and Thimann, 1956), either because of the metabolism of the substance or because of the disruption of the phloem itself (Eames, 1950). The word "downward" is somewhat misleading and is intended to mean any movement which is initially downward out of the leaf. Once out of the leaf, phloem flow may take solutes up to the stem apex as easily as down to the lower parts of the plants; this dualism caused some heated misunderstandings in the earlier days of interest in translocation (Curtis and Clark, 1950).

The movement of inorganic materials through the cell walls has been known for many years. Münch (1930) described the aqueous network through the cell walls as the *apoplast,* meaning outside the protoplast. The movement of dyes through the apoplast was clearly described by Bauer (1949), and then, using fine autoradiographs, Lüttge and Weigl (1962) established that sulfur taken up as sulfate is specifically localized in the cell-wall areas (Fig. 18-13). (A similar conclusion was made for lead; Fig. 2-6). In contrast, calcium was not localized in the apoplast, indicating differences in distribution immediately after uptake of different ions. The apoplast seems to be generally accepted as the principal region of the apparent free space, as indicated by Butler (1953).

There appear to be two forces limiting the freedom of movement of materials through the apoplast: the region of the endodermis seems to act as a screen which may limit the passage of solutes from the cortex into the stele, and the accumulation of solutes by the cells may limit the freedom of movement in the apoplast. The action of the endodermal region was clearly visualized by Priestley (1920) as a discontinuity in the aqueous continuum between the cell walls of the cortex and the stele. It is presumed that solutes must ordinarily pass through the cell protoplasts of this region before they can progress into the xylem. Experiments with radioactive ions indicate a clear pileup of solutes in the region just external to the endodermis, consistent with the idea that the endodermis is a barrier to solute passage (Fig. 18-13). Arnold (1952) believes that the passage through the endodermal cells is effected by an active-transport system.

The accumulation of solutes by cells, described many times, may clearly serve to limit the translocation of solutes through the plant. The inability of some herbicides such as 2,4-D to move

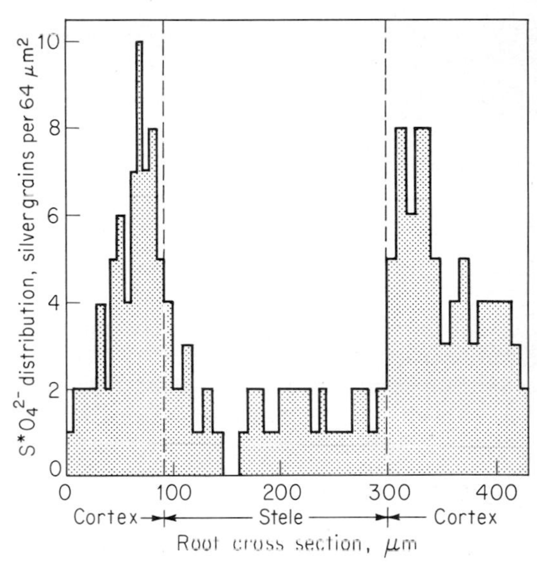

Fig. 18-13 Microautoradiography of a pea root exposed to $S^*O_4^{2+}$ for 5 min shows accumulation of radioactivity just outside the region of the endodermis and markedly less isotope inside the stele (plerome), indicating a screening action by the endodermal region of the root (Lüttge and Weigl, 1962).

out of the roots of certain plants (Crafts and Yamaguchi, 1960) is apparently a consequence of such accumulative action.

The movement of herbicides in oils is apparently a capillary flow of the oil through the cell walls (Minshall and Helson, 1949) and may show patterns of distribution not unlike those of solutes which move through the aqueous medium of the apoplast.

The rapid systemic permeation of gases and volatiles through plants indicates a ready movement through intercellular spaces, which must be very extensive in some species, e.g., rice (Barber et al., 1962).

The movement of materials through these various pathways provides some distinctive patterns of translocation. For example, materials moving in the xylem or phloem are first concentrated in the vascular bundles of the leaves and stems (Fig. 18-14). Materials which move in the apoplast tend to permeate continuous blocks of tissue rather than follow the vascular patterns, especially when applied to leaves as in Fig. 18-15

(Crafts, 1959*a*). Simazin and monuron are dramatic examples of herbicides which move only in the apoplast (Crafts, 1961).

Translocation Rates

The rates or velocities of movement of applied chemicals strongly resemble the rates of translocation in phloem and xylem, as discussed in Chap. 2. Xylem translocation of exogenous chemicals may be as fast as 9 m/h and phloem translocation between 10 and 100 cm/h (Day, 1952; Little and Blackman, 1963). Oils moving in cell walls may go from 4 cm/h (Rice and Rohrbaugh, 1953) to 24 m/h for very short periods (Hay, 1956), and gases may permeate through the plant at as much as 36 m/h (Barber et al., 1962).

Fig. 18-14 The movement of radioactivity of 2,4-D applied at the base of a bean leaf, showing concentration in the vascular strands and movement both up and down; 50 μg of 2,4-D applied 3 h before freezing the plant for autoradiography. (*Courtesy of A. S. Crafts.*)

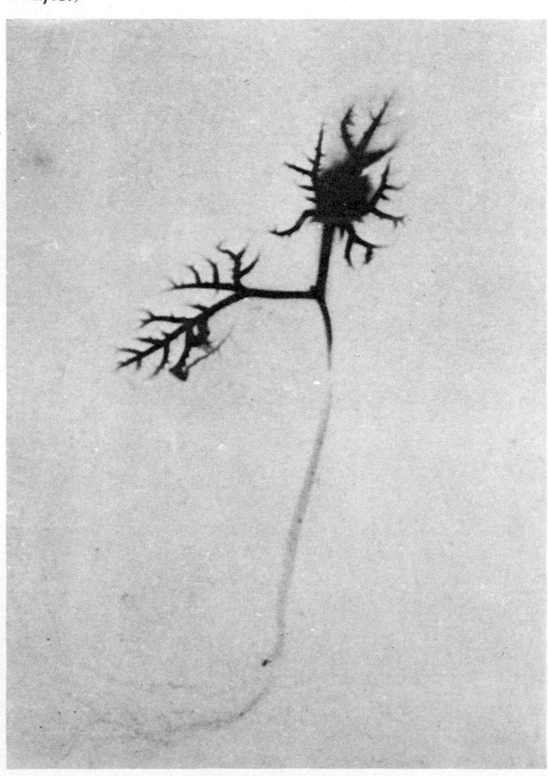

Chemical Specificity

One of the most remarkable characteristics of the translocation of exogenous chemicals is the striking specificity of movement for different chemical materials. This subject is best known to herbicide workers, who have recognized that some herbicides move only in the xylem, e.g., most triazines, substituted ureas, isopropylphenylcarbamate (IPC) and related compounds, and trichloroacetic acid. With foliar applications, some herbicides tend to remain in the phloem, e.g., 2,4-D and related auxins; and still others can move readily from one pathway to the other, e.g., maleic hydrazide. Materials which move principally in the phloem without readily crossing into the xylem may accumulate in the root tips and be released out into the soil solution, e.g., aminotriazole. As mentioned previously, some solutes are essentially nontranslocatable.

Curious changes in translocation characteristics come with altered molecular structures. For example, the movement of 2,4-dichlorophenoxyethyl sulfate is essentially restricted to xylem, but oxidation to the acid permits preferential movement in the phloem. The mobility of aminotriazole in phloem is greatly reduced when a hydroxyl group is substituted into its ring (Massini, 1959). Most striking of the chemical specificities described so far has been the dramatic increase in translocation of various aromatic acids, e.g., the auxins, and the *N*-phenyl carbamates, e.g., IPC, by a lactic acid substituent in the side chain (Mitchell and Preston, 1953; Mitchell et al., 1954).

One structural feature which can contribute to the differences in translocation abilities is the electric charge. This has been used to account for the slower movement of dyes and antibiotics with basic residual (positive) charges, on the assumption that the acidic (negative) residual charge on plant cell walls would retard the move-

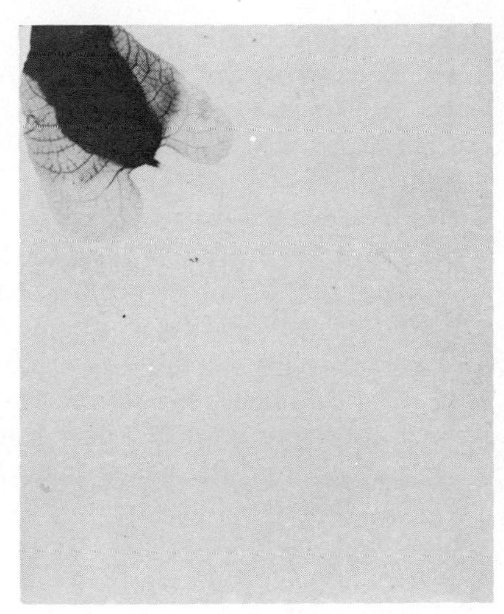

Fig. 18-15 A bean plant (*left*) treated with radioactive calcium at the base of one leaf, showing a creeping movement up the leaf tissues (*right*) without movement down the petiole to the stem; 1 μg of $^{45}Ca^{2+}$ applied 4 days before fixing for autoradiography (Crafts and Yamaguchi, 1960).

ment of such substances (Charles, 1953). But the charges cannot explain the differences in translocation of most chemicals.

Another feature affecting translocation is adsorption. The abilities of plants to hold chemicals by adsorptive means have been emphasized by Brian and Rideal (1952), and the strong adsorption of 2,4-dichlorobenzonitrile has been used to account for its slow translocation, even in the xylem (Massini, 1961). Differences in phloem translocation of inorganic ions have likewise been attributed to adsorptive differences resembling the selective partitioning in chromatography (Swanson and Whitney, 1953). The readiness with which a chemical can be adsorbed depends on complicated sets of properties, including molecular configuration, acidity, water solubility, and polarizability (Bailey and White, 1970).

Physiological Factors

For any chemical to flow in the phloem translocation system, there must be a flow of the phloem sap, and this in turn requires an osmotic gradient according to the mass-flow concept (see Chap. 2). Thus it is not surprising that applied chemicals cannot move readily out of leaves without an accumulation of photosynthates. This requirement was first reported for auxins (Mitchell and Brown, 1946), but it also applies to the movement of other organic materials such as insecticides (Thomas and Bennett, 1954) and inorganic compounds (Barrier and Loomis, 1957). Usually the movement of applied materials out of leaves can be obtained in the dark by the simple means of applying sugar to the leaves (Rohrbaugh and Rice, 1949). Applied sugars can be readily translocated out of the leaf via the phloem, and it is assumed that other applied substances can be swept along in the current created by the sugar.

Translocation differences between species have received considerable attention from herbicide physiologists, as these differences may contribute to the selectivity of some herbicides. Extensive comparisons of herbicide selectivity and translocation in various species led Weintraub et al. (1956) to suggest that differences in movement might possibly account for the selective action of 2,4-D. The suggestion has been extended to other species and to other herbicides (Ashton, 1958).

A further factor in translocation is the assimilation or accumulation of the chemical by the plant cells. The fact that some inorganic ions such as phosphate are strongly assimilated by rapidly growing tissues has been described as a mobilization effect (Chap. 2), and there is a striking contrast between the accumulation of such materials in the apex and the accumulation of a nonassimilated ion such as cobalt in the tips of leaves. Translocation of organic compounds is likewise strongly suppressed or altered by abilities of cells to assimilate the chemical, withdrawing it from the translocation system. Crafts and Yamaguchi (1958) suggested this as an explanation for the much readier translocation of maleic hydrazide than of 2,4-D, the latter being more readily assimilated. Foy (1961) attributed the movement of the herbicide dalapon to the growing points as a mobilization effect, by which the chemical would presumably be assimilated preferentially by the growing points. Such preferential assimilation may account for variations in herbicide translocation between tissues of different ages and for translocation differences between plant species (Koontz and Biddulph, 1957).

An interesting study of the ability of various herbicides to move through excised sections of plant stems was carried out by Taylor and Warren (1970), who found striking differences in the movement characteristics of several herbicides. Aminotriazole was found to move readily through the sections and with almost no delay, whereas 2,4-D and linuron showed an initial lag period followed by good movement, and amiben showed a steadfast lack of mobility.

One of the most surprising features of the translocation of exogenous chemicals is the great diversity of chemicals which can enter the phloem and be transported in it. The mass-flow concept of phloem function requires strong limitations of permeability for the maintenance of osmotic gradients, and it is perplexing to consider how the presumed metabolic loading system that introduces sugars into the phloem might also introduce such diverse chemicals as fluorescein, the auxins, aminotriazole, and maleic hydrazide. Much less surprising is the poor entry into the phloem of such materials as the triazines, the substituted ureas, and the phenols, which probably accounts for their relative inability to be translocated out of leaves (Woodford, 1957).

FATE OF CHEMICALS IN PLANTS

Despite the lack of an organ like the liver which can specifically remove and dispose of foreign chemicals, the plant organism is surprisingly effective in getting rid of applied chemicals. Unless they are applied in lethal doses, most exogenous chemicals are disposed of by the plant within a few days.

The persistence of applied chemicals varies enormously among compounds and among plants. The auxin herbicide 2,4-D persists for relatively short periods after application; Hay and Thimann (1956) found that in bean seedlings half the applied 2,4-D had been destroyed 2 days after application. Morphological evidence of persistence may continue over longer periods; the phenoxyacetic acids may cause morphological irregularities over a period of 25 days in corn (McIlrath and Ergle, 1953) or alter fruit abscission for 50 days in apples and for 7 months in *Citrus*. Periods of dormancy may extend the period of morphological response to a year (Tullis and Davis, 1950), probably because of suspended metabolic degradation.

It is not infrequent for the effects of a growth regulator to persist in the progeny of treated plants. Pridham (1947) observed the morphological responses in the progeny of bean plants which had been treated with 2,4-D. Foy (1961) observed symptoms of dalapon responses in wheat plants to persist for three generations after application. The growth retardants are often highly persistent, their effect carrying over to the progeny of treated plants (Marth et al., 1954; Riddell et al., 1962). In none of the cases reviewed, however, was it clear that the applied chemical itself persisted in the plant to the next generation.

Loss Curves

There seem to be three general types of disappearance curves for exogenously applied chemicals. If the plant has the chemical or enzymatic means for ready disposal of the material, disappearance will proceed in a roughly logarithmic curve, like curve *A* of Fig. 18-16. If the plant has only limited means of disposing of the chemical, disappearance may proceed only for a short time until adsorption or binding forces can hold the material in a steady state, as in curve *B*. If the plant must develop new enzymatic means to dispose of the chemical, there may be a lag before disappearance proceeds, as in curve *C*. (Note that curve *C* is for losses from soil.) Where adaptive enzymes are formed, the subsequent disappearance may be precipitous, as in curve *C*, rather than logarithmic, as in curve *A*, but subsequent applications of the same chemical may be disposed of without additional lag periods and at a faster rate than before (Audus, 1949). As representative examples of each of these loss curves, 2,4-D disappearance in beans or other susceptible dicotyledonous plants would proceed logarithmically (see, for example, Hay and Thimann, 1956), the urea herbicides such as monuron disappear in beans with a short-term decline followed by a steady state (Fang et al., 1955), and the triazines such as simazine disappear in corn and other susceptible species with a lag period followed by rapid disposal (Montgomery and Freed, 1961).

Adsorption of applied chemicals can often account for large amounts of the applied material. Many agricultural chemicals are strongly adsorbed by soils, especially by fine clays or organic soils (Bailey and White, 1970); likewise there was early evidence for the adsorption of auxin herbicides in plants from the work of Brian and Rideal (1952). Evidence for the specific adsorption of auxins and some related growth regulators has come from the work of Hertel et al. (1972). In their experiments, membrane particulates from plant tissues were centrifuged into a pellet, with radioactive naphthaleneacetic acid (NAA) presented for adsorption; displacement of the NAA* with unlabeled NAA demonstrated the specific adsorption of the NAA. Binding of other growth regulators, e.g., naphthylphthalamic acid, triiodobenzoic acid, and some growth retardants, to cell particulates has also been reported (Lembi et al., 1971; Thomson, 1972). Binding growth regulators to sites in the cell may in some instances be associated with biological activity; in other instances the binding may simply be the removal of the chemical from a biologically active pool (Winter and Thimann, 1966).

Some applied chemicals can be lost from the plant through volatilization. This loss is related to the vapor pressure of the compound, and sometimes there may be large differences in the volatility of even closely related chemicals. As examples, the volatilities of several triazine herbicides are shown for comparison in Fig. 18-17. Loss by volatilization is common for esters of the various acid auxins. Extrapolation from the volatilization of natural oils in plants makes it probable that the extent of such loss will depend not only on the temperature but also on the moisture levels, as native oils may be lost much more rapidly when there is abundant water, implying a loss by a type of steam distillation (cf. Biggs and Leopold, 1955).

Fig. 18-16 Some representative disappearance curves for herbicides: *A*, a logarithmic disappearance of 2,4-D in bean plants (Hay and Thimann, 1956); *B*, a short-term disappearance of CMU from bean plants (Fang et al., 1955); *C*, a delayed disappearance of 2,4-D from soil (Audus, 1949). The time intervals for *A* were multiplied by 5 for clarity.

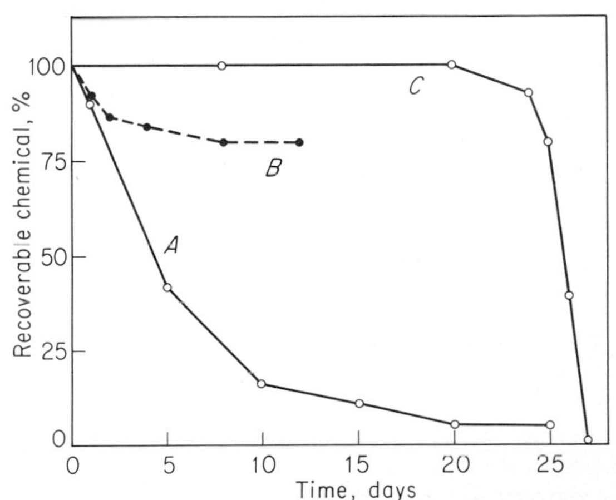

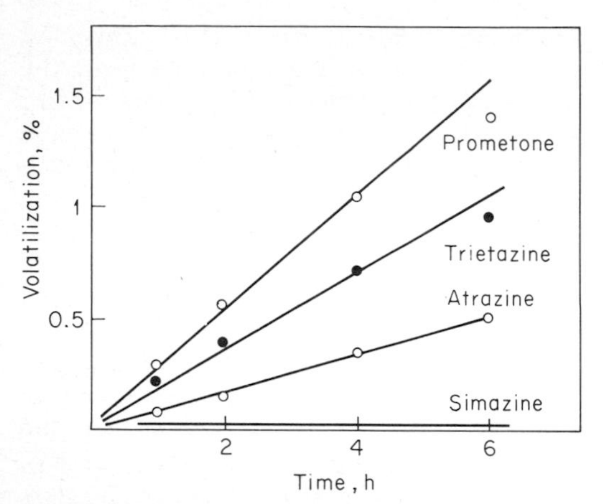

Fig. 18-17 There are differences in the volatilization of four related triazine herbicides, as illustrated by the rate of loss from steel planchets at 25°C (Kearney et al., 1964, cited by Jordan et al., 1970).

Metabolism

While it has been evident for quite a few years that applied organic chemicals can be metabolized by enzymic or nonenzymic systems in plants, the actual reactions and end products have been worked out for only a few such materials, most commonly the herbicides. In a comprehensive review of the metabolism of pesticides, Casida and Lykken (1969) have pointed out that the growth of knowledge in this area has depended largely on the availability of radioactive-labeled pesticides.

The metabolism of applied chemicals can be generally divided into two categories, metabolism by the formation of derivatives and metabolism by oxidation reactions. By far the most common among the derivatives formed are glycosides, which may be formed either for carboxyl or hydroxyl groups on the molecule. The first example of such a derivative to be found for an applied growth regulator is the glucoside of such auxins as IAA, NAA, and 2,4-D (Zenk, 1962). Glycosides are also formed with gibberellins (Fig. 5-12) and ABA (Fig. 8-8). Among the synthetic chemicals, maleic hydrazide, aminotriazole, and the benzoic acid auxin herbicides frequently form glycosides (Towers et al., 1958). Compounds with carboxyl groups can also be metabolized by the formation of peptide conjugates. The first such conjugate to be described was the aspartic acid conjugate of IAA (Andreae and Good, 1955); since then peptide or protein conjugates have been found for other auxins, such as the benzoic acids, 2,4-D, and tordon. There are early indications that esters may be formed of the acid auxins (Seeley et al., 1955), but more recent studies have not commonly reported ester products of acidic growth regulators.

The oxidative degradation of applied chemicals includes a wide variety of cases, some enzymatic and some nonenzymatic. Perhaps the first direct report of an enzymatic oxidation of a plant-growth regulator was the report of an enzymatic decarboxylation of indoleacetic acid (Tang and Bonner, 1947). Decarboxylation has been reported for each of the types of auxins with acid side chains longer than that of the benzoic acid auxins. A special kind of oxidation of the side chain is β oxidation to remove two-carbon fragments, especially from those with longer side chains (Fawcett et al., 1958). The simple decarboxylation reaction can be carried out also by nonenzymatic means such as photooxidation (Galston and Hillman, 1961). Oxidative degradation can most easily attack terminal carboxyl groups, but many other sites of oxidation can also be hit. For example, many of the herbicides applied to plants suffer removal of halides or other leaving groups with the introduction instead of hydroxyl groups. This is a part of the degradation process for the auxins, tordon, the benzoic acid series, and the triazines. Oxidation through the removal of alkyl groups is probably slightly less common but is especially evident in alkyl groups attached to nitrogen atoms, e.g., the oxidation of the triazines, trifluralin, and the substituted ureas (Casida and Lykken, 1969).

If some species are capable of degrading a herbicide better than others, this could well be the basis of selectivity. In some extensive comparisons, Luckwill and Lloyd-Jones (1960) found that 2,4-D is degraded markedly better by resistant than by sensitive varieties of currant and apple. This is probably not a general difference, however, for Weintraub et al. (1956) were unable to correlate destruction of 2,4-D with susceptibility of corn and bean, and Ragab and McCollum (1960) were unable to correlate destruction of simazine with susceptibility of corn and cucumber.

In some instances, the degradation of applied chemicals may be nonenzymatic. It has been found that a natural component of corn sap can directly detoxify simazine to the 2-hydroxytriazine (Hamilton and Moreland, 1961). The component appears to be a natural hydroxamate in corn (cf. Andersen, 1964). Photochemical degradations have been described for most herbicides, including 2,4-D (Bell, 1956), the benzoic acids (Sheets, 1963), the substituted ureas (Weldon and Timmons, 1961), triazines (Jordan et al., 1970), and trifluralin (Wright and Warren, 1965).

For some herbicides the product of degradation may be the toxic material. The use of phenoxy acid auxins with side chains of four- or even six-carbon lengths takes advantage of the tendencies of plants for β oxidation of straight-chain acids, resulting in the degradative formation of the herbicidal phenoxyacetic acid in the plant (Fawcett et al., 1958). Here additional selectivity can be achieved on the basis of the ability of some species to carry out the β oxidation. Other examples of activation within the plant may be the light-activated herbicides, e.g., triazines, phenyl ureas, acyl anilides, and quaternary dipyridyl compounds (cf. Hilton et al., 1963). Applied to plants in darkness, most of them are essentially without effect, but when treated parts of the plant are exposed to light, a toxic principle is formed, which then may have lethal effects on the plant. Most of these herbicides are inhibitors of photosynthesis, but their actual toxicity seems to reside in the light-activated formation of some toxic substance. That the herbicide itself is converted into the toxic principle seems likely for the quaternary dipyridyl herbicides (Mees, 1960) but is not certain in the other cases.

When the metabolism of auxins to peptides and glycosides was first described, it was observed that these metabolic reactions are adaptive; i.e., exposure of the plant to the applied chemical turns on the enzymes which bring about the changes (Zenk, 1962). This stimulation of the metabolism of auxins by exposure of the plant to auxin is most impressively illustrated by the work of Venis (1972), who shows that the molecular structure needed to turn on the conjugating system is exactly the same as the structural requirements for auxin activity. In the presence of inhibitors of protein synthesis, the adaptive increase in auxin metabolism does not occur.

These examples illustrate that plants have diverse means of degrading applied chemicals and that at least in some instances such degradations may regulate the effectiveness of the application in modifying plant performance.

CHEMICAL MODIFICATION OF PLANTS

Since 1950 there has been a spectacular development of the technology of chemical regulation of plant growth and development. Simultaneous with the almost exponential increase in regulatory systems and regulatory chemicals known to the physiologist and agriculturalist, there has been increasing economic pressure to offset hand labor in agricultural production, and applications of chemicals have been the salient source of change in minimizing hand labor. While herbicides have provided the most impressive economic aid to agricultural production through the use of chemicals, since about 1965 there has been a burst of development of new chemicals which can be utilized in the regulation of plants. From 1950 to 1965, the chemical-regulator scene was dominated by hormones and chemical analogs of them, but since that time the range of types of chemicals and types of chemical controls achievable has been markedly broadened. Among the most versatile of these new regulators are the growth retardants, chemicals which retard the overall growth of the plant without altering their morphology (Cathey, 1964).

Regulatory Strategies

One can divide the possibilities for achieving chemical regulatory effects into three general types: (1) chemicals which alter one or more of

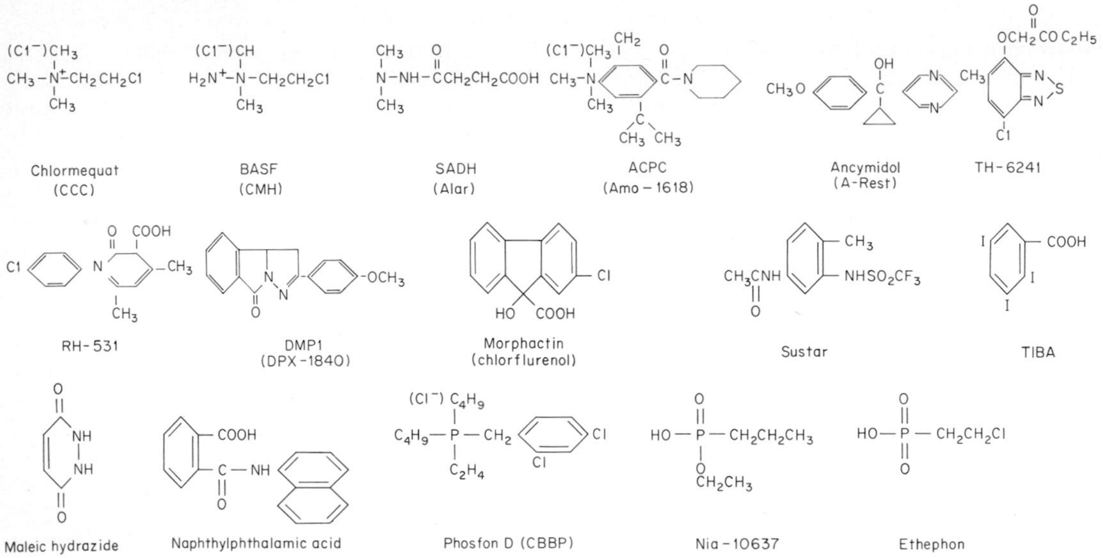

Chlormequat (CCC) | BASF (CMH) | SADH (Alar) | ACPC (Amo – 1618) | Ancymidol (A-Rest) | TH-6241

RH-531 | DMP1 (DPX-1840) | Morphactin (chlorflurenol) | Sustar | TIBA

Maleic hydrazide | Naphthylphthalamic acid | Phosfon D (CBBP) | Nia-10637 | Ethephon

Fig. 18-18 Examples of some commercially available growth regulators.

the endogenous hormone systems, (2) chemicals which exploit interactions between hormone systems, or (3) chemicals which achieve regulatory functions by acting as disruptive agents. Although examples of each of these types of regulation will be given, it should be understood at the outset that the actions of many growth regulators being developed by agricultural chemical firms do not appear to fit into any of these categories. Ordinarily, new growth regulators are discovered by empirical screening programs, and how they act may not be at all clear even when they are widely utilized in agriculture.

Altering Hormone Systems

The most direct type of regulation can be achieved by simply applying one of the plant hormones or an analog of it; we have seen numerous synthetic auxins used for this purpose (Fig. 4-20), and while the widest use of these is for herbicidal purposes, there are numerous regulatory uses as well, e.g., regulation of rooting, sprouting of potatoes, flowering of pineapple, fruit-set of tomato, and abscission of numerous fruits (Weaver, 1972). Synthetic analogs of cytokinins are likewise very diverse in structure (Figs. 6-1 and 6-7) and have found applications in a more limited range of uses, including the regulation of dormancy, and the deferral of senescence in stored vegetables (Weaver, 1972). Gibberellin is ordinarily applied as gibberellic acid (GA_3), one of the commonest endogenous hormone forms, and has found some uses in regulation of fruit growth and in the malting of barley in the beer industry. Ethylene applications are used to stimulate ripening or coloring of some fruits. ABA is not used agriculturally.

Instead of achieving regulation by the application of the hormone itself or an analog, one can apply a chemical which will alter the synthesis of the hormone, or its transport, action, or destruction. The first regulator known to inhibit a hormone synthesis was the growth retardant ACPC (AMO-1618) (Fig. 18-18); this retardant and some others, such as chlormequat and sometimes CBBP (Phosfon D), interfere with the biosynthesis of gibberellin (see review by Lang, 1970). Illustrative of the inhibition of GA biosynthesis are some data in Fig. 5-10. The regulatory effects of these retardants (and some others) are commonly relieved by applications of gibberellin to the plant.

Instead of applying a chemical which inhibits

hormone synthesis, one can apply a stimulator of hormone production; most common in this category is ethephon (Fig. 18-18), which breaks down in the plant to release ethylene (Warner and Leopold, 1967; Cooke and Randall, 1968). A part of the effect of this compound arises also out of the subsequent elevation of endogenous ethylene biosynthesis (Suzuki et al., 1971; Leopold, 1972).

An inhibition of hormone transport can be the basis of regulation by some growth retardants. The effectiveness of three such compounds in the inhibition of auxin transport is illustrated in Fig. 18-19.

Interference with hormone action itself is illustrated by the growth retardant ancymidol, shown in Fig. 18-20. The growth-stimulating action of gibberellin is erased by the retardant. A similar inhibition of gibberellin actions has been reported for RH-531 (Yih et al., 1971).

It is a reasonable presumption that hormones act at the molecular level through attachment to some binding site; TIBA has been found to compete with auxin for an attachment site on some membrane fraction, presumably the plasmalemma (Hertel et al., 1972). Naphthylphthalamic acid and morphactin become attached

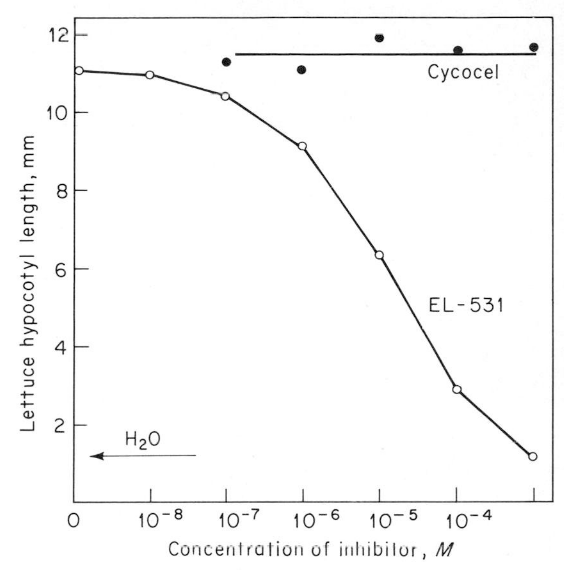

Fig. 18-20 Some growth retardants can antagonize gibberellin-stimulated growth, for example, ancymidol (EL-531); cycocel is not active in this type of test; lettuce hypocotyl test in the presence of 10 μM GA. (*Adapted from Leopold, 1971.*)

to another site on the membrane (Thomson, 1972; Thomson and Leopold, 1974). Such attachment may be involved in the interference of the action of auxin or its transport, or both.

Regulation through the chemical stimulation of hormone destruction is a reasonable possibility, but few instances of chemical regulation in this manner are known. The enzymatic oxidation of indoleacetic acid is known to be stimulated by some phenolics (Fig. 4-12), and application of *p*-coumaric acid to plants can lead to a stimulation of auxin oxidation, as illustrated in Table 8-1.

Fig. 18-19 Some growth retardants are effective in inhibiting the polar transport of auxin, including chlorofluorenol, napthylphthalamic acid, and DMPI; transport of NAA* through 10-mm sections of cotton stems, 4 h basipetal transport (Beyer, 1972).

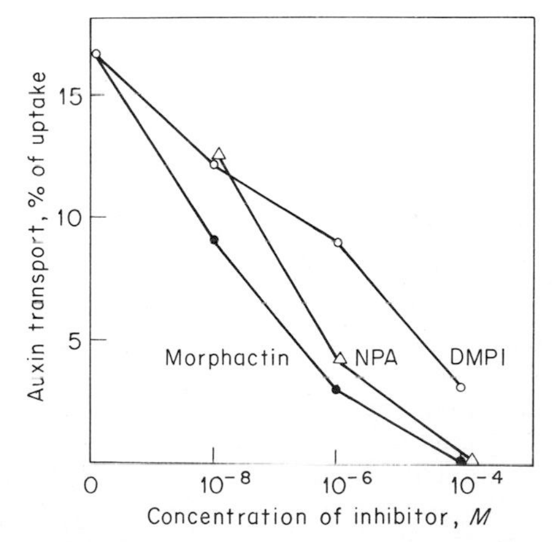

Regulation through Hormone Interaction

While achieving regulatory control by utilizing hormone interactions, e.g., combining two regulators to achieve an enhanced physiological effect or applying one type of hormone in order to bring about a feedback alteration of another, is just beginning to be exploited, its possibilities seem highly promising. To illustrate how two regulators in combination can achieve a unique or

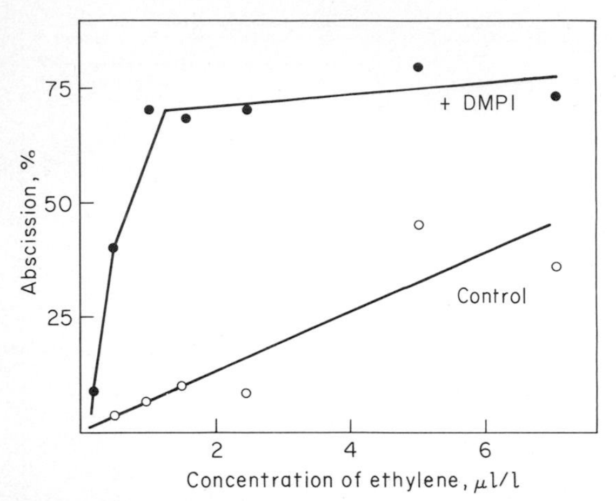

Fig. 18-21 The ethylene stimulation of cotton-leaf abscission can be greatly amplified by previous application of DMPI, an inhibitor of auxin transport (Morgan and Durham, 1972).

magnified effect, Morgan and Durham (1972) applied inhibitors of auxin transport to cotton plants, including naphthylphthalamic acid, triiodobenzoic acid, or DMPI (Fig. 18-18), in order to inhibit auxin transport, and then applied ethephon to stimulate leaf abscission. It is known that auxin serves as the natural inhibitor of abscission and ethylene as the stimulator; thus blocking auxin transport deprives the abscission zones of the natural inhibitor and they become increasingly susceptible to the stimulator of abscission (Fig. 18-21). Combinations of regulators should find numerous applications in the future.

The alteration of one hormone system by application of another is made possible by the numerous and complex types of feedback interactions between the endogenous hormone systems in plants. The first of these to be described was the mutual feedback between auxin and ethylene, as described by Burg and Burg (1966*a*), in which high concentrations of auxin lead to the biosynthesis of ethylene and ethylene leads to a suppression of auxin biosynthesis. Thus, a substantial alteration of auxin levels can be achieved by applications of ethylene, and, conversely, an enhancement of ethylene levels in plants can be achieved by applications of auxin. This was found to be the explanation of the stimulation of flowering in pineapple by auxin applications; the subsequent ethylene production was the actual cause of the flowering response (Burg and Burg, 1966*b*). An interesting case of mutual hormone alterations is reported by Mayak and Halevy (1972); ABA is consistently weak in effectiveness when applied to the foliage of plants, presumably because of inactivation during penetration, but Mayak and Halevy obtained marked increases in endogenous ABA levels in cut roses by applications of ethylene or ethephon.

In connection with chemicals which can have interactions with the hormones, hormone synergists should be mentioned. The very interesting possibility of obtaining regulatory effects by synergists has received very little attention, though occasional synergists of auxin have been described (Gorter, 1958; Thimann et al., 1962; Veldstra, 1964). It is unlikely that hormone synergists would be detected in the usual manner of screening for plant regulators, and this may account for the present shortage of interesting chemicals in this category.

Regulation through Disruptive Action

Disruption of the plant by applied chemicals can be roughly divided into cases where the plant tissue is actually desiccated by the applied chemical, e.g., some defoliants, and cases in which the tissues are only slightly disrupted but their subsequent functioning is altered. The comparative effects of these types of defoliating chemicals are compared by Weaver (1972, Chap. 9). Biggs (1957) found that the degree of stimulation of abscission by leaf desiccants may be accounted for on the basis of the amount of tissue desiccated.

Among the gentler chemical treatments which disrupt plant tissues and thus obtain an altered physiological state, the effects are often due to the ethylene production following disruption of the tissue. Mechanical bruises can cause ethylene formation and stimulate the ripening of fruits in storage (McGlasson, 1969). Some fruits stimulated to grow by ethylene can also be stimulated to grow by cutting gashes in the fruit, thus producing ethylene (Zeroni et al., 1972). In the chemical armamentorium, the abscission of citrus fruits can be hastened by numerous types of chemicals, including ascorbic acid, isoascorbic acid, fluoroacetic acid, and various cations, including copper, iron, and manganese (Cooper et al., 1968). In all except the manganese, a burst of ethylene production is associated with the stimulation of abscission. In manganese stimulation of abscission, it is likely that this cation acts through an enhancement of the oxidation of indoleacetic acid (Biggs, 1971). The abscission of citrus fruits is commercially stimulated by applications of cycloheximide, and as shown in Fig. 18-22, the concentrations which are most effective in facilitating abscission are the same concentrations as stimulate ethylene formation by the fruits. The stimulation of cotton leaf drop by ammonium thiocyanate has been described as being due to the associated destruction of auxin (Swets and Addicott, 1955). Most other cotton defoliants appear to have their effect at least in part through an enhancement of ethylene formation (Hall, 1952).

Another chemical control that might be classified as disruptive is the inhibition of suckering or lateral bud growth by various fatty acids. In the tobacco crop, the prevention of laterals is particularly important in maintaining good quality in the mature leaves, and the use of fatty acids has been successful (Tso, 1964). An overall spray of the plants with acids or alcohols of 6 or 18 carbon lengths leads to the preferential necrosis of the lateral buds (Cathey et al., 1966), in a manner suggestive of a burning or disruptive effect. Maleic hydrazide has also been widely used for the sucker control in tobacco (Zukel, 1955), and its effect may be a consequence of the readiness with which the chemical is translocated to growing meristems (Klein and Leopold, 1953), combined with inhibitory effects of the chemical on nucleic acid synthesis (Nooden, 1972).

Chemically Regulated Functions

The agricultural uses to which chemical regulators may be put have been comprehensively treated by Weaver (1972), and so only the general classes of regulations will be listed here, along with a few examples.

Growth is itself amenable to control, not only in terms of the stimulation of elongation, e.g., conversion of bush beans into pole beans by gibberellin applications (Fig. 5-2), but also the imposition of dwarf growth habit by the growth retardants. The application of gibberellins can relieve the dwarf habit of some genetic lines of corn (Fig. 5-14), and, conversely, the application of some growth retardants can convert a standard line of corn into the dwarf (Fig. 18-23).

Rooting has been stimulated commercially with auxins for many years (Fig. 9-6). More recently, ethylene-generating chemicals have been found to have marked promotive effects on the rooting of some species (Kawase, 1971). Cytokinins can stimulate the differentiation of buds (Fig. 6-8), a feature useful in some types of plant propagation (Plummer and Leopold, 1957).

Fig. 18-22 Applications of cycloheximide sprays can bring about the formation of ethylene by Valencia orange fruits, leading to a marked reduction in the force needed for their abscission. Ethylene determinations were of internal atmospheres in the fruits 4 days after spray treatments (Cooper et al., 1969).

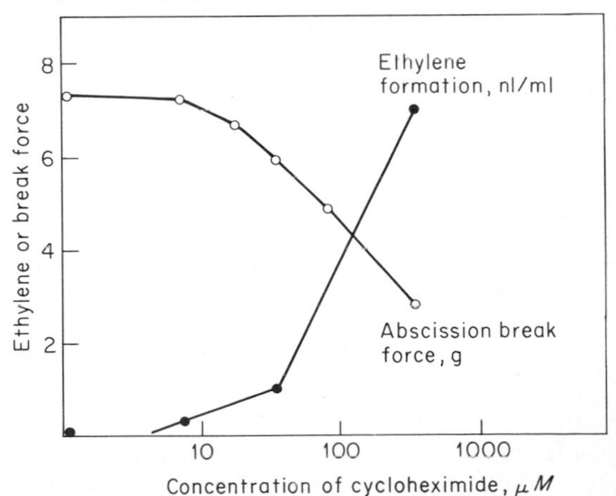

Fig. 18-23 The growth retardant ancymidol can cause the dwarfing habit in corn seedlings (*left*), and applications of gibberellin can restore normal growth to plants inhibited in this way (*center*). Untreated control plants on right; 1% talc dust of ancymidol applied to seeds (*middle and left*), and 0.1 mM GA subsequently applied (*middle*) (Leopold, 1971).

Dormancy is often broken by gibberellin applications (Fig. 10-12), although in some instances gibberelin can actually impose dormancy (Nagao and Mitsui, 1959). Ethylene is effective in breaking dormancy of some seeds (Fig. 7-8), and abscisic acid is sometimes effective in inducing dormancy (Fig. 8-1). Thiourea is one of the most useful chemicals for breaking seed dormancy, though relatively high concentrations are required to bring about the germination effect.

Flowering is susceptible to chemical control in a few crops; e.g., the pineapple (Clark and Kerns, 1942) has been commercially brought into flowering with auxins. Now ethylene-generating compounds are generally utilized to control pineapple flowering (Cooke and Randall, 1968), since the auxin stimulation was found to be a consequence of the ethylene formed in response to the applied auxin (Burg and Burg, 1966*a*). The same ethylene control is common to the bromeliads generally (Cathey and Taylor, 1970). Growth retardants are strikingly effective in inhibiting the flowering of some plants (Zeevaart, 1964*b*), but in other crops they markedly stimulate flowering (Monselise et al., 1966).

The sex of flowers can be regulated by applied chemicals, most particularly by the ethylene-generating chemicals (McMurray and Miller, 1968) and growth retardants (Robinson et al., 1971; Quebedeaux and Beyer, 1972). The induction of male sterility is another type of chemical regulation which shows great promise (Eaton, 1957).

Fruit-set is susceptible to chemical alteration in some species, though it has not been used extensively at the applied or economic level (review of Crane, 1969).

Fruit growth rates can be chemically regulated in some crops, e.g., by applications of auxins (Crane, 1952), gibberellins (Jackson, 1968), cytokinins, and ethylene or ethylene generators (Maxie and Crane, 1968).

Senescence of leaves can be altered with the applications of gibberellins (Fig. 5-8) or cytokinins (Figs. 6-3 and 11-22). This could be useful in the storage of vegetables (Dennis et al., 1967), but the effect is not used agriculturally. The ripening of fruits is of course highly susceptible to stimulation by ethylene (Fig. 7-9), and using ethylene generators makes it possible to regulate ripening for the machine harvesting of tomatoes (Fig. 18-24). Gibberellins have been useful in delaying the ripening of other fruits, especially citrus (Coggins and Lewis, 1962).

Abscission control is particularly useful in removing leaves for mechanical harvesting of cotton and soybeans and in loosening fruits for mechanical harvesting. The abscission of leaves

Fig. 18-24 The ripening of tomato fruits can be enhanced by overall sprays of ethephon. Plants were mechanically harvested at the times indicated, 2 weeks after ethephon spray (Dostal and Wilcox, 1971).

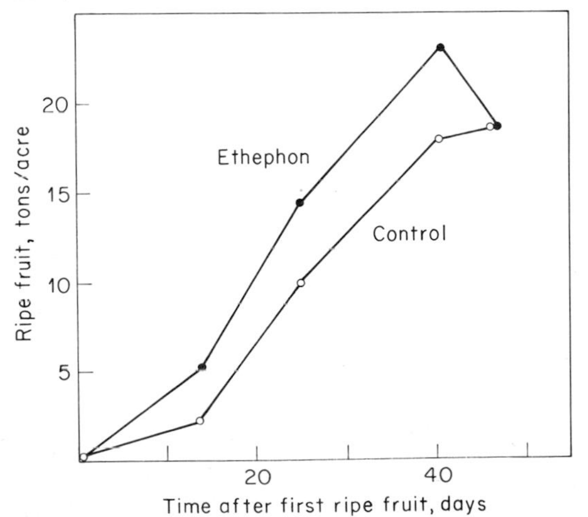

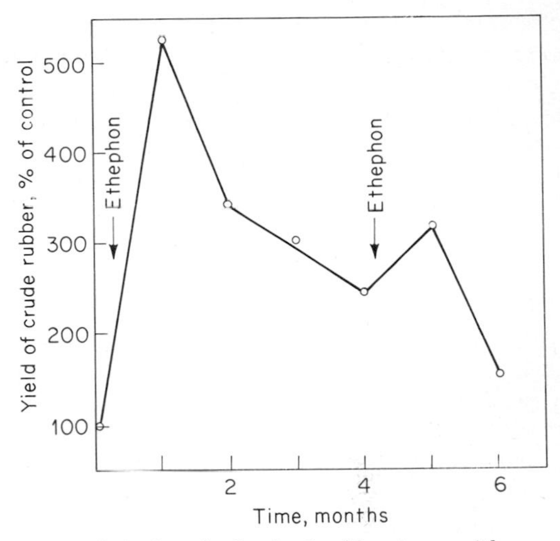

Fig. 18-25 Painting the bark of rubber trees with an ethephon solution results in a severalfold increase in rubber production over a period of several months. Retreatment after 4 months was relatively ineffective (Pakianathan, 1970).

is amenable to stimulation by ethylene-generating chemicals or by various disruptive agents (Hall, 1958). For loosening fruits for harvest, a wide range of chemical materials can be employed (Cooper et al., 1968).

Protection against environmental stress can be obtained in many species by the application of growth retardants, probably in this case due to the changes associated with growth inhibition (Halevy and Kessler, 1963). Limitations on transpiration can be achieved either with chemicals which coat the leaf in such a way that they suppress water loss (review of Gale and Hagan, 1966) or with chemicals which cause stomatal closure.

Several crops have chemical regulatory uses that are specialized rather than fitting into any of the above categories of growth and development. Most prominent is the stimulation of latex production by ethylene-generating chemicals, as in Fig. 18-25. Earlier plantation practice involved the application of auxins to obtain this effect, but, as in pineapple flowering, the auxin effectiveness was probably due to the generation of ethylene in the plant. Ethephon application is more direct and of course less damaging to the tree (Boatman, 1970). Another special crop application is chemical applications to sugarcane to increase the sucrose content of the cane (Monsanto, 1972).

When we return to the evidence cited in Table 18-1 that each of the steps in plant growth and development may be amenable to endogenous control by any of the plant-hormone systems known to the physiologist today, it is hardly surprising that the applications of chemicals to plants should not show any clear specificity for one type of chemical material in the regulation of any given growth or developmental step. To be able to achieve chemical regulation of a developmental function in any given crop, it would seem reasonable to empirically test each known type of hormone and chemical regulator, presuming that any of the hormone types may be naturally limiting or that any hormone may be altered in its level or effectiveness by the application of another (see Fig. 9-36).

The range of chemical regulations that can be exploited will certainly continue to expand. The possibilities of chemically regulating photosynthesis (Treharne and Stoddart, 1968) and of regulating photorespiration and thus increasing net photosynthesis appear to be attractive underdeveloped areas. Scattered evidence that chemicals may alter the interchange of phytochrome between the red- and the far-red-absorbing form (Mann et al., 1967) presents another highly attractive possibility. While major attention has been given to regulatory effects of organic chemicals, inorganic compounds may prove effective as well (Poovaiah and Leopold, 1973).

This brief statement of the representative types of plant functions which can be chemically manipulated illustrates the enormous range of the chemical influences on growth and developmental functions. It is consistent with the diverse evidence that chemical systems are principal regulators of growth and development of plants and that the growth and function of individual cells in the multicellular organism are to a considerable extent products of a special molecular ecology—a dynamic interaction of chemical signals between cells of the plant.

GENERAL REFERENCES

Schneider, G. 1970. Morphactins: Physiology and performance, *Annu. Rev. Plant Physiol.* 21:499–536.

Weaver, R.J. 1972. "Plant Growth Substances in Agriculture." W.H. Freeman Co., San Francisco. 594 pp.

APPENDIX

Table A-1 Abbreviations Used in the Text

Abbreviation	Meaning
ABA	Abscisic acid
ACPC	Ammonium (5-hydroxycarvacryl) trimethyl chloride piperidine carboxylate (AMO-1618)
ADP	Adenosine diphosphate
AMO-1618	*See* ACPC
AMP	Adenosine monophosphate
ATP	Adenosine triphosphate
BA	Benzyladenine
CBBP	2,4 Dichlorobenzyltributylphosphonium chloride (Phosfon D)
CCC	*See* chlormequat, Table A-2
CGR	Crop growth rate
CMP	Cytidine monophosphate
Cl-CCP	Carbonylcyanide-*m*-chlorophenyl hydrazine
CMU	*See* monuron, Table A-2
CTP	Cytidine triphosphate
2,4-D	2,4-Dichlorophenoxyacetic acid
DCMU	*See* diuron, Table A-2
DEAE	Diethylaminoethyl
DMAA	γ,γ-Dimethylallyladenine
DMPI	3,3-α-Dihydro-2-(*p*-methoxyphenyl)-8-β-pyrazolo-5, 1-α-isoindol-8-one (DPX-1840)
DNA	Deoxyribonucleic acid
DNP	2,4-Dinitrophenol
DPX-1840	*See* DMPI
EDTA	Ethylenediaminetetraacetic acid
EL-531	*See* ancymidol, Table A-2
FMN	Flavin mononucleotide
FRS	Ferredoxin-reducing substance
5-FU	5-Fluorouracil
FUDR	5-Fluorodeoxyuridine
GA	Gibberellin, gibberellic acid
GAP	Glyceraldehyde 3-phosphate
HER	High-energy reactions
IAA	Indole-3-acetic acid
IPC	Isopropylphenylcarbamate
KMB	α-Keto-γ-mercaptobutyrate
LAD	Leaf-area duration
LAI	Leaf-area index

Table A-1 (Continued)

LAF	Light-activating factor
LAR	Leaf-area ratio
LDP	Long-day plant
LSDP	Long-short-day plant
MAK	Methylated-albumin-kieselguhr (column)
MFM	Microfibrillar material
mRNA	Messenger RNA
NAA	Naphthaleneacetic acid
NAD	Nicotinamide adenine dinucleotide
NADP	Nicotinamide adenine dinucleotide phosphate
NADPH	Reduced NADP
$NADPH_2$	Reduced NADP
NAR	Net assimilation rate
OD	Optical density
OP	Osmotic pressure
P_i	Inorganic phosphate
PBA	Tetrahydrylpyranylbenzyladenine
PEP	Phosphoenol pyruvate
PGA	Phosphoglyceric acid
PhAR	Photosynthetically active radiation
PmAR	Photomorphologically active radiation
RGR	Relative growth rate
RH-531	1-(4-Chlorophenyl)-3-carboxyl-4,6-dimethyl-2-pyridine
RLGR	Relative leaf-growth rate
RNA	Ribonucleic acid
rRNA	Ribosomal RNA
RuDP	Ribulose 1,5-diphosphate
SADH	Succinic acid-2,2-dimethylhydrazine (Alar)
SDP	Short-day plant
SLDP	Short-long-day plant
SKF 7997	Tris(2-ethylaminoethyl) phosphate trihydrochloride
sRNA	Soluble RNA
2,4,5-T	2,4,5-Trichlorophenoxyacetic acid
THO	Tritiated water
TIBA	2,3,5-Triiodobenzoic acid
TP	Turgor pressure
tRNA	Transfer RNA
UDP	Uridine diphosphate

Table A-2 Systematic Names for Chemicals with Common Names

Common name	Systematic name
Alar	Succinic acid-2,2-dimethylhydrazine (SADH)
Amiben	3-Amino-2,5-dichlorobenzoic acid
Ancymidol	α-Cyclopropyl-α-(4-methoxyphenyl)-5-pyrimidine methanol (EL-531)
Chlormequat	(2-Chloroethyl)trimethyl ammonium chloride (CCC)
Chlorflurenol	2-Chloro-9-hydroxyfluorene-9-carboxylic acid (Morphactin)
Dicamba	2-Methoxy-3,6-dichlorobenzoic acid
Dactal	Dimethyl-2,3,5,6-tetrachloroterephthalate
Dalapon	Sodium 2,2-dichloropropionate
Diuron	3-(3,4-Dichlorophenyl)-1,1-dimethylurea (DCMU)
Ethephon	2-Chloroethylphosphonic acid
Linuron	1-Methoxy-3-(3,4,-dichlorophenyl)-1-methylurea
Monuron	3-(*p*-Chlorophenyl)-1,1-dimethylurea
Morphactin	*See* chlorflurenol
Simazine	2-Cloro-4,6-bis(ethylamino)-1,3,5-triazine
Tordon	4-amino - 3, 5, 6-trichloropicolinic acid
Tricamba	2-Methoxy-3,5,6-trichlorobenzoic acid
Trifluralin	*N*,*N*-Di-*n*-propyl-2,6-dinitro-4-trifluoromethylaniline

Table A-3 Common Prefixes for Powers of 10†

Value	Prefix	Abbreviation
10^{12}	tera	T
10^{9}	giga	G
10^{6}	mega	M
10^{3}	kilo	k
10^{-1}	deci	d
10^{-2}	centi	c
10^{-3}	milli	m
10^{-6}	micro	μ
10^{-9}	nano	n
10^{-12}	pico	p

†Duplicated prefixes are no longer allowed, so that mμ- becomes n- and $\mu\mu$- becomes p-.

REFERENCES

Abbott, D. L. 1970. The role of budscales in the morphogenesis and dormancy of the apple fruit bud, pp. 65–80. In L. C. Luckwill and C. V. Cutting (eds.), "Physiology of Tree Crops." Academic, London.

Abdel-Gawad, H. A., and R. J. Romani. 1967. Effects of phytohormones on maturation and postharvest behavior of fruit. *Plant Physiol.,* 42 (suppl.): 43.

Abeles, F. B. 1967. Mechanism of action of abscission accelerators. *Physiol. Plant.,* 20: 442–454.

———. 1969. Abscission: Role of cellulase. *Plant Physiol.,* 44: 447–452.

——— and H. E. Gahagan. 1968. Abscission: The role of ethylene, ethylene analogues, carbon dioxide and oxygen. *Plant Physiol.,* 43: 1255–1258.

——— and R. E. Holm. 1967. Abscission: Role of protein synthesis. In J. P. Fredrick (ed.), "Plant Growth Regulators." *Ann. N.Y. Acad. Sci.,* 144: 367–373.

———, ———, and H. E. Gahagan. 1967. Abscission: The role of aging. *Plant Physiol.,* 42: 1351–1356.

——— and G. R. Leather. 1971. Abscission: Control of cellulase secretion by ethylene. *Planta,* 97: 87–91.

——— and J. Lonski. 1969. Stimulation of lettuce seed germination by ethylene. *Plant Physiol.,* 44: 227–280.

——— and B. Rubinstein. 1964*a*. Cell-free ethylene evolution from etiolated pea seedlings. *Proc. 2d Defoliation Conf., Ft. Detrick, Md.,* pp. 21–25.

——— and ———. 1964*b*. Regulation of ethylene evolution and leaf abscission by auxin. *Plant Physiol.,* 39: 963–969.

———, J. M. Ruth, L. E. Forrence, and G. R. Leather. 1972. Mechanisms of hormone action: Use of deuterated ethylene. *Plant Physiol.,* 49: 669–671.

Åberg, B. 1958. Studies on plant growth regulators. XIV: Some indole derivatives. *Ann. R. Agric. Coll. Swed.,* 24: 375–395.

——— and I. Johansson. 1969. Studies on plant growth regulators. XXIV: Some phenolic compounds. *K. Lantbrukshogsk. Ann.,* 35: 3–27.

——— and E. Jonsson. 1955. Studies on plant growth regulators, XI. *K. Lantbrukshogsk. Ann.* 21: 401–416.

Abrams, G. J. von, and H. K. Pratt. 1968. Effect of the kinetin–naphthaleneacetic acid interaction upon total RNA and protein in senescing detached leaves. *Plant Physiol.,* 43: 1271–1278.

Abrol, Y. P., and W. J. McIlrath. 1966. Studies on the dehydration of seeds during fruit development in *Lycopersicon esculentum. Indian J. Plant Physiol.,* 9: 66–80.

Acevedo, E., T. C. Hsiao, and D. W. Henderson. 1971. Immediate and subsequent growth responses of maize leaves to changes in water status. *Plant Physiol.,* 48: 631–636.

Adams, M. S., and B. R. Strain. 1968. Photosynthesis in stems and leaves of *Cercidium floridum*: Spring and summer diurnal field response and relation to temperature. *Oecol. Plant.,* 3: 285–297 (Gauthier-Villars, Paris).

———, ———, and I. P. Ting. 1967. Photosynthesis in chlorophyllous stem tissue and leaves of *Cercidium floridum*: Accumulation and distribution of ^{14}C from $^{14}CO_2$. *Plant Physiol.,* 42: 1797–1799.

Addicott, F. T., and R. S. Lynch. 1951. Acceleration and retardation of abscission by indoleacetic acid. *Science,* 114: 688–689.

——— and J. L. Lyon. 1969. Physiology of abscisic acid and related substances. *Ann. Rev. Plant Physiol.,* 20: 139–164.

———, K. Ohkuma, O. E. Smith, and W. E. Thiessen. 1966. *Adv. Chem.,* 53: 97–105.

Ahlgren, G. E., and T. W. Sudia. 1964. Absorption of P-32 by leaves of *Glycine max* of different ages. *Bot. Gaz.,* 125: 204.

Aikman, D. P., and N. P. Anderson. 1971. A quantitative investigation of a peristaltic model for phloem translocation. *Ann. Bot.,* 35: 761–772.

Akamine, E. K. 1963. Ethylene production in fading *Vanda* orchid blossoms. *Science,* 140: 1217–1218.

——— and G. Girolami. 1959. Pollination and fruit set in the yellow passion fruit. *Hawaii Agric. Exp. Stn. Tech. Bull.,* 39: 1–45.

Albersheim, P., and J. Bonner. 1959. Metabolism and hormonal control of pectic substances. *J. Biol. Chem.,* 234: 3105–3108.

Albrigo, L. G. 1972. Ultrastructure of cuticular surfaces and stomata of developing leaves and fruit of the Valencia orange. *J. Am. Soc. Hort. Sci.,* 97: 761–765.

Allard, H. A., and W. W. Garner. 1940. Observations on responses to length of day. *USDA Tech. Bull.* 727.

Allen, L. H., C. S. Yocum, and E. R. Lemon. 1964. Photosynthesis under field conditions. VII: Radiant energy exchanges within a corn crop canopy and implications in water use efficiency. *Agron. J.,* 56: 253–259.

Amen, R. D. 1968. A model of seed dormancy. *Bot. Rev.,* 34: 1–31.

Amir, S., and L. Reinhold. 1971. Interaction between K-deficiency and light in ^{14}C-sucrose translocation in bean plants. *Physiol. Plant.,* 24: 226–231.

Andersen, K. S., J. M. Bain, D. G. Bishop, and R. M. Smillie. 1972*b*. Photosystem II activity in agranal bundle sheath chloroplasts from *Zea mays*. *Plant Physiol.,* 49: 461–466.

Andersen, R. N. 1964. Differential response of corn inbreds to simazine and atrazine. *Weeds,* 12: 60–61.

Anderson, A. S., I. Moller, and J. Hansen. 1972. 3-Methyleneoxindole and plant growth regulation. *Physiol. Plant.,* 27: 105–108.

Anderson, J. K., and K. S. Rowan. 1965. Activity of peptidase in tobacco leaf tissue in relation to senescence. *Biochem. J.,* 97: 741–746.

Anderson, M. B., and J. H. Cherry. 1969. Differences in leucyl-transfer RNAs and synthetase in soybean seedlings. *Proc. Natl. Acad. Sci. (U.S.),* 62: 202–209.

Anderson, M. C. 1964. Light relations of terrestrial plant communities and their measurement. *Biol. Rev.,* 39: 425–486.

Anderssen, F. G. 1929. Some seasonal changes in the tracheal sap of pear and apricot trees. *Plant Physiol.,* 4: 459–476.

Andreae, W. A., and N. E. Good. 1955. The formation of indoleaspartic acid in pea seedlings. *Plant Physiol.,* 30: 380–382.

——— and ———. 1957. Studies in indoleacetic acid metabolism. IV: Conjugation with aspartic acid and ammonia as processes in the metabolism of carboxylic acids. *Plant Physiol.,* 32: 566–572.

Andreae, W. A., M. A. Venis, F. Jursic, and T. Dumas. 1968. Does ethylene mediate root growth inhibition by indoleacetic acid? *Plant Physiol.,* 43: 1375–1379.

——— and M. W. van Ysselstein. 1960. Studies of indoleacetic acid metabolism. VI: Indoleacetic acid uptake and metabolism by pea roots and epicotyls. *Plant Physiol.,* 35: 225–232.

Andrews, S., and J. Levitt. 1967. The effect of cryoprotective agents on intermolecular SS formations during freezing of Thiogel. *Cryobiology,* 4: 85–89.

Anker, L. 1958. The influence of the pH on the growth and geotropism of decapitated *Avena* coleoptiles supplied either with indoleacetic acid or with indoleacetonitrile. *Acta Bot., Neerl.,* 7: 69–76.

Antoszewsky, R. 1970. Studies on the physiology of fruit growth and fruit development. *Tagungsber. Dtsch. Akad. Landwirtsch. Ber.,* 99: 73–84.

Aoki, S., and E. Hase. 1964. De- and re-generation of chloroplasts in *Chlorella* I and II. *Plant Cell Physiol.,* 5: 473–493.

Appleman, C. O., and E. V. Miller. 1926. A chemical and physiological study of maturity in potatoes. *Agric. Res.* 33: 569–578.

Arnold, A. 1952. Uber den Funktionsmechanismus der Endodermiszellen der Wurzeln. *Protoplasma,* 41: 189–211.

Arnold, W. N. 1968. The selection of sucrose as the translocate of higher plants. *J. Theoret. Biol.,* 21: 13–20.

Arnon, D. I. 1960. The chloroplast as a functional unit in photosynthesis. *Handb. Pflanzenphysiol.,* 5(1): 773–829.

Aronsson, A., and L. Eliasson. 1970. Frost hardiness in scots pine (*Pinus silvestris* L.). I: Conditions for test on hardy plant tissues and for evaluation of injuries by conductivity measurements. *Stud. For. Suec.* 77: 1–30.

Arreguin-Lozano, B., and J. Bonner. 1949. Experiments on sucrose formation by potato tubers as influenced by temperature. *Plant Physiol.,* 24: 720–738.

Asahira, T., and J. P. Nitsch. 1968. Tuberisation in vitro: *Ullucus tuberosus* et *Dioscorea*. *Bull. Soc. Bot. Fr.,* 115: 345–352.

Ascher, P. D. 1966. A gene action model to explain gametophytic self-incompatibility. *Euphytica,* 15: 179–183.

——— and S. J. Peloquin. 1966. Influence of temperature on incompatible and compatible pollen tube growth in *Lilium longiflorum*. *Can. J. Genet. Cytol.,* 8: 661–664.

Ashton, F. M. 1956. Effects of a series of cycles of low and high soil water on the rate of apparent photosynthesis in sugarcane. *Plant Physiol.,* 31: 266–274.

———. 1958. Absorption and translocation of radioactive 2, 4-D in sugarcane and bean plants. *Weeds,* 6: 257–262.

Aslander, A. 1927. Sulphuric acid as a weed spray. *J. Agric. Res.,* 34: 1065–1091.

Atsmon, D., A. Lang, and E. N. Light. 1968. Contents and recovery of gibberellins in monoecious and gynoecious cucumber plants. *Plant Physiol.,* 43: 806–810.

Audus, L. J. 1949. The biological detoxication of 2,4-D in soil. *Plant Soil,* 2: 31–35.

———. 1962. The mechanism of the perception of gravity by plants. *Symp. Soc. Exp. Biol.,* 16: 197–226.

———. 1972. "Plant Growth Substances." Leonard Hill, London.

——— and A. N. Lahiri. 1961. Studies on the geotropism of roots, III. *J. Exp. Bot.,* 12: 75–84.

Aufdemgarten, H. 1939. The so-called induction process in carbon dioxide assimilation. *Planta,* 29: 643–678.

Aung, L. H., A. A. de Hertogh, and G. Staby. 1969. Temperature regulation of endogenous gibberellin activity and development of *Tulipa gesneriana*. *Plant Physiol.,* 44: 403–406.

Avery, D. J. 1969. Comparisons of fruiting and deblossomed maiden apple trees and of non-fruiting trees on a dwarfing and an invigorating rootstock. *New Phytol.,* 68: 323–336.

———. 1970. Effects of fruiting on the growth of apple trees on four rootstock varieties. *New Phytol.,* 69: 19–30.

Badenhuizen, N. P., and R. W. Dutton. 1956. *Protoplasma,* 47: 156.

Bahr, J. T., and R. G. Jensen. 1974. Ribulose diphosphate carboxylase from freshly ruptured spinach chloroplasts having an *in vivo* K_m [CO_2]. *Plant Physiol.,* 53: 39–44.

Bailey, G. W., and J. L. White. 1970. Factors influencing the adsorption, desorption and movement of pesticides in soils. *Residue Rev.,* 32: 29–92.

Bain, J. M., and F. V. Mercer. 1964. Organization resistance and the respiration climacteric. *Aust. J. Biol. Sci.,* 17: 78–85.

——— and R. N. Robertson. 1951. Physiology of growth of apple fruits, I. *Aust. J. Sci. Res.,* 4: 75–91.

Baker, D. N., and R. B. Musgrave. 1964. Photosynthesis under field conditions. V: Further plant chamber studies of the effects of light on corn (*Zea mays* L.). *Crop Sci.,* 4: 127–131.

Baker, H. G. 1963. Evolutionary mechanisms in pollination biology. *Science,* 139: 877–883.

Bakhuyzen, H. L. S. 1926. Physiological phenomena at the time of flowering. *Proc. Soc. Exp. Biol. Med.,* 24: 143–145.

Baldev, B., and A. Lang. 1965. Control of flower formation by growth retardants and gibberellin in *Samolus parviflorus,* a long-day plant. *Am. J. Bot.,* 52: 408–417.

———, ———, and A. O. Agatep. 1965. Gibberellin production in pea seeds developing in excised pods: Effect of AMO-1618. *Science,* 147: 155–156.

Balegh, S. E., and O. Biddulph. 1970. The photosynthetic action spectrum of the bean plant. *Plant Physiol.,* 46: 1–5.

Ball, N. G., and I. J. Dyke. 1954. An endogenous 24-hour rhythm in the growth rate of the *Avena* coleoptile. *J. Exp. Bot.,* 5: 421–433.

Ballantine, D. J. 1966. Respiration of floral tissue of the daffodil treated with benzyl adenine and auxin. *Can. J. Bot.,* 44: 117–119.

Ballantine, J. E. M., and B. J. Forde. 1970. The effect of light intensity and temperature on plant growth and chloroplast ultrastructure in soybean. *Am. J. Bot.,* 57: 1150–1159.

Ballard, L. A. T., and A. H. K. Petrie. 1936. Physiological ontogeny in plants and its relation to nutrition. *Aust. J. Exp. Biol. Med. Sci.,* 14: 135–163.

Balz, H. P. 1966. Intrazellulare Lokalization and Funktion von hydrolytischen Enzymen bei Tabak. *Planta,* 70: 207–236.

Barber, D. A., M. Ebert, and N. T. S. Evans. 1962. The movement of O^{15} through barley and rice plants. *J. Exp. Bot.,* 13: 397–403.

——— and M. G. T. Shone. 1966. The absorption of silica from aqueous solutions by plants. *J. Exp. Bot.,* 17: 569–578.

Barber, H. N., and D. M. Paton. 1952. A gene-controlled flowering inhibitor in *Pisum. Nature,* 169: 592.

Barendse, G. W. M., H. Kende, and A. Lang. 1968. Fate of radioactive gibberellin A^1 in seeds of peas and Japanese morning glory. *Plant Physiol.,* 43: 815–822.

———, S. Rodriguez-Pereira, P. A. Beckers, F. N. van Eyden-Emons, and H. F. Linskens. 1970. Growth hormones in pollen. *Acta Bot. Neerl.,* 19: 175–186.

Barker, W. G. 1953. A method for *in vitro* culturing of potato tubers. *Science,* 118: 384–385.

Barkley, G. M., and A. C. Leopold. 1973. Comparative effects of hydrogen ions, carbon dioxide and auxin on pea stem segment elongation. *Plant Physiol.,* 52: 76–78.

Barlow, H. W. B. 1970. Some aspects of morphogenesis in fruit trees. Pp. 25–43 in L. C. Luckwill and C. V. Cutting (eds.), "Physiology of Tree Crops." Academic, London.

Barrier, G. E., and W. E. Loomis. 1957. Absorption and translocation of 2,4-dichlorophenoxyacetic acid and P^{32} in leaves. *Plant Physiol.,* 32: 225–231.

Barrs, H. D. 1971. Cyclic variations in stomatal aperture, transpiration, and leaf water potential under constant environmental conditions. *Annu. Rev. Plant Physiol.,* 22: 223–236.

Barton, L. V. 1947. Special studies on seed coat impermeability. *Contrib. Boyce Thompson Inst.,* 14: 355–362.

Bassham, J. A. 1964. Kinetic studies on the photosynthetic carbon reduction cycle. *Annu. Rev. Plant Physiol.,* 15: 101–120.

———. 1965. Photosynthesis: The path of carbon. Pp. 875–902 in J. Bonner and J. E. Varner (eds.), "Plant Biochemistry." Academic, New York.

——— and M. Calvin. 1957. The Path of Carbon in Photosynthesis. Prentice-Hall, Englewood Cliffs, N.J.

——— and R. G. Jensen. 1967. Photosynthesis of carbon compounds. Pp. 79–110 in A. San Pietro, F. A. Greer, and T. J. Army (eds.), "Harvesting the Sun." Academic, New York.

Bate, G. C., and D. T. Canvin. 1971. The effect of some environmental factors on the growth of young aspen trees *(Populus tremuloides)* in controlled environments. *Can. J. Bot.,* 49: 1443–1453.

Bateman, A. J. 1956. Cryptic self incompatibility in the wall flower, *Cheiranthus. Heredity,* 10: 257–261.

Bauchop, T., and S. R. Elsden. 1960. The growth of micro-organisms in relation to their energy supply. *J. Gen. Microbiol.,* 23: 457–469.

Bauer, L. 1949. Über den Wanderungsweg fluoreszierenden Farbstoffe in den Siebrohren. *Planta,* 37: 221–243.

Baur, A., and S. F. Yang. 1969. Ethylene production from propanol. *Plant Physiol.,* 44: 189–192.

———, ———, H. K. Pratt, and J. B. Biale. 1971.

Ethylene biosynthesis in fruit tissues. *Plant Physiol.,* 47: 696–699.

Bavel, C. H. M. van, L. J. Fritschen, and W. E. Lewis. 1963. Transpiration by Sudangrass as an externally controlled process. *Science,* 141: 269–270.

Bayley, S. T., and G. Setterfield. 1957. The influence of mannitol and auxin on growth of cell walls in *Avena* coleoptiles. *Ann. Bot.,* 21: 633–641.

Beams, H. W., and R. L. King. 1944*a*. Effect of ultracentrifuging on polarity in the pollen grains of *Vinca. J. Cell. Comp. Physiol.,* 24: 109–116.

——— and ———. 1944*b*. The negative group effect in the pollen grains of *Vinca. J. Cell. Comp. Physiol.,* 23: 39–46.

Becker, D., and H. Ziegler. 1973. Cyclic adenosine -3′5′-monophosphate in translocation tissues of plants. *Planta,* 110: 85–89.

Beever, J. E., and H. W. Woolhouse. 1973. Increased cytokinin from root system of *Perilla* and flower and fruit development. *Nature New Biol.,* 246: 31–32.

Beevers, H. 1961. Metabolic production of sucrose from fat. *Nature,* 191: 433–436.

———, M. L. Stiller, and V. S. Butt. 1966. Metabolism of the organic acids. Pp. 119–262 in F. C. Steward (ed.), Plant Physiology, vol. IV-B. Academic, New York.

Beinhart, G. 1962. Effects of temperature and light intensity on CO_2 uptake, respiration and growth of white clover. *Plant Physiol.,* 37: 709–715.

Beissner, L. 1888. Über Jugendformen von Pflanzen, special Coniferen. *Ber. Dtsch. Bot. Ges.,* 6: 83.

Belhanafi, A., and G. F. Collet. 1970. Modalités de l'inhibition de la croissance et de la synthèse des acides nucléiques des plantules de blé par l'acide abscissique. *Physiolog. Plant.,* 23: 859–870.

Bell, G. R. 1956. On the photochemical degradation of 2,4-D and related compounds in the presence and absence of riboflavin. *Bot. Gaz.,* 118: 133.

Benedict, C. R., K. J. McCree, and R. J. Kohel. 1972. High photosynthetic rate of chlorophyll mutant of cotton. *Plant Physiol.,* 49: 968–971.

Benes, J., K. Veres, L. Choojka, and A. Friedrich. 1965. New types of kinins and their action on fruit tree species. *Nature,* 206: 830–831.

Bennet-Clark, T. A. 1956. Salt accumulation and mode of action of auxin. Pp. 284–294 in R. L. Wain and F. Wightman (eds.), "Chemistry and Mode of Action of Plant Growth Substances." Butterworth, London.

——— and N. P. Kefford. 1953. Chromatography of the growth substances in plant extracts. *Nature,* 171: 645–647.

Bentley, J. A. 1950. Growth-regulating effect of certain organic compounds. *Nature,* 65: 449.

———, K. R. Farrar, S. Housley, G. F. Smith, and W. C. Taylor. 1956. Some chemical and physiological properties of 3-indolylpyruvic acid. *Biochem J.,* 64: 44–49.

———, and S. Housley. 1952. Studies on plant growth hormones, I. *J. Expt. Bot.,* 3: 393–405.

Ben-Yehoshua, S. 1964. Respiration and ripening of discs of the avocado fruit. *Physiolog. Plant.,* 17: 71–80.

———, R. N. Robertson, and J. B. Biale. 1963. Respiration and internal atmosphere of avocado fruit. *Plant Physiol.,* 38: 194–201.

Ben-Zioni, A., C. Itai, and Y. Vaadia. 1967. Water and salt-stresses kinetin and protein synthesis in tobacco leaves. *Plant Physiol.,* 42: 361–365.

Bergh, B. O., M. J. Garber, and C. O. Gustafson. 1966. The effect of adjacent trees of other avocado varieties on Fuerte fruit-set. *Proc. Am. Soc. Hort. Sci.,* 89: 167–174.

Bernier, G., J. Kinet, and R. Bronchard. 1967. Cellular events at the meristem during floral induction in *Sinapis alba. Physiol. Veg.,* 5: 311–324.

Berridge, M. V., R. K. Ralph, and D. S. Letham. 1972. On the significance of cytokinin binding of plant ribosomes. Pp. 248–255 in D. J. Carr (ed.), "Plant Growth Substances, 1970." Springer-Verlag, Berlin.

Berrie, A. M. M., M. R. Hendrie, W. Parker, and B. A. Knights. 1967. Induction of light sensitive dormancy in seed of lettuce by patulin. *Plant Physiol.,* 42: 889–890.

Beuerlein, J. E., and J. W. Pendelton, 1971. Photosynthetic rates and light saturation curves of individual soybean leaves under field conditions. *Crop Sci.,* 11: 217–219.

Bex, J. H. M. 1972*a*. Effects of abscisic acid on nucleic acid metabolism in maize coleoptiles. *Planta,* 103: 1–10.

———. 1972*b*. Effects of abscisic acid on the soluble RNA polymerase activity in maize coleoptiles. *Planta,* 103: 11–17.

Beyer, E. M. 1972*a*. Auxin transport: A new synthetic inhibitor. *Plant Physiol.,* 50: 322–327.

———. 1972*b*. Mechanism of ethylene action: Biological activity of deuterated ethylene. *Plant Physiol.,* 49: 672–675.

——— and P. W. Morgan. 1969. Time sequence of the effect of ethylene on transport, uptake and decarboxylation of auxin. *Plant Cell Physiol.,* 10: 787.

Bezemer-Sybrandy, S. M., and H. Veldstra. 1971. Investigations on cytokinins. IV: The metabolism of benzylaminopurine in *Lemna minor. Physiolog. Plant.,* 25: 1–7.

Bhargava, S. C. 1963. A transmissible flower bud inhibitor in the short-day plant, *Salvia occidentalis. Proc. K. Ned. Akad. Wet. Amst.,* C66: 371–376.

———. 1964. Photoperiodism, floral induction and floral inhibition in Salvia. *Meded. Landbouwhogesch. Wageningen* 255, 70 pp.

Biale, J. B. 1950. Postharvest physiology and biochemistry of fruits. *Annu. Rev. Plant Physiol.,* 1: 183–206.

———. 1960. Respiration of fruits. *Hand. Pflanzenphysiol.*, 12(-2): 536–592.

———. 1964. Growth, maturation and senescence in fruits. *Science,* 146: 880–888.

——— and F. F. Halma. 1937. The use of heteroauxin in rooting of subtropicals. *Proc. Am. Soc. Hort. Sci.*, 35: 443–447.

———, R. E. Young, and A. J. Olmstead. 1954. Fruit respiration and ethylene production. *Plant Physiol.*, 29: 168–174.

Bick, M., H. Liebke, J. H. Cherry, and B. L. Strehler. 1970. Changes in leucyl- and tyrosyl-*t*RNA of soybean cotyledons during plant growth. *Biochim. Biophys. Acta,* 204: 175–182.

Biddulph, O. 1951. The translocation of minerals in plants. Pp. 261–278 in E. Truog (ed.), "Mineral Nutrition of Plants." University of Wisconsin Press, Madison.

———, S. Biddulph, R. Cory, and H. Koontz. 1958. Circulation patterns for phosphorus, sulfur and calcium in the bean plant. *Plant Physiol.*, 33: 293–300.

——— and R. Cory. 1965. Translocation of ^{14}C metabolites in the phloem of the bean plant. *Plant Physiol.*, 40: 119–129.

——— and J. Markle. 1944. Translocation of radiophosphorus in the phloem of the cotton plant. *Am. J. Bot.*, 31: 65–70.

Bidwell, R. G. S. 1972. Products of photosynthesis by *Acetabularia* chloroplasts: Possible control mechanisms. Pp. 1927–1934 in *2d Int. Congr. Photosynthesis, Stresa, 1971.*

———, W. B. Levin, and D. C. Shepard. 1969. Photosynthesis, photorespiration and respiration of chloroplasts from *Acetabularia mediterrania. Plant Physiol.*, 44: 946–954.

——— and W. B. Turner. 1966. Effect of growth regulators on CO_2 assimilation in leaves, and its correlation with the bud break response in photosynthesis. *Plant Physiol.,* 41: 267–270.

Bieleski, R. L. 1966. Accumulation of phosphate, sulfate and sucrose by excised phloem tissues. *Plant Physiol.,* 41: 447–454.

———. 1969. Phosphorus compounds in translocating phloem. *Plant Physiol.,* 44: 497–502.

Biemann, L., S. Tunakawa, J. Sonnenbichler, H. Feldmann, G. Dütting, and H. G. Zachau. 1966. Struktur eines ungewöhnlichen Nukleosids aus Serin spezifischer Transfer-Ribonukleinsäure. *Angew. Chem.,* 78: 600.

Biggs, R. H. 1957. Physiological basis of abscission in plants. Doctoral thesis, Purdue Univ. 116 pp.

———. 1971. Citrus abscission. *HortScience,* 6: 388–392.

——— and A. C. Leopold. 1955. The effects of temperature on peppermint. *Proc. Am. Soc. Hort. Sci.,* 66: 315–321.

——— and ———. 1957. Factors influencing abscission. *Plant Physiol.,* 32: 626–632.

——— and ———. 1958. The two-phase action of auxin on abscission. *Am. J. Bot.,* 45: 547–551.

Bilderback, D. E. 1972. The effects of hormones upon the development of the excised floral buds of *Aquilegia. Am. J. Bot.,* 55: 525–529.

Billings, W. D., and R. J. Morris. 1951. Reflection of visible and infrared radiation from leaves of different ecological groups. *Am. J. Bot.*, 38: 327–331.

Biran, I., I. Gur, and A. H. Halevy. 1972. The relationship between exogenous growth inhibitors and endogenous levels of ethylene and tuberization of dahlias. *Physiol. Plant.*, 27: 226–230.

——— and A. H. Halevy. 1973. Endogenous levels of growth regulators and their relationship to the rooting of *Dahlia* cuttings. *Physiol. Plant.* 28: 436–442.

Birch, A. J., R. W. Rickards, H. Smith, A. Harris, and W. B. Whalley. 1959. Studies in relation to biosynthesis, XXI. *Tetrahedron,* 7: 241–251.

Birch-Hirschfeld, L. 1920. Untersuchungen über die Ausbreitungschwindigkeit gelöster Stoffe in der Pflanze. *Jahrb. Wiss. Bot.,* 59: 171–262.

Birth, G. S. 1960. Agricultural applications of the dual-monochromator spectrophotometer. *Agric. Eng.,* 41: 432–435.

Bisalputra, T., W. J. S. Downton, and E. B. Tregunna. 1969. The distribution and ultrastructure of chloroplasts in leaves differing in photosynthetic carbon metabolism. I: Wheat, sorghum and *Aristida gramineae. Can. J. Bot.,* 47: 15–21.

Bishop, D. G., K. S. Andersen, and R. M. Smillie. 1971. Lamellar structure and composition in relation to photochemical activity. Pp. 372–381 in M. D. Hatch, C. B. Osmond, and R. O. Slatyer (eds.), "Photosynthesis and Photorespiration." Wiley-Interscience, New York.

Biswas, V. B., and S. P. Sen. 1959. Translocation and utilisation of sulfate and phosphate in the pea plant. *Indian J. Plant. Physiol.,* 2: 1–8.

Björkman, O. 1966. Comparative studies of photosynthesis and respiration in ecological races. *Brittonia,* 18(3): 214–224.

———. 1968. Further studies on differentiation of photosynthetic properties in sun and shade ecotypes of *Solidago virgaurea. Physiol. Plant.,* 21: 84–99.

——— and P. Holmgren. 1963. Adaptability of the photosynthetic apparatus to light intensity in ecotypes from exposed and shaded habitats. *Physiol. Plant.,* 16: 889–914.

———, R. W. Pearcy, A. T. Harrison, and H. Mooney. 1972. Photosynthetic adaptation to high temperatures: A field study in Death Valley, California. *Science,* 175: 786–789.

Blaauw, A. H. 1915. Licht und Wachstum. II. *Zeit. Bot.,* 7: 465–532.

Blaauw, O. H. 1961. The influence of blue, red and

far-red light on geotropism and growth of the *Avena* coleoptile. *Acta Bot.*, 10: 397–450.

Blaauw-Jensen, G. 1954. On the light-induced transformation of chlorophyllide into a growth-inhibiting substance. *Proc. Kon. Ned. Alkad. Wet. Amsterdam,* C51: 498–506.

Black, C. C., Jr., T. M. Chen, and R. H. Brown. 1969. Biochemical basis for plant competition. *Weed Sci.,* 17: 338–344.

——— and B. C. Mayne. 1970. P700 activity and chlorophyll concentration of plants with different photosynthetic carbon dioxide fixation cycles. *Plant Physiol.*, 45: 738–741.

Black, M., and P. F. Wareing. 1954. Photoperiodic control of germination in seed of birch. *Nature,* 174: 705.

Blackman, F. F. 1905. Optima and limiting factors. *Ann. Bot.*, 19: 281–295.

——— and R. Parija. 1928. *Proc. R. Soc. (Lond.), Bio.* 3: 422–445; cited by Biale (1950).

Blackman, G. E., and J. M. Black. 1959. Physiological and ecological studies in the analysis of plant environment. XII: The role of the light factor in limiting growth. *Ann. Bot.,* n.s., 23: 131–145.

——— and G. L. Wilson. 1951. Physiological and ecological studies in the analysis of plant environment. VI: The constancy for different species of a logarithmic relationship between net assimilation rate and light intensity and its ecological significance. *Ann. Bot.,* n.s., 15: 63–94.

Blackman, V. H. 1919. The compound interest law and plant growth. *Ann. Bot.,* 33: 353–360.

Blair, D. S., M. MacArthur, and S. H. Nelson. 1956. Observations in the growth phases of fruit trees. *Proc. Am. Soc. Hort. Sci.,* 67: 75–79.

Bloch, R. 1952. Wound healing in higher plants. *Bot. Rev.*, 18: 655–679.

Blommaert, K. L. J. 1955. The significance of auxins and growth inhibiting substances in relation to winter dormancy of the peach tree. *Union S. Afr. Dept. Agric. Sci. Bull.* 368, pp. 1–23.

Blumenthal-Goldschmidt, S., and L. Rappaport. 1965. Regulation of bud rest in tubers of potato, II. *Plant Cell Physiol.,* 6: 601–608.

Boardman, N. K. 1967. Chloroplast structure and development. Pp. 211–230 in A. San Pietro, F. A. Greer, and T. J. Army (eds.), "Harvesting the Sun." Academic, New York.

——— and J. M. Anderson. 1964. Isolation from spinach chloroplasts of particles containing different proportions of chlorophyll *a* and *b* and their possible role in the light reactions of photosynthesis. *Nature,* 203: 166–167.

———, R. B. Francki, and S. G. Wildman. 1965. Protein synthesis by cell-free extracts from tobacco leaves. II: Association of activity with chloroplast ribosomes. *Biochem.,* 4: 872–876.

Boatman, S. G. 1970. Physiological aspects of the exploitation of rubber trees. Pp. 323–333 in L. C. Luckwill and C. V. Cutting (eds.), "Physiology of Tree Crops." Academic, London.

Bode, N. R. 1939. Über die Blattausscheidung des Wermuts und ihre Wirkung auf andere Pflanzen. *Planta,* 30: 567–589.

Bodlaender, K. B. A. 1963. Influence of temperature, radiation and photoperiod on development and yield. Pp. 199–210 in J. D. Ivins and F. L. Milthorpe (eds.), "The Growth of the Potato." Butterworth, London.

Bogorad, L. 1967. Chloroplast structure and development. Pp. 191–210 in A. San Pietro, F. A. Greer, and T. J. Army, "Harvesting the Sun." Academic, New York.

Böhmer, K. 1928. Die Bedeutung der Samenteile für die Lichtwirkung und die Wechselbeziehung von Licht und Sauerstoff bei der Keimung lichtempfindlicher Samen. *Jahrb. Wiss. Bot.,* 68: 549–601.

Böhning, R. H. 1949. Time course of photosynthesis in apple leaves exposed to continuous illumination. *Plant Physiol.,* 24: 222–240.

——— and C. A. Burnside. 1956. The effect of light intensity on rate of apparent photosynthesis in leaves of sun and shade plants. *Am. J. Bot.,* 43: 557–561.

Bolin, B. 1970. The carbon cycle. *Sci. Am.,* 223(3): 125–132.

Bollard, E. G. 1953. The use of tracheal sap in the study of apple tree nutrition. *J. Exp. Bot.,* 4: 363–368.

———. 1958. Nitrogenous compounds in tree xylem sap. Pp. 83–94 in K. V. Thimann (ed.), "The Physiology of Forest Trees." Ronald, New York.

———. 1960. Transport in the xylem. *Annu. Rev. Plant Physiol.,* 11: 141–166.

———. 1970. The physiology and nutrition of developing fruits. Pp. 387–427 in A. C. Hulme (ed.), "The Biochemistry of Fruits and Their Products." Academic, London.

Bonner, J. 1943. Effects of application of thiamine to *Cosmos. Bot. Gaz.,* 104: 475–479.

———. 1957. The chemical cure of climatic lesions. *Eng. Sci. Mag.,* 20: 28–30.

———. 1960. The mechanical analysis of auxin-induced growth. *Z. Schweiz. Forstv.,* 30: 141–159.

———. 1961. The biology of plant growth. Pp. 439–452 in M. X. Zarrow et al. (eds.), "Growth in Living Systems." Basic Books, N.Y.

———. 1962. The upper limit of crop yield. *Science,* 137: 11–15.

———, R. S. Bandurski, and A Millerd. 1953. Linkage of respiration to auxin-induced water uptake. *Physiol. Plant.,* 6: 511–522.

———, E. Heftmann, and J. A. D. Zeevaart. 1963. Suppression of floral induction by inhibition of steroid biosynthesis. *Plant Physiol.,* 38: 81.

Bonnett, H. T., and J. G. Torrey. 1965. Chemical control of organ formation in root segments of *Convolvulus* cultured in vitro. *Plant Physiol.,* 40: 1228–1236.

Booth, A., J. Moorby, C. R. Davies, H. Jones, and P. F.

Wareing. 1962. Effects of indolylacetic acid on the movement of nutrients within plants. *Nature,* 194: 204–205.

Borgström, G. 1939. "The Transverse Reactions of Plants," Gleerup, Lund, Sweden.

Bormann, F. H. 1953. Factors determining the role of loblolly pine and sweet gum in early old field succession in the Piedmont of North Carolina. *Ecol. Monog.* 23: 339–358.

Borthwick, H. A., S. B. Hendricks, M. W. Parker, E. H. Toole, and V. K. Toole. 1952. A reversible photoreaction controlling seed germination. *Proc. Natl. Acad. Sci. (U.S.),* 38: 662–666.

———, ———, E. H. Toole, and V. K. Toole. 1954. Action of light on lettuce seed germination. *Bot. Gaz.,* 115: 205–225.

——— and M. W. Parker. 1940. Floral initiation in Biloxi soybean influenced by age and position of leaf. *Bot. Gaz.,* 101: 806–817.

Bose, T. K., and J. P. Nitsch. 1970. Chemical alteration of sex expression in *Luffa acutangula. Physiol. Plant.,* 23: 1206.

Boswell, V. R. 1929. Studies of premature flower formation in wintered-over cabbage. *Md. Agric. Exp. Stn. Bull.* 313.

Bottger, I., and R. Wollgiehn. 1958. Untersuchungen über den Zusammenhang zwischen Nucleinsäure und Eiweisstoffwechsel in grünen Blättern. *Flora,* 146: 302–320.

Bouillenne, R. 1964. Aspects physiologique de la formation des racines. *Bull. Soc. R. Bot. Belg.,* 95: 193–204.

Bouma, D. 1967. Growth changes of subterranean clover during recovery from phosphorus and sulphur stresses. *Aust. J. Biol. Sci.,* 20: 51–66.

——— and E. J. Dowling. 1966. The physiological assessment of the nutrient status of plants. II: The effect of the nutrient status of the plant with respect to phosphorus, sulphur, potassium, calcium, or boron on the pattern of leaf area response following the transfer to different nutrient solutions. *Aust. J. Agric. Res.,* 17: 633–646.

Bowen, G. D., and A. D. Rovira. 1969. The influence of microorganisms on growth and metabolism of plant roots. Pp. 170–201 in W. J. Whittington (ed.), "Root Growth." Butterworth, London.

Boyer, J. S. 1970. Leaf enlargement and metabolic rates in corn, soybean and sunflower at various leaf water potentials. *Plant Physiol.,* 46: 233–235.

———. 1971*a*. Nonstomatal inhibition in photosynthesis in sunflower at low leaf water potentials in high light intensities. *Plant Physiol.,* 48: 532–536.

———. 1971*b*. Recovery of photosynthesis in sunflower after a period of low leaf water potential. *Plant Physiol.,* 47: 816–820.

Boysen-Jensen, P. 1932. "Die Stoffproduktion der Pflanzen." Jena.

Bradley, M. V., and W. J. Dahmen. 1972. Cytohistological effects of ethylene, 2,4-D, kinetin and carbon dioxide on peach mesocarp tissue in vitro. *Phytomorphology,* 21B: 154–163.

Brady, C. J., P. B. H. O'Connell, J. Smydzuk, and N. L. Wade. 1970. Permeability, sugar accumulation and respiration rate in ripening banana fruits. *Aust. J. Biol. Sci.,* 23: 1143–1152.

Brandon, P. C. 1967. Temperature features of enzymes affecting crassulacean acid metabolism. *Plant Physiol.,* 42: 977–984.

Brauner, L., and E. Appel. 1960. Zum Problem der Wuchsstoff-Querverschiebung bei der geotropischen Induktion. *Planta,* 55: 226–234.

——— and M. Brauner. 1947. Untersuchungen über der Mechanismus der phototropischen Reaktion der Blättfiedern von *Robinia pseudoacacia. Rev. Fac. Sci. Univ. Istanbul,* B.2: 35–79.

——— and A. Hager. 1957. Über die geotropische "Mneme." *Naturwiss.,* 15: 429–430.

——— and ———. 1958. Versuche zur Analyse der geotropischen Perzeption, I. *Planta,* 51: 115–146.

——— and A. Zipperer. 1961. Über die Anfangsphasen der geotropischen Krummungs-bewegung von *Avena* Koleoptilen. *Planta,* 57: 503–517.

Brewbaker, J. L. 1959. Biology of the angiosperm pollen grain. *Indian J. Genet. Plant Breed.,* 19: 121–133.

——— and S. K. Majumder. 1961. Cultural studies of the pollen population effect. *Am. J. Bot.,* 48: 457–464.

Brian, P. W. 1958. Role of gibberellin-like hormones in regulation of plant growth and flowering. *Nature,* 181: 1122–1123.

———. 1964. Gibberellins and their biological significance. *Proc. 10th. Bot. Congr.,* pp. 81–88.

——— and H. G. Hemming. 1955. The effect of gibberellic acid on shoot growth of pea seedlings. *Physiol. Plant.,* 8: 669–681.

———, ———, and M. Radley. 1955. A physiological comparison of gibberellic acid with some auxins. *Physiol. Plant.,* 8: 889–912.

———, J. H. Petty, and P. T. Richmond. 1959. Extended dormancy of deciduous woody plants treated in autumn with gibberellic acid. *Nature,* 184: 69.

Brian, R. C., and E. K. Rideal. 1952. On the action of plant growth regulators. *Biochim. Biophys. Acta,* 9: 1–18.

Briggs, W. R. 1960. Light dosage and phototropic responses of corn and oat coleoptiles. *Plant Physiol.,* 35: 951–962.

———. 1963. Red light, auxin relationships and the phototropic sensitivity of corn and oat coleoptiles. *Am. J. Bot..* 50: 196–207.

——— and H. V. Rice. 1972. Phytochrome: Chemical and physical properties and mechanism of action. *Annu. Rev. Plant Physiol.,* 23: 293–334.

———, R. D. Tocher, and J. F. Wilson. 1957. Phototropic auxin redistribution in corn coleoptiles. *Science,* 126: 210–212.

Brink, R. A. 1924. The physiology of pollen. *Am. J. Bot.,* 11: 417–463.

Brix, H. 1962. The effect of water stress on the rates of photosynthesis and respiration in tomato plant and loblolly pine seedlings. *Physiol. Plant.,* 15: 10–20.

Brock, T. D. 1967. Micro-organisms adapted to high temperatures. *Nature,* 214: 882–885.

Bronchard, R., and A. Nougaréde. 1970. Événements métaboliques et structuraux au niveau du méristème du *Perilla nankinensis* lors de la phase préflorale. Pp. 27–34 in G. Bernier (ed.), "Cellular and Molecular Aspects of Floral Induction." Longman, London.

Brougham, R. W. 1958. Interception of light by the foliage of pure and mixed stands of pasture plants. *Aust. J. Agric. Res.,* 9: 39–52.

Brouwer, R. 1965. Water movement across the root. *Symp. Soc. Exp. Biol.,* 19: 131–149.

Brown, H. T., and W. E. Wilson. 1905. On the thermal emissivity of a green leaf in still and moving air. *Proc. R. Soc. (Lond.),* B76: 122–137.

Brown, K. W. 1969. A model of the photosynthesizing leaf. *Physiol. Plant.,* 22: 620–637.

Brown, R. 1940. An experimental study of permeability to gases of the seed coat membranes of *Cucurbita. Ann. Bot.,* 4: 379–395.

Bruce, M. I., and J. A. Zwar. 1966. Cytokinin activity of some substituted ureas and thioureas. *Proc. R. Soc.,* B165: 245–265.

Brun, L. J., E. T. Kanemasu, and W. L. Powers. 1972. Evapotranspiration from soybean and sorghum fields. *Agron. J.,* 64: 145–148.

Buch, M. L., and O. Smith. 1959. The acidic growth inhibitor of potato tubers in relation to their dormancy. *Physiol. Plant.,* 12: 706–715.

Bukovac, M. J., F. Zucconi, F. P. Larsen, and C. D. Kesner. 1969. Chemical promotion of fruit abscission in cherries and plums with 2-chloroethylphosphonic acid. *J. Am. Soc. Hort. Sci.,* 94: 226–230.

Bünning, E. 1948. Die entwicklungsphysiologische Bedeutung der endogenen Tagesrhythmik bei den Pflanzen. In A. E. Murneek and R. O. Whyte (eds.), "Vernalization and Photoperiodism." Chronica Botanica, Waltham, Mass.

———. 1957. Polarität und inäquale Teilung des pflanzlichen Protoplasten. *Protoplasmologia,* B. 8 (Springer-Verlag, Vienna).

———. 1959. Tagesperiodische Bewegungen. *Handb. Pflanzenphysiol.,* 17: 579–656.

——— and L. Lörcher. 1957. Regulierung und Auslösung endogentagesperiodischer Blattbewegung durch verschiedene Lichtqualitäten. *Naturwiss.,* 44: 472.

Bünning, E. 1963. "Die physiologische Uhr. Springer-Verlag, Berlin. 153 pp.

———. 1961. Measurements of the velocity of translocation. *Ann. Bot.,* 25: 152–167.

Bünsow, R. 1953. Über den Einfluss der Lichtmenge auf die endogene Tagesrhythmik bei *Kalanchoë blossfeldiana. Biol. Zentralbl.,* 72: 465–477.

———, J. Penner, and R. Harder. 1958. Blütenbildung bei *Bryophyllum* durch Extrakt aus Bohnensamen. *Naturwiss.,* 45: 46–47.

Burden, R. S., R. D. Firn, R. W. P. Hiron, F. T. Taylor, and S. T. C. Wright. 1971. Induction of plant growth inhibitor xanthoxin in seedlings by red light. *Nature New Biol.,* 234: 95–96.

Burdett, A. N. 1972. Ethylene synthesis in lettuce seeds: Its physiological significance. *Plant Physiol.,* 50: 719–722.

——— and W. E. Vidaver. 1971. Synergistic action of ethylene with gibberellin on red light in germinating lettuce seeds. *Plant Physiol.,* 48: 656–657.

Burg, S. P. 1963. Studies on the formation and function of ethylene gas in plant tissues. Pp. 719–725 in J. P. Nitsch (ed.), "Régulateurs naturels de la croissance végétale." CNRS, Paris.

———. 1968. Ethylene, plant senescence and abscission. *Plant Physiol.,* 43: 1503–1511.

———, A. Apelbaum, W. Eisinger, and B. G. Kang. 1971. Physiology and mode of action of ethylene. *HortScience,* 6: 359–364.

——— and E. A. Burg. 1962. Role of ethylene in fruit ripening. *Plant Physiol.,* 37: 179–189.

——— and ———. 1964. Biosynthesis of ethylene. *Nature,* 203: 869–870.

——— and ———. 1965*a*. Ethylene action and the ripening of fruits. *Science,* 148: 1190–1195.

——— and ———. 1965*b*. Gas exchange in fruits. *Physiol. Plant.,* 18: 870–884.

——— and ———. 1965*c*. Relationships between ethylene production and ripening in bananas. *Bot. Gaz.,* 126: 200–204.

——— and ———. 1966*a*. Fruit storage at subatmospheric pressures. *Science,* 153: 314–315.

——— and ———. 1966*b*. The interaction between auxin and ethylene and its role in plant growth. *Proc. Natl. Acad. Sci. (U.S.),* 55: 262–269.

——— and ———. 1966*c*. Auxin-induced ethylene formation: Its relation to flowering in the pineapple. *Science,* 152: 1269.

——— and ———. 1967*a*. Ethylene action independent of protein synthesis. *Plant Physiol.,* 42 (suppl.): 31.

——— and ———. 1967*b*. Inhibition of polar auxin transport by ethylene. *Plant Physiol.,* 42: 1224–1228.

——— and ———. 1967*c*. Molecular requirements for the biological activity of ethylene. *Plant Physiol.,* 42: 144–152.

——— and ———. 1968*a*. Auxin stimulated ethylene formation: Its relationship to auxin inhibited growth, root geotropism and plant processes. Pp. 1275–1294 in F. Wightman and G. Setterfield (eds.), "Biochemistry and Physiology of Plant Growth Substances." Runge Press, Ottawa.

——— and ———. 1968*b*. Ethylene formation in pea seedlings: Its relation to the inhibition of bud growth caused by indoleacetic acid. *Plant Physiol.,* 43: 1069–1074.

———, ———, and R. Marks. 1964. Relationship of solute leakage to solution tonicity in fruits and other plant tissues. *Plant Physiol.,* 39: 185–195.

——— and C. O. Clagett. 1967. Conversion of methionine to ethylene in vegetative tissue and fruits. *Biochem. Biophys. Res. Commun.,* 27: 125–130.

——— and M. J. Dijkman. 1967. Ethylene and auxin participation in pollen induced fading of *Vanda* orchid blossoms. *Plant Physiol.,* 42: 1648–1650.

——— and K. V. Thimann. 1959. The physiology of ethylene formation in apples. *Proc. Natl. Acad. Sci. (U.S.),* 45: 335–344.

——— and ———. 1960. Studies on the ethylene production of apple tissue. *Plant Physiol.,* 35: 24–35.

Burk, L. G., and T. C. Tso. 1958. Effects of gibberellic acid on *Nicotiana* plants. *Nature,* 181: 1672–1673.

Burström, H. G. 1955. Zur Wirkungsweise chemischer Regulatoren des Wurzelwachstums. *Bot. Not.,* 108: 400–416.

———. 1969. Influence of the tonic effect of gravitation and auxin on cell elongation and polarity in roots. *Am. J. Bot.,* 56: 679–684.

———, I. Uhrström, and B. Olausson. 1970. Influence of auxin on Young's modulus in stems and root of *Pisum. Physiol. Plant.,* 23: 1223–1233.

———, ———, and R. Wurscher. 1967. Growth, turgor, water potential, and Young's modulus in pea internodes. *Physiol. Plant.,* 20: 213–231.

Burt, R. L. 1964. Carbohydrate utilisation as a factor in plant growth. *Aust. J. Biol. Sci.,* 17: 867–877.

Büsgen, M., E. Münch, and T. Thomson. 1929. "The Structure and Life of Forest Trees." Chapman Hall, London.

Butler, G. W. 1953. Ion uptake by young wheat plants. II: The apparent free space of wheat roots. *Physiol. Plant.,* 6: 617–635.

Butler, R. D. 1963. The effect of light intensity on stem and leaf growth in broad bean seedlings. *J. Exp. Bot.,* 14: 142–152.

Butler, W. L., and H. C. Lane. 1964. Dark transformations of phytochrome in vivo. II. *Plant Physiol.,* 40: 13–17.

———, ———, and H. W. Siegelman. 1963. Non-photo-chemical transformations of phytochrome in vivo. *Plant Physiol.,* 38: 514–519.

———, H. W. Siegelman, and C. O. Miller. 1964. Denaturation of phytochrome. *Biochemistry,* 3: 851–857.

———, K. H. Norris, H. W. Siegelman, and S. B. Hendricks. 1959. Detection, assay and purification of the pigment controlling photoresponsive development of plants. *Proc. Natl. Acad. Sci. (U.S.),* 45: 1703–1708.

Buttrose, M. S., and C. R. Hale. 1971. Effects of temperature on accumulation of starch or lipid in chloroplasts of grapevine. *Planta,* 101: 166–170.

Buy, H. G. du, and R. A. Olson. 1940. The relation between respiration, photoplasmic streaming and auxin transport in the *Avena* coleoptile. *Am. J. Bot.,* 27: 401–413.

Byrd, H. W., and J. C. Delouche. 1971. Deterioration of soybean seed in storage. *Proc. Assoc. Off. Seed Anal.,* 61: 41–57.

Cajlachjan, M. C. 1936. On the mechanism of the photoperiodic reaction. *C. R. Dokl. Acad. Sci. U.S.S.R.,* 10: 89–93.

———. 1965. Flowering in graft hybrids with both components in the vegetative state. *C. R. Dokl. Akad. Nauk SSSR,* 159: 1421–1424.

——— and R. Butenko. 1957. Movement of assimilates of leaves to shoots under differential photoperiodic conditions of leaves. *C. R. Dokl. Acad. Sci.,* 4: 450–462.

———, ———, and I. E. Lyarnarskaya. 1961. Effect of derivatives of nucleic acid metabolism on the growth and flowering of *Perilla. Fiziol. Rast.,* 8: 71–80.

Cajlachjan, N. K., and V. N. Lozhnikova. 1964. The dynamics of gibberellin-like substances in relation to photoperiodism. *Dokl. Bot. Sci.,* 157: 482–485.

——— and P. C. Marth. 1960. Effectiveness of a quaternary ammonium carbamate and a phosphonium in controlling growth of *Chrysanthemum morifolium* (Ramat.). *Am. Soc. Hort. Sci.,* 76: 609–619.

Calvert, A. 1957. Effect of the early environment on the development of flowering in the tomato. *J. Hort. Sci.,* 32: 9–17.

———. 1972. Effects of day and night temperatures and carbon dioxide enrichment on yield of glasshouse tomatoes. *J. Hort. Sci.,* 47: 231–247.

Calvin, M., C. Heidelberger, J. C. Reid, B. M. Tolbert, and P. F. Yankwich. 1949. "Isotopic Carbon." Wiley, New York.

Cameron, J. W., and R. K. Soost. 1952. Size, yield and fruit characters of orchard trees of *Citrus* propagated from young nucellar seedling lines. *Proc. Am. Soc. Hort. Sci.,* 60: 255–264.

Candolle, A. de. 1855. "Géographie botanique raisonée." Masson, Paris. 606 pp.

Canny, M. J. 1960. The rate of translocation. *Biol. Rev.,* 35: 507–532.

———. 1962*a*. The translocation profile: Sucrose and carbon dioxide. *Ann. Bot.,* n.s., 26: 181–196.

———. 1962*b*. The mechanism of translocation. *Ann. Bot.,* n.s., 26: 603–612.

———. 1971. Translocation: Mechanisms and kinetics. *Annu. Rev. Plant Physiol.,* 22: 237–260.

———. 1973. Translocation and distance. I. The

growth of the fruit of the sausage tree, *Kigelia pinnata*. *New Phytol.*, 72: 1269–1280.

——— and J. J. Askham. 1967. Physiological inferences from the evidence of translocation tracer: A caution. *Ann. Bot.*, n.s., 31: 409–416.

———, B. Nairn, and M. Harvey. 1968. The velocity of translocation in trees. *Aust. J. Bot.*, 16: 479–485.

Carfantan, N. 1970. Observations préliminaires sur le métabolisme de la fleur de tulipe. *Rev. Gen. Bot.*, 77: 313–330.

Carlson, R. F., and H. B. Tukey. 1945. Differences in afterripening requirements of several sources of peach seeds. *Proc. Am Soc. Hort. Sci.*, 46: 199–202.

Carns, H. R., F. T. Addicott, and R. S. Lynch. 1951. Effects of water and oxygen on abscission in vitro. *Plant Physiol.*, 26: 629–630.

Carpenter, W. J. G., and J. H. Cherry. 1966. Effects of benzyladenine on accumulation of P^{32} into nucleic acids. *Biochim. Biophys. Acta*, 114: 640–642.

Carr, D. J. 1952. A critical experiment on Bünning's theory of photoperiodism. *Z. Naturforsch.*, 76: 570.

———. 1959. Translocation between leaf and meristem in the flowering response of short-day plants. *9th. Int. Bot. Congr.*, 2: 11.

———. 1967. The relationship between florigen and the flowering hormones. *Ann. N.Y. Acad. Sci.*, 144: 305–312.

——— (ed.). 1972. "Plant Growth Substances, 1970." Springer-Verlag, Berlin.

——— and E. K. Ng. 1955. Experimental induction of flower formation in kikuyu grass *(Pennisetum clandestinum)*. *Aust. J. Agric. Res.*, 7: 1–6.

——— and J. S. Pate. 1967. Ageing in the whole plant. *Soc. Exp. Biol. Symp.*, 21: 559–600.

——— and D. M. Reid. 1968. The physiological significance of the synthesis of hormones in roots and of their export to the shoot system. Pp. 1169–1185 in P. Wightman and G. Setterfield (eds.), "Biochemistry and Physiology of Plant Growth Substances." Runge Press, Ottawa.

———, ———, and K. G. M. Skene. 1964. The supply of gibberellins from root to the shoot. *Planta*, 63: 382–392.

——— and K. G. M. Skene. 1961. Diauxic growth curves of seeds with references to french beans. *Aust. J. Biol. Sci.*, 14: 1–12.

Casida, J. E., and L. Lykken. 1969. Metabolism of organic pesticide chemicals in higher plants. *Annu. Rev. Plant Physiol.*, 20: 607–636.

Casperson, G. 1965. Über endogen Faktoren der Reaktionsholzbildung, I. *Planta*, 64: 225–240.

Cataldo, D. A., A. L. Christy, and C. L. Coulson. 1972*a*. Solution-flow in the phloem. II: Phloem transport of THO in *Beta vulgaris*. *Plant Physiol.*, 49: 690–695.

———, ———, ———, and J. M. Ferrier. 1972*b*. Solution-flow in the phloem. I: Theoretical considerations. *Plant Physiol.*, 49: 685–689.

Catchpole, A. H., and J. Hillman. 1969. Effect of ethylene on tuber formation in *Solanum tuberosum*. *Nature*, 223: 1387.

Cathey, H. M. 1957. Chrysanthemum temperature study. F: The effect of temperature upon the critical photoperiod necessary for the initiation and development of flowers of *Chrysanthemum morifolium*. *Proc. Am. Soc. Hort. Sci.*, 69: 485–491.

———. 1964. Physiology of growth retarding chemicals. *Ann. Rev. Plant Physiol.*, 15: 271–302.

———, G. L. Steffens, N. W. Stuart, and R. H. Zimmerman. 1966. Chemical pruning of plants. *Science*, 153: 1382–1383.

——— and R. C. Taylor. 1970. Flowering of bromeliads with spray applications of 2-chloroethanephosphonic acid, *Florist Nursery Exch.*

Chadwick, A. V., and S. P. Burg. 1967. An explanation of the inhibition of root growth caused by indoleacetic acid. *Plant Physiol.*, 42: 415–420.

——— and ———. 1970. Regulation of root growth by auxin-ethylene interaction. *Plant Physiol.*, 45: 192–200.

Challenger, S., H. J. Lacey, and B. H. Howard. 1964. The demonstration of root promoting substances in apple and plum root stocks. *Annu. Rep. E. Malling Res. Stn. 1964*, p. 124.

Chalutz, E., J. E. DeVay, and E. C. Maxie. 1969. Ethylene-induced isocoumarin formation in carrot root tissue. *Plant Physiol.*, 44: 235–241.

Chandler, W. H., M. H. Kimball, G. L. Philip, W. P. Tufts, and G. B. Weldon. 1937. Chilling requirements for opening of buds on deciduous orchard trees. *Univ. Calif. Agric. Exp. Stn. Bull.* 611.

Chang, F. H., and J. H. Troughton. 1972. Chlorophyll *a*/*b* ratios in C_3- and C_4-plants. *Photosynthetica*, 6: 57–65.

Chapman, H. W. 1958. Tuberization in the potato plant. *Physiol. Plant.*, 11: 215–224.

Charles, A. 1953. Uptake of dyes into cut leaves. *Nature*, 171: 435.

Charley, P., and P. Saltman. 1963. Chelation of calcium by lactose: Its role in transport mechanisms. *Science*, 139: 1205–1206.

———, B. Sarkar, C. F. Stitt, and P. Saltman. 1963. Chelation of iron by sugars. *Biochem. Biophys. Acta*, 69: 313–321.

Chatterjee, S. K., and A. C. Leopold. 1964. Kinetin and gibberellin actions on abscission processes. *Plant Physiol.*, 39: 334–337.

Chen, C. M., and R. H. Hall. 1969. Biosynthesis of N^6-(delta, delta isopentenyl)-adenosine in the transfer ribonucleic acid of cultured tobacco pith tissue. *Phytochem.*, 8: 1687–1695.

Chen, J. L., and S. G. Wildman. 1967. Functional chloroplast polyribosomes from tobacco leaves. *Science*, 155: 1271–1273.

Chen, T. M., R. H. Brown, and C. C. Black, Jr. 1969. Photosynthetic activity of chloroplasts isolated from bermuda grass (*Cynodon dactylon* L.) a species with a high photosynthetic capacity. *Plant Physiol.,* 44: 649–654.

Cherry, J. H. 1967*a*. Nucleic acid metabolism in aging cotyledons. *Soc. Exp. Biol. Symp.,* 21: 247–268.

———. 1967*b*. Nucleic acid biosynthesis in seed germination: Influences of auxin and growth-regulating substances. *Ann. N.Y. Acad. Sci.,* 144: 154–168.

———. 1968. Regulation of invertase in washed sugar beet tissue. Pp. 417–431 in F. Wightman and G. Setterfield (eds.), "Biochemistry and Physiology of Plant Growth Substances." Runge Press, Ottawa.

———. 1970. Nucleic acid synthesis during floral induction of *Xanthium*. Pp. 173–190 in G. Bernier (ed.), "Cellular and Molecular Aspects of Floral Induction." Longman, London.

——— and M. B. Anderson. 1972. Cytokinin-induced changes in transfer RNA species. Pp. 181–189, in D. J. Carr (ed.), "Plant Growth Substances 1970." Springer-Verlag, Berlin.

——— and R. B. von Huystee. 1965. Comparison of messenger RNA in photoperiodically induced and non-induced *Xanthium* buds. *Science,* 150: 1450–1453.

Chiang, K. S., and N. Sueoka. 1967. Replication of chloroplast DNA in *Chlamydomonas reinhardi* during vegetative cell cycle: Its mode and regulation. *Proc. Natl. Acad. Sci. (U.S.),* 57: 1506–1513.

Chibnall, A. D. 1939. "Protein Metabolism in the Plant." Yale University Press, New Haven. 306 pp.

——— and S. H. Piper. 1934. The metabolism of plant and insect waxes. *Biochem. J.,* 28: 2209–2219.

Chlor, M. A. 1940. Translocation of tritium-labelled gibberellic acid in pea stem segments and potato tuber cylinders. *Nature,* 214: 1263–1264.

Cho, S. C. 1970. Response to gibberellic acid of the sterile-cultured buds of *Begonia*. *Sci. Rep. Tohoku Univ.,* 25: 139–148.

Cholodny, N. 1924. Über die hormonale Wirkung der Organspitze bei der geotropischen Krummung. *Ber. Dtsch. Bot. Ges.,* 42: 356–362.

———. 1936. On the theory of yarovization. *C. R. Dokl. Acad. Sci.,* 3: 9.

Chouard, P. 1960. Vernalization and its relations to dormancy. *Annu. Rev. Plant Physiol.,* 11: 191–238.

Chrispeels, M. J., and J. E. Varner. 1966. Inhibition of gibberellic acid induced formation of amylase by abscisin II. *Nature,* 212: 1066–1067.

——— and ———. 1967. Gibberellic acid enhanced synthesis and release of amylase and ribonuclease by isolated barley aleurone layers. *Plant Physiol.,* 42: 398–406.

Christiansen, G. S., and K. V. Thimann. 1950. Metabolism of stem tissue during growth and its inhibition. II: Respiration and ether soluble material. *Arch. Biochem.,* 26: 248–259.

Christopher, E. P. 1954. "The Pruning Manual." Macmillian, New York. 320 pp.

Chroboczek, E. 1934. A study of some ecological factors influencing seed-stalk development in beets. *Cornell Univ. Agric. Exp. Stn. Mem.,* 154: 3–84.

Clark, H. E., and K. R. Kerns. 1942. Control of flowering with phytohormones. *Science,* 95: 536–537.

——— and ———. 1943. Effects of growth regulating substances on parthenocarpic fruit. *Bot. Gaz.,* 104: 639–644.

Clark, J. E., and O. V. S. Heath. 1959. Auxin and the bulbing of onions. *Nature,* 184: 345–347.

——— and ———. 1962. Studies in the physiology of the onion plant. V: Growth substance content. *J. Exp. Bot.,* 13: 227–249.

Cleland, C. F., and W. R. Briggs. 1969. Gibberellin and CCC effects on flowering and growth in the long-day plant *Lemma gibba*. *Plant Physiol.,* 44: 503–507.

Cleland, R. 1960*a*. Effect of auxin upon loss of calcium from cell walls. *Plant Physiol.,* 35: 581–584.

———. 1960*b*. Auxin-induced methylation in maize. *Nature,* 185: 44.

———. 1961. The relation between auxin and metabolism. *Handb. Pflanzenphysiol.,* 14: 754–783.

———. 1963. Hydroxyproline as an inhibitor of auxin-induced cell elongation. *Nature,* 200: 908–909.

———. 1973. Auxin-induced hydrogen ion excretion from *Avena* coleoptiles. *Proc. Nat. Acad. Sci.,* 70: 3092–3093.

——— and N. McCombs. 1965. Gibberellic acid: Action in barley endosperm does not require endogenous auxin. *Science,* 150: 497–498.

Clements, H. F. 1940. Movement of organic solutes in the sausage tree, *Kigelia africana*. *Plant Physiol.,* 15: 689–700.

Clutter, M. E. 1960. Hormonal induction of vascular tissue in tobacco pith *in vitro*. *Science,* 132: 548–549.

Coggins, C. W., H. Z. Hield, and M. J. Garber. 1960. The influence of gibberellin on Valencia orange trees and fruit. *Proc. Am. Soc. Hort. Sci.,* 76: 193–198.

——— and L. N. Lewis. 1962. Regreening of Valencia orange as influenced by gibberellin. *Plant Physiol.,* 37: 625–627.

Cohen, D. 1958. The mechanism of germination stimulation by alternating temperatures. *Bull. Res. Coun. Isr. Bot.* 6D: 111–117.

Cohen, S. S. 1970. Are/were mitochondria and chloroplasts microorganisms? *Am. Sci.,* 58: 281–289.

Collier, H. O. J. 1962. Kinins. *Science,* 207: 111–118.

Collins, W. T., F. B. Salisbury, and C. W. Ross. 1963. Growth regulators and flowering. III: Antimetabolites. *Planta,* 60: 131–144.

Commoner, B., S. Fogel, and W. H. Muller. 1943. The mechanism of auxin action: The effect of auxin on water absorption. *Am. J. Bot.,* 30: 23–38.

Condit, H. R., and F. Grum. 1964. Spectral energy distribution of daylight. *J. Opt. Soc. Am.,* 54: 937–944.

Considine, J. A., and P. E. Kriedemann. 1972. Fruit splitting in grapes: Determination of the critical turgor pressure. *Aust. J. Agric. Res.,* 23: 17–24.

Cook, G. D., J. L. Dixon, and A. C. Leopold. 1964. Transpirational effects on plant leaf temperature. *Science,* 144: 546–547.

Cooke, A. R. 1954. Changes in free auxin content during the photo-induction of short-day plants. *Plant Physiol.,* 29: 440–444.

——— and D. I. Randall. 1968. 2-Haloethanephosphonic acids as ethylene releasing agents for the induction of flowering in pineapples. *Nature,* 218: 974–975.

Coombe, B. G. 1960. Relationship of growth and development to changes in sugars, auxins, and gibberellins in fruit of seeded and seedless varieties of *Vitis. Plant Physiol.,* 35: 241–250.

———. 1971. GA_{32}: A polar gibberellin with high biological potency. *Science,* 172: 856–857.

———, D. Cohen, and L. G. Paleg. 1967. Barley endosperm bioassay for gibberellins, I. *Plant Physiol.,* 42: 105–112.

Cooper, M. J., J. Digby, and P. J. Cooper. 1972. Effects of plant hormones on the stomata of barley: A study of the interaction between abscisic acid and kinetin. *Planta,* 105: 43–49.

Cooper, T. G., D. Filmer, M. Wishnick, and M. D. Lane. 1969. The active species of "CO_2" utilized by ribulose diphosphate carboxylase. *J. Biol. Chem.,* 244: 1081–1083.

Cooper, W. C., A. C. Burkett, and A. Hern. 1945. Flowering of Peruvian cube induced by girdling. *Am. J. Bot.,* 32: 655–657.

——— and A. Peynado. 1958. Effect of gibberellic acid on growth and dormancy in *Citrus. Proc. Am. Soc. Hort. Sci.,* 72: 284–289.

———, G. K. Rasmussen, and D. J. Hutchinson. 1969. Promotion of abscission of orange fruits by cycloheximide as related to site of treatments. *BioScience,* 19: 443–444.

———, ———, B. J. Rogers, D. C. Reece, and W. H. Henry. 1968. Control of abscission in agricultural crops and its physiological basis. *Plant Physiol.,* 43: 1560–1576.

Coorts, G. D., J. P. McCollum, and J. B. Gartner. 1965. Effect of senescence and preservative on mitochondrial activity in flower petals of *Rosa hybrida. Proc. Am. Soc. Hort. Sci.,* 86: 791–797.

Corcoran, M. R., and B. O. Phinney. 1962. Changes in amount of gibberellin in developing seed of *Echinocystis, Lupinus* and *Phaseolus. Physiol. Plant.,* 15: 252–262.

———, C. A. West, and B. O. Phinney. 1961. Natural inhibitors of gibberellin-induced growth. In "Gibberellins." *Adv. Chem.,* 28: 152–158.

Cornforth, J. W., B. V. Milborrow, and G. Ryback. 1965*a*. Synthesis of abscisin II. *Nature,* 206: 715.

———, ———, and ———. 1966*a*. Identification and estimation of abscisin II in plant extracts by spectropolarimetry. *Nature,* 210: 627–628.

———, ———, ———, K. Rothwell, and R. L. Wain. 1966*b*. Identification of the yellow lupin growth inhibitor as abscisin II. *Nature,* 211: 742–743.

———, ———, ———, and P. F. Wareing. 1965*b*. Identity of sycamore dormin with Abscisin II. *Nature,* 205: 1269–1270.

Coster, C. 1927. Zur Anatomi und Physiologie der Zuwachszonen und Jahresringbildung in den Tropen. *Ann. Jard. Bot. Buitenzorg,* 37: 49–160.

Coulson, C. L., A. L. Christy, D. A. Cataldo, and C. A. Swanson. 1972. Carbohydrate translocation in sugar beet petioles in relation to petiolar respiration and adenosine 5′-triphosphate. *Plant Physiol.,* 49: 919–923.

Cowan, I. R., and F. L. Milthorpe. 1968. Plant factors influencing the water status of plant tissues. Pp. 137–193 in T. T. Kozlowski (ed.), "Water Deficits and Plant Growth," vol. I. Academic, New York.

Cox, E. F. 1966. Resistance to water flow through the plant. Ph.D. thesis, Univ. Nottingham, England.

Cracker, L. E., and F. B. Abeles. 1969. Abscission: Role of abscisic acid. *Plant Physiol.,* 44: 1144–1149.

Crafts, A. S. 1931. Movement of organic materials in plants. *Plant Physiol.,* 6: 1–41.

———. 1933. Sieve-tube structure and translocation in the potato. *Plant Physiol.,* 8: 81–104.

———. 1959*a*. Improvement of growth regulator formulation. Pp. 789–801 in R. M. Klein (ed.), "Plant Growth Regulation." Iowa State University Press, Ames.

———. 1959*b*. Further studies on comparative mobility of labeled herbicides. *Plant Physiol.,* 34: 613–620.

———. 1961. "The Chemistry of Mode of Action of Herbicides." Interscience, New York. 269 pp.

——— and O. A. Lorenz. 1944. Fruit growth and food transport in cucurbits. *Plant Physiol.,* 19: 131–138.

——— and ———. 1960. Gross autoradiography of solute translocation and distribution in plants. *Med. Biol. Illust.,* 10: 103–109.

——— and S. Yamaguchi. 1958. Comparative tests on the uptake and distribution of labeled herbicides by *Zebrina pendula* and *Tradescantia fulminensis. Hilgardia,* 27: 421–454.

Crane, J. C. 1965. The chemical induction of parthenocarpy in the Calimyrna fig and its physiological significance. *Plant Physiol.,* 40:606–610.

———. 1969. The role of hormones in fruit set and development. *HortScience,* 4: 108–111.

———, M. V. Bradley, and L. C. Luckwill. 1959. Auxins in parthenocarpic and non-parthenocarpic figs. *J. Hort. Sci.,* 34: 142–153.

——— and J. van Overbeek. 1965. Kinin induced parthenocarpy in the fig. *Science,* 147: 1468–1469.

———, P. Primer, and R. Campbell. 1960. Gibberel-

lin induced parthenocarpy in *Prunus*. *Proc. Am. Soc. Hort. Sci.*, 75: 129–137.
———. 1952. Ovary-wall development as influenced by growth-regulators inducing parthenocarpy in the *Calimyrna* fig. *Bot. Gaz.*, 114: 102–107.
Craniades, P. 1954. Tubercules aerines chez les pommes de terre. *Bull. Soc. Chim. Biol.*, 36: 1671–1674.
Crocker, W. 1906. Role of seed coat in delayed germination. *Bot. Gaz.*, 42: 265–291.
——— and L. V. Barton. 1957. "Physiology of Seeds." Chronica Botanica, Waltham, Mass.
——— and W. E. Davis. 1914. Delayed germination in seed of *Alisma plantago*. *Bot. Gaz.*, 58: 285–321.
———, A. E. Hitchcock, and P. W. Zimmerman. 1935. Similarities in the effects of ethylene and the plant auxins. *Contrib. Boyce Thompson Inst.*, 7: 231–248.
———, N. C. Thornton, and E. M. Schroeder. 1946. Internal pressure necessary to break shells of nuts and the role of the shells in delayed germination. *Contrib. Boyce Thompson Inst.*, 14: 173–201.
Cross, B. E. 1954. Gibberellic acid, I. *J. Chem. Soc. Lond.*, 1954: 4670–4676.
———. 1968. Biosynthesis of the gibberellins. Pp. 195–222 in L. Reinhold and Y. Liwshitz (eds.), "Progress in Phytochemistry." Interscience, London.
Crowdy, S. H., and D. Rudd-Jones. 1956. The translocation of sulphonamides in higher plants, I. *J. Exp. Bot.*, 7: 335–346.
——— and T. W. Tanton. 1970. Water pathways in higher plants, I. *J. Exp. Bot.*, 21: 102–111.
Crozier, A., and D. M. Reid. 1971. Do roots synthesize gibberellins? *Can. J. Bot.*, 49: 967–975.
Cumming, B. G. 1967. Circadian rhythmic flowering response in *Chenopodium rubrum*: Effects of glucose and sucrose. *Can. J. Bot.*, 45: 2173–2193.
Cummins, B. 1969. in L. T. Evans (ed.), "The Induction of Flowering." Cornell Univ. Press, Ithaca, N.Y.
Currier, H. B., and C. D. Dybing. 1959. Foliar penetration of herbicides: Review and present status. *Weeds*, 7: 195–213.
Curry, G. M. 1969. Phototropism. Pp. 243–273 in M. B. Wilkins (ed.), "Physiology of Plant Growth and Development." McGraw-Hill, London.
Curtis, O. F. 1935. "The Translocation of Solutes in Plants." McGraw-Hill, New York.
———. 1936. Leaf temperature and the cooling of leaves by radiation. *Plant Physiol.*, 11: 343–364.
——— and D. G. Clark. 1950. "An Introduction to Plant Physiology." McGraw-Hill, New York. 752 pp.
Czopek, M. 1962. The oligodynamic action of light on the germination of turions of *Spirodela polyrrhiza*. *Acta. Soc. Bot. Pol.*, 31: 715–722.
———. 1964. Action of kinetin and gibberellic acid and red light on the germination of *Spirodela polyrrhiza*. *Bull. Acad. Pol. Sci.*, (II)12: 117–182.
Danielli, J. F. 1954. Morphological and molecular aspects of active transport. *Soc. Exp. Biol.*, 8: 502–515.
Dankwardt-Lilliestrom, C. 1957. Kinetin induced shoot formation from isolated roots of *Isatis*. *Physiol. Plant.*, 10: 794–797.
Darbyshire, D. 1971*a*. The effect of water stress on indoleacetic acid oxidase in pea plants. *Plant Physiol.*, 47: 65–67.
———. 1971*b*. Changes in indoleacetic acid oxidase activity associated with plant water potential. *Physiol. Plant.*, 25: 80–83.
——— and B. T. Steer. 1973. Dehydration of macromolecules. I: Effect of dehydration-rehydration on indoleacetic acid oxidase, ribonuclease, ribulose diphosphate carboxylase and aldolase. *Aus. J. Biol. Sci.*, 26: 591–604.
Darrow, G. M. 1942. Rest period requirements for blueberries. *Proc. Am. Soc. Hort. Sci.*, 41: 189–194.
——— and W. H. Camp. 1945. *Vaccinium* hybrids and the development of new horticultural material. *Bull. Torrey Bot. Club*, 72: 1–21.
Darwin, C. 1897. "The Power of Movement in Plants." Appleton, N.Y.
Das, N. K., K. Patau, and F. Skoog. 1956. Initiation of mitosis and cell division by kinetin and indoleacetic acid in excised tobacco pith tissue. *Physiol. Plant.*, 9: 640–651.
Das, T. M. 1968. Physiological changes with leaf senescence: Kinins on cell ageing and organ senescence. Pp. 91–102 in S. M. Sircar (ed.), *Proc. Int. Symp. Plant Growth Substances. Calcutta Univ.*
Das, U. S. R., and P. V. Raju. 1965. Photosynthetic $^{14}CO_2$ assimilation by rice leaves under the influence of blue light. *Indian J. Plant Physiol.*, 8: 1–4.
Dashek, W. W., and W. G. Rosen. 1966. Electron microscopical localization of chemical components in the growth zone of lily pollen tubes. *Protoplasma*, 61: 192–204.
Daubenmire, R. 1954. Alpine timber lines in the Americas and their interpretation. *Butler Univ. Bot. Stud.*, 11: 119–136.
Davidson, D. 1966. The onset of mitosis and DNA synthesis in roots of germinating beans. *Am. J. Bot.*, 53: 491–495.
Davis, L. A., and F. T. Addicott. 1972. Abscisic acid: Correlations with abscission and with development in the cotton fruit. *Plant Physiol.*, 49: 644–648.
Day, B. E. 1952. The absorption and translocation of 2,4-D by bean plants. *Plant Physiol.*, 27: 143–152.
Deleuze, G. G., J. D. McChesney, and J. E. Fox. 1972. Identification of a stable cytokinin metabolite. *Biochim. Biophys. Res. Commun.*, 48: 1426–1432.
Deltour, R. 1970. Induction florale in vitro de plantes régénérées à partir de *Sinapis alba*. Pp. 416–429, in G. Bernier (ed.), "Cellular and Molecular Aspects of Floral Induction." Longman, London.
Demorest, D. M., and M. A. Stahmann. 1971. Ethy-

lene production from peptides and protein containing methionine. *Plant Physiol.*, 47: 450–451.
Denisen, E. L. 1951. Carotenoid content of tomato fruits. I: Effect of temperature and light. II: Effects of nutrients, storage, and variety. *Iowa State Coll. J. Sci.*, 25: 549–574.
Dennis, D. T., C. D. Upper, and C. A. West. 1965. An enzymic site of inhibition of gibberellin biosynthesis by AMO-1618 and other plant growth retardants. *Plant Physiol.*, 40: 948–952.
Dennis, D. T., M. Stubbs, and T. P. Coultate. 1967. The inhibition of brussels sprout leaf senescence by kinins. *Can. J. Bot.*, 45: 1019–1024.
Dennis, F. G. 1967. Apple fruit-set: Evidence for a specific role of seeds. *Science,* 156: 71–73.
——— and J. B. Nitsch. 1966. Identification of gibberellins A_4 and A_7 in immature apple seeds. *Nature,* 211: 781–782.
Denny, F. E. 1924. Hastening the coloration of lemons. *J. Agric. Res.*, 27: 757–769.
———. 1926. Hastening the sprouting of dormant potato tubers. *Am. J. Bot.*, 13: 118–125.
——— and L. P. Miller. 1935. Production of ethylene by plant tissue as indicated by the epinastic response of leaves. *Contrib. Boyce Thompson Inst.*, 7: 97–102.
De Vries, H. 1885. Über die Bedeutung der Zirkulation und der Rotation des Protoplasmas für den Stofftransport in der Pflanze. *Bot. Ztg.*, 43: 1–6, 16–26.
Dickinson, D. B. 1966. The relation between external sugars and respiration of germinating lily pollen. *Proc. Am. Soc. Hort. Sci.*, 88: 651–656.
Diemer, R. 1961. Untersuchungen des phototropischen Induktionsvorganges an *Helianthus* Keimlingen. *Planta,* 57: 111–137.
Dilley, R. R. 1969. Hormonal control of fruit ripening. *Hort. Sci.*, 4: 11–114.
Dixon, H. H. 1914. "Transpiration and the Ascent of Sap in Plants." Macmillan, London.
——— and Ball, N. G. 1922. Transport of organic substances in plants. *Nature,* 109: 236–237.
Dolk, H. E. 1929. Über die Wirkung der Schwerkraft auf Koleoptilen von *Avena sativa*. *Proc. K. Akad. Wet. Amst.*, 32: 40–47.
Dolzmann, P. 1964. Elektronenmikroskopische Untersuchungen an den Saughaaren von *Tillandsia usneoides*, I. *Planta,* 60: 461.
Domanski, R., and T. T. Kozlowski. 1968. Variations in kinetin-like activity in buds of *Betula* and *Populus* during release from dormancy. *Can. J. Bot.*, 46: 397–403.
Donald, C. M. 1961. Competition for light in crops and pastures. *Soc. Exp. Biol. Symp.*, 15: 282–313.
Doorenbos, J. 1953. Review of the literature on dormancy in buds of woody plants. *Meded. Landbouwhogesch. Wageningen,* 53: 1–24.
———. 1954. "Rejuvenation" of *Hedera helix* in graft combinations. *K. Akad. Wet. Amst. Proc. Sec. Sci.*, C57: 99–102.
Dörffling, K. 1966. Weitere Untersuchungen über korrelative Knospenhemmung. *Planta,* 70: 257–274.
Dorland, R. E., and F. W. Went. 1947. Plant growth under controlled conditions, VIII. *Am. J. Bot.*, 34: 393–401.
Dostal, H. C., and A. C. Leopold. 1967. Gibberellin delays ripening of tomatoes. *Science,* 158: 1579–1580.
——— and G. E. Wilcox. 1971. Chemical regulation of fruit ripening of field-grown tomatoes with 2-chloroethylphosphonic acid. *J. Am. Soc. Hort. Sci.*, 96: 656–660.
Dostal, R. 1950. Morphogenetic experiments with *Bryophyllum verticillatum*. *Acta. Acad. Sci. Nat. Moravo-Silesiaca,* 22: 57–98.
———. 1962. Über die korrelative Erhaltung entspreiteter Blattstiele. *Biol. Plant.*, 4: 191–202.
Downes, R. W. 1971. Relationship between evolutionary adaptation and gas exchange characteristics of diverse *Sorghum* taxa. *Aust. J. Biol. Sci.*, 24: 843–852.
Downs, R. J. 1955. Photoreversibility of leaf and hypocotyl elongation of dark grown red kidney bean seedlings. *Plant Physiol.*, 30: 468–473.
———, S. B. Hendricks, and H. A. Borthwick. 1957. Photoreversible control of elongation of pinto beans and other plants under normal conditions of growth. *Bot. Gaz.*, 118: 199–208.
Downton, W. J. S. 1970. Preferential C_4-dicarboxylic acid synthesis, the postillumination CO_2 burst, carboxyl transfer step, and grana configurations in plants with C_4-photosynthesis. *Can. J. Bot.*, 48: 1795–1800.
——— and N. A. Pyliotis. 1971. Loss of photosystem II during ontogeny of sorghum bundle sheath chloroplasts. *Can. J. Bot.*, 49: 179–180.
Dravnieks, D. E., F. Skoog, and R. H. Burris. 1969. Cytokinin activation of de novo thiamine biosynthesis in tobacco culture. *Plant Physiol.*, 44: 866–870.
Dugger, B. M. 1913. Lycopersicin, the red pigment of the tomato. *Wash. Univ. Stud.*, 1: 22–45.
Dugger, W. M., O. C. Taylor, E. Cardiff, and C. R. Thompson. 1962. Stomatal action in plants as related to damage from photochemical oxidants. *Plant Physiol.*, 37: 487–491.
Dumbroff, E. B., and D. R. Pierson. 1971. Probable sites for passive movements of ions across the endodermis. *Can. J. Bot.*, 49: 35–38.
Duncan, W. G., R. S. Loomis, W. A. Williams, and R. Hanau. 1967. A model for simulating photosynthesis in plant communities. *Hilgardia,* 38: 181–205.
Dybing, C. D., and H. B. Currier. 1961. Foliar penetration by chemicals. *Plant Physiol.*, 36: 169–174.
Eagles, C. F. 1967*a*. Apparent photosynthesis and

respiration in populations of *Lolium perenne* from contrasting climatic regions. *Nature,* 215: 100–101.
———. 1967*b*. The effect of temperature on vegetative growth in climatic races of *Dactylis glomerata* in controlled environments. *Ann. Bot.,* n.s., 31: 31–39.
——— and K. J. Treharne. 1969. Photosynthetic activity of *Dactylis glomerata* L. in different light regimes. *Photosynthetica,* 3: 29–38.
——— and P. F. Wareing. 1963. Dormancy regulators in woody plants. *Nature,* 199: 874–875.
——— and ———. 1964. Role of growth substances in the regulation of bud dormancy. *Physiol. Plant.,* 17: 697–709.
Eames, A. J. 1950. Destruction of phloem in young bean plants after treatment with 2,4-D. *Am. J. Bot.,* 37: 840–847.
Eashi, Y., and A. C. Leopold. 1969. Dormancy regulation in subterranean clover seeds by ethylene. *Plant Physiol.,* 44: 1470–1473.
Eaton, F. M., and D. R. Ergle. 1953. Relationship of seasonal trends in carbohydrate and nitrogen levels to the nutritional interpretation of boll shedding in cotton. *Plant Physiol.,* 28: 503–520.
Eaton, L. M. 1957. Selective gametocide opens way to hybrid cotton. *Science,* 126: 1174–1175.
Eberhardt, F. 1955. Der Atmungsverlauf alternder Blätter und reifender Früchte. *Planta,* 45: 57–67.
Edelman, J. 1959. *Proc. 4th Int. Cong. Biochem.,* 13: 336 (Pergamon Press, London); cited in Edelman (1963).
———. 1963. Physiological and biochemical aspects of carbohydrate metabolism during tuber growth. Pp. 135–147 in J. D. Ivins and F. L. Milthorpe (eds.), "The Growth of the Potato." Butterworth, London.
——— and M. A. Hall. 1964. Effects of growth hormones in the development of invertase associated with cell walls. *Nature,* 201: 296–297.
——— and K. Popov. 1962. *Proc. Bulg. Acad. Sci.,* 15: 627; cited in Edelman (1963).
Egle, E., and H. Fock. 1967. Pp. 79–87 in T. Goodwin (ed.), "Biochemistry of Chloroplasts." Academic, New York.
——— and ———. 1967. Light respiration: Correlations between CO_2 fixation. O_2 pressure and glycollate concentration. Pp. 79–87 in T. W. Goodwin (ed.), "Biochemistry of Chloroplasts." *NATO Adv. Stud. Inst. Aberystwyth Proc.* Academic, New York.
Eglinton, G., and R. J. Hamilton. 1967. Leaf epicuticular waxes. *Science,* 156: 1322–1335.
Eilam, Y. 1965. Permeability changes in senescing tissue. *J. Exp. Bot.,* 16: 614–627.
Eilati, S. K., S. P. Monselise, and P. Budowski. 1969. Seasonal development of external color and carotenoid content in the peel of ripening "Shamouti" oranges. *J. Am. Soc. Hort. Sci.,* 94: 346–348.
El-Antably, H. M. M., and P. F. Wareing. 1966. Stimulation of flowering in certain short-day plants by abscisin. *Nature,* 210: 328–329.
———, ———, and J. Hillman. 1967. Some physiological responses to D-L abscisin (dormin). *Planta,* 73: 74–90.
Elliott, B. B., and A. C. Leopold. 1953. An inhibitor of germination and of amylase activity in oat seeds. *Plant Physiol.,* 6: 66–78.
Elliott, F. C. 1958. "Plant Breeding and Cytogenetics." McGraw-Hill, New York. 395 pp.
Ellyard, P. W., and M. Gibbs. 1969. Inhibition of photosynthesis by oxygen in isolated spinach chloroplasts. *Plant Physiol.,* 44: 1115–1121.
Elmore, C. D., J. D. Hesketh, and H. Muramoto. 1967. A survey of rates of leaf growth, leaf aging and leaf photosynthetic rates among and within species. *J. Ariz. Acad. Sci.,* 4: 215–219.
El-Sharkawy, M., and J. Hesketh. 1965. Photosynthesis among species in relation to characteristics of leaf anatomy and CO_2 diffusion resistances. *Crop Sci.,* 5: 517–521.
Elzam, O. E., and E. Epstein. 1965. Absorption of chloride by barley roots: Kinetics and selectivity. *Plant Physiol.,* 40: 620–624.
Emerson, R., and W. Arnold. 1932. The separation of the reactions in photosynthesis by means of intermittent light. *J. Gen. Physiol.,* 15: 391–420.
———, R. V. Chalmers, and C. Cederstrand. 1957. Some factors influencing the long-wave limit of photosynthesis. *Proc. Natl. Acad. Sci. (U.S.),* 43: 133–143.
——— and C. M. Lewis. 1943. The dependence of the quantum yield of *Chlorella* photosynthesis on wavelength of light. *Am. J. Bot.,* 30: 165–178.
Emerson, S. H. 1940. Growth of incompatible pollen tubes in *Oenothera. Bot. Gaz.,* 101: 890–911.
Emmerling, A. 1880. Studien über die Eiweissbildung in der Pflanze, I. *Landw. Vers. Sta.,* 24: 113–160.
Ende, H. van den and J. A. D. Zeevaart. 1971. Influence of day length on gibberellin metabolism and stem growth in *Silene armeria. Planta,* 98: 164–176.
Engelsma, G., and G. Meijer. 1965. The influence of light on the synthesis of phenolic compounds in gherkin seedlings. *Acta Bot. Neerl.,* 14: 54–92.
Engle, R. L., and W. H. Gabelman. 1966. Inheritance and mechanism for resistance to ozone damage in onion. *Proc. Am. Soc. Hort. Sci.,* 89: 423–430.
Engelbrecht, L., and K. Mothes. 1960. Kinetin als Faktor der Hitzresistenz. *Ber. Dtsch. Bot. Ges.,* 73: 246–257.
Enns, T. 1967. Facilitation by carbonic anhydrase of carbon dioxide transport. *Science,* 155: 44–47.
Epstein, E. 1972. "Mineral Nutrition of Plants: Principles and Perspectives." Wiley, New York. 412 pp.
——— and C. E. Hagen. 1952. Kinetic study of

absorption of alkali cations by barley roots, *Plant Physiol.,* 27: 457–474.
——— and J. E. Leggett. 1954. The absorption of alkaline earth cations by barley roots: Kinetics and mechanism. *Am. J. Bot.,* 41: 785–791.
———, D. W. Rains, and W. E. Schmidt. 1962. Course of cation absorption by plant tissue. *Science,* 136: 1051–1052.
Esashi, Y. 1960. Studies on the formation and sprouting of aerial tubers in *Begonia evansiana* Andr. IV: Cutting method and tuberizing stages. *Sci. Rep. Tohoku Univ.,* 26: 239–246.
———. 1961. Studies on the formation and sprouting of aerial tubers in *Begonia evansiana* Andr. VI: Photoperiodic conditions for tuberization and sprouting in the cutting plants. *Sci. Rep. Tohoku Univ.,* 27: 101–112.
———. 1964. Studies on the formation and sprouting of aerial tubers in Begonia. X: Tuberization under long days and in darkness. *Plant Cell Physiol.,* 5: 101–117.
———, T. Eguchi, and M. Nagao. 1964*a*. The role of auxin in the photoperiodic tuberization in *Begonia evansiana. Plant Cell Physiol.,* 5: 413–427.
——— and A. C. Leopold. 1968*a*. Physical forces in dormancy and germination of *Xanthium* seeds. *Plant Physiol.,* 43: 871–876.
——— and ———. 1968*b*. Regulation of tuber development in *Begonia* by cytokinin. Pp. 923–941 in P. Wightman and G. Setterfield (eds.), "Biochemistry and Physiology of Plant Growth Substances." Runge Press, Ottawa.
——— and ———. 1969*a*. Cotyledon expansion as a bioassay for cytokinins. *Plant Physiol.,* 44: 618–620.
——— and ———. 1969*b*. Dormancy regulation in subterranean clover seeds by ethylene. *Plant Physiol.,* 44: 1470–1473.
——— and ———. 1969*c*. Regulation of the onset of dormancy in tubers of *Begonia evansiana. Plant Physiol.,* 44: 1200–1202.
——— and M. Nagao. 1958. Studies on the formation and sprouting of aerial tubers in *Begonia evansiana* Andr. I: Photoperiodic conditions for tuberization. *Sci. Rep. Tohoku Univ.,* 24: 81–88.
———, K. Ogata, and M. Nagao. 1964*b*. Studies on the formation and sprouting of aerial tubers in Begonia. XI: Temperature and tuber initiation. *Plant Cell Physiol.,* 5: 1–10.
Esau, K. 1953. "Plant Anatomy." Wiley, New York. 735 pp.
———. 1960. "Anatomy of Seed Plants." Wiley, New York.
———. 1961. "Plants, Viruses and Insects." Harvard University Press, Cambridge, Mass.
———. 1965. "Plant Anatomy," 2d ed. Wiley, New York.
———. 1967. Minor veins in *Beta* leaves: Structure related to function. *Proc. Am. Phil. Soc.,* 3: 219–233.
———, E. M. Engleman, and T. Bisalputra. 1963. What are transcellular strands? *Planta,* 59: 617–623.
Eschrich, W. 1965. Physiologie der Siebröhrencallose. *Planta,* 65: 280–300.
———, R. F. Evert, and J. H. Young. 1972. Solution flow in tubular semipermeable membranes. *Planta,* 107: 279–300.
Evans, G. C. 1956. An area survey method of investigating the distribution of light intensity in woodlands, with particular reference to sunflecks. *J. Ecol.,* 44: 391–428.
——— and A. P. Hughes. 1961. Plant growth and the aerial environment. I: Effect of artificial shading on *Impatiens parviflora. New Phytol.,* 60: 150–180.
——— and ———. 1962. Plant growth and the aerial environment. III: On the computation of unit leaf rate. *New Phytol.,* 61: 322–327.
Evans, L. T. 1960. Inflorescence initiation in *Lolium temulentum.* I: Effect of plant age and leaf area on sensitivity to photoperiodic induction. *Aust. J. Biol. Sci.,* 13: 123–131.
———. 1963. Extrapolation from controlled environments to the field. Pp. 421–437 in L. T. Evans (ed.), "Environmental Control of Plant Growth." Academic, New York.
———. 1964*a*. Inflorescence initiation in *Lolium temulentum.* VI: Effects of inhibitors of nucleic acid, protein and steroid biosynthesis. *Aust. J. Biol. Sci.,* 17: 24–35.
———. 1964*b*. Infloresence initiation in *Lolium temulentum,* V. *Aust. J. Biol. Sci.,* 17: 10–23.
———. 1966. Abscisin II: Inhibitory effect on flower induction in a long-day plant. *Science,* 151: 107–108.
———. 1971. Flower induction and the florigen concept. *Ann. Rev. Plant Physiol.,* 22: 365–394.
Evans, M. L., and R. Hokanson. 1969. Timing of the response of coleoptiles to the application and withdrawal of various auxins. *Planta,* 85: 85–90.
——— and P. M. Ray. 1969. Timing of the auxin response in coleoptiles and its implications regarding auxin action. *J. Gen. Physiol.,* 53: 1–20.
———, ———, and L. Reinhold. 1971. Induction of coleoptile elongation by carbon dioxide. *Plant Physiol.,* 47: 335–341.
Evenari, M. 1949. Germination inhibitors. *Bot. Rev.,* 15: 153–194.
———. 1952*a*. The water balance of plants in desert conditions. *Desert Res.,* 2: 1-9.
———. 1952*b*. The germination of lettuce seeds. *Palest. J. Bot.,* 5: 138–160.
———. 1957. The physiological action and biological importance of germination inhibitors. *Soc. Exp. Biol. Symp.,* 11: 21–43.
———, G. Stein, and G. Neumann. 1954. The action of light in conjunction with thiourea on germination. *Proc. 1st. Int. Photobiol. Congr. 1954,* pp. 82–86.
Everson, R. G., and C. R. Slack. 1968. Distribution of

carbonic anhydrase in relation to the C_4 pathway of photosynthesis. *Phytochemistry,* 7: 581–584.

Everett, M., and K. V. Thimann. 1968. Second positive phototropism in the *Avena* coleoptile. *Plant Physiol.,* 43: 1786–1792.

Evert, R. F., and W. F. Derr. 1964. Slime substance and strands in sieve elements. *Am. J. Bot.,* 51: 875–880.

———, W. Eschrich, and S. E. Eichhorn. 1971. Sieve-plate pores in leaf veins of *Hordeum vulgare. Planta,* 100: 262–267.

——— and L. Murmanis. 1965. Ultrastructure of the secondary phloem of *Tilia americana. Am. J. Bot.,* 52: 95–106.

———, ———, and I. B. Sachs. 1966. Another view of the ultrastructure of *Cucurbita* phloem. *Ann. Bot.,* n.s., 30: 563–585.

Evins, W. H. 1971. Enhancement of polyribosome formation and induction of tryptophan-rich proteins by gibberellic acid. *Biochemistry,* 10: 4295–4303.

——— and J. E. Varner. 1971. Hormone controlled synthesis of endoplasmic reticulum in barley aleurone cells. *Proc. Natl. Acad. Sci. (U.S.),* 68: 1631–1633.

Eyk, J. van, and H. Veldstra. 1966. A comparative investigation of kinetin and some similarly substituted purines with *Lemna. Phytochemistry,* 5: 457–462.

Faegri, K., and L. van der Pijl. 1966. "The Principles of Pollination Ecology." Pergamon, London. 248 pp.

Fan, D. F., and G. A. Maclachlan. 1966. Control of cellulase activity by indoleacetic acid. *Can. J. Bot.,* 44: 1025–1034.

——— and ———. 1967. Massive synthesis of ribonucleic acid and cellulase in the pea epicotyl in response to indoleacetic acid. *Plant Physiol.,* 42: 1114–1122.

Fang, S. C., V. H. Freed, R. H. Johnson, and D. R. Coffee. 1955. Absorption, translocation and metabolism of radioactive CMV in bean plants. *Agric. Food Chem.,* 3: 400–402.

——— and L. Y. Dalbert. 1954. The adaptive formation and physiological significance of indoleacetic acid oxidase. *Am. J. Bot.,* 41: 373–380.

Fassett, N. C. 1960. "A Manual of Aquatic Plants." University of Wisconsin Press, Madison. 405 pp.

Fawcett, C. H., J. M. A. Inhram, and R. L. Wain. 1952. β-Oxidation of phenoxyallylcarboxylic acid in the flax plant. *Nature,* 170: 887.

———, R. L. Wain, and F. Wightman. 1958. Beta-oxidation of omega indolylalkanecarboxylic acids in plant tissues. *Nature,* 181: 1387–1389.

Fekete, M. A. R. de, and C. E. Cardini. 1964. Mechanism of glucose transfer from sucrose into starch granule of sweet corn. *Arch. Biochem. Biophys.,* 104: 173–184.

Fensom, D. S. 1957. The bioelectric potentials of plants and their functional significance. I: An electro-kinetic theory of transport. *Can. J. Bot.,* 35: 537–582.

———. 1972. A theory of translocation in phloem of *Heracleum* by contractile protein microfibrillar material. *Can. J. Bot.,* 50: 479–497.

——— and H. R. Davidson. 1970. Micro-injection of ^{14}C-sucrose into single living sieve tubes of *Heracleum. Nature,* 227: 857–858.

Ferguson, J. H. A. 1957. Photothermographs, a tool for climate studies in relation to the ecology of vegetable varieties. *Euphytica,* 6: 97–105.

Fernald, M. L. 1926. The antiquity and dispersal of vascular plants. *Q. Rev. Biol.,* 1: 212–245.

Fernqvist, I. 1966. Studies on factors in adventitious root formation. *Lantbrukshogsk. Ann.,* 32: 109–244.

Filippova, L. A. 1957. Qualitative distribution of photosynthetic carbon among organic substances in a leaf. *Proc. 2d All-Union Conf. Photosynthesis,* pp. 325–329 (Acad. Sci. USSR, Moscow.)

Filner, P., and J. E. Varner. 1967. A simple and unequivocal test for de novo synthesis of enzyme: Density labeling of barley α-amylase with H_2O^{18}. *Proc. Natl. Acad. Sci. (U.S.),* 58: 1520–1526.

Fine, J. M., and L. Barton. 1958. Biochemical studies of dormancy and after-ripening in seeds, I. *Contrib. Boyce Thompson Inst.,* 19: 483–500.

Fischer, J. E., and W. E. Loomis. 1954. Auxin-florigen balance in flowering of soybean. *Science,* 119: 71–73.

Fischer, R. A., and T. C. Hsiao. 1968. Stomatal opening in isolated epidermal strips of *Vicia faba.* II: Responses to KCl concentrations and the role of potassium absorption. *Plant Physiol.,* 43: 1953–1958.

———, ———, and R. M. Hagan. 1970. After-effect of water stress on stomatal opening potential. *J. Exp. Bot.,* 21: 371–385.

Fitting, H. 1909. Die Beeinflussung der Orchideenblüten durch die Bestandung und durch anders Umstande. *Z. Bot.,* 1: 1–86.

———. 1938. Die Umkehrbarkeit der durch aussenfaktoren induzierten Dorsiventralität. *Jahrb. Wiss. Bot.,* 86: 107.

Fittler, F., and R. H. Hall. 1966. Selective modification of yeast seryl-tRNA and its effect on the acceptance and binding functions. *Biochem. Biophys. Res. Commun.,* 25: 441–446.

Fleet, D. S. van. 1961. Histochemistry and function of the endodermis. *Bot. Rev.,* 27: 165–220.

Flemion, F. 1959. Effect of temperature, light and gibberellin on stem elongation in dwarfed peach and *Rhodotypos. Contrib. Boyce Thompson Inst.,* 20: 57–70.

——— and E. Waterbury. 1945. Further studies with dwarf seedlings of non-after-ripened peach seeds. *Contrib. Boyce Thompson Inst.,* 13: 415–422.

Fletcher, F. A., and D. J. Osborne. 1965. Regulation

of protein and nucleic acid synthesis by gibberellin during leaf senescence. *Nature,* 207: 1176–1177.

Flint, L. H., and E. D. McAlister. 1937. Wave lengths of radiation in the visible spectrum promoting the germination of light-sensitive lettuce seed. *Smithson. Misc. Coll.,* 96: 1–8.

Fogg, G. E. 1947. Quantitative studies on the wetting of leaves by water. *Proc. R. Soc. Lond.,* B134: 503–522.

———. 1969. Survival of algae under adverse conditions. *Symp. Soc. Exp. Bio.,* 23: 123–142.

Ford, M. A., and G. W. Thorne. 1967. Effect of CO_2 concentration on growth of sugar-beet, barley, kale, and maize. *Ann. Bot.,* 31: 629–644.

Forrester, M. L., G. Krotkov, and C. D. Nelson. 1966. Effect of oxygen on photosynthesis, photorespiration and respiration in detached leaves. II: Corn and other monocotyledons. *Plant Physiol.,* 41: 428–431.

Fortanier, E. J. 1954. Some observations on the influence of spectral regions of light on stem elongation, flower bud elongation, flower bud opening and leaf movement in *Arachis hypogea* L. *Meded. Landbouwhogesch. Wageningen,* 54: 103–114.

Forward, D. F. 1960. The effect of temperature on respiration. *Encycl. Plant Physiol.,* 12(2): 234–254. (Springer-Verlag, Berlin).

Foster, R. J., D. H. McRae, and J. Bonner. 1952. Auxin induced growth inhibition, a natural consequence of two point attachment. *Proc. Natl. Acad. Sci. (U.S.),* 38: 1014–1022.

Fox, J. E. 1964. Indoleacetic acid–kinetin antagonism in certain tissue culture systems. *Plant Cell Physiol.,* 5: 251–254.

———. 1966. Incorporation of a kinin, *N*-6-benzyladenine, into soluble RNA. *Plant Physiol.,* 41: 75–82.

———. 1969. The cytokinins. Pp. 85–123 in M. B. Wilkins (ed.), "Physiology of Plant Growth and Development." McGraw-Hill, London.

——— and C. M. Chen. 1967. Characterization of labeled RNA from tissue grown on C^{14}-containing cytokinins. *J. Biol. Chem.,* 242: 4490–4494.

——— and J. S. Weis. 1965. Transport of the kinin, *N*-benzyladenine; non-polar or polar? *Nature,* 206: 678–679.

Foy, C. L. 1961. Absorption, distribution, and metabolism of 2,2-dichloropropionic acid in relation to phytotoxicity, II. *Plant Physiol.,* 36: 698–709.

———, W. Hurtt, and M. G. Hale. 1971. Root exudation of plant growth regulators. Pp. 75–85 in "Biochemical Interactions among Plants." National Academy of Science, Washington.

Franke, W. 1961. Ectodesmata and foliar absorption. *Am. J. Bot.,* 48: 683–690.

Frankland, B., and P. F. Wareing. 1960. Effect of gibberellic acid on hypocotyl growth of lettuce seedlings. *Nature,* 185: 255–256.

——— and ———. 1962. Changes in endogenous gibberellins in relation to chilling of dormant seeds. *Nature,* 194: 313–314.

Frattianne, D. G. 1965. The interrelationship between the flowering of dodder and the flowering of some long and short day plants. *Am. J. Bot.,* 52: 556–562.

Frazier, W. A., and J. L. Bowers. 1947. A final report on studies of tomato fruit cracking in Maryland. *Proc. Am. Soc. Hort. Sci.,* 49: 241–255.

Fredericq, H. 1964. Conditions determining effects of far-red and red radiations on flowering response in *Pharbitis nil. Plant Physiol.,* 39: 812–816.

Frenkel, C. 1972. Involvement of peroxidase and indoleacetic acid oxidase isozymes from pear, tomato and blueberry fruit in ripening. *Plant Physiol.,* 49: 757–763.

———, I. Klein, and D. R. Dilley. 1968. Protein synthesis in relation to ripening of pome fruits. *Plant Physiol.,* 43: 1146–1153.

Frost, H. B. 1952. Characteristics in the nursery of *Citrus* budlings of young nucellar seedling lines and parental old lines. *Proc. Am. Soc. Hort. Sci.,* 60: 247–254.

Fuchs, Y., and M. Lieberman. 1968. Effects of kinetin IAA and gibberellin on ethylene production and their interactions in growth of seedlings. *Plant Physiol.,* 43: 2029–2036.

Fuente, R. K. dela, and A. C. Leopold. 1966. Kinetics of polar auxin transport. *Plant Physiol.,* 41: 1481–1484.

——— and ———. 1968. Senescence processes in leaf abscission. *Plant Physiol.,* 43: 1496–1502.

——— and ———. 1969. Kinetics of abscission in the bean petiole explant. *Plant Physiol.,* 44: 251–254.

——— and ———. 1970. Time course of auxin stimulations of growth. *Plant Physiol.,* 46: 186–189.

Fujino, M. 1967. Role of adenosine triphosphate and adenosine-triphosphatase in stomatal movement. *Sci. Bull. Educ. Nagasaki Univ.,* 18: 1–47.

Fujisawa, H. 1966. Role of nucleic acid and protein metabolism in the initiation of growth at germination. *Plant Cell Physiol.,* 7: 185–198.

Fuller, G. L., and A. C. Leopold. 1972. Timing of nucleic acid changes in cucumber ovary tissue with pollination. *Plant Physiol.,* 49(suppl.): 29.

Funke, G. L. 1938. Observations on the growth of water plants. *Bot. Jaarb.,* vol. 25.

Furr, J. R., W. C. Cooper, and P. C. Reece. 1947. Flower formation in citrus trees. *Am. J. Bot.,* 34: 1–8.

Furuya, M., A. W. Galston, and B. B. Stowe. 1962. Isolation from peas of co-factors and inhibitors of indolyl-3-acetic acid oxidase. *Nature,* 193: 456–457.

——— and R. G. Thomas. 1964. Flavonoid complexes in *Pisum sativum,* II. *Plant Physiol.,* 39: 634–642.

Gaastra, P. 1958. Light energy conversion in field

crops in comparison with photosynthetic efficiency under laboratory conditions. *Meded. Landbouwhogesch. Wageningen,* 58: 1–12.

———. 1959. Photosynthesis of crop plants as influenced by light, carbon dioxide, temperature and stomatal diffusion resistance. *Meded. Landbouwhogesch. Wageningen,* 59: 1–68.

———. 1962. Photosynthesis of leaves and field crops. *Neth. J. Agric. Sci.,* 10: 311–324.

Gabelman, W. H., and D. D. F. Williams. 1962. Water relations affecting pod set of green beans. *Plant Sci. Symp., Campbell Soup Co., Camden, N.J.,* pp. 25–35.

Gabr, O. M. K., and C. G. Guttridge. 1968. Identification of abscisic acid in strawberry leaves. *Planta,* 78: 305–309.

Gaff, D. F. 1966. The sulfhydryl-disulfide hypothesis in relation to desiccation injury of cabbage leaves. *Aust. J. Biol. Sci.,* 19: 291–299.

———. 1971. Desiccation-tolerant flowering plants in southern Africa. *Science,* 174: 1033–1034.

———, T. C. Chambers, and K. Markus. 1964. Studies of extrafascicular movement of water in the leaf. *Aust. J. Biol. Sci.,* 17: 581–586.

Gahagan, H. E., R. E. Holm, and F. B. Abeles. 1968. Effect of ethylene on peroxidase activity. *Physiol. Plant.,* 21: 1270–1279.

Gale, J., and R. M. Hagan. 1966. Plant antitranspirants. *Annu. Rev. Plant Physiol.,* 17: 269–282.

———, R. Naaman, and A. Poljakoff-Mayber. 1970. Growth of *Atriplex halinus* L. in sodium chloride salinated culture solutions as affected by the relative humidity of the air. *Aust. J. Biol. Sci.,* 23: 947–952.

——— and A. Poljakoff-Mayber. 1968. Resistances to the diffusion of gas and vapour in leaves. *Physiol. Plant.,* 21: 1170–1176.

——— and ———. 1970. Interrelations between growth and photosynthesis of salt bush (*Atriplex halinus* L.) grown in saline media. *Aust. J. Biol. Sci.,* 23: 937–945.

Galil, J. 1968. Vegetative dispersal in *Oxalis cernua. Am. J. Bot.,* 55: 68–73.

Galsky, A. G., and J. A. Lippincott. 1969. Promotion and inhibition of α-amylase production in barley endosperm by cyclic AMP and adenosine diphosphate. *Plant Cell Physiol.,* 10: 607–620.

Galston, A. W. 1955. Some metabolic consequences of the administration of indoleacetic acid to plant cells. Pp. 219–233 in R. L. Wain and F. Wightman (eds.), "The Chemistry and Mode of Action of Plant Growth Substances." Butterworth, London.

———, and R. S. Baker. 1949. Studies on the physiology of light action. II: The photodynamic action of riboflavin. *Am. J. Bot.,* 36: 773–780.

——— and ———. 1951. Studies on the physiology of light action, IV. *Plant Physiol.,* 26: 311–317.

———, J. Bonner, and R. S. Baker. 1953. Flavoprotein and peroxidase as components of the indoleacetic acid oxidase system of peas. *Arch. Biochem. Biophys.,* 49: 456–470.

——— and M. E. Hand. 1949*a*. Adenine as a growth factor for etiolated peas and its relation to the thermal inactivation of growth. *Arch. Biochem.,* 22: 434–443.

——— and ———. 1949*b*. Studies on the physiology of light action, I. *Am. J. Bot.,* 36: 85–94.

——— and W. S. Hillman. 1961. The degradation of auxin. *Handb. Pflanzenphysiol.,* 14: 647–670.

Galun, E. 1959. Effects of gibberellic acid and naphthaleneacetic acid on sex expression and some morphological characters in the cucumber plant. *Phyton,* 13: 1–8.

———, J. Gressel, and A. Keynan. 1964. Suppression of floral induction by actinomycin-D, an inhibitor of messenger RNA synthesis. *Life Sci.,* 3: 911–915.

———, Y. Jung, and A. Lang. 1962. Culture and sex modification of male cucumber buds in vitro. *Nature,* 194: 596–598.

Gane, R. 1934. Production of ethylene by some ripening fruits. *Nature,* 134: 1008.

———. 1935. The formation of ethylene by plant tissues and its significance in the ripening of fruits. *J. Pomol. Hort. Sci.,* 13: 351.

———. 1937. The respiration of bananas in presence of ethylene. *New Phytol.,* 36: 170.

Gardner, F. E. 1929. The relationship between tree age and the rooting of cuttings. *Proc. Am. Soc. Hort. Sci.,* 26: 101.

Gardner, G., C. S. Pike, H. V. Rice, and W. R. Briggs. 1971. "Disaggregation" of phytochrome *in vitro*: A consequence of proteolysis. *Plant Physiol.,* 48: 686–693.

Gardner, W. R., and R. H. Nieman. 1964. Lower limit of water availability to plants. *Science,* 143: 1460–1462.

Garner, W. W., and H. A. Allard. 1920. Effect of length of day on plant growth. *J. Agric. Res.,* 18: 553–606.

——— and ———. 1931. Effect of abnormally long and short alterations of light and darkness on growth and development of plants. *J. Agric. Res.,* 42: 629–651.

Garrison, R. 1949. Origin and development of axillary buds: *Syringa vulgaris. Am. J. Bot.,* 36: 205–213.

——— and R. H. Wetmore. 1961. Studies in shoot-tips abortion: *Syringa vulgaris. Am. J. Bot.,* 48: 789–795.

Gaspar, T., and A. Xhaufflaire. 1967. Effect of kinetin on growth, auxin catabolism, peroxidase and catalase activities. *Planta,* 72: 252–257.

Gassner, G. 1918. Beiträge zur physiologischen Charakteristik Sommer und Winter annueller Gewächse insbesondere der Getreidepflanzen. *Z. Bot.,* 10: 417–430.

Gates, C. T. 1955*a*. The response of the young tomato

plant to a brief period of water shortage. I: The whole plant and its principal parts. *Aust. J. Biol. Sci.,* 8: 196–214.

———. 1955*b*. The response of the young tomato plant to a brief period of water shortage. II: The individual leaves. *Aust. J. Biol. Sci.,* 8: 215–230.

——— and J. Bonner. 1959. The response of the young tomato plant to a brief period of water shortage. IV: Effects of water stress on the ribonucleic acid metabolism of tomato leaves. *Plant Physiol.,* 34: 49–55.

Gates, D. M. 1963. The energy environment in which we live. *Am. Sci.,* 51: 327–348.

———. 1965. Radiant energy, its receipt and disposal. *Meteorol. Monogr.,* 6(28): 1–26.

———. 1968. Transpiration and leaf temperature. *Annu. Rev. Plant Physiol.,* 19: 211–238.

———. 1969. The ecology of an elfin forest in Puerto Rico, 4: Transpiration rates and temperatures of leaves in cool humid environment. *J. Arnold Arbor.,* 50: 93–98.

———, R. Alderfer, and E. Taylor. 1968. Leaf temperatures of desert plants. *Science,* 159: 994–995.

———, W. H. Hiesey, H. W. Milner, and M. A. Nobs. 1964. Temperature and environment of *Mimulus* in the Sierra Nevada Mountains. *Carnegie Inst. Wash. Year Book,* 63: 418–426.

———, H. J. Keegan, J. C. Schleter, and V. R. Weidner, 1965. Spectral properties of plants. *Appl. Opt.,* 4: 11–20.

Gauch, H. G., and W. M. Dugger, Jr. 1953. The role of boron in the translocation of sucrose. *Plant Physiol.,* 28: 457–466.

Geiger, D. R., and D. A. Cataldo. 1969. Leaf structure and translocation in sugar beet. *Plant Physiol.,* 44: 45–54.

——— and A. L. Christy. 1971. Effect of sink region anoxia on translocation rate. *Plant Physiol.,* 47: 172–174.

———, J. Malone, and D. A. Cataldo. 1971. Structural evidence for a theory of vein loading of translocate. *Am. J. Bot.,* 58: 672–675.

Geiger, R. 1957. "The Climate near the Ground," Trans., M. N. Stewart, Harvard University Press, Cambridge, Mass. 482 pp.

Geissman, T. A., A. J. Verbiscar, B. O. Phinney, and G. Cragg. 1966. Studies on the biosynthesis of gibberellins from kaurenoic acid in cultures of *Gibberella fujikuroi. Phytochemistry,* 5: 933.

Genkel, P. A., K. A. Badanoua, and I. N. Andreeva. 1967. Significance of respiration in plant cell water content under drought conditions. *Fiziol. Rast.,* 14: 418–423 (trans.).

Gerhardt, F., and E. Smith. 1946. *Proc. Wash. State Hort. Assoc., 41st Annu. Meet.;* cited in Smock and Neubert (1950).

Giaquinta, R. T., and D. R. Geiger. 1973. Mechanism of inhibition of translocation by localized chilling. *Plant Physiol.,* 51: 372–377.

Gibbs, M. 1970. Photorespiration, Warburg effect and glycolate. *Ann. N.Y. Acad. Sci.,* 168: 356–368.

———, E. Latzko, R. G. Everson, and W. Cockburn. 1967. Carbon mobilization by the green plant. Pp. 111–130 in A. San Pietro, G. A. Greer, and T. J. Army (eds.), "Harvesting the Sun." Academic, New York.

Gibbs, R. D. 1958. Patterns in the seasonal water content of trees. Pp. 43–69 in K. V. Thimann (ed.), "The Physiology of Forest Trees." Ronald, New York.

Gientka-Rychter, A., and J. H. Cherry. 1968. De novo synthesis of isocitritase in peanut cotyledons. *Plant Physiol.,* 43: 653–659.

Gifford, E. M. 1964. Developmental studies of vegetative and floral meristems. *Brookhaven Symp. Biol.,* 16: 126–137.

——— and K. D. Stewart. 1965. Ultrastructure of vegetative and reproductive apices of *Chenopodium album. Science,* 149: 75–77.

——— and H. B. Tepper. 1962. Ontogenetic and histochemical changes in the vegetative shoot tip of *Chenopodium album. Am. J. Bot.,* 49: 902–911.

Gilbert, M. L., and A. G. Galsky. 1972. The action of cyclic-AMP on GA controlled responses, III. *Plant Cell Physiol.,* 13: 867–873.

Gilder, J., and J. Cronshaw. 1973. Adenosine triphosphatase in the phloem of *Cucurbita. Planta,* 110: 189–204.

Gillespie, B., and K. V. Thimann. 1961. The lateral transport of indoleacetic acid–C^{14} in geotropism. *Experientia,* 17: 126–129.

Ginsburg, H., and B. Z. Ginzburg. 1971. Radial water and solute flow in roots of *Zea mays,* III. *J. Expt. Bot.,* 22: 337–341.

Glasziou, K. T., K. R. Gayler, and J. C. Waldron. 1968. Effects of auxin and gibberellic acid on the regulation of enzyme synthesis in sugar-cane stem tissue. Pp. 433–442 in F. Wightman and G. Setterfield (eds.), "Biochemistry and Physiology of Plant Growth Substances." The Runge Press, Ottawa.

Glinka, Z., and L. Reinhold. 1971. Abscisic acid raises the permeability of plant cells to water. *Plant Physiol.,* 48: 103–105.

——— and ———. 1972. Induced changes in permeability of plant cell membranes to water. *Plant Physiol.,* 49: 602–606.

Gmelin, R., and A. I. Virtanen. 1961. Glucobrassicin, the precursor of indolylacetylnitrile, ascorbigen and SCN in *Brassica oleracea. Suomen Kem.,* 34: 15–18.

Goebel, K. 1889. Über die Jugendzustände der Pflanzen. *Flora,* 72: 1–45.

Goeschl, J. D., H. K. Pratt, and B. A. Bonner. 1967. An effect of light on the production of ethylene and the growth of the plumular portion of etiolated pea seedlings. *Plant Physiol.,* 42: 1077–1080.

———, L. Rappaport, and H. K. Pratt. 1966. Ethylene as a factor regulating the growth of pea epicotyls subjected to physical stress. *Plant Physiol.,* 41: 877–884.

Goldacre, P. L., and W. Bottomley. 1959. A kinin in apple fruitlets. *Nature,* 184: 555–556.

———, A. W. Galston, and R. L. Weintraub. 1953. The effect of substituted phenols on the activity of the indoleacetic acid oxidase of peas. *Arch. Biochem. Biophys.,* 43: 358–373.

Goldschmidt, E. E. 1968. The auxin induced curvature on *Citrus* petals. *Plant Physiol.,* 43: 1973–1977.

Goldsmith, M. H. M. 1967. Separation of transit of auxin from uptake: Average velocity and reversible inhibition by anaerobic conditions. *Science,* 156: 661–663.

——— and K. V. Thimann. 1962. Some characteristics of movement of indoleacetic acid in coleoptiles of *Avena,* I. *Plant Physiol.,* 37: 492–505.

——— and M. B. Wilkins. 1964. Movement of auxin in coleoptiles of *Zea mays* during geotropic stimulation. *Plant Physiol.,* 39: 151–162.

Goldsworthy, A. 1968. Comparison of the kinetics of photosynthetic carbon dioxide fixation in maize, sugar cane and tobacco and its relation to photorespiration. *Nature,* 217: 62.

Goldthwaite, J. J. 1972. Further studies of hormone-regulated senescence in *Rumex* leaf tissue. Pp. 581–588 in D. J. Carr (ed.), "Plant Growth Substances, 1970." Springer-Verlag, Berlin.

——— and W. M. Laetsch. 1967. Regulation of senescence in bean leaf discs by light and chemical growth regulators. *Plant Physiol.,* 42: 1757–1762.

Goodall, D. W. 1945. Distribution of weight change in the young tomato plant. I: Dry-weight changes of the various organs. *Ann. Bot.,* n.s., 9: 101–139.

Goodchild, D. J., O. Björkman, and N. A. Pyliotis. 1972. Chloroplast ultrastructure, leaf anatomy, and content of chlorophyll and soluble protein in rainforest species. *Annu. Rep. Carnegie Inst. Wash., Stanford, Calif., 1971-1972,* pp. 102–107.

Goodwin, R. H. 1944. Inheritance of flowering time in a short-day species, *Solidago sempervirens. Genetics,* 29: 503–519.

——— and O. H. Owens. 1951. The effectiveness of the spectrum in *Avena* internode inhibition. *Bull. Torrey Bot. Club,* 78: 11–21.

Gordon, S. A. 1946. Auxin-protein complexes of the wheat grain. *Am. J. Bot.,* 33: 160–169.

———. 1956. Auxin biosynthesis: A cytoplasmic locus of radiation damage. Pp. 44–47 in "Progress in Radiobiology." Oliver & Boyd, London.

———. 1961. The biogenesis of auxin. *Handb. Pflanzenphysiol.,* 14: 620–646.

——— and R. P. Weber. 1951. Colorimetric estimation of indoleacetic acid. *Plant Physiol.,* 26: 192–195.

Goren, R., and S. P. Monselise. 1964. Survey of hesperidin and nitrogen in the developing flower of the Shamouti orange tree. *Proc. Am. Soc. Hort. Sci.,* 85: 218–223.

Gorter, C. J. 1932. Groeistofproblemen bej Wortels. Ph.D. thesis, Univ. Utrecht.

———. 1958. Syngergism of indole and indole-3-acetic acid in the root production of *Phaseolus* cuttings. *Physiol. Plant.,* 11: 1–9.

Gortner, W. A., and M. Kent. 1953. Indoleacetic acid oxidase and an inhibitor in pineapple tissue. *J. Bio. Chem.,* 204: 594–603.

———, ——— and G. K. Sutherland. 1958. Ferulic and *p*-coumaric acids in pineapple tissue as modifiers of pineapple indoleacetic acid oxidase. *Nature,* 181: 630–631.

Gott, M. B., F. G. Gregory, and O. N. Purvis. 1955. Studies in vernalisation of cereals. XIII: Photoperiodic control of stages in flowering between initiation and ear formation in vernalised and unvernalised Petkus winter rye. *Ann. Bot.,* 21: 87–126.

Gouwentak, C. A., and G. Hellings. 1935. Beobachtungen über Wurzelbildung. *Meded. Landbouwhogesch. Wageningen,* 39: 1–6.

Graebe, J. E., D. T. Dennis, C. D. Upper, and C. A. West. 1965. The biosynthesis of kaurene, kaurene-ol and *trans*-geranylgeraniol in endosperm nucellus of *Echinocystis macrosarpa. J. Biol. Chem.,* 240: 1847.

Graham, D., C. A. Atkins, M. L. Reed, B. D. Patterson, and R. M. Smillie. 1971. Carbonic anhydrase, photosynthesis, and light-induced pH changes. Pp. 267–274 in M. D. Hatch, C. B. Osmond, and R. O. Slayter (eds.), "Photosynthesis and Photorespiration." Wiley-Interscience, New York.

———, A. M. Grieve, and R. M. Smillie. 1968. Phytochrome as the primary photoregulator of the synthesis of Calvin cycle enzymes in etiolated pea seedlings. *Nature,* 218: 89–90.

———, M. D. Hatch, C. R. Slack, and R. M. Smillie. 1970. Light-induced formation of enzymes of the C_4-dicarboxylic acid pathway of photosynthesis in detached leaves. *Phytochemistry,* 9: 521–532.

Greathouse, D. C., W. M. Laetsch, and B. O. Phinney. 1971. The shoot growth rhythm of a tropical tree. *Am. J. Bot.,* 58: 281–286.

Green, B. R., and M. P. Gordon. 1966. Replication of chloroplast DNA of tobacco. *Science,* 152: 1071–1074.

Greenwood, M. S., S. Shaw, J. R. Hillman, A. Ritchie, and M. B. Wilkins. 1972. Identification of auxin from *Zea* coleoptile tips by mass spectroscopy. *Planta,* 108: 179–183.

Gregory, D. W., and E. C. Cocking. 1966. Studies on isolated protoplasts and vacuoles, II. *J. Exp. Bot.,* 17: 68–77.

Gregory, F. G. 1917. Physiological conditions in cucumber houses. *Exp. Res. Stn. Cheshunt 3d Annu. Rep.*, p. 19.

———. 1926. Effect of climatic conditions on the growth of barley. *Ann. Bot.*, 40: 1–26.

——— and O. N. Purvis. 1937. Devernalization of *spring rye* by anaerobic conditions and revernalization by low temperatures. *Nature*, 140: 547.

——— and R. S. de Ropp. 1938. Vernalization of excised embryos. *Nature*, 142: 481–482.

——— and J. A. Veale. 1957. A reassessment of the problem of apical dominance. *Soc. Exp. Biol. Symp.*, 11: 1–20.

Gregory, L. E. 1956. Some factors for tuberization in the potato plant. *Am. J. Bot.*, 43: 281–288.

Griesel, W. O. 1966. Inheritance of factors affecting floral primordia initiation in *Cestrum*. *Plant Physiol.*, 41: 111–114.

Grieshaber-Scheuber, D., and G. Fellenberg. 1971. Influence of growth substances on the binding of lysine-rich and arginine-rich histone to DNA. *Z. Pflanzenphysiol.*, 66: 106–112.

Grieve, B. J., and E. O. Hellmuth. 1968. Ecophysiological studies of Western Australian plants. *Proc. Ecol. Soc. Aust.*, 3: 46–54.

Grime, J. P. 1965. Shade tolerance in flowering plants. *Nature*, 208: 161–163.

Grunwald, C., J. Mendez, and B. B. Stowe. 1968. Substrates for the optimum gas chromagraphic separation of indolic methyl esters. Pp. 163–171 in F. Wightman and G. Setterfield (eds.), "Biochemistry and Physiology of Plant Growth Substances." Runge Press, Ottawa.

Guern, J. 1964. Remarques à propos des Méthodes d'extraction, de purification et d'identification des inhibiteurs de croissance. P. 352 in J. P. Nitsch (ed.), "Régulateurs naturels de la croissance végétale." CNRS, Paris.

———, M. Doree, and P. Sadorge. 1968. Transport, metabolism and biological activity of some cytokinins. Pp. 1155–1167 in F. Wightman and G. Setterfield (eds.), "Biochemistry and Physiology of Plant Growth Substances." Runge Press, Ottawa.

Gunckel, J. E., and K. V. Thimann. 1949. Studies of development of long shoots and short shoots of *Ginkgo biloba* L. III: Auxin production of short shoot. *Am. J. Bot.*, 36: 145–151.

Gunning, B. E. S., and W. K. Barkley. 1963. Kinin-induced directed transport and senescence in detached oat leaves. *Nature*, 199: 262–264.

——— and L. G. Briarty. 1968. Specialized "transfer cells" in minor veins of leaves and their possible significance in phloem translocation. *J. Cell Biol.*, 37: C7–C12.

——— and J. S. Pate. 1969. "Transfer cells": Plant cells with wall ingrowth, specialised in relation to short distance transport of solutes—their occurrence, structure and development. *Protoplasma*, 68: 107–133.

Gustafson, F. G. 1936. Inducement of fruit development by growth promoting chemicals. *Proc. Natl. Acad. Sci. (U.S.)*, 22: 626–636.

———. 1939. The cause of natural parthenocarpy. *Am. J. Bot.*, 26: 135–138.

Guttenberg, H. von, and H. Leike. 1958. Untersuchungen über den Wuchs- und Hemmstoffgehalt ruhender und treibender Knospen von *Syringa vulgaris*. *Planta*, 52: 96–120.

——— and I. Strutz. 1952. Zur Keimungsphysiologie von *Ustilage zeae*. *Arch. Mikrobiol.*, 14: 189–198.

Guttridge, C. D. 1959. Further evidence for a growth-promoting and flower-inhibiting hormone in strawberry. *Ann. Bot.*, 23: 612–621.

Guttman, R. 1956. Effects of kinetin on cell division with special reference to initiation and duration of mitosis. *Chromosoma*, 8: 341–350.

Haber, A. H., D. E. Foard, and S. W. Perdue. 1969. Actions of gibberellic and abscisic acids on lettuce seed germination without actions on nuclear DNA synthesis. *Plant Physiol.*, 44: 463–467.

——— and H. J. Luippold. 1960. Effects of gibberellin on gamma-irradiated wheat. *Am. J. Bot.*, 47: 140–144.

———, P. J. Thompson, P. L. Walne, and L. L. Triplett. 1969. Nonphotosynthetic retardation of chloroplast senescence by light. *Plant Physiol.*, 44: 1619–1628.

Haberlandt, G. 1884. "Physiological Plant Anatomy," 4th German ed. Trans. M. Drummond, MacMillan, London, 1914.

———. 1913. Zur Physiologie der Zellteilung. *Sitzungsber. K. Preuss. Akad. Wiss.*, 1913: 318–345.

———. 1914. "Physiological Plant Anatomy." Trans. M. Drummond. Macmillan, London. 776 pp.

———. 1921. Wundhormone als Erreger von Zellteilungen. *Beitr. Allg. Bot.*, 2: 1.

Hackett, D. P. 1952. The osmotic change during the auxin-induced water uptake by potato tissue. *Plant Physiol.*, 27: 279–284.

Hackett, W. P., and R. M. Sachs. 1968. Experimental separation of inflorescence development from initiation in *Bougainvillea*. *Proc. Am. Soc. Hort. Sci.*, 92: 615–621.

Hadfield, W. 1968. Leaf temperature, leaf pose and productivity of the tea bush. *Nature*, 219: 282–284.

Hager, A. 1970. Ausbildung von Maxima in Absorptionsspektrum von Carotinoiden in Bereich um 370 nm Folgen für die Interpretation bestimmter Wirkungsspektren. *Planta*, 91: 38–53.

———, H. Menzel, and A. Krauss. 1971. Versuche und Hypothese zur Primärwirkung des Auxins beim Streckungswachstum. *Planta*, 100: 47–75.

Hahne, I. 1961. Untersuchungen zum Problem der Beteiligung des Wuchsstoffes an der geotropischen Induktion bei Koleoptilen. *Planta*, 57: 557–582.

Hale, C. R., and R. J. Weaver. 1962. The effect of developmental stage on direction of translocation of photosynthate in *Vitis vinifera*. *Hilgardia*, 33: 89–131.

Hales, S. 1727. "Vegetable Statics." Oldbourne, London.

Halevy, A. H., A. Ashri, and Y. Ben-Tal. 1969. Peanuts: Gibberellin antagonists and genetically controlled differences in growth habit. *Science,* 164: 1397–1398.

——— and B. Kessler. 1963. Increased tolerance of bean plants to soil drought by means of growth retarding substances. *Nature,* 197: 310–311.

——— and S. H. Wittwer. 1965. Chemical regulation of leaf senescence. *Q. Bull. Mich. Agric. Exp. Stn.,* 48: 30–35.

Hall, R. H. 1967. An N^6-alkyl-adenosine in the sRNA of *Zea mays. Ann. N.Y. Acad. Sci.,* 144: 258–259.

———. 1968. Cytokinins in the transfer-RNA; their significance to the structure of tRNA. Pp. 47–56 in F. Wightman and G. Setterfield (eds.), "Biochemistry and Physiology of Plant Growth Substances." Runge Press, Ottawa.

———, L. Csonka, H. David, and B. McLennan. 1967. Cytokinins in the soluble RNA of plant tissues. *Science,* 156: 69–71.

———, N. J. Robins, L. Stasiuk, and T. Roosevelt. 1966. Isolation of N^6-γ,γ-dimethylallyladenosine from soluble ribonucleic acid. *J. Am. Chem. Soc.,* 88: 2614–2615.

———, M. Silagh, and D. A. Baker. 1972. The chemical composition of *Ricinus* phloem exudate. *Planta,* 106: 131–140.

Hall, W. C. 1951. Studies on the origin of ethylene from plant tissues. *Bot. Gaz.,* 113: 55–65.

———. 1952. Evidence on the auxin-ethylene balance hypothesis of foliar abscission. *Bot. Gaz.,* 113: 310–322.

———. 1958. Physiology and biochemistry of abscission in the cotton plant. *Tex. Agric. Exp. Stn.* MP-285.

——— and H. C. Lane. 1952. Compositional and physiological changes associated with the chemical defoliation of cotton. *Plant Physiol.,* 27: 754–768.

———, G. B. Truchelut, C. L. Leinweber, and F. A. Herrero. 1957. Ethylene production by the cotton plant and its effects under experimental and field conditions. *Physiol. Plant.,* 10: 306–317.

Hamilton, R. H., and D. E. Moreland. 1961. Simazine degradation by corn seedlings. *Science,* 135: 373–374.

Hammond, D. 1941. The expression of genes for leaf shape in *Gossypium hirsutum* and *Gossypium arboreum.* I: The expression of genes for leaf shape in *Gossypium hirsutum. Am. J. Bot.,* 28: 124–138.

Hamner, K. C. 1940. Interrelation of light and darkness in photoperiodic induction. *Bot. Gaz.,* 101: 658–687.

——— and J. Bonner. 1938. Photoperiodism in relation to hormones as factors in floral initiation. *Bot. Gaz.,* 100: 388–431.

——— and E. M. Long. 1939. Localization of photoperiod perception in *Helianthus. Bot. Gaz.,* 101: 81–90.

Hamzi, H. Q., and F. Skoog. 1964. Kinetin-like growth promoting activity of substituted adenines. *Proc. Natl. Acad. Sci. (U.S.),* 51: 76.

Hänsel, H. 1953. Vernalisation of winter rye by negative temperatures and the influence of vernalisation upon the lamina length of the first and second leaf in winter rye, spring barley, and winter barley. *Ann. Bot.,* 7: 418–431.

Hansen, E. 1942. Quantitative study of ethylene production in pears. *Bot. Gaz.,* 103: 543–548.

——— and G. D. Blanpied. 1968. Ethylene induced ripening of pears in relation to maturity and length of treatment. *Proc. Am. Soc. Hort. Sci.,* 93: 807–812.

——— and H. T. Hartmann. 1937. Effect of ethylene and metabolic gases upon respiration and ripening of pears before and after cold storage. *Plant Physiol.,* 12: 441–454.

Harada, H., and A. Lang. 1965. Effect of some 2-chloroethyl trimethylammonium chloride analogs and other growth retardants on gibberellin biosynthesis. *Plant Physiol.,* 40: 176–183.

——— and J. P. Nitsch. 1961. Isolement et propriétés physiologiques d'une substance de montaison. *Ann. Physiol. Veg.,* 3: 193–208.

Hardenburg, E. V. 1949. "Potato Production." Comstock Publishing Co., Ithaca, N.Y.

Harder, R., and O. Bode. 1943. Wirkung von Zwischenbelichtungen während der Dunkelperiode auf *Kalanchoe. Planta,* 33: 469–504.

——— and R. Bunsöw. 1956. Einfluss des Gibberellins auf die Blütenbildung bei *Kalanchoe blossfeldiana. Naturwiss.,* 23: 544.

Hardin, J. W., J. H. Cherry, D. J. Morre, and C. A. Lembi. 1972. Enhancement of RNA polymerase activity by a factor released by auxin from plasma membrane. *Proc. Natl. Acad. Sci. (U.S.),* 69: 3146–3150.

Hardwick, K., M. Wood, and H. W. Woolhouse. 1968. Photosynthesis and respiration in relation to leaf age in *Perilla frutescens* (L.) Britt. *New Phytol.,* 67: 79–86.

Hardy, P. J., and J. V. Possingham. 1969. Studies on translocation of metabolites in the xylem of grapevine shoots. *J. Exp. Bot.,* 20: 325–335.

Harley, C. P., J. R. Magness, M. P. Masure, L. A. Fletcher, and E. S. Degman. 1942. Investigations on the course and control of biennial bearing of apple trees. *USDA Tech. Bull.* 792.

Harris, C. I., D. D. Kaufman, T. J. Sheets, R. G. Nash, and P. C. Kearney. 1968. Behavior and fate of *s*-triazines in soils. *Adv. Pest Control Res.,* 8: 1–55.

Harris, H. C. 1949. Effect on the growth of peanuts of nutrient deficiencies in the root and pegging zone. *Plant Physiol.,* 24: 150–161.

Hartig, T. 1860. Beiträge zur physiologischen Forstbotanik. *Allg. Forst-Jagdztg.,* 36: 257–261.

Hartmann, H. T. 1950. Tests with growth regulators for increasing fruit set in olives. *Proc. Am. Soc. Hort. Sci.,* 55: 181–189.

Hartmann, K. M. 1966. A general hypothesis to interpret "high energy phenomena" of photomorphogenesis on the basis of phytochrome. *Photochem. Photobiol.,* 5: 349–366.

———. 1967*a*. Ein Wirkungsspektrum der Photomorphogenese unter Hochenergiebedingungen und seine Interpretation auf der Basis des Phytochroms (Hypokotylwachstumshemmung bei *Lactuca sativa* L.). *Z. Naturforsch.,* 22b: 1172–1175.

———. 1967*b*. Photoreceptor problems in photomorphogenic responses under high-energy-conditions (UV-blue-far-red). Pp. 29–32 in *Book Abstr. Eur. Photobiol. Symp., Hvar, Jugoslavia.*

Hartsema, A. M. 1961. Influence of temperatures on flower formation and flowering of bulbous and tuberous plants. *Encycl. Plant Physiol.,* 16: 123–167 (Springer-Verlag, Berlin).

———, I. Luyten, and A. H. Blaauw. 1930. The optimal temperatures from flower formation to flowering of Darwin tulips, II. *Proc. Ned. Akad. Wet.,* 27: 1–46.

Hartt, C. E. 1963. Translocation as a factor in photosynthesis. *Naturwiss.,* 21: 666–667.

———. 1965. Light and translocation of ^{14}C in detached blades of sugar cane. *Plant Physiol.,* 40: 718–724.

———. 1967. Effect of moisture supply upon translocation and storage of ^{14}C in sugar cane. *Plant Physiol.,* 42: 338–346.

———, H. P. Kortschak, and G. O. Burr. 1964. Effects of defoliation, deradication, and darkening the blade upon translocation of ^{14}C in sugar cane. *Plant Physiol.,* 39: 15–22.

Hartung, W. 1972. Die Wirkung von cyclischen AMP auf *Avena* Coleoptilzylinder. *Z. Pflanzenphysiol.,* 67: 380–382.

Hasegawa, S., V. P. Maier, H. P. Kaszychi, and J. K. Crawford. 1969. Polygalacturonase content of dates and its relation to maturity and softness. *J. Food Sci.,* 34: 527.

——— and D. C. Smolensky. 1970. Date invertase: Properties and activity associated with maturation and quality. *J. Agric. Chem.,* 18: 902–904.

——— and ———. 1971. Cellulase in dates and its role in fruit softening. *J. Food. Sci.,* 36: 966–967.

Hashimoto, T. 1958. Increase in percentage of gibberellin-induced dark germination of tobacco seeds by N-compounds. *Bot. Mag. (Tokyo),* 71: 845–846.

———, K. Hasegawa, and A. Kawarada. 1972. Batatasins: new dormancy inducing substances of yam bulbils. *Planta,* 108: 369–374.

——— and L. Rappaport. 1966. Variations in endogenous gibberellins in developing bean seeds, I and II. *Plant Physiol.,* 41: 623–628, 629–632.

Haskins, F. A., and H. J. Gorz. 1961. Assay of EIS and *o*-hydroxycinnamic acids in sweet clover extracts. *Biochem. Biophys. Res. Commun.,* 6: 298–303.

Hasman, M., N. Inanc, and C. Verter. 1961. Certain comparative studies on physiological and synergistic actions of gibberellin. *Physiol. Plant.,* 14: 290–302.

Hatakeyama, I., and J. Kato. 1965. Studies on the water relation of *Buxus* leaves. *Planta,* 65: 259–268.

Hatch, M. D. 1972. Photosynthesis and the C_4-pathway. *CSIRO Div. Plant Ind. Annu. Rep. 1971,* pp. 19–26.

———, C. B. Osmond, and R. D. Slatyer. 1971. "Photosynthesis and Photorespiration." Wiley-Interscience, New York.

——— and C. R. Slack. 1966. Photosynthesis by sugar cane leaves: A new carboxylation reaction and the pathway of sugar formation. *Biochem. J.,* 101: 103–111.

——— and ———. 1970. Photosynthetic CO_2-fixation pathways. *Ann. Rev. Plant Physiol.,* 21: 141–162.

———, ———, and T. A. Bull. 1969. Light-induced changes in the content of some enzymes of the C_4-dicarboxylic acid pathway of photosynthesis and its effect on other characteristics of photosynthesis. *Phytochemistry,* 8: 697–706.

Hatcher, E. S. J. 1959. Auxin relations of the woody shoot. *Ann. Bot.,* 23: 409–423.

Haupt, W. 1957. Die Induktion der Polarität bei der Spore von *Equisetum. Planta,* 49: 61–90.

———. 1958. Die Blütenbildung bei *Pisum sativum. Z. Bot.,* 46: 242–256.

Hawker, J. S. 1965. The sugar content of cell walls and intercellular spaces in sugar-cane stems and its relation to sugar transport. *Aust. J. Biol. Sci.,* 18: 959–969.

——— and M. D. Hatch. 1965. Mechanism of sugar storage by mature stem tissue of sugar cane. *Physiol. Plant.,* 18: 444–453.

Hawker, L. E. 1932. Experiments on the perception of gravity by roots. *New Phytol.,* 31: 321–328.

Hay, J. R. 1956. Translocation of herbicides in Marabu, II. *Weeds,* 4: 349–356.

——— and K. V. Thimann. 1956. The fate of 2,4-D in bean seedlings, I. *Plant Physiol.,* 31: 382–386.

Hayashi, T. 1940. Biochemical studies on "bakanae" fungus of rice. 6: Effect of gibberellin on the activity of amylase in germinated cereal grains. *J. Agric. Chem. Soc. Jap.,* 16: 531–538.

Hayashi, F., R. Naito, M. J. Bukovac, and H. M. Sell. 1968. Occurrence of gibberellin GA_3 in parthenocarpic apple fruit. *Plant Physiol.,* 43: 448–450.

Heath, O. V. S. 1943. Studies in the physiology of the onion plant. *Ann. Appl. Biol.,* 30: 208–220.

——— and F. G. Gregory. 1938. The constancy of mean net assimilation rate and its ecological importance. *Ann. Bot.,* n.s., 2: 811–818.

——— and M. Holdsworth. 1943. Bulb formation and flower production in onion. *Nature,* 152: 334–335.

——— and ———. 1948. Morphogenic factors as exemplified by the onion plant. *Symp. Soc. Exp. Biol.,* 2: 326–350.

——— and H. Meidner. 1961. The influence of water strain on the minimum intercellular space CO_2 concentration and stomatal movement in wheat leaves. *J. Exp. Bot.,* 12: 226–242.

——— and T. Russell. 1954. Studies of stomatal behaviour. VI: An investigation of the light responses of wheat stomata with the attempted elimination of control by the mesophyll, I. *J. Exp. Bot.* 5: 1–15.

Hecht, A. 1964. Partial inactivation of an incompatibility substance in the stigmas and styles of *Oenothera.* Pp. 237–243 in H. F. Linskens (ed.), "Pollen Physiology and Fertilization." North-Holland, Amsterdam.

Hecht, S. M., N. J. Leonard, W. J. Burrows, D. J. Armstrong, F. Skoog, and J. Occolowitz. 1969. Cytokinin of wheat germ transfer RNA: 6(4-hydroxy-3-methyl-2-butenylamine)-2-methylthio-9-*D*-ribofuranosyl purine. *Science,* 166: 1272–1274.

Heide, O. M. 1968. Auxin level and regeneration of *Begonia* leaves. *Planta,* 81: 153–159.

———. 1972. The role of cytokinin in regeneration processes. Pp. 207–219 in H. Kaldewey and Y. Vardar (eds.), "Hormonal Regulation in Plant Growth and Development." Verlag Chemie, Weinheim.

——— and F. Skoog. 1967. Cytokinin activity in *Begonia* and *Bryophyllum. Physiol. Plant.,* 20: 771–780.

Heine, R. W. 1971. Hydraulic conductivity in trees. *J. Exp. Bot.,* 22: 503–511.

Heinze, W. 1959. Über den Einfluss des Blattes auf den Bluhtermin bei der Tomate. *Naturwiss.,* 21: 609.

Hemberg, T. 1949*a.* Growth inhibiting substances in buds of *Fraxinus. Physiol. Plant.,* 2: 37–44.

———. 1949*b.* Significance of growth inhibiting substances and auxins for the rest period of potato. *Physiol. Plant.,* 2: 24–36.

———. 1955. Studies on the balance between free and bound auxin in germinating maize. *Physiol. Plant.,* 8: 418–432.

Henckel, P. A. 1970. Role of protein synthesis in drought resistance. *Can. J. Bot.,* 48: 1235–1241.

Henderson, T. H., C. G. Skinner, and R. E. Eakin. 1962. Kinetin and kinetin analogues as substrates and inhibitors of xanthine oxidase. *Plant Physiol.,* 37: 552–555.

Hendricks, S. B. 1968. How light interacts with living matter. *Sci. Am.,* 219: 174–186.

——— and H. A. Borthwick. 1954. Photoperiodism in plants. *Proc. 1st Int. Photobiol. Congr., Amst.,* pp. 23–35.

——— and ———. 1965. The physiological functions of phytochrome. pp. 405–436 in T. W. Goodwin (ed.), "Chemistry and Biochemistry of Plant Pigments." Academic, London.

——— and ———. 1967. The function of phytochrome in the regulation of plant growth. *Proc. Natl. Acad. Sci. (U.S.),* 58: 2125–2130.

———, ———, and R. J. Downs. 1956. Pigment conversion in the formative responses of plants. *Proc. Natl. Acad. Sci. (U.S.),* 42: 19–26.

Hernandez-Gil, R., and M. Schaedle. 1973. Functional and structural changes in senescing *Populus deltoides* (Partr.) chloroplasts. *Plant Physiol.,* 51: 245–249.

Herrero, F. A., and W. C. Hall. 1960. General effects of ethylene on enzyme systems in the cotton leaf. *Physiol. Plant.,* 13: 736–750.

Hertel, R., M. L. Evans, A. C. Leopold, and H. M. Sell. 1969*a.* The specificity of the auxin transport system. *Planta,* 85: 238–249.

———, R. K. dela Fuente, and A. C. Leopold. 1969*b.* Geotropism and the lateral transport of auxin in the mutant amylomaize. *Planta,* 88: 204–214.

——— and A. C. Leopold. 1962. Auxintransport and Schwerkraft. *Naturwiss.,* 49: 377–378.

——— and ———. 1963. Versuche zur Analyse des Auxintransports in der Koleoptile von *Zea mays. Planta,* 59: 535–562.

———, K. S. Thomson, and V. E. A. Russo. 1972. In vitro auxin binding to particulate cell fractions from corn coleoptiles. *Planta,* 107: 325–340.

Hesketh, J. D. 1963. Limitations to photosynthesis responsible for differences among species. *Crop Sci.,* 3: 493–496.

——— and D. N. Moss. 1963. Variation in the response of photosynthesis to light. *Crop Sci.,* 3: 107–110.

Heslop-Harrison, J. 1956. Auxin and sexuality in *Cannabis sativa. Physiol. Plant.,* 9: 588–597.

———. 1957*a.* The physiology of reproduction in *Dactylorchis,* I. *Bot. Not.,* 110: 28–49.

———. 1957*b.* The experimental modifications of sex expression in flowering plants. *Biol. Rev.,* 32: 38–90.

———. 1960. Suppressive effects of 2-thiouracil on differentiation and flowering in *Cannabis sativa. Science,* 132: 1943–1944.

———. 1963. Sex expression in flowering plants. *Brookhaven Symp. Biol.,* 16: 109–122.

——— and Y. Heslop-Harrison. 1958. Long-day and auxin induced male sterility in *Silene pendula. Port. Acta. Biol.,* 5: 79–94.

——— and ———. 1970. The state of the apex and the response to induction in *Cannabis sativa.* Pp. 3–26 in G. Bernier (ed.), "Cellular and Molecular Aspects of Floral Induction." Longman Group Ltd., London.

Hess, C. E. 1962. A physiological analysis of root initiation in easy and difficult-to-root cuttings. *Proc. XVI Int. Hort. Cong.,* pp. 375–387.

———. 1964. Naturally occurring substances which stimulate root initiation. Pp. 517–527 in J. P. Nitsch (ed.), "Régulateurs Naturels de la Croissance Végétale." CNRS, Paris.

Hess, D. 1961. Ribosenucleinsäure und Blühinduktion. *Planta,* 56: 229–232.

Hewitt, S. P., and O. F. Curtis. 1948. The effect of temperature on loss of dry matter and carbohydrate from leaves by respiration and translocation. *Am. J. Bot.,* 35: 746–755.

Heyn, A. N. J. 1931. Der Mechanismus der Zellstreckung. *Recl. Trav. Bot. Neerl.,* 28: 113–244.

Hiesey, W. M., and H. W. Milner. 1965. Physiology of ecological races and species. *Annu. Rev. Plant Physiol.,* 16: 203–216.

Highkin, H. R. 1958. Temperature-induced variability in peas. *Am. J. Bot.,* 45: 626–631.

——— and A. Lang. 1966. Residual effect of germination temperature on the growth of peas. *Planta,* 68: 94–98.

Hill, R. 1937. Oxygen evolved by isolated chloroplasts. *Nature,* 139: 881–882.

Hillman, S. K., and I. D. J. Phillips. 1970. Transport and metabolism of indoleacetic acid-C^{14} in pea roots. *J. Exp. Bot.,* 21: 959–967.

Hillman, W. S. 1956. Injury of tomato plants by continuous light and unfavorable photoperiodic cycles. *Am. J. Bot.,* 43: 89–96.

———. 1964. Phytochrome levels detectable by in vivo spectrophotometry. *Am. J. Bot.,* 51: 1102–1107.

———. 1966. Photoperiodism in *Lemna*: Reversal of night-interruption depends on color of the main photoperiod. *Science,* 154: 1360–1362.

———. 1967. The physiology of phytochrome. *Annu. Rev. Plant Physiol.,* 18: 301–324.

———. 1969. Photoperiodism and vernalization. Pp. 558–601 in M. B. Wilkins (ed.), "Physiology and Plant Growth and Development." McGraw-Hill, London.

——— and W. K. Purves. 1961. Does gibberellin act through an auxin-mediated mechanism? Pp. 589–600 in R. M. Klein (ed.), "Plant Growth Regulation." Iowa State University Press, Ames.

Hilton, J. L., L. L. Jansen, and H. M. Hull. 1963. Mechanisms of herbicide action. *Annu. Rev. Plant. Physiol.,* 14: 353–377.

Hinsvark, O. N., W. H. Houff, S. H. Wittwer, and H. M. Sell. 1954. The extraction and colorimetric estimation of indoleacetic acid and its esters in developing corn kernels. *Plant Physiol.,* 29: 107–108.

Hiroi, T., and M. Monsi. 1963. Physiological and ecological analyses of shade tolerance of plants. 3: Effect of shading on growth attributes of *Helianthus annuus. Bot. Mag. Tokyo,* 76: 121–129.

——— and ———. 1966. Dry-matter economy of *Helianthus annuus* communities grown at varying densities and light intensities. *J. Fac. Sci. Univ. Tokyo,* (III)9: 241–285.

Hitchcock, A. E., and P. W. Zimmerman. 1935. Absorption and movement of synthetic growth substances from soil as indicated by responses of aerial parts. *Contrib. Boyce Thompson Inst.,* 7: 447–476.

Hitchcock, A. S., and A. Chase. 1931. Grass, *Smithson. Sci. Ser.,* 11: 201–250.

Hoad, G. V. 1967. (+)-Abscisin II in phloem exudate of willow. *Life Sci.,* 6: 1113.

——— and M. R. Bowen. 1968. Evidence for gibberellin-like substances in phloem exudate of higher plants. *Planta,* 82: 22–32.

Hobson, G. E. 1964. Polygalacturonase in normal and abnormal tomato fruit. *Biochem. J.,* 92: 324–332.

———. 1965. The ripening of tomato fruit as affected by the injection of certain chemicals. *J. Expt. Bot.,* 16: 411–422.

Hodges, T. K., R. T. Leonard, C. E. Bracker, and T. W. Kramer. 1972. Purification of an ion-stimulated adenosine triphosphatase from plant roots: Association with plasma membranes. *Proc. Natl. Acad. Sci. (U.S.),* 69: 3307–3311.

Hodson, H. K., and K. C. Hamner. 1970. Floral inducing extract from *Xanthium. Science,* 167: 384–385.

Hofstra, G., and C. D. Nelson. 1969. A comparative study of translocation of assimilates ^{14}C from leaves of different species. *Planta,* 88: 103–112.

Holdsworth, M. 1959. The spectral sensitivity of light-induced leaf movements. *J. Exp. Bot.,* 11: 40–44.

Holly, K. 1964. Herbicide selectivity in relation to formulation and application methods. P. 423 in "The Physiology and Biochemistry of Herbicides." Academic, New York.

Holm, R. E., and F. B. Abeles. 1967. Abscission: The role of RNA synthesis. *Plant Physiol.,* 42: 1094–1102.

——— and ———. 1969. The role of ethylene in 2,4-D induced growth inhibition. *Planta,* 78: 293–304.

———, T. J. O'Brien, J. L. Key, and J. H. Cherry. 1970. The influence of auxin and ethylene on chromatin-directed ribonucleic acid synthesis in soybean hypocotyl. *Plant Physiol.,* 45: 41–45.

Holmgren, P. 1968. Leaf factors affecting light-saturated photosynthesis in ecotypes of *Solidago virgaurea* from exposed and shaded habitats. *Physiol. Plant.,* 21: 676–698.

———, P. G. Jarvis, and M. S. Jarvis. 1965. Resistances to carbon dioxide and water vapour transfer in leaves of different plant species. *Physiol. Plant.,* 18: 557–573.

Homann, P. H., and G. H. Schmid. 1967. Photosynthetic reactions of chloroplasts with unusual structures. *Plant Physiol.,* 42: 1619–1632.

Hope, A. B., and R. N. Robertson. 1956. Initial absorption of ions by plant tissue. *Nature,* 177: 43–44.

——— and P. G. Stevens. 1952. Electrical potential differences in bean roots and their relation to salt uptake. *Aust. J. Sci. Res.,* 5: 335.

Horowitz, S. B., and I. R. Fenichel. 1965. Diffusion and the transport of organic nonelectrolytes in cells. *Ann. N.Y. Acad. Sci.,* 125: 572–594.

Horton, R. F., and D. J. Osborne. 1967. Senescence, abscission and cellulase activity in *Phaseolus vulgaris. Nature,* 214: 1086–1088.

Horvath, I., and I. V. Feher. 1965. Influence of the spectral composition on carbohydrate metabolism. I: Quantity and proportion of the carbohydrates. *Acta. Bot. Acad. Hung.,* 11: 159–164.

Horwitz, L. 1958. Some simplified mathematical treatments of translocation in plants. *Plant Physiol.,* 33: 81–93.

Howard, F. D., and M. Yamaguchi. 1957. Respiration and oxidative activity of particulate fractions from pepper fruits. *Plant Physiol.,* 32: 418–428.

Howard, J. A. 1966. Spectral energy relations of isobilateral leaves. *Aust. J. Biol. Sci.,* 19: 757–766.

Howlett, F. S. 1936. The effect of carbohydrate and of nitrogen deficiency upon microsporogenesis and the development of the male gametophyte in the tomato, *Lycopersicum esculentum. Ann. Bot.,* 50: 767–804.

Hsiang, T. T. 1951. Physiological and biochemical changes accompanying pollination in orchid flowers, I. *Plant Physiol.,* 26: 441–455.

Hsiao, T. C., E. Acevedo, and D. W. Henderson. 1970. Maize leaf elongation: Continuous measurements and close dependence on plant water status. *Science,* 168: 590–591.

Huang, A. H. C., and H. Beevers. 1972. Microbody enzymes and carboxylases in sequential extracts from C_4 and C_3 leaves. *Plant Physiol.,* 50: 242–248.

Huber, B. 1956. Die Gefässleitung. *Handb. Pflanzenphysiol.,* 3: 541.

Huber, B., and E. Schmidt. 1937. Eine Kompensationsmethode zur Thermoelektrischen Messung langsames Saftströmen. *Ber. Dtsch. Bot. Ges.,* 55: 514–529.

———, ———, and H. Jahnel. 1937. Untersuchungen über den Assimilatström, I. *Tharandt. Forstl. Jahrb.,* 88: 1017–1050.

Huck, M. G., B. Klepper, and H. M. Taylor. 1970. Diurnal variation in root diameter. *Plant Physiol.,* 45: 529–530.

Hudock, G. A., G. C. McLeod, J. Moravkooa-Kiely, and L. P. Levine. 1964. The relation of oxygen evolution to chlorophyll and protein. *Plant Physiol.,* 39: 898–903.

Hughes, A. P. 1965. The importance of light compared with other factors affecting plant growth. Pp. 121–146 in "Light as an Ecological Factor," *Br. Ecol. Soc., Symp.* 6.

——— and K. E. Cockshull. 1971. The variation in response to light intensity and carbon dioxide concentration shown by two cultivars of *Chrysanthemum morifolium* grown in controlled environments at two times of year. *Ann. Bot.,* 35: 933–945.

Hull, H. H. 1970. Leaf structure as related to absorption of pesticides. *Residue Rev.,* 31: 1–150.

Hulme, A. C. 1939. The nitrogen metabolism of the apple fruit in relation to its growth and respiration. *Rep. Food Invest. Board 1939,* p. 55.

———. 1954. The climacteric rise in respiration in relation to changes in the equilibrium between protein synthesis and breakdown. *J. Exp. Bot.,* 5: 159–173.

——— and G. E. Neal. 1957. A new factor in the respiration climacteric of apple fruits. *Nature,* 179: 1192–1193.

———, M. J. C. Rhodes, and L. S. Wooltorton. 1971. The relationship between ethylene and the synthesis of RNA and protein in ripening apples. *Phytochemistry,* 10: 749–756.

Humble, G. D., and T. C. Hsiao. 1970. Light dependent influx and efflux of potassium of guard cells during stomatal opening and closing. *Plant Physiol.,* 46:483–487.

——— and K. Raschke. 1971. Stomatal opening quantitatively related to potassium transport (evidence from electron probe analysis). *Plant Physiol.,* 48: 447–453.

Humphries, E. C. 1960. Inhibition of root development on petioles and hypocotyls of dwarf bean. *Physiol. Plant.,* 13: 659–663.

——— and A. W. Wheeler. 1964. Cell division and growth substances in leaves. P. 505 in J. P. Nitsch (ed.), "Régulateurs Naturels de la Croissance Végétale." CNRS, Paris.

Huxley, J. S. 1932. "Problems of Relative Growth." Dial, New York. 276 pp.

———. 1949. Soviet genetics and world science. "Lysenko and the Meaning of Heredity." Chatto & Windus, London. 245 pp.

Huxley, P. A. 1969. The effect of fluctuating light intensity on plant growth. *J. Appl. Ecol.,* 6: 273–276.

Hyde, E. O. C. 1954. The function of the hilum in some Papilionaceae in relation to the ripening of the seed and permeability of the testa. *Ann. Bot.,* 18: 241–256.

Hylmo, B. 1953. Transpiration and ion absorption. *Physiol. Plant.,* 6: 333–405.

Ikuma, H., and K. V. Thimann. 1959. Photosensitive site in lettuce seeds. *Science,* 130: 568–569.

——— and ———. 1963. Action of kinetin on photosensitive germination of lettuce seed as compared with that of gibberellic acid. *Plant Cell Physiol.,* 4: 113–128.

Iljin, W. S. 1957. Drought resistance in plants and physiological processes. *Annu. Rev. Plant Physiol.,* 8: 257–274.

Imamura, S. 1953. Photoperiodic initiation of flower primordia in Japanese morning glory, *Pharbitis nil. Proc. Jap. Acad.,* 29: 368–373.

——— and A. Takimoto. 1955. Transmission rate of photoperiodic stimulus in *Pharbitis nil. Bot. Mag.,* 68: 260–266.

Imber, D., and M. Tal. 1970. Phenotypic reversion of *flacca,* a wilty mutant of tomato, by abscisic acid. *Science,* 169: 592–593.

Ingen-Housz, J. 1779. "Experiments upon Vegetables, Discovering Their Great Power of Purifying

Common Air in the Sunshine and of Injuring in the Shade and at Night." London.
Ingersoll, R. B., and O. E. Smith. 1971. Transport of abscisic acid. *Plant Cell Physiol.,* 12: 301–309.
Ino, Y. 1970. The effect of fluctuating light on photosynthesis. Pp. 68–70 in "Photosynthesis and Utilisation of Solar Energy." *Level III. Rep. 1969, Jap. Nat. Subcomm. PP IBP.*
Irvine, V. C. 1938. Studies in growth promoting substances related to x-radiation and photoperiodism. *Univ. Colo. Stud.,* 26: 69–70.
Irving, R. M. 1969. Characterization and role of an endogenous inhibitor in the induction of cold hardiness in *Acer negundo. Plant Physiol.,* 44: 801–805.
Isely, D. (ed.). 1965. *Rules for Testing Seeds.* Association Official Seed Analysis, Lake Mills, Iowa. 112 pp.
Isikawa, S. 1954. Light-sensitivity against the germination. I: "Photoperiodism" of seeds. *Bot. Mag.,* 67: 789–790.
——— and R. Fujii. 1961. Photocontrol and temperature dependence of germination of *Rumex* seeds. *Plant Cell Physiol.,* 2: 51–62.
Itai, C., and Y. Vaadia. 1971. Cytokinin activity in water-stressed shoots. *Plant Physiol.,* 47: 87–90.
Iversen, T. H. 1969. Elimination of geotropic responses in roots of cress by removal of statolith starch. *Physiol. Plant.,* 22: 1251–1262.
Ivins, J. D., and F. L. Milthorpe. 1963. "The Growth of the Potato." Butterworth, London.
Iwaki, H. 1959. Ecological studies on interspecific competition in a plant community. I: An analysis of growth of competing plants in mixed stands of buckwheat and green-grains. *Jap. J. Bot.,* 17: 120–138.
Iwanami, Y. 1959. Physiological studies of pollen. *J. Yokohama Munic. Univ.,* 116: 1–137.
———. 1960. On the movement of the stamen of *Portulaca grandiflora. J. Yokohama Munic. Univ.,* 121: 1–25.
——— and I. Hoshino. 1963. The opening and closing movement of the flower of *Portulaca. Bot. Mag. (Tokyo),* 76: 108–114.
———. 1957. Physiological researches of pollen. XIII: Growth inhibition of the pollen tube of *Camellia japonica. Bot. Mag. (Tokyo),* 70: 144–149.
Izhar, S., and D. H. Wallace. 1967. Studies of the physiological basis for yield differences. III: Genetic variation in photosynthetic efficiency of *Phaseolus vulgaris* L. *Crop Sci.,* 7: 457–460.
Jablonski, J. R., and F. Skoog. 1954. Cell enlargement and cell division in excised tobacco pith tissue. *Physiol. Plant.,* 7: 16–24.
Jachymczyk, W. J., and J. H. Cherry. 1968. Studies on messenger RNA from peanut plants: In vitro polyribosome formation and protein synthesis. *Biochem. Biophys. Acta,* 157: 368–377.
Jackson, D. I. 1968. Gibberellin and the growth of peach and apricot fruits. *Aust. J. Biol. Sci.,* 21: 209–215.
——— and B. G. Coombe. 1966. Gibberellin-like substances in the developing apricot fruit. *Science,* 154: 277–278.
Jackson, G. A., and D. M. V. Prosser. 1959. The induction of parthenocarpic development in *Rosa* by auxins and gibberellic acid. *Naturwiss.,* 12: 407–408.
Jackson, L. W. R. 1967. Effect of shade on leaf structure of deciduous tree species. *Ecology,* 48: 498–499.
Jackson, W. A., and R. J. Volk. 1970. Photorespiration. *Annu. Rev. Plant Physiol.,* 21: 385–432.
Jacobs, W. P. 1950*a*. Control of elongation in bean hypocotyl by the ability of the hypocotyl tip to transport auxin. *Am. J. Bot.,* 37: 551–555.
———. 1952. Role of auxin in differentiation of xylem around a wound. *Am. J. Bot.,* 39: 301–309.
———. 1954. Acropetal auxin transport and xylem regeneration: A quantitative study. *Am. Nat.,* 88: 327–337.
———. 1961. The polar movement of auxin in the shoots of higher plants. Pp. 397–409 in R. M. Klein (ed.), "Plant Growth Regulation." Iowa State University Press, Ames.
——— and B. Bullwinkel. 1953. Compensatory growth in *Coleus* shoots. *Am. J. Bot.,* 40: 385–392.
——— and H. Kaldewey. 1970. Polar movement of gibberellic acid through young *Coleus* petioles. *Plant Physiol.,* 45: 539–541.
———, J. Danielson, V. Hurst, and P. Adams. 1959. What substance normally controls a given biological process? II: The relation of auxin to apical dominance. Pp. 534–554 in "Developmental Biology." Academic, New York.
——— and I. B. Morrow. 1957. A quantitative study of xylem development in the vegetative shoot apex of *Coleus. Am. J. Bot.,* 44: 823–842.
——— and ———. 1958. Quantitative relations between stages of leaf development and differentiation of sieve tubes. *Science,* 128: 1084–1085.
———. 1950*b*. Auxin transport in the hypocotyl of *Phaseolus vulgaris. Am. J. Bot.,* 37: 248–254.
Jacoby, B., and A. H. Halevy. 1970. Participation of light and temperature fluctuations in the induction of contractile roots of *Gladiolus. Bot. Gaz.,* 131: 74–77.
Jacqmard, A. 1968. Early effects of gibberellic acid on mitotic activity and DNA synthesis in the apical bud of *Rudbeckia bicolor. Physiol. Veg.,* 6: 409–416.
Jaffe, L. F. 1968. Localization in the developing *Fucus* egg and the general role of localizing current. *Adv. Morphogen.,* 7: 295–328.
Jaffe, M. J. 1970. Physiological studies of pea tendrils. VII: Evaluation of a technique for asymmetrical application of ethylene. *Plant Physiol.,* 45: 631–633.
——— and A. W. Galston. 1966. Physiological studies on pea tendrils, I. *Plant Physiol.,* 41: 1014–1025.
——— and ———. 1967*a*. Phytochrome control of

rapid nyctinastic movements and membrane permeability in *Albizzia*. *Planta,* 77: 135–141.

——— and ———. 1967*b*. Physiological studies on pea tendrils, III and IV. *Plant Physiol.,* 42: 845–850.

——— and ———. 1968*a*. Physiological studies on pea tendrils. V: Membrane changes and water movement associated with contact coiling. *Plant Physiol.,* 43: 537–542.

——— and ———. 1968*b*. The physiology of tendrils. *Ann. Rev. Plant Physiol.,* 19: 417–434.

Janick, J., and E. C. Stevenson. 1955. Environmental influences on sex expression in monecious lines of spinach. *Proc. Am. Soc. Hort. Sci.,* 65: 416–422.

Jankiewicz, L. S. 1956. The effect of auxins on crotch angles in apple trees. *Acad. Pol. Sci.,* 4: 173–178.

———. 1972. A cybernetic model of growth correlation in young apple trees. *Biol. Plant.,* 14: 52–61.

Jansen, E. F. 1965. Ethylene and polyacetylenes. Pp. 641–664 in J. Bonner and J. E. Varner (eds.), "Plant Biochemistry." Academic, New York.

Jarvis, B. C., B. Frankland, and J. H. Cherry. 1968*a*. Increased DNA template and RNA polymerase associated with the breaking of seed dormancy. *Plant Physiol.,* 43: 1734–1736.

———, ———, and ———. 1968*b*. Increased nucleic acid synthesis in relation to the breaking of dormancy of hazel seed by gibberellic acid. *Planta,* 83: 257–266.

Jarvis, P., and R. Thaine. 1971. Strands in sections of sieve elements cut in a cryostat. *Nature New Biol.,* 232: 236–237.

Jarvis, P. G., and M. S. Jarvis. 1963. The water relations of tree seedlings. I: Growth and water use in relation to soil water potential. *Physiol. Plant.,* 16: 215–235.

——— and ———. 1964. Growth rates of woody plants. *Physiol. Plant.,* 17: 654–666.

Jefford, T. G., and J. Edelman. 1960. Changes in content and composition of the fructose polymers in tubers of *Helianthus tuberosus*. *J. Exp. Bot.,* 12: 177–187.

Jenny, H. 1961. Two-phase studies on availability of iron in calcareous soils, V. *Agrochimica,* 5: 281–289.

Jensen, R. D., S. A. Taylor, and H. H. Wiebe. 1961. Negative transport and resistance to water flow through plants. *Plant Physiol.,* 36: 633–638.

Johnson, K. D., and H. Kende. 1971. Hormonal control of lecithin synthesis in barley aleurone cells: Regulation of the CDP-choline pathway by gibberellin. *Proc. Natl. Acad. Sci., (U.S.),* 68: 2674–2677.

Johnson, M., and J. Bonner. 1956. The uptake of auxin by plant tissues. *Physiol. Plant.,* 9: 102–118.

Johnsson, A. 1965. Investigations of the reciprocity rule by means of geotropic and geoelectric measurements. *Physiol. Plant.,* 18: 945–967.

Johri, M. M., and J. E. Varner. 1968. Enhancement of RNA synthesis in isolated pea nuclei by gibberellic acid. *Proc. Natl. Acad. Sci. (U.S.),* 59: 269–276.

Jones, C. M. 1965. Effects of benzyladenine on fruit set in muskmelon. *Proc. Am. Soc. Hort. Sci.,* 87: 335–340.

Jones, E. R. H., H. R. Henbest, G. F. Smith, and J. A. Bentley. 1952. 3-Indolylacetonitrile: A natural occurring plant growth hormone. *Nature,* 169: 485.

Jones, H. C., H. S. Black, and A. M. Altschul. 1967. Comparison of the effects of gibberellic acid and aflatoxin in germinating seeds. *Nature,* 214: 171–172.

Jones, M. B., and J. V. Enzie. 1961. Identification of a cyanogenetic growth-inhibiting substance in extracts from peach flower buds. *Science,* 134: 284.

Jones, R. J., and T. A. Mansfield. 1970. Suppression of stomatal opening in leaves treated with abscisic acid. *J. Exp. Bot.,* 21: 714–719.

Jones, R. L. 1969. Ethylene enhanced release of α-amylase from barley endosperm. *Plant Physiol.,* 43: 442–444.

———. 1969. The fine structure of barley aleurone cells. *Planta,* 85: 359–375.

——— and A. Lang. 1968. Extractable and diffusible gibberellins from light and dark grown pea seedlings. *Plant Physiol.,* 43: 629–634.

——— and I. D. J. Phillips. 1966. Organs of gibberellin synthesis in light-grown sunflower plants. *Plant Physiol.,* 41: 1381–1386.

——— and J. E. Varner. 1967. The bioassay of gibberellin. *Planta,* 72: 155–161.

Jones, W. W., and J. H. Beaumont. 1937. Carbohydrate accumulation in relation to vegetative propagation of the *Litchi*. *Science,* 86: 313.

Jordan, L. S., W. J. Farmer, J. R. Goodin, and B. E. Day. 1970. Nonbiological detoxication of the *s*-triazine herbicides. *Residue Rev.,* 32: 267–286.

Jost, L. 1891*b*. Über Dichenwachsthum und Jahresringbildung. *Bot. Ztg.,* 49: 485–630.

———. 1894. Über den Einfluss des Lichtes auf das Knospentreiben der Rothbucke. *Ber. Dtsch. Bot. Ges.,* 12: 188–197.

———, S. Groves, B. L. Schachar, and L. J. Audus. 1966. Root cap and the perception of gravity. *Nature,* 209: 93–94.

Juniper, B. E. 1959. The surfaces of plants. *Endeavour,* 18: 20–25.

Jyung, W. H., and S. H. Wittwer. 1964. Foliar absorption: An active uptake process. *Am. J. Bot.,* 51: 437–444.

Kahn, A. J., A. Goss, and D. E. Smith. 1957. Effect of gibberellin on germination of lettuce seed. *Science,* 125: 645–646.

Kaldeway, H. 1957. Wachstumsverlauf, Wuchsstoffbildung und Nutations-bewegungen von *Fritillaria Meleagris* L. im Laufe der Vegetations-periode. *Planta,* 49: 300–344.

Kamerbeek, G. A., and A. L. Verlind. 1972. Characteristic stimulation by ethylene of respiration in Dutch iris bulbs. *Physiol. Plant.,* 27: 5–8.

Kamisaka, S., and Y. Masuda. 1970. Stimulation of auxin-induced cell elongation by cyclic AMP. *Naturwiss.,* 57: 546.

———, H. Sano, M. Katsumi, and Y. Masuda. 1972. Effects of cyclic AMP and gibberellic acid on lettuce hypocotyl elongation. *Plant Cell Physiol.,* 13: 167–173.

Kang, B. G., and S. P. Burg. 1973. Role of ethylene in photochrome-induced anthocyanin synthesis. *Planta,* 110: 227–235.

———, W. Newcomb, and S. P. Burg. 1971. Mechanism of auxin-induced ethylene production. *Plant Physiol.,* 47: 504–509.

———, C. S. Yokum, S. P. Burg, and P. M. Ray. 1967. Ethylene and carbon-dioxide: Mediation of hypocotyl hook opening response. *Science,* 156: 958–959.

Kanta, K., and P. Maheshwari. 1964. Intraovarian pollination in some Papaveraceae. *Phytomorphology,* 13: 215–229.

Karpilov, Y. S. 1970. "The Photosynthesis of Xerophytes." Academy of Sciences, Moldavian SSR; Kishinov.

Kasperbauer, M. J. 1971. Spectral distribution of light in a tobacco canopy and effect of end-of-day light quality on growth and development. *Plant Physiol.,* 47: 775–778.

Kato, J. 1953. Studies on the physiological effect of gibberellin. I: On the differential activity between gibberellin and auxin. *Mem. Coll. Sci. Univ. Kyoto,* B20: 190–193.

———. 1958. Nonpolar transport of gibberellin through pea stem and a method for its determination. *Science,* 128: 1008–1009.

——— and M. Katsumi. 1959. Studies on the physiological effect of gibberellin. V: Effect of gibberellic acid and gibberellin A on the activity of indoleacetic acid oxidase. *Mem. Coll. Sci. Univ. Kyoto,* 26: 53–60.

——— and ———. 1967. Pseudo gibberellin A as an inhibitor of the GA_3 induced growth of rice seedlings. *Planta,* 74: 194–196.

Katrushenko, I. V. 1965. On the photosynthetic adaptation to the light of the perennial needles of spruce *(Picea abies)* saplings. (In Russian.) *Bot. Zh. (Mos.),* 50(8): 1119–1120.

Katsumi, M. 1962. Physiological effects of kinetin on the thickening of etiolated pea stem sections. *Physiol. Plant.,* 15: 115–121.

———, B. O. Phinney, P. R. Jeffries, and C. A. Henrick. 1964. Growth response of the *d*-5 and *an*-1 mutants of maize to some kaurene derivatives. *Science,* 144: 849–850.

———, ———, and W. K. Purves. 1965. The roles of gibberellin and auxin in cucumber hypocotyl growth. *Physiol. Plant.,* 18: 462–473.

Katunsky, V. M. 1936. On the causes of pre- and post-floral movements of peduncles and scapes. *C. R. Acad. Sci.,* 12: 343–346.

Kaufman, P. B., L. B. Petering, C. S. Yocum, and D. Baic. 1970. Ultrastructural studies on stomata development in internodes of *Avena sativa. Am. J. Bot.,* 57: 33–49.

Kaufmann, M. R. 1968. Water relations of pine seedlings in relation to root and shoot growth. *Plant Physiol.,* 43: 281–288.

———. 1969. Relation of slice gap width in oranges and plant water stress. *J. Am. Soc. Hort. Sci.,* 94: 161–163.

Kaul, R. 1966. Effect of water stress on respiration of wheat. *Can. J. Bot.,* 44: 623–632.

Kawase, M. 1961. Growth substances related to dormancy in *Betula. Proc. Am. Soc. Hort. Sci.,* 78: 532–544.

———. 1971. Causes of centrifugal root promotion. *Physiol. Plant.,* 25: 64–70.

Kearney, P. C., T. J. Sheets, and J. W. Smith. 1964. Volatility of seven *s*-triazines. *Weeds,* 12: 83–86.

Kefeli, V. I., and C. S. Kadyrov. 1971. Natural growth inhibitors, their chemical and physiological properties. *Ann. Rev. Plant Physiol.,* 22: 185–196.

Kefford, N. P. 1962. Auxin-gibberellin interactions in rice coleoptile elongation. *Plant Physiol.,* 37: 380–386.

———, J. Brockwell, and J. A. Zwar. 1960. The symbiotic synthesis of auxin by legumes and nodule bacteria and its role in nodule development. *Aust. J. Biol. Sci.,* 13: 456–467.

——— and O. H. Caso. 1966. A potent auxin with unique chemical structure: 4-amino-3,5,6-trichloropicolinic acid. *Bot. Gaz.,* 127: 139–163.

——— and P. L. Goldacre. 1961. The changing concept of auxin. *Am. J. Bot.,* 48: 643–650.

———, J. A. Zwar, and M. I. Bruce. 1965. Enhancement of lettuce seed germination by some urea derivatives. *Planta,* 67: 103–106.

———, ———, and ———. 1968. Antagonism of purine and urea cytokinin activities by derivatives of benzylurea. Pp. 61–69 in F. Wightman and G. Setterfield (eds.), "Biochemistry and Physiology of Plant Growth Substances." Runge Press, Ottawa.

Kende, H. 1965. Kinetin-like factors in the root exudate of sunflowers. *Proc. Natl. Acad. Sci. (U.S.),* 53: 1302–1307.

———. 1967. Preparation of radioactive gibberellin A and its metabolism in dwarf peas. *Plant Physiol.,* 42: 1612–1618.

———. 1971. The cytokinins. *Int. Rev. Cytol.,* 31: 301–338.

———, H. Ninnemann, and A. Lang. 1963. Inhibition of gibberellic acid biosynthesis in Fusarium by Ammo 1618 and CCC. *Naturwiss.,* 50: 599–600.

——— and J. E. Tavares. 1968. On the significance of cytokinin incorporation into RNA. *Plant Physiol.,* 43: 1244–1248.

Kendrick, R. E., and C. J. P. Spruit. 1973. Phytochrome properties and the molecular environment. *Plant Physiol.* 52: 327–331.

Kent, M., and W. A. Gortner. 1951. Effect of pre-illumination on response of split pea stems to growth substances. *Bot. Gaz.,* 112: 307–311.

Kentzer, T. 1966*a*. Gibberellin-like substances and growth inhibitors in relation to the dormancy and after-ripening of ash seeds. *Acta Soc. Bot. Pol.,* 35: 575–585.

———. 1966*b*. The dynamics of gibberellin-like and growth inhibiting substances during seed development of *Fraxinus excelsior*. *Acta Soc. Bot. Pol.,* 35: 447–484.

Kerk, G. J. M. van der, M. H. Raalte, A. S. Kaars, and R. van der Veen. 1955. A new type of plant growth-regulating substances. *Nature,* 176: 308.

Kerner, A. 1895. "The Natural History of Plants." Trans. F. W. Oliver. Holt, New York; cited by Meeuse (1961).

Kessler, B. 1961. Nucleic acids as factors in drought resistance in higher plants. *Recent Adv. Bot.,* 2: 1153–1159.

——— and J. Frank-Tishel. 1962. Dehydration induced synthesis of nucleic acids and changing RNA composition: A possible protective reaction in drought resistance. *Nature,* 196: 542–543.

Ketring, D. L., and P. W. Morgan. 1969. Ethylene as a component of the emanations from germinating peanut seeds and its effect on dormant Virginia-type seeds. *Plant Physiol.,* 44: 326–330.

Key, J. L. 1964. Ribonucleic acid and protein synthesis as essential processes for cell elongation. *Plant Physiol.,* 39: 365–370.

———, N. M. Barnett, and C. Y. Lin. 1967. RNA and protein biosynthesis and the regulation of cell elongation by auxin. *Ann. N.Y. Acad. Sci.,* 144: 49–62.

——— and J. Ingle. 1964. Requirement for the synthesis of DNA-like RNA for growth of excised plant tissue. *Proc. Natl. Acad. Sci. (U.S.),* 52: 1382–1388.

——— and J. C. Shannon. 1964. Enhancement by auxin of ribonucleic acid synthesis in excised soybean hypocotyl tissue. *Plant Physiol.,* 39: 360–364.

Khan, A. A. 1966. Breaking of dormancy in *Xanthium* seeds by kinetin mediated by light and DNA dependent RNA synthesis. *Physiol. Plant.,* 19: 869–874.

———. 1968. Inhibition of gibberellic acid induced germination by abscisic acid and reversal by cytokinins. *Plant Physiol.,* 43: 1463–1465.

———. 1971. Cytokinins: permissive role in seed germination. *Science,* 171: 853–859.

Kharin, N. G. 1965. The reflective capacity of some plants and plant communities. (In Russian.) *Bot. Zh. (Mos.),* 50(8): 1115–1119.

Kidd, F. 1934. *Proc. R. Inst. G. B.,* 1934: 1–33; cited by Biale (1950).

——— and C. West. 1930. Physiology of fruit, I. *Proc. R. Soc. (Lond.),* B106: 93–109.

——— and ———. 1933. The influence of the composition of the atmosphere upon the incidence of the climacteric in apples. *Rep. Food Invest. Board.,* pp. 51–57.

——— and ———. 1937. The keeping quality of apples in relation to their maturity when gathered. *Sci. Hort.,* 5: 78.

———, ———, and G. E. Briggs. 1920. What is the significance of the efficiency index of plant growth? *New Phytol.,* 19: 88–96.

Kimura, K. 1961. Effect of temperature and nutrients on flower initiation of *Raphanus sativus* L. in total darkness. *Bot. Mag. (Tokyo),* 74: 361–368.

King, R. W., and L. T. Evans. 1967. Photosynthesis in artificial communities of wheat, lucerne, subterranean clover plants. *Aust. J. Biol. Sci.,* 20: 623–635.

———, ———, and I. F. Wardlaw. 1968. Translocation of the floral stimulus in *Pharbitis nil* in relation to that of assimilates. *Z. Pflanzenphysiol.,* 59: 377–388.

Kirk, J. T. O., and R. A. E. Tilney-Bassett. 1967. "The Plastids: Their Chemistry, Structure, Growth and Inheritance." Freeman, San Francisco.

Kirk, S. C., and W. P. Jacobs. 1968. Polar movement of indoleacetic acid ^{14}C in roots of *Lens* and *Phaseolus*. *Plant Physiol.,* 43: 675–682.

Kislyakova, T. E. 1960. On the twenty-four hour photosynthesis of plants in the far north. *Fiziol. Rast.,* 7: 45–47.

Klebs, G. 1914. Über das Treiben der einheimischen Baume speziell der Buche. *Abh. Heidelb. Akad. Wiss.,* 3: 116.

———. 1918. Über die Blutenbildung von *Sempervivum*. *Flora,* 11: 128–151.

Klein, S. 1960. The effect of low temperature on the development of the lamellar system in chloroplasts. *J. Biophys. Biochem. Cytol.,* 8: 529–538.

——— and B. M. Pollock. 1968. Cell fine structure of developing lima bean seeds related to seed desiccation. *Am. J. Bot.,* 55: 658–672.

Klein, W. H., and A. C. Leopold. 1953. Effect of maleic hydrazide on flower initiation. *Plant Physiol.,* 28: 293–298.

———, J. L. Edwards, and W. Shropshire. 1967. Spectrophotometric measurements of phytochrome in vivo and their correlation with photomorphogenic responses in *Phaseolus*. *Plant Physiol.,* 42: 264–270.

———, L. Price and K. Mitrakos. 1963. Light stimulated starch degradation in plastids and leaf morphogenesis. *Photochem. Photobiol.,* 2: 233–240.

Kleinkopf, G. E., R. C. Huffaker, and A. Matheson. 1970. Light-induced *de novo* synthesis of ribulose 1,5-diphosphate carboxylase in greening leaves of barley. *Plant Physiol.,* 46: 416–418.

Kliewer, W. M. 1964. Influence of environment on metabolism of organic acids and carbohydrates in *Vitis vinifera*. I: Temperature. *Plant Physiol.,* 39: 869–880.

Knecht, E., and J. Bruinsma. 1973. A rapid, sensitive

and accurate determination of indoleacetic acid. *Phytochemistry,* 12: 753–756.
Knudson, L. 1920. The secretion of invertase by plant roots. *Am. J. Bot.,* 7: 371–379.
Köckemann, A. 1934. Über eine keimungshemmende Substanz in fleischigen Fruchten. *Ber. Dtsch. Bot. Ges.,* 52: 523–526.
Koepfli, J. B., K. V. Thimann, and F. W. Went. 1938. Phytohormones: Structure and physiological activity. *J. Biol. Chem.,* 122: 763–780.
Kögl, F., A. J. Haagen Smit, and H. Erxleben. 1934. Über ein neues Auxin ("Heteroauxin") aus Horn, XI. *Z. Physiol. Chem.*, 228: 90–103.
——— and D. G. F. R. Kostermans. 1934. Heteroauxin als Stoff-wechselprodukt niederer pflanzlicher Organismen Isolierung aus Hefe, XIII. *Z. Physiol. Chem.,* 228: 113–121.
Köhler, D. 1964. Kinetische Unter suchungen über die Wirkung eines aus unreifen Samen von *Vicia* isolierten Antigibberellin. *Planta,* 63: 326–343.
——— and A. Lang. 1963. Evidence for substances in higher plants interfering with gibberellin responses. *Plant Physiol.,* 38: 555–560.
Kojima, H., and S. Maeda. 1958. Promotion of flower-initiation by restraining the vegetative growth in the Japanese radish. *Bot. Mag. (Tokyo),* 71: 841–842.
Kok, B. 1956*a*. On the inhibition of photosynthesis by intense light. *Biochim. Biophys. Acta,* 21: 234–244.
———. 1956*b*. Photosynthesis in flashing light. *Biochim. Biophys. Acta,* 21: 245–258.
Koller, D. 1957. Germination-regulating mechanisms in some desert seeds. IV: *Atriplex dimorphostegia. Ecology,* 38: 1–13.
——— and H. R. Highkin. 1960. Environmental control of reproductive development in *Hordeum bulbosum,* a perennial pasture grass. *Am. J. Bot.,* 47: 843–847.
———, A. Poljakoff-Mayber, A. Berg, and T. Diskin. 1963. Germination regulating mechanisms in *Citrullus colocynthis. Am. J. Bot.,* 50: 597–603.
Könings, H. 1968. The significance of the root cap for geotropism. *Acta Bot. Neerl.,* 17: 203–211.
Konishi, M. 1956. Studies on development of flowering stalks in long-day plants in relation to auxin metabolism. *Mem. Coll. Agric. Kyoto Univ., Bot. Ser.,* 3: 1–70.
Könitz, W. 1958. Blühemmung bei Kurztagpflanzen durch Hellrot- und Dunkelrotlicht in der Photo- und Skotophilen Phase. *Planta,* 51: 1–29.
Koontz, H., and O. Biddulph. 1957. Factors affecting absorption and translocation of foliar applied phosphorus. *Plant Physiol.,* 32: 463–470.
Korn, E. D. 1966. Structure of biological membranes. *Science,* 153: 1491–1498.
Kortschak, H. P.,and C. E. Hartt. 1966. The effects of varied conditions on carbon dioxide fixation in sugar-cane leaves. *Naturwiss.,* 10: 253.
———, ———, and G. O. Burr. 1965. Carbon dioxide fixation in sugar-cane leaves. *Plant Physiol.,* 40: 209–213.
Koshimizu, K., H. Fukui, T. Mitsui, and Y. Ogawa. 1966. Identity of lupine inhibitor with abscisin II and its biological activity on growth of rice seedlings. *Agric. Biol. Chem.,* 30: 941–943.
———, M. Inui, H. Fukui, and T. Mitsui. 1968. Isolation of abscisyl- β-*D*-glucopyranoside from immature fruit of *Lupinus luteus. Agric. Biol. Chem.,* 32: 789–791.
Kozlowski, T. T. 1957. Effect of continuous light intensity on photosynthesis of forest tree seedlings. *For. Sci.,* 3: 220–224.
———. 1971. "Growth and Development of Trees," vol. 1. Academic, New York. 443 pp.
——— and T. Keller. 1966. Food relations of woody plants. *Bot. Rev.,* 32: 293–382.
Kramer, P. J. 1937. The relation between rate of transpiration and rate of absorption of water in plants. *Am. J. Bot.,* 24: 10–15.
———. 1956. Physical and physiological aspects of water absorption. Pp. 124–159 in W. Ruhland (ed.), "Encyclopedia of Plant Physiology," vol. 3. Springer-Verlag, Berlin.
———. 1957. Outer space in plants: Some possible implications of the concept. *Science,* 125: 663–635.
Kraus, E. J., and H. R. Kraybill. 1918. Vegetation and reproduction with special reference to the tomato. *Oreg. Agric. Exp. Stn. Bull.* 149, pp. 1–90.
Krelle, E. 1967. Untersuchungen über die Wirkung des Antigibberellins 2-Chloro-9-fluorenol-9-carbonsäure Methylester. Pp. 691–692 in "Wachstumsregulatoren bei Pflanzen." Fischer-Verlag, Jena.
Kriedemann, P. E. 1968*a*. An effect of kinetin on the translocation of ^{14}C labelled photosynthate in citrus. *Aust. J. Biol. Sci.,* 21: 569–571.
———. 1968*b*. Observations on gas exchange in the developing sultana berry. *Aust. J. Biol. Sci.,* 21: 907–916.
———. 1968*c*. Some photosynthetic characteristics of citrus leaves. *Aust. J. Biol. Sci.,* 21: 895–905.
———. 1969. Sugar uptake by the grape berry: A note on the absorption pathway. *Planta,* 85: 111–117.
———. 1971. Photosynthesis and transpiration as a function of gaseous diffusive resistances in orange leaves. *Physiol. Plant.,* 24: 218–225.
——— and H. Beevers. 1967*a*. Sugar uptake and translocation in the castor bean seedling. I: Characteristics of transfer in intact and excised seedlings. *Plant Physiol.,* 42: 161–173.
——— and ———. 1967*b*. Sugar uptake and translocation in the castor bean seedling. II: Sugar transformations during uptake. *Plant Physiol.,* 42: 174–180.
——— and R. L. Canterford. 1971. The photosynthet-

ic activity of pear leaves (*Pyrus communis* L.). *Aust. J. Biol. Sci.,* 24: 197–205.
———, W. M. Kliewer, and J. M. Harris. 1970. Leaf age and photosynthesis in *Vitis vinifera* L. *Vitis,* 9: 97–104.
———, B. R. Loveys, G. L. Fuller, and A. C. Leopold. 1972*a*. Abscisic acid and stomatal regulation. *Plant Physiol.,* 49: 842–847.
——— and T. F. Neales. 1963. Studies on the use of cetyl alcohol as a transpiration suppressor. *Aust. J. Biol. Sci.,* 16: 743–750.
———, ———, and D. H. Ashton. 1964. Photosynthesis in relation to leaf orientation and light interception. *Aust. J. Biol. Sci.,* 17: 591–600.
——— and R. E. Smart. 1971. Effects of irradiance, temperature and leaf water potential on photosynthesis of vine leaves. *Photosynthetica,* 5: 6–15.
———, E. Törökfalvy, and R. E. Smart. 1972*b*. Natural occurrence of sunflecks by grapevine leaves. *Photosynthetica,* 7:
Krizek, D. T., H. H. Klueter, and W. A. Bailey. 1972. Effects of day and night temperature and type of container on the growth of F_1 hybrid annuals in controlled environments. *Am. J. Bot.,* 59: 284–289.
Kroeger, G. S. 1941. Dormancy in seeds of *Impatiens balsamina*. *Contrib. Boyce Thompson Inst.,* 12: 203–212.
Kroh, M. 1964. An electron microscopic study of the behavior of Cruciferae pollen after pollination. Pp. 221–224 in H. F. Linskins (ed.), "Pollen Physiology and Fertilization." North-Holland, Amsterdam.
Krotokov, G. 1963. Effect of light on respiration. In "Photosynthetic Mechanisms in Green Plants", *Publ.* 1145, National Academy of Sciences, National Research Council, Washington, D.C.
Krumweide, D. 1960. Über die Wirkung von Stark- und Schwachlichtkombinationen auf das Blühen von *Kalanchoe blossfeldiana*. *Biol. Zentralbl.,* 79: 258–278.
Ku, H. S., and A. C. Leopold. 1970. Ethylene formation from peptides of methionine. *Biochem. Biophys. Res. Commun.,* 41: 1155–1160.
——— and H. K. Pratt. 1968. Active mitochondria do not produce ethylene. *Plant Physiol.,* 43: 999–1001.
———, ———, A. R. Spurr, and W. M. Harris. 1968. Isolation of active mitochondria from tomato fruit. *Plant Physiol.,* 43: 883–887.
———, H. Suge, L. Rappaport, and H. K. Pratt. 1970. Stimulation of rice coleoptile growth by ethylene. *Planta,* 90: 333–339.
———, S. F. Yang, and H. K. Pratt. 1969. Ethylene formation from α-keto-γ-methylbutyrate by tomato fruit extracts. *Phytochemistry,* 8: 567–573.
Kuiper, P. J. C. 1972. Water transport across membranes. *Annu. Rev. Plant. Physiol.,* 23: 157–172.
Kumamoto, J., H. H. A. Dollwet, and J. M. Lyons. 1969. Evidence and hypothesis for a Taube bridge electron transfer propagating to a remote site through bonding: The formation of ethylene from monoethyl sulfate. *J. Am. Chem. Soc.,* 91: 1207–1210.
Kuraishi, S. 1968. The effect of kinetin on protein level of *Brassica* leaf discs. *Physiol. Plant.,* 21: 78–73.
——— and R. M. Muir. 1964*a*. The relationship of gibberellin and auxin in plant growth. *Plant Cell Physiol.,* 5: 61–69.
——— and ———. 1964*b*. The mechanism of gibberellin action in the dwarf pea. *Plant Cell Physiol.,* 5: 259.
———, T. Tezuka, T. Ushijima, and T. Tazaki. 1966. Effect of cytokinins on frost hardiness. *Plant Cell Physiol.,* 7: 705–706.
——— and T. Yamaki. 1967. Cytokinin-like activities of 4-benzylaminobenzimidazole. *Physiol. Plant.,* 20: 208–212.
Kuroiwa, S., T. Hiroi, K. Takada, and M. Monsi. 1964. Distribution ratio of net photosynthate to photosynthetic and non-photosynthetic systems in shaded plants. *Bot. Mag. (Tokyo),* 77: 37–42.
Kurosawa, E. 1926. Experimental studies on the secretion of *Fusarium heterosporum* on rice plants. *Trans. Nat. Hist. Soc. Formosa,* 16: 213–227.
Kursanov, A. L. 1961. The transport of organic substances in plants. *Endeavour,* 20: 19–26.
Kushimitzu, K., T. Kusaki, T. Mitsui, and S. Matsubara. 1967. Isolation of a cytokinin, dihydrozeatin, from immature seeds of *Lupinus luteus*. *Tetrahedron Lett.,* 14: 1317.
Kutacek, M., and Z. Prochazka. 1964. Méthodes de détermination et d'isolement des composés indoliques chez les cruciferes. Pp. 445–456 in J. P. Nitsch (ed.), "Régulateurs Naturels de la Croissance Végétale." CNRS, Paris.
———, Z. Prochazka, and D. Grunberger. 1960. Biosynthesis of ascorbigen, indoleacetonitrile, indole carboxylic acid from tryptophan-3C^{14} in *Brassica oleracea*. *Nature,* 187: 61–62.
Kwack, B. H. 1965. Stylar culture of pollen and physiological studies of self-incompatibility in *Oenothera*. *Physiol. Plant.,* 18: 297–305.
Kylin, A., and B. Hylmo. 1957. Uptake and transport of sulphate in wheat; active and passive components. *Physiol. Plant.,* 10: 467–484.
Laan, P. A. van der. 1934. Der Einfluss von Aethylen auf die Wuchsstoffibildung bei *Avena* und *Vicia*. *Recl. Trav. Bot. Neerl.,* 31: 691–742.
Labarca, C., P. B. Nicholls, and R. S. Bandurski. 1965. A partial characterization of indoleacetylinositols from *Zea mays*. *Biochem. Biophys. Res. Commun.,* 20: 641–646.
Lachman, W. H., and E. F. Upham. 1954. Effect of warm storages on the bolting of onions grown from sets: A preliminary report. *Proc. Am. Soc. Hort. Sci.,* 63: 342–346.
Laetsch, W. M. 1968. Chloroplast specialization in

dicotyledons possessing the C_4-dicarboxylic acid pathway of photosynthetic CO_2 fixation. *Am. J. Bot.,* 55: 875–883.

Lagarde, J. 1971. Influence de divers traitements photothermopériodiques sur le développement chez le crosne du Japon dormant. *Soc. Bot. Fr. Mem.,* 1971: 221–228.

Laibach, F. 1929. Ectogenesis in plants. *J. Hered.,* 20: 201–208.

———. 1932. Pollenhormon und Wuchsstoff. *Ber. Dtsch. Bot. Ges.,* 50: 383–390.

———. 1933. Versuche mit Wuchsstoffpaste. *Ber. Dtsch. Bot. Ges.,* 51: 386–392.

——— and F. J. Kribben. 1950. Einfluss von Wuchsstoff auf die Blütenbildung der Gurke. *Z. Naturforsch.,* 56: 160.

——— and W. Troll. 1955. Uber die Ursachen des vorzeitigen Abwurfes von Blutenteilen bei *Coleus. Bietr. Biol. Pflanzen,* 31: 15–26.

Lake, J. V. 1967. Respiration of leaves during photosynthesis. I: Estimates from electrical analogue. *Aust. J. Biol. Sci.,* 20: 487–493.

Lam, S. L. 1965. Movement of the flower stimulus in *Xanthium. Am. J. Bot.,* 52: 924–928.

——— and H. B. Cordner. 1955. Flowering hormone in relation to blooming in sweetpotatoes. *Science,* 121: 140–141.

——— and A. C. Leopold. 1960. Reversion from flowering to the vegetative state in *Xanthium. Am. J. Bot.,* 47: 256–260.

——— and ———. 1961. Reversion and reinduction of flowering in *Perilla. Am. J. Bot.,* 48: 306–310.

——— and ———. 1966. Role of leaves in phototropism. *Plant Physiol.,* 41: 847–851.

Lambertz, P. 1954. Untersuchungen über das Vorkommen von Plasmodesmen in den Epidermisaussenwanden. *Planta,* 44: 147–190.

Lang, A. 1948. Beiträge zur Genetik des Photoperiodismus. Pp. 175–189 in A. E. Murneek and R. O. Whyte (eds.), "Vernalization and Photoperiodism." Chronica Botanica, Waltham, Mass.

———. 1952. Physiology of flowering. *Annu. Rev. Plant Physiol.,* 3: 265–306.

———. 1956. Gibberellin and flower formation. *Naturwiss.,* 23: 544.

———. 1957. The effect of gibberellin upon flower formation. *Proc. Natl. Acad. Sci. (U.S.),* 43: 709–504.

———. 1960. Gibberellin-like substances in photoinduced and vegetative *Hyoscyamus* plants. *Planta,* 54: 498–504.

———. 1970. Gibberellins: Structure and metabolism. *Ann. Rev. Plant Physiol.,* 21: 537–570.

——— and G. Melchers. 1943. Die Photoperiodische Reaktion von *Hyoscyamus niger. Planta,* 33: 653–702.

——— and E. Reinhard. 1961. Gibberellins and flower formation. *Adv. Chem.,* 28: 71–79.

———, J. A. Sandoval, and A. Bedri. 1957. Induction of bolting and flowering in *Hyoscyamus* and *Samolus* by a gibberellin-like material from a seed plant. *Proc. Natl. Acad. Sci. (U.S.),* 43: 960–964.

Lang, A. R. G., B. Klepper, and M. J. Cumming. 1969. Leaf water balance during oscillation of stomatal aperture. *Plant Physiol.,* 44: 826–830.

Lange, H., I. Bienger, and H. Mohr. 1967. Eine neue Beweisführung für die Hypothese einer differentiellen Genaktivierung durch Phytochrom 730. *Planta,* 76: 359–366.

——— and H. Mohr. 1965. Die Hemmung der phytochrominduzierten Anthocyansynthese durch Actinomycin D und Puromycin. *Planta,* 67: 107–121.

Lange, O. C., and B. Schwemmle. 1960. Untersuchungen zur Hitzeresistenz vegetativer und blühender Pflanzen von *Kalanchôe blossfeldiana. Planta,* 55: 208–225.

Langridge, J., and B. Griffing. 1959. A study of high temperature lesions in *Arabidopsis thaliana. Aust. J. Biol. Sci.,* 12: 117–135.

Larcher, W. 1969. The effect of environmental and physiological variables on the carbon dioxide gas exchange of trees. *Photosynthetica,* 3: 167–198.

Larsen, P. 1944. 3-Indole acetaldehyde as a growth hormone in higher plants. *Dan. Bot. Ark.,* 11: 1–132.

———. 1951. Enzymatic conversion of indole acetaldehyde and naphthalene acetaldehyde to auxins. *Plant Physiol.,* 26: 697–707.

———. 1962. Orthogeotropism in roots. *Handb. Pflanzenphysiol.,* 17(2): 153–199.

———. 1971. The susception of gravity by higher plants. Pp. 73–88 in S. A. Gordon and M. J. Cohen (eds.), "Gravity and the Organism." University of Chicago Press, Chicago.

———, A. Harbo, S. Klungsoyr, and T. Aasheim. 1962. On the biogenesis of some indole compounds in *Acetobacter xylinin. Physiol. Plant.,* 15: 552–565.

LaRue, C. D. 1936. Effect of auxin on abscission of petioles. *Proc. Natl. Acad. Sci. (U.S.),* 22: 254–259.

——— and S. Narayanswami. 1957. Auxin inhibition in the liverwort *Lunularia. New Phytol.,* 56: 61–70.

Läuger, P. 1972. Carrier-mediated ion transport. *Science,* 178: 24–30.

Leach, R. W. A., and P. F. Wareing. 1967. Distribution of auxin in horizontal woody stems in relation to gravimorphism. *Nature,* 214: 1025–1027.

Ledbetter, M. C., P. W. Zimmerman, and A. E. Hitchcock. 1960. The histopathological effects of ozone on plant foliage. *Contrib. Boyce Thompson Inst.,* 20: 275–282.

Lehenbauer, P. A. 1914. Growth of maize seedlings in relation to temperature. *Physiol. Res.,* 1: 247–288.

Leitsch, J. 1916. Some experiments on the influence of temperature on the rate of growth in *Pisum sativum. Ann. Bot. (Lond.),* 30: 25–46.

Lek, H. A. A. van der. 1925. Over de wortelvorming van houtige stekken. Doctoral diss. Utrecht; cited by Went and Thimann (1937).

———. 1934. Over den invloed der knappen op de wortelvorming der stekken. *Meded. Landbouwhogesch. Wageningen,* 38: 1–95.

Leloir, L. F., C. E. Cardini, and E. Cabib. 1960. Utilization of free energy for the biosynthesis of saccharides. Pp. 97–138 in M. Florkin and H. S. Mason (eds.), "Comparative Biochemistry," vol. 2. Academic, New York.

———, M. A. R. de Fekete, and C. E. Cardins. 1961. Starch and oligosaccharide synthesis from uridine diphosphate glucose. *J. Biol. Chem.,* 236: 636–641.

Leman, V. M., and I. I. Bogacheva. 1957. Accumulation of dry matter in intermittently illuminated plants. Pp. 980–987 in "Problems of Photosynthesis." *Rep. 2d All-Union Conf. Photosynthesis, Moscow, Jan. 21–26, 1957.*

Lembi, C., D. J. Morré, K. S. Thomson, and R. Hertel. 1971. *N*-1-naphthylphthalamic acid binding activity of a plasma membrane rich fraction from maize coleoptiles. *Planta,* 99: 37.

Lemon, E. R. 1963. Energy and water balance of plant communities. Pp. 55–77 in L. T. Evans (ed.), "Environmental Control of Plant Growth." Academic, New York.

———. 1968. The measurement of height distribution of plant community activity using the energy and momentum balance approaches. Pp. 381–389 in "Plant Environment and Efficient Water Use." *UNESCO Symp. Proc. Copenhagen, July 1965.*

———. 1970. Mass and energy exchange between plant stands and environment. Pp. 199–205 in "Prediction and Measurement of Photosynthetic Productivity." *Proc. Int. Biol. Progr. Sec. PP Tech. Meet. Trebon, Sept., 14–21, 1969.* (Pudoc, Wageningen, Netherlands).

———, D. W. Stewart, and R. W. Shawcroft. 1971. The sun's work in a corn field. *Science,* 174: 371–378.

———, J. L. Wright, and G. M. Drake. 1969. Photosynthesis under field conditions XB: Origins of short-time CO_2 fluctuations in a corn field. *Agron. J.,* 61: 411–413.

Lenton, J. R., M. R. Bowen, and P. F. Saunders. 1968. Detection of abscisic acid in the xylem sap of willow by gas chromatography. *Nature,* 220: 86–87.

———, V. M. Perry, and P. F. Saunders. 1972. Endogenous abscisic acid in relation to photoperiodically induced bud dormancy. *Planta,* 106: 13–22.

Leonard, E. R., and C. W. Wardlaw. 1941. Studies in tropical fruits. XII: Respiration of bananas. *Ann. Bot.,* 5: 379–423.

Leonard, N., S. Achmatowixz, R. Loeppky, K. Carraway, W. Grimm, A. Szweykowska, H. Hamzi, and F. Skoog. 1966. Development of cytokinin activity by rearrangement of 1-substituted adenines to 6-substituted aminopurines. *Proc. Natl. Acad. Sci. (U.S.),* 56: 709–716.

———, S. M. Hecht, F. S. Skoog, and R. Y. Schnitz. 1968. Cytokinins: Synthesis of 6(3-methyl-3-butenylamino)-9-8-D-ribofuranosylpurine and the effect of side-chain unsaturation on the biological activity of isopentylaminopurines and their ribosides. *Proc. Natl. Acad. Sci. (U.S.),* 59: 15–21.

Leopold, A. C. 1961. Senescence in plant development. *Science,* 134: 1727–1732.

———. 1971. Antagonism of some gibberellin actions by a substituted pyrimidine. *Plant Physiol.,* 48: 537–540.

———. 1972. Ethylene as a plant hormone. Pp. 245–262 in H. Kaldeway and Y. Vardar (eds.), "Hormone Regulation in Plant Growth and Development." Verlag Chemie, Weinheim.

———, K. M. Brown, and F. H. Emerson. 1972. Ethylene in the wood of stressed trees. *Hort Science,* 7: 175.

——— and R. K. dela Fuente. 1968. A view of polar auxin transport. Pp. 24–47 in Y. Vardar (ed.), "The Transport of Plant Hormones." North-Holland, Amsterdam.

——— and O. F. Hall. 1966. A mathematical model of polar auxin transport in plants. *Plant Physiol.,* 41: 1467–1480.

——— and M. Kawase. 1964. Benzyladenine effects on bean leaf growth and senescence. *Am. J. Bot.,* 51: 294–298.

———, E. Niedergang-Kamien, and J. Janick. 1959. Experimental modification of plant senescence. *Plant Physiol.,* 34:570–573.

——— and F. I. Scott. 1952. Physiological factors in tomato fruit-set. *Am. J. Bot.,* 39: 310–217.

Leshem, Y. 1971. Abscisic acid as a ribonuclease promoter. *Physiol. Plant.,* 24: 85–89.

——— and L. Schwarz. 1972. The selective effect of abscisic acid on ribonucleic acid components. *Physiol. Plant.,* 26: 328–331.

Letham, D. S. 1963*a*. Regulators of cell division in plant tissue, I: N. Z. J. Sci., 1: 336–349.

———. 1963*b*. Zeatin a factor inducing cell division isolated from *Zea mays. Life Sci.,* 8: 569–573.

———. 1967. Chemistry and physiology of kinetin-like compounds. *Annu. Rev. Plant Physiol.,* 18: 349–364.

———. 1968. A new cytokinin bioassay and the naturally occurring cytokinins. Pp. 19–31 in F. Wightman and G. Setterfield (eds.), "Biochemistry and Physiology of Plant Growth Substances." Runge Press, Ottawa.

———, J. S. Shannon, and I. R. McDonald. 1964. The structure of zeatin, a factor inducing cell division. *Proc. Chem. Soc. (Lond.),* 1964: 230.

——— and M. W. Williams. 1969. Regulators of cell division in plant tissues, VIII. *Physiol. Plant.,* 22: 925–936.

Levering, C. A., and W. W. Thomson. 1971. The ultrastructure of the salt gland of *Spartina foliosa. Planta,* 97: 183–196.

Levine, R. P. 1969. The mechanism of photosynthesis. *Sci. Am.,* 221: 58–70.

Levitt, J. 1954. Steady state versus equilibrium thermodynamics in the concept of "active" water absorption. *Physiol. Plant.,* 7: 592–594.

———. 1956. "The Hardiness of Plants," Academic, New York. 278 pp.

———. 1957*a*. The moment of frost injury. *Protoplasma,* 48: 289–303.

———. 1957*b*. The role of cell sap concentration in frost hardiness. *Plant Physiol.,* 32: 237–239.

———. 1959*a*. Effects of artificial increase in sugar content on frost hardiness. *Plant Physiol.,* 34: 401–402.

———. 1959*b*. Bound water and frost hardiness. *Plant Physiol.,* 34: 674–677.

———. 1962. A sulfhydryl-disulfide hypothesis of frost injury and resistance in plants. *J. Theoret. Biol.,* 3: 355–391.

———. 1966. Resistance to water transport in plants: A misconception. *Nature,* 212: 527.

———. 1967. The mechanism of stomatal action. *Planta,* 74: 101–118.

———. 1969. Growth and survival of plants at extremes of temperature: A unified concept. *Symp. Soc. Exp. Biol.,* 23: 395–448.

Lewington, R. J., M. Talbot, and E. W. Simon. 1967. The yellowing of attached and detached cucumber cotyledons. *J. Exp. Bot.,* 18: 526–534.

Lewis, D. 1949. Incompatibility in flowering plants. *Biol. Rev.,* 24: 472.

———. 1952. Serological reactions of pollen incompatibility substance. *Proc. R. Soc. (Lond.),* B140: 127–135.

———. 1953. Some factors affecting flower production in the tomato. *J. Hort. Sci.,* 23: 207–220.

———. 1964. Incompatibility: A protein dimer hypothesis. *Proc. XI Int. Congr. Genet.,* 2: 29 (Pergamon Press, The Hague).

Lewis, L. N., and J. E. Varner. 1970. Synthesis of cellulase during abscission of *Phaseolus vulgaris* leaf explants. *Plant Physiol.,* 46: 194–199.

Libbert, E., S. Wichner, U. Schiewer, H. Risch, and W. Kaiser. 1966. The influence of epiphytic bacteria on auxin metabolism. *Planta,* 68: 327–334.

Lieberman, M., and C. C. Craft. 1961. Ethylene production by cytoplasmic particles from apples and tomato fruits in the presence of thiomalic and thioglycolic acid. *Nature,* pp. 189–243.

——— and A. T. Kunishi. 1967. Propanal may be a precursor of ethylene in metabolism. *Science,* 158: 938.

——— and ———. 1969. Stimulation of ethylene production in tomato tissue by propionic acid. *Plant Physiol.,* 44: 1446–1450.

——— and ———. 1974. Ethylene-forming systems in etiolated pea seedling epicotyl segments and in apple tissue. *Plant Physiol.* 53: supp. 17.

———, ———, L. W. Mapson, and D. A. Wardale. 1966. Stimulation of ethylene production in apple tissue slices by methionine. *Plant Physiol.,* 41: 376–382.

——— and L. W. Mapson. 1962*a*. Fatty acid control of ethane production by subcel ular particles from apples and its possible relation to ethylene biosynthesis. *Nature,* 195: 1016–1017.

——— and ———. 1962*b*. Inhibition of the evolution of ethylene and ripening of fruit by ethylene oxide. *Nature,* 196: 660–661.

Lin, P. P., and J. E. Varner. 1972. Cyclic nucleotide phosphodiesterase in pea seedlings. *Biochem. Biophys. Acta,* 276: 454–474.

Lincoln, R. G., D. L. Mayfield, and A. Cunningham. 1961. Preparation of a floral initiating extract from *Xanthium. Science,* 133: 756.

———, K. A. Raven, and K. C. Hamner. 1956. Certain factors influencing expression of the flowering stimulus in *Xanthium. I:* Translocation and inhibition of the flowering stimulus. *Bot. Gaz.,* 117: 194–206.

Linder, P. J., J. C. Craig, and T. R. Walton. 1957. Movement of taggedα-methoxyphenylacetic acid out of roots. *Plant Physiol.,* 32: 572–575.

Lindner, R. C. 1940. Factors affecting regeneration of the horseradish root. *Plant Physiol.,* 15: 161–181.

Linschitz, H., V. Kasche, W. L. Butler, and H. W. Siegelman. 1966. The kinetics of the phytochrome conversion. *J. Biol. Chem.,* 241: 3395–3403.

Lipe, J. A., and P. W. Morgan. 1972. Ethylene role in fruit abscission and dehiscence processes. *Plant Physiol.,* 50: 759–764.

Lips, S. H., and N. Roth-Bejerano. 1969. Light and hormones: Interchangeability in the induction of nitrate reductase. *Science,* 166: 109–110.

Little, E. C. S., and G. E. Blackman. 1963. The movement of growth regulators in plants, III. *New Phytol.,* 62: 173–197.

Liverman, J. L. 1955. The physiology of flowering. *Ann. Rev. Plant Physiol.,* 6: 177–210.

——— and J. Bonner. 1953. The interaction of auxin and light in the growth responses of plants. *Proc. Natl. Acad. Sci. (U.S.),* 39: 905–916.

Loach, K. 1967. Shade tolerance in tree seedlings. I: Leaf photosynthesis and respiration in plants raised under artificial shade. *New Phytol.,* 66: 607–621.

Lockhart, J. A. 1957. Studies of the organ of production of the natural gibberellin factor in higher plants. *Plant Physiol.,* 32: 204–207.

——— and V. Gottschall. 1961. Fruit-induced and apical senescence in *Pisum sativum. Plant Physiol.,* 36: 389–398.

Loeffler, J. E., and J. van Overbeek. 1964. Kinin activity in coconut milk. Pp. 77–82 in J. P. Nitsch (ed.), "Régulateurs Naturels de la Croissance Végétale." CNRS, Paris.

Loening, U. E., and J. Ingle. 1967. Diversity of RNA

components in green plant tissues. *Nature,* 215: 363–367.
Loesecke, H. W. von. 1950. "Bananas," Interscience, New York, 189 pp.
Lona, F. 1946. Sui fenomeni di induzione, post-effeto e localizzazione fotoperiodica. *Nuovo Giorn. Bot. Ital.,* 53: 548–575.
———. 1962. La brevidiurna *Impatiens balsamina* stimolata alla fiotura dalle gibberelline ed inibità dall'acido indolacetico. *Ateneo Parmense,* 34: 1–4.
———. 1964. The action of morphogenins, especially gibberellins, on flowering. *Proc. 10th Int. Bot. Congr.,* pp. 189.
——— and A. Bocchi. 1956. La distensione caulinare nella canapa incrementata dall'acido gibberellico. *Riv. Int. Agric.,* 7:
Longman, K. A. 1961. Factors affecting flower initiation in certain conifers. *Proc. Linn. Soc. Lond.,* 172: 124–127.
———. 1969. The dormancy and survival of plants in the humid tropics. *Soc. Exp. Biol. Symp.,* 23: 471–488.
Loomis, R. S., and J. G. Torrey. 1964. Chemical control of vascular cambium initiation in isolated radish roots. *Proc. Natl. Acad. Sci. (U.S.),* 52: 3–11
——— and W. A. Williams. 1963. Maximum crop productivity: An estimate. *Crop Sci.,* 3: 67–72.
———, ———, and A. E. Hall. 1971. Agricultural Productivity. *Annu. Rev. Plant Physiol.,* 22: 431–468.
Loomis, W. E. 1965. Absorption of radiant energy by leaves. *Ecology,* 46: 14–17.
——— and M. M. Evans. 1928. Experiments in breaking the rest period of corms and bulbs. *Proc. Am. Soc. Hort. Sci.,* 65: 73–79.
Looney, N. E., and M. E. Patterson. 1967. Chlorophyllase activity in apples and bananas during the climacteric phase. *Nature,* 214: 1245–1246.
Lovell, P. H. 1971. Translocation of photosynthates in tall and dwarf varieties of pea, *Pisum sativum. Physiol. Plant.,* 25: 382–385.
Loveys, B. R. 1970. The phytochrome and hormonal control of unrolling in etiolated wheat leaves. Ph.D. thesis, Univ. Wales, Aberystwyth. 240 pp.
——— and P. E. Kriedemann. 1973. Rapid changes in ABA-like inhibitors following alterations in vine leaf water potential. *Physiol. Plant.,* 28: 476–479.
———, A. C. Leopold, and P. E. Kriedemann. 1974. Abscisic acid metabolism and stomatal physiology in *Betula lutea* following alteration in photoperiod.
——— and P. F. Wareing. 1971*a*. The red light controlled production of gibberellin in etiolated wheat leaves. *Planta,* 98: 109–116.
——— and ———. 1971*b*. The hormonal control of wheat leaf unrolling. *Planta,* 98: 117–127.
Luckwill, L. C. 1949. Fruit drop in the apple in relation to seed development. *Ann. Appl. Biol.,* 36: 567–568.
———. 1948. Hormone content of the seed in relation to endosperm development and fruit drop in apple. *J. Hort. Sci.,* 24: 32–44.
———. 1957. Hormonal aspects of fruit development in higher plants. *Symp. Soc. Exp. Biol.,* 11: 63–85.
———. 1952. Growth-inhibiting and growth-promoting substances in relation to the dormancy of apple seeds. *J. Hort. Sci.,* 27: 53–67.
———. 1953. Studies of fruit development in relation to plant hormones, I. *J. Hort. Sci.,* 28: 14–24.
———. 1959. Factors controlling the growth and form of fruits. *J. Linn. Soc. Lond. Bot.,* 56: 294–302.
———. 1970. Control of growth and fruitfulness of apple trees. ' Pp. 237–254 in L. C. Luckwill and C. V. Cutting (eds.), "Physiology of Tree Crops." Academic, London.
——— and C. Lloyd-Jones. 1960. Metabolism of plant growth regulators, II. *Ann. Appl. Biol.,* 48: 626–636.
Ludlow, M. M., and G. L. Wilson. 1971. Photosynthesis of tropical pasture plants. I: Illuminance, carbon dioxide concentration, leaf temperature and leaf-air vapor pressure difference. *Aust. J. Biol. Sci.,* 24: 449–470.
Ludwig, L. J., T. Saeki, and L. T. Evans. 1965. Photosynthesis in artificial communities of cotton plants in relation to leaf area. I: Experiments with progressive defoliation of mature plants. *Aust. J. Biol. Sci.,* 18: 1103–1118.
Lund, H. 1956. Growth hormones in the styles and ovaries of tobacco responsible for fruit development. *Am. J. Bot.,* 43: 562–568.
Lüttge, V., and J. Weigl. 1962. Mikroautoradiographische Untersuchungen der Aufnahme und des Transportes von $S^{35}O_4$ und Ca^{45} in Keimwurzeln von *Zea mays* und *Pisum sativum. Planta,* 58: 113–126.
———, and ———. 1965. Zur Mikroautoradiographie wasserlöslicher Substanzen. *Planta,* 64: 28–36.
Lyons, J. M., W. B. McGlasson, and I. R. Pratt. 1962. Ethylene production, respiration and internal gas concentration in cantaloupe fruits at various stages of maturity. *Plant Physiol.,* 37: 31–36.
——— and H. K. Pratt. 1964*a*. An effect of ethylene on swelling of isolated mitochondria. *Arch. Biochem. Biophys.,* 104: 318–324.
——— and ———. 1964*b*. Effect of stage of maturity and ethylene treatment on respiration and ripening of tomato fruits. *Proc. Am. Soc. Hort. Sci.,* 84: 491–500.
——— and J. K. Raison. 1970. Oxidative activity of mitochondria isolated from plant tissue sensitive and resistant to chilling injury. *Plant Physiol.,* 45: 386–389.
———, T. A. Wheaton, and H. K. Pratt. 1964. Relationship between the physical nature of mitochondrial membranes and chilling sensitivity in plants. *Plant Physiol.,* 39: 262–268.
Lysenko, T. D. 1928. *Tr. Azerb. Op. Stat.,* 3: 168.
MacDougal, D. T., and J. Dufrenoy. 1944. Mycorrhi-

zal symbiosis in *Aplectrum, Corallorhiza,* and *Pinus. Plant Physiol.,* 19: 440–465.

Mackay, J. P. G., and P. E. Weatherley. 1973. The effects of transverse cuts through the stems of transpiring woody plants on water transport and stress in leaves. *J. Exp. Bot.,* 24: 15–28.

Macklon, A. E. S., and P. E. Weatherley. 1965. Controlled environment studies of the nature and origins of water deficits in plants. *New Phytol.,* 64: 414–427.

MacRobbie, E. A. C. 1971. Phloem translocation: Facts and mechanisms. A comparative survey. *Biol. Rev.,* 46: 429–481.

Madec, P. 1963. Tuber-forming substances in the potato. Pp. 121–130 in J. D. Ivins and P. L. Milthorpe (eds.), "The Growth of the Potato." Butterworth, London.

——— and P. Perennec. 1962. Les Relations entre l'induction de la tuberisation el la craissance chez la plante de pomme de terre. *Ann. Phys. Veg.,* 4: 5–84.

Maggs, D. H. 1965. Growth rates in relation to assimilate supply and demand, II. *J. Exp. Bot.,* 16: 387–404.

——— and D. McE. Alexander. 1969. The quantitative growth of young seedlings of the citrus rootstocks *Citrus sinensis. Ann. Bot.,* 33: 981–990.

Magness, J. R., L. A. Fletcher, and W. W. Aldrich. 1933*a*. Time during which fruitbud formation in apples may be influenced. *Proc. Am. Soc. Hort. Sci.,* 30: 313–318.

———, L. O. Regeimbal, and E. S. Degman. 1933*b*. Accumulation of carbohydrates in apple foliage as influenced by moisture supply. *Proc. Am. Soc. Hort. Sci.,* 29: 246–252.

Magruder, R., and H. A. Allard. 1937. Bulb formation in onions and length of day. *J. Agric. Res.,* 54: 715–752.

Maheshwari, S. C., and R. Venkataraman. 1966. Induction of flowering in duckweed by a new kinin, zeatin. *Planta,* 70: 304–306.

Mai, G. 1934. Korrelationsuntersuchungen an entspreiteten Blattstielen mit lebender Orchideenpollinien als Wuchstoffquelle. *Jahrb. Wiss. Bot.,* 79: 581–713.

Major, W., and E. H. Roberts. 1968. Dormancy in cereal seeds. I: The effects of oxygen and respiratory inhibitors. *J. Exp. Bot.,* 19: 77–89.

Makinen, Y. L. A., and D. Lewis. 1962. Immunological analysis of incompatibility proteins and of cross-reacting material in a self-compatible mutant of *Oenothera. Genet. Res.,* 3: 352–363.

Malpighi, M. 1674. "Anatomia plantarum idea."

Maltzahn, K. E. von. 1959. Interaction between kinetin and indoleacetic acid in the control of bud reactivation in *Splachnum ampullacreum. Nature,* 183: 60–61.

Mann, J. D., H. Haid, L. S. Jordan, and B. E. Day. 1967. Inhibition of the action of phytochrome by the herbicide CIPC. *Nature,* 213: 420–421.

Mansfield, T. A., and O. V. S. Heath. 1963. Photoperiodic effects on rhythmic phenomena in stomata of *Xanthium pennsylvanicum. J. Exp. Bot.,* 14: 334–352.

——— and R. J. Jones. 1971. Effects of abscisic acid on potassium uptake and starch content of stomatal guard cells. *Planta,* 101: 147–158.

——— and H. Meidner. 1966. Stomatal opening in light of different wavelengths: Effects of blue light independent of carbon dioxide concentration. *J. Exp. Bot.,* 17: 510–521.

Mansour, B. M. M. 1968. Effects of temperature and light on growth, flowering and corm formation in *Freesia. Meded. Landbouwhogesch. Wageningen* 68–80,76 pp.

Mapson, L. W., and D. A. Wardale. 1971. Enzymes involved in the synthesis of ethylene from methionine or its derivatives in tomatoes. *Phytochemistry,* 10: 29–39.

Marcus, A., and J. Feeley. 1964. Activation of protein synthesis in the imbibition phase of seed germination. *Proc. Natl. Acad. Sci. (U.S.),* 51: 1075–1079.

Marei, N., and R. Romani. 1971. Ethylene-stimulated synthesis of ribosomes, ribonucleic acid and protein in developing fig fruits. *Plant Physiol.,* 48: 806–808.

Marks, J. O., R. Bernlohr, and J. E. Varner. 1957. Oxidative phosphorylation in ripening fruit. *Plant Physiol.,* 32: 259–262.

Marrè, E. 1946. Histophysiological aspects of the action of the stamen on the pistil. *Boll. Soc. Ital. Biol. Sper.,* 22: 1208–1209.

———. 1948*a*. Regolazione ormonale del ricambio dell'amido nella pianta. I: Funzione amilofissatrice del seme. *Boll. Soc. Ital. Biol. Sper.,* 24: 1–4.

———. 1948*b*. Hormone regulation of starch metabolism in the higher plant. II: The starch fixing and starch mobilizing action of the seed in the competition between neighboring fruits. *Boll. Soc. Ital. Biol. Sper.,* 24: 602–605.

———. 1949. Growth hormones and carbohydrate metabolism in the higher plant. I: Starch fixing action of auxin after castration or removal of the developing seeds. *Boll. Soc. Ital. Biol. Sper.,* 25: 331–334.

Marre, E., R. Colombo, P. Lado, and F. Rasi-Caldogno. 1974. Correlation between proton extrusion and stimulation of cell enlargement. *Plant Sci. Letters,* 2: 139–150.

———, P. Lado, F. R. Caldogno and R. Colombo. 1973. Correlation between cell enlargement in pea internode segments and decrease in the pH of the medium. I. *Plant Sci. Letters,* 1: 179–184.

Marshall, C., and I. F. Wardlaw. 1973. A comparative study of the distribution and speed of movement of ^{14}C assimilates and foliar-applied ^{32}P-labeled phosphate in wheat. *Aust. J. Biol. Sci.,* 26: 1–13.

Marth, P. C., W. H. Preston, and J. W. Mitchell. 1954. Growth controlling effects of some quaternary ammonium compounds on various species of plants. *Bot. Gaz.*, 115: 200–204.

Martin, C., and K. V. Thimann. 1972. The role of protein synthesis in the senescence of leaves. I: The formation of protease. *Plant Physiol.*, 49: 64–71.

Martin, D., T. L. Lewis, J. Cerny, and A. Grassia. 1967. Interrelationships of cropping level, chemical composition and disorder susceptibility of Jonathan apples within and between seasons. *Field Stn. Rec. CSIRO Div. Plant Ind.*, 6(2): 29–36.

Martin, E. V. 1943. Studies of evaporation and transpiration under controlled conditions. *Carnegie Inst. Wash. Pub.* 550. 48 pp.

Marushige, K., and Y. Marushige. 1966. Effects of light on the appearance of photoperiodic sensitivity of etiolated *Pharbitis*. *Bot. Mag.*, 79: 397–403.

Marx, G. A. 1968. Influence of genotype and environment on senescence in peas. *Bioscience*, 18: 505–506.

Mascarenhas, J. P. 1966. Pollen tube growth and ribonucleic acid synthesis by vegetative and generative nuclei of *Tradescantia*. *Am. J. Bot.*, 53: 563–569.

——— and E. Bell. 1969. Protein synthesis during germination of pollen. *Biochim. Biophys. Acta*, 179: 199–203.

——— and L. Machlis. 1962. The pollen tube chemotropic factor from *Antirrhinum*. *Am. J. Bot.*, 49: 482–489.

——— and ———. 1964. Chemotropic response of the pollen of *Antirrhinum* to calcium. *Plant Physiol.*, 39: 70–77.

Mason, S. C. 1925. The inhibitive effect of direct sunlight on the growth of the date palm. *J. Agric. Res.*, 31: 455–468.

Mason, T. G., and C. J. Lewin. 1924. On the rate of carbohydrate transport in the greater yam, *Dioscorea alata*. *Sci. Proc. Roy. Dublin Soc.*, 18: 203–205.

——— and E. J. Maskell. 1928*a*. Studies of the transport of carbohydrate in the cotton plant. I: A study of diurnal variation in the carbohydrates of leaf, bark and wood, and the effect of ringing. *Ann. Bot.*, 42: 189–253.

———, ———, and E. Phyllis. 1936. Further studies on transport in the cotton plant. III. *Ann. Botany (London)*, 50: 23–58.

——— and ———. 1928*b*. Studies of the transport of carbohydrate in the cotton plant. II: The factors determining the rate and the direction of movement of sugars. *Ann. Bot.*, 42: 571–636.

———. 1961. Movement of 2,6-dichlorobenzonitrile in soils and plants in relation to its physical properties. *Weed Res.*, 1: 142–146.

Massini, P. 1959. Uptake and translocation of 3-amino and 3-hydroxy-1,2,4,-triazole in plants. *Proc. 2d Int. Conf. Atomic Energy.* Pergamon Press, London, pp. 58–62.

Masuda, Y. 1968. Role of cell-wall degrading enzymes in cell-wall loosening in oat coleoptiles. *Planta*, 83: 171–184.

——— and E. Tanimoto. 1967. Effect of auxin and antiauxin on the growth and RNA synthesis of etiolated pea internodes. *Plant Cell Physiol.*, 8: 459–465.

Matson, D. T., and W. R. Jarvis. 1970. Post-harvest ripening of strawberries. *Hort. Res.*, 10: 125–132.

Matthysse, A. J., and M. Abrams. 1970. A factor mediating interaction of kinins with the genetic material. *Biochim. Biophys. Acta*, 199: 511–518.

Maxie, E. C., and C. E. Baker. 1954. Air filtration studies in a commercial type apple storage. *Proc. Am. Soc. Hort. Sci.*, 64: 235–247.

——— and J. C. Crane. 1968. Effect of ethylene on growth and maturation of the fig. *Proc. Am. Soc. Hort. Sci.*, 92: 255–267.

Mayak, S., and H. Halevy. 1972. Interrelationships of ethylene and abscisic acid in the control of rose petal senescence. *Plant Physiol.*, 50: 341–346.

Mayer, A. M. 1964. Germination inhibitory action of ramulosin. *Isr. J. Bot.*, 13: 41.

——— and A. Poljakoff-Mayber. 1957. The influence of gibberellic acid and kinetin on germination and seedling growth of lettuce. *Bull. Res. Counc. Isr.*, 6D: 65–72.

——— and ———. 1963. "The Germination of Seeds." Pergamon, Oxford. 236 pp.

Mayer, J. R. 1845. "Die organische Bewegung in ihrem Zusammenhange mit dem Stoffwechsel." Heilbronn.

McAlister, E. D., and J. Myers. 1940. The time course of photosynthesis and fluorescense observed simultaneously. *Smithson. Mis. Coll.* 99(6): 1–37.

McCalla, D. R., D. J. Morré, and D. Osborne. 1962. The metabolism of a kinin, benzyladenine. *Biochim. Biophys. Acta*, 55: 522–528.

McComb, A. J. 1961. Bound gibberellin in mature runner bean seeds. *Nature*, 192: 575–576.

———. 1964. The stability and movement of gibberellic acid in pea seedlings. *Ann. Bot.*, 28: 669–687.

——— and D. J. Carr. 1958. Evidence from a dwarf pea bioassay for naturally occurring gibberellins in the growing plant. *Nature*, 181: 1548–1549.

McComb, A. J., and J. A. McComb. 1970. Growth substances and the relation between phenotype and genotype in *Pisum sativum*. *Planta*, 91: 235–245.

McCready, C. C. 1963. Movement of growth regulators in plants. I: Polar transport of 2,4-D. *New Phytol.*, 62: 3–18.

McCree, K. J. 1970. An equation for the rate of respiration of white clover plants grown under controlled conditions. Pp. 221–229 in "Prediction and Measurement of Photosynthetic Productivity." *Proc. IBP/PP Tech. Meet, Trebon. Sept., 14–21, 1969.* (Pudoc, Wageningen, Netherlands).

———. 1971. Significance of enhancement for calculations based on the action spectrum for photosynthesis. *Plant Physiol.*, 49: 704–706.

——— and R. S. Loomis. 1969. Photosynthesis in fluctuating light. *Ecology,* 50: 422–428.

——— and J. H. Troughton. 1966*a*. Prediction of growth rates at different light levels from measured photosynthesis and respiration rates. *Plant Physiol.,* 41: 559–566.

——— and ———. 1966*b*. Non-existence of an optimum leaf area index for the production rate of white clover grown under constant conditions. *Plant Physiol.,* 41: 1615–1622.

McDonough, W. T., and H. G. Gauch. 1959. The contribution of the awns to the development of the kernels of bearded wheat. *Md. Agric. Exp. Stn. Bull.* A-103. 15 pp.

McGlasson, W. G. 1969. Ethylene production by slices of green banana fruit and potato tuber tissue during the development of induced respiration. *Aust. J. Biol. Sci.,* 22: 489–491.

——— and H. K. Pratt. 1964. Effects of wounding on respiration and ethylene production by cantaloupe fruit tissue. *Plant Physiol.,* 39: 128–132.

McIlrath, W. J., Y. P. Abrol, and F. Heiligman. 1963. Dehydration of seeds in intact tomato fruits. *Science,* 142: 1681–1682.

——— and D. R. Ergle. 1953. Further evidence of the persistence of the 2,4-D stimulus in cotton. *Plant Physiol.,* 28: 693–702.

McKee, H. S., L. Nestel, and R. N. Robertson. 1955. Physiology of pea fruits, II. *Aust. J. Sci.,* 8: 467–475.

McMurchie, E. J., W. B. McGlasson, and I. L. Eaks. 1972. Treatment of fruit with propylene gives information about the biogenesis of ethylene. *Nature,* 237: 235–236.

McMurray, A. L., and C. H. Miller. 1968. Cucumber sex expression modified by 2-chloroethylphosphonic acid. *Science,* 162: 1397–1398.

McNairn, R. B., and H. B. Currier. 1968. Translocation blockage by sieve plate callose. *Planta,* 82: 369–380.

McNaughton, S. J. 1967. Photosynthetic system. II: Racial differentiation in *Typha latifolia. Science,* 156: 1363.

McNitt, R. E., L. Glassner, and J. Shen-Miller. 1974. Spectral effects on corn root geotropism. *Plant Physiol.,* 53: supp. 46.

McRae, D. H., and J. Bonner. 1953. Chemical structure and anti-auxin activity. *Physiol. Plant.,* 6: 485–510.

———, R. J. Foster, and J. Bonner. 1953. Kinetics of auxin interaction. *Plant Physiol.,* 28: 343–355.

McWilliam, J. R., and A. W. Naylor. 1967. Temperature and plant adaption. I: Interaction of temperature and light in the synthesis of chlorophyll in corn. *Plant Physiol.,* 42: 1711–1715.

Medawar, P. B. 1957. "The Uniqueness of the Individual." Basic Books, New York. 191 pp.

Medcalf, J. 1949. *Pineapple Res. Inst. Bull.,* Honolulu.

Mees, G. C. 1960. Experiments on the herbicidal action of 1,1′-ethlyene-2,2′-dipyridylium dibromide. *Ann. Appl. Biol.,* 48: 601–612.

Meeuse, B. J. D. 1961. "The Story of Pollination." Ronald, New York. 243 pp.

Mehard, C. W., and J. M. Lyons. 1970. A lack of specificity for ethylene-induced mitochondrial changes. *Plant Physiol.,* 46: 36–39.

———, ———, and J. Kumamoto. 1970. Utilization of model membranes in a test for the mechanism of ethylene action. *J. Membr. Biol.,* 3: 173–179.

Meheriuk, M., and M. Spencer. 1964. Ethylene production during germination of oat seeds and *Penicillium* spores. *Can. J. Bot.,* 42: 337–340.

——— and ———. 1967. Studies on ethylene production by a subcellular reaction from ripening tomatoes, I. *Phytochemistry,* 6: 535.

Meidner, H. 1965. Stomatal control of transpirational water loss. Soc. Exp. Biol. Symp. 19: 185–204.

———. 1967. Effect of kinetin on stomatal opening and the rate of intake of carbon dioxide in mature primary leaves of barley. *J. Exp. Bot.,* 18: 556–561.

———. 1969. "Rate limiting" resistances and photosynthesis. *Nature,* 222: 876–877.

——— and T. A. Mansfield. 1968. "Physiology of Stomata." McGraw-Hill, Maidenhead, England. 179 pp.

Meigh, D. F. 1959. Nature of the olefins produced by apples. *Nature,* 184: 1072–1073.

———, J. D. Jones, and A. C. Hulme. 1967. The respiration climacteric in the apple: Production of ethylene and fatty acids in fruits attached to and detached from the tree. *Phytochemistry,* 6: 1507–1515.

———, K. H. Norris, C. C. Craft, and M. Lieberman. 1960. Ethylene production by tomato and apple fruits. *Nature,* 186: 902–903.

Meijer, G. 1968. Rapid growth inhibition of gherkin hypocotyls in blue light. *Acta Bot. Neerl.,* 17: 9–14.

Melchers, G. 1937. Die Wirkung von Genen, tiefen Temperaturen und blühenden Propfpartnern auf die Blühreife von *Hyoscyamus niger. Biol. Zentralbl.,* 57: 568–614.

——— and A. Lang. 1941. Weitere Untersuchungen zur Frage der Blühhormone. *Biol. Zentralbl.,* 61: 16–39.

Menke, W. 1962. Structure and chemistry of plastids. *Annu. Rev. Plant Physiol.,* 13: 27–44.

Mes, M. G. 1957. Studies on the flowering of *Coffea Arabica* L. II: Breaking the dormancy of coffee flower buds. *Port. Acta Biol.,* 4: 342–354.

——— and I. Menge. 1954. Potato shoot and tuber cultures *in vitro. Physiol. Plant.,* 7: 637–649.

Meyer, A. 1883. "Das Chlorophyllkorn in chemischer morphologischer und biologischer Beiziehung." Leipzig, 1883.

Meyer, B. S., and D. B. Anderson. 1952. "Plant Physiology." Van Nostrand, Princeton, N.J. 784 pp.

Meyer, W. H. 1938. Yield of even-aged stands of Ponderosa pine. USDA Tech. Bull. 630. 59 pp.

Michener, H. D. 1938. The action of ethylene on plant growth. *Am. J. Bot.,* 25: 711–720.

———. 1942. Dormancy and apical dominance in potato tubers. *Am. J. Bot.,* 29: 558–562.

Michniewicz, M. 1967. The dynamics of gibberellin-like substances and growth inhibitors in ontogeny of conifers. *Wiss. Z. Univ. Rostock,* 16: 577–583.

——— and A. Kamienska. 1965. Flower formation induced by kinetin and vitamin treatment in a long-day plant, *Arabidopsis thaliana. Naturwiss.,* 52: 1–2.

——— and J. Kopcewicz. 1966. The dynamics of growth substances during germination of larch and pine. *Roszniki Nauk Roln.,* 90: 689–698.

Miedema, P. 1973. A physiological study of adventitious bud formation in potato. Ph.D. thesis, Groningen. 66 pp.

Mikkelsen, D. S. 1966. Germination inhibitors as a possible factor in rice dormancy. *Int. Rice Comm. News Lett., Tokyo,* pp. 132–145.

Milborrow, B. V. 1968. Identification and measurement of abscisic acid in plants. Pp. 1531 1545 in F. Wightman and G. Setterfield (eds.), "Biochemistry and Physiology of Plant Growth Substances." Runge Press, Ottawa.

———. 1970. The metabolism of abscisic acid. *J. Exp. Bot.,* 21: 17–29.

———. 1972. The biosynthesis and degradation of abscisic acid. Pp. 281–290 in D. J. Carr (ed.), "Plant Growth Substances, 1970." Springer Verlag, Berlin.

———. 1974. Biosynthesis of abscisic acid by a cell-free system. *Phytochemistry,* 13: 131–136.

——— and R. C. Noddle. 1970. Conversion of 5-(1,2-epoxy-2,6,6-trimethyl-cyclohexyl)-3-methyl penta-*cis*-2-*trans*-4-dienoic acid into abscisic acid in plants. *Biochem. J.,* 119: 729–734.

Miller, C. O. 1951. Promoting effect of cobaltous and nickelous ions on expansion of etiolated bean leaf disks. *Arch. Biochem. Biophys.,* 32: 216–218.

———. 1956. Similarity of some kinetin and red light effects. *Plant Physiol.,* 31: 318–319.

———. 1961. Kinetin and related compounds in plant growth. *Annu. Rev. Plant Physiol.,* 12: 395–408.

———. 1963. Kinetin and kinetin-like compounds. Pp. 194–202 in K. Paech and M. V. Tracey (eds.), "Modern Methods of Plant Analysis," vol. 6. Springer-Verlag, Berlin.

———. 1965. Evidence for the natural occurrence of zeatin and derivatives: Compounds from maize which promote cell division. *Proc. Natl. Acad. Sci. (U.S.),* 54: 1052–1058.

———. 1967. Cytokinins in *Zea mays. Ann. N.Y. Acad. Sci.,* 144: 251–257.

———. 1968. Naturally-occurring cytokinins. Pp. 33–45 in F. Wightman and G. Setterfield (eds.), "Biochemistry and Physiology of Plant Growth Substances." Runge Press, Ottawa.

———, F. Skoog, F. S. Okumura, M. H. von Saltza, and F. M. Strong. 1955. Structure and synthesis of kinetin. *J. Am. Chem. Soc.,* 77: 2662–2663.

———, ———, ———, ———, and ———. 1956. Isolation, structure and synthesis of kinetin, a substance promoting cell division. *J. Am. Chem. Soc.,* 78: 1375–1380.

Miller, E. S., and G. O. Burr. 1935. Carbon dioxide balance at high light intensities. *Plant Physiol.,* 10: 93–114.

Miller, E. V., J. R. Winston, and H. A. Schomer. 1940. Physiological studies of plastid pigments in rinds of maturing oranges. *J. Agric. Res.,* 60: 259–267.

Miller, J. C. 1929. Seedstalk development in cabbage. *Cornell Univ. Agric. Exp. Stn. Bull.* 488.

Miller, P. M., H. C. Sweet, and J. H. Miller. 1970. Growth regulation by ethylene in fern gametophytes. *Am. J. Bot.,* 57: 212–217.

Millerd, A., J. Bonner, and J. B. Biale. 1953. The climacteric rise in fruit respiration as controlled by phosphorylative coupling. *Plant Physiol.,* 28: 521–531.

Millington, W. F., and E. L. Fisk. 1956. Shoot development in *Xanthium pennsylvanicum,* I. *Am. J. Bot.,* 43: 655–665.

Milthorpe, F. L. (ed.). 1956. "The Growth of Leaves." Butterworth, London.

———. 1959. Studies on the expansion of the leaf surface. I: The influence of temperature. *J. Exp. Bot.,* 10: 233–249.

——— and J. Moorby. 1969. Vascular transport and its significance in plant growth. *Ann. Rev. Plant Physiol.,* 20: 117–138.

Minshall, W., and V. A. Helson. 1949. The herbicidal action of oils. *Proc. Am. Soc. Hort. Sci.,* 53: 294–298.

Mirov, N. T. 1941. Distribution of growth hormone in shoots of two species of pine. *J. For.,* 39: 457–464.

Mitchell, J. W., and J. W. Brown. 1946. Movement of 2,4-D stimulus and its relation to the translocation of organic food materials in plants. *Bot. Gaz.,* 107: 393–407.

——— and C. L. Hamner. 1944. Polyethylene glycols as carriers for growth regulating substances. *Bot. Gas.,* 105: 474–483.

———, N. Mandava, J. R. Plimmer, J. F. Worley, and M. E. Drown. 1969. Plant growth properties of some acetone condensation products. *Nature,* 223: 1386–1387.

——— and W. A. Preston. 1953. Secondary galls and other plant growth modifying effects induced by translocated α-methoxyphenylacetic acid. *Science,* 118: 518–519.

———, ———, C. F. Krewson, C. H. Neufeld, T. F. Drake, and T. D. Fontaine. 1954. Synthetic plant-growth modifiers, IV. *Weeds,* 3: 28–37.

———, B. C. Smale, and R. L. Metcalf. 1960. Absorption and translocation of regulators and compounds used to control plant diseases and insects. *Adv. Pest. Control. Res.,* 3: 359–436.

Mitchell, R. C., and T. A. Villiers. 1972. Polysome formation in light controlled dormancy. *Plant Physiol.,* 50: 671–674.

Mitchell, W. D. 1971. 1-Arylbiurets as plant growth regulants. U.S. Patent 3,556,733. Jan. 19, 1971.

——— and S. H. Wittwer. 1962. Chemical regulation of flower sex expression and vegetative growth in *Cucumis sativa. Science,* 136: 880–881.

Mitra, R., and S. P. Sen. 1966. The dependence of flowering on nucleic acid and protein synthesis. *Plant Cell Physiol.,* 7: 167–169.

Mittelheuser, C. J., and R. F. M. van Steveninck. 1969. Stomatal closure and inhibition of transpiration by RS-abscisic acid. *Nature,* 221: 281–282.

Mittler, T. E. 1958. Studies on the feeding and nutrition of *Tuberolachnus. II:* Nitrogen and sugar composition of phloem sap and honeydew. *J. Exp. Biol.,* 35: 74–84.

Miura, G. A., and C. O. Miller. 1969. 6-γ,γ-Dimethylallylamino purine as a precursor of zeatin. *Plant Physiol.,* 44: 372–376.

Mizrahi, Y., A. Blumenfeld, S. Bittner, and A. E. Richmond. 1971. Abscisic acid and cytokinin contents of leaves in relation to salinity and relative humidity. *Plant Physiol.,* 48: 752–755.

———, ———, and A. E. Richmond. 1970. Abscisic acid and transpiration in leaves in relation to osmotic root stress. *Plant Physiol.,* 46: 169–171.

———, ———, and ———. 1972. The role of abscisic acid and salination in the adaptive response of plants to reduced aeration. *Plant Cell Physiol.,* 13: 15–21.

——— and A. E. Richmond. 1972. Hormonal modification of plant response to water stress. *Aust. J. Biol. Sci.,* 25: 437–442.

——— and A. E. Richmond. 1972. Abscisic acid in relation to mineral deprivation. *Plant Physiol.,* 50: 667–670.

Mohl, H. von. 1837. Untersuchungen über die anatomischen Verhältnisse des Chlorophylls. Diss. Univ. Tübingen.

Mohr, H. 1959. Der Lichteinfluss auf die Haarbildung am Hypokotyl von *Sinapis alba. Planta,* 53: 109–124.

———. 1960. The effects of long visible and near infrared radiation on plants. Pp. 44–49 in "Progress in Photobiology." Elsevier, Amsterdam.

———. 1966*a*. Untersuchungen zur phytochrominduzierten Photomorphogenese des Senfkeimlings (*Sinapis alba* L.). *Z. Pflanzenphysiol.,* 54: 63–83.

———. 1966*b*. Differential gene activation as a mode of action of phytochrome 730. *Photochem. Photobiol.,* 5: 469–483.

——— and A. Noble. 1960. Die Steurung der Schliessung und Öffnung des Plumul-hakens bei Keimlingen von *Lactuca sativa* durch sichtbare Strahlung. *Planta,* 55: 327–342.

——— and K. Ohlenroth. 1962. Photosynthese und Photomorphogenese bei Farnovorkeimen von *Drypopteris felix-mas. Planta,* 57: 656–664.

——— and M. Wehrung. 1960. Die Steuerung des Hypocotylwachstums bei den Keimlingen von *Lactuca sativa* durch sichtbare Strahlung. *Planta,* 55: 438–450.

Molisch, H. 1909. Warmbad und Pflanzentreiberei. *Oesterr. Gartenztg.,* 4: 17–23.

———. 1938. "The Longevity of Plants." Science Press, Lancaster, Pa. 226 pp.

Monsanto Chemical Company. 1972. *Monsanto Tech. Bull* 845.

Monselise, S. P. 1951. Growth analysis of citrus seedlings. I: Growth of sweet lime seedlings in dependence upon illumination. *Palest. J. Bot. (Rehovot Ser.),* 8: 54–75.

———, R. Goren, and A. H. Halevy. 1966. Effects of B-nine, cycocel and benzothiazole oxyacetate on flower bud induction of lemon trees. *Proc. Am. Soc. Hort. Sci.,* 89: 195–200.

Monsi, M. 1960. Dry-matter production in plants. I: Schemata of dry-matter reproduction. *Bot. Mag. (Tokyo),* 73: 81–90.

———, H. Iwaki, S. Kuraishi, T. Saeki, and N. Nomoto. 1962. Physiological and ecological analyses of shade tolerance of plants. 2: Growth of dark-treated greengram under varying light intensities. *Bot. Mag. (Tokyo),* 75: 185–194.

——— and T. Saeki. 1953. Über den Lichtfaktor in den Pflanzengesellschaften und seine Bedeutung für die Stoff produktion *Jap. J. Bot.,* 14: 22–52.

———. 1966. The photosynthesis and transpiration of crops. *Exp. Agric.,* 2: 1–14.

Monteith, J. L. 1965. Light distribution and photosynthesis in field crops. *Ann. Bot.,* 29: 17–38.

Montgomery, M., and V. H. Freed. 1961. The uptake, translocation and metabolism of simazine and atrazine by corn plants. *Weeds,* 9: 231–237.

Moorby, J., M. Ebert, and N. T. Evans. 1963. The translocation of ^{11}C labelled photosynthate in the soybean. *J. Exp. Bot.,* 14: 210–220.

Moore, J. F. 1959. Male sterility induced in tomato by sodium dichloroisobutyrate. *Science,* 129: 1738–1740.

Moore, P. H., H. B. Reid, and K. C. Hamner. 1967. Flowering responses of *Xanthium pennsylvanicum* to long dark periods. *Plant Physiol.,* 42: 503–509.

Moore, T. C., and E. K. Bonde. 1958. Interaction of gibberellic acid and vernalization in the dwarf telephone pea. *Physiol. Plant.,* 11: 752–759.

——— and ———. 1962. Physiology of flowering in peas. *Plant Physiol.,* 37: 149–153.

Morey, P. R., and J. Cronshaw. 1968. Developmental changes in the secondary xylem of *Acer rubrum* induced by various auxins and triiodobenzoic acid. *Protoplasma,* 65: 287–313.

Morgan, D. G. and H. Soding. 1958. Über die Wirkungsweise von Phthalsäuremono-α-naphthylamid

auf das Wachstum der Haferkoleoptile. *Planta,* 52: 235–249.
Morgan, P. W., and J. I. Durham. 1972. Abscission: Potentiating action of auxin transport inhibitors. *Plant Physiol.,* 50: 313–318.
——— and H. W. Gausman. 1966. Effects of ethylene on auxin transport. *Plant Physiol.,* 41: 45–52.
——— and W. C. Hall. 1962. Effect of 2,4-dichlorophenoxyacetic acid on the production of ethylene by cotton and grain sorghum. *Physiol. Plant.,* 15: 420–427.
Morinaga, T. 1926. Effect of alternating temperatures upon the germination of seeds. *Am. J. Bot.,* 13: 141–158.
Morré, D. J. 1968. Cell wall dissolution and enzyme secretion during leaf abscission. *Plant Physiol.,* 43: 1545–1559.
——— and A. C. Olson. 1965. An analysis of *Avena* coleoptile pectin fractions. *Can. J. Bot.,* 43: 1083–1095.
Mortimer, D. C. 1965. Translocation of the products of photosynthesis in sugar beet petioles. *Can. J. Bot.,* 43: 269–280.
Moser, B. C., and C. E. Hess. 1968. The physiology of tuberous root development in *Dahlia. Proc. Am. Hort. Sci.,* 93: 595–603.
Moskov, B. S. 1936. Die photoperiodische Reaktion der Blätter und die Möglichkeite einer Ausnützung derselben bei Pfropfungen. *Bull. Appl. Bot. Genet. Plant Breed.,* ser. A., 17: 25–30.
———. 1941. On the photoperiodic after-effects. *C. R. Dokl. Acad. Sci. URSS,* 31: 699–701.
Moss, D. N., R. B. Musgrave, and E. R. Lemon. 1961. Photosynthesis under field conditions. III: Some effects of light, carbon dioxide, temperature and soil moisture on photosynthesis, respiration, and transpiration of corn. *Crop Sci.,* 1: 83–87.
——— and H. P. Rasmussen. 1969. Cellular localization of CO^2 fixation and translocation of metabolites. *Plant Physiol.,* 44: 1063–1068.
Mothes, K. 1928. Die Wirkung des Wassermangels auf den Eiweissumsatz in höheren Pflanzen. *Ber. Dtsch. Bot. Ges.,* 46: 59–67.
———. 1959. Bemerkungen über isolierten Blättern. *Colloq. Ges. Physiol. Chem.,* 1959: 72–81.
———. 1960. Über das Altern der Blätter und die Möglichkeit ihrer Wiederverjungung. *Naturwiss.,* 47: 337–350.
———. 1961. Kinetin induced directed transport of substances in excised leaves in the dark. *Phytochemistry,* 1: 58–62.
———. 1964. The role of kinetin in plant regulation. Pp. 131–140 in "Régulateurs Naturels de la Croissance Végétale." CNRS, Paris.
——— and W. Baudisch. 1958. Untersuchungen über die Reversibilität der Ausbleichung grüner Blätter. *Flora,* 146: 521–531.
———, L. Engelbrecht, and O. Kulajewa. 1959. Über die Wirkung des Kinetins auf Stickstoffverteilung und Eiweiβsynthese in isolierten Blättern. *Flora (Jena),* 147: 445–464.
———, ———, and H. R. Schütte. 1961. Über die Akkumulation von α-Aminoisobuttersäure im Blattgewebe unter dem Einfluss von Kinetin. *Physiol. Plant.,* 14: 72–75.
Mousseron-Canet, M., J. C. Mani, J. P. Dalle, and J. L. Olive. 1966. Photoxydation sensibilisée de quelques composés apparentes à la dihydro-β-ionone, synthese de l'ester methylique de la abscisine. *Bull. Soc. Chim. Fr.,* 12: 3874–3878.
Muckadell, M. S. de. 1954. Juvenile stages in woody plants. *Physiol. Plant.,* 7: 782–796.
———. 1956. Experiments on development in *Fagus silvatica* by means of herbaceous grafting. *Physiol. Plant.,* 9: 396–400.
———. 1959. Investigations on aging of apical meristems in woody plants and its importance in sivliculture. Doctoral thesis, R. Vet. Agric. Coll., Copenhagen.
Muir, R. M. 1942. Growth hormones as related to the setting and development of fruit in *Nicotiana tabacum. Am. J. Bot.,* 29: 716–720.
———. 1947. The relationship of growth hormones and fruit development. *Proc. Natl. Acad. Sci. (U.S.),* 33: 303–312.
———. 1951. The growth hormone mechanism in fruit development. Pp. 357–364 in F. Skoog (ed.), "Plant Growth Substances." University of Wisconsin Press, Madison.
——— and B. P. Lantican. 1968. Purification and properties of the enzyme system forming indoleacetic acid. Pp. 259–272 in F. Wightman and G. Setterfield (eds.), "Biochemistry and Physiology of Plant Growth Substances." Runge Press, Ottawa.
Müller, B., and H. Ziegler. 1969. Die lichtinduzierte Aktivitätssteigerung der $NDAP^+$ abhängigen Glycerinaldehyd - 3 - Phosphat - Dehydrogenase. IX: Die Reaktion in isolierten Chloroplasten. *Planta,* 85: 96–104.
Muller, C. H., W. H. Muller, and B. L. Haines. 1964. Volatile growth inhibitors produced by aromatic shrubs. *Science,* 143: 471–473.
Müller, K., and A. C. Leopold. 1966. Correlative aging and transport of P^{32} in corn leaves under the influence of kinetin. *Planta,* 68: 167–185.
Müller-Thurgau, H. 1880. Über das Gefrieren und Erfrieren der Pflanze. *Landwirtsch. Jahrb. Schweiz.,* 9: 133–189.
———. 1882. Über Zuckeranhäufung in Pflanzentheilen in Folge niederer Temperatur. *Landwirtsch. Jahrb. Schweiz,* 11: 751–828.
Mullins, M. G. 1970. Hormone-directed transport of assimilates in decapitated internodes of *Phaseolus vulgaris* L. *Ann. Bot.,* 34: 897–909.
Mumford, F. E., D. H. Smith, and J. E. Castle. 1961. An inhibitor of indoleacetic acid oxidase from pea tips. *Plant Physiol.,* 36: 752–756.
———, H. M. Stark, and D. H. Smith. 1963. 4-

Hydroxybenzylalcohol, a naturally occurring cofactor of indoleacetic oxidase. *Phytochemistry,* 2: 215–220.
Münch, E. 1930. "Die Stoffbewegungen in der Pflanze." Gustav Fischer, Jena. 234 pp.
Murabaa, A. I. M. el. 1957. Factors affecting seed set in brussels sprouts, radish and cyclamen. *Meded. Landbouwhogesch. Wageningen,* 57: 1–33.
Murakami, Y. 1961. Paper-chromatographic studies on change in gibberellins during seed development and germination in *Pharbitis nil. Bot. Mag.,* 74: 241–247.
Murneek, A. E. 1926. Effects or correlation between vegetative and reproductive functions in the tomato. *Plant Physiol.,* 1: 3–56.
———. 1927. Physiology of reproduction in horticultural plants, *II. Mo. Agric. Exp. Stn. Res. Bull.* 106.
Musgrave, A., M. B. Jackson, and E. Ling. 1972. *Callitriche* stem elongation is controlled by ethylene and gibberellin. *Nature New Biol.,* 238: 93–96.
Muzik, T. J., and H. J. Cruzado. 1958. Transmission of juvenile rooting ability from seedlings to adults of *Hevea brasiliensis. Nature,* 181: 1288.
Myers, R. M. 1940. Effect of growth substances on the abscission layer in leaves of *Coleus. Bot. Gaz.,* 102: 323–338.
Nagao, M., Y. Esashi, T. Tanako, T. Kumagai, and S. Funkumoto. 1959. Effects of photoperiod and gibberellin on germination seeds of *Begonia. Plant Cell Physiol.,* 1: 39–48.
——— and E. Mitsui. 1959. Studies on the formation and sprouting of aerial tubers in *Begonia evansiana* Andr., III. *Sci. Rep. Tohoku Univ.,* 25: 199–205.
——— and Y. Ohwaki. 1968. Auxin transport in the elongation zone of *Vicia* roots. *Bot. Mag.,* 81: 44–45.
——— and N. Okagami. 1966. Effect of 2-chloroethyltrimethyl-ammonium chloride on the formation and dormancy of aerial tubers of *Begonia. Bot. Mag.,* 79: 687–692.
Nair, P. M., and L. C. Vining. 1965. Cinnamic acid hydroxylase in spinach. *Phytochemistry,* 4: 161–168.
Nakamura, T., T. Yamada, and N. Takahashi. 1966. Effect of gibberellic acid on the growth of the plumular hook section of etiolated pea seedling. *Bot. Mag.,* 79: 404–413.
Nakanishi, T., Y. Esashi, and K. Hinata. 1969. Control of self-incompatibility by CO_2 gas in *Brassica. Plant Cell Physiol.,* 10: 925–927.
Nakata, S. 1955. Floral initiation and fruit set in lychee, with special reference to the effect of sodium naphthaleneacetate. *Bot. Gaz.,* 117: 126–134.
——— and A. C. Leopold. 1967. Radioautographic study of translocation in bean leaves. *Am. J. Bot.,* 54: 769–772.
Nakayama, S. 1958. Photoreversible control of flowering at the start of inductive dark period in *Pharbitis nil. Ecol. Rev.,* 14: 325–326.
———, H. Tobita, and F. S. Okumura. 1962. Antagonism of kinetin and far-red light or indoleacetic acid in the flowering of *Pharbitis* seedlings. *Phyton,* 19: 43–48.
Napp-Zinn, K. 1957. Die Abhängigkeit des Vernalizationseffektes bei *Arabidopsis* von der Vorquellung der Samen sowie von Alter der Pflanzen. *Z. Bot.,* 45: 379–394.
———. 1960. Vernalisation, Licht und Alter bei *Arabidopsis thaliana,* I. *Planta,* 54: 409–444.
Nasr, T. A. S., and P. F. Wareing. 1961. Studies on flower initiation in black currant, I. *J. Hort. Sci.,* 36: 1–10.
Naylor, J. M., and G. M. Simpson. 1961. Dormancy studies in seed of *Avena fatua,* 2. *Can. J. Bot.,* 39: 281–295.
Neales, T. F. 1959. Effect of boron supply on the sugars soluble in 80% ethanol in flax seedlings. *Nature,* 183: 483.
———. 1970. Effect of ambient carbon dioxide concentration on the rate of transpiration of *Agave americana* in the dark. *Nature,* 228: 880–882.
——— and L. D. Incoll. 1968. The control of leaf photosynthesis rate by the level of assimilate concentration in the leaf: A review of the hypothesis. *Bot. Rev.,* 34: 107–125.
———, A. A. Patterson, and V. J. Hartney. 1968. Physiological adaptation to drought in the carbon assimilation and water loss of xerophytes. *Nature,* 219: 469–472.
———, K. J. Treharne, and P. F. Wareing. 1971. A relationship between net photosynthesis, diffusive resistance, and carboxylating enzyme activity in bean leaves. Pp. 89–96 in M. D. Hatch, C. B. Osmond and R. O. Slatyer (eds.), "Photosynthesis and Photorespiration." Wiley-Interscience, New York.
Necessany, V. 1958. Effect of β-indoleacetic acid on the formation of reaction wood. *Phyton,* 11: 117–127.
Negi, S., and P. Olmo. 1966. Sex conversion in a male *Vitis vinifera* by kinin. *Science,* 152: 1624–1625.
Neish, A. C. 1965. Coumarins, phenylpropanes and lignin. Pp. 581–617 in J. Bonner and J. Varner (eds.), "Plant Biochemistry." Academic, New York.
Neljubow, D. 1901. Uber die horizontale Nutation der Stengel von *Pisum sativum* und einigen anderen Pflanzen. *Beih. Bot. Zentralbe.,* 10: 128–138.
Nelson, C. D. 1962. The translocation of organic compounds in plants. *Can. J. Bot.,* 40: 757–770.
———. 1963. Effect of climate on the distribution and translocation of assimilates. Pp. 149–174 in L. T. Evans (ed.), "Environmental Control of Plant Growth." Academic, New York.
———. 1964. The production and translocation of photosynthate ^{14}C in conifers. Pp. 243–257 in M. H. Zimmermann (ed.), "The Formation of Wood in Forest Trees." Academic, New York.
———, H. J. Perkins, and P. R. Gorham. 1959*a*. Note on a rapid translocation of photosynthetically assimilated ^{14}C out of the primary leaf of the young soybean plant. *Can. J. Biochem. Physiol.,* 36: 1277–1279.

———, ———, and ———. 1959*b*. Evidence for different kinds of concurrent translocation of photosynthetically assimilated ^{14}C in the soybean plant. *Can. J. Bot.*, 37: 1181–1189.

Nelson, L. E. 1967. Effect of root temperature variation on growth and transpiration of cotton (*Gossypium hirsutum* L.) seedlings. *Agron. J.*, 59: 391–395.

Nelson, P. E., and J. E. Hoff. 1968. Food volatiles: Gas chromatographic determination of partition coefficents in water-lipid systems. *J. Food Sci.*, 33: 479–482.

Nelson, R. C., and R. B. Harvey. 1935. The presence in self-blanching celery of unsaturated compounds similar to ethylene. *Science,* 82: 133–134.

Nemec, B. 1900. Über die Art der Wahrnehmung des Schwerkraftreizes bei den Pflanzen. *Ber. Dtsch. Bot. Ges.,* 18: 241–245.

Niedergang, E., and F. Skoog. 1956. Studies on polarity and auxin transport in plants. I: Modification of polarity and auxin transport by triodobenzoic acid. *Physiol. Plant.,* 9: 60–73.

Niedergang-Kamien, E., and A. C. Leopold. 1957. Inhibitors of polar transport. *Physiol. Plant.,* 10: 29–38.

Nielsen, K. F., and E. C. Humphries. 1966. Effects of root temperature on plant growth. *Soils Fert.*, 29: 1–7.

Nir, I., and A. Poljakoff-Mayber. 1967. Effect of water stress on the photochemical activity of chloroplasts. *Nature,* 213: 418–419.

Nissl, D., and M. Zenk. 1969. Evidence against induction of protein synthesis during auxin-induced initial elongation of *Avena coleoptiles. Planta,* 80: 323–341.

Nitsan, J. 1962. Accumulation of distinct macromolecular components in rye embryos as a result of vernalization. *Nature,* 194: 400–401.

——— and A. Lang. 1965. Inhibition of cell division and elongation in higher plants by inhibitors of DNA synthesis. *Dev. Biol.,* 12: 358–376.

——— and ———. 1966. DNA synthesis in the elongating nondividing cells of the lentil epicotyl and its promotion by gibberellin. *Plant Physiol.,* 41: 965.

Nitsch, C. 1968. Effects of growth substances on the induction of flowering of a short-day plant in vitro. Pp. 1385–1398 in F. Wightman and G. Setterfield (eds.), "Biochemistry and Physiology of Plant Growth Substances." Runge Press, Ottawa.

——— and J. P. Nitsch. 1963. Étude du mode d'acide gibbérellique au moyen de gibbérelline Aè marquée, I. *Bull. Soc. Bot. Frc.* 110: 7–17.

Nitsch, J. P. 1950. Growth and morphogenesis of the strawberry as related to auxin. *Am. J. Bot.,* 37: 211–215.

———. 1952. Plant hormones in the development of fruits. *Q. Rev. Biol.,* 27: 33–57.

———. 1957. Growth response of woody plants to photoperiodic stimuli. *Proc. Am. Soc. Hort. Sci.,* 70: 512–525.

———. 1963. Fruit development. Pp. 361–394 in P. Maheshwari (ed.), "Recent Advances in the Embryology of Angrosperms." International Society of Plant Morphology, Delhi.

———. 1966. Photoperiodisme et tubérisation. *Bull. Soc. Fr. Phys. Veget.,* 12: 233–246.

———. 1970. Hormonal factors in growth and development. Pp. 427–472 in A. C. Hulme (ed.), "The Biochemistry of Fruits and Their Products." Academic, London.

———, E. B. Kurtz, J. L. Liverman, and F. W. Went. 1952. The development of sex expression in cucurbit flowers. *Am. J. Bot.,* 39: 32–43.

——— and C. Nitsch. 1961. Growth factors in the tomato fruit. Pp. 687–705 in R. M. Klein (ed.), "Plant Growth Regulation." Iowa State University Press, Ames.

——— and ———. 1962. Composés phénoliques et croissance végétale. *Ann. Physiol. Veg.,* 4: 211–225.

———, C. Pratt, C. Nitsch, and N. J. Shaulis. 1960. Natural growth substances in Concord and Concord Seedless grapes in relation to berry development. *Am. J. Bot.,* 47: 566–576.

Njoku, E. 1958. Effect of gibberellic acid on leaf form. *Nature,* 182: 1097–1098.

Nobel, P. S. 1970. Relation of light-dependent potassium uptake by pea leaf fragmanets to the p*K* of the accompanying organic acid. *Plant Physiol.,* 46: 491–493.

Nooden, L. D. 1972. Inhibition of nucleic acid synthesis by maleic hydrazide. *Plant Cell Physiol.,* 13: 609–621.

——— and K. V. Thimann. 1965. Inhibition of protein synthesis and auxin-induced growth by chloramphenicol. *Plant Physiol.,* 40: 193–201.

Norman, J. M., and C. B. Tanner. 1969. Transient light measurements in plant canopies. *Agron. J.,* 61: 847–849.

Norris, R. F., and M. J. Bukovac. 1969. Some physical-kinetic considerations in penetration of naphthaleneacetic acid. *Physiol. Plant.,* 22: 701–712.

Nutile, G. E. 1945. Inducing dormancy in lettuce seed with coumarin. *Plant Physiol.,* 20: 433–442.

Nutman, F. G. 1937. Studies on the physiology of *Coffea arabica. II: Stomatal movements in relation to photosynthesis under natural conditions. Ann. Bot., n.s., 1: 681–694.*

Nuttonson, M. Y. 1948. Some preliminary observations of phenological data. Pp. 129–145 in A. E. Murneek and R. O. Whyte (eds.), "Vernalization and Photoperiodism." Chronica Botanica, Waltham, Mass.

O'Brien, D. G., and E. G. Prentice. 1930. An eelworm disease of potatoes caused by *Heterodera schachtii. Scot. J. Agric.,* 13: 413–432; cited by Foy et al. (1971.).

O'Brien, T. J., B. C. Jarvis, J. H. Cherry, and J. B. Hanson. 1968. Enhancement by 2,4-dichlorophenoxyacetic acid of chromatin RNA polymerase in soybean hypocotyl tissue. *Biochem. Biophys. Acta,* 169: 35–43.

O'Brien, T. P., S. Zee, and J. G. Swift. 1970. The occurrence of transfer cells in the vascular tissues of the coleoptilar node of wheat. *Aust. J. Biol. Sci.,* 23: 709–712.

Oda, Y. 1962. Effect of light quality on flowering of *Lemna perpuisilla* 6746. *Plant Cell Physiol.,* 3: 415–417.

Ogawa, Y. 1960. Über die Auslösung der Blütenbildung von *Pharbitis nil* durch niedere Temperatur. *Bot. Mag.,* 73: 334–335.

———. 1961. Über die Wirkung von Kinetin auf die Blütenbildung von *Pharbitis nil. Plant Cell Physiol.,* 2: 343–359.

———. 1962. Weitere Untersuchungen über die Wirkung von gibberellinähnlichen Substanzen auf die Blütenbildung von *Pharbitis nil. Plant Cell Physiol.,* 3: 5–21.

———. 1963. Changes in the content of gibberellin-like substances in ripening seed of *Lupinus. Plant Cell Physiol.,* 4: 85–94.

——— and S. Imamura. 1958. Über die fordernde Wirkung von Samendiffusat auf die Blütenbildung von *Pharbitis nil. Proc. Jap. Acad.,* 34: 631–632.

Ohkuma, K., F. T. Addicott, O. E. Smith, and W. E. Thiesson. 1965. The structure of abscissin, II. *Tetrahedron Lett.,* 29: 2529–2535.

Okazawa, Y. 1960. Studies on the relation between the tuber formation of potato and its natural gibberellin content. *Proc. Crop Sci. Soc. Jap.,* 29: 121–124.

———, N. Katsura, and T. Tagawa. 1967. Effects of auxin and kinetin on the development and differentiation of potato tissue cultured in vitro. *Physiol. Plant.,* 20: 862–869.

O'Kelley, J. C. 1955. External carbohydrates in growth and respiration of pollen tubes. *Am. J. Bot.,* 42: 322–327.

Olson, A. O., and M. Spencer. 1968. Studies on the mechanism of action of ethylene, II. *Can. J. Biochem.,* 46: 283–288.

Ondok, J. P. 1971. Calculation of mean leaf area ratio in growth analysis. *Photosynthetica,* 5: 269–271.

——— and J. Kvet. 1971. Integral and differential formulae in growth analysis. *Photosynthetica,* 5: 358–363.

Ordin, L., and J. Bonner. 1957. Effect of galactose on growth and metabolism on *Avena* coleoptile sections. *Plant Physiol.,* 32: 212–215.

———, R. Cleland, and J. Bonner. 1955. Influence of auxin on cell wall metabolism. *Proc. Natl. Acad. Sci. (U.S.),* 41: 1023–1029.

Osborne, D. J. 1958. Changes in the distribution of pectin methylesterase across leaf abscission zones of *Phaseolus vulgaris. J. Exp. Bot.,* 9: 446–457.

——— 1962*a*. Effect of kinetin on protein and nucleic acid metabolism in *Xanthium* leaves during senescence. *Plant Physiol.,* 37: 595–602.

——— and M. K. Black. 1964. Polar transport of a kinin, benzyladenine. *Nature,* 201: 97.

———, G. E. Blackman, S. Novoa, F. Sudzuke, and R. G. Powell. 1955. The physiological activity of 2,6-substituted phenoxyacetic acids. *J. Exp. Bot.,* 6: 392–408.

——— and M. Halloway. 1964. The auxin 2,4-dichlorophenoxyacetic acid as a regulator of protein synthesis and senescence in detached leaves of *Prunus. New Phytol.,* 63: 334–347.

——— and D. R. McCalla. 1961. Rapid bioassay for kinetin and kinins using senescing leaf tissue. *Plant Physiol.,* 36: 219–221.

——— and F. W. Went. 1953. Climatic factors influencing parthenocarpy and normal fruit set in tomatoes. *Bot. Gaz.,* 114: 312–322.

Osipova, O. P., and N. I. Ashur. 1965. Structure of chloroplasts of leaves of corn cultivated under different conditions of illumination. *Fiziol. Rast.,* 12(2): 257–262.

Osman, A. M. 1971. Dry-matter production of a wheat crop in relation to light interception and photosynthetic capacity of leaves. *Ann. Bot.,* 35: 1017–1035.

——— and F. L. Milthorpe. 1971*a*. Photosynthesis of wheat leaves in relation to age, illuminance and nutrient supply. I: Techniques. *Photosynthetica,* 5: 55–60.

——— and ———. 1971*b*. Photosynthesis of wheat leaves in relation to age, illuminance and nutrient supply. II: Results. *Photosynthetica,* 5: 61–70.

Ota, J. 1925. Continuous respiration studies of dormant seeds of *Xanthium. Bot. Gaz.,* 80: 288–299.

Overbeek, J. van. 1956. Absorption and translocation of plant regulators. *Annu. Rev. Plant Physiol.,* 7: 355–372.

———. 1946. Control of flower formation and fruit size in the pineapple. *Bot. Gaz.,* 108: 64–73.

———. 1966. Plant hormones and regulators. *Science,* 152: 721.

——— and R. Blondeau. 1954. Mode of action of phytotoxic oils. *Weeds,* 3: 55–65.

———, ———, and V. Horne. 1951. Transcinnamic acids as an anti-auxin. *Am. J. Bot.,* 38: 589–595.

———, M. E. Conklin, and A. F. Blakeslee. 1942. Cultivation *in vitro* of small *Datura* embryos. *Am. J. Bot.,* 29: 472–477.

———, S. A. Gordon, and L. E. Gregory. 1946. An analysis of the function of the leaf in the process of root formation in cuttings. *Am. J. Bot.,* 33: 100–107.

———, J. E. Loeffler, and M. I. R. Mason. 1967. Dormin, inhibitor of plant DNA synthesis? *Science,* 156: 1497–1499.

———, D. Olivio, and E. M. S. de Vasquez. 1945. Rapid extraction method for free auxin and its application in geotropic reactions of bean and sugarcane. *Bot. Gaz.,* 106: 440–451.

Overland, L. 1960. Endogenous rhythm in opening

and odor of flowers of *Cestrum nocturnum*. *Am. J. Bot.*, 47: 378–382.
Paál, A. 1919. Über phototropische Reizleitung. *Jahrb. Wiss. Bot.*, 58: 406–458.
Pakianathan, S. W. 1970. The search for new stimulants. *Rubber Res. Inst. Malaya Plant. Conf.*, preprint E.
Paleg, L. G. 1960. Physiological effects of gibberellic acid, II. *Plant Physiol.*, 35: 902–906.
———. 1961. Physiological effects of gibberellic acid, III. *Plant Physiol.*, 36: 829–837.
——— and D. Aspinall. 1970. Field control of plant growth and development through the laser activation of phytochrome. *Nature,* 228: 970–973.
———, H. Kende, H. Ninnemann, and A. Lang. 1965. Physiological effects of gibberellic acid. VIII: *Plant Physiol.*, 40: 165–169.
——— and R. M. Muir. 1959. Neutral and acidic auxins in developing tobacco fruit. *Aust. J. Biol. Sci.*, 12: 340–343.
———, D. H. B. Sparrow, and A. Jennings. 1962. Physiological effects of gibberellic acid, IV. *Plant Physiol.*, 37: 579–583.
Pallas, J. E., Jr., and H. H. Mollenhauer. 1972. Physiological implications of *Vicia faba* and *Nicotiana tabacum* guardcell ultrastructure. *Am. J. Bot.*, 59: 504–514.
Palmer, C. E., and O. E. Smith. 1969. Effect of abscisic acid on elongation and kinetin-induced tuberization of isolated stolons of *Solanum tuberosum*. *Plant Cell Physiol.*, 10: 657–664.
Palmer, J. M. 1966. The influence of growth regulating substances on the development of enhanced metabolic rates in thin slices of beetroot. *Plant Physiol.*, 41: 1173–1178.
Palmer, R. L., L. N. Lewis, H. Z. Hield, and J. Kumamoto. 1967. Abscission induced by conversion of betahydroxyethylhydrazine to ethylene. *Nature,* 126: 1216–1217.
Pandy, K. K. 1967. Origin of genetic variability: Combinations of peroxidase isozymes determine multiple allelism of the s gene. *Nature,* 213: 669–672.
Pardee, A. B. 1968. Membrane transport proteins. *Science,* 162: 632–637.
Park, R. B. 1965. Substructure of chloroplast lamellae. *J. Cell Biol.*, 27: 151–161.
Parker, C. W., and D. S. Letham. 1973. Regulators of cell division in plant tissues. XVI. *Planta,* 114: 199–218.
Parker, M. W. and H. A. Borthwick. 1950. Influence of light on plant growth. *Ann. Rev. Plant Physiol.*, 1: 43–58.
———, S. B. Hendricks, H. A. Borthwick, and N. J. Scully. 1946. Action spectrum for the photoperiodic control of floral initiation of short day plants. *Bot. Gaz.*, 108: 1–26.
———, ———, ———, and F. W. Went. 1949. Spectral sensitivities for leaf and stem growth of etiolated pea seedlings. *Am. J. Bot.*, 36: 194–204.
Parkinson, K. J. 1970. The effects of silicon coatings on leaves. *J. Exp. Bot.*, 21: 566–579.
——— and H. L. Penman. 1970. A possible source of error in the estimation of stomatal resistance. *J. Exp. Bot.*, 21: 405–409.
Parlange, J., and P. E. Waggoner. 1972. Boundary layer resistance and temperature distribution on still and flapping leaves. II: Field experiments. *Plant Physiol.*, 50: 60–63.
Partheir, B., and R. Wollgiehn. 1961. Über den Einfluss des kinetin auf den Eiweiss- und Nukleinsīure Stoffwechsel in isolierten Tabaksblattern. *Ber. Dtsch. Bot. Ges.*, 74: 47–51.
Passecker, F. 1949. Zur Frage der Jungendformen der Apfel. *Züchter*, 19: 311.
Patau, K., N. K. Das, and F. Skoog. 1957. Induction of DNA synthesis by kinetin and indoleacetic acid in excised tobacco pith tissue. *Physiol. Plant.*, 10: 949–966.
Pate, J. S., and B. E. S. Gunning. 1969. Vascular transfer cells in angiosperm leaves: A taxonomic and morphological survey. *Protoplasma,* 68: 135–156.
Patefield, W. M., and R. B. Austin. 1971. A model for the simulation of the growth of *Beta vulgaris* L. *Ann. Bot. (Lond.),* 35: 1227–1250.
Patil, S. S., and C. S. Tang. 1974. Inhibition of ethylene evolution in papaya pulp tissue by benzyl isothiocyanate. *Plant Physiol.*, 53: 585–588.
Pawar, S. S., and H. C. Thompson. 1950. The effect of age and size of plant at the time of exposure of low temperature on reproductive growth in celery. *Proc. Am. Soc. Hort. Sci.*, 55: 367–371.
Pearce, H. L. 1943. The effect of nutrition and phytohormones on the rooting of *Vinca* cuttings. *Ann. Bot.*, n.s., 7: 123–132.
Pearce, R. B., R. H. Brown, and R. E. Blaser. 1967. Photosynthesis in plant communities as influenced by leaf angle. *Crop Sci.*, 7: 321–324.
Pearman, G. I., and J. R. Garratt. 1972. Global aspects of carbon dioxide. *Search,* 3: 67–73.
Pearson, A., and P. F. Wareing. 1969. Effects of abscisic acid on activity of chromatin. *Nature,* 221: 672–673.
Pearson, J. A., and R. N. Robertson. 1952. The climacteric rise in respiration of fruit. *Aust. J. Sci.*, 15: 99–100.
Peel, A. J. 1970. Further evidence for the relative immobility of water in sieve tubes of willow. *Physiol. Plant.*, 23: 667–672.
———, R. J. Field, C. L. Coulson, and D. C. Gardner. 1969. Movement of water and solutes in sieve tubes of willow in response to puncture by aphid stylets: Evidence against a mass flow of solution. *Physiol. Plant.* 22: 768–775.
Penman, H. L. 1942. Theory of porometers used in the study of stomatal movement in leaves. *Proc. Roy. Soc.*, B130: 416–433.
———. 1970. The water cycle. *Sci. Am.*, 1970 (Sept.): 99–108.

Penner, D., and F. M. Ashton. 1966. Proteolytic enzyme control in squash cotyledons. *Nature,* 212: 935.

——— and ———. 1967. Hormonal control of proteinase activity in squash cotyledons. *Plant Physiol.,* 42: 791–796.

Penny, P. 1971. Growth-limiting proteins in relation to auxin-induced elongation in lupin hypocotyls. *Plant Physiol.,* 48: 720–723.

Peterkovsky, A. 1968. The incorporation of mevalonic acid into the isopentenyl adenosine of transfer RNA in *Lactobacillus. Biochemistry,* 7: 42.

Peterson, L. W., G. E. Kleinkopf, and F. C. Huffaker. 1973. Evidence for a lack of turnover of ribulose 1,5-diphosphate carboxylase in barley leaves. *Plant Physiol.,* 51: 1042–1045.

Peterson, P. A. 1955. Dual cycle of avocado flowers. *Calif. Agric.,* 1955 (Oct.): 6–13.

Petrie, A. H. K., R. Watson, and E. D. Ward. 1939. Physiological ontogeny in the tobacco plant, I. *Aust. J. Exp. Biol. Med. Sci.,* 17: 93–122.

Pfeffer, W. 1873. "Physiologische Untersuchungen." Leipzig; cited by Bünning (1959).

———. 1904. "Pflanzenphysiologie," vol. II, "Kraftwechsel," 2d ed. Pp. 247–278. Leipzig.

———. 1906. "The Physiology of Plants," vol. III. Trans. A. J. Ewart. Clarendon Press, Oxford.

———. 1915. Beitrage zur Kenntnis der Entstehung der Schlafbewegungen. *Ahb. K. Sachs Ges. Wiss. Math. Phys.,* 34: 1–154; cited by Bünning (1959).

Pharis, R. P. 1972. Flowering of *Chrysanthemum* under noninductive long days by gibberellins and benzyladenine. *Planta,* 105: 205–212.

———, H. Hellmers, and E. Schnurmans. 1967. Kinetics of the daily rate of photosynthesis at low temperatures for two conifers. *Plant Physiol.,* 42: 525–531.

——— and W. Morf. 1968. Physiology of gibberellin-induced flowering in conifers. Pp. 1341–1356 in F. Wightman and G. Setterfield (eds.), "Biochemistry and Physiology of Plant Growth Substances." Runge Press, Ottawa.

Pharr, D. M., and A. A. Kattan. 1971. Effects of air flow rate, temperature and maturity on respiration and ripening of tomato fruits. *Plant Physiol.,* 48: 53–55.

Philip, J. R. 1958. Propagation of turgor and other properties through cell aggregations. *Plant Physiol.,* 33: 271–274.

Phillips, I. D. J. 1961. Induction of light requirement for germination of lettuce seed by naringenin and its removal by gibberellic acid. *Nature,* 192: 240–241.

——— and P. F. Wareing. 1958. Studies in dormancy of sycamore, I. *J. Exp. Bot.,* 9: 350–364.

——— and ———. 1959. Studies in dormancy of sycamore. II. *J. Exp. Bot.,* 10: 504–514.

Phinney, B. O. 1956. Growth response of single-gene dwarf mutants in maize to gibberellic acid. *Proc. Natl. Acad. Sci. (U.S.),* 42: 185–189.

———, West, C. A., M. Ritzel, and P. M. Neely. 1957. Evidence for "gibberellin-like" substances from flowering plants. *Proc. Natl. Acad. Sci. (U.S.),* 43: 398–404.

Pickard, B. G. 1969. Second positive phototropic response patterns of the oat coleoptile. *Planta,* 88: 1–33.

——— and K. V. Thimann. 1966. Geotropic response of wheat coleoptiles in absence of amyloplast starch. *J. Gen. Physiol.,* 49: 1065–1086.

Pieniazek, J., and L. S. Jankiewicz. 1965. Acropetal and basipetal transmission of 6-benzylaminopurine effect in dormant apple seedlings. *Bull. Acad. Pol. Sci.,* 13: 607–609.

Pijl, L. van der. 1969. "Principles of Dispersal in Higher Plants." Springer-Verlag, Berlin. 153 pp.

Pilet, P. E. 1951*a*. Contribution a l'etude des hormones de croissance (auxines) dans la racine de *Lens culinaris. Mem. Soc. Vaud. Sci. Nat.,* 10: 137–244.

———. 1951*b*. Etude de la circulation des auxines dans la racine de *Lens culinaris. Bull. Soc. Bot. Suisse,* 61: 410–424.

———. 1957. Action des gibberellines sur l'activité auxines-oxydasique de tissus cultivés *in vitro. C. R. Acad. Sci. (Paris),* 245: 1327–1328.

———. 1958. Action de l'indole sur la destruction des auxines en relation avec la senescense cellulaire. *C. R. Acad. Sci. (Paris),* 246: 1896–1898.

Pilet, P. E. 1960. Le catabolisme auxinique. *Bull. Soc. Fr. Physiol. Veg.,* 6: 119–137.

———. 1972. Geoperception et georéaction rasinaires. *Physiol. Veg.,* 10: 347–367.

——— and S. Meylan. 1953. Polarité electrique, auxines et physiologie des racines du *Lens culinaris. Bull. Soc. Bot. Suisse,* 63: 430–465.

Plaisted, P. H. 1957. Growth of the potato tuber. *Plant Physiol.,* 32: 445–453.

Plaut, Z., and L. Reinhold. 1965. The effect of water stress on (^{14}C) sucrose transport in bean plants. *Aust. J. Biol. Sci.,* 18: 1143–1155.

Plumb, R. C., and W. B. Bridgman. 1972. Ascent of sap in trees. *Science,* 176: 1129–1131.

Plummer, T. H., and A. C. Leopold. 1957. Chemical treatment for bud formation in *Saintpaulia. Proc. Am. Soc. Hort. Sci.,* 70: 442–444.

Poapst, P. A., A. B. Durkee, W. A. McGugan, and F. B. Johnston. 1968. Identification of ethylene in gibberellic acid treated potatoes. *J. Sci. Food Agric.,* 19: 325.

Poidevic, N. 1965. Inhibition of the germination of mustard seeds by unsaturated fatty acids. *Phytochemistry,* 4: 525.

Poljakoff-Mayber, A., S. Goldschmidt Blumenthal, and M. Evenari. 1957. The growth substance content of germinating lettuce seeds. *Physiol. Plant.,* 10: 14–19.

Pollard, C. J., and B. N. Singh. 1968. Early effects of gibberellic acid on barley aleurone layers. *Biochem. Biophys. Res. Commun.*, 33: 321–326.

Pollock, B. M. 1953. The respiration of *Acer* buds in relation to the inception and termination of winter rest. *Physiol. Plant.*, 6: 47–64.

———. 1960. Studies of rest period, III. *Plant Physiol.*, 35: 975–977.

———. 1962. Temperature control of physiological dwarfing in peach seedlings. *Plant Physiol.*, 37: 190–197.

——— and J. R. Manolo. 1970. Simulated mechanical damage to garden beans during germination. *J. Am. Soc. Hort. Sci.*, 95: 415–417.

Pollock, J. R. A., B. H. Kirsop, and R. E. Essery. 1955. Studies in barley and malt, IV. *Inst. Brew. J.*, 61: 301–307.

Poovaiah, B. W., and A. C. Leopold. 1973. Deferral of leaf senescence with calcium. *Plant Physiol.*, 52: 236–239.

Porter, H. K. 1966. Leaves as collecting and distributing agents of carbon. *Aust. J. Sci.*, 29: 31–40.

——— and W. R. Rees. 1954. Some effects of ethanol extracts of potatoes on the activity of phosphorylase. *Plant Physiol.*, 29: 514–520.

Porter, W. L., and K. V. Thimann. 1965. Molecular requirements for auxin action, I. *Phytochemistry*, 4: 229–243.

Possingham, J. V., and W. Saurer. 1969. Changes in chloroplast number per cell during leaf development in spinach *(Spinacea oleracea)*. *Planta*, 86: 186–194.

Post, K. 1935. Some effects of temperature and light upon the flower bud formation and leaf character of stock. *Proc. Am. Soc. Hort. Sci.*, 33: 649–652.

Postlethwait, S. N., and B. Rogers. 1958. Tracing the path of the transpiration stream in trees by the use of radioactive isotopes. *Am. J. Bot.*, 45: 753–757.

Pratt, H. K. 1954. Direct chemical proof of ethylene production by detached leaves. *Plant Physiol.*, 29: 16–18.

——— and J. D. Goeschl. 1969. Physiological roles of ethylene in plants. *Annu. Rev. Plant Physiol.*, 20: 541–584.

Pratt, L. H., and W. R. Briggs. 1966. Photochemical and non-photochemical reactions of phytochrome in vivo. *Plant Physiol.*, 41: 467–474.

Preiss, J., and T. Kosuge. 1970. Regulation of enzyme activity in photosynthetic systems. *Annu. Rev. Plant Physiol.*, 21: 433–466.

Prendeville, G. N. 1968. Shoot zone uptake of soil-applied herbicides. *Weed Res.*, 8: 106–114.

Presley, H. J., and L. Fowden. 1965. Acid phosphatase and isocitritase production during seed germination. *Phytochemistry*, 4: 169–176.

Pressey, R., and R. Shaw. 1966. Effect of temperature on invertase, invertase inhibitor and sugars in potato tubers. *Plant Physiol.*, 41: 1657–1661.

Preston, R. D. 1958. The ascent of sap and the movement of soluble carbohydrates in stems of higher plants. Pp. 366–382 in D. H. Everett and F. S. Stone (eds.), "The Structure and Properties of Porous Materials." Butterworth, London.

Pridham, A. M. S. 1947. Effect of 2,4-D on bean progeny seedlings. *Science*, 105: 412.

Pridham, J. B. 1965. Low molecular weight phenols in higher plants. *Annu. Rev. Plant Physiol.*, 16: 13–36.

Priestley, J. 1776. "Experiments and Observations on Different Kinds of Air." London.

Priestly, J. H. 1920. The mechanism of root pressure. *New Phytol.*, 19: 189–200.

Puckridge, D. W. 1972. Photosynthesis of wheat under field conditions, V. *Aust. J. Agric. Res.*, 23: 397–404.

Purvis, O. N. 1948. Studies in vernalization, XI. *Ann. Bot.*, 12: 183–206.

———. 1960. Effect of gibberellin on the flower initiation and stem extension in petkus winter rye. *Nature*, 185: 479.

——— and F. G. Gregory. 1937. Studies in vernalization of cereals, I. *Ann. Bot.*, 1: 569–592.

——— and ———. 1945. Devernalization by high temperature. *Nature*, 155: 113.

Quatrano, R. S. 1968. Rhizoid formation in *Fucus* zygotes; dependence on protein and ribonucleic acid synthesis. *Science*, 162: 468–470.

Quebedeaux, B., and E. M. Beyer. 1972. Chemically induced parthenocarpy in cucumber by a new inhibitor of auxin transport. *HortScience*, 7: 474–476.

Quinlan, J. D., and R. J. Weaver. 1969. Influence of benzyladenine, leaf darkening and ringing on movement of ^{14}C-labelled assimilates into expanding leaves of *Vitis vinifera*. *Plant Physiol.*, 44: 1247–1252.

Qureshi, F. A., and D. C. Spanner. 1973*a*. The effect of nitrogen on the movement of tracers down the stolon of *Saxifraga sarmentosa*, with some observations on the influence of light. *Planta*, 110: 131–144.

——— and ———. 1973*b*. Movement of (^{14}C) sucrose along the stolon of *Saxifraga sarmentosa*. *Planta*, 110: 145–152.

——— and ———. 1973*c*. The influence of dinitrophenol on phloem transport along the stolon of *Saxifraga sarmentosa*. *Planta*, 111: 1–12.

Rabideau, G. S., C. S. French, and A. S. Holt. 1946. The absorption and reflection spectra of leaves, chloroplast suspensions and chloroplast fragments as measured in Ulbricht sphere. *Am. J. Bot.*, 33: 769–777.

Rabinowitch, E. I. 1951*a*. "Photosynthesis and Related Processes." Interscience, New York.

———. 1951*b*. "Photosynthesis," vol. 2, pt. 1. Interscience, New York.

———. 1956. "Photosynthesis and Related Processes," 2d ed. Interscience, New York.

——— and Govindjee. 1969. "Photosynthesis." Wiley, New York.

Racusen, D. 1955. Formation of indole-3-aldehyde by indoleacetic oxidase. *Arch. Biochem.,* 58: 508–509.
Radin, J. W., and R. S. Loomis. 1969. Ethylene and carbon dioxide in the growth and development of cultured radish roots. *Plant Physiol.,* 44: 1584–1589.
Radley, M. 1963. Gibberellin content of spinach in relation to photoperiod. *Ann. Bot.,* 27: 373–377.
Ragab, M. T. H., and J. P. McCollum. 1960. Degradation of C^{14}-labeled simazine by plants and soil microorganisms. *Weeds,* 9: 72–84.
Railton, I. D., and P. F. Wareing. 1973. Effects of abscisic acid on the levels of endogenous gibberellin-like substances in *Solanum andigena. Planta,* 112: 65–70.
——— and ———. 1973. Effects of daylength on endogenous gibberellins in leaves of *Solanum.* I, II. *Physiol. Plant,* 28: 88–94, 127–131.
Raju, P. V., and V. S. R. Das. 1968. Natural and GA-induced changes in levels of gibberellin-like substances and nucleic acids during growth of the pepper fruit. *Z. Pflanzenphysiol.,* 58: 266–276.
Rakitin, Y. V., A. V. Krylov, and K. G. Garaeva. 1957. Concerning the distribution and transformation of the methyl ester of naphthalene acetic acid in potato tubers. *Dokl. Biol. Sci. Sec.,* 116: 696.
Rappaport, L., S. Blumenthal-Goldschmidt, M. Clegg, and O. E. Smith. 1965. Regulation of bud rest in tubers of potato, I. *Plant Cell Physiol.,* 6: 587–600.
——— and M. Sachs. 1967. Wound-induced gibberellins. *Nature,* 214: 1149–1150.
——— and N. Wolf. 1969. The problem of dormancy in potato tubers and related structures. *Soc. Exp. Biol. Symp.,* 23: 219–240.
Raschke, K. 1956. Über die physikalischen Beziehungen zwischen Wärmeübergangszahl, Strahlungsaustausch, Temperatur und Transpiration eines Blattes. *Planta,* 48: 200–238.
———. 1967. Der Einfluss von Rot and Blaulicht auf die Öffnungs und Schliess-Geschwindigkeit der Stomata von *Zea mays. Naturwiss.,* 54: 72–73.
———. 1970. Stomatal responses to pressure changes and interruptions in the water supply of detached leaves of *Zea mays* L. *Plant Physiol.,* 45: 415–423.
Ray, P. M. 1962. Destruction of indoleacetic acid. IV: Kinetics of enzymic oxidation. *Arch. Biochim. Biophys.,* 96: 199–209.
———, P. B. Green, and R. Cleland. 1972. Role of turgor in plant cell growth. *Nature (Lond.),* 239: 163–164.
——— and A. W. Ruesink. 1962. Kinetic experiments on the nature of the growth mechanism in oat coleoptile cells. *Dev. Biol.,* 4: 377–397.
——— and K. V. Thimann. 1955. Steps in the oxidation of indoleacetic acid. *Science,* 122: 187–188.
Rayle, D. L., and R. Cleland. 1970. Enhancement of wall loosening and elongation by acid solutions. *Plant Physiol.,* 46: 250–253.
——— and W. K. Purves. 1967*a.* Conversion of indole-ethanol to indoleacetic acid in cucumber seedling shoots. *Plant Physiol.,* 42: 1091–1093.
——— and ———. 1967*b.* Isolation and identification of indole-3-ethanol from cucumber seedlings. *Plant Physiol.,* 42: 520–524.
Redemann, C. T., L. Rappaport, and R. H. Thompson. 1968. Phaseolic acid, a new plant growth regulator from bean seeds. Pp. 109–124 in F. Wightman and G. Setterfield (eds.), "Biochemistry and Physiology of Plant Growth Substances." Runge Press, Ottawa.
———, S. H. Wittwer, and H. M. Sell. 1951. Characterization of indoleacetic acid and its esters. *J. Am. Chem. Soc.,* 73: 2957.
Redshaw, A. J., and H. Meidner. 1972. Effects of water stress on the resistance to uptake of carbon dioxide in tobacco. *J. Exp. Bot.,* 23: 229–240.
Reeve, R. M. 1959. Histological and histochemical changes in developing and ripening peaches. I: The catechol tannins. *Am. J. Bot.,* 46: 210–217.
Reger, B. J., and R. W. Krauss. 1970. The photosynthetic response to a shift in chlorophyll *a* to chlorophyll *b* ratio of *Chlorella. Plant Physiol.,* 46: 568–575.
Reid, D. M., and J. B. Clements. 1968. RNA and protein synthesis: Prerequisites of red light induced gibberellin synthesis. *Nature,* 219: 607–609.
———, ———, and D. J. Carr. 1968. Red light induction of gibberellin synthesis in leaves. *Nature,* 217: 580–582.
Reinhold, L. 1967. Induction of coiling in tendrils by auxin and carbondioxide. *Science,* 158: 791–793.
Rhodes, M. J. C., T. Galliard, L. S. C. Wooltorton, and A. C. Hulme. 1968. The development of a malate decarboxylation system during the ageing of apple peel discs. *Phytochemistry,* 7: 405–408.
Rice, E. L. 1948. Absorption and translocation of ammonium 2,4-dichlorophenoxyacetate by bean plants. *Bot. Gaz.,* 109: 301–314.
——— and L. M. Rohrbaugh. 1953. Effect of kerosene on movement of 2,4-D and some derivatives through destarched bean leaves in darkness. *Bot. Gaz.,* 115: 76–81.
Richardson, S. D. 1956. Studies of root growth in *Acer saccharinum* L. IV: The effect of differential shoot and root temperature on root growth. *Proc. K. Ned. Akad. Wet.,* 59: 428–438.
Richmond, A., and J. B. Biale. 1966. Protein and nucleic acid metabolism in fruits, I. *Plant Physiol.,* 41: 1247–1253.
——— and A. Lang. 1957. Effect of kinetin on protein content and survival of detached *Xanthium* leaves. *Science,* 125: 650–651.
Rick, C. M., and R. I. Bowman. 1961. Galapagos tomatoes and tortoises. *Evolution,* 15: 407–417.
Riddell, J. A., H. A. Hagerman, C. M. J. Anthony, and W. L. Hubbard. 1962. Retardation of plant growth by a new group of chemicals. *Science,* 136: 391.

Ridge, J., and D. J. Osborne. 1969. Cell growth and cellulase: Regulation by ethylene and indoleacetic acid. *Nature,* 223: 318–319.

——— and ———. 1971. Role of peroxidase when hydroxyproline-rich protein in plant cell walls is increased by ethylene. *Nature,* 229: 205–208.

Rijven, A. H. G. C., and V. Parkash. 1970. Cytokinin induced growth responses by fenugreek cotyledons. *Plant Physiol.,* 45: 638–640.

——— and ———. 1971. Action of kinetin on cotyledons of fenugreek. *Plant Physiol.,* 47: 59–64.

Riov, J., S. P. Monselise, and R. S. Kahan. 1969. Ethylene controlled induction of phenylalanine ammonia lyase in citrus fruit peel. *Plant Physiol.,* 44: 631–635.

Robbins, W. J. 1957. Gibberellic acid and the reversal of adult *Hedera* to a juvenile state. *Am. J. Bot.,* 44: 743–746.

Roberts, E. H. 1969. Seed dormancy and oxidation processes. *Soc. Exp. Biol. Symp.,* 23: 161–192.

———. 1972. "Viability of Seeds." Chapman and Hall, London. 448 pp.

Roberts, L. W. 1960. Experiments on xylem regeneration in stem wound responses in *Coleus. Bot. Gaz.,* 121: 201–208.

Roberts, R. H., and B. E. Struckmeyer. 1944. Use of sprays to set greenhouse tomatoes. *Proc. Am. Soc. Hort. Sci.,* 44: 417–427.

Robinson, D. R., and G. Ryback. 1969. Incorporation of tritium from (4R)-4-^{3}H-mevalonic acid into abscisic acid. *Biochem. J.,* 113: 825–827.

Robinson, N. 1966. "Solar Radiation," vol. XII. Elsevier, Amsterdam. 347 pp.

Robinson, P. M., and P. F. Wareing. 1964. Chemical nature and biological properties of the inhibitor varying with photoperiod in sycamore. *Physiol. Plant.,* 17: 314–323.

Robinson, R. W., D. J. Cantliff, and S. Shannon. 1971. Morphactin-induced parthenocarpy in the cucumber. *Science,* 171: 1251–1252.

———, S. Shannon, and M. D. dela Guardia. 1969. Regulation of sex expression in the cucumber. *Bioscience,* 19: 141–142.

Robitaille, H. 1973. Effects of bruising on ethylene formation in apple fruits. *HortScience.*

Rodriguez, A. B. 1932. Smoke and ethylene in fruiting of pineapple. *J. Dept. Agric. P. R.,* 26: 5–18.

Rohrbaugh, L. M., and E. L. Rice. 1949. Effect of application of sugar on the translocation of sodium 2,4-D by bean plants in the dark. *Bot. Gaz.,* 111: 85–89.

——— and ———. 1956. Relation of phosphorus nutrition to the translocation of 2,4-D in tomato plants. *Plant Physiol.,* 31: 196–199.

Romani, R. J., and J. B. Biale. 1957. Metabolic processes in cytoplasmic particles of the avocado fruit, IV. *Plant Physiol.,* 32: 662–668.

Romberg, L. D. 1944. Some characteristics of the juvenile and the bearing pecan tree. *Proc. Am. Soc. Hort. Sci.,* 44: 255–259.

Roos, E. E., and B. M. Pollock. 1971. Soaking injury in lima beans. *Crop Sci.,* 11: 78–81.

Ropp, R. S. de. 1947. Studies in the physiology of leaf growth, IV. *Ann. Bot.,* 11: 439–447.

Rosa, J. T. 1925. Shortening the rest period of potatoes with ethylene gas. *Potato News Bull.,* 2: 363–365.

Rosado-Alberio, J., T. E. Weier, and C. R. Stocking. 1968. Continuity of the chloroplast membrane systems in *Zea mays. Plant Physiol.,* 43: 1325–1331.

Rosen, W. G., S. R. Gawlik, W. V. Dashek, and K. A. Siegesmond. 1964. Fine structure and cytochemistry of *Lilium* pollen tubes. *Am. J. Bot.,* 51: 61–71.

Ross, C. W. 1962. Nucleotide composition of ribonucleic acid from vegetative and flowering cocklebur-shoot tips. *Biochim. Biophys. Acta,* 55: 387–388.

Ross, J. D., and J. W. Bradbeer. 1968. Concentrations of gibberellin in chilled hazel seeds. *Nature,* 220: 85–86.

Rothwell, K., and S. T. C. Wright. 1967. Phytokinin activity in some new 6-substituted purines. *Proc. R. Soc.,* B167: 202–223.

Rovira, A. D., and G. D. Bowen. 1968. Anion uptake by the apical region of seminal wheat roots. *Nature (Lond.),* 218: 685–686.

Rowan, K. S., W. B. McGlasson, and H. K. Pratt. 1969. Changes in adenosine pyrophosphates in cantaloupe fruit ripening normally and after treatment with ethylene. *J. Exp. Bot.,* 20: 145–155.

Rowsell, E. V., and L. J. Goad. 1962. Latent β-amylase of wheat: Its mode of attachment to glutenin and its release. *Biochem. J.,* 84: 73–74.

Roychoudhury, R., A. Datta, and S. P. Sen. 1965. The mechanism of action of plant growth substances: The role of nuclear *RNA* in growth substance action. *Biochim. Biophys. Acta,* 107: 346–351.

——— and S. P. Sen. 1964. Studies on the mechanism of auxin action. *Physiol. Plant.,* 17: 352–362.

Rubinstein, B., and A. C. Leopold. 1962. Effects of amino acids on bean leaf abscission. *Plant Physiol.,* 37: 398–401.

——— and ———. 1963. Analysis of the auxin control of bean leaf abscission. *Plant Physiol.,* 38: 262–267.

Ruddat, M., A. Lang, and E. Mosettig. 1963. Gibberellin activity of steviol, a plant terpenoid. *Naturwiss.,* 50: 23.

Rudich, J., A. H. Halevy, and N. Kedar. 1972. Ethylene evolution from cucumber plants as related to sex expression. *Plant Physiol.,* 49: 998–999.

———, ———, and ———. 1972. The level of phytohormones in monoecious and gynoecious cucumbers as affected by photoperiod and ethephon. *Plant Physiol.,* 50: 585–590.

Rudnicki, R., J. Pieniazek, and N. Pieniazek. 1968. Abscisin II in strawberry plants at two different stages of growth. *Bull. Acad. Pol. Sci. Biol.,* 16: 127–130.

Ruesink, A. W. 1969. Polysaccharidases and the control of cell wall elongation. *Planta,* 89: 95–107.

Rufelt, H. 1957. The course of the geotropic reaction of wheat roots. *Physiol. Plant.,* 10: 231–247.

Ruge, U. 1957. Zur Wirkstoff-Analyse des Rhizokalin-Komplexes, I. *Z. Bot.,* 45: 273–296.

Russell, M. B., and J. T. Woolley. 1961. Transport processes in the soil-plant system. Pp. 695–722 in M. X. Zarrow (ed.), "Growth in Living Systems." Basic Books, New York.

Russo, L., H. C. Dostal, and A. C. Leopold. 1968. Chemical stimulation of fruit ripening. *BioScience,* 18: 109.

Rutten-Pekelharing, C. J. 1909. Untersuchungen über die Perzeption des Schwerkraftreizes. *Rec. Trav. Bot. Neerl.,* 7: 241–348.

Ryan, C. A., and O. C. Housman. 1967. Chymotrypsin inhibitor I from potatoes. *Nature,* 1047–1049.

Rylski, I., and A. H. Halevy. 1972. Factors controlling the readiness to flower of buds along the main axis of pepper. *J. Am Soc. Hort. Sci.,* 97: 309–312.

Ryther, J. H. 1970. Is the world's oxygen supply threatened? *Nature,* 227: 374–375.

Sacher, J. A. 1957. Relationship between auxin and membrane-integrity in tissue senescence and abscission. *Science,* 125: 1199 - 1200.

———. 1959. Studies of auxin-membrane permeability relations in fruit and leaf tissues. *Plant Physiol.,* 34: 365–372.

———. 1966*a*. Permeability characteristics and amino acid incorporation during senescence of banana tissue. *Plant Physiol.,* 41: 701–708.

———. 1966*b*. The regulation of sugar uptake and accumulation in bean pod tissue. *Plant Physiol.,* 41: 181–189.

Sachs, J. 1862. Uber den Einfluss des Lichtes auf die Bildung des Amylums in den Chlorophyllkörnern. *Bot. Ztg.,* 20: 365–373.

———. 1864. Uber die obere Temperatur-Gränze der Vegetation. *Flora (Jena),* 47: 5–12.

———. 1872. Uber den Einfluss der Lufttemperatur und des Tageslichts auf die stundlichen und taglichen Anderungen des langenwachsthums (Streckung) der Internodien. *Arb. Bot. Inst. Wurtzburg,* 1: 99–192.

———. 1882. Stoff und Form der Pflanzenorgane. *Arb. Bot. Inst. Wurzburg,* (II)3: 452–488.

———. 1887. Lecture on the Physiology of Plants. Trans. H. Marshall Wared. Oxford.

Sachs, R. M., C. Bretz, and A. Lang. 1958. Cell division and gibberellic acid. *Exp. Cell Res.,* 18: 230–244.

———, ———, and ———. 1959. Shoot histogenesis: The early effects of gibberellin upon stem elongation in two rosette plants. *Am. J. Bot.,* 46: 376–384.

Sachs, T., and K. V. Thimann. 1964. Release of lateral buds from apical dominance. *Nature,* 201: 939–940.

Saeki, T. 1960. Interrelationship between leaf amount, light distribution and total photosynthesis in a plant community. *Bot. Mag. Tokyo,* 73: 55–63.

Sakai, A. 1956. Survival of plant tissues at super low temperature. *Low Temp. Sci.,* B14: 17–23; cited by Levitt (1956).

Sakai, S. and H. Imaseki. 1973. A proteinaceous inhibitor of ethylene biosynthesis by etiolated mungbean hypococtyl sections. *Planta,* 113: 115–128.

——— and ———. 1973. Properties of the proteinaceous inhibitor of ethylene synthesis: Action of ethylene production and indoleacetyl-aspartate formation. *Plant Cell Physiol.* 14: 881–892.

Sakurai, A., and S. Tamura. 1965. Syntheses of several compounds related to helminthosporol and the plant growth regulating activities. *Agric. Biol. Chem.,* 29: 407–411.

Salisbury, F. B. 1955. The dual role of auxin flowering. *Plant Physiol.,* 30: 327–334.

——— and ———. 1956. The reactions of the photoinductive dark period. *Plant Physiol.,* 31: 141–147.

——— and ———. 1960. Inhibition of photoperiodic induction by 5-flourouracil. *Plant Physiol.,* 35: 173–177.

Salter, P. J., and J. E. Goode. 1967. "Crop Responses to Water at Different Stages of Growth." Published Commonwealth Agricultural Bureau, *Res. Rev.* 2 *Commonw. Bur. Hort. Plant. Crops, East Malling, Kent,* Farnham Royal, Bucks., England.

Samish, R. M. 1954. Dormancy in woody plants. *Annu. Rev. Plant Physiol.,* 5: 183–204.

Sanchez-Diaz, M. F., and P. J. Kramer. 1971. Behaviour of corn and sorghum under water stress and during recovery. *Plant Physiol.,* 48: 613–616.

Sano, H., and M. Nagao. 1970. Change in the indoleactic acid oxidase levels in leaves of *Begonia evansiana* under short-day conditions. *Plant Cell Physiol.,* 11: 849–856.

Sargent, J. A., and G. E. Blackman. 1962. Studies of foliar penetration, I. *J. Exp. Bot.,* 13: 348–368.

Satter, R. L., and A. W. Galston. 1971. Potassium flux: A common feature of *Albizzia* leaflet movement controlled by phytochrome or endogenous rhythm. *Science,* 174: 518–520.

Saunders, P. F., C. F. Jenner, and G. E. Blackman. 1966. The uptake of growth substances, V. *J. Exp. Bot.,* 17: 241–269.

Saussure, T. de. 1804. Recherches chimiques sur la végétation. Ayon, Paris.

Sawhney, B. L., and I. Zelitch. 1969. Direct determination of potassium ion accumulation in guard cells in relation to stomatal opening in light. *Plant Physiol.,* 44: 1350–1354.

Sawyer, J. S. 1972. Man-made carbon dioxide and the "greenhouse" effect. *Nature (Lond.),* 239: 23–26.

Sax, K., and H. J. Sax. 1962. Effects of x-ray on the aging of seeds. *Nature,* 194: 459–460.

Scarth, G. W. 1932. Mechanism of the action of light and other factors on stomatal movement. *Plant Physiol.,* 7: 481–504.

——— and M. Shaw. 1951. Stomatal movement and photosynthesis in Pelargonium, II. *Plant Physiol.,* 26: 581–597.

Schaeffer, G. W., and H. H. Smith. 1963. Auxin-kinetin interaction in tissue cultures of *Nicotiana* species and tumor-conditioned hybrids. *Plant Physiol.,* 38: 291–297.

Schenk, R. V. 1961. Development of the peanut fruit. *Ga. Agric. Exp. Stn. Tech. Bull.* 22. 53 pp.

Scherf, H., and M. H. Zenk. 1967. Der Einfluss des Lichtes auf die Flavonoidsynthese und die Enzyminduktion bei *Fagopyrum esculentum. Z. Pflanzenphysiol.,* 57: 401–418.

Schieferstein, R. H., and W. E. Loomis. 1959. Development of the cuticular layers in angiosperm leaves. *Am. J. Bot.,* 46: 625–635.

Schlee, D., H. Reinbothe, and K. Mothes. 1966. Wirkungen von Kinetin auf den Adeninabbau in chlorophyll-defecten Blttern von Pelargonium. *Z. Pflanzenphysiol.,* 54: 223–236.

Schmitz, H. 1933. Über Wuchsstoff und Geotropismus bei Grasern. *Planta,* 19: 614–635.

Schmucker, T. 1933. Zur Blutenbiologie tropischer *Nymphaa arten.* II: Bor als entscheidener Faktor. *Planta,* 18: 641–650.

Schneider, E. A., R. A. Gibson, and F. Wightman. 1972. Pathways of auxin biosynthesis in shoots of higher plants. Pp. 82–90 in D. J. Carr (ed.), "Plant Growth Substances, 1970." Springer-Verlag, Berlin.

Schneider, G. W., and N. F. Childers. 1941. Influence of soil moisture on photosynthesis, respiration and transpiration of apple leaves. *Plant Physiol.,* 16: 565–583.

Scholander, P. F. 1972. Tensile water. *Am. Sci.,* 60: 584–590.

———, H. T. Hammel, E. D. Bradstreet, and E. A. Hemmingsen. 1965. Sap pressure in vascular plants. *Science,* 148: 339–346.

———, ———, E. Hemmingsen, and W. Carey. 1962. Salt balance in mangroves. *Plant Physiol.,* 37: 722–729.

———, E. Hemmingsen, and W. Garey. 1961. Cohesive lift of sap in the rattan vine. *Science,* 134: 1835–1838.

———, W. E. Love, and J. W. Kanwisher. 1955. The rise of sap in tall grapevines. *Plant Physiol.,* 30: 93–104.

———, B. Ruud, and H. Leivestad. 1957. The rise of sap in a tropical liana. *Plant Physiol.,* 32: 1–6.

Schopfer, P., and H. Mohr. 1972. Phytochrome-mediated induction of phenylalanine ammonia-lyase in mustard seedlings: A contribution to eliminate some misconceptions. *Plant Physiol.,* 49: 8–10.

Schramm, R. 1912. Über die anatomischen Jugendformen der Blätter einheimischen Holzpflanzen. *Flora,* 104: 225–292.

Schrank, A. R. 1956. Ethionine inhibition of elongation and geotropic curvature of *Avena* coleoptiles. *Arch. Biochem. Biophys.,* 61: 348–355.

Schreven, D. A. van. 1949. Premature tuber formation in early potatoes. *Tijdschr. Plantenziekten,* 55: 290–308.

———. 1956. On the physiology of tuber formation in potatoes, I and II. *Plant Soil,* 8: 49–86.

Schroeder, C. A. 1963. Induced temperature tolerance in plant tissue *in vitro. Nature,* 200: 1301–1302.

Schumacher, W. 1933. Untersuchungen über die Wanderung des Fluoresgeng in den Siebrohren. *Jahrb. Wiss Bot.,* 77: 685–732.

Schwabe, W. W. 1954. Acceleration of flowering in non-vernalized chrysanthemums by the removal of apical sections of the stem. *Nature,* 174: 1022.

———. 1959. Studies of long-day inhibition in short-day plants. *J. Exp. Bot.,* 10: 317–329.

Schwarz, K., and A. A. Bitancourt. 1957. Paper chromatography of unstable substances. *Science,* 126: 607–608.

Schwertner, H. A., and P. W. Morgan. 1966. Role of IAA oxidase in abscissin control in cotton. *Plant Physiol.,* 41: 1513–1519.

Scott, D., P. H. Menalda, and J. A. Rowley. 1970. CO_2 exchange of plants. I: Technique, and response of seven species to light intensity. *N.Z. J. Bot.,* 8: 82–90.

Scott, F. M. 1950. Internal suberization of tissues. *Bot. Gaz.,* 111: 378–394.

Scott, L., and J. H. Priestley. 1928. The root as an absorbing organ. *New Phytol.,* 27: 125–140.

Scott, P. C., and A. C. Leopold. 1967. Opposing effects of gibberellin and ethylene. *Plant Physiol.,* 42: 1021–1022.

Scott, T. K. 1972. Auxins and roots. *Ann. Rev. Plant Physiol.,* 23: 235–258.

——— and W. R. Briggs. 1960. Auxin relationships in the Alaska pea. *Am. J. Bot.,* 47: 492–499.

———, D. B. Case, and W. P. Jacobs. 1967. Auxin-gibberellin interaction in apical dominance. *Plant Physiol.,* 42: 1329–1333.

Scurfield, G. 1973. Reaction wood: Its structure and function. *Science,* 179: 647–655.

Sebanek, J. 1965. Die Interaktion endogener Gibberelline in der Korrelation zwischen Wurzel und Epikotyl bei *Pisum* Keimlingen. *Flora,* 156: 303–311.

———. 1966. Interaction of indoleacetic acid with synthetic and native growth regulators during transfer of ^{32}P into epicotyls of etiolated pea seedlings. *Biol. Plant.,* 8: 213–219.

Seeley, R. C., C. H. Fawcett, R. L. Wain, and F. Wightman. 1955 Chromatographic investigations on the metabolism of certain indole derivatives in plant tissues. Pp. 234–247 in R. L. Wain and F.

Wightman (eds.), "The Chemistry and Mode of Action of Plant Growth Substances." Academic, London.

Selman, I. W., and S. Kulasegaram. 1967. Development of the stem tuber in kohlrabi. *J. Exp. Bot.,* 18: 471–490.

Sembdner, G., G. Schneider, J. Wieland, and K. Schreiber. 1964. Über ein gebundenes Gibberellin aus *Phaseolus coccineus. Experientia,* 20: 1–4.

——— and J. Wieland. 1968. Pp. 70–86 in "Plant Growth Regulators." *Soc. Chem. Ind. Monogr.* 31.

Sequiera, L., and L. Mineo. 1966. Partial purification and kinetics of indoleacetic acid oxidase from tobacco roots. *Plant Physiol.,* 41: 1200–1208.

——— and T. A. Steeves. 1954. Auxin inactivation and its relation to leaf drop caused by the fungus, *Omphalia flavida. Plant Physiol.,* 29: 11–16.

Sestak, Z., and J. Catsky. 1962. Intensity of photosynthesis and chlorophyll content as related to leaf age in *Nicotiana. Biol. Plant.,* 4: 131–140.

Seth, A. K., C. R. Davies, and P. F. Wareing. 1966. Auxin effects on the mobility of kinetin in the plant. *Science,* 151: 587–588.

——— and P. F. Wareing. 1964. Interaction between auxins, gibberellins and kinins in hormone-directed transport. *Life Sci.,* 3: 1483–1486.

Sfakiotakis, E. M., D. H. Simons, and D. R. Dilley. 1972. Pollen germination and tube growth: Dependence on carbon dioxide and independence of ethylene. *Plant Physiol.,* 49: 963–967.

Shain, Y., and A. M. Mayer. 1968. Activation of enzymes during germination: amylopectin-1, 6-glucosidase in peas. *Physiol. Plant.,* 21: 765–766.

Shantz, E. M., and F. C. Steward. 1955. The identification of compound A from coconut milk as 1,3-diphenylurea. *J. Am. Chem. Soc.,* 77: 6351–6353.

——— and ———. 1957. The growth stimulating substances in extracts of immature corn grain: A progress report. *Plant Physiol.,* 32: vii.

Shaw, M., P. K. Bhattacharya, and W. A. Quick. 1965. Chlorophyll, protein and nucleic acid levels in detached senescing wheat leaves. *Can. J. Bot.,* 43: 739–746.

——— and A. R. Hawkins. 1958. The physiology of host-parasite relations. *Can. J. Bot.,* 36: 1–16.

——— and M. S. Manocha. 1965. Fine structure in detached, senescing wheat leaves. *Can. J. Bot.,* 43: 747.

Shaw, R. H. 1954. Leaf and air temperatures under freezing conditions. *Plant Physiol.,* 29: 102–104.

Sheets, T. J. 1963. Photochemical alteration and inactivation of amiben. *Weeds,* 111: 186–190.

Shen-Miller, J. 1970. Reciprocity in the activation of geotropism in oat coleoptiles grown on clinostats. *Planta,* 92:152–163.

——— and C. Miller. 1972. Distribution and activation of the Golgi apparatus in geotropism. *Plant Physiol.,* 49: 634–639.

Shephard, D. C., W. B. Levin, and R. G. S. Bidwell. 1968. Normal Photosynthesis by isolated chloroplasts. *Biochim. Biophys. Res. Commun.,* 32(3): 413–420.

Sheriff, D. W. 1972. A new apparatus for the measurement of sap flux in small shoots with the magnethohydrodynamic technique. *J. Exp. Bot.,* 23: 1086–1095.

Shibaoka, H., and T. Yamaki. 1959. Studies on the growth movement of sunflower plant. *Sci. Pap. Coll. Gen. Educ. Univ. Tokyo.,* 9: 105–126.

Shifriss, O. 1961. Gibberellin as a sex regulator in *Ricinus communis. Science,* 133: 2061–2062.

Shimokawa, A., and Z. Kasai. 1967. Ethylene formation from pyruvate by subcellular particles of apple tissue. *Plant Cell Phys.,* 8: 227–230.

——— and ———. 1968. A possible incorporation of ethylene into RNA in Japanese morning glory seedlings. *Agric. Biol. Chem.,* 32: 680–682.

Shindy, W. W., W. M. Kliener, and R. J. Weaver. 1973. Benzyladenine-induced movement of ^{14}C-labeled photosynthate into roots of *Vitis vinifera. Plant Physiol.,* 51: 345–349.

Shone, M. G. T., and D. A. Barber. 1966. The initial uptake of ions by barley roots, I. *J. Exp. Bot.,* 17: 78–88.

Short, K. C., and J. G. Torrey. 1972. Cytokinins in seedling roots of pea. *Plant Physiol.,* 49: 155–160.

Shropshire, W., W. H. Klein, and J. L. Edwards. 1964. Photomorphogenesis induced by flavin mononucleotide fluorescence. *Physiol. Plant.,* 17: 676–682.

Shropshire, W., Jr., and R. B. Withrow. 1958. Action spectrum of phototropic tip-curvature of *Avena. Plant Physiol.,* 33: 360–365.

Shul'gin, I. A., and A. F. Kleshnin. 1959. Correlation between optical properties of plant leaves and their chlorophyll content. *Dokl. Akad. Nauk,* 125 (1–6): 119–121.

Shull, C. A. 1916. Measurement of the surface forces in soils. *Bot. Gaz.,* 62: 1–31.

Sibaoka, T. 1969. Physiology of rapid movements in higher plants. *Annu. Rev. Plant Physiol.,* 20: 165–184.

Sideris, C. P. 1925. Observations on the development of the root system of *Allium. Am. J. Bot.,* 12: 255–258.

Siegel, B. Z., and A. W. Galston. 1967. The isoperoxidases of *Pisum sativum. Plant Physiol.,* 42: 221–226.

Siegel, S. M. 1950. Effects of exposures of seeds to various physical agents, I. *Bot. Gaz.,* 112: 57–70.

Siegelman, H. W., and W. L. Butler. 1965. Properties of phytochrome. *Annu. Rev. Plant Physiol.,* 16: 383–392.

———, C. T. Chow, and J. B. Biale. 1958. Respiration of developing rose petals. *Plant Physiol.,* 33: 403–409.

——— and S. B. Hendricks. 1958. Photocontrol of alcohol, aldehyde, and anthocyanin production in apple skins. *Plant Physiol.,* 33: 409–413.

Sievers, A., and D. Volkmann. 1972. Verursacht

differentieller Druck der Amyloplasten auf die komplexes Endomembransystem die Geoperzeption in Wurzeln? *Planta,* 102: 160–172.

Sij, J. W., and C. A. Swanson. 1973. Effect of petiole anoxia on phloem transport in squash. *Plant Physiol.,* 51: 368–371.

Silberger, J., and F. Skoog. 1953. Changes induced by indoleacetic acid in nucleic acid contents of tobacco pith tissue. *Science,* 118: 443–444.

Siminovitch, D., and D. R. Briggs. 1954. Studies on the chemistry of living bark in relation to frost hardiness, VII. *Plant Physiol.,* 29: 331–337.

———, C. M. Wilson, and D. R. Briggs. 1953. Studies on the chemistry of living bark in relation to frost hardiness, V. *Plant Physiol.,* 28: 383–400.

Singh, B. N., and K. N. Lal. 1935. Investigations of the effect of age on assimilation of leaves. *Ann. Bot.,* 49: 291–307.

Singh, L. B. 1960. "The Mango." Leonard Hill, London.

Singh, M., D. B. Peters, and J. W. Pendleton. 1968. Net and spectral radiation in soybean canopies. *Agron. J.,* 60: 542–545.

Sinnott, E. W. 1952. Reaction wood and the regulation of tree form. *Am. J. Bot.,* 39: 69–78.

———. 1960. "Plant Morphogenesis." McGraw-Hill, New York.

——— and R. Bloch. 1944. Visible expression of cytoplasmic pattern in the differentation of xylem strands. *Proc. Natl. Acad. Sci. (U.S.),* 30: 388–392.

Sironval, C. 1963. Chlorophyll metabolism on the leaf content in some other tetrapyrrole pigments. *Photochem. Photobiol.,* 2: 207–221.

Sitton, D., C. Itai, and H. Kende. 1967. Decreased cytokinin production in the roots as a factor in shoot senescence. *Planta,* 73: 296–300.

Skene, K. G. M., and D. J. Carr. 1961. A quantitative study of gibberellin of seeds of *Phaseolus. Aust. J. Biol. Sci.,* 14: 13–25.

——— and G. H. Kerridge. 1967. Effect of root temperature on cytokinin activity in root exudate of *Vitis vinifera* L. *Plant Physiol.,* 42: 1131–1139.

Skoog, F., and D. J. Armstrong. 1970. Cytokinins. *Ann. Rev. Plant Physiol.* 21: 359–284.

———, ———, J. D. Cherayil, A. E. Hampel, and R. M. Bock. 1966. Cytokinin activity: Localization in transfer RNA preparations. *Science,* 154: 1354–1356.

———, ———, and D. J. Goodchild. 1969. Distribution of enzymes in mesophyll and parenchyma sheath chloroplasts of maize leaves in relation to the C_4 pathway of photosynthesis. *Biochem. J.,* 114: 489–498.

———, H. Q. Hamzi, A. M. Szweykowska, N. J. Leonard, K. L. Caraway, T. Fujii, J. P. Helgeson, and R. N. Loeppky. 1967. Cytokinins: Structure/activity relationships. *Phytochemistry,* 6: 1169–1192.

——— and N. J. Leonard. 1968. Sources and structure-activity relationships of cytokinins. Pp. 1–18 in F. Wightman and G. Setterfield (eds.), "Biochemistry and Physiology of Plant Growth Substances." Runge Press, Ottowa.

——— and C. O. Miller. 1957. Chemical regulation of growth and organ formation in plant tissues cultured in vivo. *Symp. Soc. Exp. Biol.,* 11: 118–131.

———, R. Y. Schmitz, R. M. Bock, and S. M. Hecht. 1973. Cytokinin antagonists: Synthesis and physiological effects of 7-substituted 3-methylpyrazolo(4,3-d)pyrimidines. *Phytochemistry,* 12: 25–37.

———, C. L. Schneider, and P. Malan. 1942. Interactions of auxins in growth and inhibition. *Am. J. Bot.,* 29: 568–576.

——— and C. Tsui. 1948. Chemical control of growth and bud formation on tobacco stem and callus cultured *in vitro. Am. J. Bot.,* 35: 782–787.

Skoss, J. D. 1955. Structure and composition of plant cuticle in relation to environmental factors and permeability. *Bot. Gaz.,* 117: 55–72.

Slack, C. R., and D. J. Goodchild. 1969. Distribution of enzymes in mesophyll and parenchyma sheath chloroplasts of maize leaves in relation to the C_4-dicarboxylic acid pathway of photosynthesis. *Biochem. J.,* 114: 489–498.

——— and M. D. Hatch. 1967. Comparative studies on the activity of carboxylases and other enzymes in relation to the new pathway of photosynthetic carbon dioxide fixation in tropical grasses. *Biochem. J.,* 103: 660–665.

Slatyer, R. O. 1957. The influence of progressive increases in total soil moisture stress on transpiration, growth, and internal water relationships of plants. *Aust. J. Biol. Sci.,* 10: 320–336.

———. 1960*a*. Absorption of water by plants. *Bot. Rev.,* 26: 331–392.

———. 1960*b*. Aspects of the tissue water relationships of an important arid zone species (*Acacia aneura* F. Muell) in comparison with two mesophytes. *Bull. Res. Coun. Isr.,* sec D, 8D: 159–168.

———. 1967. "Plant-Water Relationships." Academic Press, London. 366 pp.

——— and J. F. Bierhuizen. 1964*a*. Transpiration from cotton leaves under a range of environmental conditions in relation to internal and external diffusive resistances. *Aust. J. Biol. Sci.,* 17: 115–130.

——— and ———. 1964*b*. The influence of several transpiration suppressants on transpiration, photosynthesis, and water-use efficiency of cotton leaves. *Aust. J. Biol. Sci.,* 17: 131–146.

——— and I. C. McIlroy. 1961. "Practical Microclimatology." UNESCO, Paris.

Smillie, R. M. 1962. Photosynthetic and respiratory activities of growing pea leaves. *Plant Physiol.,* 37: 716–721.

——— and G. Krotkov. 1961. Changes in the dry weight, protein, nucleic acid and chlorophyll contents of growing pea leaves. *Can. J. Bot.,* 39: 891–900.

Smith, A. E., J. W. Zukel, G. M. Stone, and J. A.

Riddell. 1959. Factors affecting the performance of maleic hydrazide. *Agric. Food Chem.,* 7: 341–344.

Smith, F. H. 1930. The corn and contractile roots of *Brodiaea lactea. Am. J. Bot.,* 17: 916–927.

——— and Q. D. Clarkson. 1956. Cytological studies of interspecific hydridization in *Iris. Am. J. Bot.,* 43: 582–588.

Smith, H., and P. F. Wareing. 1964. Gravimorphism in trees, III. *Ann. Bot.,* 28: 297–309.

Smith, I. K., and L. Fowden. 1966. A study of mimosine toxicity in plants. *J. Exp. Bot.,* 17: 750–761.

Smith, O. 1932. Relation of temperature to anthesis and blossom drop of the tomato, together with histological study of the pistils. *J. Agric. Res.,* 44: 183–190.

Smith, O. E., and L. Rappaport. 1961. Endogenous gibberellins in resting and sprouting potato tubers. *Adv. Chem.,* 28: 42–48.

Smith, W. H., and J. C. Parker. 1966. Prevention of ethylene injury to carnations by carbon dioxide. *Nature,* 211: 100–101.

Smock, R. M. 1958. Controlled atmosphere storages of apples. *Cornell Ext. Bull.* 750.

———. 1972. Influence of detachment from the tree on the respiration of apples. *J. Am. Soc. Hort. Sci.,* 97: 509–511.

——— and G. D. Blanpied. 1958. A comparison of controlled atmosphere storage and film liners for the storage of apples. *Proc. Am. Soc. Hort. Sci.,* 71: 36–44.

——— and A. M. Neubert. 1950. "Apples and Apple Products." Interscience, New York.

Snow, R. 1932. Growth-regulators in plants. *New Phytol.,* 31: 336–353.

———. 1935. Activation of cambial growth by pure hormones. *Nature,* 135: 876.

Sondheimer, E., E. C. Galson, Y. P. Chang, and D. C. Walton. 1971. Asymmetry: Its importance to the action and metabolism of abscisic acid. *Science,* 174: 829–831.

———, D. S. Tzou, and E. C. Galson. 1968. Abscisic acid levels and seed dormancy. *Plant Physiol.,* 43: 1443–1447.

Sorokin, C. 1960. Kinetic studies of temperature effects on the cellular level. *Biochim. Biophys. Acta,* 38: 197–204.

——— and R. W. Krauss. 1961. Relative efficiency of photosynthesis in the course of cell development. *Biochim. Biophys. Acta,* 48: 314–319.

——— and ———. 1962. Effects of temperature and illuminance on *Chlorella* growth uncoupled from cell division. *Plant Physiol.,* 37: 37–42.

Sorokin, H. P., and K. V. Thimann. 1964. The histological basis for inhibition of axillary buds in *Pisum. Protoplasma,* 59: 326–350.

Soudain, P. 1965. Diffusion de l'oxygéne et du gaz carbonique à travers un fruit en cours de sa croissance. *Physiol. Veg.,* 3: 91–105.

Spanner, D. C. 1952. The action potential of plant cells and some related topics. *Ann. Bot.,* n.s., 16: 379.

———. 1958. The translocation of sugar in sieve tubes. *J. Exp. Bot.,* 9: 332–342.

Sparling, J. H. 1967. Assimilation rates of some woodland herbs in Ontario. *Bot. Gaz.,* 128: 160–168.

Spence, J. A., and E. C. Humphries. 1972. Effect of moisture supply, root temperature and growth regulators on photosynthesis of isolated rooted leaves of sweet potato *(Ipomea batatas). Ann. Bot.,* 36: 115–121.

Spencer, J. L., and W. C. Kennard. 1955. Studies on mango fruit set in Puerto Rico. *Trop. Agric.,* 32: 323–330.

Spencer, M. 1969. Ethylene in nature. *Fortschr. Chem. Org. Naturs.,* 27: 31–80.

——— and A. O. Olson. 1965. Ethylene production and lipid mobilization during germination of castor beans. *Nature,* 205: 699–700.

Spomer, G. G. 1968. Sensors monitor tensions in transpiration stream of trees. *Science,* 161: 484–485.

——— and F. B. Salisbury. 1968. Eco-physiology of *Geum turbinatum* and implications concerning alpine environments. *Bot. Gaz.,* 129: 33–49.

Spragg, S. P., and E. W. Yemm. 1959. Respiratory mechanisms and changes of glutathione and ascorbic acid in germinating peas. *J. Exp. Bot.,* 10: 409–425.

Spurr, A. R., and W. M. Harris. 1968. Ultrastructure of chloroplasts and chromoplasts in *Capsicum annuum.* I: Thylakoid membrane changes during fruit ripening. *Am. J. Bot.,* 55: 1210–1224.

Srivastava, B. I. S. 1968. Acceleration of senescence and of the increase of chromatin associated nucleases in barley leaves by abscisin and its reversal by kinetin. *Biochim. Biophys. Acta,* 169: 534–536.

——— and W. O. S. Meredith. 1962. Mechanism of action of gibberellic acid. *Can. J. Bot.,* 40: 1257.

——— and G. Ware. 1965. The effect of kinetin on nucleic acids and nucleases of excised barley leaves. *Plant Physiol.,* 40: 62–73.

Staden, J. van, and P. F. Wareing. 1972. The effect of light on endogenous cytokinin levels in seeds of *Rumex obtusifolius. Planta (Ber.),* 104: 126–133.

Stahl, A. L., and A. F. Camp. 1936. Cold storage studies of Florida citrus fruits, I. *Fla. Agric. Exp. Sta. Bull.* 303. 67 pp.

Stahl, E. 1909. "Zur Biologie des Chlorophylls"; cited by Molisch (1938).

Stahly, E. A., and A. H. Thompson. 1959. Auxin levels of developing Halehaven peach ovules. *Md. Agric. Exp. Stn. Bull.* A-104. 22 pp.

Stahmann, M. A., B. G. Clare, and W. Woodbury. 1966. Increased disease resistance and enzyme activity induced by ethylene and ethylene production by black rot infected sweet potato tissue. *Plant Physiol.,* 41: 1505–1512.

Stanhill, G. 1960. The relationship between climate and the transpiration and growth of pastures. *Proc. 8th Int. Grassl. Cong.*, (British Grassland Society, 1961). pp. 293–296.
———, G. J. Hofstede, and J. D. Kalma. 1966. Radiation balance of natural and agricultural vegetation. *R. Meteorol. Soc. Q. J.*, 92: 128–140.
Stanley, R. G., and H. F. Linskens. 1964. Enzyme activation in germinating *Petunia* pollen. *Nature,* 203: 542–544.
Stebbins, G. L. 1957. "Variation and Evolution in Plants." Columbia University Press, N.Y. 643 pp.
Steemann-Nielsen, E. 1952. Experimental carbon dioxide curves in photosynthesis. *Physiol. Plant.*, 5: 145–159.
———. 1957. The chlorophyll content and the light utilization in communities of plankton algae and terrestrial higher plants. *Physiol. Plant.*, 10: 1009–1021.
Steeves, T. A., and W. R. Briggs. 1960. Morphogenetic studies on *Osmunda cinnamonea* L.: The auxin relationship of expanding fronds. *J. Exp. Bot.*, 11: 45–67.
Steinberg, R. A. 1952. Premature blossoming: Effects of vernalization, seedling age and environment on subsequent growth and flowering of transplanted tobacco. *Plant Physiol.*, 27: 745–753.
Steiner, M. 1933. Zum Chemismus der osmotischen Jahresschwankungen einiger immergrüner Holzgewächse. *Jahrb. Wiss. Bot.*, 78: 564–622.
Steinhart, C. E., J. D. Mann, and S. H. Mudd. 1964. Alkaloids and plant metabolism. VII: The kinetin produced elevation in tyramine methylpherase levels. *Plant Physiol.*, 39: 1030–1038.
Stephen, W. P. 1958. Pear pollination studies in Oregon. *Oreg. Agric. Tech. Bull.* 43.
Stern, A. I., J. A. Schiff, and H. T. Epstein. 1964. Studies of chloroplast development in Euglena. V: Pigment biosynthesis, photosynthetic oxygen evolution and carbon dioxide fixation during chloroplast development. *Plant Physiol.*, 39: 220–226.
Stern, W. R., and C. M. Donald. 1961. Relationship of radiation, leaf area index and crop growth rate. *Nature,* 189: 597–598.
Stetler, D. A., and W. M. Laetsch. 1965. Kinetin induced chloroplast maturation in cultures of tobacco tissue. *Science,* 149: 1387–1388.
Steveninck, R. F. M. van. 1957. Factors affecting abscission of reproductive organs of lupine. *J. Exp. Bot.*, 8: 373–381.
———. 1959. Abscission accelerators in lupine. *Nature,* 183: 1246–1248.
Steward, F. C., and S. M. Caplin. 1952. Investigations on growth and metabolism of plant cells. *Ann. Bot.*, 16: 478–489.
———, M. O. Mapes, and K. Mears. 1958. Growth and organized development of cultured cells. II: Organization in cultures grown from freely suspended cells. *Am. J. Bot.*, 45: 705–708.
——— and E. M. Shantz. 1956. The chemical induction of growth in plant tissue cultures. Pp. 165–186 in R. L. Wain and F. Wightman (eds.), "The Chemistry and Mode of Action of Plant Growth Substances." Academic Press, N.Y.
——— and N. W. Simmonds. 1954. Growth-promoting substances in the ovary and immature fruit of the banana. *Nature,* 173: 1083.
Stewart, E. R., and H. T. Freebairn. 1969. Ethylene, seed germination and epinasty. *Plant Physiol.*, 44: 955–958.
Stewart, G. R., and J. A. Lee. 1972. Desiccation injury in mosses. II: The effect of moisture stress on enzyme levels. *New Phytol.*, 71: 461–466.
Stewart, L. 1963. Chelation in the absorption and translocation of mineral elements. *Ann. Rev. Plant Physiol.*, 14: 295–310.
Still, C. C., C. C. Olivier, and H. S. Moyed. 1965. Inhibitory oxidation products of indoleacetic acid: Enzymic formation and detoxification by pea seedlings. *Science,* 149: 1249–1251.
Stiller, M. 1962. The path of carbon in photosynthesis. *Annu. Rev. Plant Physiol.*, 13: 151–170.
Stocking, C. R. 1959. Chloroplast isolation in nonaqueous media. *Plant Physiol.*, 34: 56–61.
Stoddart, J. L. 1966. Studies on the relationship between gibberellin metabolism in normal and non-flowering red clover. *J. Exp. Bot.*, 17: 96–107.
Stokes, P. 1952. A physiological study of embryo development in *Heracleum sphondylium*. *Ann. Bot.*, 16: 442–447.
——— and K. Verkerk. 1951. Flower formation in brussels sprouts. *Meded.* Landbouwhogesch. Wageningen, 50: 141–160.
Stolwijk, J. A. J., and J. A. D. Zeevaart. 1955. Wave length dependence of light reactions governing flowering in *Hyoscyamus niger*. *Proc. K. Akad. Wet. Amst. Proc. Sec. Sci.*, C58: 386–396.
Stosser, R. 1971. Localization of RNA and protein synthesis in the developing abscission layer in fruit of *Prunus cerasus*. *Z. Pflanzenphysiol.*, 64: 328–334.
Stout, M. 1945. Translocation of the reproductive stimulus in sugar beets. *Bot. Gaz.*, 107: 86–95.
———. 1946. Relation of temperature to reproduction in sugar beets. *J. Agric. Res.*, 72: 49–68.
Stout, P. R., and D. R. Hoagland. 1939. Upward and lateral movement of salt in certain plants as indicated by radioactive isotopes of potassium, sodium and phosphorus absorbed by roots. *Am. J. Bot.*, 26: 320–324.
Stoutemyer, V. T., and O. K. Britt. 1961. Effect of temperature and grafting on vegetative growth phases of Algerian ivy. *Nature,* 189: 854–855.
——— and ———. 1965. The behavior of tissue cultures from English and Algerian ivy in different growth phases. *Am. J. Bot.*, 52: 805–810.
Straub, J. 1946. Zur Entwicklungsphysiologie der

Selbsterlität von *Petunia*. *Z. Naturforsch.*, 1: 287–291.
——. 1947. Zur Entwicklungsphysiologie der Selbsterlität von *Petunia*, I. *Z. Naturforsch.*, 2: 433–444.
Strugger, S. 1938–1939. Die lumineszenzmikroskopische Analyse des Transpirationsstromes in Parenchymen. *Flora*, N.F., 33: 56–68.
——. 1949. "Praktikum der Zell- und Gewebephysiologie der Pflanze." Springer-Verlag, Berlin.
Stuart, N. W., S. Asen, and C. J. Gould. 1966. Accelerated flowering of bulbous iris after exposure to ethylene. *Hort. Sci.*, 1: 19–20.
Suda, S. 1960. On the physiological properties of mimosine. *Bot. Mag. (Tokyo)*, 73: 142–147.
Suge, H. 1971. Chloroethylphosphonic acid as ethylene releasing agent for the stimulation of rice. *Proc. Crop Sci. Soc. Jap.*, 40: 127–131.
——. 1972. Mesocotyl elongation in japonica rice: Effect of high temperature pre-treatment and ethylene. *Plant Cell Physiol.*, 13: 401–405.
—— and A. Osada. 1966. Inhibitory effect of growth retardants on the induction of flowering in winter wheat. *Plant Cell Physiol.*, 7: 617–629.
—— and N. Yamada. 1963. Chemical control of plant growth and development, 4. *Proc. Crop Sci. Soc. Jap.*, 32: 77–80.
—— and ——. 1965. Effect of nucleic acid and its antimetabolite on induction of flowering in winter cereals. *Proc. Crop Sci. Soc. Jap.*, 33: 324.
Sugiura, M., K. Umemura, and K. Oota. 1962. The effect of kinetin on protein level of tobacco leaf disks. *Physiol. Plant.*, 15: 457–464.
Sussex, I. M., and M. E. Clutter. 1960. A study of the effect of externally supplied sucrose on the morphology of excised fern leaves *in vitro*. *Phytomorphology*, 10: 87–96.
Suzuki, Y., A. C. Leopold, and H. S. Ku. 1971. The stimulation of ethylene biosynthesis by ethephon. *Plant Physiol.*, 47(suppl.): 15.
—— and N. Takahaski. 1968. Effects of after-ripening and gibberellic acid on the thermoinduction of seed germination in *Solanum melongena*. *Plant Cell Phys.*, 9: 653–660.
Swain, R. R., and E. E. Dekker. 1969. Seed Germination studies, III. *Plant Physiol.*, 44: 319–325.
Swanson, C. A., and R. H. Böhning. 1951. The effect of petiole temperature on the translocation of carbohydrate from bean leaves. *Plant Physiol.*, 26: 557–564.
—— and D. R. Geiger. 1967. Time course of low temperature inhibiton of sucrose translocation in sugar beets. *Plant Physiol.*, 42: 751–756.
—— and J. B. Whitney. 1953. Studies on the translocation of foliar applied ^{32}P and other radioisotopes in bean plants. *Am. J. Bot.*, 40: 816–823.
Sweet, G. B., and P. F. Wareing. 1966. Role of plant growth in regulating photosynthesis. *Nature*, 210(5031): 77–79.
Swets, W. A., and F. T. Addicott. 1955. Experiments on the physiology of defoliation. *Proc. Am. Soc. Hort. Sci.*, 65: 291–295.
Swingle, W. T. 1928. Metaxenia in the date palm. *J. Hered.*, 19: 257–268.
——. 1932. Recapitulation of seedling characters by nucellar buds developing in the embryo sac of *Citrus*. *Proc. Int. Congr. Genet.*, 2: 196–197.
Tagawa, T., and J. Bonner. 1957. Mechanical properties of the *Avena* coleoptile as related to auxin and to ionic interactions. *Plant Physiol.*, 32: 207–212.
Tageeva, S. V., and A. B. Brandt. 1961. Study of optical properties of leaves depending on the angle of light incidence. Pp. 163–169 in "Progress in Photobiology." *Proc. 3d Int. Congr. Photobiol., Copenh, 1960. (Elsevier, London).*
Tager, J. M., and B. Clark. 1961. Replacement of alternating temperature requirement for germination by gibberellic acid. *Nature*, 192: 83–84.
Takahashi, K. 1972. Abscisic acid as a stimulator for rice mesocotyl growth. *Nature New Biol.*, 238: 92–93.
Takeo, T., and M. Lieberman. 1969. 3-Methylthiopropionaldehyde peroxidase from apples: An ethylene-forming enzyme. *Biochim. Biophys. Acta*, 178: 235–247.
Takeyosi, H., and M. Fujii. 1961. On the growth substance economy before and after flowering in each organ of *Portulaca grandiflora*. *Bot. Mag. (Tokyo)*, 74: 357–360.
Takimoto, A. 1960. Effect of sucrose on flower initiation of *Pharbitis*. *Plant Cell Physiol.*, 1: 241–246.
—— and K. C. Hamner. 1964. Effect of temperature and preconditioning on photoperiodic response of *Pharbitis*. *Plant Physiol.*, 39: 1024–1030.
—— and ——. 1965. Studies on red light interruption in relation to timing mechanisms involved in the photoperiodic response of *Pharbitis*. *Plant Physiol.*, 40: 852–854.
—— and Y. Naito. 1962. Studies on the light controlling flower initiation of *Pharbitis, IX*. *Bot. Mag. (Tokyo)*, 75: 205–211.
Tal, M., and D. Imber. 1970. Abnormal stomatal behaviour and hormonal inbalance in *flacca*, a wilty mutant of tomato. II: Auxin- and abscisic acid-like activity. *Plant Physiol.*, 46: 373–376.
—— and ——. 1971. Abnormal stomatal behaviour and hormonal imbalance in *flacca*, a wilty mutant of tomato. III: Hormonal effects on the water status in the plant. *Plant Physiol.*, 47: 849–850.
—— and ——. 1972. The effect of abscisic acid on stomatal behaviour in *flacca*, a wilty mutant of tomato, in darkness. *New Phytol.*, 71: 81–84.
——, ——, and C. Itai. 1970. Abnormal stomatal behaviour and hormonal imbalance in *flacca*, a wilty mutant of tomato. I: Root effect and kinetin-like activity. *Plant Physiol.*, 46: 367–372.
Tamura, S., and M. Nagao. 1969. 5-(1,2-Epoxy-2,6,6-trimethyl-1-cyclohexyl)-3-methyl-*cis*-*trans*-2,4-

pentadienoic acid and its esters: New Plant Growth inhibitors structurally related to abscisic acid. *Planta,* 85: 209–212.
——— and A. Sakurai. 1967. Helminthosporal. *Protein, nucleic acid, enzyme,* 12: 24.
———, ———, K. Kainuma, and M. Takai. 1965. Isolation of helminthosporal as a natural plant growth regulator. *Agric. Biol. Chem.,* 29: 216–221.
———, N. Takahashi, T. Yokota, and N. Murofushi. 1968. Isolation of water-soluble gibberellins from immature seeds of *Pharbitis nil. Planta,* 78: 208–212.
Tanada, T. 1968. A rapid photoreversible response by barley root tips in the presence of IAA. *Proc. Natl. Acad. Sci. (U.S.),* 59: 376–380.
———. 1972. Phytochrome control of another phytochrome mediated process. *Plant Physiol.,* 49: 560–562.
———. 1973. Indoleacetic acid and abscisic acid antagonism. II: On the phytochrome mediated attachment of barley root tips on glass. *Plant Physiol.,* 51:
Tang, Y. W., and J. Bonner. 1947. The enzymatic inactivation of indoleacetic acid, I. *Arch. Biochem. Biophys.,* 3: 11–25.
Tanimoto, E., and Y. Masuda. 1968. Effect of auxin on cell wall degrading enzymes. *Physiol. Plant.,* 21: 820–826.
Tanner, C. B., and E. R. Lemon. 1962. Radiant energy utilised in evapotranspiration. *Agron. J.,* 54: 207–212.
Tavares, J., and H. Kende. 1970. The effect of benzylaminopurine on protein metabolism in senescing corn leaves. *Phytochemistry,* 9: 1763–1770.
Taylor, A. O., and A. S. Craig. 1971. Plants under climatic stress. II: Low Temperature, high light effects on chloroplast ultrastructure. *Plant Physiol.,* 47: 719–725.
———, N. M. Jepsen, and J. T. Christeller. 1972. Plants under climatic stress. III: Low temperature, high light effects on photosynthetic products. *Plant Physiol.,* 49: 798–802.
——— and J. A. Rowley. 1971. Plants under climatic stress. I: Low temperature, high light effects on photosynthesis. *Plant Physiol.,* 47: 713–718.
Taylor, H. F., and R. S. Burden. 1970. Identification of plant growth inhibitors produced by photolysis of violaxanthin. *Phytochemistry,* 9: 2217–2223.
Taylor, H. M. 1969. The rhizotron at Auburn, Alabama: A plant root observation laboratory. *Agric. Exp. Stn. Auburn Univ. Circ.* 171.
Taylor, T. D., and G. F. Warren. 1970. The effect of metabolic inhibitors on herbicide movement in plants. *Weed Sci.,* 18: 68–74.
Taylorson, R. B., and S. B. Hendricks. 1972. Phytochrome control of germination of *Rumex crispus* seeds induced by temperature shifts. *Plant Physiol.,* 50: 645–648.
Tepper, H. B., C. A. Hollis, E. C. Galson, and E. Sondheimer. 1967. Germination of excised *Praxinus* embryos with and without phleomycin. *Plant Physiol.,* 42: 1493–1496.
Teraoka, H. 1967. Proteins of wheat embryos in the period of vernalization. *Plant Cell Physiol.,* 8: 87–95.
Teubner, F. G., and S. H. Wittwer. 1955. Effect of *N-m*-tolylphthalmic acid on tomato flower formation. *Science,* 122: 74–75.
Tewari, K. K., and S. G. Wildman. 1966. Chloroplast DNA from tobacco leaves. *Science,* 153: 1269–1271.
Tezuka, T., and Yukio Yamamoto. 1969. NAD kinase and phytochrome. *Bot. Mag. (Tokyo),* 82: 130–133.
——— and ———. 1974. Kinetics of activation of nicotinamide adenine dinucleotide kinase by phytochrome–far–red–absorbing form. *Plant Physiol.,* 53:717–722.
Thaine, R. 1961. Transcellular strands and particle movement in mature sieve tubes. *Nature,* 192: 772–773.
———. 1964. The protoplasmic-streaming theory of phloem transport. *J. Exp. Bot.,* 15: 470–484.
———. 1969. Movement of sugars through plants by cytoplasmic pumping. *Nature (Lond.),* 22: 873–875.
———, S. L. Ovenden, and J. S. Turner. 1959. Translocation of labelled assimilates in soya bean. *Aust. J. Biol. Sci.,* 12: 349–372.
Thimann, K. V. 1935. On the plant growth hormone produced by *Rhizopus sinuis. J. Biol. Chem.,* 109: 279–291.
———. 1936. Auxins and the growth of roots. *Am. J. Bot.,* 23: 561–569.
———. 1937. On the nature of inhibitons caused by auxin. *Am. J. Bot.,* 24: 407–412.
———. 1952. "The Action of Hormones in Plants and Invertebrates." Academic, New York.
———. 1958. Auxin activity of some indole derivatives. *Plant Physiol.,* 33: 311–321.
———. 1963. Plant growth substances: Past, present and future. *Annu. Rev. Plant Physiol.,* 14: 1–18.
——— and Y. Edmondson. 1949. The biogensis of the anthocyamins, I. *Arch. Biochem.,* 22: 33–53.
——— and M. Grochowska. 1968. The role of tryptophan and tryptamine as IAA precursors. Pp. 231–242 in F. Wightman and G. Setterfield (eds.), "Biochemistry and Physiology of Plant Growth Substances." Runge Press, Ottawa.
——— and S. Mahadevan. 1958. Enzymatic hydrolysis of indoleacetonitrile. *Nature,* 181: 1466–1467.
——— and E. Skoog. 1934. Inhibiton of bud development and other functions of growth substances in *Vicia faba. Proc. R. Soc. Lond.,* B114: 317–339.
——— and ———. 1940. The extraction of auxin from plant tissues. *Am. J. Bot.,* 27: 951–960.
———, M. Tomaszewski, and W. L. Porter. 1962. Growth-promoting activity of caffeic acid. *Nature,* 183: 1203.
——— and F. W. Went. 1934. On the chemical nature of the root forming hormone. *Proc. K. Akad. Wet. Amst.,* 37: 456–459.
——— and M. Wickson. 1958. The antagonism of

auxin and kinetin in apical dominance. *Physiol. Plant.,* 11: 62–74.
Thoday, J. M., and R. J. Davey. 1932. Mechanism of root contraction in *Oxalis incarnata. Ann. Bot.,* 46: 993–1006.
Thomas, H. D., and G. R. Hill. 1937. The continuous measurement of photosynthesis respiration and transpiration of alfalfa and wheat growing under field conditions. *Plant Physiol.,* 12: 285–307.
Thomas, M. D. 1965. Photosynthesis (carbon assimilation): Environmental and metabolic relationships. Pp. 9–202 in F. C. Steward (ed.), "Plant Physiology," vol. IVA. Academic, New York.
Thomas, T. H., P. F. Wareing, and P. M. Robinson. 1965. Action of the sycamore dormin as a gibberellin antagonist. *Nature,* 205: 1270–1272.
Thomas, W. D. E., and S. H. Bennett. 1954. The absorption translocation and breakdown of schradan applied to leaves, using ^{32}P-labelled material, III. *Ann. Appl. Biol.,* 41: 501–519.
Thompson, H. C. 1929. Premature seeding of celery. *Cornell Univ. Agric. Exp. Stn. Bull.* 480.
Thompson, P. A. 1961. Evidence for a factor which prevents the development of parthenocarpic fruits in the strawberry. *J. Exp. Bot.,* 12: 199–206.
———. 1969. Germination of *Lycopus europaeus* in response to fluctuating temperatures and light. *J. Exp. Bot.,* 20: 1–11.
Thompson, R. C., and W. F. Kosar. 1939. Stimulation of germination of dormant lettuce seed by sulphur compounds. *Plant Physiol.,* 14: 567–573.
——— and ———. 1972. Effect of light and gibberellin on RNA species of pea stem tissues as studied by DNA-RNA hybridization. *Plant Physiol.,* 50: 289–292.
Thomson, K. S. 1972. The binding of naphthylphthalamic acid, an inhibitor of auxin transport, to particulate fractions of corn coleoptiles. Pp. 83–88 in H. Kaldewey and Y. Vardar (eds.), "Hormonal Regulation of Plant Growth and Development." Verlag Chemie, Wienheim.
——— and A. C. Leopold. 1974. In vitro binding of morphactins and naphthylphthalamic acid in corn coleoptiles and their effects on auxin transport. *Planta,* 115: 259–270.
Thomson, W. W. 1966. Ultrastructural development of chromoplasts in Valencia oranges. *Bot. Gaz.,* 127: 133–139.
———, W. L. Berry, and L. L. Liu. 1969. Localization and secretion of salt by the salt glands of *Tamarix aphylla. Proc. Natl. Acad. Sci. (U.S.),* 63: 310–317.
Thorne, Gillian N. 1960. Variation with age in NAR and other growth attirbutes of sugar-beet, potato and barley in a controlled environment. *Ann. Bot.,* n.s., 24: 356–371.
——— and A. F. Evans. 1964. Influence of tops and roots on net assimilation rate of sugar-beet and spinach beet and grafts between them. *Ann. Bot.,* 28: 499–508.
Thornley, J. H. M. 1972*a*. A model to describe the partitioning of photosynthate during vegetative plant growth. *Ann. Bot.,* 36: 419–430.
———. 1972*b*. A balanced quantitative model for root:shoot ratios in vegetative plants. *Ann. Bot.,* 36: 431–441.
Thrower, S. L. 1962. Translocation of labelled assimilates in soybean. II: The patterns of translocation in intact and defoliated plants. *Aust. J. Biol. Sci.,* 15: 629–649.
———. 1964. Translocation of labelled assimilates in soybean. III: Translocation and other factors affecting leaf growth. *Aust. J. Biol. Sci.,* 17: 412–426.
Thurlow, V., and J. Bonner. 1947. Inhibiton of photoperiodic induction in *Xanthium. Am. J. Bot.,* 34: 603–604.
Tiffin, L. O., and J. C. Brown. 1962. Iron chelates in soybeans exudate. *Science,* 135: 311–313.
Ting, I. P., and C. B. Osmond. 1973*a*. Photosynthetic phosphoenolpyruvate carboxylases: Characteristics of alloenzymes from leaves of C^2 and C^4 plants. *Plant Physiol.,* 51: 439–447.
——— and ———. 1973*b*. Multiple forms of plant phosphoenolpyruvate carboxylase associated with different metabolic pathways. *Plant Physiol.,* 51: 448–453.
Titus, E., H. Weissbach, R. E. Peterson, and S. Udenfriend. 1956. Biogenesis and metabolism of 5-hydrozyindole compounds. *J. Biol. Chem.,* 219: 335–344.
Tizio, R. 1966. Présence de kinines dans le périderme de tubercules de pomme de terre. *C. R. Acad. Sci. Paris,* 262: 868–869.
Tognoni, F., A. H. Halvey, and S. H. Wittwer. 1967. Growth of bean and tomato plants as affected by root absorbed growth substances and atmospheric carbon dioxide. *Planta,* 73: 43–52.
Tolbert, N. E., A. Oeser, R. K. Yamazake, R. H. Hageman, and T. Kisaki. 1969. A survey of plants for leaf peroxisomes. *Plant Physiol.,* 44: 135–147.
Tomaszewski, M. 1964. The mechanism of synergistic effects between auxin and some phenolic substances. Pp. 335–351 in J. P. Nitsch (ed.), "Régulateurs naturels de la croissance végétale." CNRS, Paris.
——— and K. V. Thimann. 1966. Interactions of phenolic acids, metallic ions and chelating agents on auxin-induced growth. *Plant Physiol.,* 41: 1443–1454.
Tomita, T. 1959. The fractions of diffusate obtained from vernalized winter rye and their effect on flowering of annual meadow grass. *Tohoku J. Agric. Res.,* 10: 1–6.
———. 1961. A vernalization-like phenomenon in winter wheat treated with flower-promoting substances. *Proc. Crop Sci. Soc. Jap.,* 30: 83–88.
———. 1964. Studies on vernalization and flowering substances, IV. *Tohoku J. Agric. Res.,* 15: 1–11.
Ton, L. D., and A. H. Krezdorn. 1966. Growth of

pollen tubes in three incompatible varieties of *Citrus. Proc. Am. Soc. Hort. Sci.*, 89: 211–215.

Toole, E. H. 1959. Effect of light on the germination of seeds. Pp. 88–99 in R. B. Withrow (ed.), "Photoperiodism." AAAS, Washington, D.C.

———, V. K. Toole, H. A. Borthwick, and S. B. Hendricks. 1955. Photocontrol of *Lepidium* seed germination. *Plant Physiol.*, 30: 15–21.

Toole, V. K., W. K. Bailey, and E. H. Toole. 1964. Factors influencing dormancy of peanut seeds. *Plant Physiol.*, 39: 822–832.

——— and H. M. Cathey. 1961. Responses to gibberellin of light-requiring seeds of lettuce and *Lepidium virginicum. Plant Physiol.*, 36: 663–671.

Toriyama, H. 1967. On the relation between tannin vacuoles and protoplasm in the motor cell of *Mimosa pudica. Proc. Jap. Acad.*, 43: 777–782.

Torrey, J. G. 1955. On the determination of vascular patterns during tissue differentiation in excised pea roots. *Am. J. Bot.*, 42: 183–198.

———. 1956. Physiology of root elongation. *Ann. Rev. Plant Physiol.*, 7: 237–266.

———. 1957. Auxin control of vascular pattern formation in regenerating pea root meristems growth *in vitro. Am. J. Bot.*, 44: 859–870.

———. 1958. Endogenous bud and root formation by isolated roots of *Convolvulus* grown *in vitro. Plant Physiol.*, 33: 258–263.

———. 1961. Kinetin as trigger for mitosis in mature endomitotic plant cells. *Exp. Cell Res.*, 23: 281–299.

Towers, G. H. N., A. Hutchinson, and W. A. Andreae. 1958. Formation of a glycoside of maleic hydrazide in plants. *Nature,* 181: 1535–1536.

Tregunna, E. B., G. Krotokov, and C. D. Nelson. 1966. Effect of oxygen on the rate of photorespiration in detached tobacco leaves. *Physiol. Plant.*, 19: 723–733.

Treharne, K. J., and J. L. Stoddart. 1968. Effects of gibberellin on photosynthesis in red clover. *Nature,* 220: 457–458.

———, ———, J. Pughe, K. Paranjothy, and P. F. Wareing. 1970. Effects of gibberellin and cytokinins on the activity of photosynthetic enzymes and plastid ribosomal RNA synthesis in *Phaseolus vulgaris* L. *Nature,* 228: 129–131.

Trotter, R. J. 1971. Advances in photosynthesis: Priestley to the present. *Sci. News.* 99(22): 372–374.

Troughton, J. H., and R. O. Slatyer. 1969. Plant water status, leaf temperature, and the calculated mesophyll resistance to carbon dioxide of cotton leaves. *Aust. J. Biol. Sci.*, 22: 815–827.

Tso, T. C. 1964. Plant growth inhibition by some fatty acids and their analogues. *Nature,* 202: 511.

Tuan, D. Y. H., and J. Bonner. 1964. Dormancy associated with repression of genetic activity. *Plant Physiol.*, 39: 768–772.

Tucker, D. J., and T. A. Mansfield. 1971. A simple bioassay for detecting antitranspirant activity of naturally occurring compounds such as abscisic acid. *Planta,* 98: 157–163.

Tukey, H. B. 1933. Embryo abortion in early ripening varieties of *Prunus avium. Bot. Gaz.*, 94: 433–468.

——— and R. F. Carlson. 1945*a*. Breaking the dormancy of peach seed by treatment with thiourea. *Plant Physiol.*, 20: 505–516.

——— and ———. 1945*b*. Morphological changes in peach seedlings following after-ripening treatments of seeds. *Bot. Gaz.*, 106: 431–440.

——— and J. O. Young. 1939. Histological study of the developing fruit of the sour cherry. *Bot. Gaz.*, 100: 723–749.

Tukey, H. B., Jr. 1970. The leaching of substances from plants. *Ann. Rev. Plant Physiol.*, 21: 305–324.

———. 1934. Growth of the embryo, seed and pericarp of the sour cherry in relation to season of fruit ripening. *Proc. Am. Soc. Hort. Sci.*, 31: 125–144.

Tuli, V., and H. S. Moyed. 1969. The role of 3-methyleneoxindole in auxin action. *J. Biol. Chem.*, 244: 4914–4920.

Tullis, E. C., and W. C. Davis. 1950. Persistence of 2,4-D in plant tissue. *Science,* 111: 90.

Tupy, J. 1962. Radiorespirometric study of the utilization of exogenous sucrose, glucose and fructose by germinating apple pollen. *Biol. Plant.*, 4: 69–84.

——— and N. S. Rangaswamy. 1973. The investigation of the effect of pollination on ribosomal RNA, transfer RNA and DNA contents in styles of *Nicotiana alata. Biol Plant.*, 15: 95–101

Turrell, F. M. 1947. Citrus leaf stomata: Structure, composition and pore size in relation to penetration of liquids. *Bot. Gaz.*, 108: 476–483.

———. 1965. Internal surface-intercellular space relationships and the dynamics of humidity maintainance in leaves. Pp. 39–53 in E. J. Andur (ed.), *Int. Symp. Humidity Moisture,* vol. 2. Reinhold, New York.

Uchijima, Z. 1970. Carbon dioxide environment and flux within a corn crop canopy. Pp. 179–196 in "Prediction and Measurement of Photosynthetic Productivity." Pudoc, Wageningen, Netherlands.

Udenfriend, S., W. Lowenberg, and W. Sjoerlsma. 1959. Physiologically active amines in common fruits and vegetables. *Arch. Biochem. Biophys.*, 85: 487–490.

Ulrich, A. 1952. The influence of temperature and light factors on the growth and development of sugar beets. *Agron. J.*, 44: 66–73.

Ulrich, R. 1958. Postharvest physiology of fruits. *Ann. Rev. Plant Physiol.*, 9: 385–416.

Umemoto, T. 1971. Effect of chlorogenic acid on flower production in long-day duckweed, *Lemna gibba. Plant Cell Physiol.*, 12: 165–169.

Umemura, K., and Y. Oota. 1965. Effects of nucleic acid and protein antimetabolites on frond and flower production in *Lemna. Plant Cell Physiol.*, 6: 73.

Uphof, J. C. T. 1938. Cleistogamous flowers. Bot. Rev. 4:21-49.

Urata, V. 1954. Pollination requirements of *Macadamia*. *Tech. Bull. Hawaii Agric. Exp. Stn.*, pp. 1–40.
Vacha, G. A., and R. B. Harvey. 1927. The use of ethylene, propylene and similar compounds in breaking the rest period of tubers, bulbs, cuttings and seeds. *Plant Physiol.*, 2: 187–194.
Van Den Honert, T. H. 1930. Carbon dioxide assimilation and limiting factors. *Recl. Trav. bot. neerl.*, 27: 149–286
Vanderhoef, L. N., and J. L. Key. 1968. Inhibition by kenetin of cell elongation and RNA synthesis in excised soybean hypocotyl. *Plant Cell Physiol.*, 9: 343–351.
Vanderkooi, G., and D. E. Green. 1971. New insights into biological membrane structure. *BioScience*, 21: 409.
Van Niel, C. B. 1949. The comparative biochemistry of photosynthesis. In J. Franck and W. E. Loomis (eds.), "Photosynthesis in Plants." Iowa State College Press, Ames.
Vardar, Y. 1953. A study of the auxin factor in epinastic and hyponastic movements. *Rev. Fac. Sci. Univ. Istanbul.*, 18: 318–352.
———. 1955. A study on the apical bud inhibition upon the lateral branches. *Rev. Fac. Sci. Univ. Istanbul*, 20: 245–256.
——— and B. Tözün. 1958. Role played by decapitation in growth and differentiation of *Lens culinaris* roots. *Am. J. Bot.*, 45: 714–718.
Vardjan, M., and J. P. Nitsch. 1961. La Régénération chez *Chicorium*: étude des auxines et des kinines endogénes. *Bull. Soc. Bot. Fr.*, 108: 363–374.
Varga, M. 1957. Examination of growth-inhibiting substances separated by paper chromatography in fleshy fruits, III. *Acta Biol. Szeged.*, 3: 225–232.
Varner, J. E. 1964. Gibberellic acid controlled synthesis of α-amylase in barley endosperm. *Plant Physiol.*, 39: 413–415.
———, V. Balce, and R. C. Huang. 1963. Senescence of cotyledons of germinating peas: Influence of axis tissue. *Plant Physiol.*, 38: 89–92.
——— and G. R. Chandra. 1964. Hormonal control of enzyme synthesis in barley endosperm. *Proc. Natl. Acad. Sci. (U.S.)*, 52: 100–106.
———, ———, and M. J. Chrispeels. 1965. Gibberellic acid and controlled synthesis of α-amylase in barley endosperm. *J. Cell. Comp. Phys.*, 66: 55–68.
Vasil, I. K. 1957. Effect of kinetin and gibberellic acid on excised anthers of *Allium cepa*. *Science*, 126: 1294–1295.
———. 1960. Pollen germination in some graminae: *Pennisetum typhoideum*. *Nature*, 187: 1134–1135.
Vegis, A. 1964. Dormancy in higher plants. *Annu. Rev. Plant Physiol.*, 15: 185–224.
Veldstra, H. 1956. On form and function of plant growth substances. Pp. 117–133 in R. L. Wain and F. Wightman (eds.), "The Chemistry and Mode of Action of Plant Growth Substances." Academic, N.Y.
———. 1964. Synergism of drugs. Pp. 42–51 in *Proc. III Int. Congr. Chemother.* Thieme-Verlag, Stuttgart.
Vendrell, M. 1969. Reversion of senescence: Effects of 2,4-D and indoleacetic acid on respiration, ethylene production and ripening of bananas. *Aust. J. Biol. Sci.*, 22: 601–610.
Venis, M. A. 1972. Auxin-induced conjugation systems in peas. *Plant Physiol.*, 49: 24–27.
——— and G. E. Blackman. 1966. The uptake of growth substances, VII. *J. Exp. Bot.*, 17: 270–282.
Venkataraman, R., and P. de Leo. 1972. Changes in leucyl-t-RNA species during aging of detached soybean cotyledons. *Phytochemistry*, 11: 923–927.
Verkerk, D. 1957. The pollination of tomatoes. *J. Agric. Sci. Neth.* 5: 37–54.
Vickery, H. B., G. W. Pucher, A. J. Wakeman, and C. S. Leavenworth. 1937. Chemical investigations of the tobacco plant. VI: Chemical changes in light and darkness. *Conn. Agric. Exp. Stn. New Haven Bull.* 399; pp. 757–828.
Villiers, T. A., and P. F. Wareing. 1965. The growth substance content of dormant fruits of *Fraxunus*. *J. Exp. Bot.*, 16: 533–544.
Vince, D., and R. Grill. 1966. The photoreceptors involved in anthocyanin synthesis. *Photochem. Photobiol.*, 5: 407–411.
Virgin, H. I. 1962. Light induced unfolding of the grass leaf. *Physiol. Plant.*, 15: 380–390.
Virtanen, A. J., T. Laine, and S. von Hausen. 1936. Excretion of amino acids from the root nodules and their chemical nature. *Suom. Kemistil.*, B9: 1.
Visser, T. 1955. Germination and storage of pollen. *Meded. Landbouwhogesch. Wageningen*, 55: 1–68.
———. 1964. Juvenile phase and growth of apple and pear seedlings. *Euphytica*, 13: 119–129.
———. 1965. On the inheritance of the juvenile period in apple. *Euphytica*, 14: 125–134.
——— and D. P. DeVries. 1970. Precocity and productivity of propagated apple and pear seedlings as dependent on the juvenile period. *Euphytica*, 19: 141–144.
Vöchting, H. 1878. "Organbildung in Pflanzenreich." Cohen, Bonn. 258 pp.
Vogel, S. 1970. Convective cooling at low airspeeds and the shapes of broad leaves. *J. Exp. Bot.*, 21: 91–101.
Voskresenskaya, N. P. 1950. The effect of the light wavelength on the synthesis of carbohydrates and proteins in a leaf. *Dok. Akad. Naus. SSSR*, 72(1): 173–176.
———. 1953. The effect of illumination on the nature of photosynthetic products. *Tr. Inst. Fiziol. Rast. Akad. Nauk SSSR (Proc. Timiriazev Inst. Plant Physiol.)*, 8(1): 42–56.
———. 1972. Blue light and carbon metabolism. *Annu. Rev. Plant Physiol.*, 23: 219–234.
Wada, S., E. Tanimoto, and Y. Masuda. 1968. Cell

elongation and metabolic turnover of the cell wall as affected by auxin and cell wall degrading enzymes. *Plant Cell Physiol.,* 9: 369–376.

Wade, N. L., and C. J. Brady. 1971. Effects of kinetin on respiration, ethylene production and ripening of banana fruit slices. *Austral. J. Biol. Sci.,* 24: 165–167.

Wadhi, M., and H. Y. M. Ram. 1967. Shortening the juvenile phase for flowering in *Kalanchoe. Planta,* 73: 28–36.

Wagenknecht, A. C., and R. H. Burris. 1950. IAA inactivating enzymes from bean roots and pea seedlings. *Arch. Biochem. Biophys.,* 25: 30–53.

Wagné, C. 1964. The distribution of the light effect in partly irradiated grass leaves. *Physiol. Plant.,* 17: 751–756.

Wald, G. 1960. Letter to the editor. *Sci. Am.,* 202: 13.

Walker, G. W. R., and J. Dietrich. 1961. Kinetin induced meiotic prophase acceleration in *Tradescantia* anthers. *Nature,* 192: 889–890.

Walker, T. S., and R. Thaine. 1971. Protein and fine structural components in exudate from sieve tubes in *Cucurbita pepo* stems. *Ann. Bot.,* 35: 773–790.

Walter, H., and E. Stadelmann. 1968. The physiological prerequisites for the transition of autotrophic plants from water to terrestrial life. *Bioscience,* 18: 694–701.

Walton, D. C. 1968. Phenylalanine ammonia lyase activity during germination of *Phaseolus vulgaris. Plant Physiol.,* 43: 1120–1124.

——— and E. Sondheimer. 1968. Effects of abscisin II on phenylalanine ammonia lyase activity in excised bean axes. *Plant Physiol.,* 43: 467–469.

——— and ———. 1972. Activity and metabolism of ^{14}C-abscisic acid derivatives. *Plant Physiol.,* 49: 290–292.

——— and G. S. Soofi. 1969. Germination of *Phaseolum vulgaris,* III. *Plant Cell Physiol.,* 10: 307–315.

———, ———, and E. Sondheimer. 1970. The effects of abscisic acid on growth and nucleic acid synthesis in excised embryonic bean axes. *Plant Physiol.,* 45: 37–40.

Wanner, H. 1958. Mobilisierung der Kohlenhydrate bei der Keimung. *Handb. Pflanzenphysiol.,* 6: 935–951.

Warburg, O. 1920. Über die Geschwindigkeit der photochemischen Kohlensäurezersetzung in lebenden Zellen. *Biochem. J.,* 103: 188–217.

———, G. Krippahl, and A. Lehman. 1969. Chlorophyll catalysis and Einstein's law of photochemical equivalence in photosynthesis. *Am. J. Bot.,* 56: 961–971.

——— and E. Negelein. 1922. Uber den Energieumsatz bei der Kohlensäureassimilation. *Z. Phys. Chem.,* 102: 235–266.

Wardlaw, C. W., and E. R. Leonard. 1936. Studies in tropical fruits, I. *Ann. Bot.,* 50: 622–653.

Wardlaw, I. F. 1967. The effect of water stress on translocation in relation to photosynthesis and growth. I: Effect during grain development in wheat. *Aust. J. Biol. Sci.,* 20: 25–39.

———. 1968. The control and pattern of movement of carbohydrates in plants. *Bot. Rev.,* 34(1): 79–105.

———. 1969. The effect of water stress on translocation in relation to photosynthesis and growth. II: Effect during leaf development in *Lolium temulentum* L. *Aust. J. Biol. Sci.,* 22: 1–16.

Wardrop, A. B. 1956. The nature of reaction wood, V. *Aust. J. Bot.,* 4: 152–166.

Wareing, P. F. 1953. A new photoperiodic phenomenon in short-day plants. *Nature,* 181: 614.

———. 1954. Growth studies in woody species, VI. *Physiol. Plant.,* 7: 261–277.

———. 1958*a*. Interaction between indoleacetic acid and gibberellic acid in cambial activity. *Nature,* 181: 1744–1745.

———. 1958*b*. Reproductive development in *Pinus sylvestris.* Pp. 643–654 in K. V. Thimann (ed.), *"The Physiology of Forest Trees."* Ronald Press, New York.

———. 1961. Juvenility and induction of flowering. Pp. 1652–1654 in "Recent Advances in Botany," University of Toronto Press, Toronto.

———. 1969. The control of bud dormancy in seed plants. *Soc. Exp. Biol. Symp.,* 23: 241–262.

——— and H. M. M. El-Antably. 1970. The possible role of endogenous growth inhibitors in the control of flowering. Pp. 285–300 in G. Bernier (ed.), "Cellular and Molecular Aspects of Floral Induction." Longman, London.

——— and H. A. Foda. 1957. Growth inhibitors and dormancy in *Xanthium* seed. *Physiol. Plant.,* 10: 266–280.

———, M. M. Khalifa, and K. J. Treharne. 1968. Rate-limiting processes in photosynthesis at saturating light intensities. *Nature,* 220: 453–457.

——— and D. L. Roberts. 1956. Photoperiodic control of cambial activity in *Robinia pseudoacacia. New Phytol.,* 55: 289–388.

——— and P. F. Saunders. 1971. Hormones and dormancy. *Annu. Rev. Plant Physiol.,* 22: 261–288.

——— and A. K. Seth. 1967. Ageing and senescence in the whole plant. *Soc. Exp. Biol. Symp.,* 21: 543–558.

——— and T. A. Villers. 1961. Growth substance and inhibitor changes in buds and seeds in response to chilling. Pp. 95–107 in R. M. Klein (ed.), "Plant Growth Regulation." Iowa State University Press., Ames.

———. 1970. Growth and its coordination in trees. Pp. 1–21 in L. C. Luckwill and C. V. Cutting (eds.), "Physiology of Tree Crops." Academic, London.

Warmke, H. E., and G. L. Warmke. 1950. Role of auxin in differentiation of root and shoot of *Taraxacum* and *Cichorium. Am. J. Bot.,* 37: 272–280.

Warner, H. L., and A. C. Leopold. 1967. Plant growth

regulation by stimulation of ethylene production. *BioScience,* 17: 722.
——— and ———. 1969. Timing of plant hormone responses in etiolated pea seedlings. *Plant Physiol.,* 44: 528.
——— and ———. 1971. Timing of growth regulator responses in peas. *Biochem. Biophys. Res. Commun.,* 44: 989–994.
Warren-Wilson, J. 1959. Analysis of the distribution of foliage area in grassland. Pp. 51–61 in "The Measurement of Grassland Productivity." *Proc. Univ. Nottingham, 6th Easter Sch. Agric. Sci.* (Butterworth, London).
———. 1960. Observations on net assimilation rates in arctic environments. *Ann. Bot.,* n.s., 24: 372–381.
———. 1966. Effect of temperature on net assimilation rate. *Ann. Bot.,* n.s., 30: 753–761.
Wassink, E. C. 1946. Experiments on photosynthesis of horticultural plants with the aid of the Warburg method. *Enzymologia,* 12: 33–55.
———. 1953. Starch conversion in leaves of *Helianthus tuberosus* and *H. annuus:* Preliminary observations. *Acta Botan. Neerl.,* 2: 327–348.
———. 1959. Efficiency of light energy conversion in plant growth. *Plant Physiol.,* 34: 356–361.
———, S. D. Richardson, and G. A. Pieters. 1956. Photosynthetic adaptation to light intensity in leaves of *Acer pseudoplatanus. Acta Botan. Neerl.,* 5: 247–256.
Watanabe, K. 1961. Studies on the germination of grass pollen, II. *Bot. Mag. (Tokyo),* 74: 131–137.
Watson, D. J. 1947. Comparative physiological studies on the growth of field crops. I: Variation in net assimilation rate and leaf area between species and varieties, and within and between years. *Ann. Bot.,* n.s., 11: 41–76.
———. 1952. The physiological basis of variation in yield. *Adv. Agron.,* 4: 101–145.
———. 1956. Leaf growth in relation to crop yield. Pp. 178–191 in F. L. Milthorpe (ed.), "Growth of Leaves." Butterworth, London.
———. 1958*a*. Factors limiting production. In "The Biological Productivity of Britain." Institute of Biology.
———. 1958*b*. The dependence of net assimilation rate on leaf-area index. *Ann. Bot.,* n.s., 22: 37–54.
Watson, R. W. 1942. Mechanism of elongation in palisade cells. *New Phytol.,* 41: 206–221.
——— and A. H. K. Petrie. 1940. Physiological ontogeny in the tobacco plant, IV. *Aust. J. Exp. Biol. Med. Sci.,* 18: 313–339.
Waygood, E. R., A. Oaks, and G. A. Maclachlan. 1956. On the mechanism of indoleacetic acid oxidation by wheat leaf enzymes. *Can. J. Bot.,* 34: 54–59.
Weatherley, P. E., and R. P. C. Johnson. 1968. The form and function of the sieve tube: A problem in reconciliation. *Int. Rev. Cytol.,* 24: 149–192 (Academic, New York).
Weaver, R. J. 1958. Effect of gibberellic acid on fruit set and berry enlargement in seedless grapes of *Vitis vinifera. Nature,* 181: 851–852.
———. 1959. Prolonging dormancy in *Vitis vinifera* with gibberellin. *Nature,* 183: 1198.
———. 1972. "Plant Growth Substances in Agriculture." Freeman, San Francisco. 594 pp.
——— and H. R. DeRose. 1946. Absorption and translocation of 2,4-D. *Bot. Gaz.,* 107: 509–521.
——— and S. B. McCune. 1958. Gibberellin tested on grapes. *Calif. Agric.,* 12: 6–15.
——— and J. van Overbeek. 1963. Kinins stimulate grape growth. *Calif. Agric.,* 17(9): 12.
———, ———, and R. M. Pool. 1966. Effect of kinins on fruit set and development in *Vitis vinifera. Hilgardia,* 37: 181–201.
——— and R. M. Pool. 1965. Gibberellin-like activity in seeded fruit of *Vitis. Naturwiss.,* 52: 111–112.
Webb, J. A., and P. R. Gorham. 1965. Radial movement of ^{14}C translocates from squash phloem. *Can. J. Bot.,* 43: 97–103.
Webster, B. D. 1968. Anatomical aspects of abscission. *Plant Physiol.,* 43: 1512–1544.
Weigl, J. 1969. Wechselwirkung pflanzlicher Wachstumshormone mit Membranen. *Z. Naturforsch.,* 24b: 1046–1052.
Weij, H. G. van der. 1932. Der Mechanismus des Wuchsstafftransportes. *Recl. Trav. Bot. Neerl.,* 29: 379–496.
Weintraub, M. 1952. Leaf movements in *Mimosa pudica. New Phytol.,* 50: 357–382.
Weintraub, R. L., and E. D. McAlister. 1942. Developmental physiology of the grass seedling. I: Inhibition of the mesocotyl of *Avena sativa* by continuous exposure to light of low intensities. *Smithson. Inst. Misc. Coll.,* 101: 1–10.
———, J. H. Reinhart, and R. A. Scherff. 1956. Role of entry, translocation and metabolism in specificity of 2,4-D and related compounds. *AEC Rep.* TID-7512, pp. 203–208.
Weiser, C. J. 1970. Cold resistance and injury in woody plants. *Science,* 169: 1269–1278.
Weisner, W., and H. Molisch. 1890. Untersuchungen über die Gasbewegung in der Pflanzen. *Sitzber. Akad. Wien,* 98: 534.
Weiss, C., and Y. Vaadia. 1965. Kinetin activity in root apices of sunflower plants. *Life Sci.,* 4: 1323.
Weldon, L. W., and F. L. Timmons. 1961. Photochemical degradation of Diuron and Monuron. *Weeds,* 9: 111–116.
Wellensiek, S. J. 1929. The physiology of tuber formation in *Solanum tuberosum* L. *Meded. Landbouwhogesch. Wageningen,* 33: 6–42.
———. 1952. Rejuvenation of woody plants by formation of sphaeroblasts. *Proc. Ned. K. Akad. Wet. Amst.,* C55: 567–573.
———. 1958. Vernalization and age in *Lunaria. Proc. K. Ned. Akad. Wet. Amst.,* C61: 561–571.

———. 1959. The inhibitory action of light on the floral induction of *Perilla*. *Proc. K. Ned. Acad. Wet. Amst.*, 62: 195–203.

———. 1961. De Fysiologie der Bloemknopvorming (The physiology of flower bud formation). *Med. K. Vlaamse Acad. Wet. Belg.*, 23: 1–15.

———. 1962. Dividing cells as the locus for vernalization. *Nature,* 195: 307–308.

———. 1966. The flower forming stimulus in *Silene armeria*. *Z. Pflanzenphysiol.*, 55: 1.

———, J. Doorenbos, and D. de Zeeuw. 1954. The mechanism of photoperiodism. *Proc. VIII Bot. Congr.*, 12: 307–315.

——— and F. A. Hakkaart. 1955. Vernalization and age. *Proc. K. Ned. Akad. Wet. Amst. Proc. Sec. Sci.*, C58: 16–21.

Went, F. W. 1928. Wuchsstoff und Wachstum. *Recl. Trav. Bot. Neerl.*, 25: 1–116.

———. 1942. Growth, auxin and tropisms in decapitated *Avena* coleoptiles. *Plant Physiol.*, 17: 236–249.

———. 1945. Plant growth under controlled conditions. V: Relation between age, light, variety, and thermoperiodicity of tomatoes. *Am. J. Bot.*, 32: 469–479.

———. 1948. Thermoperiodicity. Pp. 145–157 in A. E. Murneek and R. O. Whyte (eds.), "Vernalization and Photoperiodism." Chronica Botanica, Waltham, Mass.

———. 1949. Ecology of desert plant. II: Effect of rain and temperature on germination and growth. *Ecology,* 30: 1–13.

———. 1953. The effect of temperature on plant growth. *Annu. Rev. Plant Physiol.*, 4: 347–362.

———. 1957. "The Experimental Control of Plant Growth." Chronica Botanica, Waltham, Mass. 343 pp.

———. 1958. The physiology of photosynthesis in higher plants. *Preslia,* 30: 225–240.

———. 1959. Effects of environment of parent and grandparent generations on tuber production by potatoes. *Am. J. Bot.*, 46: 277–282.

———. 1961. Temperature. Pp. 1–23 in "Encyclopedia of Plant Physiology," vol. 16. Springer-Verlag, Berlin.

———. 1962. Phytotronics. *Plant Sci. Symp. Campbell Soup Co., Camden, N.J.*, pp. 149–162.

——— and K. V. Thimann. 1937. "Phytohormones." MacMillan, New York. 294 pp.

Werner, H. O. 1935. The effect of temperature, photoperiod and nitrogen upon tuberization in the potato. *Am. Potato J.,* 12: 274–280.

Wershing, H. F., and I. W. Bailey. 1942. Seedlings as experimental material in the study of redwood in conifers. *J. For.*, 40: 411–414.

West, C., G. E. Briggs, and F. Kidd. 1920. Methods and significant relations in the quantitative analysis of plant growth. *New Phytol.*, 19: 200–207.

West, C. A., and R. R. Fall. 1972. Gibberellin biosynthesis and its regulation. Pp. 133–142 in D. J. Carr (ed.), "Plant Growth Substances, 1970." Springer-Verlag, Berlin.

Westergaard, M. 1948. The relation between chromosome constitution and sex in the off-spring of triploid melandriom. *Hereditas,* 34: 257–279.

Wetmore, R. H. 1954. The use of in vitro cultures in the investigation of growth and differentiation in vascular plants. *Brookhaven Symp. Biol.*, 6: 22–40.

———. 1955. Differentiation of xylem in plants. *Science,* 121: 626–627.

———. 1959. Morphogensis in plants: A new approach. *Am. Sci.*, 47: 326–340.

——— and W. P. Jacobs. 1953. Studies on abscission: The inhibiting effect of auxin. *Am. J. Bot.*, 40: 272–276.

——— and S. Sorokin. 1955. On the differentiation of xylem. *J. Arnold Arbor.*, 36: 305–317.

Wettstein, D. von. 1953. Beeinflussung der Polarität und undifferenzierte Gewebebildung aus Moossporen. Z. Bot., 41: 199–226.

———. 1967. Chloroplast structure and genetics. Pp. 153–190 in A. San Pietro, F. A. Greer, and T. J. Army (eds.), "Harvesting the Sun." Academic, New York.

Wheeler, A. W. 1960. Changes in leaf growth substance in cotyledons and primary leaves during the growth of dwarf bean seedlings. *J. Exp. Bot.*, 11: 217–226.

Whitehead, F. H., and P. J. Myerscough. 1962. Growth analysis of plants: The ratio of mean relative growth rate to mean relative rate of leaf area increase. *New Phytol.*, 61: 314–321.

Whiteman, P. C., and D. Koller. 1964. Saturation deficit of the mesophyll evaporating surfaces in a desert halophytes. *Science,* 146: 1320–1321.

Whittingham, C. P., and G. G. Pritchard. 1963. The production of glycolate during photosynthesis in *Chlorella*. *Proc. R. Soc. Lond.*, B157: 366–382.

Whittle, C. M. 1964. Translocation and temperature. *Ann. Bot.*, 28: 339–344.

———. 1971. The behavior of ^{14}C profiles in *Helianthus* seedlings. *Planta,* 98: 136–149.

Whyte, P., and L. C. Luckwill. 1966. A sensitive bioassay for gibberellins based on retardation of leaf senescence. *Nature,* 210: 1360–1361.

Wickson, M., and K. V. Thimann. 1958. The antagonism of auxin and kinetin in apical dominance. *Physiol. Plant.*, 11: 62–74.

Wiebe, H. H., and P. J. Kramer. 1954. Translocation of radioactive isotopes from various regions of roots of barley seedlings. *Plant Physiol.*, 29: 342–348.

Wiesner, J. 1892. "Ein Elementarstruktur und das Wachstum der lebender Substanz." Hölder, Vienna.

Wiley, L., and F. M. Ashton. 1967. Influence of the embryonic axis on protein hydrolysis in cotyledons of *Cucurbita maxima*. *Physiol. Plant.*, 20: 688–696.

Wilkins, M. B., and T. K. Scott. 1968. Auxin transport in roots. *Nature,* 219: 1388–1389.
Williams, R. F. 1936. Physiological ontogeny in plants and its relation to nutrition. *Aust. J. Exp. Biol. Med. Sci.,* 14: 165–185.
———. 1938. Physiological ontogeny in plants and its relation to nutrition. IV: The effect of phosphorus supply on the total-protein-, and soluble-nitrogen contents, and water content of leaves and other plant parts. *Aust. J. Exp. Biol. Med. Sci.,* 16: 65–83.
———. 1946. The physiology of plant growth with special reference to the concept of net assimilation rate. *Ann. Bot.,* n.s., 10: 41–72.
———. 1955. Redistribution of mineral elements during development. *Annu. Rev. Plant Physiol.,* 6: 25–42.
Williams, R. J., and H. T. Meryman. 1970. Freezing injury and resistance in spinach chloroplast grana. *Plant Physiol.,* 45: 752–755.
Williamson, R. E. 1972. An investigation of the contractile protein hypothesis of phloem translocation. *Planta,* 106: 149–157.
Willstätter, R., and A. Stoll. 1913. "Untersuchungen über Chlorophyll." Springer-Verlag, Berlin. English trans. F. M. Schertz and A. R. Mertz, 1928.
Wilson, D., and J. P. Cooper. 1969. Effects of light intensity and CO_2 on apparent photosynthesis in genotypes of *Lolium perenne. New Phytol.,* 68: 627–644.
Winter, A., and K. V. Thimann. 1966. Bound indoleacetic acid in *Avena* coleoptiles. *Plant Physiol.,* 41: 335–342.
Wiskich, J. T., R. E. Young, and J. B. Biale. 1964. Metabolic processes in cytoplasmic particles of the avocado fruit, VI. *Plant Physiol.,* 39: 312–322.
Wit, C. T. de. 1965. "Photosynthesis of Leaf Canopies." *Agric. Res. Rep.* 663, Centre for Agricultural Publications and Documentation, Wageningen, Netherlands.
Withrow, R. B. 1941. Response of seedlings to various wavebands of low intensity irradiation. *Plant Physiol.,* 16: 241–256.
———, W. H. Klein and V. Elstad. 1957. Action spectra of photomorphogenic induction and its photoinactivation. *Plant Physiol.,* 32: 453–462.
———, ———, L. Price and V. Elstad. 1953. Influence of visible and near infrared radiant energy on organ development and pigment synthesis in bean and corn. *Plant Physiol.,* 28: 1–14.
Wittwer, S. H., and M. J. Bukovac. 1957. Gibberellins: New Chemicals for Crop production. *Mich. Agric. Exp. Stn. Q. Bull.,* 39: 469–494.
———, ———, H. M. Sell, and L. E. Weller. 1957. Some effects of flowering and fruit setting. *Plant Physiol.,* 32: 39–41.
Wohlpart, A., and T. J. Mabry. 1968. On the light requirement for betalin biosynthesis. *Plant Physiol.,* 43: 457–459.
Wolf, F. T. 1952. The production of indoleacetic acid by *Ustilago zeae* and its possible significance in tumor formation. *Proc. Natl. Acad. Sci. (U.S.),* 38: 106–111.
Wollgiehn, R. 1961. Untersuchungen über den Zusammenhang zwischen Nukleinsäure und Eiweisssstoffwechsel in grünen Blättern. *Flora,* 150: 117–127.
———. 1965. Kinetin and Nucleinsäurestoffwechsel. *Flora,* 156: 291–302.
——— and B. Parthier. 1964. Der Einfluss des Kinetins auf den RNS- und Protein-Stoffwechsel in abgeschnittenen mit hemmstoffen behandelten tabakblättern. *Phytochemistry,* 3: 241–248.
Wolpert, A. 1962. Heat transfer analysis of factors affecting plant leaf temperature: Significance of leaf hairs. *Plant Physiol.,* 37: 113–120.
Woo, K. C., N. A. Pyliotis, and W. J. S. Downton. 1971. Thylakoid aggregation and chlorophyll a/chlorophyll b ratio in C_4-plants. *Zeit. für Pflanzenphysiologie,* 64: 400–413.
Wood, A., and L. G. Paleg. 1972. The influence of gibberellic acid on the permeability of model membrane systesm. *Plant Physiol.,* 50: 103–108.
Wood, H. N., and A. C. Braun. 1961. Studies on the regulation of certain essential leiosynthetic systems in crown-gall tumor cells. *Proc. Natl. Acad. Sci. (U.S.),* 47: 1907–1913.
Woodford, E. K. 1957. The toxic action of herbicides. *Outlook Agric.,* 1: 145–154.
———. 1958. How a selective herbicide works. *World Crops, 1–4.*
Woodham, R. C., and D. McE. Alexander. 1966. The effect of root temperature on development of small fruiting Sultana vines, *Vitis. Ber. Rebenforsch.,* 5: 345–350.
Wooding, F. B. P. 1969. P Protein and microtubular systems in *Nicotiana* callus. *Planta,* 85: 284–298.
Woodwell, G. M. 1970. The energy cycle of the biosphere. *Sci. Am.,* 223: 64–74.
Woolhouse, H. W. 1967. The nature of senescense in plants. In H. W. Woolhouse (ed.), "Aspects of the Biology of Aging." *Symp. Soc. Exp. Biol.,* 21: 179–213.
Woolley, J. T. 1971. Reflectance and transmittance of light by leaves. *Plant Physiol.,* 47: 656–662.
Wooltorton, L. S. C., J. D. Jones, and A. C. Hulme. 1965. Genesis of ethylene in apples. *Nature,* 207: 999–1000.
Workman, M. 1963. Color and pigment changes in Golden Delicious and Grimes Golden apples. *Proc. Am. Soc. Hort. Sci.,* 83: 149–161.
Worley, J. F. 1966. Injection of foreign particles and their intercellular translocation in living phloem fibres. *Planta,* 68: 286–291.
———. 1973. Evidence in support of "open" sieve tube pores. *Protoplasma,* 76: 129–132.
Wright, R. D., D. T. N. Pillay., and J. H. Cherry. 1973. Changes in leucyl tRNA species of pea leaves during

senescence and after zeatin treatment. *Mech. Aging Dev.*, 1: 403–412.

Wright, S. T. C. 1956. Studies of fruit development in relation to plant hormones, III. *J. Hort. Sci.*, 31: 196–211.

———. 1961. A sequential growth response to gibberellic acid kinetin, and indolyacetic acid in the wheat coleoptile. *Nature*, 190: 699–700.

——— and R. W. P. Hiron. 1969. Abscisic acid, the growth inhibitor induced in detached wheat leaves by a period of wilting. *Nature*, 224: 719–720.

——— and ———. 1972. The accumulation of abscisic acid in plants during wilting and under stress conditions. Pp. 291–298 in D. J. Carr (ed.), "Plant Growth Substances, 1970." Springer-Verlag, Berlin.

Wright, W. L., and G. F. Warren. 1965. Photochemical decomposition of trifluralin. *Weeds*, 13: 329.

Yabuki, K., M. Aoki, and K. Hamotani. 1971. The effect of wind speed on the photosynthesis of rice field. 2: Photosynthesis of rice field in relation to wind speed and solar radiation. Pp. 7–9 in "Photosynthesis and Utilisation of Solar Energy." Level III Experiments. IBP-Japanese National Subcommittee for PP.

Yabuta, T. 1935. Biochemistry of the "bakanae" fungus of rice. *Agric. Hort. (Tokyo)*, 10: 17–22.

Yamada, Y., H. P. Rasmussen, M. J. Bukovac, and S. H. Wittwer. 1966. Binding sites for inorganic ions and urea on isolated cuticular membrane surfaces. *Am. J. Bot.*, 53: 170–172.

———, *S. H. Wittwer, and M. J. Bukovac. 1965.* Penetration of organic compounds through isolated cuticular membranes with special reference to urea. *Plant Physiol.*, 40: 170–175.

Yamaki, T., and K. Nakamura. 1952. Formation of indoleacetic acid in maize embryo. *Sci. Pap. Coll. Gen. Educ. Univ. Tokyo.*, 2: 81–98.

Yang, S. F. 1967. Biosynthesis of ethylene: Ethylene formation from methional by horseradish peroxidase. *Arch. Biochem. Biophys.*, 122: 481–487.

———. 1968. Biosynthesis of ethylene. Pp. 1217–1228 in F. Wightman and G. Setterfield (eds.), "Biochemistry and Physiology of Plant Growth Regulators." Runge Press, Ottawa.

———, H. S. Ku, and H. K. Pratt. 1966. Ethylene production from methionine as mediated by flavin mononucleotide and light. *Biochem. Biophys. Res. Commun.*, 24: 739–743.

———, ———, and ———. 1967. Photochemical production of ethylene from methionine and its analogues in the presence of flavin mononucleotide. *J. Biol. Chem.*, 242: 5274–5280.

Yarwood, C. E. 1961. Translocated heat injury. *Plant Physiol.*, 36: 721–726.

Yasuda, S. 1934. Parthenocarpy caused by the stimulus of pollination in some plants of Solanaceae. *Agric. and Hort.*, 9: 647–656.

Yemm, E. W. 1956. The metabolism of senescent leaves. In G. E. W. Wolstemholme (ed.), *CIBA Colloq. Aging*, 2: 207–214.

Yeung, E. C., and R. L. Peterson. 1972. Xylem transfer cells in the rosette plant *Hieracium floribundum*. *Planta*, 107: 183–188.

Yih, R. Y. 1966. Inhibition of stem growth and flower formation in *Pharbitis nil* with *N,N*-dimethylammosuccinamic acid (B-995). *Planta*, 71: 68–80.

———, P. L. McNulty, M. C. Seidel, and K. L. Viste. 1971. Plant growth regulating properties of 3-carboxy-2-pyridones. *HortScience*, 6: 460–461.

Yin, H. C. 1938. Diaphototropic movements of the leaves of *Malva neglecta*. *Am. J. Bot.*, 25: 1–6.

Yokota, T., N. Takahashi, N. Murofushi, and S. Tamura. 1969. Isolation of gibberellins A_{26} and their glucosides from immature seeds of *Pharbitis*. *Planta*, 87: 180–184.

Yoshida, K., K. Umemura, K. Yoshinaga, and Y. Oota. 1967. Specific RNA from photoperiodically induced cotyledons of *Pharbitis nil*. *Plant Cell Physiol.*, 8: 97–108.

Young, L. C. T., and R. G. Stanley. 1963. Incorporation of tritiated nucleosides in nuclei of germinating pine pollen. *Nucleus*, 6: 83–90.

Young, R. E. 1965. Effect of ionizing radiation on respiration and ethylene production of avocado fruit. *Nature*, 205: 1113.

Younis, A. F. 1954. Experiments on the growth and geotropism of roots. I: Technique for achieving regular growth, and a study of the effects of decapitation and reheading on the growth of *Vicia faba* roots. *J. Exp. Bot.*, 5: 357–372.

Zaerr, J. B., and J. W. Mitchell. 1967. Polar transport related to mobilization of plant constituents. *Plant Physiol.*, 42: 863–874.

Zafar, M. A. 1955. Application of certain hormones to prevent flower abscission in two potato varieties. *Am. Potato J.*, 32: 283–292.

Zauberman, G., and M. Schiffmann-Nadel. 1972. Pectin methylesterase and polygalacturonase in avocado fruit at various stages of development. *Plant Physiol.*, 49: 864–865.

Zeeuw, D. de. 1953. Flower initiation and light intensity in *Perilla*. *Proc. R. Acad. Sci.*, C56: 418–422.

———. 1954. De invloed van het blad op de bloei. *Meded. Landbouwhogesch. Wageningen*, 54: 1–44.

———. 1956. Leaf induced inhibition of flowering in tomato. *Proc. K. Ned. Akad. Wet. Amst.*, 59: 535–540.

———. 1957. Flowering of *Xanthium* under long-day conditions. *Nature*, 180: 588.

——— and A. C. Leopold. 1955. Altering juvenility with auxin. *Science*, 122: 925–926.

Zeevaart, J. A. D. 1957. Studies on flowering by means of grafting. II: Photoperiodic treatment of detached *Perilla* and *Xanthium* leaves. *Proc. K. Ned. Acad. Wet. Amst.*, 6D: 332–337.

———. 1958. Flower formation as studied by grafting. *Med. Landbouwhogesch. Wageningen,* 58: 1–88.
———. 1962. Physiology of flowering. *Science,* 137: 723–731.
———. 1964*a*. Chemical basis of induction. Pp. 343–347, in J. Bonner and P. Tso (eds.), "The Nucleohistones." Holden-Day, San Francisco.
———. 1964*b*. Effects of the growth retardant CCC on floral initiation and growth in *Pharbitis nil. Plant Physiol.,* 39: 402–408.
———. 1964*c*. Inhibition of stem growth and flower formation in *Pharbitis nil* with *N,N*-dimethylaminosuccinamic acid (B-995). *Planta,* 71: 68–80.
———. 1969. In L. T. Evans (ed.), "The Induction of Flowering." Cornell Univ. Press, Ithaca, N.Y. 488 pp.
———. 1971. Abscisic acid content of spinach in relation to photoperiod and water stress. *Plant Physiol.,* 48: 86–90.
———. 1971. Effects of photoperiod on growth rate and endogenous gibberellins in the long-day rosette plant spinach. *Plant Physiol.,* 47: 821–827.
———. 1974. Levels of abscisic acid and xanthoxin in spinach under different environmental conditions. *Plant Physiol.,* 53: 644–648.
——— and A. Lang. 1962. The relationship between gibberellin and floral stimulus in *Bryophyllum daigremontianum. Planta,* 58: 531–542.
Zenk, M. H. 1961. Indoleacetyl glucose, a new compound in the metabolism of indoleacetic acid in plants. *Nature,* 191: 493–494.
———. 1962. Aufnahme und Stoffwechsel von Naphthylessigsäure durch Erbsenepicotyle. *Planta,* 58: 75–94.
———. 1963. Isolation, biosynthesis and function of indoleacetic acid conjugates. Pp. 241–249 in J. P. Nitsch (ed.), "Régulateurs naturels de la croissance végétale." CNRS, Paris.
———. 1964. Über Primareffekte der phototropischen Perzeption. *Ber. deutsch. bot. Ges.,* 74: 328.
———. 1968. The action of light on the metabolism of auxin in relation to phototropism. Pp. 1109–1128 in F. Wightman and G. Setterfield (eds.), "Biochemistry and Physiology of Plant Growth Substances." Runge Press, Ottawa.
——— and G. Müller. 1963. In vivo destruction of exogenously applied indoleacetic acid as influenced by phenolic acids. *Nature,* 200: 761–763.
——— and ———. 1964. Biosynthese von 2,6 -Hydroxybenzoesäure and anderen Benzoesäuren in höheren Pflanzen. *Z. Naturforsch.,* 19: 398–405.
Zeroni, M., S. Ben-Yehoshua, and J. Galil. 1972. Relationship between ethylene and the growth of *Ficus sycomorus. Plant Physiol.,* 50: 378–381.
Ziegler, H. 1953. Über die Bildung und Lokalizierung der Formazan der Pflanzenzelle. *Naturwiss.,* 40: 144.
———. 1956. Untersuchungen über die Leitung und Sekretion der Assimilate. *Planta,* 47: 447–500.
———, D. Kohler, and B. Streitz. 1966. Ist 2-Chloro-9-fluorenol-9-carbonsäure ein Gibberellinantagonist? *Z. Pflanzenphysiol.,* 54: 118–124.
——— and U. Lüttge. 1966. The salt-glands of *Limonium vulgare.* I: The fine structure. *Planta,* 70: 193–206.
——— and ———. 1967. The salt-glands of *Limonium vulgare.* II: The localisation of chloride. *Planta,* 74: 1–17.
——— and G. H. Vieweg. 1961. Der experimentelle Nachweis einer Massenströmung im Phloem von *Heracleum mantigazzianum* (somm. et Lev.). *Planta,* 56: 402–408.
——— and Irmgard Ziegler. 1966. Die lichtinduzierte Synthese der NADP$^+$-abhängigen Glycerinaldehyd-3-Phosphat-Dehydrogenase. III: Die Bedeutung des Chlorophylls und der Einfluss von Stoffwechselinhibitoren. *Planta,* 69: 111–123.
———, and ———, and H. J. Schmidt-Clausen. 1965. The influence of light intensity and light quality on the increase in activity of the NADP$^+$-dependent glyceraldehyde-3-phosphate dehydrogenase. *Planta,* 67: 344–356.
Zimmerman, P. W. 1935. Anaesthetic properties of carbon monoxide and other gases in relation to plants, insects and centipedes. *Contrib. Boyce Thompson Inst.,* 7: 147–155.
——— and A. E. Hitchcock. 1933. Initiation and stimulation of adventitious roots caused by unsaturated hydrocarbon gases. *Contrib. Boyce Thompson Inst.,* 5: 351–369.
——— and ———. 1962. Substituted phenoxy and benzoic acid growth substances and the relation of structure to physiological activity. *Contrib. Boyce Thompson Inst.,* 12: 321–363.
———, ———, and W. Crocker. 1931. The effect of ethylene and illuminating gas on roses. *Contrib. Boyce Thompson Inst.,* 3: 459–481.
———, ———, and F. Wilcoxon. 1936. Several esters as plant hormones. *Contrib. Boyce Thompson Inst.,* 8: 105–112.
——— and F. Wilcoxon. 1935. Several chemical growth substances which cause initiation of roots and other responses in plants. *Contrib. Boyce Thompson Inst.,* 7: 209–229.
——— and ———. 1942. Substituted phenoxy and benzoic acid growth substances and the relation of structure to physiological activity. *Contrib. Boyce Thompson Inst.,* 12: 321–343.
Zimmerman, W. 1929. Die Schlafbewegungen der Laubblätter. *Tueb. Naturwiss. Abh.,* 12: 163–6; cited by Bünning (1959).
Zimmermann, M. H. 1957*a*. Translocation of organic substances in trees. I: The nature of the sugars in the sieve tube exudate of trees. *Plant Physiol.,* 32: 288–291.

———. 1957*b*. Translocation of organic substances in trees. II: On the translocation mechanism in the phloem of white ash (*Fraxinus americana* L.). *Plant Physiol.,* 32: 399–404.

———. 1958. Translocation of organic substances in trees. III: The removal of sugars from sieve tubes in the white ash. *Plant Physiol.,* 33: 213–217.

———. 1961. Movement of organic substances in trees. *Science,* 133: 73–79.

———. 1964. Effect of low temperature on ascent of sap in trees. *Plant Physiol.,* 39: 568–572.

———. 1969. Translocation velocity and specific mass transfer in the sieve tubes of *Fraxinus americana* L. *Planta,* 84: 272–278.

———, A. B. Wardrop,and P. B. Tomlinson. 1968. Tension wood in aerial roots of *Ficus benjamina. Wood Sci. Tech.,* 2: 95–104.

Zinsmeister, H. D. 1960. Das phototropische Verhalten der Blütenstiele von *Cylamen persicum. Planta,* 55: 647–668.

——— and W. Hollmuller. 1964. Gerbstoffe und Wachstum, II. *Planta,* 63: 133–145.

Zobel, R. W. 1973. Some physiological characteristics of the ethylene-requiring tomato mutant, diageotropica. *Plant Physiol.,* 52: 385–389.

Zucker, M. 1965. Induction of phenylalanine deaminase by light and its relation to chlorogenic acid synthesis in potato tuber tissue. *Plant Physiol.,* 40: 779–784.

———. 1969. Induction of phenylalanine ammonia lyase in *Xanthium* leaf discs. *Plant Physiol.,* 44: 912–922.

———. 1972. Light and enzymes. *Annu. Rev. Plant Physiol.,* 23: 133–156.

Zukel, J. W. 1955. "Literature Summary on Maleic Hydrazide." Naugatuck Chemical Co., MHIS no. 6C.

Zurzycki, J. 1953. Arrangement of chloroplasts and light absorption in plant cells. *Acta Soc. Bot. Pol.,* 22: 299–320.

INDEX